Lecture Notes in Computer Science 2724
Edited by G. Goos, J. Hartmanis, and J. van Leeuwen

Springer-Verlag
Berlin Heidelberg GmbH

Erick Cantú-Paz James A. Foster
Kalyanmoy Deb Lawrence David Davis
Rajkumar Roy Una-May O'Reilly
Hans-Georg Beyer Russell Standish
Graham Kendall Stewart Wilson
Mark Harman Joachim Wegener
Dipankar Dasgupta Mitch A. Potter
Alan C. Schultz Kathryn A. Dowsland
Natasha Jonoska Julian Miller (Eds.)

Genetic and Evolutionary Computation – GECCO 2003

Genetic and Evolutionary Computation Conference
Chicago, IL, USA, July 12-16, 2003
Proceedings, Part II

Springer

Series Editors

Gerhard Goos, Karlsruhe University, Germany
Juris Hartmanis, Cornell University, NY, USA
Jan van Leeuwen, Utrecht University, The Netherlands

Main Editor

Erick Cantú-Paz
Center for Applied Scientific Computing (CASC)
Lawrence Livermore National Laboratory
7000 East Avenue, L-561, Livermore, CA 94550, USA
E-mail: cantupaz@llnl.gov

Cataloging-in-Publication Data applied for

A catalog record for this book is available from the Library of Congress

Bibliographic information published by Die Deutsche Bibliothek
Die Deutsche Bibliothek lists this publication in the Deutsche Nationalbibliografie;
detailed bibliographic data is available in the Internet at <http://dnb.ddb.de>.

CR Subject Classification (1998): F.1-2, D.1.3, C.1.2, I.2.6, I.2.8, I.2.11, J.3

ISSN 0302-9743
ISBN 978-3-540-40603-7 ISBN 978-3-540-45110-5 (eBook)
DOI 10.1007/978-3-540-45110-5

This work is subject to copyright. All rights are reserved, whether the whole or part of the material is
concerned, specifically the rights of translation, reprinting, re-use of illustrations, recitation, broadcasting,
reproduction on microfilms or in any other way, and storage in data banks. Duplication of this publication
or parts thereof is permitted only under the provisions of the German Copyright Law of September 9, 1965,
in its current version, and permission for use must always be obtained from Springer-Verlag. Violations are
liable for prosecution under the German Copyright Law.

http://www.springer.de

© Springer-Verlag Berlin Heidelberg 2003
Originally published by Springer-Verlag Berlin Heidelberg New York in 2003.

Typesetting: Camera-ready by author, data conversion by PTP Berlin GmbH
Printed on acid-free paper   SPIN 10929001   06/3142   5 4 3 2 1 0

Volume Editors

James A. Foster
University of Idaho
Department of Computer Science
P.O. Box 441010
Moscow, ID 83844-1010, USA
E-mail: foster@uidaho.edu

Kalyanmoy Deb
Kanpur Genetic Algorithms Laboratory
Department of Mechanical Engineering
Indian Institute of Technology
Kanpur, PIN 208 016, India
E-mail: deb@iitk.ac.in

Lawrence David Davis
NuTech Solutions, Inc.
28 Green Street, Newbury, MA 01951, USA
E-mail: david.davis@nutechsolutions.com

Rajkumar Roy
Enterprise Integration
Cranfield University
Bedford MK43 0AL, UK
E-mail: r.roy@cranfield.ac.uk.

Una-May O'Reilly
Artificial Intelligence Lab.
MIT 20
200 Technology SQ, Rm 933
Cambridge, MA 02139, USA
E-mail: unamay@ai.mit.edu

Hans-Georg Beyer
University of Dortmund
Dept. of Computer Science XI
44221 Dortmund, Germany
E-mail: beyer@Ls11.cs.uni-dortmund.de

Russell Standish
School of Mathematics
University of New South Wales
Sydney 2052, Australia
E-mail: R.Standish@unsw.edu.au

Graham Kendall
University of Nottingham
School of Computer Science and IT
Jubilee Campus, Wollaton Road
Nottingham NG8 1BB, UK
E-mail: gxk@cs.nott.ac.uk

Stewart W. Wilson
Prediction Dynamics
Concord, MA 01742, USA
E-mail: wilson@prediction-dynamics.com

Mark Harman
Brunel University
Dept. of Information Systems
and Computing
Uxbridge, Middlesex, 11B8 3PH, UK
E-mail: Mark.Harman@brunel.ac.uk

Joachim Wegener
DaimlerChrysler AG
Research and Technology
Alt-Mohabit 96 A, 10559 Berlin
Germany
E-mail: joachim.wegener@daimlerchrysler.com

Dipankar Dasgupta
University of Memphis
Division of Computer Science
Memphis, TN 38152, USA
E-mail: dasgupta@memphis.edu

Mitchell A. Potter
Naval Research Laboratory
Washington, DC, USA
E-mail: mpotter@aic.nrl.navy.mil

Alan C. Schultz
Navy Center for Applied Research in
Artificial Intelligence
Naval Research Laboratory
4555 Overlook Ave. S.W., Washington
DC 20375
E-mail: schultz@aic.nrl.navy.mil

Kathryn A. Dowsland
Gower Optimal Algorithms Ltd
5 Whitstone Lane, Newton
Swansea SA3 4UH, UK
E-mail: k.a.dowsland@btconnect.com

Natasha Jonoska
University of South Florida
Department of Mathematics
4202 E. Fowler Av. PHY 114
Tampa FL 33559, USA
E-mail: jonoska@math.usf.edu

Julian F. Miller
School of Computer Science
University of Birmingham
Birmingham, B15 2TT, UK
E-mail: j.miller@cs.bham.ac.uk

Preface

These proceedings contain the papers presented at the 5th Annual Genetic and Evolutionary Computation Conference (GECCO 2003). The conference was held in Chicago, USA, July 12–16, 2003.

A total of 417 papers were submitted to GECCO 2003. After a rigorous doubleblind reviewing process, 194 papers were accepted for full publication and oral presentation at the conference, resulting in an acceptance rate of 46.5%. An additional 92 submissions were accepted as posters with two-page extended abstracts included in these proceedings.

This edition of GECCO was the union of the 8th Annual Genetic Programming Conference (which has met annually since 1996) and the 12th International Conference on Genetic Algorithms (which, with its first meeting in 1985, is the longest running conference in the field). Since 1999, these conferences have merged to produce a single large meeting that welcomes an increasingly wide array of topics related to genetic and evolutionary computation.

Possibly the most visible innovation in GECCO 2003 was the publication of the proceedings with Springer-Verlag as part of their Lecture Notes in Computer Science series. This will make the proceedings available in many libraries as well as online, widening the dissemination of the research presented at the conference.

Other innovations included a new track on Coevolution and Artificial Immune Systems and the expansion of the DNA and Molecular Computing track to include quantum computation.

In addition to the presentation of the papers contained in these proceedings, the conference included 13 workshops, 32 tutorials by leading specialists, and presentation of late-breaking papers.

GECCO is sponsored by the International Society for Genetic and Evolutionary Computation (ISGEC). The ISGEC by-laws contain explicit guidance on the organization of the conference, including the following principles:

(i) GECCO should be a broad-based conference encompassing the whole field of genetic and evolutionary computation.

(ii) Papers will be published and presented as part of the main conference proceedings only after being peer-reviewed. No invited papers shall be published (except for those of up to three invited plenary speakers).

(iii) The peer-review process shall be conducted consistently with the principle of division of powers performed by a multiplicity of independent program committees, each with expertise in the area of the paper being reviewed.

(iv) The determination of the policy for the peer-review process for each of the conference's independent program committees and the reviewing of papers for each program committee shall be performed by persons who occupy their positions by virtue of meeting objective and explicitly stated qualifications based on their previous research activity.

(v) Emerging areas within the field of genetic and evolutionary computation shall be actively encouraged and incorporated in the activities of the conference by providing a semiautomatic method for their inclusion (with some procedural flexibility extended to such emerging new areas).

(vi) The percentage of submitted papers that are accepted as regular full-length papers (i.e., not posters) shall not exceed 50%.

These principles help ensure that GECCO maintains high quality across the diverse range of topics it includes.

Besides sponsoring the conference, ISGEC supports the field in other ways. ISGEC sponsors the biennial Foundations of Genetic Algorithms workshop on theoretical aspects of all evolutionary algorithms. The journals *Evolutionary Computation* and *Genetic Programming and Evolvable Machines* are also supported by ISGEC. All ISGEC members (including students) receive subscriptions to these journals as part of their membership. ISGEC membership also includes discounts on GECCO and FOGA registration rates as well as discounts on other journals. More details on ISGEC can be found online at http://www.isgec.org.

Many people volunteered their time and energy to make this conference a success. The following people in particular deserve the gratitude of the entire community for their outstanding contributions to GECCO:

James A. Foster, the General Chair of GECCO for his tireless efforts in organizing every aspect of the conference.
David E. Goldberg and John Koza, members of the Business Committee, for their guidance and financial oversight.
Alwyn Barry, for coordinating the workshops.
Bart Rylander, for editing the late-breaking papers.
Past conference organizers, William B. Langdon, Erik Goodman, and Darrell Whitley, for their advice.
Elizabeth Ericson, Carol Hamilton, Ann Stolberg, and the rest of the AAAI staff for their outstanding efforts administering the conference.
Gerardo Valencia and Gabriela Coronado, for Web programming and design.
Jennifer Ballentine, Lee Ballentine and the staff of Professional Book Center, for assisting in the production of the proceedings.
Alfred Hofmann and Ursula Barth of Springer-Verlag for helping to ease the transition to a new publisher.

Sponsors who made generous contributions to support student travel grants:

Air Force Office of Scientific Research
DaimlerChrysler
National Science Foundation
Naval Research Laboratory
New Light Industries
Philips Research
Sun Microsystems

The track chairs deserve special thanks. Their efforts in recruiting program committees, assigning papers to reviewers, and making difficult acceptance decisions in relatively short times, were critical to the success of the conference:

A-Life, Adaptive Behavior, Agents, and Ant Colony Optimization,
    Russell Standish
Artificial Immune Systems, Dipankar Dasgupta
Coevolution, Graham Kendall
DNA, Molecular, and Quantum Computing, Natasha Jonoska
Evolution Strategies, Evolutionary Programming, Hans-Georg Beyer
Evolutionary Robotics, Alan Schultz, Mitch Potter
Evolutionary Scheduling and Routing, Kathryn A. Dowsland
Evolvable Hardware, Julian Miller
Genetic Algorithms, Kalyanmoy Deb
Genetic Programming, Una-May O'Reilly
Learning Classifier Systems, Stewart Wilson
Real-World Applications, David Davis, Rajkumar Roy
Search-Based Software Engineering, Mark Harman, Joachim Wegener

The conference was held in cooperation and/or affiliation with:

American Association for Artificial Intelligence (AAAI)
Evonet: the Network of Excellence in Evolutionary Computation
5th NASA/DoD Workshop on Evolvable Hardware
Evolutionary Computation
Genetic Programming and Evolvable Machines
Journal of Scheduling
Journal of Hydroinformatics
Applied Soft Computing

Of course, special thanks are due to the numerous researchers who submitted their best work to GECCO, reviewed the work of others, presented a tutorial, organized a workshop, or volunteered their time in any other way. I am sure you will be proud of the results of your efforts.

May 2003

Erick Cantú-Paz
Editor-in-Chief GECCO 2003
Center for Applied Scientific Computing
Lawrence Livermore National Laboratory

GECCO 2003 Conference Organization

Conference Committee

General Chair: James A. Foster
Proceedings Editor-in-Chief: Erick Cantú-Paz
Business Committee: David E. Goldberg, John Koza, J.A. Foster
Chairs of Program Policy Committees:
    A-Life, Adaptive Behavior, Agents, and Ant Colony Optimization, Russell Standish
    Artificial Immune Systems, Dipankar Dasgupta
    Coevolution, Graham Kendall
    DNA, Molecular, and Quantum Computing, Natasha Jonoska
    Evolution Strategies, Evolutionary Programming, Hans-Georg Beyer
    Evolutionary Robotics, Mitchell A. Potter and Alan C. Schultz
    Evolutionary Scheduling and Routing, Kathryn A. Dowsland
    Evolvable Hardware, Julian Miller
    Genetic Algorithms, Kalyanmoy Deb
    Genetic Programming, Una-May O'Reilly
    Learning Classifier Systems, Stewart Wilson
    Real-World Applications, David Davis, Rajkumar Roy
    Search-Based Software Engineering, Mark Harman and Joachim Wegener
Workshops Chair: Alwyn Barry
Late-Breaking Papers Chair: Bart Rylander

Workshop Organizers

Biological Applications for Genetic and Evolutionary Computation (Bio GEC 2003), Wolfgang Banzhaf, James A. Foster
Application of Hybrid Evolutionary Algorithms to NP-Complete Problems, Francisco Baptista Pereira, Ernesto Costa, Günther Raidl
Evolutionary Algorithms for Dynamic Optimization Problems, Jürgen Branke
Hardware Evolutionary Algorithms and Evolvable Hardware (HEAEH 2003), John C. Gallagher
Graduate Student Workshop, Maarten Keijzer, Sean Luke, Terry Riopka
Workshop on Memetic Algorithms 2003 (WOMA-IV), Peter Merz, William E. Hart, Natalio Krasnogor, Jim E. Smith
Undergraduate Student Workshop, Mark M. Meysenburg
Learning, Adaptation, and Approximation in Evolutionary Computation, Sibylle Mueller, Petros Koumoutsakos, Marc Schoenauer, Yaochu Jin, Sushil Louis, Khaled Rasheed
Grammatical Evolution Workshop (GEWS 2003), Michael O'Neill, Conor Ryan
Interactive Evolutionary Search and Exploration Systems, Ian Parmee

Analysis and Design of Representations and Operators (ADoRo 2003), Franz Rothlauf, Dirk Thierens

Challenges in Real-World Optimisation Using Evolutionary Computing, Rajkumar Roy, Ashutosh Tiwari

International Workshop on Learning Classifier Systems, Wolfgang Stolzmann, Pier-Luca Lanzi, Stewart Wilson

Tutorial Speakers

Parallel Genetic Algorithms, Erick Cantú-Paz
Using Appropriate Statistics, Steffan Christiensen
Multiobjective Optimization with EC, Carlos Coello
Making a Living with EC, Yuval Davidor
A Unified Approach to EC, Ken DeJong
Evolutionary Robotics, Dario Floreano
Immune System Computing, Stephanie Forrest
The Design of Innovation & Competent GAs, David E. Goldberg
Genetic Algorithms, Robert Heckendorn
Evolvable Hardware Applications, Tetsuya Higuchi
Bioinformatics with EC, Daniel Howard
Visualization in Evolutionary Computation, Christian Jacob
Data Mining and Machine Learning, Hillol Kargupta
Evolvable Hardware, Didier Keymeulen
Genetic Programming, John Koza
Genetic Programming Theory I & II, William B. Langdon, Riccardo Poli
Ant Colony Optimization, Martin Middendorf
Bionics: Building on Biological Evolution, Ingo Rechenberg
Grammatical Evolution, C. Ryan, M. O'Neill
Evolution Strategies, Hans-Paul Schwefel
Quantum Computing, Lee Spector
Anticipatory Classifier Systems, Wolfgang Stolzmann
Mathematical Theory of EC, Michael Vose
Computational Complexity and EC, Ingo Wegener
Software Testing via EC, J. Wegener, M. Harman
Testing & Evaluating EC Algorithms, Darrell Whitley
Learning Classifier Systems, Stewart Wilson
Evolving Neural Network Ensembles, Xin Yao
Neutral Evolution in EC, Tina Yu
Genetics, Annie S. Wu

Keynote Speakers

John Holland, "The Future of Genetic Algorithms"
Richard Lenski, "How the Digital Leopard Got His Spots: Thinking About Evolution Inside the Box"

Members of the Program Committee

Hussein Abbass
Adam Adamopoulos
Alexandru Agapie
José Aguilar
Jesús Aguilar
Hernán Aguirre
Chang Wook Ahn
Uwe Aickelin
Enrique Alba
Javier Alcaraz Soria
Dirk Arnold
Tughrul Arslan
Atif Azad
Meghna Babbar
Vladan Babovic
B.V. Babu
Thomas Bäck
Julio Banga
Francisco Baptista Pereira
Alwyn Barry
Cem Baydar
Thomas Beielstein
Theodore Belding
Fevzi Belli
Ester Bernado-Mansilla
Tom Bersano-Begey
Hugues Bersini
Hans-Georg Beyer
Filipic Bogdan
Andrea Bonarini
Lashon Booker
Peter Bosman
Terry Bossomaier
Klaus Bothe
Leonardo Bottaci
Jürgen Branke
Wilker Bruce
Peter Brucker
Anthony Bucci
Dirk Bueche
Magdalena Bugajska
Larry Bull
Edmund Burke
Martin Butz

Stefano Cagnoni
Xiaoqiang Cai
Erick Cantú-Paz
Uday Chakraborty
Weng-Tat Chan
Alastair Channon
Ying-Ping Chen
Shu-Heng Chen
Junghuei Chen
Prabhas Chongstitvatana
John Clark
Lattaud Claude
Manuel Clergue
Carlos Coello Coello
David Coley
Philippe Collard
Pierre Collet
Clare Bates Congdon
David Corne
Ernesto Costa
Peter Cowling
Bart Craenen
Jose Cristóbal Riquelme Santos
Keshav Dahal
Paul Darwen
Dipankar Dasgupta
Lawrence Davis
Anthony Deakin
Kalyanmoy Deb
Ivanoe De Falco
Hugo De Garis
Antonio Della Cioppa
A. Santos Del Riego
Brahma Deo
Dirk Devogelaere
Der-Rong Din
Phillip Dixon
Jose Dolado Cosin
Marco Dorigo
Keith Downing
Kathryn Dowsland
Gerry Dozier
Rolf Drechsler

Stefan Droste
Marc Ebner
R. Timothy Edwards
Norberto Eiji Nawa
Aniko Ekart
Christos Emmanouilidis
Hector Erives
Felipe Espinoza
Matthew Evett
Zhun Fan
Marco Farina
Robert Feldt
Francisco Fernández
Sevan Ficici
Peter John Fleming
Stuart Flockton
Dario Floreano
Cyril Fonlupt
Carlos Fonseca
Stephanie Forrest
Alex Freitas
Clemens Frey
Chunsheng Fu
Christian Gagne
M. Gargano
Ivan Garibay
Josep Maria Garrell i Guiu
Alessio Gaspar
Michel Gendreau
Zhou Gengui
Pierre Gérard
Andreas Geyer-Schulz
Tushar Goel
Fabio Gonzalez
Jens Gottlieb
Kendall Graham
Buster Greene
John Grefenstette
Darko Grundler
Dongbing Gu
Steven Gustafson
Charles Guthrie
Pauline Haddow
Hani Hagras

Hisashi Handa
Georges Harik
Mark Harman
Emma Hart
William Hart
Inman Harvey
Michael Herdy
Jeffrey Hermann
Arturo Hernández
  Aguirre
Francisco Herrera
Jürgen Hesser
Robert Hierons
Mika Hirvensalo
John Holmes
Tadashi Horiuchi
Daniel Howard
William Hsu
Jianjun Hu
Jacob Hurst
Hitoshi Iba
Kosuke Imamura
Iñnaki Inza
Christian Jacob
Thomas Jansen
Segovia Javier
Yaochu Jin
Bryan Jones
Natasha Jonoska
Hugues Juille
Bryant Julstrom
Mahmoud Kaboudan
Charles Karr
Balakrishnan Karthik
Sanza Kazadi
Maarten Keijzer
Graham Kendall
Didier Keymeulen
Michael Kirley
Joshua Knowles
Gabriella Kokai
Arthur Kordon
Bogdan Korel
Erkan Korkmaz
Tim Kovacs
Natalio Krasnogor

Kalmanje Krishnakumar
Renato Krohling
Sam Kwong
Gary Lamont
William Langdon
Pedro Larrañnaga
Jesper Larse
Marco Laumanns
Paul Layzell
Martin Lefley
Claude Le Pape
Kwong Sak Leung
Warren Liao
Derek Linden
Michael Littman
Xavier Llora
Fernando Lobo
Jason Lohn
Michael Lones
Sushil Louis
Manuel Lozano
Jose Antonio Lozano
Jose Lozano
Pier Luca Lanzi Sean Luke
John Lusth
Evelyne Lutton
Nicholas Macias
Ana Madureira
Spiros Mancoridis
Martin Martin
Pete Martin
Arita Masanori
Iwata Masaya
Keith Mathias
Dirk Mattfeld
Giancarlo Mauri
David Mayer
Jon McCormack
Robert McKay
Nicholas McPhee
Lisa Meeden
Jörn Mehnen
Karlheinz Meier
Ole Mengshoel
Mark Meysenburg
Zbigniew Michalewicz

Martin Middendorf
Risto Miikkulainen
Julian Miller
Brian Mitchell
Chilukuri Mohan
David Montana
Byung-Ro Moon
Frank Moore
Alberto Moraglio
Manuel Moreno
Yunjun Mu
Sibylle Mueller
Masaharu Munetomo
Kazuyuki Murase
William Mydlowec
Zensho Nakao
Tomoharu Nakashima
Olfa Nasraoui
Bart Naudts
Mark Neal
Chrystopher Nehaniv
David Newth
Miguel Nicolau
Nikolay Nikolaev
Fernando Nino
Stefano Nolfi
Peter Nordin
Bryan Norman
Wim Nuijten
Leandro Nunes De Castro
Gabriela Ochoa
Victor Oduguwa
Charles Ofria
Gustavo Olague
Markus Olhofer
Michael O'Neill
Una-May O'Reilly
Franz Oppacher
Jim Ouimette
Charles Palmer
Liviu Panait
Gary Parker
Anil Patel
Witold Pedrycz
Martin Pelikan
Marek Perkowski

Sanja Petrovic
Hartmut Pohlheim
Riccardo Poli
Tom Portegys
Reid Porter
Marie-Claude Portmann
Mitchell A. Potter
Walter Potter
Jean-Yves Potvin
Dilip Pratihar
Alexander Pretschner
Adam Prügel-Bennett
William Punch
Günther Raidl
Khaled Rasheed
Tom Ray
Tapabrata Ray
Victor Rayward-Smith
Patrick Reed
John Reif
Andreas Reinholz
Rick Riolo
Jose Riquelme
Denis Robilliard
Katya Rodriguez-Vazquez
Marc Roper
Brian Ross
Franz Rothlauf
Jon Rowe
Rajkumar Roy
Günter Rudolph
Thomas Runarsson
Conor Ryan
Bart Rylander
Kazuhiro Saitou
Ralf Salomon
Eugene Santos
Kumara Sastry
Yuji Sato
David Schaffer
Martin Schmidt
Thorsten Schnier
Marc Schoenauer
Sonia Schulenburg
Alan C. Schultz

Sandip Sen
Bernhard Sendhoff
Kisung Seo
Franciszek Seredynski
Jane Shaw
Martin Shepperd
Alaa Sheta
Robert Shipman
Olivier Sigaud
Anabela Simões
Mark Sinclair
Abhishek Singh
Andre Skusa
Jim Smith
Robert Smith
Donald Sofge
Alan Soper
Terence Soule
Lee Spector
Andreas Spillner
Russell Standish
Harmen Sthamer
Adrian Stoica
Wolfgang Stolzmann
Matthew Streeter
V. Sundararajan
Gil Syswerda
Walter Tackett
Keiki Takadama
Uwe Tangen
Alexander Tarakanov
Ernesto Tarantino
Gianluca Tempesti
Hugo Terashima-Marin
Sam Thangiah
Scott Thayer
Lothar Thiele
Dirk Thierens
Adrian Thompson
Jonathan Thompson
Jonathan Timmis
Ashutosh Tiwari
Marco Tomassini
Andy Tomlinson
Jim Torresen

Paolo Toth
Michael Trick
Shigeyoshi Tsutsui
Andy Tyrrell
Jano Van Hemert
Clarissa Van Hoyweghen
Leonardo Vanneschi
David Van Veldhuizen
Robert Vanyi
Manuel Vazquez-
 Outomuro
Oswaldo Vélez-Langs
Hans-Michael Voigt
Roger Wainwright
Matthew Wall
Jean-Paul Watson
Ingo Wegener
Joachim Wegener
Karsten Weicker
Peter Whigham
Ronald While
Darrell Whitley
R. Paul Wiegand
Kay Wiese
Dirk Wiesmann
Janet Wile
Janet Wiles
Wendy Williams
Stewart Wilson
Mark Wineberg
Alden Wright
Annie Wu
Zheng Wu
Chia-Hsuan Yeh
Ayse Yilmaz
Tian-Li Yu
Tina Yu
Hongnian Yu
Ricardo Zebulum
Andreas Zell
Byoung-Tak Zhang
Lyudmila A. Zinchenko

A Word from the Chair of ISGEC

You may have just picked up your proceedings, in hard copy and CD-ROM, at GECCO 2003, or purchased it after the conference. You've doubtless already noticed the new format – publishing our proceedings as part of Springer's Lecture Notes in Computer Science (LNCS) series will make them available in many more libraries, broadening the impact of the GECCO conference dramatically!

If you attended GECCO 2003, we, the organizers, hope your experience was memorable and productive, and you have found these proceedings to be of continuing value. The opportunity for first-hand interaction among authors and other participants in GECCO is a big part of what makes it exciting, and we all hope you came away with many new insights and ideas.

If you were unable to come to GECCO 2003 in person, I hope you'll find many stimulating ideas from the world's leading researchers in evolutionary computation reported in these proceedings, and that you'll be able to participate in a future GECCO – for example, next year, in Seattle!

The International Society for Genetic and Evolutionary Computation, the sponsoring organization of the annual GECCO conferences, is a young organization, formed through the merger of the International Society for Genetic Algorithms (sponsor of the ICGA conferences) and the organization responsible for the annual Genetic Programming conferences. It depends strongly on the voluntary efforts of many of its members. It is designed to promote not only the exchange of ideas among innovators and practitioners of well-known methods such as genetic algorithms, genetic programming, evolution strategies, evolutionary programming, learning classifier systems, etc., but also the growth of newer areas such as artificial immune systems, evolvable hardware, agentbased search, and others. One of the founding principles is that ISGEC operates as a confederation of groups with related but distinct approaches and interests, and their mutual prosperity is assured by their representation in the program committees, editorial boards, etc., of the conferences and journals with which ISGEC is associated. This also insures that ISGEC and its functions continue to improve and evolve with the diversity of innovation that has characterized our field.

ISGEC has seen many changes this year, in addition to its growth in membership. We have completed the formalities for recognition as a tax-exempt charitable organization. We have created the new designations of Fellow and Senior Fellow of ISGEC to recognize the achievements of leaders in the field, and by the time you read this, we expect to have elected the first cohort. Additional Fellows and Senior Fellows will be added annually. GECCO continues to be subject to dynamic development – the many new tutorials, workshop topics, and tracks will evolve again next year, seeking to follow and encourage the developments of the many fields represented at GECCO. The best paper awards were presented for the second time at GECCO 2003, and we hope many of you participated in the balloting. This year, for the first time, most presentations at GECCO were electronic, displayed with the LCD projectors that ISGEC has recently purchased. Our journals, *Evolutionary Computation* and *Genetic Programming and Evolvable Machines*, continue to prosper, and we are exploring ways to make them even more widely available. The inclusion of the proceedings in Springer's Lecture Notes in Computer Science series, making them available in many more libraries worldwide, should have a strong positive impact on our field.

ISGEC is your society, and we urge you to become involved or continue your involvement in its activities, to the mutual benefit of the whole evolutionary computation community. Three members were elected to new five-year terms on the Executive Board at GECCO 2002 – Wolfgang Banzhaf, Marco Dorigo, and Annie Wu.

Since that time, ISGEC has been active on many issues, through actions of the Board and the three Councils – the Council of Authors, Council of Editors, and Council of Conferences.

The organizers of GECCO 2003 are listed in this frontmatter, but special thanks are due to James Foster, General Chair, and Erick Cantú-Paz, Editor-in-Chief of the Proceedings, as well as to John Koza and Dave Goldberg, the Business Committee. All of the changes this year, particularly in the publication of the proceedings, have meant a lot of additional work for this excellent team, and we owe them our thanks for a job well done.

Of course, we all owe a great debt to those who chaired or served on the various core and special program committees that reviewed all of the papers for GECCO 2003. Without their effort it would not have been possible to put on a meeting of this quality.

Another group also deserves the thanks of GECCO participants and ISGEC members – the members of the ISGEC Executive Board and Councils, who are listed below. I am particularly indebted to them for their thoughtful contributions to the organization and their continuing demonstrations of concern for the welfare of ISGEC.

I invite you to communicate with me (goodman@egr.msu.edu) if you have questions or suggestions for ways ISGEC can be of greater service to its members, or if you would like to get more involved in ISGEC and its functions.

Don't forget about the 8th Foundations of Genetic Algorithms (FOGA) workshop, also sponsored by ISGEC, the biennial event that brings together the world's leading theorists on evolutionary computation, which will be held in 2004.

Finally, I hope you will join us at GECCO 2004 in Seattle. Get your ideas to Ricardo Poli, the General Chair of GECCO 2004, when you see him at GECCO 2003, and please check the ISGEC Website, www.isgec.org, regularly for details as the planning for GECCO 2004 continues.

Erik D. Goodman

ISGEC Executive Board

Erik D. Goodman (Chair)
David Andre
Wolfgang Banzhaf
Kalyanmoy Deb
Kenneth De Jong
Marco Dorigo
David E. Goldberg
John H. Holland
John R. Koza
Una-May O.'Reilly
Ingo Rechenberg
Marc Schoenauer
Lee Spector
Darrell Whitley
Annie S. Wu

Council of Authors

Erick Cantú-Paz (chair), Lawrence Livermore National Laboratory
David Andre, University of California – Berkeley
Plamen P. Angelov, Loughborough University
Vladan Babovic, Danish Hydraulic Institute
Wolfgang Banzhaf, University of Dortmund
Forrest H. Bennett III, FX Palo Alto Laboratory, Inc.
Hans-Georg Beyer, University of Dortmund
Jergen Branke, University of Karlsruhe
Martin Butz, University of Illinois at Urbana-Champaign
Runwei Cheng, Ashikaga Institute of Technology
David A. Coley, University of Exeter
Marco Dorigo, IRIDIA, Université Libre de Bruxelles
Rolf Drechsler, University of Freiburg
Emanuel Falkenauer, Optimal Design and Brussels University (ULB)
Stephanie Forrest, University of New Mexico
Mitsuo Gen, Ashikaga Institute of Technology
Andreas Geyer-Schulz, Abteilung fuer Informationswirtschaft
David E. Goldberg, University of Illinois at Urbana-Champaign
Jens Gottlieb, SAP, AG
Wolfgang A. Halang, Fernuniversitaet
John H. Holland, University of Michigan and Sante Fe Institute
Hitoshi Iba, University of Tokyo
Christian Jacob, University of Calgary
Robert E. Keller, University of Dortmund
Dimitri Knjazew, SAP, AG

John R. Koza, Stanford University
Sam Kwong, City University of Hong Kong
William B. Langdon, University College, London
Dirk C. Mattfeld, University of Bremen
Pinaki Mazumder, University of Michigan
Zbigniew Michalewicz, University of North Carolina at Charlotte
Melanie Mitchell, Oregon Health and Science University
Ian Parmee, University of North Carolina at Charlotte
Frederick E. Petry, University of North Carolina at Charlotte
Riccardo Poli, University of Essex
Moshe Sipper, Swiss Federal Institute of Technology
William M. Spears, University of Wyoming
Wallace K.S. Tang, Swiss Federal Institute of Technology
Adrian Thompson, University of Sussex
Michael D. Vose, University of Tennessee
Man Leung Wong, Lingnan University

Council of Editors

Erick Cantú-Paz (chair), Lawrence Livermore National Laboratory
Karthik Balakrishnan, Fireman's Fund Insurance Company
Wolfgang Banzhaf, University of Dortmund
Peter Bentley, University College, London
Lance D. Chambers, Western Australian Department of Transport
Dipankar Dasgupta, University of Memphis
Kenneth De Jong, George Mason University
Francisco Herrera, University of Granada
William B. Langdon, University College, London
Pinaki Mazumder, University of Michigan
Eric Michielssen, University of Illinois at Urbana-Champaign
Witold Pedrycz, University of Alberta
Rajkumar Roy, Cranfield University
Elizabeth M. Rudnick, University of Illinois at Urbana-Champaign
Marc Schoenauer, INRIA Rocquencourt
Lee Spector, Hampshire College
Jose L. Verdegay, University of Granada, Spain

Council of Conferences, Riccardo Poli (Chair)

The purpose of the Council of Conferences is to provide information about the numerous conferences that are available to researchers in the field of Genetic and Evolutionary Computation, and to encourage them to coordinate their meetings to maximize our collective impact on science.

ACDM, Adaptive Computing in Design and Manufacture, 2004, Ian Parmee (Ian.Parmee@uwe.ac.uk)

EuroGP, European Conference on Genetic Programming, Portugal, April 2004, Ernesto Costa (ernesto@dei.uc.pt)

EvoWorkshops, European Evolutionary Computing Workshops, Portugal, April 2004, Stefano Cagnoni (cagnoni@ce.unipr.it)

FOGA, Foundations of Genetic Algorithms Workshop, 2004

GECCO 2004, Genetic and Evolutionary Computation Conference, Seattle, June 2004, Riccardo Poli (rpoli@essex.ac.uk)

INTROS, INtroductory TutoRials in Optimization, Search and Decision Support Methodologies, August 12, 2003, Nottingham, UK, Edmund Burke (ekb@cs.nott.ac.uk)

MISTA, 1st Multidisciplinary International Conference on Scheduling: Theory and Applications August 8-12, 2003, Nottingham, UK, Graham Kendall (gxk@cs.nott.ac.uk)

PATAT 2004, 5th International Conference on the Practice and Theory of Automated Timetabling, Pittsburgh, USA, August 18–20, 2004, Edmund Burke (ekb@cs.nott.ac.uk)

WSC8, 8th Online World Conference on Soft Computing in Industrial Applications, September 29th - October 10th, 2003, Internet (hosted by University of Dortmund), Frank Hoffmann (hoffmann@esr.e-technik.uni-dortmund.de)

An up-to-date roster of the Council of Conferences is available online at http://www.isgec.org/conferences.html.

Please contact the COC chair Riccardo Poli (rpoli@essex.ac.uk) for additions to this list.

Papers Nominated for Best Paper Awards

In 2002, ISGEC created a best paper award for GECCO. As part of the double blind peer review, the reviewers were asked to nominate papers for best paper awards. The chairs of core and special program committees selected the papers that received the most nominations for consideration by the conference. One winner for each program track was chosen by secret ballot of the GECCO attendees after the papers were presented in Chicago. The titles and authors of the winning papers are available at the GECCO 2003 website (www.isgec.org/GECCO-2003).

Finite Population Models of Co-evolution and Their Application to Haploidy versus Diploidy, Anthony M.L. Liekens, Huub M.M. ten Eikelder, and Peter A.J. Hilbers

A Game-Theoretic Memory Mechanism for Coevolution, Sevan G. Ficici and Jordan B. Pollack

A Non-dominated Sorting Particle Swarm Optimizer for Multiobjective Optimization, Xiaodong Li

Emergence of Collective Behavior in Evolving Populations of Flying Agents, Lee Spector, Jon Klein, Chris Perry, and Mark Feinstein

Immune Inspired Somatic Contiguous Hypermutation for Function Optimisation, Johnny Kelsey and Jon Timmis

Efficiency and Reliability of DNA-Based Memories, Max H. Garzon, Andrew Neel, and Hui Chen

Hardware Evolution of Analog Speed Controllers for a DC Motor, D.A. Gwaltney and M.I. Ferguson

Integration of Genetic Programming and Reinforcement Learning for Real Robots, Shotaro Kamio, Hideyuki Mitshuhashi, and Hitoshi Iba

Co-evolving Task-Dependent Visual Morphologies in Predator-Prey Experiments, Gunnar Buason and Tom Ziemke

The Steady State Behavior of $(\mu/\mu_I, \lambda)$-ES on Ellipsoidal Fitness Models Disturbed by Noise, Hans-Georg Beyer and Dirk V. Arnold

On the Optimization of Monotone Polynomials by the (1+1) EA and Randomized Local Search, Ingo Wegener and Carsten Witt

Ruin and Recreate Principle Based Approach for the Quadratic Assignment Problem, Alfonsas Misevicius

Evolutionary Computing as a tool for Grammar Development, Guy De Pauw

Adaptive Elitist-Population Based Genetic Algorithm for Multimodal Function Optimization, Kwong-Sak Leung and Yong Liang

Scalability of Selectorecombinative Genetic Algorithms for Problems with Tight Linkage, Kumara Sastry and David E. Goldberg

Effective Use of Directional Information in Multi-objective Evolutionary Computation, Martin Brown and R.E. Smith

Are Multiple Runs of Genetic Algorithms Better Than One? Erick Cantú-Paz and David E. Goldberg

Selection in the Presence of Noise, Jürgen Branke and Christian Schmidt

Difficulty of Unimodal and Multimodal Landscapes in Genetic Programming, Leonardo Vanneschi, Marco Tomassini, Manuel Clergue, and Philippe Collard

Dynamic Maximum Tree Depth: a Simple Technique for Avoiding Bloat in Tree-Based GP, Sara Silva and Jonas Almeida

Generative Representations for Evolving Families of Designs, Gregory S. Hornby

Identifying Structural Mechanisms in Standard Genetic Programming, Jason M. Daida and Adam M. Hilss

Visualizing Tree Structures in Genetic Programming, Jason M. Daida, Adam M. Hilss, David J. Ward, and Stephen L. Long

Methods for Evolving Robust Programs, Liviu Panait and Sean Luke

Population Implosion in Genetic Programming, Sean Luke, Gabriel Catalin Balan, and Liviu Panait

Designing Efficient Exploration with MACS: Modules and Function Approximation, Pierre Gérard and Olivier Sigaud

Tournament Selection: Stable Fitness Pressure in XCS, Martin V. Butz, Kumara Sastry, and David E. Goldberg

Towards Building Block Propagation in XCS: a Negative Result and Its Implications, Kurian K. Tharakunnel, Martin V. Butz, and David E. Goldberg

Quantum-Inspired Evolutionary Algorithm-Based Face Verification, Jun-Su Jang, Kuk-Hyun Han, and Jong-Hwan Kim

Mining Comprehensive Clustering Rules with an Evolutionary Algorithm, Ioannis Sarafis, Phil Trinder and Ali Zalzala

System-Level Synthesis of MEMS via Genetic Programming and Bond Graphs, Zhun Fan, Kisung Seo, Jianjun Hu, Ronald C. Rosenberg, and Erik D. Goodman

Active Guidance for a Finless Rocket Using Neuroevolution, Faustino J. Gomez and Risto Miikkulainen

Extracting Test Sequences from a Markov Software Usage Model by ACO, Karl Doerner and Walter J. Gutjahr

Modeling the Search Landscape of Metaheuristic Software Clustering Algorithms, Brian S. Mitchell and Spiros Mancoridis

Table of Contents

Volume II

Genetic Algorithms (continued)

Design of Multithreaded Estimation of Distribution Algorithms 1247
   Jiri Ocenasek, Josef Schwarz, Martin Pelikan

Reinforcement Learning Estimation of Distribution Algorithm 1259
   Topon Kumar Paul, Hitoshi Iba

Hierarchical BOA Solves Ising Spin Glasses and MAXSAT 1271
   Martin Pelikan, David E. Goldberg

ERA: An Algorithm for Reducing the Epistasis of SAT Problems 1283
   Eduardo Rodriguez-Tello, Jose Torres-Jimenez

Learning a Procedure That Can Solve Hard Bin-Packing Problems:
A New GA-Based Approach to Hyper-heuristics 1295
   Peter Ross, Javier G. Marín-Blázquez, Sonia Schulenburg, Emma Hart

Population Sizing for the Redundant Trivial Voting Mapping 1307
   Franz Rothlauf

Non-stationary Function Optimization Using Polygenic Inheritance 1320
   Conor Ryan, J.J. Collins, David Wallin

Scalability of Selectorecombinative Genetic Algorithms for
Problems with Tight Linkage 1332
   Kumara Sastry, David E. Goldberg

New Entropy-Based Measures of Gene Significance and Epistasis 1345
   Dong-Il Seo, Yong-Hyuk Kim, Byung-Ro Moon

A Survey on Chromosomal Structures and Operators for Exploiting
Topological Linkages of Genes 1357
   Dong-Il Seo, Byung-Ro Moon

Cellular Programming and Symmetric Key Cryptography Systems 1369
   Franciszek Seredyński, Pascal Bouvry, Albert Y. Zomaya

Mating Restriction and Niching Pressure: Results from Agents and
Implications for General EC 1382
   R.E. Smith, Claudio Bonacina

EC Theory: A Unified Viewpoint 1394
    Christopher R. Stephens, Adolfo Zamora

Real Royal Road Functions for Constant Population Size 1406
    Tobias Storch, Ingo Wegener

Two Broad Classes of Functions for Which a No Free Lunch Result
Does Not Hold 1418
    Matthew J. Streeter

Dimensionality Reduction via Genetic Value Clustering 1431
    Alexander Topchy, William Punch III

The Structure of Evolutionary Exploration: On Crossover,
Buildings Blocks, and Estimation-of-Distribution Algorithms 1444
    Marc Toussaint

The Virtual Gene Genetic Algorithm 1457
    Manuel Valenzuela-Rendón

Quad Search and Hybrid Genetic Algorithms 1469
    Darrell Whitley, Deon Garrett, Jean-Paul Watson

Distance between Populations 1481
    Mark Wineberg, Franz Oppacher

The Underlying Similarity of Diversity Measures Used in
Evolutionary Computation 1493
    Mark Wineberg, Franz Oppacher

Implicit Parallelism 1505
    Alden H. Wright, Michael D. Vose, Jonathan E. Rowe

Finding Building Blocks through Eigenstructure Adaptation 1518
    Danica Wyatt, Hod Lipson

A Specialized Island Model and Its Application in
Multiobjective Optimization 1530
    Ningchuan Xiao, Marc P. Armstrong

Adaptation of Length in a Nonstationary Environment 1541
    Han Yu, Annie S. Wu, Kuo-Chi Lin, Guy Schiavone

Optimal Sampling and Speed-Up for Genetic Algorithms on the
Sampled OneMax Problem 1554
    Tian-Li Yu, David E. Goldberg, Kumara Sastry

Building-Block Identification by Simultaneity Matrix 1566
    Chatchawit Aporntewan, Prabhas Chongstitvatana

A Unified Framework for Metaheuristics 1568
  Jürgen Branke, Michael Stein, Hartmut Schmeck

The Hitting Set Problem and Evolutionary Algorithmic Techniques
with ad-hoc Viruses (HEAT-V) 1570
  Vincenzo Cutello, Francesco Pappalardo

The Spatially-Dispersed Genetic Algorithm 1572
  Grant Dick

Non-universal Suffrage Selection Operators Favor Population
Diversity in Genetic Algorithms 1574
  Federico Divina, Maarten Keijzer, Elena Marchiori

Uniform Crossover Revisited: Maximum Disruption in
Real-Coded GAs 1576
  Stephen Drake

The Master-Slave Architecture for Evolutionary Computations
Revisited 1578
  Christian Gagné, Marc Parizeau, Marc Dubreuil

Genetic Algorithms – Posters

Using Adaptive Operators in Genetic Search 1580
  Jonatan Gómez, Dipankar Dasgupta, Fabio González

A Kernighan-Lin Local Improvement Heuristic That Solves Some Hard
Problems in Genetic Algorithms 1582
  William A. Greene

GA-Hardness Revisited 1584
  Haipeng Guo, William H. Hsu

Barrier Trees For Search Analysis 1586
  Jonathan Hallam, Adam Prügel-Bennett

A Genetic Algorithm as a Learning Method Based on
Geometric Representations 1588
  Gregory A. Holifield, Annie S. Wu

Solving Mastermind Using Genetic Algorithms 1590
  Tom Kalisker, Doug Camens

Evolutionary Multimodal Optimization Revisited 1592
  Rajeev Kumar, Peter Rockett

Integrated Genetic Algorithm with Hill Climbing for Bandwidth
Minimization Problem 1594
  Andrew Lim, Brian Rodrigues, Fei Xiao

A Fixed-Length Subset Genetic Algorithm for the p-Median Problem 1596
    Andrew Lim, Zhou Xu

Performance Evaluation of a Parameter-Free Genetic Algorithm for
Job-Shop Scheduling Problems 1598
    Shouichi Matsui, Isamu Watanabe, Ken-ichi Tokoro

SEPA: Structure Evolution and Parameter Adaptation in
Feed-Forward Neural Networks 1600
    Paulito P. Palmes, Taichi Hayasaka, Shiro Usui

Real-Coded Genetic Algorithm to Reveal Biological Significant
Sites of Remotely Homologous Proteins 1602
    Sung-Joon Park, Masayuki Yamamura

Understanding EA Dynamics via Population Fitness Distributions 1604
    Elena Popovici, Kenneth De Jong

Evolutionary Feature Space Transformation Using Type-Restricted
Generators 1606
    Oliver Ritthoff, Ralf Klinkenberg

On the Locality of Representations 1608
    Franz Rothlauf

New Subtour-Based Crossover Operator for the TSP 1610
    Sang-Moon Soak, Byung-Ha Ahn

Is a Self-Adaptive Pareto Approach Beneficial for Controlling
Embodied Virtual Robots? 1612
    Jason Teo, Hussein A. Abbass

A Genetic Algorithm for Energy Efficient Device Scheduling in
Real-Time Systems 1614
    Lirong Tian, Tughrul Arslan

Metropolitan Area Network Design Using GA Based on Hierarchical
Linkage Identification 1616
    Miwako Tsuji, Masaharu Munetomo, Kiyoshi Akama

Statistics-Based Adaptive Non-uniform Mutation for Genetic
Algorithms 1618
    Shengxiang Yang

Genetic Algorithm Design Inspired by Organizational Theory:
Pilot Study of a Dependency Structure Matrix Driven
Genetic Algorithm 1620
    Tian-Li Yu, David E. Goldberg, Ali Yassine, Ying-Ping Chen

Are the "Best" Solutions to a Real Optimization Problem Always
Found in the Noninferior Set? Evolutionary Algorithm for Generating
Alternatives (EAGA) 1622
   Emily M. Zechman, S. Ranji Ranjithan

Population Sizing Based on Landscape Feature 1624
   Jian Zhang, Xiaohui Yuan, Bill P. Buckles

Genetic Programming

Structural Emergence with Order Independent Representations 1626
   R. Muhammad Atif Azad, Conor Ryan

Identifying Structural Mechanisms in Standard Genetic Programming ... 1639
   Jason M. Daida, Adam M. Hilss

Visualizing Tree Structures in Genetic Programming 1652
   Jason M. Daida, Adam M. Hilss, David J. Ward, Stephen L. Long

What Makes a Problem GP-Hard? Validating a Hypothesis of
Structural Causes 1665
   Jason M. Daida, Hsiaolei Li, Ricky Tang, Adam M. Hilss

Generative Representations for Evolving Families of Designs 1678
   Gregory S. Hornby

Evolutionary Computation Method for Promoter Site Prediction
in DNA 1690
   Daniel Howard, Karl Benson

Convergence of Program Fitness Landscapes 1702
   W.B. Langdon

Multi-agent Learning of Heterogeneous Robots by
Evolutionary Subsumption 1715
   Hongwei Liu, Hitoshi Iba

Population Implosion in Genetic Programming 1729
   Sean Luke, Gabriel Catalin Balan, Liviu Panait

Methods for Evolving Robust Programs 1740
   Liviu Panait, Sean Luke

On the Avoidance of Fruitless Wraps in Grammatical Evolution 1752
   Conor Ryan, Maarten Keijzer, Miguel Nicolau

Dense and Switched Modular Primitives for Bond Graph Model Design .. 1764
   *Kisung Seo, Zhun Fan, Jianjun Hu, Erik D. Goodman,
   Ronald C. Rosenberg*

Dynamic Maximum Tree Depth 1776
  Sara Silva, Jonas Almeida

Difficulty of Unimodal and Multimodal Landscapes in
Genetic Programming 1788
  Leonardo Vanneschi, Marco Tomassini, Manuel Clergue,
  Philippe Collard

Genetic Programming – Posters

Ramped Half-n-Half Initialisation Bias in GP 1800
  Edmund Burke, Steven Gustafson, Graham Kendall

Improving Evolvability of Genetic Parallel Programming Using
Dynamic Sample Weighting 1802
  Sin Man Cheang, Kin Hong Lee, Kwong Sak Leung

Enhancing the Performance of GP Using an Ancestry-Based Mate
Selection Scheme 1804
  Rodney Fry, Andy Tyrrell

A General Approach to Automatic Programming Using Occam's Razor,
Compression, and Self-Inspection 1806
  Peter Galos, Peter Nordin, Joel Olsén, Kristofer Sundén Ringnér

Building Decision Tree Software Quality Classification Models
Using Genetic Programming 1808
  Yi Liu, Taghi M. Khoshgoftaar

Evolving Petri Nets with a Genetic Algorithm 1810
  Holger Mauch

Diversity in Multipopulation Genetic Programming 1812
  Marco Tomassini, Leonardo Vanneschi, Francisco Fernández,
  Germán Galeano

An Encoding Scheme for Generating λ-Expressions in
Genetic Programming 1814
  Kazuto Tominaga, Tomoya Suzuki, Kazuhiro Oka

AVICE: Evolving Avatar's Movernent 1816
  Hiromi Wakaki, Hitoshi Iba

Learning Classifier Systems

Evolving Multiple Discretizations with Adaptive Intervals for a
Pittsburgh Rule-Based Learning Classifier System 1818
  Jaume Bacardit, Josep Maria Garrell

Limits in Long Path Learning with XCS 1832
  Alwyn Barry

Bounding the Population Size in XCS to Ensure
Reproductive Opportunities 1844
  Martin V. Butz, David E. Goldberg

Tournament Selection: Stable Fitness Pressure in XCS 1857
  Martin V. Butz, Kumara Sastry, David E. Goldberg

Improving Performance in Size-Constrained Extended
Classifier Systems 1870
  Devon Dawson

Designing Efficient Exploration with MACS: Modules and Function
Approximation 1882
  Pierre Gérard, Olivier Sigaud

Estimating Classifier Generalization and Action's Effect:
A Minimalist Approach 1894
  Pier Luca Lanzi

Towards Building Block Propagation in XCS: A Negative Result and
Its Implications 1906
  Kurian K. Tharakunnel, Martin V. Butz, David E. Goldberg

Learning Classifier Systems – Posters

Data Classification Using Genetic Parallel Programming 1918
  Sin Man Cheang, Kin Hong Lee, Kwong Sak Leung

Dynamic Strategies in a Real-Time Strategy Game 1920
  William Joseph Falke II, Peter Ross

Using Raw Accuracy to Estimate Classifier Fitness in XCS 1922
  Pier Luca Lanzi

Towards Learning Classifier Systems for Continuous-Valued
Online Environments 1924
  Christopher Stone, Larry Bull

Real World Applications

Artificial Immune System for Classification of Gene Expression Data 1926
  Shin Ando, Hitoshi Iba

Automatic Design Synthesis and Optimization of Component-Based
Systems by Evolutionary Algorithms 1938
  P.P. Angelov, Y. Zhang, J.A. Wright, V.I. Hanby, R.A. Buswell

Studying the Advantages of a Messy Evolutionary Algorithm for
Natural Language Tagging 1951
    Lourdes Araujo

Optimal Elevator Group Control by Evolution Strategies 1963
    Thomas Beielstein, Claus-Peter Ewald, Sandor Markon

A Methodology for Combining Symbolic Regression and Design of
Experiments to Improve Empirical Model Building 1975
    Flor Castillo, Kenric Marshall, James Green, Arthur Kordon

The General Yard Allocation Problem 1986
    Ping Chen, Zhaohui Fu, Andrew Lim, Brian Rodrigues

Connection Network and Optimization of Interest Metric for
One-to-One Marketing 1998
    Sung-Soon Choi, Byung-Ro Moon

Parameter Optimization by a Genetic Algorithm for a Pitch
Tracking System 2010
    Yoon-Seok Choi, Byung-Ro Moon

Secret Agents Leave Big Footprints: How to Plant a Cryptographic
Trapdoor, and Why You Might Not Get Away with It 2022
    John A. Clark, Jeremy L. Jacob, Susan Stepney

GenTree: An Interactive Genetic Algorithms System for Designing
3D Polygonal Tree Models 2034
    Clare Bates Congdon, Raymond H. Mazza

Optimisation of Reaction Mechanisms for Aviation Fuels Using a
Multi-objective Genetic Algorithm 2046
    Lionel Elliott, Derek B. Ingham, Adrian G. Kyne, Nicolae S. Mera,
    Mohamed Pourkashanian, Christopher W. Wilson

System-Level Synthesis of MEMS via Genetic Programming and
Bond Graphs 2058
    Zhun Fan, Kisung Seo, Jianjun Hu, Ronald C. Rosenberg,
    Erik D. Goodman

Congressional Districting Using a TSP-Based Genetic Algorithm 2072
    Sean L. Forman, Yading Yue

Active Guidance for a Finless Rocket Using Neuroevolution 2084
    Faustino J. Gomez, Risto Miikkulainen

Simultaneous Assembly Planning and Assembly System Design Using
Multi-objective Genetic Algorithms 2096
    Karim Hamza, Juan F. Reyes-Luna, Kazuhiro Saitou

Multi-FPGA Systems Synthesis by Means of
Evolutionary Computation 2109
  J.I. Hidalgo, F. Fernández, J. Lanchares, J.M. Sánchez, R. Hermida,
  M. Tomassini, R. Baraglia, R. Perego, O. Garnica

Genetic Algorithm Optimized Feature Transformation –
A Comparison with Different Classifiers............................. 2121
  Zhijian Huang, Min Pei, Erik Goodman, Yong Huang, Gaoping Li

Web-Page Color Modification for Barrier-Free Color Vision with
Genetic Algorithm 2134
  Manabu Ichikawa, Kiyoshi Tanaka, Shoji Kondo, Koji Hiroshima,
  Kazuo Ichikawa, Shoko Tanabe, Kiichiro Fukami

Quantum-Inspired Evolutionary Algorithm-Based Face Verification 2147
  Jun-Su Jang, Kuk-Hyun Han, Jong-Hwan Kim

Minimization of Sonic Boom on Supersonic Aircraft Using an
Evolutionary Algorithm... 2157
  Charles L. Karr, Rodney Bowersox, Vishnu Singh

Optimizing the Order of Taxon Addition in Phylogenetic Tree
Construction Using Genetic Algorithm.............................. 2168
  Yong-Hyuk Kim, Seung-Kyu Lee, Byung-Ro Moon

Multicriteria Network Design Using Evolutionary Algorithm 2179
  Rajeev Kumar, Nilanjan Banerjee

Control of a Flexible Manipulator Using a Sliding Mode Controller
with Genetic Algorithm Tuned Manipulator Dimension 2191
  N.M. Kwok, S. Kwong

Daily Stock Prediction Using Neuro-genetic Hybrids 2203
  Yung-Keun Kwon, Byung-Ro Moon

Finding the Optimal Gene Order in Displaying Microarray Data 2215
  Seung-Kyu Lee, Yong-Hyuk Kim, Byung-Ro Moon

Learning Features for Object Recognition 2227
  Yingqiang Lin, Bir Bhanu

An Efficient Hybrid Genetic Algorithm for a Fixed Channel
Assignment Problem with Limited Bandwidth 2240
  Shouichi Matsui, Isamu Watanabe, Ken-ichi Tokoro

Using Genetic Algorithms for Data Mining Optimization in an
Educational Web-Based System.................................... 2252
  Behrouz Minaei-Bidgoli, William F. Punch III

Improved Image Halftoning Technique Using GAs with Concurrent
Inter-block Evaluation 2264
    Emi Myodo, Hernán Aguirre, Kiyoshi Tanaka

Complex Function Sets Improve Symbolic Discriminant Analysis of
Microarray Data 2277
    David M. Reif, Bill C. White, Nancy Olsen, Thomas Aune,
    Jason H. Moore

GA-Based Inference of Euler Angles for Single Particle Analysis 2288
    Shusuke Saeki, Kiyoshi Asai, Katsutoshi Takahashi, Yutaka Ueno,
    Katsunori Isono, Hitoshi Iba

Mining Comprehensible Clustering Rules with an
Evolutionary Algorithm 2301
    Ioannis Sarafis, Phil Trinder, Ali Zalzala

Evolving Consensus Sequence for Multiple Sequence Alignment with
a Genetic Algorithm 2313
    Conrad Shyu, James A. Foster

A Linear Genetic Programming Approach to Intrusion Detection 2325
    Dong Song, Malcolm I. Heywood, A. Nur Zincir-Heywood

Genetic Algorithm for Supply Planning Optimization under
Uncertain Demand 2337
    Tezuka Masaru, Hiji Masahiro

Genetic Algorithms: A Fundamental Component of an Optimization
Toolkit for Improved Engineering Designs 2347
    Siu Tong, David J. Powell

Spatial Operators for Evolving Dynamic Bayesian Networks from
Spatio-temporal Data 2360
    Allan Tucker, Xiaohui Liu, David Garway-Heath

An Evolutionary Approach for Molecular Docking 2372
    Jinn-Moon Yang

Evolving Sensor Suites for Enemy Radar Detection 2384
    Ayse S. Yilmaz, Brian N. McQuay, Han Yu, Annie S. Wu,
    John C. Sciortino, Jr.

Real World Applications – Posters

Optimization of Spare Capacity in Survivable WDM Networks 2396
    H.W. Chong, Sam Kwong

Partner Selection in Virtual Enterprises by Using Ant
Colony Optimization in Combination with the Analytical
Hierarchy Process 2398
   Marco Fischer, Hendrik Jähn, Tobias Teich

Quadrilateral Mesh Smoothing Using a Steady State
Genetic Algorithm 2400
   Mike Holder, Charles L. Karr

Evolutionary Algorithms for Two Problems from the Calculus of
Variations... 2402
   Bryant A. Julstrom

Genetic Algorithm Frequency Domain Optimization of an
Anti-Resonant Electromechanical Controller 2404
   Charles L. Karr, Douglas A. Scott

Genetic Algorithm Optimization of a Filament Winding Process 2406
   Charles L. Karr, Eric Wilson, Sherri Messimer

Circuit Bipartitioning Using Genetic Algorithm 2408
   Jong-Pil Kim, Byung-Ro Moon

Multi-campaign Assignment Problem and Optimizing
Lagrange Multipliers 2410
   Yong-Hyuk Kim, Byung-Ro Moon

Grammatical Evolution for the Discovery of Petri Net Models of
Complex Genetic Systems.. 2412
   Jason H. Moore, Lance W. Hahn

Evaluation of Parameter Sensitivity for Portable Embedded Systems
through Evolutionary Techniques 2414
   James Northern III, Michael Shanblatt

An Evolutionary Algorithm for the Joint Replenishment of
Inventory with Interdependent Ordering Costs 2416
   Anne Olsen

Benefits of Implicit Redundant Genetic Algorithms for Structural
Damage Detection in Noisy Environments............................ 2418
   Anne Raich, Tamás Liszkai

Multi-objective Traffic Signal Timing Optimization Using
Non-dominated Sorting Genetic Algorithm II 2420
   Dazhi Sun, Rahim F. Benekohal, S. Travis Waller

Exploration of a Two Sided Rendezvous Search Problem Using
Genetic Algorithms 2422
    T.Q.S. Truong, A. Stacey

Taming a Flood with a T-CUP – Designing Flood-Control Structures
with a Genetic Algorithm 2424
    Jeff Wallace, Sushil J. Louis

Assignment Copy Detection Using Neuro-genetic Hybrids 2426
    Seung-Jin Yang, Yong-Geon Kim, Yung-Keun Kwon, Byung-Ro Moon

Search Based Software Engineering

Structural and Functional Sequence Test of Dynamic and
State-Based Software with Evolutionary Algorithms 2428
    André Baresel, Hartmut Pohlheim, Sadegh Sadeghipour

Evolutionary Testing of Flag Conditions 2442
    Andr#e Baresel, Harmen Sthamer

Predicate Expression Cost Functions to Guide Evolutionary Search
for Test Data 2455
    Leonardo Bottaci

Extracting Test Sequences from a Markov Software Usage Model
by ACO 2465
    Karl Doerner, Walter J. Gutjahr

Using Genetic Programming to Improve Software Effort Estimation
Based on General Data Sets 2477
    Martin Lefley, Martin J. Shepperd

The State Problem for Evolutionary Testing 2488
    Phil McMinn, Mike Holcombe

Modeling the Search Landscape of Metaheuristic Software
Clustering Algorithms 2499
    Brian S. Mitchell, Spiros Mancoridis

Search Based Software Engineering – Posters

Search Based Transformations 2511
    Deji Fatiregun, Mark Harman, Robert Hierons

Finding Building Blocks for Software Clustering 2513
    Kiarash Mahdavi, Mark Harman, Robert Hierons

Author Index

Volume I

A-Life, Adaptive Behavior, Agents, and Ant Colony Optimization

Swarms in Dynamic Environments 1
   T.M. Blackwell

The Effect of Natural Selection on Phylogeny
Reconstruction Algorithms 13
   Dehua Hang, Charles Ofria, Thomas M. Schmidt, Eric Torng

AntClust: Ant Clustering and Web Usage Mining 25
   Nicolas Labroche, Nicolas Monmarché, Gilles Venturini

A Non-dominated Sorting Particle Swarm Optimizer for
Multiobjective Optimization 37
   Xiaodong Li

The Influence of Run-Time Limits on Choosing Ant
System Parameters 49
   Krzysztof Socha

Emergence of Collective Behavior in Evolving Populations of
Flying Agents 61
   Lee Spector, Jon Klein, Chris Perry, Mark Feinstein

On Role of Implicit Interaction and Explicit Communications in
Emergence of Social Behavior in Continuous Predators-Prey
Pursuit Problem 74
   Ivan Tanev, Katsunori Shimohara

Demonstrating the Evolution of Complex Genetic Representations:
An Evolution of Artificial Plants 86
   Marc Toussaint

Sexual Selection of Co-operation 98
   M. Afzal Upal

Optimization Using Particle Swarms with Near Neighbor Interactions ... 110
   *Kalyan Veeramachaneni, Thanmaya Peram, Chilukuri Mohan,
   Lisa Ann Osadciw*

Revisiting Elitism in Ant Colony Optimization 122
   Tony White, Simon Kaegi, Terri Oda

A New Approach to Improve Particle Swarm Optimization 134
   Liping Zhang, Huanjun Yu, Shangxu Hu

A-Life, Adaptive Behavior, Agents, and Ant Colony Optimization – Posters

Clustering and Dynamic Data Visualization with Artificial
Flying Insect 140
    S. Aupetit, N. Monmarché, M. Slimane, C. Guinot, G. Venturini

Ant Colony Programming for Approximation Problems 142
    Mariusz Boryczka, Zbigniew J. Czech, Wojciech Wieczorek

Long-Term Competition for Light in Plant Simulation 144
    Claude Lattaud

Using Ants to Attack a Classical Cipher 146
    Matthew Russell, John A. Clark, Susan Stepney

Comparison of Genetic Algorithm and Particle Swarm Optimizer When
Evolving a Recurrent Neural Network 148
    Matthew Settles, Brandon Rodebaugh, Terence Soule

Adaptation and Ruggedness in an Evolvability Landscape 150
    Terry Van Belle, David H. Ackley

Study Diploid System by a Hamiltonian Cycle Problem Algorithm 152
    Dong Xianghui, Dai Ruwei

A Possible Mechanism of Repressing Cheating Mutants
in Myxobacteria 154
    Ying Xiao, Winfried Just

Tour Jeté, Pirouette: Dance Choreographing by Computers 156
    Tina Yu, Paul Johnson

Multiobjective Optimization Using Ideas from the Clonal
Selection Principle 158
    Nareli Cruz Cortés, Carlos A. Coello Coello

Artificial Immune Systems

A Hybrid Immune Algorithm with Information Gain for the Graph
Coloring Problem 171
    Vincenzo Cutello, Giuseppe Nicosia, Mario Pavone

MILA – Multilevel Immune Learning Algorithm 183
    Dipankar Dasgupta, Senhua Yu, Nivedita Sumi Majumdar

The Effect of Binary Matching Rules in Negative Selection 195
    Fabio González, Dipankar Dasgupta, Jonatan Gómez

Immune Inspired Somatic Contiguous Hypermutation for
Function Optimisation... 207
    Johnny Kelsey, Jon Timmis

A Scalable Artificial Immune System Model for Dynamic
Unsupervised Learning 219
    *Olfa Nasraoui, Fabio Gonzalez, Cesar Cardona, Carlos Rojas,
    Dipankar Dasgupta*

Developing an Immunity to Spam 231
    Terri Oda, Tony White

Artificial Immune Systems – Posters

A Novel Immune Anomaly Detection Technique Based on
Negative Selection 243
    F. Niño, D. Gómez, R. Vejar

Visualization of Topic Distribution Based on Immune Network Model ... 246
    Yasufumi Takama

Spatial Formal Immune Network 248
    Alexander O. Tarakanov

Coevolution

Focusing versus Intransitivity (Geometrical Aspects of Co-evolution) 250
    Anthony Bucci, Jordan B. Pollack

Representation Development from Pareto-Coevolution 262
    Edwin D. de Jong

Learning the Ideal Evaluation Function 274
    Edwin D. de Jong, Jordan B. Pollack

A Game-Theoretic Memory Mechanism for Coevolution................ 286
    Sevan G. Ficici, Jordan B. Pollack

The Paradox of the Plankton: Oscillations and Chaos in
Multispecies Evolution... 298
    Jeffrey Horn, James Cattron

Exploring the Explorative Advantage of the Cooperative
Coevolutionary (1+1) EA 310
    Thomas Jansen, R. Paul Wiegand

PalmPrints: A Novel Co-evolutionary Algorithm for Clustering
Finger Images 322
    Nawwaf Kharma, Ching Y. Suen, Pei F. Guo

Coevolution and Linear Genetic Programming for Visual Learning 332
  Krzysztof Krawiec, Bir Bhanu

Finite Population Models of Co-evolution and Their Application to
Haploidy versus Diploidy 344
  Anthony M.L. Liekens, Huub M.M. ten Eikelder, Peter A.J. Hilbers

Evolving Keepaway Soccer Players through Task Decomposition 356
  Shimon Whiteson, Nate Kohl, Risto Miikkulainen, Peter Stone

Coevolution – Posters

A New Method of Multilayer Perceptron Encoding 369
  Emmanuel Blindauer, Jerzy Korczak

An Incremental and Non-generational Coevolutionary Algorithm 371
  Ramón Alfonso Palacios-Durazo, Manuel Valenzuela-Rendón

Coevolutionary Convergence to Global Optima 373
  Lothar M. Schmitt

Generalized Extremal Optimization for Solving Complex Optimal
Design Problems 375
  Fabiano Luis de Sousa, Valeri Vlassov, Fernando Manuel Ramos

Coevolving Communication and Cooperation for Lattice
Formation Tasks 377
  *Jekanthan Thangavelautham, Timothy D. Barfoot,
  Gabriele M.T. D'Eleuterio*

DNA, Molecular, and Quantum Computing

Efficiency and Reliability of DNA-Based Memories 379
  Max H. Garzon, Andrew Neel, Hui Chen

Evolving Hogg's Quantum Algorithm Using Linear-Tree GP 390
  André Leier, Wolfgang Banzhaf

Hybrid Networks of Evolutionary Processors 401
  *Carlos Martín-Vide, Victor Mitrana, Mario J. Pérez-Jiménez,
  Fernando Sancho-Caparrini*

DNA-Like Genomes for Evolution *in silico* 413
  Michael West, Max H. Garzon, Derrel Blain

DNA, Molecular, and Quantum Computing – Posters

String Binding-Blocking Automata 425
  M. Sakthi Balan

On Setting the Parameters of QEA for Practical Applications: Some
Guidelines Based on Empirical Evidence 427
  Kuk-Hyun Han, Jong-Hwan Kim

Evolutionary Two-Dimensional DNA Sequence Alignment............. 429
  Edgar E. Vallejo, Fernando Ramos

Evolvable Hardware

Active Control of Thermoacoustic Instability in a Model Combustor
with Neuromorphic Evolvable Hardware 431
  John C. Gallagher, Saranyan Vigraham

Hardware Evolution of Analog Speed Controllers for a DC Motor 442
  David A. Gwaltney, Michael I. Ferguson

Evolvable Hardware – Posters

An Examination of Hypermutation and Random Immigrant Variants of
mrCGA for Dynamic Environments 454
  Gregory R. Kramer, John C. Gallagher

Inherent Fault Tolerance in Evolved Sorting Networks 456
  Rob Shepherd, James Foster

Evolutionary Robotics

Co-evolving Task-Dependent Visual Morphologies in Predator-Prey
Experiments.. 458
  Gunnar Buason, Tom Ziemke

Integration of Genetic Programming and Reinforcement Learning for
Real Robots... 470
  Shotaro Kamio, Hideyuki Mitsuhashi, Hitoshi Iba

Multi-objectivity as a Tool for Constructing Hierarchical Complexity 483
  Jason Teo, Minh Ha Nguyen, Hussein A. Abbass

Learning Biped Locomotion from First Principles on a Simulated
Humanoid Robot Using Linear Genetic Programming.................. 495
  Krister Wolff, Peter Nordin

Evolutionary Robotics – Posters

An Evolutionary Approach to Automatic Construction of
the Structure in Hierarchical Reinforcement Learning 507
  Stefan Elfwing, Eiji Uchibe, Kenji Doya

Fractional Order Dynamical Phenomena in a GA 510
  E.J. Solteiro Pires, J.A. Tenreiro Machado, P.B. de Moura Oliveira

Evolution Strategies/Evolutionary Programming

Dimension-Independent Convergence Rate for
Non-isotropic $(1, \lambda) - ES$ 512
  Anne Auger, Claude Le Bris, Marc Schoenauer

The Steady State Behavior of $(\mu/\mu_I, \lambda)$-ES on Ellipsoidal Fitness
Models Disturbed by Noise 525
  Hans-Georg Beyer, Dirk V. Arnold

Theoretical Analysis of Simple Evolution Strategies in Quickly
Changing Environments 537
  Jürgen Branke, Wei Wang

Evolutionary Computing as a Tool for Grammar Development 549
  Guy De Pauw

Solving Distributed Asymmetric Constraint Satisfaction Problems
Using an Evolutionary Society of Hill-Climbers 561
  Gerry Dozier

Use of Multiobjective Optimization Concepts to Handle Constraints
in Single-Objective Optimization 573
  *Arturo Hernández Aguirre, Salvador Botello Rionda,
  Carlos A. Coello Coello, Giovanni Lizárraga Lizárraga*

Evolution Strategies with Exclusion-Based Selection Operators
and a Fourier Series Auxiliary Function 585
  Kwong-Sak Leung, Yong Liang

Ruin and Recreate Principle Based Approach for the Quadratic
Assignment Problem 598
  Alfonsas Misevicius

Model-Assisted Steady-State Evolution Strategies 610
  Holger Ulmer, Felix Streichert, Andreas Zell

On the Optimization of Monotone Polynomials by the (1+1) EA and
Randomized Local Search 622
  Ingo Wegener, Carsten Witt

Evolution Strategies/Evolutionary Programming – Posters

A Forest Representation for Evolutionary Algorithms Applied to
Network Design 634
  A.C.B. Delbem, Andre de Carvalho

Solving Three-Objective Optimization Problems Using Evolutionary
Dynamic Weighted Aggregation: Results and Analysis 636
   Yaochu Jin, Tatsuya Okabe, Bernhard Sendhoff

The Principle of Maximum Entropy-Based Two-Phase Optimization of
Fuzzy Controller by Evolutionary Programming 638
   Chi-Ho Lee, Ming Yuchi, Hyun Myung, Jong-Hwan Kim

A Simple Evolution Strategy to Solve Constrained
Optimization Problems 640
   Efrén Mezura-Montes, Carlos A. Coello Coello

Effective Search of the Energy Landscape for Protein Folding 642
   Eugene Santos Jr., Keum Joo Kim, Eunice E. Santos

A Clustering Based Niching Method for Evolutionary Algorithms 644
   Felix Streichert, Gunnar Stein, Holger Ulmer, Andreas Zell

Evolutionary Scheduling Routing

A Hybrid Genetic Algorithm for the Capacitated Vehicle
Routing Problem 646
   Jean Berger, Mohamed Barkaoui

An Evolutionary Approach to Capacitated Resource Distribution by
a Multiple-agent Team... 657
   Mudassar Hussain, Bahram Kimiaghalam, Abdollah Homaifar,
   Albert Esterline, Bijan Sayyarodsari

A Hybrid Genetic Algorithm Based on Complete Graph
Representation for the Sequential Ordering Problem 669
   Dong-Il Seo, Byung-Ro Moon

An Optimization Solution for Packet Scheduling: A Pipeline-Based
Genetic Algorithm Accelerator....................................... 681
   Shiann-Tsong Sheu, Yue-Ru Chuang, Yu-Hung Chen, Eugene Lai

Evolutionary Scheduling Routing – Posters

Generation and Optimization of Train Timetables Using Coevolution 693
   Paavan Mistry, Raymond S.K. Kwan

Genetic Algorithms

Chromosome Reuse in Genetic Algorithms 695
   Adnan Acan, Yüce Tekol

Real-Parameter Genetic Algorithms for Finding Multiple Optimal
Solutions in Multi-modal Optimization 706
   Pedro J. Ballester, Jonathan N. Carter

An Adaptive Penalty Scheme for Steady-State Genetic Algorithms 718
  Helio J.C. Barbosa, Afonso C.C. Lemonge

Asynchronous Genetic Algorithms for Heterogeneous Networks
Using Coarse-Grained Dataflow 730
  John W. Baugh Jr., Sujay V. Kumar

A Generalized Feedforward Neural Network Architecture and Its
Training Using Two Stochastic Search Methods 742
  Abdesselam Bouzerdoum, Rainer Mueller

Ant-Based Crossover for Permutation Problems 754
  Jürgen Branke, Christiane Barz, Ivesa Behrens

Selection in the Presence of Noise 766
  Jürgen Branke, Christian Schmidt

Effective Use of Directional Information in Multi-objective
Evolutionary Computation 778
  Martin Brown, R.E. Smith

Pruning Neural Networks with Distribution Estimation Algorithms 790
  Erick Cantú-Paz

Are Multiple Runs of Genetic Algorithms Better than One? 801
  Erick Cantú-Paz, David E. Goldberg

Constrained Multi-objective Optimization Using Steady State
Genetic Algorithms 813
  Deepti Chafekar, Jiang Xuan, Khaled Rasheed

An Analysis of a Reordering Operator with Tournament Selection on
a GA-Hard Problem 825
  Ying-Ping Chen, David E. Goldberg

Tightness Time for the Linkage Learning Genetic Algorithm 837
  Ying-Ping Chen, David E. Goldberg

A Hybrid Genetic Algorithm for the Hexagonal Tortoise Problem 850
  Heemahn Choe, Sung-Soon Choi, Byung-Ro Moon

Normalization in Genetic Algorithms 862
  Sung-Soon Choi and Byung-Ro Moon

Coarse-Graining in Genetic Algorithms: Some Issues and Examples 874
  Andrés Aguilar Contreras, Jonathan E. Rowe,
  Christopher R. Stephens

Building a GA from Design Principles for Learning Bayesian Networks ... 886
  Steven van Dijk, Dirk Thierens, Linda C. van der Gaag

A Method for Handling Numerical Attributes in GA-Based Inductive
Concept Learners 898
    Federico Divina, Maarten Keijzer, Elena Marchiori

Analysis of the (1+1) EA for a Dynamically Bitwise
Changing ONEMAX 909
    Stefan Droste

Performance Evaluation and Population Reduction for a Self
Adaptive Hybrid Genetic Algorithm (SAHGA) 922
    Felipe P. Espinoza, Barbara S. Minsker, David E. Goldberg

Schema Analysis of Average Fitness in Multiplicative Landscape 934
    Hiroshi Furutani

On the Treewidth of NK Landscapes 948
    Yong Gao, Joseph Culberson

Selection Intensity in Asynchronous Cellular Evolutionary Algorithms ... 955
    Mario Giacobini, Enrique Alba, Marco Tomassini

A Case for Codons in Evolutionary Algorithms 967
    Joshua Gilbert, Maggie Eppstein

Natural Coding: A More Efficient Representation for
Evolutionary Learning 979
    Raúl Giráldez, Jesús S. Aguilar-Ruiz, José C. Riquelme

Hybridization of Estimation of Distribution Algorithms with a
Repair Method for Solving Constraint Satisfaction Problems 991
    Hisashi Handa

Efficient Linkage Discovery by Limited Probing 1003
    Robert B. Heckendorn, Alden H. Wright

Distributed Probabilistic Model-Building Genetic Algorithm 1015
    Tomoyuki Hiroyasu, Mitsunori Miki, Masaki Sano, Hisashi Shimosaka,
    Shigeyoshi Tsutsui, Jack Dongarra

HEMO: A Sustainable Multi-objective Evolutionary
Optimization Framework 1029
    Jianjun Hu, Kisung Seo, Zhun Fan, Ronald C. Rosenberg,
    Erik D. Goodman

Using an Immune System Model to Explore Mate Selection in
Genetic Algorithms 1041
    Chien-Feng Huang

Designing A Hybrid Genetic Algorithm for the Linear
Ordering Problem.. 1053
  Gaofeng Huang, Andrew Lim

A Similarity-Based Mating Scheme for Evolutionary
Multiobjective Optimization... 1065
  Hisao Ishibuchi, Youhei Shibata

Evolutionary Multiobjective Optimization for Generating an
Ensemble of Fuzzy Rule-Based Classifiers 1077
  Hisao Ishibuchi, Takashi Yamamoto

Voronoi Diagrams Based Function Identification 1089
  Carlos Kavka, Marc Schoenauer

New Usage of SOM for Genetic Algorithms............................ 1101
  Jung-Hwan Kim, Byung-Ro Moon

Problem-Independent Schema Synthesis for Genetic Algorithms 1112
  Yong-Hyuk Kim, Yung-Keun Kwon, Byung-Ro Moon

Investigation of the Fitness Landscapes and Multi-parent
Crossover for Graph Bipartitioning 1123
  Yong-Hyuk Kim, Byung-Ro Moon

New Usage of Sammon's Mapping for Genetic Visualization 1136
  Yong-Hyuk Kim, Byung-Ro Moon

Exploring a Two-Population Genetic Algorithm....................... 1148
  Steven Orla Kimbrough, Ming Lu, David Harlan Wood, D.J. Wu

Adaptive Elitist-Population Based Genetic Algorithm for
Multimodal Function Optimization 1160
  Kwong-Sak Leung, Yong Liang

Wise Breeding GA via Machine Learning Techniques for
Function Optimization... 1172
  Xavier Llorà, David E. Goldberg

Facts and Fallacies in Using Genetic Algorithms for Learning
Clauses in First-Order Logic.. 1184
  Flaviu Adrian Mărginean

Comparing Evolutionary Computation Techniques via
Their Representation 1196
  Boris Mitavskiy

Dispersion-Based Population Initialization 1210
  Ronald W. Morrison

A Parallel Genetic Algorithm Based on Linkage Identification 1222
  Masaharu Munetomo, Naoya Murao, Kiyoshi Akama

Generalization of Dominance Relation-Based Replacement Rules for
Memetic EMO Algorithms 1234
  Tadahiko Murata, Shiori Kaige, Hisao Ishibuchi

Author Index

Design of Multithreaded Estimation of Distribution Algorithms

Jiri Ocenasek[1], Josef Schwarz[2], and Martin Pelikan[1]

[1] Computational Laboratory (CoLab), Swiss Federal Institute of Technology ETH
Hirschengraben 84, 8092 Zürich, Switzerland
jirio@inf.ethz.ch,pelikanm@inf.ethz.ch
[2] Faculty of Information Technology, Brno University of Technology
Bozetechova 2, 612 66 Brno, Czech Republic
schwarz@fit.vutbr.cz

Abstract. Estimation of Distribution Algorithms (EDAs) use a probabilistic model of promising solutions found so far to obtain new candidate solutions of an optimization problem. This paper focuses on the design of parallel EDAs. More specifically, the paper describes a method for parallel construction of Bayesian networks with local structures in form of decision trees in the Mixed Bayesian Optimization Algorithm. The proposed Multithreaded Mixed Bayesian Optimization Algorithm (MMBOA) is intended for implementation on a cluster of workstations that communicate by Message Passing Interface (MPI). Communication latencies between workstations are eliminated by multithreaded processing, so in each workstation the high-priority model-building thread, which is communication demanding, can be overlapped by low-priority model sampling thread when necessary. High performance of MMBOA is verified via simulation in TRANSIM tool.

1 Introduction

Estimation of Distribution Algorithms (EDAs) [1], also called Probabilistic Model-Building Genetic Algorithms (PBMGAs) [2] and Iterated Density Estimation Evolutionary Algorithms (IDEAs) [3], have been recently proposed as a new evolutionary technique to allow effective processing of information represented by sets of high-quality solutions. EDAs incorporate methods for automated learning of linkage between genes of the encoded solutions and incorporate this linkage into a graphical probabilistic model. The process of sampling new individuals from a probabilistic model respects these mutual dependencies, so that the combinations of values of decision variables remain correlated in high-quality solutions. The quality of new offspring is affected neither by the ordering of genes in the chromosome nor the fitness function epistasis. It is beyond the scope of this paper to give a thorough introduction to EDAs, for a survey see [2,1].

Recent work on parallel Estimation of Distribution Algorithms concentrated on the parallelization of the construction of a graphical probabilistic model, mainly on the learning of structure of a Bayesian network. The Parallel Bayesian

Optimization Algorithm (PBOA) [4] was designed for pipelined parallel architecture, the Distributed Bayesian Optimization Algorithm (DBOA) [5] was designed for a cluster of workstations and the Parallel Estimation of Bayesian Network Algorithm (Parallel EBNA$_{BIC}$) [6] was designed for MIMD architecture with shared memory, also called PRAM.

Exponential size of the tabular representation of local conditional probability distributions emerged as a major problem in learning Bayesian networks. The most advanced implementations of EDAs use decision trees to capture local distributions of a Bayesian network more effectively. Nevertheless, parallelization of EDAs using Bayesian networks with decision trees has not been tackled yet. This paper – although applicable to parallel Bayesian network learning in general – deals especially with the parallel learning of decision trees and proposes advanced multithreaded techniques to accomplish it effectively.

The following section introduces the principles of Estimation of Distribution Algorithms. We identify the disadvantages of Bayesian network with tabular representation of local structures and motivate the usage of decision trees. Section 3 provides an overview of present parallel Estimation of Distribution Algorithms. We identify the main differences between them, analyze the way these algorithms learn dependency graphs and derive design guidelines. In Sect. 4 a new Multithreaded Mixed Bayesian Optimization Algorithm is proposed. We develop an original technique for parallel construction of Bayesian networks with decision trees using a cluster of workstations. Additional efficiency is achieved on a multithreaded platform, where the communication latencies between workstations are overlapped by switching between model-building thread and model-sampling thread. The efficiency and scalability of the proposed algorithm is verified via simulation using TRANSIM tool and demonstrated in Sect. 5.

2 Estimation of Distribution Algorithms

2.1 Basic Principles

The general procedure of EDAs is similar to that of genetic algorithms (GAs), but the traditional recombination operators are replaced by (1) estimating the probability distribution of selected solutions and (2) sampling new points according to this estimate:

```
Set t := 0;
Randomly generate initial population P(0);
while termination criteria are not satisfied do begin
   Select a set of promising parent solutions D(t) from P(t);
   Estimate the probability distribution of the selected set D(t);
   Generate a set of new strings O(t) according to the estimate;
   Create a new pop. P(t+1) by replacing part of P(t) by O(t);
   Set t := t+1;
end
```

2.2 EDAs with Bayesian Network Model

First EDAs differed mainly in the complexity of the used probabilistic models. At the present time Bayesian networks have become most popular for representing discrete distributions in EDAs because of their generality. For the domain of possible chromosomes $\mathbf{X} = (X_0, \ldots, X_{n-1})$ the Bayesian network represents a joint probability distribution over $\mathbf{X}$. This representation consists of two parts – a dependency graph and a set of local conditional probability distributions.

The first part, the dependency graph, encodes the dependencies between values of genes throughout the population of chromosomes. Each gene corresponds to one node in the graph. If the probability of the value of a certain gene X_i is affected by a value of another gene X_j, then we say that "X_i depends on X_j" or "X_j is a parent variable of X_i". This assertion is expressed by the existence of an edge (j,i) in the dependency graph. A set of all parent variables of X_i, denoted Π_i, corresponds to the set of all starting nodes of edges ending in X_i. The second part, the set of local conditional probability distributions $p(X_i|\Pi_i)$, is usually expressed in the tabular form.

The well known EDA implementations with the Bayesian network model are the Bayesian Optimization Algorithm (BOA) [7], the Estimation of Bayesian Network Algorithm (EBNA) [8] or the Learning Factorized Distribution Algorithm (LFDA) [9].

2.3 EDAs with Decision Tree Models

For each possible assignment of values to the parents Π_i, we need to specify a distribution over the values X_i can take. In its most naive form, local conditional probability distributions $p(X_i|\Pi_i)$ are encoded using tabular representation that is exponential in the number of parents of a variable X_i. More effective representation can be obtained by encoding local probability distributions using n decision trees. Since decision trees usually require less parameters, frequency estimation is more robust for given population size N.

The implementation of the Bayesian Optimization Algorithm with decision trees appeared in [10] and an extension of this idea for continuous variables resulted in the Mixed Bayesian Optimization Algorithm (MBOA) [11]. See also [12] for the multiobjective version of MBOA with epsilon-archiving approach.

3 Parallel Estimation of Distribution Algorithms

3.1 Performance Issues

The analysis of BOA complexity identifies two potential performance bottlenecks – the construction of a probabilistic model and the evaluation of solutions. It depends on the nature of final application which part of MMBOA should be parallelized.

For problems with computationally cheap fitness evaluation – such as spin glass optimization with complexity $O(n*N)$ – great speedup can be achieved by parallelizing model construction, which can take $O(n^2 * N * log(N))$ steps in each generation. On the other hand, for problems with expensive fitness evaluation – such as ground water management optimization [13] – fitness evaluation should be also parallelized to achieve maximum efficiency.

In this paper we focus only on the parallelization of model building because it is not straightforward and needs special effort to be done effectively whereas the parallel evaluation of solutions can be done in obvious way and you can find many papers on this topic. For further analysis of parallelization of fitness evaluation see Cantú-Paz [14].

3.2 Parallel Learning of Dependency Graph

All present parallel EDAs use greedy algorithms driven by statistical metrics to obtain the dependency graph. The so-called BIC score in EBNA [8] algorithm and also the Bayesian-Dirichlet metric in PBOA [4] and DBOA [5] algorithms are decomposable and can be written as a product of n factors, where the i-th factor expresses the influence of edges ending in the variable X_i. Thus, for each variable X_i the metrics gain for the set of parent variables Π_i can be computed independently. It is possible to utilize up to n processors, each processor corresponds to one variable X_i and it examines only edges leading to this variable.

The addition of edges is parallel, so an additional mechanism has to be used to keep the dependency graph acyclic. Parallel EBNA uses master-slave architecture, where the addition of edges is synchronous and controlled by the master.

PBOA and DBOA use predefined topological ordering of nodes to avoid cycles in advance. At the beginning of each generation, a random permutation of numbers $\{0, 1, ..., n-1\}$ is created and stored in an array $\mathbf{o} = (o_0, o_1, ..., o_{n-1})$. The direction of all edges in the network should be consistent with the ordering, so the addition of an edge from X_{o_j} to X_{o_i} is allowed if only $j < i$. The advantage is that each processor can create its part of the probabilistic model asynchronously and independently of other processors. No communication among processors is necessary because acyclicity is implicitly satisfied. The network causality might be violated by this additional assumption, but according to our empirical experience the quality of generated network is comparable to the quality of sequentially constructed network (see [4]).

3.3 Parallel Offspring Generation

The difference between various parallel EDAs arises more if we analyze how the offspring is generated.

PBOA was proposed for fine-grained type of parallelism with tightly connected communication channels. It can generate offspring in a linear pipeline way, because in the fine-grained architecture there are negligible communication latencies. It takes n cycles to generate the whole chromosome, but $n + N$ cycles to generate the whole population. For example let us consider the generation of first offspring chromosome. Its o_0-th bit is generated independently in processor number 0 at time t. Then, its o_1-th bit is generated in processor number 1 at time

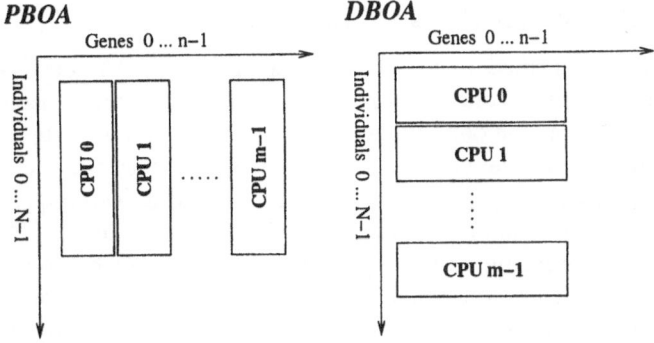

Fig. 1. Comparison of offspring generation. PBOA distributes the work between processors "horizontally" and DBOA "vertically".

$t+1$ conditionally on the o_0-th bit received from neighbouring processor number 0, etc. Generally, X_{o_i} is generated in CPU number i at time $t+i$ conditionally on $\Pi_{o_i} \subseteq \{X_{o_0}, ..., X_{o_{i-1}}\}$. The advantage is that each conditional probability distribution is sampled locally at the place where it has been estimated.

In DBOA case communication delays are too long to use pipelined generation of offspring. Thus, DBOA uses distributed generation of offspring, each processor generates one complete subpopulation of chromosomes. See Fig. 1 for comparison of both types of offspring generation. Note that for this kind of offspring generation each processor needs to use a complete probabilistic model, which is constructed piecewise. Thus, gathering of local parts of model is a necessary step between model estimation and offspring generation, when each processor exchanges its part of model with the other processors.

In the case of parallel EBNA the target architecture is aimed to be the shared-memory MIMD, so the problem with the distribution of new population does not occur, the shared memory architecture allows for both "horizontal" as well as for "vertical" layout.

4 Multithreaded Mixed Bayesian Optimization Algorithm

The Mixed Bayesian Optimization Algorithm (MBOA) [11] uses a set of decision trees to express the local structures of Bayesian network. Its implementation was described in detail in [11]. It is suitable for both continuous and/or discrete optimization problems. In this paper we propose its parallel version. The target architecture for parallel MBOA is the cluster of workstations connected with the hardware switch.

4.1 Parallel Decision Trees Construction

Some principles identified in Sect. 3 are useful for MBOA too. Namely in the model learning step the concept of restricted ordering of nodes from DBOA can be used again to keep the dependencies acyclic and to remove the need for communication during parallel construction of the probabilistic model. An ordering permutation $(o_0, o_1, ..., o_{n-1})$ means that only the variables $\{X_{o_0}, ..., X_{o_{i-1}}\}$ can serve as splits in the binary decision tree of target variable X_{o_i}. The drawback of predetermined ordering of nodes is that each tree can be constructed independently of the other trees, which allows for efficient parallelization. Moreover, it significantly reduces memory requirements for temporary data structures because only one decision tree is constructed at each time. In binary case the simplified code for building one decision tree is:

```
function BuildTree(Population Pop, TreeIndex i,
                   Permutation o): DecisionTreeNode;
begin
    for j:=0 to i-1 do
      if X[o[j]] has not been used as the split in i-th tree yet
        Evaluate the gain of X[o[j]] split for X[o[i]] target;
    Pick the split X[o[j']] with the highest metrics gain;
    if model complexity penalty >  gain of X[o[j']]
        return new UnivariateProbabilityLeaf(X[o[i]],Pop);
    Pop1 := SelectIndividuals (Pop,"X[o[j']] = 0");
    Pop2 := SelectIndividuals (Pop,"X[o[j']] = 1");
    return new SplitNode(new SplitVariable(X[o[j']]),
            BuildTree(Pop1,i,o), BuildTree(Pop2,i,o));
end
```

4.2 Parallel Decision Trees Sampling

In MBOA the decision trees are used also for continuous or mixed domains, so the decision nodes in the trees can have parameters of various types: real numbers for splitting on real-coded parameters, sequences of integers for splitting on categorical parameters, etc. Also the leaf nodes in the trees can contain various structured parameters: a list of Gaussian kernels, a list of Gaussian network coefficients, probability bias for binary leaves, etc.

It is a very time consuming task for each process to communicate the whole decision tree. It would be far more natural if each decision tree could be used exactly at the same workstation where it has been built. Thus, the goal of our design is to propose the parallel MBOA that uses the "horizontal" layout for offspring generation (see Fig. 1 left), but we would like parallel MBOA to be suited for coarse-grained platform, where the long communication latencies force us to use the "vertical" layout. To solve this problem, we interlace the model building task with the offspring generation task.

Each workstation uses two threads – a high priority thread for building the decision tree and a low priority thread for sampling previously built decision trees. The model building thread has to communicate with the master workstation to get the job and the model sampling threads have to communicate with other workstations to exchange the generated parts of offspring. By switching the threads it is possible to avoid most of the communication delays.

4.3 Multithreaded Processing and TRANSIM Tool

To switch the threads automatically and effectively, we presume workstations with multithreading capabilities. In simultaneous multithreading (SMT) one or more additional threads can take up the slack when the processor cannot continue execution for any reason. Since this is done by switching the active register set, without the need to copy the registers off to slower memory, the switch can be very fast. SMT capabilities, while not giving the performance of multiple processors, are very inexpensive to incorporate into the processor, because the execution units need not be replicated.

Although these days SMT becomes very popular – especially with the new Intel's Hyper-Threading(HT) technology – the current implementation of MPI standard does not support multithreaded processing to implement Multithreaded MBOA. Another possibility was to use the multitasking nature of Unix system, but we were looking for a concept at higher level of abstraction.

Finally, we decided to simulate the multithreaded MBOA using the well known TRANSIM tool. TRANSIM is a CASE tool used in the design of parallel algorithms. Its major function is the prediction of time performance early in the life cycle before coding has commenced. The TRANSIM language is a subset of transputers' Occam language with various extensions. It is not intended as a general purpose programming language, but to provide a control structure whereby the performance properties of an application may be expressed. Parallel execution, alternation, channel communication, time-slicing, priorities, interruption, concurrent operation of links and the effects of external memory are taken into account.

4.4 Multithreaded MBOA Design

The whole architecture of multithreaded MBOA is shown in Fig. 2. The *farmer* thread is responsible for dynamic workload assignment, *builder* threads are responsible for building decision trees and *generator* threads are responsible for generating new genes. *Buffer* threads are used only for buffering dataflow between builders and generators, because TRANSIM does not implicitly support buffered channels. Threads $builder_i$, $buffer_i$ and $generator_i$ are mapped to the same i-th workstation.

The processes in TRANSIM communicate through pairwise unidirectional channels. The array of external channels $ch.master.in$ is used by *farmer* to receive the job requests and the array of external channels $ch.master.out$ is used to send the jobs to *builders*. A two-dimensional array of external channels $ch.network$ is used to transfer the partially generated population between *generators*. The internal channels $ch.buffer.in$, $ch.buffer.req$ and $ch.buffer.out$ are used for exchange of local parts of model (decision tree) between *builder* and *generator*.

The principles of the most important threads *builder* and *generator* will be illustrated by fragments of TRANSIM input file. For those readers non-familiar with the statements of TRANSIM input language the non-intuitive parts will be re-explained afterwards. The complete code can be downloaded from web page http://jiri.ocenasek.com/.

First see the code fragment of the *builder* thread:

```
PLACED PAR i=0 FOR NUMWORKERS
  INT prev,next,j,prefetch.prev,prefetch.next,prefetch.j:
  SEQ | builder
    ch.farmer.in[i]    ! i | REQ.LENGTH
    ch.farmer.out[i]   ? prefetch.j
    ch.farmer.out[i]   ? prefetch.prev
    ch.farmer.out[i]   ? prefetch.next
    WHILE prefetch.j < n
      SEQ
        j    := prefetch.j
        next := prefetch.next
        gene := prefetch.gene
        PRI PAR
          SEQ | getjob
            ch.farmer.in[i]    ! i | REQ.LENGTH
            ch.farmer.out[i]   ? prefetch.j
            ch.farmer.out[i]   ? prefetch.prev
            ch.farmer.out[i]   ? prefetch.next
          SEQ | build
            ch.buffer.in[i] ! prev | WORK.LENGTH
            SERV(j*TIMESLICE)
            ch.buffer.in[i] ! j    | WORK.LENGTH
            ch.buffer.in[i] ! next | WORK.LENGTH
```

Each *builder* requests for a job via *ch.farmer.in[i]* and it receives job info from *ch.farmer.out[i]*. More precisely, it receives the job number j, the number of workstation working on job $j-1$ and the number of workstation requesting the farmer for job $j+1$.

The useful computation is simulated by a *SERV()* command. You see that for fixed population size the time for building the decision tree for gene X_{o_j} is proportional to j and is scaled by *TIMESLICE* constant, according to empirical observations on sequential MBOA. The building of the decision tree is overlapped by high priority communication (see the sequence named "*getjob*") which serves for prefetching the next job from *farmer*.

Now see the code of *generator* thread. This fragment is very simplified, ignoring the initial stage of generating independent genes and the final stage of broadcasting the full population:

```
PLACED PAR i=0 FOR NUMWORKERS
  INT from,to,population,j:
  SEQ | generator
    WHILE TRUE
      SEQ
        ch.buffer.req[i] ! i | WORK.LENGTH
        ch.buffer.out[i] ? from
        PAR
          SEQ | recvpred
            ch.network[from][i] ? population
          SEQ | recvmodel
            ch.buffer.out[i] ? j
            ch.buffer.out[i] ? to
        SERV(GENSLICE)
        ch.network[i][to] ! population | j * N
```

Each *generator* receives from *builder* (via buffered channels) the number of workstation from which it then receives the population with genes $\{X_{o_0},...,X_{o_{j-1}}\}$ generated. Simultaneously it also receives from *builder* the decision tree for gene X_{o_j} and the number of consecutive workstation. Then the *SERV()* command simulated the sampling of the decision tree. According to empirical observations the sampling time is almost constant for fixed population size, independent of j. At the end, the population with generated $\{X_{o_0},...,X_{o_j}\}$ is sent to a consecutive workstation. To lower the communication demands further, it is sufficient to communicate only genes $\{X_{o_{k+1}},...,X_{o_j}\}$ if it exists some X_{o_k} that was previously generated in that consecutive workstation.

Note that the priority of threads is crucial. The *generator* thread should have high priority and the workstation processor should use the *builder* threads to overlap the communication latency between generators.

5 Simulation Results

The simulation parameters were set according to the parameters measured on the real computing environment - a cluster of Athlon-600 computers connected by the hardware multiport router Summit48 (from Extreme Networks). The external communication is supposed to be the MPI (Message Passing Interface) which provides the well known standard for message passing communication.

TRANSIM is able to simulate the capacity of real communication channels, so the speed of external channels was set to 100Mb/s (like the speed of Summit48 Ethernet switch) and their communication latency was set to 0.5 ms (like the average software latency in MPI). The speed of internal channels was set to maximum, because in the real implementation the *builder* and *generator* would communicate via shared memory space. According to our empirical experience with sequential MBOA on Athlon-600 processor, the TIMESLICE constant was set for about 7.5 ms and the GENSLICE was set for 10 ms. The job assignment

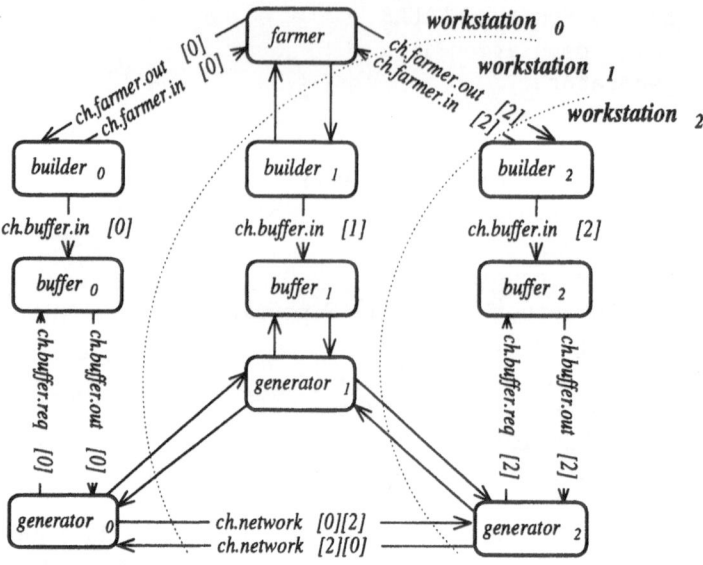

Fig. 2. Architecture of multithreaded MBOA in TRANSIM. An example with 3 workstations. The dashed curves separate processes mapped to different hardware.

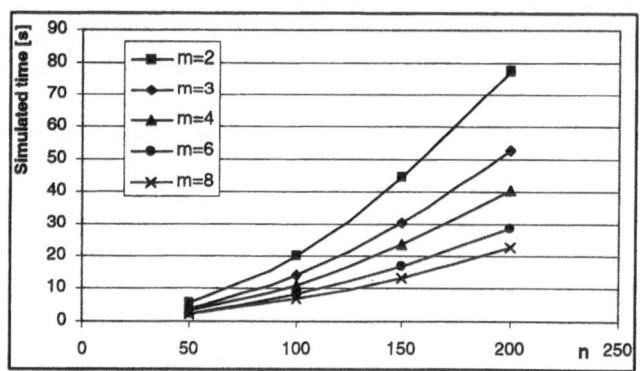

Fig. 3. The total simulated time of Multithreaded MBOA for varying problem size n and varying number of processors m for fixed population size $L = 2200$.

from *farmer* thread does not need extra communication capacity, so the message-length constants WORK.LENGTH and REQ.LENGTH were set to 1.

Figure 3 depicts the execution time of parallel MBOA with respect to the problem size and number of processors. In Fig. 4 the results indicate that the achieved speedup slightly differs from ideal case because it is impossible to remove all communication latencies completely (for example there remain no de-

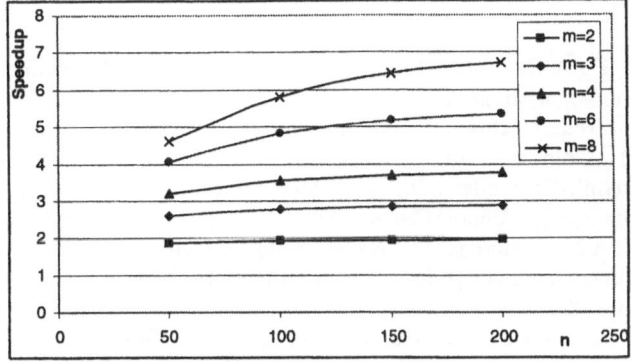

Fig. 4. The speedup of simulated multithreaded MBOA for varying problem size n and varying number of processors m for fixed population size $L = 2200$.

cision trees to build when generating the last gene $X_{o_{n-1}}$). Fortunately, this difference becomes negligible with increasing problem size, so our approach is linearly scalable.

6 Summary and Conclusions

We proposed and simulated the Multithreaded Mixed Bayesian Optimization Algorithm (MMBOA). We focused mainly on the effective construction of Bayesian networks with decision trees in the distributed environment. MMBOA is especially suitable for running on a cluster of workstations. Where available, additional efficiency can be achieved by utilizing multithreaded processors.

The results of simulation indicate that our algorithm is efficient for a large number of processors and its scalability improves with an increasing problem size. Moreover, the workload balance of each distributed process is adapted according to its computational resources when different types of workstations are used.

The use of the proposed parallelization of decision tree construction is not limited to BOA or other EDAs; it can be used in machine learning, data mining, and other areas of artificial intelligence. The use of the proposed technique outside the scope of EC is a topic for future research.

Acknowledgement. The authors would like to thank the members of the Computational Laboratory ETH Zürich for valuable comments and discussions. This research had been partially supported by the Grant Agency of Czech Republic from research grant GA 102/02/0503 "Parallel system performance prediction and tuning".

References

1. Larrañaga, P.: A Review on Estimation of Distribution Algorithms. Estimation of Distribution Algorithms. A new Tool for Evolutionary Computation. P. Larrañaga, J.A. Lozano (eds.). Kluwer Academic Publishers, pp. 57–100, 2001.
2. Pelikan, M., Goldberg, D.E., Lobo, F.: A survey of optimization by building and using probabilistic models, IlliGAL Report No. 99018, University of Illinois at Urbana-Champaign, Illinois Genetic Algorithms Laboratory, Urbana, Illinois, 1999.
3. Bosman, P.A.N., Thierens, D.: An algorithmic framework for density estimation based evolutionary algorithms. Utrecht University Technical Report UU-CS-1999-46, Utrecht, 1999.
4. Ocenasek, J., Schwarz, J.: The Parallel Bayesian Optimization Algorithm, In: Proceedings of the European Symposium on Computational Inteligence, Physica-Verlag, Kosice, Slovak Republic, pp. 61–67, 2000.
5. Ocenasek, J., Schwarz, J.: The Distributed Bayesian Optimization Algorithm for combinatorial optimization, EUROGEN 2001 – Evolutionary Methods for Design, Optimisation and Control, Athens, Greece, CIMNE, pp. 115–120, 2001.
6. Lozano, J. A., Sagarna, R., Larrañaga, P.: Parallel Estimation of Distribution Algorithms. Estimation of Distribution Algorithms. A new Tool for Evolutionary Computation. P. Larrañaga, J.A. Lozano (eds.). Kluwer Academic Publishers, pp. 129–145, 2001.
7. Pelikan, M., Goldberg, D.E., Cantú-Paz, E.: BOA: The Bayesian optimization algorithm. Proceedings of the Genetic and Evolutionary Computation Conference GECCO-99, vol. I, Orlando, FL, Morgan Kaufmann Publishers, pp. 525–532, 1999.
8. Etxeberria, R., Larrañaga, P.: Global optimization using Bayesian networks. Second Symposium on Artificial Intelligence (CIMAF-99), Habana, Cuba, pp 332–339, 1999.
9. Mühlenbein, H., Mahnig, T.: FDA - a scalable evolutionary algorithm for the optimization of additively decomposed functions. Evolutionary Computation, 7(4), pp. 353–376, 1999.
10. Pelikan, M., Goldberg, D.E., Sastry, K.: Bayesian Optimization Algorithm, Decision Graphs, and Occam's Razor, IlliGAL Report No. 2000020, University of Illinois at Urbana-Champaign, Illinois Genetic Algorithms Laboratory, Urbana, IL, 2000.
11. Ocenasek, J., Schwarz, J.: Estimation of Distribution Algorithm for mixed continuous discrete optimization problems, In: 2nd Euro-International Symposium on Computational Intelligence, Kosice, Slovakia, IOS Press, pp. 227–232, 2002.
12. Laumanns, M., Ocenasek, J.: Bayesian Optimization Algorithms for multi-objective optimization, In: Parallel Problem Solving from Nature – PPSN VII, Springer-Verlag, pp. 298–307, 2002.
13. Arst. R., Minsker, B.S., Goldberg, D.E.: Comparing Advanced Genetic Algorithms and Simple Genetic Algorithms for Groundwater Management. 2002 Water Resources Planning & Management Conference, Roanoke, VA, 2002.
14. Cantú-Paz, E.: Efficient and Accurate Parallel Genetic Algorithms. Boston, MA: Kluwer Academic Publishers. 2000.

Reinforcement Learning Estimation of Distribution Algorithm

Topon Kumar Paul and Hitoshi Iba

Graduate School of Frontier Sciences, The University of Tokyo
Hongo 7-3-1, Bunkyo-ku, Tokyo 113-8656, Japan
{topon,iba}@miv.t.u-tokyo.ac.jp

Abstract. This paper proposes an algorithm for combinatorial optimizations that uses reinforcement learning and estimation of joint probability distribution of promising solutions to generate a new population of solutions. We call it Reinforcement Learning Estimation of Distribution Algorithm (RELEDA). For the estimation of the joint probability distribution we consider each variable as univariate. Then we update the probability of each variable by applying reinforcement learning method. Though we consider variables independent of one another, the proposed method can solve problems of highly correlated variables. To compare the efficiency of our proposed algorithm with other Estimation of Distribution Algorithms (EDAs) we provide the experimental results of the two problems: four peaks problem and bipolar function.

1 Introduction

After the introduction of Genetic Algorithm (GA) it has been widely used for the optimization of problems because it is easy to use and can be applied in a problem with little prior knowledge. The heart of GA is its selection and recombination of promising solutions; however, crossover and mutation of GA are problem specific and hence success depends on the choice of these parameters. They do not consider problem specific interactions among the variables called *linkage information*. As a result, the prediction of the movements of the populations in the search space is difficult in GA. Holland[12] introduced the idea that linkage information would be beneficial for Genetic Algorithms. Following his idea, Goldberg et al.[7], Harik [9]and Kargupta[13] extended GA by either changing the representation of the solutions or evolving recombination operators among individual solutions to process the building blocks. Recently, there have been proposed evolutionary algorithms based on the probabilistic model where selection and recombination of building blocks of Genetic Algorithm are replaced by generating new solutions by sampling the probability distribution which is calculated from the selected promising solutions. These algorithms are called Estimation of Distribution Algorithms (EDAs) [17] or Probabilistic Model-Building Genetic Algorithms (PMBGAs) [20]. As the EDAs try to capture the structure of the problem, EDAs are thought to be more efficient than GAs.

The purpose of this paper is to introduce a new Estimation of Distribution Algorithm (EDA) that uses Reinforcement Learning (RL) method and marginal probability distribution of selected individuals in order to generate new solutions. We call this algorithm Reinforcement Learning Estimation of Distribution Algorithm (RELEDA). The proposed method extends existing EDAs to solve difficult classes of problems more efficiently and accurately. We consider variables as univariate and calculate their marginal distributions accordingly. We then update the probability vector of the variables by reinforcement learning method using these marginal distributions and the best fit individual of a generation. New individuals (solutions) are generated by sampling this probability vector. The experiments were done with the four peaks problem and bipolar function which are very difficult for some EDAs to optimize. The experimental results show that our proposed algorithm is able to solve the tested problems more efficiently.

In Sects. 2 and 3, we provide the background needed to understand the motivation of our algorithm. Section 4 describes the proposed algorithm. In Sect- 5, a short review of different existing EDA approaches are presented. Section 6 describes the test functions and presents the experimental results on these problems. Section 7 addresses our future work. The summary and conclusion are provided in Sect. 8.

2 Optimizations by Free Energy Minimization

Before we proceed, we need to give some mathematical notations. The variable X denotes the set of n discrete binary variables $\{X_1, X_2, \ldots, X_n\}$ and x denotes the set of values of these variables. The search space is $S = \{0, 1\}^n$. $p(x)$ is the joint probability of X and $f(x)$ denotes the function to be maximized (minimized).

The Gibbs distribution of a function $f(x)$ at given temperature T is

$$p(x) = \frac{e^{-\frac{f(x)}{T}}}{Z} \qquad (1)$$

where $Z = \sum_{y \in S} e^{-\frac{f(y)}{T}}$ is the partition function.

When $T \to 0$ the Boltzmann distribution converges uniformly to one of the global optima. The Metropolis algorithm, whose stationary distribution is precisely Boltmann distribution, samples from $p(x)$ to generate solution for optimization task. When it is not possible to calculate $p(x)$ precisely one approximates it with a target distribution $p_T(x)$ using Kullback Leibler (KL) divergence. The KL divergence is

$$D(p_T(x), p(x)) = \ln Z + \frac{1}{T}(E - TS) \qquad (2)$$

where E and S are internal energy and entropy of the system and $F = E - TS$ is the free energy. Here F is a function of $p(x)$. So by minimizing F one can minimize KL divergence. In order to minimize the F we need to know all probability distributions of the variables in the search space. Instead, the search to probability distributions can be reduced to a small number of parameters denoted by $\theta = \{\theta_1, \theta_2, \ldots, \theta_v\}$ where $\theta_i \in \mathbb{R}$ is a parameter related to the probability of the variable X_i through a function and the joint probability is denoted by $p(x, \theta)$ which is differentiable with respect to its parameters for each $x \in S$. Therefore, F is differentiable with respect to θ. Thus the problem has been transformed from the search in discrete domain to continuous domain. The update rule of the parameter θ can be found by defining a dynamical system by the ordinary differential equation: $\frac{d\theta}{dt} + \frac{\delta F}{\delta \theta} = 0$. After inserting the values of F we obtain by some calculations:

$$\frac{d\theta}{dt} + \sum_{x \in S}(f(x) + T(1 + \ln p(x, \theta)))\frac{\delta p(x, \theta)}{\delta \theta} = 0. \quad (3)$$

Then the basic update rule proposed by Berny[4] for gradient descent is as follows:

$$\Delta\theta = -\alpha(f(x) + T(1 + \ln p(x, \theta)))\frac{\delta p(x, \theta)}{\delta \theta} \quad (4)$$

where $\alpha > 0$ is a small constant called learning rate.

3 Reinforcement Learning

In reinforcement learning an agent is connected to its environment through perception and action. At every step of interactions the agent gets some input signals and indications of current state of the environment, and then the agent chooses some actions to produce output. The action changes the state of the environment, and the value of this transition is sent back to the agent as a scalar reinforcement signal. Here learning is unsupervised, and the agent modified its behavior to maximize or minimize the reinforcement signal. Williams[22] in his REINFORCEMENT Algorithm for learning with associative networks has used the following incremental rule for the updating of the weight of the connection from input i to unit $j(w_{ij})$:

$$\Delta w_{ij} = \alpha_{ij}(r - b_{ij})\frac{\delta \ln g_i}{\delta w_{ij}} \quad (5)$$

where α_{ij} is the learning rate, r is the reinforcement signal, b_{ij} is a reinforcement baseline and g_i is the probability of the output of unit i given the input and the weights to this unit. For combinatorial optimizations Berny[4] has used cost function (fitness) as reinforcement signal and low cost binary string as agent's output. His reinforcement algorithm is:

$$\Delta\theta_i = \alpha(f(x) - b_i)\frac{\delta \ln p(x, \theta)}{\delta \theta_i} \quad (6)$$

where $f(x)$ is the cost function, $p(x, \theta)$ the joint probability distribution and θ_i is a parameter related to the probability of ith variable ($p_i(x_i)$), b_i is the baseline and α is the learning rate.

The baseline (b) is updated by the recursive formula:

$$b(t+1) = \gamma b(t) + (1-\gamma)f(x(t)) \tag{7}$$

where $0 < \gamma < 1$ is the baseline factor. He (Berny) has called this baseline update rule as expectation strategy where, if a randomly generated solution x has cost $f(x)$ lower than average cost, the agent takes the solution more likely.

4 Reinforcement Learning Estimation of Distribution Algorithm (RELEDA)

In RELEDA we consider no interactions among the variables. Then the joint probability($p(x,\theta)$) becomes:

$$p(x,\theta) = \prod_{i=1}^{n} p_i(x_i) \tag{8}$$

where $p_i(x_i)$ is the probability of $X_i = 1$.

The correlation between $p_i(x_i)$ and θ_i is expressed through the sigmoid function:

$$p_i(x_i) = \frac{1}{2}(1 + \tanh(\beta\theta_i)) \tag{9}$$

where β is the sigmoid gain. Now

$$\frac{\delta \ln p(x,\theta)}{\delta \theta_i} = \frac{1}{p_i(x_i)} \frac{\delta p_i(x_i)}{\delta \theta_i}$$
$$= 2\beta(1 - p_i(x_i)) \ .$$

Inserting this value into (6) we get:

$$\Delta\theta_i = 2\alpha\beta(f(x) - b_i)(1 - p_i(x_i)) \ . \tag{10}$$

To incorporate EDA we change these equations according to our needs. We shall decompose the equations into variable levels, i.e. we replace the cost function with the value of the variable of that position in the best fit individual. We rewrite the two equations (10) and (7) as follows:

$$\Delta\theta_i = \alpha(b_i - p_i(x_i))(1 - d_i) \tag{11}$$

[where we have put $\alpha = 2\alpha\beta$] and

$$b_i(t+1) = \gamma b_i(t) + (1-\gamma)x_i \tag{12}$$

where d_i is the marginal distribution of the variable X_i and x_i is the value of the variable X_i in the best individual in that generation. Marginal distribution d_i is calculated from the selected individuals in a generation by the formula:

$$d_i = \frac{\sum_{j=1}^{N} \delta_j(X_i = 1) + 1}{N+2} \tag{13}$$

where

$$\delta_j(X_i = 1) = 1 \text{ if in } j^{th} \text{ individual } X_i = 1$$
$$= 0 \text{ otherwise}.$$

During calculation of marginal distributions we have used Laplace correction [8] in equation (13) to make sure that no probability is 0.

We update each θ_i by using equation (11) as follows:

$$\theta_i = \theta_i + \Delta\theta_i . \tag{14}$$

According to equation (11) each θ_i is updated by a small amount if either the marginal distribution is closing to 1 or the value of the the variable X_i is 0. Since the baseline is dependent on X_i, the reinforcement signal (baseline) b_i plays a role in controlling the directions of the probability of each variable.

4.1 Algorithm

Our resulting algorithm now is a combination of reinforcement learning and EDA. Here is the algorithm:

1. Initialize θ_i and b_i ($p_i(x_i)$ will be calculated according to equation(9)).
2. Generate initial population.
3. Select N promising individuals.
4. For $i = 1$ to n do
   a) Calculate marginal distribution (d_i) according to equation(13).
   b) Update b_i according to equation(12).
   c) Update θ_i by equation (14) and find $p_i(x_i)$ by equation (9).
5. Generate M offspring by sampling the probabilities calculated in the previous step.
6. Create new population by replacing some old individuals with offspring (M).
7. If termination criteria not met, go to step (3).

The initial parameter $p_i(x_i)$ should chosen such that $p(x, \theta)$ is uniform. The algorithm runs until either an optimal solution is found or the allowed maximum no. of generations have passed.

5 Other EDA Approaches

The two main steps of EDAs are to estimate the probability distribution of selected individuals (promising solutions) and generate new population by sampling this probability distribution. There is no simple and general approach to do these efficiently. Different EDAs use different models for the estimation of probability distribution.

Univariate Marginal Distribution Algorithm (UMDA) [15], Population Based Incremental Learning (PBIL)[1] and Compact Genetic Algorithm (CGA) [11] treat variables in a problem as independent of one another. As a result, n-dimensional joint probability distribution factorizes as a product of n univariate and independent probability distributions. PBIL and CGA use a probability vector while UMDA uses both a probability vector and a population. The update rules of probability vector of PBIL and CGA are different.

Mutual Information Maximizing Input Clustering Algorithm (MIMIC) [5], Bivariate Marginal Distribution Algorithm (BMDA) [21] and Combining Optimizers with Mutual Information Trees (COMIT) [3] consider pairwise interactions among variables. These algorithms learn parameters as well as structure of the problems. They can mix building blocks of order one and two successfully.

Bayesian Optimization Algorithm (BOA) [19], Estimation of Bayesian Networks Algorithm (EBNA) [14], Factorized Distribution Algorithm (FDA) [16] and Extended Compact Genetic Algorithm (ECGA) [10] use models that can cover multiple interactions among variables. BOA and EBNA both use Bayesian Network as structure learning but different score metrics. BOA uses Bayesian Dirichlet equivalence (BDe) scores while EBNA uses K2+Penalization and Bayesian Information Criterion (BIC) scores to measure the likelihood of a model. FDA can be applied to additively decomposable problems. It uses Boltzmann selection for Boltzmann distribution. In ECGA the variables of a problem are grouped into disjoint sets and marginal distributions of these sets are used to compress the population of selected individuals. This algorithm uses the Minimum Description Length (MDL) approach to measure the goodness of a structure. All these multivariate algorithms can produce a good solution with the sacrifice of computation time.

A detailed overview of different EDA approaches in both discrete and continuous domains can be found in [14]. For applications of UMDA with Laplace corrections in binary and permutation domains, see [18].

6 Experiments

The experiments have been done for functions taken from [19] and [2]. In these functions of unitation variables are highly correlated. The following section describes the function and presents the results of the experiments.

6.1 Test Functions

Before we define our test functions, we need to define elementary function like deceptive function taken from [19].

A deceptive function of order 3 is defined as

$$f^3_{dec}(X) = \begin{cases} 0.9 & \text{if } u = 0 \\ 0.8 & \text{if } u = 1 \\ 0 & \text{if } u = 2 \\ 1 & \text{if } u = 3 \end{cases} \qquad (15)$$

where X is a vector of order 3 and u is the sum of input variables.

A bipolar deceptive function of order 6 is defined with the use of deceptive function as follows:
$$f^6_{bipolar} = f^3_{dec}(|3-u|) . \qquad (16)$$
Bipolar function has global solutions of either $(1,1,1,1,1,1)$ or $(0,0,0,0,0,0)$.

So non overlapping bipolar function for the vector $X = (X_1, X_2, \ldots, X_n)$ is defined as:
$$f_{bipolar}(X) = \sum_{i=1}^{n/6} f^6_{bipolar}(S_i) \qquad (17)$$
where $S_i = (X_{6i-5}, X_{6i-4}, \ldots, X_{6i})$. Bipolar function has total $2^{\frac{n}{6}}$ global optima and $\binom{6}{3}^{\frac{n}{6}}$ local optima, making it highly multimodal. According to Deb and Goldberg [6] all the schema of order less than 6 misleads Simple Genetic Algorithm (SGA) away from the global optimum into a local one for this bipolar function.

Another test function is the four peaks problem taken from [2]. The function is defined as
$$f_{4\_peak}(X) = max(z(X), o(X)) + R(X) \qquad (18)$$
where $z(X)$ is number of trailing 0s in the vector $X = (X_1, X_2, \ldots, X_n)$ and $o(X)$ is the number of leading 1s in X. $R(X)$ is a conditional reward defined with the threshold $s = \frac{n}{10}$ as
$$R(X) = \begin{cases} n & \text{if } z(X) > s \text{ and } o(X) > s \\ 0 & \text{otherwise.} \end{cases} \qquad (19)$$

The function has global optimum of the form: first and last $(s+1)$ bits are all 1 and 0 respectively and the remaining $(n-2s-2)$ bits in the middle are either all 0s or all 1s. As n increases, local optima increase exponentially while global optima decrease in the same rate, making local search methods trapped into local optima [2].

6.2 Experimental Results

Here we present the experimental results running the algorithm on a computer with 1133 MHZ Intel Pentimum (III)^{TM} Processor and 256 MB of RAM and in Borland C++ builder 6.0 environment. For all the problems, the experimental results of our algorithm of 50 independent runs are provided.

For all the problems, the algorithm runs until either an optimal solution is found or the maximum no. of generations has passed. This is somehow different from [19] where a population is said to be converged when the portion of some values in each position reaches 95%. In each run of the algorithm we apply truncation selection: the best half of the population is selected for the estimation of marginal probability distribution. Then we update the probability vector of the variables by reinforcement learning method using these marginal distributions and the best fit individual of a generation. For the bipolar function and

Table 1. No. of fitness evaluations required until an optimal solution of the bipolar function is found

Problem	Average			Standard Deviation		
Size	RELEDA	PBIL	UMDA	RELEDA	PBIL	UMDA
12	750.00	897.60	784.80	532.98	645.31	569.83
24	13608.00	26940.00	31920.00	6502.86	14959.58	15607.80
36	77918.40	123577.20	121096.80	53929.71	36238.56	38886.05
48	256089.60	488745.60	512227.20	109425.70	132108.30	136246.98
60	529278.00	1521948.00	1418778.00	250798.44	548318.17	508102.96
72	1014084.00	3718238.40	3802471.20	621080.42	1310214.60	1268969.22
84	1855072.80	9372510.00	9164064.00	774195.88	869822.56	1432628.32
96	2366457.60	NA	NA	765031.96	NA	NA

four peaks problem, we set the baseline factor $\gamma = 0.1$, learning rate $\alpha = 0.9$, elite=50%, maximum no. of generations=30000 and population size=10*n where n is the size of the problem. We set initial probability of each variable $p_i(x_i) = \frac{1}{2}$; hence $\theta_i = 0$. The baseline is initialized by setting each $b_i = 1$. Our replacement strategy is elitism so that the best individual of a generation is not lost, and the algorithm performs gradient ascent in search space.

To compare the experimental results of our proposed algorithm with those of PBIL and UMDA, we apply these algorithms to the bipolar function and only PBIL to the four peaks problem. We set the population size, maximum no. of generations and the rate of elitism for PBIL and UMDA with the same values as those for RELEDA. The best half of the population is selected for the calculation of marginal probabilities of the variables. We apply Laplace corrections during calculation of marginal probabilities. We also use the same selection and replacement strategy as those of RELEDA for PBIL and UMDA .

For the bipolar function, we set sigmoid gain $\beta = 0.1$ and learning rate for PBIL=0.9. In table 1 we show no. of fitness evaluations required for bipolar function by RELEDA, PBIL and UMDA until an optimal solution is found. In addition to convergence to a solution, RELEDA can discover a number of different solutions out of total $2^{\frac{n}{6}}$ global optima. For problem sizes of 12 and 24, it finds all the global optima only in 50 runs. For higher size problems, it discovers different global optima. So our algorithm can be used to find multiple solutions of a problem. For the problem size of 84, we provide results of 8 independent runs of PBIL and 15 of UMDA. For the problem size of 96, neither PBIL nor UMDA converged to any solution at all within the maximum no. of generations. We have indicated it in the table by inserting 'Not Available (NA)'.

Figures 1 and 2 show graphical views of average no. of fitness evaluations required by RELEDA vs PBIL and RELEDA vs UMDA respectively for the bipolar function. We show the experimental results in two graphs because in one graph it is very difficult to distinguish between the graph of PBIL and that of UMDA. Both PBIL and UMDA require almost the same no. of fitness evaluations to find an optimal solution of the bipolar function.

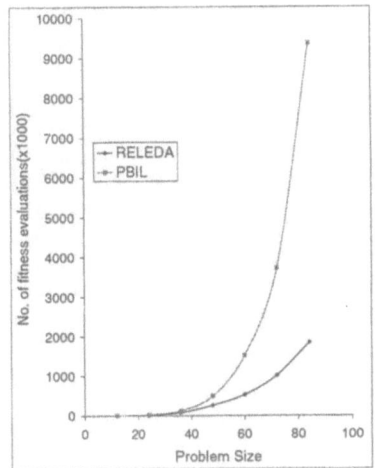 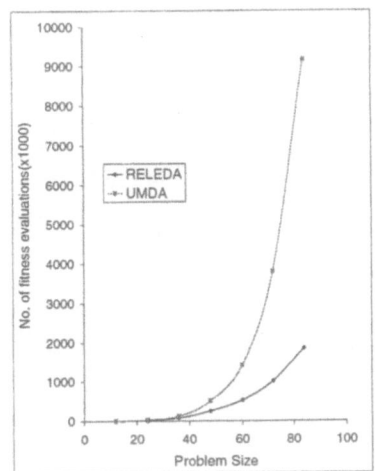

Fig. 1. Average no. of fitness evaluations required until an optimal solution of the bipolar function is found(RELEDA vs PBIL)

Fig. 2. Average no. of fitness evaluations required until an optimal solution of the bipolar function is found(RELEDA vs UMDA)

Table 2. No. of fitness evaluations required until an optimal solution of the four peak problem is found

Problem Size	Average RELEDA	PBIL	Standard Deviation RELEDA	PBIL
10	286	395	138.15	336.20
20	2366	38350	1027.30	8889.35
30	9219	189390	4765.65	63131.02
40	24904	462660	14858.31	75257.90
50	84085	643950	59619.60	52987.26
60	181224	1204290	159526.53	154901.78
70	316764	1337735	264597.58	92907.18
80	512280	1832760	425668.28	217245.50
90	715248	2283975	750726.15	262165.18

For the four peaks problem, we set sigmoid gain $\beta = 0.5$ and learning rate for PBIL=0.005 (as used in [2]). The results are shown in table 2. Graphical representation of average no. of fitness evaluations required by RELEDA and PBIL for the four peaks problem is shown in Fig. 3.

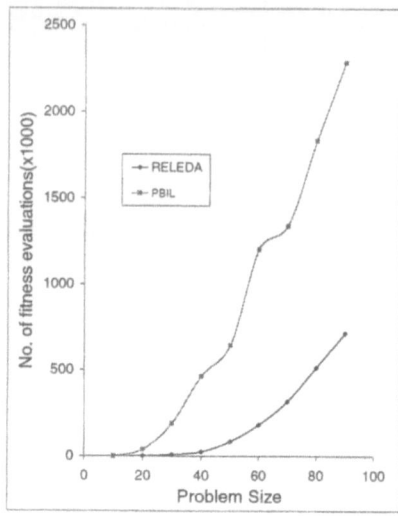

Fig. 3. Average no. fitness evaluations required until an optimal solution of the four peaks problem is found

6.3 Discussion on Experimental Results

The proposed algorithm finds optimal solutions of those problems which are very hard for simple GA (SGA) to solve. Deb and Goldberg [6] have said that in fully deceptive function with subfunctions of order k all the schema of order less than k mislead SGA away from the global optimum into a local one. In similar fashion, these deceptive functions are very hard for some EDAs. RELEDA finds optimal solutions of those problems in fewer no. of fitness evaluations because it involves a nonlinear function of hyperbolic tangent. Though the marginal distribution of a variable calculated from the selected individuals may be zero, the probability of that variable in RELEDA will not be zero. But in UMDA it will be zero. RELEDA actually shifts the probability of a variable in either directions depending on the values of the parameters by a small amount. Our algorithm has some biases to the best fit individual of a generation. As we start with the initial probability of a variable 0.5, if that variable in the best individual of the first generation is 0, the algorithm tends to produce 0 for that variable. That is why the algorithm finds optimal solutions of more 0s than 1s of the problems stated above quickly and sometimes terminates with no solutions at all within allowed maximum no. of generations.

BOA seems to be a promising algorithm for the functions of unitation, but it takes more time to find a good Bayesian network for the selected individuals.

And the learning of Bayesian Network is an NP-hard problem, but our proposed algorithm can quickly converge to an optimal solution.

7 Future Works

Our proposed algorithm is dependent on many parameters such as learning rate, sigmoid gain, baseline factor, population size etc. We are now investigating how to control these parameters for quick convergence and lesser fitness evaluations. Another question is how it can be applied to multiary search space. In this paper we have considered all variables as binary, but real world variables can take values from multiary alphabet as well as from continuous domains. We have to find some methods to make our algorithm generally applicable to all problems.

RELEDA assumes no dependency among variables, but in real life problems variables are highly correlated. Problems in permutation domains are very difficult to optimize. There are very few EDAs for combinatorial optimization in permutation domains. We shall investigate an algorithm for such kind of problems. And finally, we shall try to apply this algorithm to real world problems.

8 Summary and Conclusions

In this paper we propose the Reinforcement Learning Estimation of Distribution Algorithm (RELEDA). Our algorithm is a combination of the Estimation of Distribution Algorithm and Reinforcement Learning. We calculate marginal probability distribution of each variable using the selected individuals of a generation. Then the probability of a variable is updated using this marginal probability and the best fit individual of that generation. By sampling the probability vector of the variables, new solutions are generated.

The RELEDA is designed to solve problems in binary search space. It can solve problems with some correlated variables. With appropriate values of different parameters, it can solve problems of diverse directions. In the experiments we have shown that it can outperform some existing EDAs in terms of no. of fitness evaluations required to produce an optimal solution.

References

1. Baluja, S.: Population based incremental learning: A method for integrating genetic search based function optimization and competitive learning. Technical Report No. CMU-CS-94-163, Carnegie Mellon University, Pittsburgh, Pennsylvania, USA (1994).
2. Baluja, S. and Caruana,R.: Removing the genetics from standard genetic algorithm. In A. Prieditis and S. Russell, editors, Proceedings of the International Conference on Machine Learning, Morgan Kaufmann, (1995) 38–46 .
3. Baluja, S. and Davies, S.: Using optimal dependency trees for combinatorial optimization: Learning the structure of search space. Technical Report No. CMU-CS-97-107, Carnegie Mellon University, Pittsburgh, Pennsylvania, USA (1997).

4. Berny, A.: Statistical Machine Learning and Combinatorial Optimization. In Kallel, L., Naudts, B. and Rogers, A., editors, Theoretical Aspects of Evolutionary Computing, Springer (2001).
5. De Bonet, J.S., Isbell, C.L. and Viola, P.: MIMIC: Finding Optima by estimating probability densities. Advances in Neural Information Processing Systems, **9** (1997).
6. Deb, K. and Goldberg, D.E.: Sufficient conditions for deceptive and easy binary functions. Annals of Mathematics and Artificial Intelligence, **10** (1994), 385–408.
7. Goldberg, D.E., Korb, B. and Deb, K.: Messy genetic algorithms: Motivation, analysis and first results. Complex Systems **3(5)** (1989) 493–530.
8. González, C., Lozano, J.A. and Larrañaga, P.: Mathematical modeling of discrete estimation of distribution algorithms. In P. Larrañaga and J.A. Lozano, editors, Estimation of Distribution Algorithms: A New Tool for Evolutionary Optimization. Kluwer Academic Publishers, Boston(2001).
9. Harik, G.: Learning gene linkage to efficiently solve problems of bounded difficulty using genetic algorithms. IlliGAL Report No. 97005, Illinois Genetic Algorithms Laboratory, University of Illinois, Urbana, Illinois, USA (1997).
10. Harik, G.: Linkage learning via probabilistic modeling in the ECGA. Illigal Report No. 99010, Illinois Genetic Algorithm Laboratory, University of Illinois, Urbana, Illinois, USA (1999).
11. Harik, G.R., Lobo, F.G. and Goldberg, D.E.: The compact genetic algorithm. In Proceedings of the IEEE Conference on Evolutionary Computation,(1998) 523–528
12. Holland, J.H.: Adaptation in Natural and Artificial Systems. The University of Michigan Press (1975).
13. Kargupta, H.: Revisiting the GEMGA: Scalable evolutionary optimization through linkage learning. In Proceedings of 1998 IEEE International Conference on Evolutionary Computation,IEEE Press, Piscataway, New Jersey, USA (1998) 603–608.
14. Larrañaga, P. and Lozano, J.A.: Estimation of Distribution Algorithms: A New Tool for Evolutionary Optimization. Kluwer Academic Publishers, Boston, (2001).
15. Mühlenbein, H.: The equation for response to selection and its use for prediction. Evolutionary Computation, **5(3)** (1998) 303–346.
16. Mühlenbein, H. and Mahnig, T.: The Factorized Distribution Algorithm for additively decomposed functions. Proceedings of the 1999 Congress on Evolutionary Computation, IEEE press (1999) 752–759.
17. Mühlenbein, H. and Paaß, G.: From recombination of genes to the estimation of distributions I. Binary parameters. In Lecture Notes in Computer Science 1411: Parallel Problem Solving from Nature-PPSN IV, (1996) 178–187.
18. Paul,T.K. and Iba, H.: Linear and Combinatorial Optimizations by Estimation of Distribution Algorithms. 9th MPS Symposium on Evolutionary Computation, IPSJ Symposium **2003**,Japan (2002),99–106.
19. Pelikan, M., Goldberg, D.E. and Cantú-Paz, E.: Linkage Problem, Distribution Estimation and Bayesian Networks. Evolutionary Computation, **8(3)** (2000) 311–340.
20. Pelikan, M., Goldberg, D.E. and Lobo, F.G.: A survey of optimization by building and using probabilistic models. Technical Report, Illigal Report No. 99018, University of Illinois at Urbana-Champaign, USA (1999).
21. Pelikan, M. and Mühlenbein, H.: The bivariate marginal distribution algorithm. Advances in Soft Computing-Engineering Design and Manufacturing,(1999) 521–535.
22. Williams, R.J.: Simple statistical gradient-following algorithms for connectionist reinforcement learning. Machine Learning **8** (1992) 229–256.

Hierarchical BOA Solves Ising Spin Glasses and MAXSAT

Martin Pelikan[1,2] and David E. Goldberg[2]

[1] Computational Laboratory (Colab)
Swiss Federal Institute of Technology (ETH)
Hirschengraben 84
8092 Zürich, Switzerland
[2] Illinois Genetic Algorithms Laboratory (IlliGAL)
University of Illinois at Urbana-Champaign
104 S. Mathews Ave.
Urbana, IL 61801
{pelikan,deg}@illigal.ge.uiuc.edu

Abstract. Theoretical and empirical evidence exists that the hierarchical Bayesian optimization algorithm (hBOA) can solve challenging hierarchical problems and anything easier. This paper applies hBOA to two important classes of real-world problems: Ising spin-glass systems and maximum satisfiability (MAXSAT). The paper shows how easy it is to apply hBOA to real-world optimization problems—in most cases hBOA can be applied without any prior problem analysis, it can acquire enough problem-specific knowledge automatically. The results indicate that hBOA is capable of solving enormously difficult problems that cannot be solved by other optimizers and still provide competitive or better performance than problem-specific approaches on other problems. The results thus confirm that hBOA is a practical, robust, and scalable technique for solving challenging real-world problems.

1 Introduction

Recently, the hierarchical Bayesian optimization algorithm (hBOA) has been proposed to solve hierarchical and nearly decomposable problems [1,2,3]. The success in designing a competent hierarchical optimizer has two important consequences. First, many complex real-world systems can be decomposed into a hierarchy [4], so we can expect hBOA to provide robust and scalable solution to many real-world problems. Second, many difficult hierarchical problems are intractable by any other algorithm [2] and thus hBOA should allow us to solve problems that could not be solved before.

This paper applies hBOA to two important classes of real-world problems to confirm that hierarchical decomposition is a useful concept in solving real-world problems. More specifically, two classes of problems are considered: (1) two- and three-dimensional Ising spin glass systems with periodic boundary conditions, and (2) maximum satisfiability of predicate calculus formulas in conjunctive normal form (MAXSAT). The paper shows how easy it is to apply hBOA to combinatorial problems and achieve competitive or better performance than problem-specific approaches without the need for much problem-specific knowledge in advance.

The paper starts with a brief description of hBOA. Section 3 defines the problem of finding ground states of an Ising spin glass system and presents the results of applying hBOA with and without local search to this class of problems. Section 4 defines MAXSAT and discusses its difficulty. hBOA is again combined with local search and applied to several benchmark instances of MAXSAT. For both problem classes, the performance of hBOA is compared to that of problem-specific approaches. Finally, Sect. 5 summarizes and concludes the paper.

2 Hierarchical Bayesian Optimization Algorithm (hBOA)

Probabilistic model-building genetic algorithms (PMBGAs) [5] replace traditional variation operators of genetic and evolutionary algorithms—such as crossover and mutation—by building and sampling a probabilistic model of promising solutions. PMBGAs are sometimes referred to as estimation of distribution algorithms (EDAs) [6], or iterated density estimation algorithms (IDEAs) [7]. For an overview of PMBGAs, see [2,5,8].

One of the most advanced PMBGAs for discrete representations is the hierarchical Bayesian optimization algorithm (hBOA) [1,2,3]. hBOA evolves a population of candidate solutions to the given optimization problem starting with a random population. In each iteration, hBOA updates the population in the following four steps. hBOA first selects a population of promising solutions from the current population using one of the popular selection operators, such as tournament or truncation selection. Next, hBOA builds a Bayesian network with local structures as a probabilistic model of promising solutions. New candidate solutions are then generated by sampling the learned network. Finally, restricted tournament replacement (RTR) [9,1] is used to incorporate the new solutions into the current population.

hBOA can learn and exploit hierarchical problem decomposition, which simplifies the problem via decomposition over one or more levels. Exploitation of hierarchical decomposition enables exploration of the space of candidate solutions by juxtaposing high-quality partial solutions, starting with short-order partial solutions and ending with those of high order. This enables hBOA to solve challenging hierarchical problems that are practically unsolvable by traditional black-box optimizers such as simulated annealing or traditional genetic algorithms (GAs). Of course, hBOA can solve anything easier. It is beyond the scope of this paper to provide a detailed description of hBOA; please see [1] and [2] for more information.

3 Ising Spin Glasses

The task of finding ground states of Ising spin-glass systems is a well known problem of statistical physics. In context of GAs, Ising spin-glass systems are usually studied because of their interesting properties, such as symmetry and a large number of plateaus [10,11,12,13,14].

The physical state of an Ising spin-glass system is defined by (1) a set of spins $(\sigma_0, \sigma_1, \ldots, \sigma_{n-1})$, where each spin σ_i can obtain a value from $\{+1, -1\}$, and (2) a set of coupling constants J_{ij} relating pairs of spins σ_i and σ_j. A Hamiltonian specifies the energy of the system as $H(\sigma) = -\sum_{i,j=0}^{n} \sigma_i J_{ij} \sigma_j$. The task is to find a state of spins called the *ground state* for given coupling constants J_{ij} that *minimizes* the energy of the system. There are at least two ground states of each Ising spin-glass system (the energy of a system doesn't change if one inverts all the spins). In practice the number of ground states is even greater.

The problem of finding *any* ground state of an Ising spin glass is equivalent to a well known combinatorial problem called *minimum-weight cut* (MIN-CUT). Since MIN-CUT is NP-complete [15], the task of finding a ground state of an unconstrained Ising spin-glass system is NP-complete—that means that there exists no algorithm for solving general Ising spin glasses in polynomial time and it is believed that it's impossible to do this.

Here we consider a special case, where the spins are arranged on a two- or three-dimensional grid and each spin interacts with only its nearest neighbors in the grid. Periodic boundary conditions are used to approximate the behavior of a large-scale system. Therefore, spins are arranged on a two- or three-dimensional toroid. Additionally, we constrain coupling constants to contain only two values, $J_{ij} \in \{+1, -1\}$. In the two-dimensional case, several algorithms exist that can solve the restricted class of spin glasses in polynomial time. We will compare the best known algorithms to hBOA later in this section. However, none of these methods is applicable in the three-dimensional case.

3.1 Methodology

In hBOA each state of the system is represented by an n-bit binary string, where n is the total number of spins. Each bit in a solution string determines the state of the corresponding spin: 0 denotes the state -1, 1 denotes the state $+1$. To estimate the scalability of hBOA on two-dimensional spin-glass systems, we tested the algorithm on a number of two-dimensional spin-glass systems of size from $n = 10 \times 10 = 100$ spins to $n = 16 \times 16 = 256$ spins. For each problem size, we generated 8 random problem instances with uniform distribution over all problem instances. To ensure that a correct ground state was found for each system, we verified the results using the Spin Glass Ground State Server provided by the group of Prof. Michael Jünger (http://www.informatik.uni-koeln.de/ls_juenger/projects/sgs.html).

For each problem instance, 30 independent runs are performed and hBOA is required to find the optimum in all the 30 runs. The performance of hBOA is measured by the average number of evaluations until the optimum is found. The population size for each problem instance is determined empirically as the minimal population size for the algorithm to find the optimum in all the runs. A parameter-less population sizing scheme [16] could be used to eliminate the need for specifying the population size in advance, which could increase the total

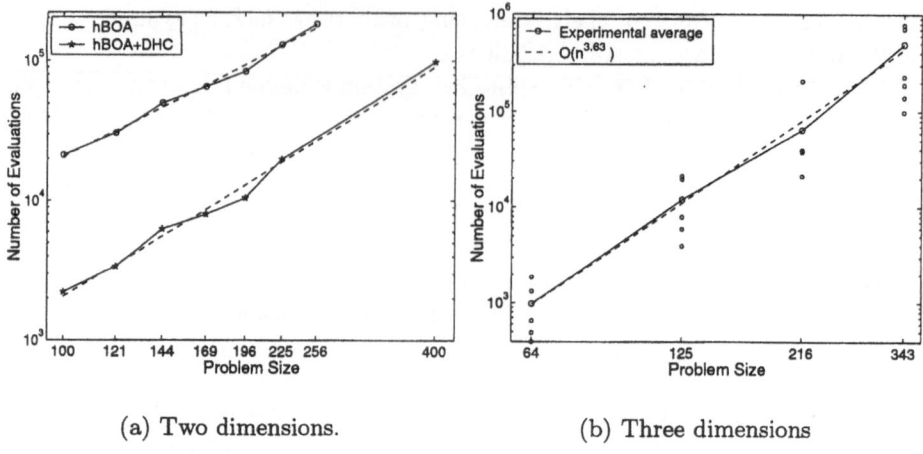

Fig. 1. hBOA on 2D and 3D Ising spin glasses

number of evaluations by at most a logarithmic factor [17]. Binary tournament selection with replacement is used in all experiments and the window size for RTR is set to the number of bits (spins) in a problem. Bayesian networks with decision graphs are used and K2 metric with the term penalizing complex models is used to measure the quality of each candidate model as described in [18].

The performance of hBOA is compared to that of hBOA combined with local search referred to as the *discrete hill climber* (DHC). DHC is applied prior to evaluation of each solution by flipping a bit that improves the solution the most; this is repeated until no more improvement is possible. For most constraint satisfaction problems including Ising spin glasses, DHC increases the computational cost of each evaluation at most $n \log n$ times; in practice, the increase in computational complexity is still significantly lower because only a few bits are flipped on average.

Using local search often improves the performance of selectorecombinative search, because the search can focus on local optima, which reveal more information about the problem than randomly generated solutions do. Furthermore, selectorecombinative search can focus its exploration on basins of attraction (peaks around each local optimum) as opposed to individual solutions. On the other hand, in some cases local search may cause premature convergence; nonetheless, we believe that this is rarely going to be the case with advanced algorithms such as BOA and hBOA. hBOA with DHC was applied to systems of size up to $n = 20 \times 20 = 400$.

3.2 Results

Figure 1a shows the average number of evaluations for both hBOA alone and hBOA with DHC. The number of evaluations is averaged over all runs for problem instances of the same size. Note that each data point in the plot corresponds to $8 \times 30 = 240$ runs. Additionally, the figure shows the best-fit polynomial approximating the growth of the number of evaluations until the optimum has been found. The total number of evaluations appears to grow polynomially as $O(n^{2.25})$ for hBOA, whereas it is $O(n^{2.73})$ for hBOA with DHC. Despite that the asymptotic complexity grows a little faster for hBOA with DHC, the population sizes and the running times decrease significantly (approximately tenfold) with DHC—for instance, hBOA with DHC can solve a 400-bit problem in *less* evaluations than hBOA needs to solve only a 225-bit problem.

3.3 Discussion

The results presented in Fig. 1a indicate that hBOA is capable of solving Ising spin-glasses in low order polynomial time. This is good news, but how does the performance of hBOA compare to that of other algorithms for solving the same subclass of spin glasses?

To answer the above question, let us first compute the overall computational complexity of hBOA without DHC. Each evaluation of a spin-glass state can be bounded by $O(n)$ trivial operations, so the overall time spent in fitness evaluation grows as $O(n^{3.25})$. However, in this case the computational complexity of model building dominates other factors. Using an asymptotic complexity bound computed by Pelikan et al. [10] and the empirical results, the overall time complexity of hBOA can be bounded by $O(n^{4.25})$. If the model was updated incrementally in each generation, the time complexity can be expected to decrease to somewhere between $O(n^{3.25})$ and $O(n^{4.25})$. The complexity of hBOA with DHC can be computed analogically to the case with hBOA alone, yielding a conservative bound of $O(n^{4.73})$.

There are several problem-specific algorithms that attempt to solve the above special case of two-dimensional spin glasses (e.g., [19,20,21,22]). Most recently, Galluccio and Loebl [21,22] proposed an algorithm for solving spin glasses in $O(n^{3.5})$ for all graphs with bounded genus (two-dimensional toroids are a special case of graphs with bounded genus). So, the overall time complexity of the best currently known algorithm for the considered class of spin glasses is $O(n^{3.5})$.

The above results indicate that the asymptotic complexity of hBOA is slightly worse than that of the best problem-specific algorithm; in particular, hBOA requires $O(n^{4.25})$ steps without DHC and $O(n^{4.73})$ steps with DHC, whereas the algorithm of Galluccio and Loebl requires only $O(n^{3.5})$ steps. However, hBOA does not use any problem-specific knowledge except for the evaluation of suggested states of the system, whereas the method of Galluccio and Loebl fully relies on the knowledge of the problem structure and its properties. Even without requiring any problem-specific information in advance, hBOA is capable of competing with the state-of-the-art methods in the field. Using DHC leads to a speed up that results in running times that are *better* than those reported by Galluccio and Loebl. For instance, using hBOA with DHC to solve 400-bit instances took on average 9.7 minutes per instance on a Pentium II/400MHz (the worst case took about 21 minutes, the best case about 2.92 minutes), while Galluccio and Loebl reported times of about 25 minutes on an Athlon/500MHz.

Therefore, hBOA+DHC is capable of finding the optimum significantly faster despite that hBOA+DHC does not assume any particular structure of the problem. Nonetheless, it can be expected that the situation will change for larger systems and the algorithm of Galluccio and Loebl will become superior.

Another important point in favor of hBOA is that hBOA does not explicitly restrict the interaction structure of a problem; consequently, hBOA is applicable to spin glasses in more than two dimensions and other spin glasses that fall outside the scope of the method of Galluccio and Loebl.

3.4 From 2D to 3D

Despite that competent methods exist for solving two-dimensional spin glasses, none of these methods is directly applicable in the three-dimensional case. In fact, finding a ground state of three-dimensional spin glasses is NP-complete even for coupling constants restricted to $\{-1, 0, +1\}$ [23]. Since in our case zero coupling constants are not allowed, instances studied here might be solvable in polynomial time although there is no algorithm that is proven to do that. Nonetheless, since hBOA does not explicitly use the dimensionality of the underlying spin-glass problem, it is straightforward to apply hBOA+DHC to three-dimensional spin glasses.

To test the scalability of hBOA with DHC, eight random spin-glass systems on a three-dimensional cube with periodic boundary conditions were generated for systems of size from $n = 4 \times 4 \times 4 = 64$ to $n = 7 \times 7 \times 7 = 343$ spins. Since no other method exists to verify whether the found state actually represents the ground state, hBOA with DHC was first run on each instance with an extremely large population of orders of magnitude larger than the expected one. After a number of generations, the best solution found was assumed to represent the ground state.

Figure 1b shows the number of evaluations until hBOA with DHC found the ground state of the tested three-dimensional Ising spin-glass instances. The overall number of evaluations appears to grow polynomially as $O(n^{3.65})$. That means that increasing the dimensionality of spin-glass systems increases the complexity of solving these systems; however, efficient performance is retained even in three dimensions.

4 MAXSAT

The task of finding an interpretation of predicates that maximizes the number of satisfied clauses of a given predicate logic formula expressed in conjunctive normal form (MAXSAT) is an important problem of complexity theory and artificial intelligence. Since MAXSAT is NP-complete in its general form, there is no known algorithm that can solve MAXSAT in worst-case polynomial time.

In the context of GAs MAXSAT is usually used as an example class of problems that *cannot* be efficiently solved using selectorecombinative search [24], although some positive results were reported with adaptive fitness [25]. The reason for poor GA performance appears to be that short-order partial solutions lead away from the optimum (sometimes in as many as 30% of predicate variables) as hypothesized by Rana and Whitley [24]. We know that hBOA outperforms traditional GAs on challenging hierarchical problems; will hBOA do better in the MAXSAT domain as well?

Here we consider the case where each formula is given in conjunctive normal form with clauses of length at most k; formulas in this form are called k-CNFs. A CNF formula is a *logical and* of clauses, where each clause is a *logical or* of k or less literals. Each literal is either a predicate or a negation of a predicate. An example 3-CNF formula over predicates (variables) X_1 to X_5 is $(X_5 \vee X_1 \vee \neg X_3) \wedge (X_2 \vee X_1) \wedge (\neg X_4 \vee X_1 \vee X_5)$.

An interpretation of predicates assigns each predicate either true or false; for example, $(X_1 = \text{true}, X_2 = \text{true}, X_3 = \text{false}, X_4 = \text{false}, X_4 = \text{true})$ is an interpretation of X_1 to X_5. The task in MAXSAT is to find an interpretation that maximizes the number of satisfied clauses in the given formula. For example, the assignment $(X_1 = \text{true}, X_2 = \text{true}, X_3 = \text{true}, X_4 = \text{true}, X_5 = \text{true})$ satisfies all the clauses in the above formula, and is therefore one of the optima of the corresponding MAXSAT problem. MAXSAT is NP complete for k-CNF if $k \geq 2$.

4.1 Methodology

In hBOA each candidate solution represents an interpretation of predicates in the problem. Each bit in a solution string corresponds to one predicate; true is represented by 1, false is represented by 0. The fitness of a solution is equal to the number of satisfied clauses given the interpretation encoded by the solution. Similarly as earlier, DHC is incorporated into hBOA to improve its performance. DHC for MAXSAT is often called GSAT in the machine learning community [26]. Similarly as for spin glasses, GSAT enhances the efficiency of hBOA, although GSAT alone is incapable of solving most tested instances. All other parameters except for the population size are the same as for spin glasses (see Sect. 3.1). For each problem instance, a minimal population size required for reliable convergence to the optimum in 30 independent runs was used.

hBOA with GSAT is compared to two methods for solving MAXSAT: GSAT, and WalkSAT. GSAT [26] starts with a random solution. In each iteration, GSAT applies a 1-bit flip that improves the current solution the most until no more improvement is possible. WalkSAT extends GSAT to incorporate random changes. In each iteration, WalkSAT performs the greedy step of GSAT with the probability p; otherwise, one of the predicates that are included in some unsatisfied clause is randomly selected and its interpretation is changed. Best results have been obtained with $p = 0.5$, although the optimal choice of p changes from application to application.

4.2 Tested Instances

Two types of MAXSAT instances are tested: (1) random satisfiable 3-CNF formulas, and (2) instances of combined-graph coloring translated into MAXSAT.

All tested instances have been downloaded from the Satisfiability Library SATLIB (http://www.satlib.org/).

Instances of the first type are randomly generated satisfiable 3-CNF formulas. All instances belong to the phase transition region [27], where the number of clauses is equal to $4.3n$ (n is the number of predicates). *Random* problems in the phase transition are known to be the most difficult ones for most MAXSAT heuristics [27]. Satisfiability of randomly generated formulas is not forced by restricting the generation procedure itself, but a complete algorithm for verifying satisfiability such as Satz [28] is used to filter out unsatisfied instances. This results in generation of more difficult problems.

Instances of the second type were generated by translating graph-coloring instances to MAXSAT. In graph coloring, the task is to color the vertices of a given graph using a specified number of colors so that no connected vertices share the same color. Every graph-coloring instance can be mapped into a MAXSAT instance by introducing one predicate for each pair (color, vertex), and creating a formula that is satisfiable if and only if exactly one color is chosen for each vertex, and the colors of the vertices corresponding to each edge are different.

Here graph-coloring instances are generated by combining regular ring lattices and random graphs with a fixed number of neighbors [29]. Combining two graphs consists of selecting (1) all edges that overlap in the two graphs, (2) a random fraction $(1-p)$ of the remaining edges from the first graph, and (3) a random fraction p of the remaining edges from the second graph. By combining regular graphs with random ones, the amount of structure in the resulting graph can be controlled; the smaller the p, the more regular the graphs are. For small values of p (from about 0.003 to 0.03), MAXSAT instances of the second type are extremely difficult for WalkSAT and other methods based on local search. Here all instances are created from graphs of 100 vertices and 400 edges that are colorable using 5 colors, and each coloring is encoded using 500 binary variables (predicates).

4.3 Results on Random 3-CNFs

Figure 2 compares the performance of hBOA with GSAT, GSAT alone, and WalkSAT. Ten instances are tested for each problem size. More specifically, the first ten instances from the archives downloaded from SATLIB are used.

How does the performance of hBOA+GSAT compare to that of GSAT alone and WalkSAT? GSAT is capable of solving only the simplest instances of up to $n = 75$ variables, because the computational time requirements of GSAT grow extremely fast. Already for instances of $n = 100$ variables, GSAT could not find the optimal interpretation even after days of computation. This leads us to a conclusion that GSAT alone cannot solve the problem efficiently, although it improves the efficiency of hBOA when used in the hybrid hBOA+GSAT. The results also indicate that the performance of WalkSAT is slightly better than that of hBOA+GSAT, although performance of the two approaches is comparable. Thus, both selectorecombinative search and randomized local search can tackle random 3-CNFs quite efficiently.

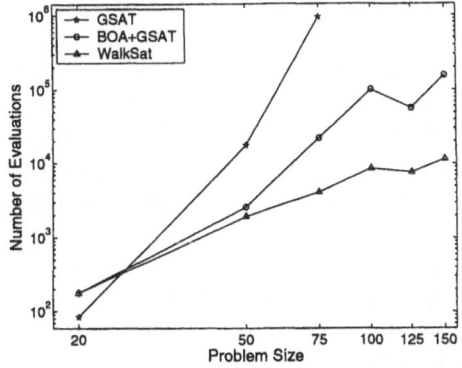

Fig. 2. hBOA+GSAT, GSAT, and WalkSAT on random 3-CNF from phase transition

Table 1. hBOA with GSAT on WalkSAT-hard MAXSAT instances. WalkSAT could not solve any of these instances even with more than 40,000,000 evaluations

Instance	p	hBOA Evals.	Instance	p	hBOA Evals.
SW100-8-5/sw100-1.cnf	2^{-5}	1,262,018	SW100-8-7/sw100-2.cnf	2^{-7}	1,558,891
SW100-8-5/sw100-2.cnf	2^{-5}	1,099,761	SW100-8-7/sw100-6.cnf	2^{-7}	1,966,648
SW100-8-5/sw100-3.cnf	2^{-5}	1,123,012	SW100-8-7/sw100-7.cnf	2^{-7}	1,222,615
SW100-8-6/sw100-1.cnf	2^{-6}	1,183,518	SW100-8-8/sw100-1.cnf	2^{-8}	1,219,675
SW100-8-6/sw100-2.cnf	2^{-6}	1,324,857	SW100-8-8/sw100-2.cnf	2^{-8}	1,537,094
SW100-8-6/sw100-3.cnf	2^{-6}	1,629,295	SW100-8-8/sw100-6.cnf	2^{-8}	1,650,568
SW100-8-7/sw100-1.cnf	2^{-7}	1,732,697	SW100-8-8/sw100-7.cnf	2^{-8}	1,287,180

4.4 Results on Graph Coloring MAXSAT

We've seen that both hBOA with GSAT and WalkSAT performed relatively well on randomly generated 3-CNF instances. Nonetheless, real-world problems are not random, most real-world problems contain a considerable amount of regularities. Combined-graph coloring described in Sect. 4.2 provides a class of problems that combines regularity with randomness. By controlling the relative amounts of structure and randomness, interesting classes of problems can be generated.

Although regular ring lattices ($p = 0$) can be solved by WalkSAT efficiently [29], introducing even a slight perturbation to the regular graph by combining it with a random graph severely affects WalkSAT's performance. More specifically, WalkSAT is practically unable to solve any instances with $p \leq 2^{-5}$. It's no surprise that for these problem instances GSAT is also intractable. On the other hand, hBOA+GSAT is capable of solving all these instances. Table 1 shows the performance of hBOA on several instances that are practically unsolvable by WalkSAT. WalkSAT is not able to solve any of these instances even when allowed to check over 40 million interpretations!

4.5 Discussion

hBOA+GSAT outperformed GSAT alone on all problem instances; not surprisingly, hBOA is capable of supplying much better starting points for GSAT than random restarts do. However, on the class of randomly generated 3-CNF, the hybrid hBOA+GSAT is outperformed by a randomized GSAT called WalkSAT. Nonetheless, the performance of hBOA+GSAT is competitive with that of WalkSAT.

On the other hand, for those problem instances that are practically unsolvable using local search (WalkSAT and GSAT), hBOA+GSAT retains efficient performance. In particular, MAXSAT instances obtained by translating graph coloring of graphs with a large amount of structure and a little amount of randomness cannot be solved by GSAT or WalkSAT even after tens of millions of evaluations, whereas hBOA+GSAT is capable of solving all these problems in fewer than two million evaluations. Therefore, hBOA+GSAT can solve those instances that are easy for local search (random 3-CNF), but it is not limited to those instances—it can solve also problems that are practically unsolvable by local search alone.

To summarize, hBOA is able to provide robust solution to two different classes of MAXSAT instances—instances that are completely random and instances that contain a significant amount of regularities and little randomness. This result is made even more important by the fact that hBOA is not given any problem specific knowledge in advance and learns how to solve the problem automatically using only the evaluation procedure.

5 Summary and Conclusions

The results presented in this paper confirmed the assumption that decomposition and hierarchical decomposition is an inherent part of many real-world problems, and that effective discovery and exploitation of single-level and hierarchical decomposition enable robust and scalable solution to a broad range of optimization problems. hBOA was capable of solving randomly generated Ising spin glass problem instances and two types of MAXSAT problems with competitive or better performance than problem specific approaches.

hBOA was told nothing about the semantics of the problem; initially it didn't know whether it was trying to solve a spin glass, MAXSAT, or any other problem. All problem-specific knowledge was acquired automatically without any interaction with the user. Recently, hBOA was successfully applied to other classes of problems also without any knowledge of the semantics of the problem; these problems included onemax, composed traps, exponentially scaled deceptive problems, and hierarchical traps. Despite the lack of prior problem-specific knowledge, hBOA was capable of automatic discovery and exploitation of problem regularities that was effective enough to solve the broad range of challenging problems in a robust and scalable manner. This adds a new piece of evidence that hBOA is indeed a robust and scalable optimization technique that should certainly make a difference in current computational optimization practice.

Acknowledgments. The authors would like to thank Martin Butz, Erick Cantú-Paz, Fernando Lobo, Jiri Ocenasek, Franz Rothlauf, Samarth Swarup, and Dav Zimak for valuable comments to earlier versions of this work and fruitful discussions.

This work was sponsored by the Air Force Office of Scientific Research, Air Force Materiel Command, USAF, under grant F49620-00-1-0163. Research funding for this work was also provided by the National Science Foundation under grant DMI-9908252. Support was also provided by a grant from the U.S. Army Research Laboratory under the Federated Laboratory Program, Cooperative Agreement DAAL01-96-2-0003. The U.S. Government is authorized to reproduce and distribute reprints for Government purposes notwithstanding any copyright notation thereon.

The views and conclusions contained herein are those of the authors and should not be interpreted as necessarily representing the official policies or endorsements, either expressed or implied, of the Air Force Office of Scientific Research, the National Science Foundation, the U.S. Army, or the U.S. Government.

References

1. Pelikan, M., Goldberg, D.E.: Escaping hierarchical traps with competent genetic algorithms. Proceedings of the Genetic and Evolutionary Computation Conference (GECCO-2001) (2001) 511–518 Also IlliGAL Report No. 2000020.
2. Pelikan, M.: Bayesian optimization algorithm: From single level to hierarchy. PhD thesis, University of Illinois at Urbana-Champaign, Urbana, IL (2002) Also IlliGAL Report No. 2002023.
3. Goldberg, D.E.: The design of innovation: Lessons from and for competent genetic algorithms. Volume 7 of Genetic Algorithms and Evolutionary Computation. Kluwer Academic Publishers (2002)
4. Simon, H.A.: The Sciences of the Artificial. The MIT Press, Cambridge, MA (1968)
5. Pelikan, M., Goldberg, D.E., Lobo, F.: A survey of optimization by building and using probabilistic models. Computational Optimization and Applications **21** (2002) 5–20 Also IlliGAL Report No. 99018.
6. Mühlenbein, H., Paaß, G.: From recombination of genes to the estimation of distributions I. Binary parameters. Parallel Problem Solving from Nature (1996) 178–187
7. Bosman, P.A., Thierens, D.: Continuous iterated density estimation evolutionary algorithms within the IDEA framework. Workshop Proceedings of the Genetic and Evolutionary Computation Conference (GECCO-2000) (2000) 197–200
8. Larranaga, P., Lozano, J.A., eds.: Estimation of Distribution Algorithms: A New Tool for Evolutionary Computation. Kluwer, Boston, MA (2002)
9. Harik, G.R.: Finding multiple solutions in problems of bounded difficulty. IlliGAL Report No. 94002, University of Illinois at Urbana-Champaign, Urbana, IL (1994)
10. Pelikan, M., Goldberg, D.E., Cantú-Paz, E.: Linkage problem, distribution estimation, and Bayesian networks. Evolutionary Computation **8** (2000) 311–341 Also IlliGAL Report No. 98013.

11. Pelikan, M., Mühlenbein, H.: The bivariate marginal distribution algorithm. Advances in Soft Computing—Engineering Design and Manufacturing (1999) 521–535
12. Naudts, B., Naudts, J.: The effect of spin-flip symmetry on the performance of the simple GA. Parallel Problem Solving from Nature (1998) 67–76
13. Van Hoyweghen, C.: Detecting spin-flip symmetry in optimization problems. In Kallel, L., et al., eds.: Theoretical Aspects of Evolutionary Computing. Springer, Berlin (2001) 423–437
14. Mühlenbein, H., Mahnig, T., Rodriguez, A.O.: Schemata, distributions and graphical models in evolutionary optimization. Journal of Heuristics **5** (1999) 215–247
15. Monien, B., Sudborough, I.H.: Min cut is NP-complete for edge weighted trees. Theoretical Computer Science **58** (1988) 209–229
16. Harik, G., Lobo, F.: A parameter-less genetic algorithm. Proceedings of the Genetic and Evolutionary Computation Conference (GECCO-99) **I** (1999) 258–265
17. Pelikan, M., Lobo, F.G.: Parameter-less genetic algorithm: A worst-case time and space complexity analysis. IlliGAL Report No. 99014, University of Illinois at Urbana-Champaign, Illinois Genetic Algorithms Laboratory, Urbana, IL (1999)
18. Pelikan, M., Goldberg, D.E., Sastry, K.: Bayesian optimization algorithm, decision graphs, and Occam's razor. Proceedings of the Genetic and Evolutionary Computation Conference (GECCO-2001) (2001) 519–526 Also IlliGAL Report No. 2000020.
19. Kardar, M., Saul, L.: The 2D +/−J Ising spin glass: Exact partition functions in polynomial time. Nucl. Phys. B **432** (1994) 641–667
20. De Simone, C., Diehl, M., Jünger, M., Reinelt, G.: Exact ground states of two-dimensional +−J Ising spin glasses. Journal of Statistical Physics **84** (1996) 1363–1371
21. Galluccio, A., Loebl, M.: A theory of Pfaffian orientations. I. Perfect matchings and permanents. Electronic Journal of Combinatorics **6** (1999) Research Paper 6.
22. Galluccio, A., Loebl, M.: A theory of Pfaffian orientations. II. T-joins, k-cuts, and duality of enumeration. Electronic Journal of Combinatorics **6** (1999) Research Paper 7.
23. Barahona, F.: On the computational complexity of ising spin glass models. Journal of Physics A: Mathematical, Nuclear and General **15** (1982) 3241–3253
24. Rana, S., Whitley, D.L.: Genetic algorithm behavior in the MAXSAT domain. Parallel Problem Solving from Nature (1998) 785–794
25. Gottlieb, J., Marchiori, E., Rossi, C.: Evolutionary algorithms for the satisfiability problem. Evolutionary Computation **10** (2002) 35–50
26. Selman, B., Levesque, H.J., Mitchell, D.: A new method for solving hard satisfiability problems. Proceedings of the National Conference on Artificial Intelligence (AAAI-92) (1992) 440–446
27. Cheeseman, P., Kanefsky, B., Taylor, W.M.: Where the really hard problems are. Proceedings of the International Joint Conference on Artificial Intelligence (IJCAI-91) (1991) 331–337
28. Li, C.M., Anbulagan: Heuristics based on unit propagation for satisfiability problems. Proceedings of the International Joint Conference on Artificial Intelligence (IJCAI-97) (1997) 366–371
29. Gent, I., Hoos, H.H., Prosser, P., Walsh, T.: Morphing: Combining structure and randomness. Proceedings of the American Association of Artificial Intelligence (AAAI-99) (1999) 654–660

ERA: An Algorithm for Reducing the Epistasis of SAT Problems

Eduardo Rodriguez-Tello and Jose Torres-Jimenez

ITESM Campus Cuernavaca, Computer Science Department
Av. Paseo de la Reforma 182-A. Lomas de Cuernavaca
62589 Temixco Morelos, Mexico
{ertello, jtj}@itesm.mx

Abstract. A novel method, for solving satisfiability (SAT) instances is presented. It is based on two components: a) An Epistasis Reducer Algorithm (ERA) that produces a more suited representation (with lower epistasis) for a Genetic Algorithm (GA) by preprocessing the original SAT problem; and b) A Genetic Algorithm that solves the preprocesed instances.
ERA is implemented by a simulated annealing algorithm (SA), which transforms the original SAT problem by rearranging the variables to satisfy the condition that the most related ones are in closer positions inside the chromosome.
Results of experimentation demonstrated that the proposed combined approach outperforms GA in all the tests accomplished.

1 Introduction

A genetic algorithm (GA) is based on three elements: an internal representation, an external evaluation function and an evolutionary mechanism. It is well known that in any GA the choice of the internal representation greatly conditions its performance [4] [14]. The representation must be designed in such a way that the gene-interaction (*epistasis*) is kept as low as possible. It is also advantageous to arrange the coding so that *building blocks* may form to aid the convergence process towards the global optimum.

We are specially interested in applying GAs to the satisfiability (SAT) problem. The most general statement of the SAT problem is very simple. Given a well-formed boolean formula F in its conjunctive normal form (CNF)[1], is there a truth assignment for the literals that satisfies it?. SAT, is of great importance in computer science both in theory and in practice. In theory SAT is one of the basic core NP-complete problems. In practice, it has become increasingly popular in different research fields, given that several problems can be easily encoded into propositional logic formulae: Planning [10], formal verification [1], and knowledge representation [6]; to mention only some.

---

[1] A CNF formula F is a conjunction of clauses C, each clause being a disjunction of literals, each literal being either a positive (x_i) or a negative ($\sim x_i$) propositional variable.

Even though the SAT problem is NP-complete, many methods have been developed to solve it. These methods can be classified into two large categories: complete and incomplete methods. Theoretically, complete methods are able to find the solution, or to prove that no solution exists provided that no time constraint is present. However, the combinatorial nature of the problem makes these complete methods impractical when the size of the problem increases. In contrast incomplete methods are based on efficient heuristics, which help to find sub-optimal solutions when applied to large optimization problems. This category includes such methods as simulated annealing [15], local search [13], and genetic algorithms (GAs) [8].

SAT has a natural representation in GAs: binary strings of length n in which the i-th bit represents the truth value of the i-th boolean variable present in the SAT formula. It is hard to imagine a better representation for this problem. Surprisingly, no efficient genetic algorithm has been found yet using this representation. We have strong reasons to think that this poor performance is caused by the effect of epistasis. It is reported in [7] that generation of building blocks will not occur when epistasis in the representation of a problem is high, in other words if the related bits in the representation are not in closer positions inside the chromosome.

In this paper a preprocessing method is proposed, called ERA (Epistasis Reducer Algorithm), to reduce the epistasis of the representation for a SAT problem, and in this way to improve the performance of a simple genetic algorithm (using classical crossover) when used to solve SAT formulae. The principle of ERA is to transform the original problem by rearranging the variables to satisfy the condition that the most related ones are in closer positions inside the chromosome.

The rest of this paper is organized as follows: Sect. 2, presents a description of the epistasis problem. In Sect. 3, a detailed description of the ERA procedure is presented. Section 4 focuses on the genetic algorithm used. In Sect. 5, the results of experiments and comparisons are presented and in last section, conclusions are discussed.

2 Tackling Epistasis

In genetics, a gene or gene pair is said to be epistatic to a gene at another locus[2], if it masks the phenotypical expression of the second one. In this way, epistasis expresses links between separate genes in a chromosome. The analogous notion in the context of GAs was introduced by Rawlins [12], who defines minimal epistasis to correspond to the situation where every gene is independent of every other gene, whereas maximal epistasis arises when no proper subset of genes is independent of any other gene. In this case the generation of building blocks will not happen.

The problems of epistasis may be tackled in two ways: as a coding problem, or as a GA theory problem. If treated as a coding problem, the solution is to

---

[2] The locus is the position within the chromosome.

find a different representation and a decoding method which does not exhibit epistasis. This will then allow conventional GA to be used. In [18] it is shown that in principle any problem can be coded in such a way that it is as simple as the "counting ones task". Similarly, any coding can be made simple for a GA by using appropriately designed crossover and mutation operators. So it is always possible to represent any problem with little or no epistasis. However, for hard problems, the effort involved in devising such a coding could be considerable.

In next section an algorithm called ERA, which is designed to tackle the epistasis of the representation in SAT instances, is presented.

3 The ERA Method

3.1 Some Definitions

The SAT problem can be represented using hypergraphs [17], where each clause is represented by a hyperedge and each variable is represented by a node. Given the following SAT problem in CNF:

$$(\sim A \vee B \vee \sim C) \wedge (A \vee B \vee \sim E) \wedge (A \vee \sim E \vee \sim F) \wedge (\sim B \vee D \vee \sim G)$$

The resulting hypergraph is shown in Fig. 1. A weighted graph can be obtained from the hypergraph SAT representation, where the weights represent how many times the variable i is related with the variable j (see Fig. 2 corresponding to the hypergraph in Fig. 1). Under this particular representation, the problem of rearranging the variables to satisfy the condition that the most related variables are in closer positions inside the chromosome, is equivalent to solve the bandwidth minimization problem for the graph.

The bandwidth minimization problem for graphs (BMPG) can be defined as finding a labeling for the vertices of a graph, where the maximum absolute difference between labels of each pair of connected vertices is minimum [3].

Formally, Let $G = (V, E)$ be a finite undirected graph, where V defines the set of vertices (labeled from 1 to N) and E is the set of edges. And a linear layout

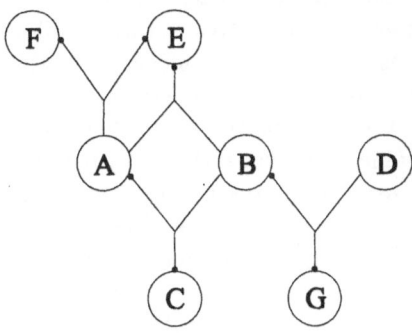

Fig. 1. Hypergraph representing a SAT problem

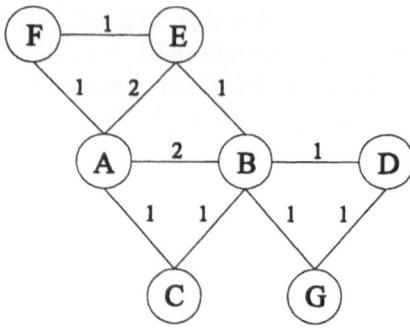

Fig. 2. Weighted graph representing a SAT problem

$\tau = \{\tau_1, \tau_2, ..., \tau_N\}$ of G is a permutation over $\{1, 2, ...N\}$, where τ_i denotes the label of the vertex that originally was identified with the label i. The bandwidth β of G for a layout τ is:

$$\beta_\tau(G) = Max_{\{u,v\} \in E} |\tau(u) - \tau(v)|$$

Then the BMPG can be defined as finding a layout τ for which $\beta_\tau(G)$ is minimum.

3.2 The Method

The ERA method is based on a simulated annealing algorithm previously reported in [16], which demonstrated to have competitive results for many classes of graphs. It will approximate the bandwidth of a graph G by examining randomly generated layouts τ of G. These new layouts are generated by interchanging a pair of distinct labels of the set of vertices V. This interchanging operation is called a *move*.

ERA begins initializing some parameters as the temperature, T; the maximum number of accepted moves at each temperature, *max_moves*; the maximum number of moves to be attempted at each temperature, *max_attempted_moves*; *max_frozen* is the number of consecutive iterations allowed for which the number of accepted moves is less than *max_moves*; and the cooling rate *cool_rate*. The algorithm continues by randomly generating a move and then calculating the change in the cost function for the new labelling of the graph. If the cost decreases then the move is accepted. Otherwise, it is accepted with probability $P(\Delta C) = e^{-\Delta C/T}$ where T is the temperature and ΔC is the increase in cost that would result from a particular move. The next temperature is obtained by using the relation $T_n = T_{n-1} *$ *cool_rate*. The minimum bandwidth of the labellings generated by the algorithm up that point in time is *min_band*. The number of accepted moves is counted and if it falls below a given limit then the system is *frozen*.

Next, the ERA procedure is presented:

Procedure ERA(G, best_map)
  $T = 0.00004;$ cool_rate $= 0.85;$
  map = best_map = random labeling;
  sa_band = Bandwidth(G, best_map);
  max_moves $= 5 * |E|;$
  max_attempted_moves $= 2*$ max_moves;
  max_frozen $= 10;$ frozen $= 0;$
  While (frozen $\leq$ max_frozen)
    moves = attempted_moves $= 0;$
    While ((moves $\leq$ max_moves) **And**
         (attempted_moves $\leq$ max_attempted_moves))
      attempted_moves $++;$
      a random move is generated, map_ran;
      If (bandwidth decreases **Or** random_number() $< e^{-\Delta Bandwidth/T}$)
        map = map_ran; moves $++;$
        If (sa_band $<$ Bandwidth(G, map))
          best_map = map;
          sa_band = Bandwidth(G, map);
        End If
      End If
    End While
    $T = T * $ cool_rate;
    If (attempted_moves $>$ max_attempted_moves)
      frozen$++;$
    Else
      frozen $= 0;$
    End If
  End While
End ERA

The parameters of the ERA algorithm were chosen taking into account our experience, and some related work reported in [11] and [15]. It is important to remark that the value of *max_moves* depends directly on the number of edges of the graph, because more moves are required for denser graphs; the value of *max_attempted_moves* is set to a large number (5 * *max_moves*), because few moves will result in bigger bandwidths. The *max_frozen* parameter that controls the external loop of our algorithm is set to 10. By modifying these three parameters one can obtain results more quickly, but probably they will not be as close to $\beta(G)$. According the experiment results the above values give a good balance between the quality of the results and the computational effort required.

4 GA for Solving SAT Problems

In order to demonstrate the benefits to preprocess SAT instances using ERA a simple genetic algorithm (using classical genetic operators) was implemented.

The basic assumption in a GA is the following: a random population of chromosomes (strings of genes that characterize particular solutions) is initially created and exposed to an environment represented by a fitness function; this fitness function is evaluated for each chromosome and only the best-fitted individuals survive and become parents for the next generation. Reproduction is then allowed by splitting the parent chromosomes to create new ones with different genetic information. This procedure is repeated and the population evolves during a given number of generations. Holland [9], introduced the theory of schemas in which he shown that above average-fit schemata get an exponentially increasing number of matches in the population as generations pass.

In the next subsections the main implementation details of the GA used in the present work is described.

4.1 Chromosome Definition

For the particular case of the SAT problem, given that it consists in a search over n boolean variables, this results in a search space of size 2^n, the most natural internal representation is: binary strings of length n in which the i-th bit represents the truth value of the i-th boolean variable.

This chromosome definition has the following advantages: It is fixed length, binary, context independent in the sense that the meaning of one bit is unaffected by changing the value of other bits, and permits the use of classic genetic operators defined by Holland [9], which have strong mathematical foundations.

4.2 Fitness Function

The choice of the fitness function is an important aspect of any genetic optimization procedure. Firstly, in order to efficiently test each individual and determine if it is able to survive, the fitness function must be as simple as possible. Secondly, it must be sensitive enough to locate promising search regions on the space of solutions. Finally, the fitness function must be consistent: a solution that is better than others must have better fitness value.

In SAT problem the fitness function used is the simplest and most intuitive one, the fraction of the clauses that are satisfied by the assignment. More formally:

$$f(chromosome) = \frac{1}{C} \sum_{c=i}^{c} f(c_i)$$

Where the contribution of each clause $f(c_i)$ is 1 if the clause is satisfied and 0 otherwise.

4.3 Operators

Recombination is done using the two-point crossover operator, and mutation is the standard bit flipping operator. The former has been applied at a 70% rate, and the later at a 0.01% rate (these parameters values were adjusted through experimentation).

Selection operator is similar to the tournament selection reported in [2]; randomly three elements of the population are selected and the one with the better fitness is chosen.

4.4 Population Size

In all cases the population size has been held fixed at: $\lfloor 1.6 * N \rfloor$, where N is the number of variables in the problem.

4.5 Termitation Criteria

The termination condition is used to determine the end of the algorithm. This GA is terminated when either a solution that satisfies all the clauses in the SAT problem is found, or the maximum number of generations is reached (300 generations).

5 Computational Results

The ERA method and the GA have been implemented in C programming language and ran into a Pentium 4 1.7 Ghz. with 256 MB of RAM. To test the algorithms described above, several SAT instances as those of the second DIMACS challenge and flat graph coloring instances were used, since they are widely-known and easily available from the SATLIB benchmarks[3].

The experimentation methodology used consistently throughout this work was as follows: For each of the selected SAT instances 20 independent runs were executed, 10 using the ERA preprocessing algorithm plus a GA (ERA+GA), and 10 solving the problem solely with the use of the same GA. All the results reported here, are data averaged over the 10 corresponding runs.

Results from ERA algorithm on the selected SAT instances are presented in Table 1. Column titled N indicates the number of nodes in the weighted graph, which represents the original SAT problem. Columns β_i and β_f represent the initial and final bandwidth obtained with the ERA algorithm and the last column presents the CPU time used by ERA measured in seconds.

As can be seen in Table 1, the ERA algorithm provides a balance between solution quality and computational effort. ERA allowed to reduce significantly the bandwidth of the weighted graphs in 5.2 seconds or less, given that the number of vertices was at most 150.

Table 2 presents the comparison between ERA+GA and GA. Data included in this comparison are: name of the formula; number of variables N; and number of clauses M. Additionally, for each approach it is presented: the number of satisfied clauses S, and the CPU time in seconds. It is important to remark that times presented for the ERA+GA approach take into account also the CPU time used for the preprocessing algorithm.

---

[3] http://www.satlib.org/benchm.html

Table 1. Results from the ERA algorithm

Formulas	N	β_i	β_f	T_β
aim100-1_6y	100	94	37	3.1
aim100-2_0y	100	96	42	5.2
dubois28	84	77	9	1.0
dubois29	87	81	9	1.0
dubois30	90	87	9	1.0
dubois50	150	144	10	2.1
flat30-1	90	85	26	2.2
flat30-2	90	81	24	2.2
pret60_75	60	56	10	1.0
pret150_40	150	143	24	2.1
pret150_60	150	140	27	3.1
pret150_75	150	142	23	3.0

Table 2. Comparative results between ERA+GA and GA

Formulas	N	M	ERA+GA S	ERA+GA T	GA S	GA T
aim100-1_6y⋆	100	160	160	8.05	159	16.81
aim100-2_0y⋆	100	200	200	9.56	198	10.11
dubois28	84	224	223	3.08	221	7.25
dubois29	87	232	231	2.09	231	19.89
dubois30	90	240	239	14.35	237	27.81
dubois50	150	400	397	4.03	395	39.94
flat30-1⋆	90	300	300	18.81	299	27.41
flat30-2⋆	90	300	300	13.64	300	17.82
pret60_75	60	160	159	2.75	159	14.67
pret150_40	150	400	399	35.73	396	23.84
pret150_60	150	400	399	15.85	397	21.31
pret150_75	150	400	397	38.82	397	65.76

The results of experimentation presented showed that ERA+GA outperforms GA in the selected SAT instances, not only in solution quality but also in computing time. The set of four satisfiable instances[4] used in the experiments were solved by ERA+GA while only one of them could be solved by GA. Better quality solutions for the ERA+GA is possible thanks to the representation used by the GA ran after ERA, which has smaller epistasis. Smaller computing time for the ERA+GA is possible because the ERA algorithm is able to find good solutions in reasonable time and the GA takes less CPU time.

Additionally it has been observed during the experimentation, that as the size of the SAT problem increases, the advantage to use the ERA algorithm also increases. It allows to conclude that it pays to make a preprocessing of SAT problems to reduce the epistasis.

---

[4] In Table 2 a ⋆ indicates a satisfiable formula.

In Figs. 3, 4, 5, and 6 is clearly appreciated the advantages to use the proposed preprocessing algorithm when SAT problems are solved. They show the convergence process of the compared algorithms. The X axis represents CPU time in seconds used by the algorithms, while the Y axis indicates the number of satisfied clauses (the results showed are the average of the 10 independent runs). It is important to point out that the ERA+GA curve initiates at the time T_β, which is the CPU time in seconds consumed in the preprocessing stage.

6 Conclusions

In this paper a novel preprocessing algorithm was introduced, called ERA, which improves the performance of GAs when used to solve SAT problems. This approach has been compared versus a simple GA without preprocessing using a set of SAT instances. The ERA+GA outperforms GA in all the selected SAT instances.

We think that the ERA+GA method is very promising and worthy to more research. In particular it will be interesting and important to identify the classes of SAT instances where is appropriate to apply our method.

Taking into account that NP-complete problems can be transformed into an equivalent SAT problem in polynomial time [5], it opens the possibility to solve through ERA+GA many NP-complete problems which do not have effective representations in GAs.

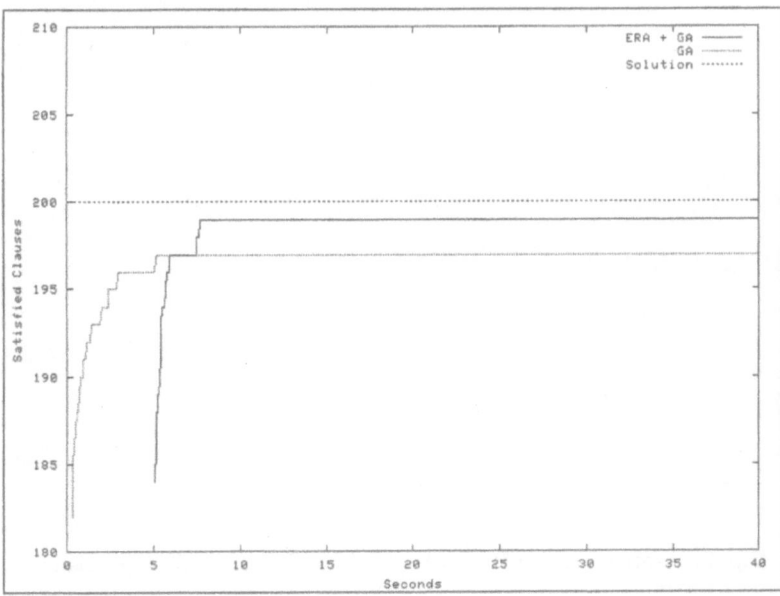

Fig. 3. Average curves for ERA+GA and GA on aim100-2_0y problem

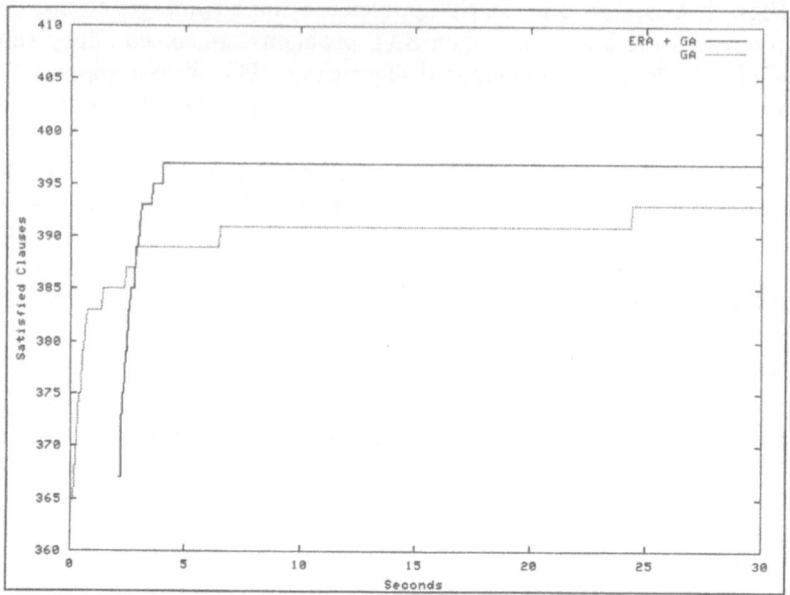

Fig. 4. Average curves for ERA+GA and GA on dubois50 problem

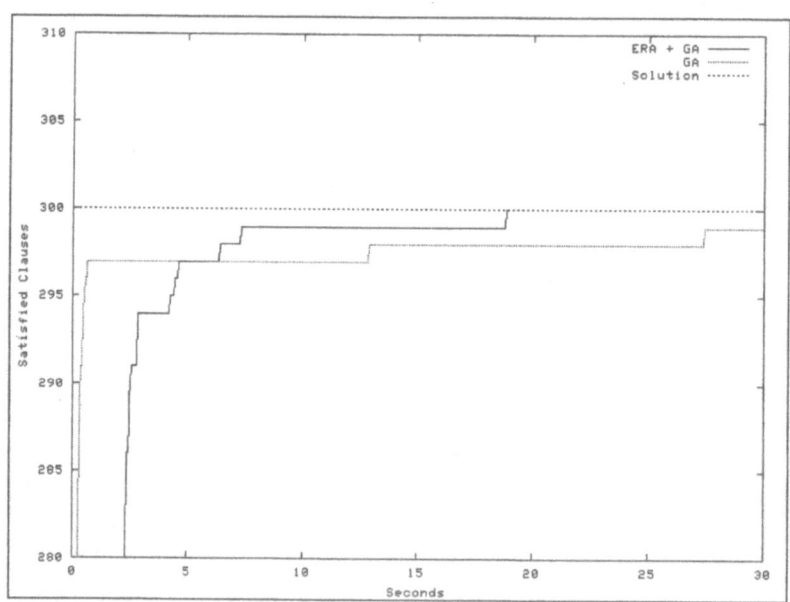

Fig. 5. Average curves for ERA+GA and GA on flat30-1 problem

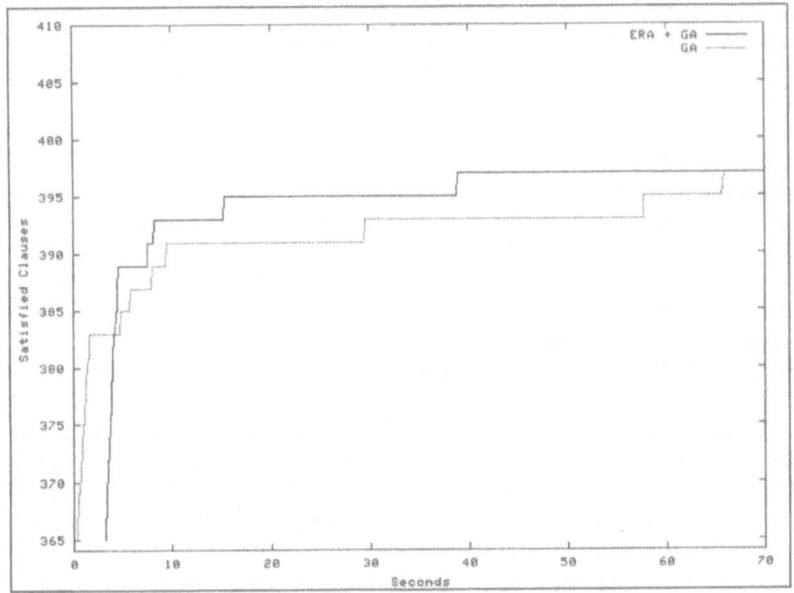

Fig. 6. Average curves for ERA+GA and GA on pret150_75 problem

References

1. Armin Biere, Alessandro Cimatti, Edmund M. Clarke, and Y. Zhu, *Symbolic Model Checking without BDDs*, Proceedings of Tools and Algorithms for the Analysis and Construction of Systems (TACAS'99), Number 1579 in LNCS, Springer Verlag, 1999, pp. 193–207.
2. T. Blickle and L. Thiele, *A mathematical analysis of tournament selection*, Proceedings of the Sixth ICGA, Morgan Kaufmann Publishers, San Francisco, Ca., 1995, pp. 9–16.
3. E. Cutchill and J. McKee, *Reducing the bandwidth of sparse symmetric matrices*, Proceedings 24th National of the ACM (1969), 157–172.
4. Y. Davidor, *Epistasis Variance: A Viewpoint of GA-Hardness*, Proceedings of the Second Foundations of Genetic Algorithms Workshop, Morgan Kaufmann, 1991, pp. 23–35.
5. M.R. Garey and D.S. Johnson, *Computers and intractability: A guide to the theory of np-completeness*, W.H. Freeman and Company, New York, 1979.
6. E. Giunchiglia, F. Giunchiglia, and A. Tacchella, *SAT-Based Decision Procedures for Classical Modal Logics*, Highlights of Satisfiability Research in the Year 2000, 2000.
7. David E. Goldberg, *Genetic Algorithms in Search, Optimization and Machine Learning*, Addison-Wesley Publishing Company, Inc., 1989.
8. J.K. Hao, *A Clausal Genetic Representation and its Evolutionary Procedures for Satisfiability Problems*, Proceedings of the International Conference on Artificial Neural Networks and Genetic Algorithms (France), April 1995.
9. J. Holland, *Adaptation in natural and artificial systems*, Ann Arbor: The University of Michigan Press, 1975.

10. Henry Kautz and Bart Selman, *Planning as Satisfiability*, Proceedings of the 10th European Conference on Artificial Intelligence (ECAI 92), 1992, pp. 359–363.
11. S. Kirkpatrick, C.D. Gelatt, and M.P. Vecchi, *Optimization by simulated annealing*, Science **220** (1983), 671–680.
12. G.J.E. Rawlins, *Foundations of Genetic Algorithms*, Morgan Kaufmann Publishers, San Mateo, 1991.
13. B. Selman, H. Levesque, and D. Mitchell, *A New Method for Solving Hard Satisfiability Problems*, Proceedings of the Tenth National Conference on Artificial Intelligence (San Jose CA), July 1992, pp. 440–446.
14. Jim Smith, *On Appropriate Adaptation Levels for the Learning of Gene Linkage*, Journal of Genetic Programming and Evolvable Machines **3** (2002), 129–155.
15. William M. Spears, *Simulated Annealing for Hard Satisfiability Problems*, Tech. Report AIC-93-015, AI Center, Naval Research Laboratory, Washington, DC 20375, 1993.
16. Jose Torres-Jimenez and Eduardo Rodriguez-Tello, *A New Measure for the Bandwidth Minimization Problem*, Proceedings of the IBERAMIA-SBIA 2000, Number 1952 in LNAI (Antibaia SP, Brazil), Springer-Verlag, November 2000, pp. 477–486.
17. Isaac Vazquez-Moran and Jose Torres-Jimenez, *A SAT Instances Construction Based on Hypergraphs*, WSEAS Transactions on Systems **1** (2002), no. 2, 244–247.
18. M. Vose and G. Liepins, *Schema Disruption*, Proceedings of the Fourth International Conference on Genetic Algorithms, Morgan Kaufmann, 1991, pp. 237–242.

Learning a Procedure That Can Solve Hard Bin-Packing Problems: A New GA-Based Approach to Hyper-heuristics

Peter Ross, Javier G. Marín-Blázquez, Sonia Schulenburg, and Emma Hart

School of Computing, Napier University
Edinburgh EH10 5DT
{P.Ross,S.Schulenburg,E.Hart}@napier.ac.uk
javierg@dai.ed.ac.uk

Abstract. The idea underlying hyper-heuristics is to discover some combination of familiar, straightforward heuristics that performs very well across a whole range of problems. To be worthwhile, such a combination should outperform all of the constituent heuristics. In this paper we describe a novel messy-GA-based approach that learns such a heuristic combination for solving one-dimensional bin-packing problems. When applied to a large set of benchmark problems, the learned procedure finds an optimal solution for nearly 80% of them, and for the rest produces an answer very close to optimal. When compared with its own constituent heuristics, it ranks first in 98% of the problems.

1 Introduction

A frequent criticism of evolutionary algorithms (EAs) is that the solutions they find can be fragile. Although the EA may solve a given problem very well, simply re-running the EA or changing the problem slightly may produce very different and/or worse results. For this reason, users may prefer to stick with using familiar and simpler but less effective heuristics. The idea of *hyper-heuristics* is to use a search process such as an EA, not to solve a given problem but to discover a combination of straightforward heuristics that can perform well on a whole range of problems.

In this paper we consider one-dimensional bin-packing problems, described below. These are simple to understand but the class is NP-hard, and they are worth studying because they appear as a factor in many other kinds of optimization problem. A very simplified representation of the possible state of any problem is used, and a messy GA is used to associate a heuristic with various specific instances of this representation. The problem solution procedure is: given a state P of the problem, find the nearest instance I and apply the associated heuristic $H(I)$. This transforms the problem state, say to P'. Repeat this until the problem has been solved.

The task of the GA is to choose the problem-state instances and to associate one of a small set of heuristics with each of them. Any given heuristic might be associated with several instances, or with none.

An example of a hyper-heuristic approach applied to the one-dimensional bin-packing problem has been reported in [12] and [13], where an accuracy-based Learning Classifier System (XCS) [14] was used to learn a set of rules that associated characteristics of the current state of a problem with eight different algorithm heuristics. In this previous work a classifier system was trained on a set of example problems and showed good generalisation to unseen problems. This represented a useful step towards the concept of using EAs to generate strong solution processes rather than merely using them to find good individual solutions. However, some questions arose from that work: could performance be improved further, and was a classifier system necessary to the approach? XCS focuses rewards on single actions [12] or on short groups of actions [13] – what if rewards are given only by the final outcome?

This paper answers these questions. It presents an alternative approach tested using a large set of benchmark one-dimensional bin-packing problems and a small set of eight heuristics. No single one of the heuristics used is capable of finding the optimal solution of more than a very few of the problems; however, the evolved rule-set was able to produce an optimal solution for nearly 80% of them, and in the rest it produced a solution very close to optimal. When compared with the solutions given by its own constituent heuristics, it ranks first in 98%.

2 One-Dimensional Bin-Packing

In the one-dimensional bin packing problem, there is an unlimited supply of bins, indexed by k, each with capacity $c > 0$. A set of n items is to be packed into the bins, the size of item $i \in \{1, \ldots, n\}$ is $s_i > 0$, and:

$$\forall k : \sum_{i \in \text{bin}(k)} s_i \leq c \qquad (1)$$

The task is to minimise the total number of bins used. Despite its simplicity, this is an NP-hard problem. If M is the minimal number of bins needed, then clearly:

$$M \geq \lceil (\sum_{i=1}^{n} s_i)/c \rceil \qquad (2)$$

and for any algorithm that does not start new bins unnecessarily, $M \leq$ bins used $< 2M$ because otherwise there would be two bins less than half filled.

Many results are known about specific algorithms. For example, a commonly-used algorithm is Largest-Fit-Decreasing (LFD): items are taken in order of size, largest first, and put in the first bin where they will fit (a new bin is opened if necessary, and effectively all bins stay open). It is known [9] that this uses no more than $11M/9 + 4$ bins. A good survey of such results can be found in [3]. A good introduction to bin-packing algorithms can be found in [11], which also introduced a widely-used heuristic algorithm, the Martello-Toth Reduction Procedure (MTRP). This simply tries to repeatedly reduce the problem to a simpler one by finding a combination of 1-3 items that provably does better than anything else (not just any combination of 1-3 items) at filling a bin, and if so packing them. This may halt with some items still unpacked; the remainder are packed using LFD. The reader may wonder whether it pays to try to fill each bin as much as possible, but [12] shows a simple counter-example.

Various authors have applied EAs to bin-packing, notably Falkenauer's grouping GA [5,7,6]; see also [10] for a different approach. Falkenauer also produced two of several sets of benchmark problems. In one of these, the so called *triplet problems*, every bin contains three items; they were generated by first constructing a solution which filled every bin exactly, and then randomly shrinking items a little so that the total shrinkage was less than the bin capacity so that the number of bins needed is unaffected. [8] raised a question about whether these problems are genuinely hard or not. It is certainly possible to solve many of them very quickly by backtracking if you exploit the knowledge that there is an optimal solution in which every bin contains three items.

3 Bin-Packing Benchmark Problems

Problems from several sources have been used in this investigation. One collection, available from Beasley's OR-Library [1], contains problems of two kinds that were generated and largely studied by Falkenauer [6]. The first kind, 80 problems named uN_M, involve bins of capacity 150. N items are generated with sizes chosen randomly from the interval 20-100. For each N in the set $(120, 250, 500, 1000)$ there are twenty problems, thus M ranges from 00 to 19. The second kind, 80 problems named tN_M, are the triplet problems mentioned earlier. The bins have capacity 1000. The number of items N is one of 60, 120, 249, 501 (all divisible by three), and as before there are twenty problems per value of N. Item sizes range from 250 to 499; the problem generation process was described earlier.

A second collection of problems studied in this paper comes from the Operational Research Library [2] at the *Technische Universität Darmstadt*: we use the 'bpp1-1' and the very hard 'bpp1-3' sets in this paper. In the bpp1-1 set problems are named NxCyWz_a where x is 1 (50 items), 2 (100 items), 3 (200 items) or 4 (500 items); y is 1 (capacity 100), 2 (capacity 120) or 3 (capacity 150); z is 1 (sizes in 1...100), 2 (sizes in 20...100) or 4 (sizes in 30...100); and a is a letter in A ... T indexing the twenty problems per parameter set. (Martello and Toth [11] also used a set with sizes drawn from 50...100, but these are far too easy.) In the hard bpp1-3 set there are just ten problems, each with 200 items and bin capacity 100,000; item sizes are drawn from the range 20,000...35,000.

Finally, a group of random problems has also been created in an attempt to be as fair as possible. The above benchmark problems do not use small item sizes, because small items could be used as sand (so to speak) to fill up boxes. A procedure that works well on such benchmark problems might still perform badly *on problems that include small items*. We generate some problems in which the bins have capacity of 100 or 500, and there are 100, 250 or 500 items whose sizes are chosen at uniform random from 1 up to the bin capacity. In all, therefore, 1016 benchmark problems are used. These were divided in two groups, a training set with 763 problems and a test set with 253.

4 The Set of Heuristics Used

Following [12], we used four basic heuristics:

- LFD, described in Sect. 2 above. This was the best of the fourteen heuristics in over 81% of the bpp1-1 problems, but was never the winner in the bpp1-3 problems.
- Next-Fit-Decreasing (NFD): an item is placed in the current bin if possible, or else a new bin is opened and becomes the current bin and the item is put in there. This is usually very poor.
- Djang and Finch's algorithm (DJD) [4]. This puts items into a bin, taking items largest-first, until that bin is at least one third full. It then tries to find one, or two, or three items that completely fill the bin. If there is no such combination it tries again, but looking instead for a combination that fills the bin to within 1 of its capacity. If that fails, it tries to find such a combination that fills the bin to within 2 of its capacity; and so on. This of course gets excellent results on, for example, Falkenauer's problems; it was the best performer on just over 79% of those problems but was never the winner on the hard bpp1-3 problems.
- DJT (Djang and Finch, more tuples): a modified form of DJD considering combinations of up to five items rather than three items. In the Falkenauer problems, DJT performs exactly like DJD, as could be expected because of the way Falkenauer problems are generated; in the bpp1-1 problems it is a little better than DJD.

As in [12] these algorithms were each also coupled with a 'filler' process that tried to find any item at all to pack in any open bins rather than moving on to a new bin, therefore creating four more heuristics. This might, for example, make a difference in DJD if a bin could be better filled by using more than three items once the bin was one-third full. So, for example, the heuristic 'Filler+LFD' first tries to fill any open bins as much as possible — this process terminates if the filling action successfully inserts at least one item. If no insertion was possible, 'Filler+LFD' invokes LFD. This forces progress: at least one gets put in a bin. Without this, endless looping might be possible. Thus, in all eight heuristics have been used. For convenience they are numbered 0..7.

5 Representing the Problem State

The problem state is defined by five real numbers. The first four numbers give the ratio R of items remaining to be packed that fall into each of four categories: huge ($c/2 < s_i$), large ($c/3 < s_i \leq c/2$), medium ($c/4 < s_i \leq c/3$) and small

($s_i \leq c/4$) These intervals are, in a sense, natural choices since at most one huge item will fit in a bin, at most two large items will fit a bin, and so on. The fifth number represents the fraction of the original number of items that remain to be packed; in a sense, it indicates the degree of progress through the problem.

It is worth noting that this representation throws away a lot of detail. For each of the benchmark problems, we tested which individual heuristics were winners on that problem. It turned out that the task of associating the initial problem state with the winning heuristics was not a linearly separable one.

6 The Genetic Algorithm

6.1 Representation

A chromosome is composed of blocks, and each block j contains six numbers $h_j, l_j, m_j, s_j, i_j, a_j$. The first five essentially represent an instance of the problem state. Here h_j corresponds to the proportion of huge items that remain to be packed, and similarly l_j, m_j and s_j refer to large medium and small items, and i_j corresponds to the proportion of items remaining to be packed. The sixth number, a_j, is an integer in the range $0\cdots 7$ indicating which heuristic is associated with this instance. An example of a set of 12 rules obtained with the GA can be seen in Fig. 1.

```
0.70 -2.16 -1.10  1.55  1.81 --> 1 |  2.34  0.67  0.19  1.93  2.75 --> 1
0.12  1.37 -0.54  1.12  0.58 --> 6 | -1.93 -2.64 -1.89  2.17 -1.46 --> 3
0.13  1.43 -1.27  0.13 -2.18 --> 2 | -1.30  0.11  2.00 -1.85  0.84 --> 4
1.87 -0.91  1.30 -1.34  1.93 --> 3 |  0.32  1.94  2.24  0.99 -0.53 --> 0
2.60  1.30 -0.54  1.12  0.58 --> 6 |  0.58  0.87  0.23 -2.11  0.47 --> 1
0.25  2.09 -1.50 -1.46 -2.56 --> 0 |  1.21  0.11  2.00  0.09  0.84 --> 4
```

Fig. 1. Example of a final set with 12 rules

For a given problem, the first five numbers would all lie in the range 0.0 to 1.0; that is, the actual problem state is a point inside a five-dimensional unit cube. The blocks in the chromosome represent a number of points, and at each step the solution process applies the heuristic associated with the point in the chromosome that is closest to the actual problem state. However, we permit the points defined in the chromosome to lie outside the unit cube, by allowing the first five numbers in a block to be in the range $-3\cdots 3$. This means that if, say, the initial problem state is $(0,0,0,1,1)$ (that is, all items are small), the nearest instance is not compelled to be an interior point of the unit cube.

It is unclear at the outset how many blocks are needed. It would be possible to use a fixed-length chromosome by guessing an upper limit on the number of blocks and, for each block, having a bit to say whether that block is to be used or not. However, we use a variable-length chromosome instead.

6.2 The Operators

We wish to keep things simple, to avoid any accusation that success rises from excessive complexity of approach. So, two very simple crossovers and three mutations have been implemented.

The first crossover is a standard two point crossover modified to work with a variable length representation and to avoid changing the meaning of any numbers. For any fixed-length GA, two point crossover uses the same two cross points for both parents and thus maintains the length of the child chromosomes. But here, each parent may choose the interchange points independently, with the caveat that the exchange points should fall inside a block in the same position for both parents. This ensures that at the end of the exchange process, the blocks are complete. The operator is implemented by choosing for each parent, a block in which to start the exchange (potentially different in each parent) and then an offset inside the block (the same in both parents). The same is done for the end point of exchange. The blocks and the points are chosen using a uniform distribution. The genetic material between these block points is then exchanged in the children.

The second crossover works at block level. It exchanges 10% of blocks between parents, that is, each block of let us say, the first parent, has a 90% chance of being passed to the first child and a 10% of being passed to the second child and vice-versa. Thus this operator merely shuffles blocks. Each crossover has an equal probability of being chosen.

The three mutations are: an add-block mutation, a remove-block mutation and a normal-mutation. The first one generates a new block and adds it to the chromosome. The first five numbers of a block are selected from a normal distribution with mean 0.5, and truncated to lie in $-3\cdots 3$; the sixth number (the heuristic, an integer in $0\cdots 7$) is chosen at uniform random. The remove-block mutation removes a random block from the chromosome. The normal-mutation changes a locus (at a rate of about one per four blocks) either adding a normally distributed random number in the case of the real values or, in the case of the heuristic, by choosing a value at uniform random.

A child has a 10% chance of being mutated. If it is to be mutated, then one type is chosen: add-block and remove-block each have a 25% chance of occurring, and normal-mutation has a 50% chance of occurring. These values were suggested by some early testing.

6.3 The Fitness Function

The fundamental ingredient in fitness is how well a chromosome solves a specified problem. But how is this to be measured? We did consider a ranking process, comparing the number of bins used with the best result so far for that problem. A chromosome's fitness would then be a weighted average of the ranking it achieved (compared to any other chromosome's result) on each of the problems it had seen so far; the weighting would be proportional to the number of time that any chromosome had tackled that problem. Being the best of only three attempts to solve a problem is not worth much.

In the end we opted for a simpler, cheaper measure: how much better was the chromosome's result on the specified problem than any single heuristic had been able to achieve (latter referred in the tables as best of four heuristics or BFH). The single-heuristic results could be prepared in advance of running the GA.

We want a chromosome to be good at solving many problems. When a chromosome is created, it is applied to just five randomly-chosen problems in order to obtain an initial fitness. Then, at each cycle of the algorithm, a new problem is given to each of the chromosomes to solve. This procedure is a trade-off between accuracy and speed; newly-created chromosomes may have not tackled many problems but they are at least built from chromosomes that had experienced many of those problems before.

The longer a chromosome survives in a population the more accurate its fitness becomes, which raises the question of how to rate a young chromosome (with therefore a poor estimate) against an older chromosome (with an improved estimate) if the fitness of the younger one is better? Several methods were tried to attempt to give higher credit to older and therefore more accurate chromosomes:

- Fixed Weighted Sum: The age and the excess/rank are combined (after normalization) with fixed weights, age weight being rather low.
- Variable Weight Sum: The weights change with the progress, age being more and more important but always using relatively low values.
- Age Projection: The search keeps track of the average degradation with age and, when comparing a chromosome with an older chromosome the fitness is updated with the expected degradation given how good is the young for its age.
- Improved Age Projection: The degradation is decomposed year by year and the updating is incremental for each year of difference.

However, preliminary trials rather surprisingly indicated that age of chromosome was not an important factor; we discarded these attempts to make any special concessions to chromosome oldsters.

6.4 The Genetic Engine

The GA uses a steady-state paradigm. From a population two parents are chosen: one by tournament size 2, the other randomly. From these two parents two children are created. The children are given a subset of four training problems to solve and obtain an initial fitness. The two children replace (if better) the two worst members of the population. Then the whole population solves a new training problem (randomly selected for each chromosome) and fitnesses are updated. The process is repeated a fixed number of times.

7 Experimental Details

The GA described above was run five times with different seeds and the results averaged. The size of the population is 40, the GA is run for 500 iterations.

That gives 24160 problems solved, 40 problems (one each member of the population) plus 2 children times 4 (initialization) problems each generation plus 40 times 4 problems for initial initialization. That means that, on average, each training problem has been seen about 31 times.

For training, all available bin-packing problems were divided into a training and a test set. In each case the training set contained 75% of the problems; every fourth problem was placed in the test set. Since the problems come in groups of twenty for each set of parameters, the different sorts of problem were well represented in both training and test sets.

At the end of training, the fittest chromosome is used on every problem in the training set to assess how well it works. It is also applied to every problem in the test set.

The initial chromosomes are composed of 16 randomly created blocks. It is of course possible to inject a chromosome that represents a single heuristic simply by specifying that heuristic in every block, and so test how your favourite heuristic fares compared to evolved hyper-heuristics. It can be argued that the imprecision regarding the fitness can kill these individual heuristics but they would have first to be the worst of the population to be in a situation were they could be removed. Being the worst of a population, even with imprecise estimations of quality, hardly qualifies as a potential winner.

8 Results

Tables 1 and 2 show cumulative percentages of problems solved with a particular number of extra bins when compared with the results of, respectively, the best of the four heuristics (BFH) and the literature reported optimal value. HH-Methods stand for Hyper-heuristic methods, XCSs and XCSm for the results of the XCS in its single [12] and multi-step [13] versions. Finally, Trn and Tst stand for results obtained with the training and testing set of problems respectively.

Cumulative percentages means that a cell is read as percentage of problems solved with x **or fewer** bins than the comparison reference (BFH or reported).

Although the main focus and achievement of this work is to be able to improve results by finding smart combinations of individual heuristics, it also seems interesting to see the performance versus known optimal results (even if the heuristics to be combined can not achieve them). Table 2 gives this information. Not all problems used to train and test the GA (and the other methods) have reported optimal values and therefore Table 2 uses a smaller number of problems, but in particular including the bpp1-1, bpp1-3 and Falkenauer problems.

Table 1 shows that there is little room for improvement over the best of the four heuristics (DJT being the best in about 95% of the problems) but the HH-Methods have achieved better performance. GA solves about 98% of the problems as well or better than the best of four heuristics (getting a 3% improvement). Even more, in about 5% of cases it outperforms its own constituent heuristics by packing all items in fewer bins than the best single heuristic, with an interesting maximum improvement of 4 bins less in 1% of the cases. XCS's

Table 1. Extra bins compared to best of four heuristics (BFH)

	HH Methods						Heuristics							
	GA		XCSs		XCSm		LFD		NFD		DJD		DJT	
Bins	Trn	Tst	Trn	Tst	Trn	Tst	Trn	Tst	Trn	Tst	Trn	Tst	Trn	Tst
-4		0.4												
-3	0.3	0.8												
-2	1.3	1.2	0.3	0.5	0.3	0.9								
-1	4.2	5.5	2.7	2.2	2.3	3.6								
0	98.3	97.6	98.3	97.3	98.8	97.3	71.1	73.9			91.2	91.7	95.4	94.1
1	100	100	100	100	100	100	83.8	82.6	0.1		97.3	97.6	99.7	99.6
2							88.9	88.5	0.1		98	98.4	100	100
3							91.9	92.5	1.1	2	99.6	98.8		
4							93.8	93.3	3.7	4	100	99.6		
5							95.8	96.1	7.2	5.9		100		
10							97.4	96.8	25.3	24.5				
20							99.7	99.6	48.1	47.8				
30							100	100	61.1	60.5				

Table 2. Extra bins compared versus optimal reported

	HH Methods						Heuristics							
	GA		XCSs		XCSm		LFD		NFD		DJD		DJT	
Bins	Trn	Tst	Trn	Tst	Trn	Tst	Trn	Tst	Trn	Tst	Trn	Tst	Trn	Tst
0	78.8	78.8	78.7	76.2	79	76.7	61.5	64.4			70	69.4	74.3	71.6
1	96	95.4	94.9	94.6	94	94.6	78.1	77.9	0.2		90.7	91.4	93.6	94.1
2	98.7	99.4	97.6	97.8	97.6	98.2	82.5	82	0.2		94.9	95.5	97.3	97.3
3	100	100	98.8	98.7	98.8	98.7	88.6	88.7	0.3	0.9	98	97.3	98.8	98.7
4			99.4	99.1	99.4	99.1	91.6	91.9	2.8	3.6	99.4	98.7	99.4	99.1
5				100	99.5	100	99.5	93.7	92.3	7.9	6.3	100	99.5	100	99.5
6					99.5			99.5	95.2	95.5	12.2	11.3		99.5		99.5
7					100			100	95.5	95.5	16.7	16.2		100		100
10							95.5	95.5	27.5	27				
20							97.8	97.8	50.2	50				
30							100	100	61.5	61.7				

overall performance is much the same than the GA, however, the percentage of saved bins (negative bins) is smaller than the GA, just about 2.5%. But to be fair, it is important to remember that XCS works on binary state representations with only 2 bits to encode the range $[0\dots1]$.

As Table 2 shows, while DJT reaches optimality in 73% of the cases, the HH methods achieve up to 79%. It is important to note the excellent performance of the GA that, at worst, would use three extra bins (and just in 1% of the cases) while the other methods have worst cases of seven extra bins. This shows the robustness that hyper-heuristics can achieve.

Table 3. Distribution of actions performed when solving problems

Action	Heur	Averaged			Single best rule set		
		GA	XCSs	XCSm	GA	XCSs	XCSm
0	LFD	0.1	3.09	14.20		29.18	
1	NFD		0.37				
2	DJD	15	17.47	22.66		24.46	18.67
3	DJT	3.7	25.39	20.76		4.58	45.05
4	LFD+F	4.3	18.46	5.949	7.02		10.91
5	NFD+F	1.7					
6	DJD+F	33.4	18.42	20.07	92.98	38.87	25.37
7	DJT+F	41.9	16.80	16.35		2.91	

In addition, the GA is able to generalise well. Results with training problems are very close to results using the new test cases. This means that particular details learned (a structure of some kind) during the adaptive phase can be reproduced with completely new data (unseen problems taken from the test sets).

Table 3 shows the distribution of actions performed when all the problems are solved using the hyper-heuristic methods. The averaged column refers to the distributions taking all the different sets of rules obtained using several seeds. The single best rule set column shows the distribution for just one set of rules, in particular the one that gives the best results.

Heuristic NFD has a very poor performance as Tables 1 and 2 demonstrate (never reaching being the best of the four heuristic in any presented problem) and the HH methods learn that fact (very low proportion of actions of that heuristic). However, very interestingly, sometimes it is useful as it is executed, in up to 2% of the actions, by rules obtained by the GA (although not in the best sets of rules) that is the winner of all bin packing methods.

XCS learns to use DJs heuristics in about 80% of the cases and LFD the other 20% of the time, with an even distribution between using or not using the filler. On the other hand, the GA learns to use DJs in about 95% of the cases and clearly favours the use of the filler. The examples of particular sets of rules seems to fit more or less into these distributions of actions.

It has to be stated that the resultant sets of rules were **never** composed of a single heuristic and that no chromosome with one block only survived in any of the tested populations. This also happened in the experiments done to tune the GA. This supports that Hyper-heuristics seems then to be the path to follow.

9 Conclusions

This paper represents another step towards developing the concept of hyper-heuristics: using EAs to find powerful combinations of more familiar heuristics.

From the experiments shown it is also interesting to note that:

1. It was shown in [12,13] that XCS is able to create and develop feasible hyper-heuristics that performed well on a large collection of benchmark data sets found in literature, and better than any individual heuristic. This work consolidates such conclusion from another approach, using a GA, and using a different reward schema.
2. The system always performed better than the best of the algorithms involved, and in fact produced results that were either optimal (in the large majority of cases) or at most one extra bin (in just 1.5% of the cases).
3. In about 5% of the cases the hyper-heuristics are able to find a combination of heuristics that **improve** the individual application of each of them.
4. The worst case of the GA compared versus the reported optimal value uses three extra bins (and in just 1% of the problems) while the other methods have worst cases of seven extra bins.

The current rules can be difficult to interpret (see Fig. 1). The distance between the antecedents and the problem state is a measure of how close are the concepts they represent. Long distances mean that the concepts are rather dissimilar. In the current implementation only being closer than any other rule is taken into account. No attempt to reduce as much as possible the average distance between the rules and the problem states is performed. Introducing a bias towards sets of rules with reduced distances would make rules much more interpretable.

On the other hand, among the particular advantages of the use of the GA it is interesting to mention that:

1. A particular set of rules can be tested for optimality by generating a chromosome containing them and injecting it into the population. This feature was used to show that individual heuristics were not optimal.
2. Variable chromosome length and encoding the action allows for a heuristic to be used in as many different contexts as necessary.
3. The use of real encoding allows for better precision to define the contexts where an action is preferred than if binary encoding of the problem state where used.
4. Rewards are given by final results. This allows for learning long chain of actions that locally can be seen as suboptimal but globally performs very well (like opening and underfilling some bins to fill them in latter stages). XCSs, for example, focuses in locally optimal actions (an action is rewarded regarding how well it filled a bin).
5. Fitness functions allows for other objectives to be taken into account (as compact sets of rules, minimized distances between rules antecedents and states, etc.).

In summary, HH methods performance is substantially better than the best heuristics alone with the GA being slightly better than the XCS method.

Acknowledgments. This work has been supported by UK EPSRC research grant number GR/N36660.

References

1. http://www.ms.ic.ac.uk/info.html.
2. http://www.bwl.tu-darmstadt.de/bwl3/forsch/projekte/binpp/.
3. E.G. Coffman, M.R. Garey, and D.S. Johnson. Approximation algorithms for bin packing: a survey. In D. Hochbaum, editor, *Approximation algorithms for NP-hard problems*, pages 46–93. PWS Publishing, Boston, 1996.
4. Philipp A. Djang and Paul R. Finch. Solving One Dimensional Bin Packing Problems. 1998.
5. Emanuel Falkenauer. A new representation and operators for genetic algorithms applied to grouping problems. *Evolutionary Computation*, 2(2):123–144, 1994.
6. Emanuel Falkenauer. A hybrid grouping genetic algorithm for bin packing. *Journal of Heuristics*, 2:5–30, 1996.
7. Emanuele Falkenauer. A Hybrid Grouping Genetic Algorithm for Bin Packing. Working Paper IDSIA-06-99, CRIF Industrial Management and Automation, CP 106 - P4, 50 av. F.D.Roosevelt, B-1050 Brussels, Belgium, 1994.
8. I.P. Gent. Heuristic Solution of Open Bin Packing Problems. *Journal of Heuristics*, 3(4):299–304, 1998.
9. D.S. Johnson. *Near-optimal bin-packing algorithms*. PhD thesis, MIT Department of Mathematics, 1973.
10. Sami Khuri, Martin Schutz, and Jörg Heitkötter. Evolutionary Heuristics for the Bin Packing Problem. In D.W. Pearson, N.C. Steele, and R.F. Albrecht, editors, *Artificial Neural Nets and Genetic Algorithms: Proceedings of the International Conference in Ales, France, 1995*, 1995.
11. Silvano Martello and Paolo Toth. *Knapsack Problems. Algorithms and Computer Implementations*. John Wiley & Sons, 1990.
12. Peter Ross, Sonia Schulenburg, Javier G. Marín-Blázquez, and Emma Hart. Hyperheuristics: learning to combine simple heuristics in bin packing problems. In *Genetic and Evolutionary Computation Conference*, New York, NY, 2002. Winner of the Best Paper Award in the Learning Classifier Systems Category of GECCO 2002.
13. Sonia Schulenburg, Peter Ross, Javier G. Marín-Blázquez, and Emma Hart. A Hyper-Heuristic Approach to Single and Multiple Step Environments in Bin-Packing Problems. In Pier Luca Lanzi, Wolfgang Stolzmann, and Stewart W. Wilson, editors, *Learning Classifier Systems: From Foundations to Applications*, Lecture Notes in Artificial Intelligence. Springer-Verlag, Berlin, 2003. In press.
14. Stewart W. Wilson. Classifier Systems Based on Accuracy. *Evolutionary Computation*, 3(2):149–175, 1995.

Population Sizing for the Redundant Trivial Voting Mapping

Franz Rothlauf

Department of Information Systems 1
University of Mannheim
68131 Mannheim, Germany
franz@rothlauf.com

Abstract. This paper investigates how the use of the trivial voting (TV) mapping influences the performance of genetic algorithms (GAs). The TV mapping is a redundant representation for binary phenotypes. A population sizing model is presented that quantitatively predicts the influence of the TV mapping and variants of this encoding on the performance of GAs. The results indicate that when using this encoding GA performance depends on the influence of the representation on the initial supply of building blocks. Therefore, GA performance remains unchanged if the TV mapping is uniformly redundant that means on average a phenotype is represented by the same number of genotypes. If the optimal solution is overrepresented, GA performance increases, whereas it decreases if the optimal solution is underrepresented. The results show that redundant representations like the TV mapping do not increase GA performance in general. Higher performance can only be achieved if there is specific knowledge about the structure of the optimal solution that can beneficially be used by the redundant representation.

1 Introduction

Over the last few years there has been an increased interest in using redundant representations for evolutionary algorithms (EAs) (Banzhaf, 1994; Dasgupta, 1995; Barnett, 1997; Shipman, 1999; Shackleton et al., 2000; Yu & Miller, 2001; Toussaint & Igel, 2002). It was recognized that redundant representations increase the evolvability of EAs (Shackleton et al., 2000; Ebner et al., 2001; Smith et al., 2001; Yu & Miller, 2001) and there is hope that such representations can increase the performance of evolutionary search. However, recent work (Knowles & Watson, 2002) indicated that redundant representations do not increase EA performance. More so, in most of the problems investigated (NK-landscapes, H-IFF, and MAX-SAT) redundant representations appeared to reduce EA performance.

The goal of this paper is to investigate how the redundancy of the trivial voting (TV) mapping, that is a redundant representation for binary phenotypes, influences the performance of genetic algorithms (GA). The developed population sizing model for the TV mapping is based on previous work (Rothlauf, 2002, sect. 3.1) that shows that the population size N that is necessary to find the optimal solution is proportional to $O(2^{k_r}/r)$, where k_r is the order of redundancy and r is the number of genotypic building blocks (BBs) that represent the optimal phenotypic BB. The results show that uniformly redundant representations do not result in a better performance of GAs. Only if the good solutions are overrepresented by the TV mapping does GA performance increase. In contrast, if the good solutions are underrepresented GA performance decreases. Therefore, the redundant TV mapping can only be used beneficially for GA search if knowledge about the structure of the optimal solution exists.

The paper is structured as follows. In the following section we review some results of previous work (Rothlauf, 2002) that presented a model on how redundant representations influence GA performance. Section 3 describes the TV mapping. We discuss the properties of the representation and formulate a population sizing model. In Sect. 4 the paper presents experimental results for one-max and deceptive trap problems. The paper ends with concluding remarks.

2 Redundant Representations

The following section reviews the population sizing model presented in Rothlauf (2002, Sect. 3.1). In particular, Sect. 2.1 reviews characteristics of redundant representations and Sect. 2.2 presents the population sizing model for redundant representations.

2.1 Characteristics of Redundant Representations

In this subsection we introduce some characteristics of redundant representations based on Rothlauf (2002, Sect. 3.1).

In general, a representation f_g assigns genotypes $x_g \in \Phi_g$ to phenotypes $x_p \in \Phi_p$. Φ_g, respectively Φ_p are the genotypic and phenotypic search spaces. A representation is redundant if the number of genotypes $|\Phi_g|$ exceeds the number of phenotypes $|\Phi_p|$. The order of redundancy k_r is defined as $\log(|\Phi_g|)/\log(|\Phi_p|)$ and measures the amount of redundant information in the encoding. When using binary genotypes and binary phenotypes, the order of redundancy can be calculated as

$$k_r = \frac{\log(2^{l_g})}{\log(2^{l_p})}, \quad (1)$$

where l_g is the length of the binary genotype and l_p is the length of the binary phenotype. When using a non-redundant representation, the number of genotypes equals the number of phenotypes and $k_r = 1$.

Furthermore, we have to describe how a representation over- or underrepresents specific phenotypes. Therefore, we introduce r as the number of genotypes that represent the one phenotype that has the highest fitness (we assume that there is only one global optimal solution). When using non-redundant representations, every phenotype is assigned to exactly one genotype and $r = 1$. However, in general, $1 \leq r \leq |\Phi_g| - |\Phi_p| + 1$.

The population sizing model presented in Rothlauf (2002) is valid for selectorecombinative GAs. Selectorecombinative GAs use crossover as the main search operator and mutation only serves as a background operator. When using selectorecombinative GAs we implicitly assume that there are building blocks (BBs) and that the GA process schemata. Consequently, we must define how k_r and r depend on the properties of the BBs.

In general, when looking at BBs of size k there are 2^k different phenotypic BBs which are represented by 2^{kk_r} different genotypic BBs. Therefore,

$$k_r = \frac{k_g}{k_p}, \qquad (2)$$

where k_g denotes the genotypic size of a BB and k_p the size of the corresponding phenotypic BB. As before, a representation is redundant if $k_r > 1$. The size of the genotypic BBs is k_r times larger than the size of the phenotypic BB. Furthermore, r is defined as the number of genotypic BBs of length kk_r that represent the best phenotypic BB of size k. Therefore, in general,

$$r \in \{1, 2, \ldots, 2^{kk_r} - 2^k + 1\}.$$

In contrast to k_r which is determined by the representation used, r depends not only on the used representation, but also on the specific problem that should be solved. Different instances of a problem result in different values of r. If we assume that k_r is an integer (each phenotypic allele is represented by k_r genotypic alleles) the possible values of the number of genotypic BBs that represent the optimal phenotypic BB can be calculated as

$$r = i^k, \text{ with } i \in [1, 2, \ldots, 2^{k_r} - 1]. \qquad (3)$$

A representation is uniformly redundant if all phenotypes are represented by the same number of different genotypes. Therefore, when using an uniformly redundant representation every phenotypic BB of size $k = k_p$ is represented by

$$r = 2^{k(k_r-1)} \qquad (4)$$

different genotypic BBs.

2.2 Influence of Redundant Representations on GA Performance

This subsection reviews the population sizing model presented in Rothlauf (2002). This population sizing model assumes that the redundancy of a representation influences the initial supply of BBs.

Earlier work (Harik, Cantú-Paz, Goldberg, & Miller, 1999) has presented a population sizing model for selectorecombinative GAs. The probability of failure α of a GA can be calculated as

$$\alpha = 1 - \frac{1 - (q/p)^{x_0}}{1 - (q/p)^N}. \qquad (5)$$

where x_0 is the expected number of copies of the best BB in the randomly initialized population, $q = 1 - p$, and p is the probability of making the right choice between a single sample of each BB

$$p = N\left(\frac{d}{\sqrt{2m'}\sigma_{BB}}\right). \quad (6)$$

d is the signal difference between the best BB and its strongest competitor, $m' = m - 1$ with m is the number of BBs in the problem, σ_{BB}^2 is the variance of a BB, and $q = 1 - p$ is the probability of making the wrong decision between two competing BBs. Therefore, we get for the population size that is necessary to solve a problem with probability $1 - \alpha$:

$$N = \frac{\log\left(1 - \left(\frac{1-(q/p)^{x_0}}{1-\alpha}\right)\right)}{\log(q/p)}.$$

Rothlauf (2002) extended the work presented in Harik, Cantú-Paz, Goldberg, and Miller (1999) and assumed that redundant representations change the initial supply of BBs:

$$x_0 = N\frac{r}{2^{kk_r}}. \quad (7)$$

After some approximations (compare Rothlauf (2002, Sect. 3.1)) we get for the population size N:

$$N \approx -\frac{2^{k_r k - 1}}{r}\ln(\alpha)\frac{\sigma_{BB}\sqrt{\pi m'}}{d}. \quad (8)$$

The population size N goes with $O\left(\frac{2^{k_r}}{r}\right)$. With increasing r the number of individuals that are necessary to solve a problem decreases. Using a uniformly redundant representation, where $r = 2^{k(k_r-1)}$, does not change the population size N in comparison to non-redundant representations.

3 The Trivial Voting Mapping

In the following subsection we give a short introduction into the trivial voting (TV) mapping.

When using the TV mapping, a set of mostly consecutive, genotypic alleles is relevant for the value of one allele in the phenotype. Each allele in the genotype can only influence the value of one allele in the phenotype. The value of the phenotypic allele is determined by the majority of the values in the genotypic alleles. In general, the different sets of alleles in the genotype defining one phenotypic allele have the same size. Furthermore, all genotypes that represent the same phenotype are very similar to one another. A mutation in a genotype results either in the same corresponding phenotype, or in one of its neighbors.

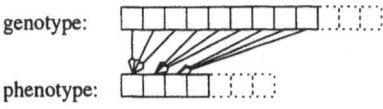

Fig. 1. The trivial voting mapping

This is an important aspect of the TV mapping as the population sizing model presented in Rothlauf (2002) is only valid for such representations.

The TV mapping can easily be characterized using the representation parameters defined in Sect. 2.1. The order of redundancy k_r is simply the number of genotypic alleles that determine the value of one phenotypic allele. As the representation is uniformly redundant, $r = 2^{k(k_r-1)}$ (Eq. (4)). Figure 1 gives an example for the TV mapping.

Shackleton, Shipman, and Ebner (2000) applied the TV mapping to binary strings illustrating that the use of redundant representations increases the evolvability of EAs. When used for binary strings, binary genotypes $x^g \in \mathbb{B}^{l_g}$ are assigned to binary phenotypes $x^p \in \mathbb{B}^{l_p}$. The length of a genotype is larger than the length of a phenotype, $l_g > l_p$. The value of one phenotypic bit is determined by the majority of the values in the corresponding genotypic bits (majority vote). However, if k_r is even then the number of ones could equal the number of zeros. Therefore, half the cases that result in a tie should encode a one in the corresponding phenotypic allele, and half the cases should represent a zero. For example, for $k_r = 4$ the genotypic BBs 1100, 1010, and 1001 represent a 1 and the phenotypic BBs 0011, 0101, 0110 represent a zero.

Because the majority of the votes determines the values of the corresponding phenotypic allele, the TV mapping is a uniformly redundant representation. Each phenotypic BB is represented by the same number of genotypic BBs which is $2^{k(k_r-1)}$, where k is the size of the phenotypic BB.

As we are not only interested in uniformly redundant representations, but also want to know how non-uniformly redundant representations influence GA performance, we extend the TV mapping to allow the encoding to overrepresent some individuals. Therefore, we want to assume that if the number of ones in the k_r genotypic alleles $x^g_{k_r i+j}$, where $j \in \{0, \ldots, k_r - 1\}$, is larger or equal than a constant u then the value of the phenotypic allele x^p_i is set to one ($i \in \{0, \ldots, l_p - 1\}$). Similarly, the phenotypic allele x^p_i is set to zero if less than u of the corresponding genotypic alleles are set to one. x^g_i respectively x^p_i denotes the ith allele of the genotype respectively phenotype. Therefore,

$$x^p_i = \begin{cases} 0 \text{ if } & \sum_{j=0}^{k_r-1} x^g_{k_r i+j} < u \\ 1 \text{ if } & \sum_{j=0}^{k_r-1} x^g_{k_r i+j} \geq u, \end{cases}$$

where $u \in \{1, \ldots, k_r\}$. u can be interpreted as the number of genotypic alleles that must be set to one to encode a one in the corresponding phenotypic allele. We denote this representation the *extended trivial voting (eTV) mapping*. For $u = (k_r + 1)/2$ (k_r must be odd) we get the original TV mapping. Extending the TV mapping in the proposed way allows us to investigate how non-uniform redundancy influences the performance of GAs.

When using the eTV mapping, the number r of genotypic BBs that can represent the optimal phenotypic BB depends on the number of ones in the genotypic alleles that determine the value of the corresponding phenotypic allele. Considering Eq. (3) we get

$$r = \left(\sum_{j=u}^{k_r} \binom{k_r}{j} \right)^k, \tag{9}$$

where $u \in \{1, \ldots, k_r\}$. k denotes the size of the phenotypic BB. We want to give a short illustration. We use a redundant representation with $k_r = 3$, $k = 1$, and the optimal BB is $x_i^p = 1$ (compare Fig. 1). Because $u \in \{1, \ldots, k_r\}$ there are three different values possible for r. For $u = 1$ the phenotypic allele x_i^p is set to one if at least one of the three corresponding genotypic alleles $x_{ik_r}^g$, $x_{ik_r+1}^g$, or $x_{ik_r+2}^g$ is set to one. Therefore, a one in the phenotype is represented by $r = \sum_{j=1}^{3} \binom{k_r}{j} = 7$ different genotypic BBs (111, 110, 101, 011, 100, 010, and 001). For $u = 2$, the optimal phenotypic BB $x_i^p = 1$ is represented by $r = \sum_{j=2}^{3} \binom{k_r}{j} = 4$ different genotypic BBs (111, 110, 101, and 011) and the representation is uniformly redundant. For $u = 2$ we get the original TV mapping. For $u = 3$, the optimal phenotypic BB is represented only by one genotypic BB (111).

Finally, we can formulate the population sizing model for the eTV mapping combining Eq. (5), 7, and 9. The probability of failure can be calculated as

$$\alpha = 1 - \frac{1 - (q/p)^{\left(\frac{N}{2^{kk_r}} \left(\sum_{j=u}^{k_r} \binom{k_r}{j} \right)^k \right)}}{1 - (q/p)^N}, \tag{10}$$

where k_r is the number of genotypic bit that represent one phenotypic bit, k is the size of the phenotypic BBs, $q = 1 - p$, and p is the probability of making the right choice between a single sample of each BB (Eq. (6)).

4 Experiments and Empirical Results

We present empirical results when using the TV and eTV mapping for the one-max problem and the concatenated deceptive trap problem.

4.1 One-Max Problem

The first test example for our empirical investigation is the one-max problem. This problem is very easy to solve for GAs as the fitness of an individual is simply the number of ones in the binary phenotype. To ensure that recombination results in a proper mixing of the BBs, we use uniform crossover for all experiments with the one-max problem. As we focus on selectorecombinative GAs we use no mutation. Furthermore, in all runs we use tournament selection without

Table 1. The trivial voting mapping for $k_r = 3$

x_i^p	$x_{3i}^g x_{3i+1}^g x_{3i+2}^g$ (with $k_r = 3$)			original TV
	extended TV			
	$u=1$	$u=2$	$u=3$	$u=2$
	$r=7$	$r=4$	$r=1$	$r=4$
0	000	001, 010, 100, 000	111, 110, 101, 011, 001, 010, 100	001, 010, 100, 000
1	111, 110, 101, 011, 001, 010, 100	111, 110, 101, 011	000	111, 110, 101, 011

replacement and a tournament size of 2. For the one-max function the signal difference d equals 1, the size k of the building blocks is 1, and the variance of a building block $\sigma_{BB}^2 = 0.25$.

When using the binary TV mapping for the one-max problem each bit of a phenotype x^p is represented by k_r bits of the genotype x^g. The string length of a genotype x^g is $l_g = k_r * l_p$ and the size of the genotypic search space is $|\Phi_g| = 2^{k_r l_p}$. Table 1 illustrates for $k_r = 3$ the three possibilities of assigning genotypic BBs $\{000, 001, 010, 100, 110, 101, 011, 111\}$ to one of the phenotypic BBs $\{0, 1\}$ when using the extended TV mapping described in the previous paragraphs. With denoting x_i^p the value of the ith bit in the phenotype, the $3i$th, $(3i+1)$th, and $(3i+2)$th bit of a genotype determine x_i^p. Because the size of the BBs $k=1$, the number of genotypic BBs that represent the optimal phenotypic BB is either $r=1$, $r=4$, or $r=7$ (compare Eq. (9)).

In Fig. 2(a) ($k_r = 2$), 2(b) ($k_r = 3$), and 2(c) ($k_r = 4$) the proportion of correct BBs at the end of a run for a 150 bit one-max problem using the TV and eTV mapping is shown. For this problem 2^{150} different phenotypes are represented by either 2^{300} ($k_r = 2$), 2^{450} ($k_r = 3$), or 2^{600} ($k_r = 4$) different genotypes. If we use the eTV mapping (indicated in the plots as eTVM) we can set u either to 1 or 2 ($k_r = 2$) or to 1, 2, or 3 ($k_r = 3$), or to 1, 2, 3, or 4 ($k_r = 4$). The corresponding values for r which can be calculated according to Eq. (9) as well as x_0/N are shown in Table 2. x_0 is the expected number of copies of the best BB in the initial population and N is the population size. Furthermore, the figures show the results when using the original, uniformly redundant TV mapping, and when using the non-redundant representation with $k_r = 1$.

The lines without line points show the theoretical predictions from Eq. (10), and the lines with line points show the empirical results which are averaged over 250 runs. The error bars indicate the standard deviation.

The results show that for the uniformly redundant TV mapping, $r = 2$ ($k_r = 2$), $r = 4$ ($k_r = 3$), or $r = 8$ ($k_r = 4$) we get the same performance as for using the non-redundant representation ($k_r = 1$). As in the original model proposed by Harik, Cantú-Paz, Goldberg, and Miller (1999) the theoretical model slightly underestimates GA performance. As predicted by the model described in Sect. 2.2, GA performance does not change when using a uniformly redundant representation. Furthermore, we can see that if the optimal BB is underrepresented GA performance decreases. Equation (10) gives us a good prediction for

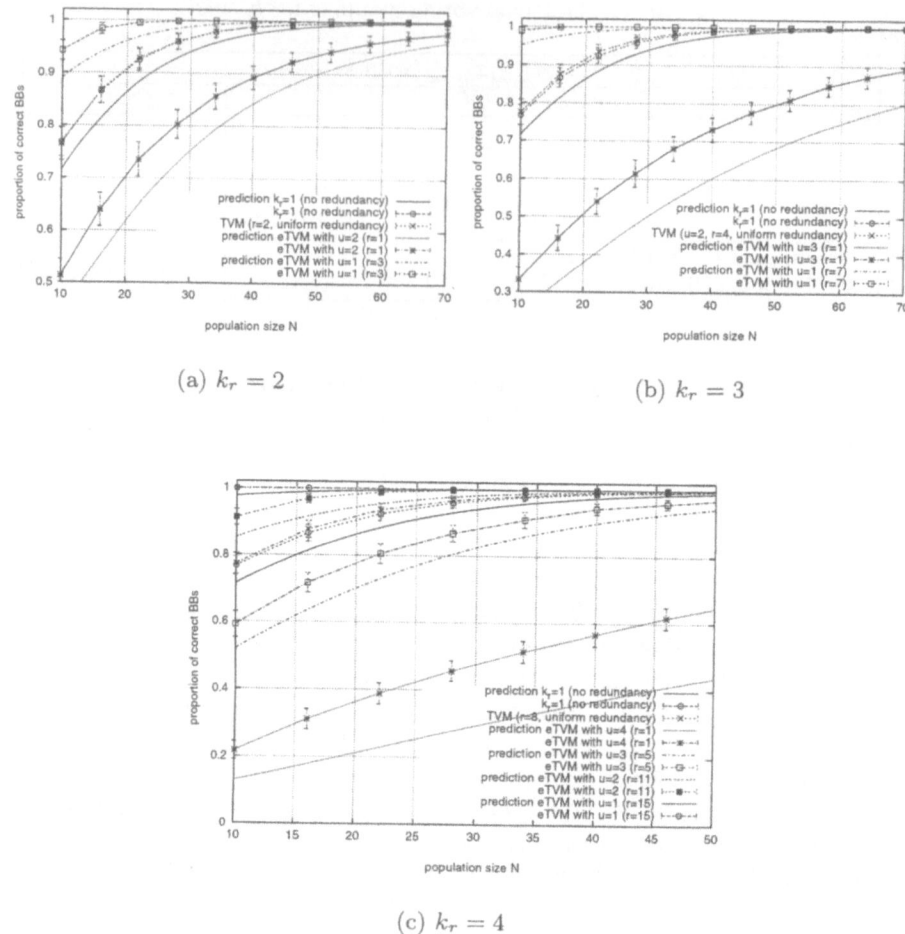

Fig. 2. Experimental and theoretical results of the proportion of correct BBs on a 150-bit one-max problem using the trivial voting mapping for $k_r = 2$ (a), $k_r = 3$ (b), and $k_r = 4$ (c). The lines without line points show the theoretical predictions. When using non-uniformly redundant representations, GA performance is changed with respect to the overrepresentation or underrepresentation of the high-quality BBs.

the expected solution quality if we consider that the non-uniform redundancy of the representation changes the initial BB supply according to Eq. (7). If the optimal solution is overrepresented GA performance increases. Again the theoretical models give a good prediction for the expected proportion of correct BBs.

Summarizing the results, we can see that using the uniformly redundant TV mapping does not change GA performance in comparison to using the non-redundant representation. Only if we overrepresent the optimal phenotypic BB, does GA performance increase; likewise, if we underrepresent the optimal BB,

Table 2. Properties of the different TV mappings for the one-max problem ($k = 1$)

		extended TV mapping				original TV mapping
		$u=1$	$u=2$	$u=3$	$u=4$	
$k_r = 2$	r	3	1	-	-	2
	x_0/N	3/4	1/4	-	-	$2/4 = 1/2$
$k_r = 3$	r	7	4	1	-	4
	x_0/N	7/8	$4/8 = 1/2$	1/8	-	$4/8 = 1/2$
$k_r = 4$	r	15	11	5	1	8
	x_0/N	15/16	11/16	5/16	1/16	$8/16 = 1/2$

GA performance drops. The derived model is able to make accurate predictions for the expected solution quality.

4.2 Concatenated Deceptive Trap Problem

Our second test example uses deceptive trap functions.

Traps were first used by Ackley (1987) and investigations into the deceptive character of these functions were provided by Deb and Goldberg (1993). Figure 3 depicts a 3-bit deceptive trap problem where the size of a BB is $k = 3$. The fitness value of a phenotype x^p depends on the number of ones v in the string of length l. The best BB is a string of l ones which has fitness l. Standard EAs are misled to the deceptive attractor which has fitness $l-1$. For the 3-bit deceptive trap the signal difference d is 1, and the fitness variance equals $\sigma_{BB}^2 = 0.75$. We construct a test problem for our investigation by concatenating $m = 10$ of the 3-bit traps so we get a 30-bit problem. The fitness of an individual x is calculated as $f(x) = \sum_{i=0}^{m-1} f_i(v)$, where $f_i(v)$ is the fitness of the ith 3-bit trap function from Fig. 3. Although this function is difficult for GAs it can be solved with proper population size N.

For deceptive traps of size $k = 3$ we can calculate the number r of genotypic BBs that represent the optimal genotypic BBs according to Eq. (9). Table 3 summarizes for the modified TV mapping how r and x_0/N depends on u, that describes how many of the genotypic alleles must be set to 1 to encode a 1 in the phenotype. x_0 is the expected number of copies of the best BB in the initial

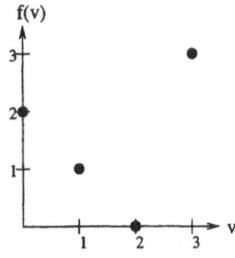

Fig. 3. A 3-bit deceptive trap problem

Table 3. Properties of the different TV mappings for the deceptive trap of size $k = 3$

		extended TV mapping				original TV mapping
		$u = 1$	$u = 2$	$u = 3$	$u = 4$	
$k_r = 2$	r	$3^3 = 27$	$1^3 = 1$	-	-	$2^3 = 8$
	x_0/N	$27/64$	$1/64$	-	-	$8/64 = 1/8$
$k_r = 3$	r	$7^3 = 343$	$4^3 = 64$	$1^3 = 1$	-	$4^3 = 64$
	x_0/N	$343/512$	$64/512 = 1/8$	$1/512$	-	$64/512 = 1/8$
$k_r = 4$	r	$15^3 = 3375$	$11^3 = 1331$	$5^3 = 125$	$1^3 = 1$	$8^3 = 2048$
	x_0/N	$3375/4096$	$1331/4096$	$125/4096$	$1/4096$	$512/4096 = 1/8$

population and N is the population size. Furthermore, we list the properties of the original uniformly redundant TV mapping.

By analogy to the previous paragraphs, in Fig. 4(a) ($k_r = 2$), Fig. 4(b) ($k_r = 3$), and Fig. 4(c) ($k_r = 4$) we show the proportion of correct BBs at the end of a run over different population sizes for ten concatenated 3-bit deceptive trap problems. In this problem, 2^{30} different phenotypes are represented by either 2^{60} ($k_r = 2$), 2^{90} ($k_r = 3$), or 2^{120} ($k_r = 4$) different genotypes. As before, we use tournament selection of size two without replacement. In contrast to the one-max problem, two-point crossover was chosen for recombination. Uniform crossover would result in an improper mixing of the BBs because the genotypic BBs are either of length $l_g = k_r l_p = 6$ ($k_r = 2$), of length $l_g = 9$ ($k_r = 3$), or of length $l_g = 12$ ($k_r = 4$). Again, the lines without line points show the predictions of the proposed model for different r. Furthermore, empirical results that are averaged over 250 runs, are shown for various values of r. The results show that for the uniformly redundant TV mapping we get the same performance as when using the non-redundant representation ($k_r = 1$). As in the experiments for the one-max problem the proposed model predicts the experimental results well if the eTV mapping is used and some BBs are underrepresented or overrepresented.

The presented results show that the influence of the redundant TV and eTV mapping on the performance of GAs can be explained well by the influence of the representation on the initial supply of high-quality BBs. If the eTV mapping favors high-quality BBs then the performance of GAs is increased. If good BBs are underrepresented the performance is reduced. If the representation is uniformly redundant, GAs show the same performance as when using the non-redundant encoding.

5 Summary and Conclusions

This paper presented a population sizing model for the trivial voting mapping and variants of this representation. The trivial voting mapping is a redundant representation for binary phenotypes. The presented population sizing model is based on previous work (Rothlauf, 2002) and assumes that redundant representations affect the initial supply of building blocks. The model was adapted to

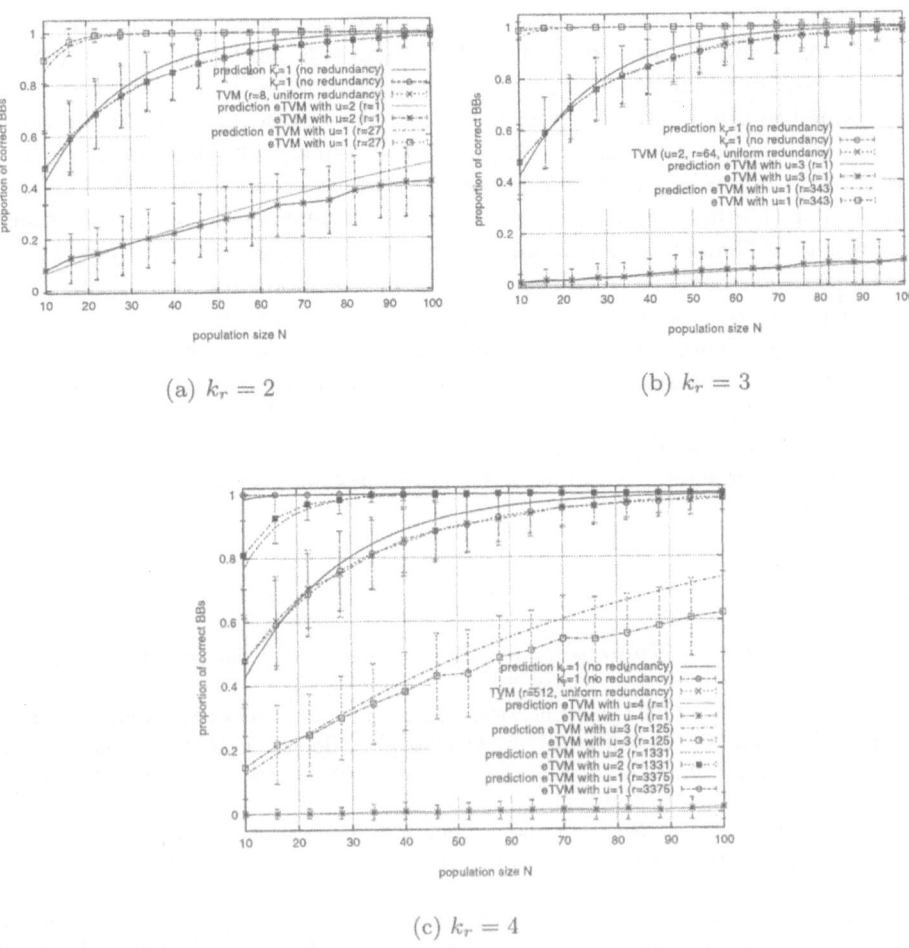

Fig. 4. Experimental and theoretical results of the proportion of correct BBs for ten concatenated 3-bit deceptive traps. We show results for different variants of the TV mapping and $k_r = 2$ (a), $k_r = 3$ (b), and $k_r = 4$ (c). The lines without line points show the theoretical predictions. As predicted, GA performance sharply decreases if the eTV mapping underrepresents the optimal BB.

the TV mapping and used for predicting the performance of genetic algorithms for one-max problems and deceptive trap problems.

The results show that the proposed population sizing model allows an accurate prediction of the influence of the redundant TV representation on GA performance. GA performance remains unchanged if the TV mapping is uniformly redundant that means each phenotype is represented on average by the same number of genotypes. Furthermore, the proposed population sizing model is able to give accurate quantitative predictions on the expected solution quality if variants of the TV mapping either overrepresent or underrepresent the optimal solution. If the optimal BB is overrepresented GA performance increases, if it is underrepresented it decreases. The results reveal that in general the redundant TV mapping and variants of it do not increase GA performance. GA performance can only be increased if there is specific knowledge about the structure of the optimal solution and if the representation overrepresents the good solutions.

Previous work (for example Shipman et al. (2000), or Ebner et al. (2001)) noticed that redundant representations increase the evolvability of EAs and assumed that this may lead to increased EA performance. However, the results we present indicate that redundant representations like the TV mapping do not increase EA performance. The influence of redundant representation on the initial supply seems to be more relevant for EA performance than an increase of evolvability. Therefore, we encourage researchers to focus on the question of how redundant representations influence EA performance and to neglect their influence on evolvability. The influence on the evolvability of EAs might be an interesting partial aspect but more relevant is the question of whether we are able to construct EAs that allow us to solve relevant problems efficiently, fast, and reliably.

References

Ackley, D. H. (1987). *A connectionist machine for genetic hill climbing.* Boston: Kluwer Academic.
Banzhaf, W. (1994). Genotype-phenotype-mapping and neutral variation – A case study in genetic programming. In Davidor, Y., Schwefel, H.-P., & Männer, R. (Eds.), *Parallel Problem Solving from Nature- PPSN III* (pp. 322–332). Berlin: Springer.
Barnett, L. (1997). *Tangled webs: Evolutionary dynamics on fitness landscapes with neutrality.* Master's thesis, School of Cognitive Sciences, University of East Sussex, Brighton.
Dasgupta, D. (1995). Incorporating redundancy and gene activation mechanisms in genetic search for adapting to non-stationary environments. In Chambers, L. (Ed.), *Practical Handbook of Genetic Algorithms* (Chapter 13, pp. 303–316). CRC Press.
Deb, K., & Goldberg, D.E. (1993). Analyzing deception in trap functions. In Whitley, L.D. (Ed.), *Foundations of Genetic Algorithms 2* (pp. 93–108). San Mateo, CA: Morgan Kaufmann.
Ebner, M., Langguth, P., Albert, J., Shackleton, M., & Shipman, R. (2001, 27-30 May). On neutral networks and evolvability. In *Proceedings of the 2001 Congress on Evolutionary Computation CEC2001* (pp. 1–8). COEX, World Trade Center, 159 Samseong-dong, Gangnam-gu, Seoul, Korea: IEEE Press.
Harik, G., Cantú-Paz, E., Goldberg, D.E., & Miller, B.L. (1999). The gambler's ruin problem, genetic algorithms, and the sizing of populations. *Evolutionary Computation, 7*(3), 231–253.
Knowles, J.D., & Watson, R.A. (2002). On the utility of redundant encodings in mutation-based evolutionary search. In Merelo, J.J., Adamidis, P., Beyer, H.-G., Fernandez-Villacanas, J.-L., & Schwefel, H.-P. (Eds.), *Parallel Problem Solving from Nature, PPSN VII* (pp. 88–98). Berlin: Springer-Verlag.
Rothlauf, F. (2002). *Representations for genetic and evolutionary algorithms.* Studies on Soft Computing and Fuzziness. Berlin: Springer Verlag.

Shackleton, M., Shipman, R., & Ebner, M. (2000, 6-9 July). An investigation of redundant genotype-phenotype mappings and their role in evolutionary search. In *Proceedings of the 2000 Congress on Evolutionary Computation CEC00* (pp. 493–500). La Jolla Marriott Hotel La Jolla, California, USA: IEEE Press.

Shipman, R. (1999). Genetic redundancy: Desirable or problematic for evolutionary adaptation? In *Proceedings of the 4th International Conference on Artificial Neural Networks and Genetic Algorithms (ICANNGA)* (pp. 1–11). Springer Verlag.

Shipman, R., Shackleton, M., & Harvey, L. (2000). The use of neutral genotype-phenotype mappings for improved evoutionary search. *British Telecom Technology Journal, 18*(4), 103–111.

Smith, T., Husbands, P., & M., O. (2001). Neutral networks and evolvability with complex genotype-phenotype mapping. In *Proceedings of the European Converence on Artificial Life: ECAL2001* (pp. 272–281).

Toussaint, M., & Igel, C. (2002). Neutrality: A necessity for self-adaptation. In Fogel, D.B., El-Sharkawi, M.A., Yao, X., Greenwood, G., Iba, H., Marrow, P., & Shackleton, M. (Eds.), *Proceedings of the 2002 Congress on Evolutionary Computation CEC2002* (pp. 1354–1359). IEEE Press.

Yu, T., & Miller, J. (2001). Neutrality and evolvability of Boolean function landscapes. In *Proceedings of the 4th European Conference on Genetic Programming (EuroGP)*, Volume LNCS 2038 (pp. 204–217). Springer.

Non-stationary Function Optimization Using Polygenic Inheritance

Conor Ryan, J.J. Collins, and David Wallin

CSIS, University of Limerick
Limerick, Ireland
{conor.ryan,j.j.collins,david.wallin}@ul.ie

Abstract. Non-stationary function optimization has proved a difficult area for Genetic Algorithms. Standard haploid populations find it difficult to track a moving target, and tend to converge on a local optimum that appears early in a run.
It is generally accepted that diploid GAs can cope with these problems because they have a *genetic memory*, that is, genes that may be required in the future are maintained in the current population. This paper describes a haploid GA that appears to have this property, through the use of Polygenic Inheritance. Polygenic inheritance differs from most implementations of GAs in that several genes contribute to each phenotypic trait.
Two non-stationary function optimization problems from the literature are described, and a number of comparisons performed. We show that Polygenic inheritance enjoys all the advantages normally associated with diploid structures, with none of the usual costs, such as complex crossover mechanisms, huge mutation rates or ambiguity in the mapping process.

1 Introduction

Natural Evolution is characterized by the adaptation of life forms to ever changing and sometimes hostile environments. Artificial Evolution, on the other hand, is usually applied to static environments, with fixed training and test cases. While Artificial Evolution has enjoyed much success with these fixed environments, there is a school of thought [6,1,4] that if Artificial Evolution were successfully applied to changing environemnts, the search power could improve.

This kind of behaviour could be useful in Genetic Algorithms for a number of reasons. Some problems exhibit shifting target functions [5,7,10], often with a periodic nature. If using a haploid genotype, it is generally better to start evolution from scratch after encountering one of these changes, but if one could force the existing population to adapt, a satisfactory solution could be found much more quickly. This would be particularly useful if the change had a cyclic nature, as the population could be expected to react more quickly to subsequent changes.

If a population could adapt to environmental changes, it would also be possible to use on-line GAs, where the fitness function could vary over time as conditions change. Indeed, it is possible that the population could adapt to previously unforeseen circumstances, as these populations tend to contain considerably more diversity than standard ones, even after lengthy periods of stability.

The "Red Queen Hypothesis", which was observed by Hillis [6], states that evolution is more powerful when the species evolving is involved in an "evolutionary arms race", that is, the problem being tackled gets progressively more difficult, as both problem and species evolve. Another reason why one might wish to employ a capricious environment could be if one was tackling a problem with an infeasibly large training set. Much time could be saved by training the population on various segments of it at a time, and one could reasonably hope that important lessons learnt on one segment will not be forgotten once it has been replaced with another set.

Most higher life forms [3], from plants up through animals to humans, use diploid genetic structures, that is, chromosomes which contain two genes at each locus. Indeed, many crops, and in particular, commercially developed crops, have polyploid structures - that is, three or more genes at each locus. It is thought that these structures have enabled populations to react to changes in the environment [3,5,10], in particular changes that bring about situations that existed in the species history. Species often exhibit a "genetic memory" [10], where it appears that genes that were once useful can be kept in the population - although rarely apparent in the phenotype - until they are needed again, and this allows the population as a whole to react more quickly to changes in the environment.

Historically, most algorithms used in these types of problems have employed a diploid representation scheme, in which two copies of each gene is maintained. However, as described in Sect. 2.1, there are some issues with this type of representation. This paper describes a haploid scheme which shares many of the characteristics of diploid schemes, but with few of the costs. It then compares this scheme, known as *Shades*, to a well known diploidy representation.

The layout of the paper is as follows. Section 2 gives a brief introduction to diploidy, and some of the more well known systems, before illustrating some of the issues with their implementation. The proposed Shades scheme is described in Sect. 3. In Sect. 4 we describe the two problems, Ošmera's dynamic problem and the constrained knapsack problem, on which our analysis are based. We then proceed with an empirical analysis of the results of the Shades scheme and compare its performance against a haploid and two diploid schemes in Sect. 5.

2 Background

Diploid GAs maintain two copies of each gene, known as alleles, typically choosing one of them to be expressed as the phenotype. In Mendel's experiments, two alleles were described for each gene, one *recessive* and the other *dominant*. In the case of a homozygous location, i.e. both forms the same, the expressed allele is selected at random. However, in the case of a heterozygous location, where there are two different forms of the gene, the dominant form is expressed, as in Fig. 1. Diploidy can shield genes from selection by holding them in abeyance, i.e.

Fig. 1. A simple dominance scheme. Upper case letters represent dominant genes, lower case recessive ones. On the left are a pair of chromosomes, while the right illustrates the expressed alleles

the gene is present but not expressed in the phenotype. During crossover, one of the genes from each locus is passed on so that in this manner, even though a gene may not have been expressed, it can still be inherited by an offspring. This endows the population with a "genetic memory" [10], where genes that were once useful can be kept in the population until they are needed again, and this allows the population as a whole to react more quickly to changes in the environment.

2.1 Previous Work

Goldberg [5] reported on a number of early diploid schemes. Unfortunately, each of these contains an inherent bias, as the resolution of heterozygous pair requires one gene to be chosen over the other.

An alternative to this scheme was proposed by Ng & Wong [10], the Dominance Change Mechanism, in which the dominance relationship between alleles could change over time. Although Lewis et al. [8] showed that it was able to track changes in an environment even after enormously long static periods, the system is dependent on hyper mutation, which must be triggered by a user-specified change in fitness. Typically, this change in fitness is relatively large, in the order of 20%, which means that not only must one be familiar with the environment being examined, but that the system is not suitable for domains in which the changes are relatively small.

2.2 Diploidy without Dominance

A different approach to diploidy avoids the use of dominance by randomly choosing which chromosome to choose a bit from [11]. The intention in this scheme is to let a bias evolve, so, in a locus in which a 1 contributes to the overall fitness, it is likely that a homozygous 11 pair will appear, and similar for the case where a 0 is required. In the case where the most desirable bit varies from generation to generation, a heterozygous pair is expected. Thus, the scheme can be summarised as

$$f(x) = rand(f(x'), f(x''))$$

where $f(x)$ is the phenotype, $f(x')$ the first chromosome and $f(x'')$ the second chromosome. This system is known as Random Diploidy [11]. A second proposal offered by Ošmera [11] was to perform a logical XOR on the pair of chromosomes. In this scheme, a heterozygous pair produce a 1, while a homozygous pair produce a 0. This scheme is interesting in that it has two representations for each phenotype, e.g. in the case of 1, the two are 10 and 01. This is useful because it means it is possible to mate two individuals of identical phenotypes producing an offspring of a different phenotype, a property which is crucial if a diploidy scheme is to be able to adapt. The XOR scheme can be summarised as below

$$f(x) = f(x') \bigoplus f(x'')$$

This scheme was tested on an objective function which varied with time, which is described in Sect. 4.

3 Polygenic Inheritance – The Shades Scheme

There is no particular reason why any trait should be controlled by a single gene, or gene pair. The first instance of this in natural biology was discovered in 1909 by Nilsson-Ehle [12] when he showed that the kernel colour in wheat, an additive trait, was in fact managed by two pairs of genes. Inheritance of genes of this type is known as polygenic inheritance.

An example of this additive effect comes from Pai [12], who used Four o'clock flowers to illustrate the point. In these flowers, the colour is controlled by a gene pair, and there are two alleles, G_r which corresponds to a red gene, and G_w which corresponds to a white gene. If the pair is $G_r G_r$ the flower is red, while the flower is white if the pair is $G_w G_w$. If, however, a white parent and a red parent are mated, all the children will have the heterozygous $G_r G_w$ pair. Unlike the other schemes where a conflict is resolved by choosing one of the alleles, *both* contribute to the colour, resulting in an intermediate form, pink.

Using two genes to control each phenotypic trait, where each gene contributes a "shade" of 1 to the trait. We are, depending on the threshold level, likely to get a phenotype of value 0, if the composite shade of the genes is a light shade, or, if the composite shade is a darker shade of 1, the phenotype will likely be 1. We refer to this style of additive polygenic inheritance as *Shades*.

Figure 2 illustrates the calculation of the *shade of one* for different genotypic combinations. In this system, there are three alleles, namely A, B and C, with shades of 0, 1 and 2 respectively. The shade for a locus is simply the sum of its alleles. A sum less than 2 gives a phenotype of 0, while a sum greater than

AA	AB	AC	BB	BC	CC
0	1	2	2	3	4

The Degree of Oneness ⟶

Fig. 2. Varying shades of oneness

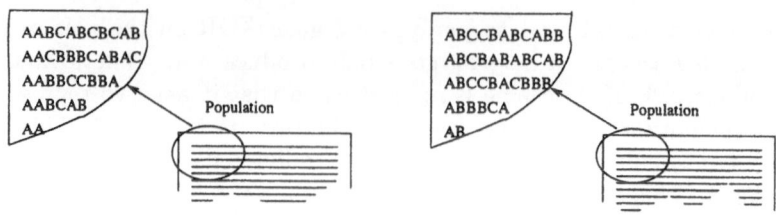

Fig. 3. Traits in a population being fixed due to lack of diversity in homozygous pair (left) and heterozygous pair (right)

2 results in a phenotype of 1. When the total is exactly two, we say that the phenotype is in a "grey area" and the value is chosen randomly.

There is no limit as to the number of genes could be involved in generating a trait, and the system can also represent more than two phenotypes. To distinguish between the different implementations of Shades, we append the number of genes involved with a dash, giving us Shades–2 and Shades–3.

Using the polygenic system, the population can still fixate on certain alleles. Figure 3 illustrates the problem using Shades–2 as an example. Once every individual in the population has a homozygous pair for a particular trait, it cannot change without mutation. In this case, all individuals have a homozygous AA pair controlling the first trait. The pairs that can cause this problem are AA and CC, as these are at the extreme of the phenotypic space.

A second problem also occurs at a pair-level. Even the heterozygous pairs can fixate, due to each pair having its genes in the same order, also illustrated in Fig. 3. This is a particularly interesting problem, as a simple examination of the gene distribution would suggest that there are ample numbers of each gene to permit a change of phenotype. Again, taking the example in Fig. 3, the gene pair controlling the first trait (AB) is identical throughout the population. Despite being heterozygous, the phenotype cannot change, due to the positions of the alleles. If two parents, both of the form AB mated, one would expect that the possible offspring would be AA, AB and BB. However, the fixed order of haploid crossover prevents this from happening. All children will have A in the first position and a B in the second, which effectively stops evolution of that pair. Both these problems were solved by the introduction of simple operators.

3.1 The Force Operator

Despite the continuous phenotypic space introduced by the additive effect of polygenic inheritance, only those shades near the threshold or grey area have the ability to quickly adapt to an environmental change. To alleviate this, the **force** operator is introduced. **Force** is applied to any homozygous pair that occurs at an extreme of the phenotypic space, and forces the mutation of one of the alleles. The mutation is always such that while the phenotype does *not* change, but if the pair is crossed over with an identical pair, the parents can produce offspring of a different phenotype. It is still possible for them to produce offspring of the same phenotype as themselves. **Force** is applied to every occurrence of extreme homozygotes in Shades scheme. This is similar to the phenomena of hyper mutation in bacteria, where the level of mutation of certain areas of their chromosome can effectively be marked to be mutated than other areas.

3.2 The Perturb Operator

Perturb operates in a similar manner to the rather unpopular inversion operator, but at a much more limited scale of disruption, hence its gentler title. **Perturb** swaps a pair that control a single trait, and therefore doesn't affect the phenotype. **Perturb** is applied randomly to possible candidate locations, and helps alleviate the fixed position problem. Neither the **force** or the **perturb** operator modify the phenotype of the individual affected.

3.3 Shades−3

Using three alleles gives us a situation as in Table 1. Again we have the grey area, and extreme values as in Shades−2.

In this case, **force** operates slightly differently. Phenotypes at the extremities have their shades adjusted up or down by 2, those near the extremities are adjusted up or down by 1. This ensures that if the individual subsequently mates with another individual of the same phenotype, they can produce an offspring of a different appearance. Table 2 shows the action taken on each genotype in the Shades−3 scheme. Notice that the order for **force** is not important.

For **perturb**, two of the three genes controlling a trait are selected at random and swapped. Notice that no attempt is made to tune the level of mutation, nor is it related the level of change in the environment.

Table 1. Genotype to phenotype mapping using 3 alleles, the order of alleles doesn't affect the shade

Combination	AAA	AAB	AAC	ABB	ABC	BBB	ACC	BBC	BCC	CCC
Shade	0	1	2	2	3	3	4	4	5	6
Phenotype	0	0	0	0			1	1	1	1

Table 2. Using the force operator on three alleles. For the extreme values, two mutations take place

Genotype	After Force
AAA	ABB or AAC
AAB	ABB or AAC
BCC	BBC or ACC
CCC	BBC or ACC

4 Experiments

In previous work, Shades−3 has been shown to outperform both the triallelic and the Dominance Change Mechanism schemes for diploidy [16]. It was quicker to react to a change in the environment and consistently found new optima following a change. We now proceed to empirically analyze the performance of haploid and the two diploid schemes from Sect. 2.2 against the Shades genotypes using two dynamic problem domains.

The first problem is taken from Ošmera [11] which involves trying to track a constantly moving target, while the second is Goldberg's classic constrained knapsack problem [5] in which the goal periodically shifts. The working hypothesis to be verified or invalidated is that Shades will yield similar performance to diploidy without incurring domain specific overheads such as selection of genotype to phenotype mapping or a dominance change mechanism.

4.1 Ošmera's Dynamic Problem Domain I

Ošmera [11] presented two dynamic problems, the first of which is specified by the following function:

$$g_1(x,t) = 1 - e^{-200(x-c(t))^2} \quad \text{where} \quad c(t) = 0.04(\lfloor t/20 \rfloor) \tag{1}$$

$x \in \{0.000, \ldots, 2.000\}$ and $t \in \{0, 1000\}$, where t is the time step, each of which is equal to one generation. Figure 4 shows a three dimensional view of the domain, with x varying linearly with t. Solutions to the problem are derived by finding a value of x at time t that minimizes $g(x,t)$. Examination of the problem shows that it displays quite a predictable pattern, with the possibility of certain loci becoming fixed at 1 without incurring any penalties.

A gray coded bit string of length 31 was used to represent the parameter, and normalized to yield a value in the range $\{0,2\}$. A generation gap of 1.0 was set as the replacement policy with test parameters of $P_{mut} = 0.01, P_{cross} = 0.7$, and population size $p = 400$. Remainder stochastic sampling without replacement as the selection scheme.

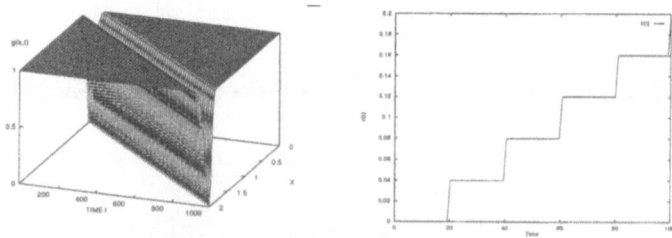

Fig. 4. Ošmera's dynamic problem domain (left), and behaviour of $c(t)$ over time (right)

4.2 The Knapsack Problem

The constrained 17-Object 0-1 knapsack problem is taken from Goldberg & Smith [5], and the weighting is used as specified therein. The population consisted of 150 individuals, made up of 17 pairs of 1 bit genes. A uniform crossover operator was used with a probability of $P_{cross} = 0.5$ for both shades−2 and shades−3, as were $P_{perturb} = 0.175$, while mutation was implemented at a rate of 0.001. Remainder stochastic sampling without replacement was also used here.

All experiments were run for 400 generations, with a period of 15 generations, and the fitness varied between 87 and 71, depending on the oscillation.

5 Results

Results are presented in two different ways. An overview is given in the following tables, which indicate the average error, while more detailed graphs give more detailed view. In both experiments, the best performer was Shades, although, surprisingly, Shades−2 outperforms Shades−3 in the Knapsack problem.

5.1 Ošmera's Function

Table 3 shows the performance of each of the approaches on Ošmera's test function. One surprising aspect is the good performance shown by the haploid scheme, however, this is probably due to the monotonically increasing nature of the problem, which means that the changes are not only regular, but also relatively small.

Figures 5, 6, and 7 indicate the manner in which each of the approaches track the error. Notice how similar the behaviour of haploid and random diploid are, with the greatest difference being the number of large errors made by the latter, giving it a larger standard deviation than the haploid system. XOR diploid, on the other hand, is consistently better across the entire run.

The performances of the two shades systems are shown in Fig. 7. In this case, we can see that shades-2 performs similarly to random diploid, but with a greater number of relatively large errors. The best performer on this problem was shades−3.

5.2 Knapsack Function

Table 4 shows the relative performances on the Knapsack problem. In this case, the haploid approach reverts to the expected, and fails to track the environment, generating an average error of almost 50%. The two diploid perform considerably better than it, but, again, Shades shows the best peformance. Curiously, in this case, Shades−2 actually outperforms Shades−3. The mean tracking error for each method are plotted in Figs. 8, 9, and 10.

In this case, the haploid scheme fails to map to the lower weight because, even by the first oscillation, it doesn't contain enough diversity to make the

Table 3. Mean tracking error

Ošmera Test Function I		
Genotype	Diploid Mapping	% Error
Haploid	–	0.043
Diploid	Random	0.026
	XOR	0.173
	Random	0.079
	XOR	0.030
Shades–2	–	0.025
Shades–3	–	0.016

Table 4. Mean tracking error

Constrained Knapsack Problem			
Genotype	Diploid Mapping	Mean Fit.	% Error
Haploid	–	44.67	49.38
Diploid	Random	77.89	3.15
	XOR	73.44	7.61
Shades–2	–	78.910	1.746
Shades–3	–	77.867	3.074

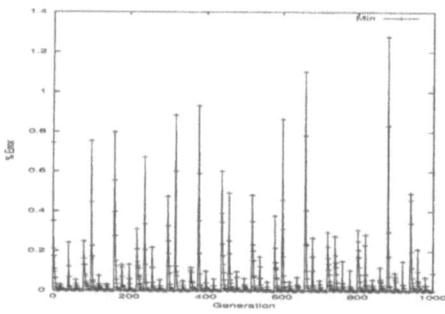

Fig. 5. Ošmera's dynamic problem. Error plotted against generation for haploid

change. The relative performances of the XOR and random diploid systems are inverted with this problem, with random performing substantially better, and successfully tracking the problem after the first few oscillations. XOR, on the other hand, never manages to track the *higher* weight. It does, however, evolve towards it, but the problem changes too quickly for it to get there.

The relative performances of shades–2 and shades–3 are also inverted, although both succeed in finding the target on each oscillation. As indicated in Table 4, the two shades methods still perform better than the other methods, although there is no statistically significant difference between random diploid and shades-3 on this problem.

Non-stationary Function Optimization Using Polygenic Inheritance

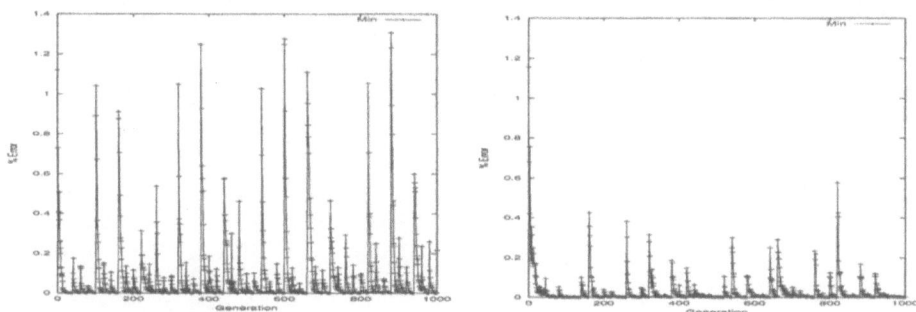

Fig. 6. Ošmera's dynamic problem. Error plotted against generation for random diploid (left) and XOR diploid (right)

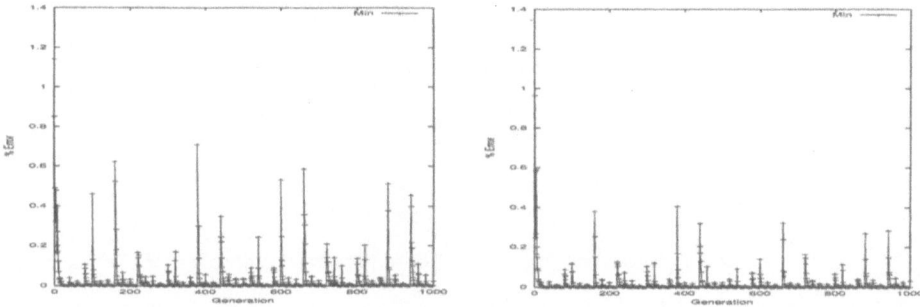

Fig. 7. Ošmera's dynamic problem. Error plotted against generation for shades−2 (left) and shades−3 (right)

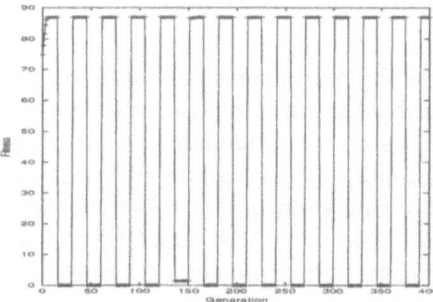

Fig. 8. The Knapsack problem. Fitness plotted against generation for haploid. Notice the different scale required compared to the plots for diploid & shades

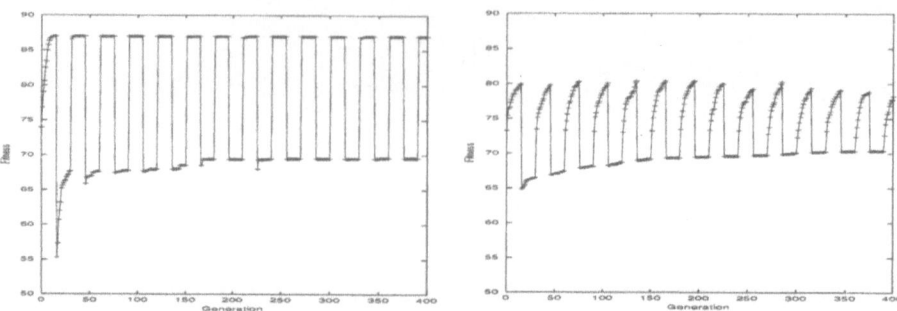

Fig. 9. The Knapsack problem. Fitness plotted against generation for random diploid (left) and XOR diploid (right)

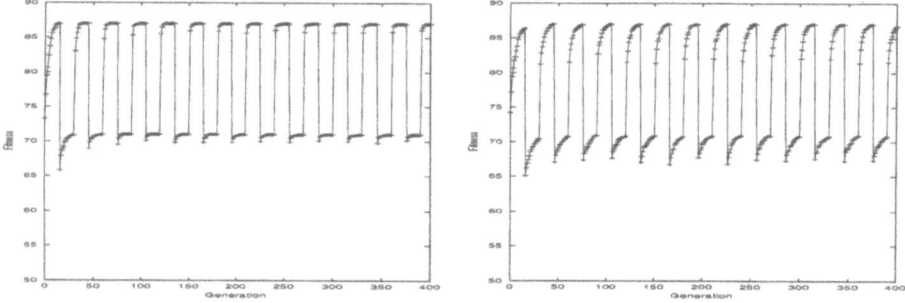

Fig. 10. The Knapsack problem. Fitness plotted against generation for shades–2 (left) and shades–3 (right)

6 Conclusions and Future Work

Shades, a new version of haploidy, coarsely modeled on naturally occurring phenomena, has been described, as have its associated functions. We have tested it and two diploidy methods on two standard benchmark problems, and demonstrated that it outperforms standard haploid and two diploid schemes on these. Perhaps somewhat surprisingly, on one of these problems, a standard haploid scheme performed better than one of the standard diploid schemes. It has been shown that a diploid genetic structure is not a prerequisite for a population to survive in a dynamic problem space.

Despite the fact that Shades–2 outperformed Shades–3 on the Knapsack problem, our belief is that the more genes involved in the phenotype mapping increases the performance on dynamic problem domains. Although there probably is a law of diminishing returns applicable to this, it should be investigated further.

The potential benefits of using dynamic environments have yet to be realized. By providing an unbiased scheme that can quickly react to environmental changes, we hope to be able to convert static problems to dynamic ones.

References

1. Jürgen Branke, "Evolutionary Approaches to Dynamic Optimization", In *Evolutionary Algorithms for Dynamic Optimization Problems*, 2001, pp. 27–30.
2. J.J. Collins and M. Eaton, "Genocodes for genetic algorithms", In *Procs. of Mendel'97*, 1997, pp. 23–30.
3. G. Elseth and K. Baumgardner, Principles of Modern Genetics, West Publishing Company, 1995.
4. A. Ghosh, S. Tsutsui and H. Tanaka, "Function Optimization in Nonstationary Environment using Steady State Genetic Algorithms with Aging of Individuals", In *Proc. of the 1998 IEEE International Conference on Evolutionary Computation*.
5. D. Goldberg and R.E. Smith, "Nonstationary function optimisation with dominance and diploidy", In *Procs. of ICGA2*, 1987, pp. 59–68.
6. D. Hillis, "Coevolving parasites improves simulated evolution as an optmisation procedure", In *Proceedings of ALife II*, 1989.
7. R.B. Hollstein, "Artificial genetic adaptation in computer control systems", PhD Dissertation, University of Michigan, 1971.
8. J. Lewis, E. Hart, and G. Ritchie, "A Comparison of Dominance Mechanisms and Simple Mutation on Non-stationary Problems", In *Proc. of Parallel Problem Solving from Nature – PPSN V*, 1998, pp. 139–148.
9. K.E. Mathias and L.D. Whitley, "Transforming the search space with gray coding", In *Proc. of IEEE Int. Conf. on Evolutionary Computing*.
10. K. Ng and K. Wong, "A new diploid scheme and dominance change mechanism for non-stationary function optimisation", In *Proc. of ICGA-5*, 1995.
11. P. Ošmera, V. Kvasnička and J. Pospíchal, "Genetic algorithms with diploid chromosomes", In *Proc. of Mendel '97*, 1997, pp. 111–116.
12. A. Pai, Foundations of Genetics : A Science for Society. McGraw-Hill, 1989.
13. M. Ridley, The Red Queen: Sex and the Evolution of Human Nature, Viking London, 1993.
14. C. Ryan, "The Degree of Oneness", In *Proceedings of the 2nd Workshop on Soft Computing*, 1996.
15. C. Ryan, "Reducing Premature Convergence in Evolutionary Algorithms", PhD Dissertation, University College Cork, Ireland, 1996.
16. C. Ryan, "Shades : A Polygenic Inheritance Scheme". In *Proceedings of Mendel '97*, 1997, pp. 140–147.
17. C. Ryan, and J.J. Collins. "Polygenic Inheritance – A Haploid Scheme that Can Outperform Diploidy", In *Proc. of Parallel Problem Solving from Nature – PPSN V*, 1998, pp. 178–187.

Scalability of Selectorecombinative Genetic Algorithms for Problems with Tight Linkage

Kumara Sastry[1,2] and David E. Goldberg[1,3]

[1] Illinois Genetic Algorithms Laboratory (IlliGAL)
[2] Department of Material Science & Engineering
[3] Department of General Engineering
University of Illinois at Urbana-Champaign, Urbana, IL
{ksastry,deg}@uiuc.edu

Abstract. Ensuring building-block (BB) mixing is critical to the success of genetic and evolutionary algorithms. This study develops facetwise models to predict the BB mixing time and the population sizing dictated by BB mixing for single-point crossover. The population-sizing model suggests that for moderate-to-large problems, BB mixing – instead of BB decision making and BB supply – bounds the population size required to obtain a solution of constant quality. Furthermore, the population sizing for single-point crossover scales as $O\left(2^k m^{1.5}\right)$, where k is the BB size, and m is the number of BBs.

1 Introduction

Since the inception of genetic algorithms (GAs), the importance of building blocks (BBs) has been recognized [1,2]. Based on Holland's notion of BBs, Goldberg proposed a design decomposition method for a successful design of GAs [3]. This design decomposition currently consists of sevens steps [4] and can be stated as follows: (1) Know what GAs process – building blocks (BBs), (2) solve problems that are of bounded BB difficulty, (3) ensure an adequate supply of raw BBs, (4) ensure increased market share for superior BBs, (5) know BB takeover and convergence times, (6) ensure that BB decisions are well made, and (7) ensure a good mixing of BBs. Significant progress has been made in developing facetwise models for many of the above decomposition steps and the interested reader should consult [4] for further details.

However, researchers have often overlooked the issues of BB identification and mixing or exchange, even though studies on selectorecombinative GAs have indicated that effective *identification* and *exchange* of BBs is critical to innovative success. Furthermore, existing facetwise models such as convergence-time and population-sizing models *assume* tight linkage. That is, alleles of a BB are assumed to be close to one another, and crossover operators are assumed to ensure necessary exchange of BBs with a high probability. Even though the tight-linkage assumption isolates the phenomenon of interest while bracketing the linkage problem, in real-world problems this is not the case, as we don't know which alleles contribute to which BBs.

It is therefore critical to understand mixing capability of popular recombination operators used in genetic and evolutionary algorithms. Dimensional models for BB mixing have been developed for uniform crossover [5,6] and similar analysis is yet to be done on single-point and two-point crossovers [7]. For problems with loosely linked BBs, mixing behavior of multi-point crossover is bounded by the mixing model of uniform crossover. On the other hand, for problems with tightly linked BBs, mixing behavior of multi-point crossovers can be different from that of uniform crossover and separate dimensional models have to be developed. Therefore the objective of this study is to develop a facetwise model to predict the mixing behavior of single-point crossover and to utilize this model to predict the mixing time and population sizing dictated by BB mixing.

This paper is organized as follows: The next section presents a brief literature review. Section 3 defines the mixing problem and states the assumptions used in developing the facetwise models. Section 4 develops mixing models for problems with two building blocks, which is extended for a more general m-BB case in Sect. 5. Finally, the paper is summarized and key conclusions are stated.

2 Literature Review

Many researchers have analyzed different aspects of fixed point crossovers and the efforts can be broadly classified into four categories [7]:

Recombination Modeled as Schema Disrupter: This class of models is motivated from the schema theorem [1]. These studies model schemata disruption of different crossover operators and are useful in comparing those operators [8,9,10]. However, as shown elsewhere, these models do not address the mixing problem [11].

Recombination Modeled as Mixer: Models belonging to this class are motivated from quantitative genetics and quantify properties such as *linkage disequilibrium* [12,13], or crossover-induced bias such as *positional bias, distributional bias, recombinative bias* and *schema bias* [14,13,15], and *relaxation time* [16]. However, these models also do not address the mixing issue.

All-in-One Models: This class of models combines all the facets of recombination operators using tools such as difference equations, Markov chains, and statistical mechanics [17,18,19,20]. Such models are more accurate and contain all the information about BB mixing, however, it is very hard to extract the BB-mixing information out of such complex models.

Recombination Modeled as an Innovator: This class of models [5,6,21] address the BB mixing issue in a more direct manner than the above mentioned class of models. These models predict *mixing* or *innovation* time, which is defined as the expected number of generations to obtain an instance of the target solution. These models have been compared with other facetwise models using dimensional arguments and *control maps* have been constructed. Such a control map identifies different competing forces affecting the genetic search. However, these models have been developed only for uniform crossover. While the mixing model for uniform crossover bounds the behavior of multi-point crossover for problems with loosely linked BBs, their mixing behavior can be significantly different from that of uniform crossover for problems with tightly linked BBs. Therefore, we analyze the mixing behavior of one-point crossover for problems with tight linkage.

3 Problem Definition

The section describes the mixing problem and the assumptions made to facilitate its analysis. Before doing so, it should be noted that there are two ways a crossover can increase the number of BBs on a particular string. One possibility is that the BB at a certain position is *created*. However, the likelihood of BBs being created decreases as the BB size increases. The other possibility is that the BBs are *mixed*. That is, the crossover operator transfers BBs at different positions from both parents to one of the offspring. For instance, a crossover can combine the following two strings, bb#### and ##b###, to yield following two offspring, bbb##, and ######. Here b refers to a BB and # refers to schemata other than the BB.

This aspect of crossover – its ability to recombine BBs in order to better solutions – is of particular interest to us. Specifically, the rate at which a recombination operator exchanges BBs dictates the success of a GA run. Therefore, we model the mixing rate of single-point crossover on search problems with tightly linked BBs. Specifically, we answer the following question: Given that the individuals in the population have m_c BBs, how long – in terms of number of generations – will it take to obtain individuals with $m_c + 1$ BBs. This time is defined as the *mixing time* and is denoted by t_x.

To ease the analytical burden, we consider generation-wise selectorecombinative GAs with non-overlapping population of fixed size. The decision variables are assumed to be encoded into a binary string of fixed length. Furthermore, we consider the class of search problems with uniformly-scaled BBs. Uniform scaling implies that the contribution of BBs from different partitions to the overall fitness of the individual is the same. Specifically, we consider fully-deceptive trap functions [22] to validate the mixing models. However, the results should apply to additively decomposable stationary fitness functions of bounded order [4].

Using these assumptions, the following section develops a dimensional model for mixing time. The mixing-time model, and a convergence-time model will be used to develop a population-sizing model dictated by BB mixing. Our approach is methodologically similar to that of Thierens and Goldberg [6], in that we consider the mixing of a simple two building block problem and later extending it to a more general m building block case. Interestingly, the similar approach leads to radically different models as should be expected. In the original study, exponential times were required to solve uniformly mixed deceptive problems. Here, polynomial time models are predicted and observed for tight linkage with single-point crossover. The ability to derive such different models accurately is testimony to the analytical power of the approach.

4 Building-Block Mixing Models: Two BBs

This section presents dimensional models to predict the mixing behavior of single-point crossover on a problem with two building blocks, each of size k. Here, we assume that individuals in the initial population have exactly one BB. Therefore, in the two-BB case, a single mixing event mixing two BBs from different partitions to yield the global optimum. That is, if the following two individuals are selected for crossover: b# and #b, and the crossover site is chosen as k, then one of the offspring is bb, which is the global optimum. Recognizing that a mixing event can occur in two scenarios (b#+#b, or #b+b#) out of the possible four crossover scenarios and if the crossover site is chosen between the BBs (crossover site is k), we can write probability of a mixing event as

$$p_{\text{mix}} = \left(\frac{2}{4}\right)\left(\frac{1}{2k-1}\right). \tag{1}$$

Equation (1) assumes that every individual in the population contains either a BB (all ones) or a deceptive attractor (all zeros). However, in the actual case, the probability of having a BB in a string is 2^{-k}, and not having a BB is $2^{-k}(2^k - 1)$. Therefore, the proportion of recombinations that can result in a mixing event is given by $2^{-2k}(2^k - 1)$. In such a case probability of mixing is given by

$$p'_{\text{mix}} = 2 \cdot 2^{-2k}(2^k - 1)\left(\frac{1}{2k-1}\right) \approx \frac{2^{-(k-1)}}{2k-1}. \tag{2}$$

Assuming a population size, n and a crossover probability, p_c, there are $\frac{n}{2} \cdot p_c$ recombinations in one generation. Therefore, the mixing time, t_x is

$$t_x = \frac{1}{\frac{n}{2}p_c p'_{\text{mix}}} = \frac{2^k(2k-1)}{np_c}. \tag{3}$$

The above mixing time model is compared with the following convergence-time model [23,24,25] to interrelate recombination time with selection time:

$$t_{\text{conv}} = \frac{c_c\sqrt{\ell}}{I} \tag{4}$$

where, I is the selection intensity, which is a function of tournament size, and c_c is a constant. Recognizing that for innovative success the mixing time has to be less than the convergence time (recombination has to create the global optimum before the population converges to the local optimum), we can write, $t_x < t_{\text{conv}}$.

For the two-BB case, $\ell = 2k$, and the above equation can be written as

$$n > \frac{I}{c_c p_c}\left(\frac{2^k(2k-1)}{\sqrt{2k}}\right) \approx c_{x2}\frac{I}{p_c}\left(2^k\sqrt{k}\right), \tag{5}$$

where $c_{x2} = 2(c_c)^{-1}$ is a constant. Equation (5) predicts the population size dictated by BB mixing. It is verified with empirical results in Fig. 1. The empirical

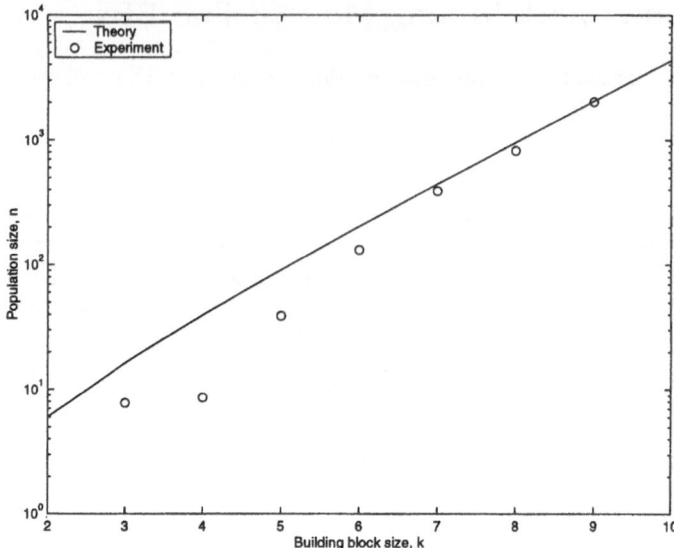

Fig. 1. Empirical validation of the population-sizing model dictated by BB mixing (Eq. (5)). Minimum population size required for a GA success is plotted as a function of BB size, k. The empirical population size is determined by a bisection method and are averaged over 25 independent bisection runs

results are obtained as follows: A binary tournament selection and single-point crossover with $p_c = 1$ is used. A GA run is terminated when all the individuals in the population converge to the same fitness value. The average number of BBs correctly converged are computed over 50 independent runs. The minimum population required such that both BBs converge to the correct value is determined by a bisection method. The results shown in Fig. 1 are averaged over 25 such bisection runs.

5 Building-Block Mixing Models: m BBs

To facilitate the analysis to m building blocks, we blend empirical observations with theory similar to the approach followed by Thierens and Goldberg [6]. The empirical results on m k-bit deceptive traps indicates two key behaviors of single-point crossover:

1. All the building block configurations at a certain *mixing level* have to be discovered before a higher mixing level can be reached. Here a mixing level denotes the number of BBs present in an individual string. For instance, individuals b#b##, and b#bb# are at mixing levels 2 and 3, respectively. This *ladder-climbing* phenomenon is illustrated in Fig. 2, which plots the proportion of all possible BB configurations at different mixing levels as a function of time. It can be clearly seen that when a certain mixing level is at

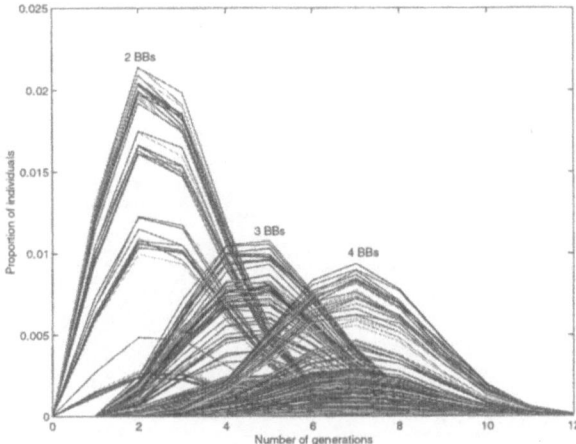

Fig. 2. Proportion of individuals with correct BBs as a function of time. Empirical results indicate that mixing at a higher level ($m_c + 1$) takes over only after BBs of lower level (m_c) are covered. The results are for ten 4-bit deceptive trap functions

its maximum value – when the proportion of all the BB configurations at a certain mixing level in the population is highest – proportion of individuals at higher mixing level is negligible, and hence, we can use the *ladder-climbing* model [6].

2. The proportion of individuals having good BBs at their two ends is higher than those at other positions. In fact, the proportion gradually reduces along the position and reaches a minimum at the center, as illustrated in Fig. 3. The figure plots the proportion of individuals having a good BB as a function of its position in the string and time. This is because the probability of mixing is much higher when mixing two BBs at the extreme points in the string. For example, the mixing probability when recombining b#···## and ##···#b to yield b#···#b is $((m-2)k+1)/(mk-1)$ as opposed to recombining b##···# and #b#···# to yield bb#···#, in which case the mixing probability is $1/(mk-1)$.

This property of single-point crossover is called the *length-reduction* phenomenon and is what makes the mixing behavior of one-point crossover different from that of uniform crossover. An immediate consequence of the length-reduction phenomenon is that an increase in mixing level leads to a decrease in problem size. For example, when we are at mixing level 3, the problem size reduces by two. This is because the BBs at the two ends get fixed for majority of the individuals.

The combined effect of *ladder-climbing* and *length reduction* illustrated in Fig. 4. The length-reduction phenomenon suggests that we can consider mixing probability as a function of number of BBs that are not converged at a given time and the BB size. Furthermore, we assume that among the BBs that are not converged the mixing level is always one.

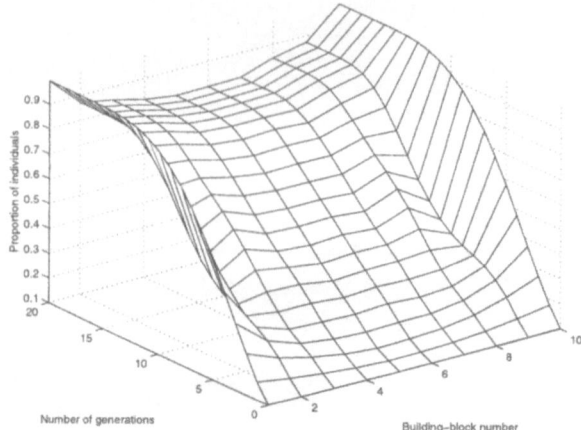

Fig. 3. Proportion of individuals with correct BBs as a function of BB position and time. Mixing of one-point crossover is dependent on the BB position. It is highest on the string ends and lowest in the middle of the string. The results shown are for a 10 4-bit deceptive trap functions

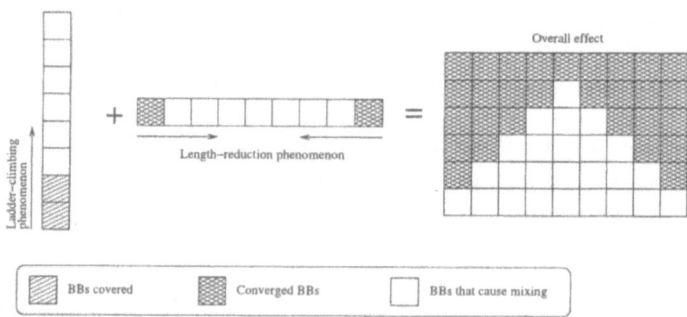

Fig. 4. Illustration of the *ladder-climbing* and *length-reduction* phenomena and their combined effect. Ladder-climbing phenomenon suggests that we increase the mixing level one step at a time. Length-reduction phenomenon suggests that as the mixing level increases, the problem size dictating mixing shrinks. It also suggests that the string ends converge faster than the BBs at the middle of the string

Given that a problem has m BBs and we are currently at mixing level one, that is, every individual has one BB at some position. Out of a total of m^2 recombination scenarios, the following result in mixing:

Recombination scenario	b##···## #b#···##	b###···# ##b#···#	...	b#··· ##···# ##···b#···#	...	b#··· ## ##···#b
Possible events	$2(m-1)$	$2(m-2)$	...	$2(m-i)$	...	2
Mixing Probability	$\frac{1}{mk-1}$	$\frac{k+1}{mk-1}$	...	$\frac{(i-1)k+1}{mk-1}$	...	$\frac{(m-2)k+1}{mk-1}$

Therefore the overall mixing probability can be written as

$$p_{\text{mix}}(m) = \frac{2}{m^2} \sum_{i=1}^{m-1} \frac{(m-i)\,((i-1)k+1)}{mk-1}, \qquad (6)$$

which can be simplified as follows:

$$p_{\text{mix}}(m) = \frac{2}{3} \cdot \frac{(m-1)\,[(m-2)k+3]}{m(mk-1)} \qquad (7)$$

From the length-reduction phenomenon, we know that, at a given mixing level, m_c, the mixing probability is given by

$$p_{\text{mix}}|_{m_c} = p_{\text{mix}}\,(m - m_c + 1) = \frac{2}{3}\left[\frac{(m-m_c)\,((m-m_c-1)k+3)}{(m-m_c+1)\,((m-m_c+1)k-1)}\right]. \qquad (8)$$

For moderate-to-large problems, the mixing probability can be assumed to be constant with respect to the mixing level and can be approximated as follows:

$$p_{\text{mix}} \approx p_{\text{mix}}\,(m_c = 1) = \frac{2}{3}\left[\frac{(m-1)\,((m-2)k+3)}{m(mk-1)}\right]. \qquad (9)$$

Now the question remains as to how we reach mixing level m_c+1 when we are at mixing level m_c. Assume that after n_x mixing events, we have n_x individuals at mixing level $m_c + 1$. We have to find out how many mixing events are needed to have all BBs covered at mixing level $m_c + 1$. Since all the n_x strings have m_c+1 BBs, it can be expected that $m\left[1 - \left(1 - \frac{m_c+1}{m}\right)^{n_x}\right]$ BBs are covered. The probability that all BBs are covered approaches 1, and the number of mixing events, n_x is increased.:

$$m\left[1 - \left(1 - \frac{m_c+1}{m}\right)^{n_x}\right] > m(1 - \alpha), \qquad (10)$$

where $\alpha < 1/m$. Rearranging the above equation and using the approximation, $\ln\left(1 - \frac{m_c+1}{m}\right) \approx -\frac{m_c+1}{m}$, we get

$$n_x > c_{mx} \frac{m}{m_c+1}, \qquad (11)$$

where $c_{mx} = -\ln \alpha$ is a constant.

The number of mixing events required to climb one step of the ladder is therefore proportional to the number of building blocks. Similar to the mixing probability, we assume n_x to be constant with respect to the mixing level:

$$n_x \approx n_x\,(m_c = 1) = \frac{1}{2}c_{mx}m. \qquad (12)$$

This approximation is slightly conservative, but bounds the mixing behavior quite accurately as will become clear in the following paragraphs. Recognizing

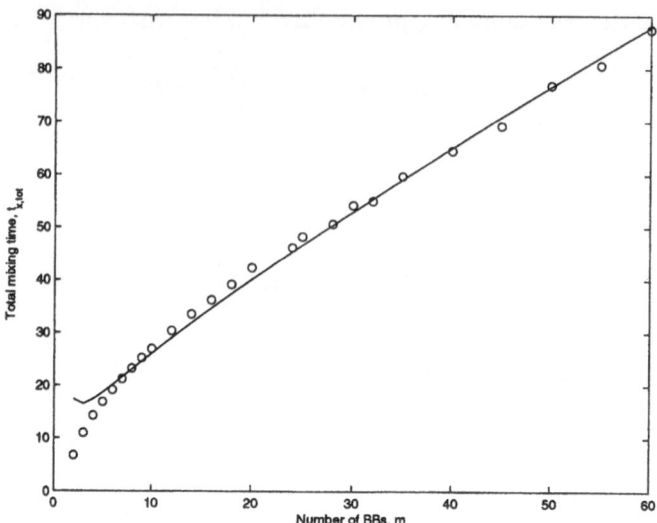

Fig. 5. Comparison of total mixing time, $t_{x,\text{tot}}$ predicted by Eq. (14) with empirical results. The total mixing time is plotted as a function of number of BBs. The empirical results are for 4-bit trap function and are averaged over 500 independent runs

that in a generation, we have $\frac{n}{2}p_c$ recombinations, and using Eqs. (9) and (12), we can calculate the mixing time, t_x:

$$t_x = \frac{n_x}{\frac{n}{2}p_c p_{\text{mix}}} = c_{x,m}\frac{m}{np_c}\left[\frac{(m-1)\left((m-2)k+3\right)}{m(mk-1)}\right], \qquad (13)$$

where $c_{x,m} = \frac{2}{3}c_{mx}$ is a constant. The above mixing time refers to the time – in number of generations – required to climb one step of the mixing ladder. However, we are interested in the total mixing time, $t_{x,\text{tot}}$, that is, the time required to go from mixing level 1 to mixing level m. Note that this is the time required to climb $m-1$ steps of the mixing ladder. Approximating, $m-1 \approx m$, we get

$$t_{x,\text{tot}} = mt_x = c_{x,m}\frac{m^2}{np_c}. \qquad (14)$$

The model suggests that the total mixing time grows quadratically with the number of building blocks and is inversely proportional to the population size and crossover probability. The mixing-time model (Eq. (14)) is compared to empirical results in Fig. 5, where the total mixing time is plotted as function of number of BBs, m. The results indicate that the agreement between theoretical and experimental results gets better with the number of BBs.

In the analysis presented above, we have assumed that every individual in the population contains either a BB (all ones) or a deceptive attractor (all zeros). However, in the actual case, the probability of having a BB in a string is 2^{-k}.

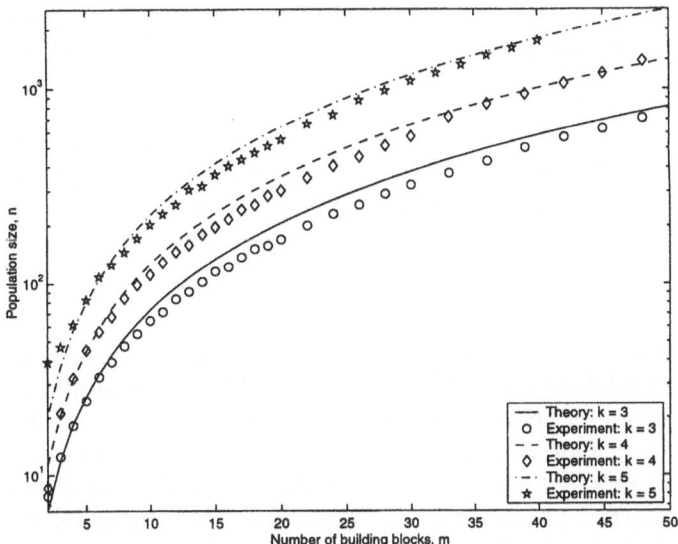

Fig. 6. Empirical validation of the population-sizing model dictated by BB mixing (Eq. (15)). Minimum population size required for a GA success is plotted as a function of number of BBs m. The minimum population size is determined by a bisection method and the results are averaged over 25 independent bisection runs

Therefore, the proportion of recombinations that can result in a mixing event is given by $2^{-mk}\left(2^{(m-1)k}-1\right) \approx 2^k$. Multiplying this term as a with the mixing-time model, yields us the total mixing time when 2^k schemata are present.

As mentioned earlier, to ensure innovative success, we need $t_x < t_{\text{conv}}$. From Eqs. (14) and (4), we get

$$n > c_x \frac{I}{p_c} 2^k m \sqrt{\frac{m}{k}} \qquad (15)$$

where $c_x = c_{x,m}/c_c$. Equation (15) suggests that the population size grows as $O(2^k m \sqrt{m})$ as opposed $O(2^k \sqrt{m})$, predicted by the gambler's ruin model [26].

The BB-mixing based population-sizing model (Eq. (15)) is verified with empirical results in Fig. 6. The figure plots the minimum population size required to obtain a solution of desired quality as a function of number of BBs. Population sizes for three different BB sizes, $k = 3, 4$, and 5 are compared. The empirical results are obtained as described in Sect. 4. The figure shows that the empirical results show good agreement with the model prediction.

These results not only demonstrate the utility of building facetwise models, but also brings forth two key behavior of one-point crossover:

1. BB supply and decision making [26] bounds the population size for small problems, and BB mixing governs population sizing for larger problems.

2. The range – in terms of number of BBs – in which BB decision-making and supply bound population sizing is dependent on the BB size. The range increases with the BB size.

Therefore, there is a boundary which segregates the problem space into two regions. One where BB decision-making governs the population sizing, and the other where BB-mixing governs the population sizing.

6 Summary and Conclusions

This study analyzed the mixing behavior of single-point crossover on boundedly decomposable additive fitness functions with tight linkage. Models for predicting the time required for achieving innovative mixing – first for a two building block problem, and then for a general m building block case – are developed. Two key features of single-point crossover, namely the *ladder-climbing* phenomenon, and the *length-reduction* phenomenon are observed empirically and successfully used in developing the mixing time model for m-BB case.

The resulting mixing time models are compared with existing convergence-time model to investigate scenarios which result in *innovative success*. Using such a comparison, population-sizing model dictated by BB mixing is derived. All the models derived in this study are compared with empirical results. Empirical results show good agreement with theory, thus validating the facetwise models.

Results indicate that the population-sizing required for a GA success is dictated by building-block mixing rather than building-block-wise decision-making or building-block supply for moderate-to-large problems. The minimum population size scales as $O\left(2^k m\sqrt{m}\right)$, as opposed to $O\left(2^k \sqrt{m}\right)$ which is the case for advanced operators that adapt linkage. This suggests that even under the best case scenario – that is, assumption of tight linkage – fixed crossover operators are less efficient in comparison to operators that adapt linkage.

Acknowledgments. This work was sponsored by the Air Force Office of Scientific Research, Air Force Materiel Command, USAF, under grant F49620-00-0163, the National Science Foundation under grant DMI-9908252, and the CSE fellowship, UIUC. The U.S. Government is authorized to reproduce and distribute reprints for government purposes notwithstanding any copyright notation thereon. The views and conclusions contained herein are those of the authors and should not be construed as necessarily representing the official policies or endorsements, either expressed or implied, of the Air Force Office of Scientific Research, the National Science Foundation, or the U.S. Government.

We thank Martin Butz, Martin Pelikan, and Dirk Thierens for their comments.

References

1. Holland, J.H.: *Adaptation in natural and artificial systems.* University of Michigan Press, Ann Arbor, MI (1975)
2. Goldberg, D.E.: *Genetic algorithms in search optimization and machine learning.* Addison-Wesley, Reading, MA (1989)
3. Goldberg, D.E., Deb, K., Clark, J.H.: Genetic algorithms, noise, and the sizing of populations. *Complex Systems* **6** (1992) 333–362 (Also IlliGAL Report No. 91010).
4. Goldberg, D.E.: *Design of innovation: Lessons from and for competent genetic algorithms.* Kluwer Academic Publishers, Boston, MA (2002)
5. Goldberg, D.E., Thierens, D., Deb, K.: Toward a better understanding of mixing in genetic algorithms. *Journal of the Society of Instrument and Control Engineers* **32** (1993) 10–16 (Also IlliGAL Report No. 92009).
6. Thierens, D., Goldberg, D.E.: Mixing in genetic algorithms. *Proceedings of the Fifth International Conference On Genetic Algorithms* (1993) 38–45.
7. Sastry, K., Goldberg, D.E.: Analysis of mixing in genetic algorithms: A survey. IlliGAL Report No. 2002012, University of Illinois at Urbana-Champaign, Urbana, IL (2002)
8. De Jong, K.A.: An analysis of the behavior of a class of genetic adaptive systems. PhD thesis, University of Michigan, Ann-Arbor, MI (1975)
9. Syswerda, G.: Uniform crossover in genetic algorithms. *Proceedings of the Third International Conference on Genetic Algorithms* (1989) 2–9.
10. De Jong, K.A., Spears, W.M.: A formal analysis of the role of multi-point crossover in genetic algorithms. *Annals of Mathematics and Artificial Intelligence* **5** (1992) 1–26.
11. Goldberg, D.E., Sastry, K.: A practical schema theorem for genetic algorithm design and tuning. *Proceedings of the Genetic and Evolutionary Computation Conference* (2001) 328–335 (Also IlliGAL Report No. 2001017).
12. Christiansen, F.B.: The effect of population subdivision on multiple loci without selection. In Feldman, M.W., ed.: *Mathematical Evolutionary Theory.* Princeton University Press, Princeton, NJ (1989) 71–85.
13. Booker, L.B.: Recombination distributions for genetic algorithms. *Foundations of Genetic Algorithms* **2** (1993) 29–44.
14. Eshelman, L.J., Caruana, R.A., Schaffer, J.D.: Biases in the crossover landscape. *Proceedings of the Third International Conference on Genetic Algorithms* (1989) 10–19.
15. Eshelman, L.J., Schaffer, J.D.: Productive recombination and propagating and preserving schemata. *Foundations of Genetic Algorithms* **3** (1995) 299–313.
16. Rabani, Y., Rabinovich, Y., Sinclair, A.: A computational view of population genetics. *Random Structures & Algorithms* **12** (1998) 313–334.
17. Bridges, C.L., Goldberg, D.E.: An analysis of reproduction and crossover in binary-coded genetic algorithm. *Proceedings of the Second International Conference on Genetic Algorithms* (1987) 9–13.
18. Nix, A.E., Vose, M.D.: Modeling genetic algorithms with Markov chains. *Annals of Mathematics and Artificial Intelligence* **5** (1992) 79–88.
19. Prügel-Bennett, A., Shapiro, J.L.: An analysis of genetic algorithms using statistical mechanics. *Physical Review Letters* **72** (1994) 1305–1309.
20. Stephens, C., Waelbroeck, H.: Schemata evolution and building blocks. *Evolutionary computation* **7** (1999) 109–124.
21. Thierens, D.: Scalability problems of simple genetic algorithms. *Evolutionary Computation* **7** (1999) 331–352.

22. Deb, K., Goldberg, D.E.: Analyzing deception in trap functions. *Foundations of Genetic Algorithms* **2** (1993) 93–108 (Also IlliGAL Report No. 91009).
23. Mühlenbein, H., Schlierkamp-Voosen, D.: Predictive models for the breeder genetic algorithm: I. continuous parameter optimization. *Evolutionary Computation* **1** (1993) 25–49.
24. Bäck, T.: Selective pressure in evolutionary algorithms: A characterization of selection mechanisms. *Proceedings of the First IEEE Conference on Evolutionary Computation* (1994) 57–62.
25. Thierens, D., Goldberg, D.E.: Convergence models of genetic algorithm selection schemes. *Parallel Problem Solving from Nature* **3** (1994) 116–121.
26. Harik, G., Cantú-Paz, E., Goldberg, D.E., Miller, B.L.: The gambler's ruin problem, genetic algorithms, and the sizing of populations. Evolutionary Computation **7** (1999) 231–253 (Also IlliGAL Report No. 96004).

New Entropy-Based Measures of Gene Significance and Epistasis

Dong-Il Seo, Yong-Hyuk Kim, and Byung-Ro Moon

School of Computer Science & Engineering, Seoul National University
Sillim-dong, Kwanak-gu, Seoul, 151-742 Korea
{diseo,yhdfly,moon}@soar.snu.ac.kr
http://soar.snu.ac.kr/~{diseo,yhdfly,moon}/

Abstract. A new framework to formulate and quantify the epistasis of a problem is proposed. It is based on Shannon's information theory. With the framework, we suggest three epistasis-related measures: gene significance, gene epistasis, and problem epistasis. The measures are believed to be helpful to investigate both the individual epistasis of a gene group and the overall epistasis that a problem has. The experimental results on various well-known problems support it.

1 Introduction

In the context of genetic algorithms, the difficulty of an optimization problem is explained in various aspects. The aspects are categorized into *deception* [1], *multimodality* [2], *noise* [3], *epistasis* [4,5], and so on. Among them, the epistasis is observed in most GA-hard problems. In biology, we refer to the suppression of gene expression by one or more other genes as epistasis. But, in the community of evolutionary algorithms, the term has a wider meaning; it means the interaction between genes.

In addition to the concepts to explain the problem difficulty, various measures quantifying the difficulty have been proposed recently. The epistasis variance, suggested by Davidor [5], is a measure quantifying the epistasis of a problem. He interpreted the epistasis as the nonlinearity embedded in the fitness landscape of the problem. The measure was explained more formally by Reeves and Wright [6] from the viewpoint of experimental design. The measures are, however, somewhat "macroscopic," i.e., they concern the epistasis merely as a factor of GA-hardness of a problem. In fact, the epistasis of a problem consists of many individual epistases between small groups of genes. This idea already affected various branches of evolutionary algorithms such as probabilistic model-building genetic algorithms (PMBGAs) [7,8], also called estimation-of-distribution algorithms (EDAs), and topological linkage-based genetic algorithms (TLBGAs) [9]. The epistases are estimated algorithmically or heuristically in the algorithms.

In this paper, we propose new algorithm-independent "microscopic" measures of epistases. We suggest a new framework for the formulation and quantification of the epistases. The framework is based on Shannon's information theory [10,

11]. We propose three measures: *gene significance*, *gene epistasis*, and *problem epistasis*. They are helpful to investigate both the individual epistasis of a gene group and the overall epistasis that a problem has.

The rest of this paper is organized as follows. The basic concepts of Shannon's entropy are introduced in Sect. 2. We establish a probability model and define new epistasis measures in Sect. 3. We provide the results of experiments on various well-known problems in Sect. 4. Finally, the conclusions are given in Sect. 5.

2 Shannon's Entropy

Shannon's information theory [10,11] provides manners to quantify and formulate the properties of random variables. According to the theory, the amount of information contained in a message notifying an event is defined to be the number of digits being required to describe the event. That is, the amount of information contained in a message notifying an event of probability p is defined to be $\log \frac{1}{p}$. The log is to the base 2 and the value is measured in bits. The lower the probability of the event is, the larger amount of information the message contains. The average amount of information contained in events is the amount of uncertainty of the random variable on the events. Thus, the uncertainty of a random variable is defined as

$$H(X) = -\sum_{x \in \mathcal{X}} p(x) \log p(x) \tag{1}$$

where $\mathcal{X}$ and $p(x)$ are the alphabet and the probability mass function (pmf), respectively. The quantity is called the *entropy* of X. It means the average number of bits being required to describe a random variable. The convention $0 \log 0 = 0$ is used in the equation, which is easily justified by continuity since $x \log x \to 0$ as $x \to 0$. Entropy is always nonnegative. Similarly, the *joint entropy* of two random variables is defined as

$$H(X,Y) = -\sum_{x \in \mathcal{X}} \sum_{y \in \mathcal{Y}} p(x,y) \log p(x,y) \tag{2}$$

where $\mathcal{X}$ and $\mathcal{Y}$ are the alphabets of random variables X and Y, respectively, and $p(x,y)$ is the joint pmf of the random variables. The *conditional entropy* of X given Y is defined as

$$H(X|Y) = -\sum_{x \in \mathcal{X}} \sum_{y \in \mathcal{Y}} p(x,y) \log p(x|y). \tag{3}$$

It means the average uncertainty of X when the value of Y is known. The conditioning reduces entropy, i.e., $H(X|Y) \leq H(X)$. The average amount of uncertainty of X reduced by knowing the value of Y is the amount of information about X contained in Y. The quantity is called *mutual information* between X

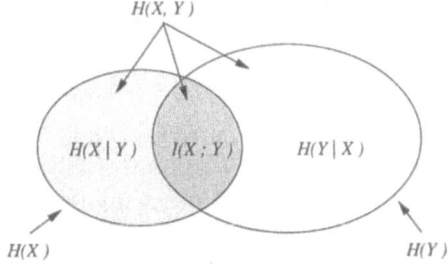

Fig. 1. Relationship between entropy and mutual information

and Y and is formally written as

$$I(X;Y) = H(X) - H(X|Y) \tag{4}$$

$$= \sum_{x \in \mathcal{X}} \sum_{y \in \mathcal{Y}} p(x,y) \log \frac{p(x,y)}{p(x)p(y)}. \tag{5}$$

Mutual information is symmetric and nonnegative. Two random variables are mutually independent if and only if the mutual information between them is zero. The Equation (4) can be rewritten as

$$I(X;Y) = H(X) + H(Y) - H(X,Y). \tag{6}$$

It is deduced from the equation that the random variables are independent if and only if the joint entropy of them is equal to the summation of the two marginal entropies. The relationship between entropy and mutual information is illustrated in Fig. 1. Table 1 shows examples of joint random variables. Table 1(a) is an example of mutually independent random variables and Table 1(b) is an example of mutually dependent random variables.

Table 1. Example joint pmf's of two pairs of random variables whose alphabets $\mathcal{X}$ and $\mathcal{Y}$ are $\{0,1\}$. (a) Two mutually independent random variables. $H(X) = 1$, $H(Y) = 0.81$, $H(X,Y) = 1.81$, and $I(X;Y) = 0$. (b) Two mutually dependent random variables. $H(X) = 1$, $H(Y) = 0.81$, $H(X,Y) = 1.75$, and $I(X;Y) = 0.06$

(a)

X	Y = 0	Y = 1	
0	1/8	3/8	1/2
1	1/8	3/8	1/2
	1/4	3/4	

(b)

X	Y = 0	Y = 1	
0	1/16	7/16	1/2
1	3/16	5/16	1/2
	1/4	3/4	

3 Probability Model and Epistasis Measures

3.1 Probability Model

Assume that a problem is encoded into n genes, and let the fitness function of the problem be $f : \mathcal{U} \to \mathbb{R}$ where $\mathcal{U}$ is the set of all feasible[1] solutions, called *universe* of the problem. When we do random sampling on $\mathcal{U}$, the probability that a feasible solution $(x_1, x_2, \ldots, x_n)$ will be chosen, is $1/|\mathcal{U}|$. By the probability model, random variables for the genes and the fitness are defined. Let the random variable for gene i be X_i and the random variable for the fitness be Y, and the set of allele values of gene i and the set of all possible fitness values be A_i and F, respectively, then the probability mass function is defined as

$$p(x_1, x_2, \ldots, x_n, y) = \begin{cases} 1/|\mathcal{U}| & \text{if } (x_1, x_2, \ldots, x_n) \in \mathcal{U} \\ & \text{and } y = f(x_1, x_2, \ldots, x_n) \\ 0 & \text{otherwise} \end{cases} \quad (7)$$

for $x_i \in A_i$, $i \in \{1, 2, \ldots, n\}$ and $y \in F$. It is practical to use a set of sampled solutions in the Equation (7) instead of the universe $\mathcal{U}$ for large-sized problems because of the spatial or computational limitations. But, in the case, the size of the set must be not too small for getting results of low levels of distortion.

3.2 Epistasis Measures

Three epistasis-related measures are proposed in this section. They are based on the probability model described in Sect. 3.1. They quantify *gene significance*, *gene epistasis*, and *problem epistasis*, respectively.

The significance of a gene i is defined to be the amount of its contribution to the fitness. It could be understood as the amount of information contained in X_i about Y, i.e., $I(X_i; Y)$. Since the minimum and the maximum of the mutual information are 0 and $H(Y)$, respectively, a normalization could be done by dividing $I(X_i; Y)$ by $H(Y)$. As a result, the significance ξ_i of a gene i is defined as

$$\xi_i = \frac{I(X_i; Y)}{H(Y)}. \quad (8)$$

It ranges from 0 to 1; if the value is zero, the gene has no contribution to the fitness and if the value is one, the gene wholly determines the fitness value.

The epistasis (often referred to as interaction) between genes means the dependence of a gene's contribution to the fitness upon the value of other genes. The contribution of gene i and gene j to the fitness are quantified as $I(X_i; Y)$ and $I(X_j; Y)$, respectively. And the contribution of the gene pair (i, j) to the fitness is quantified as $I(X_i, X_j; Y)$. Therefore, the epistasis between the two genes could be written as $I(X_i, X_j; Y) - I(X_i; Y) - I(X_j; Y)$. A normalization

---

[1] A solution is feasible if the fitness function is defined on it.

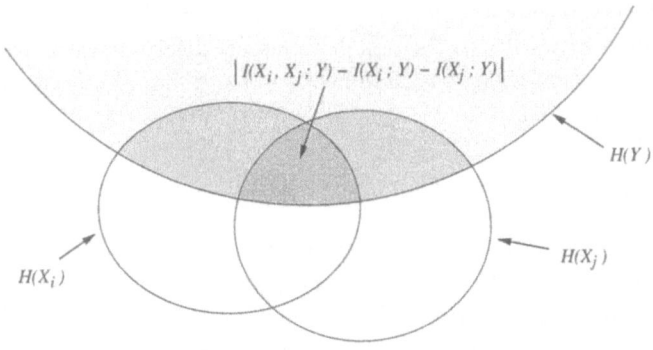

Fig. 2. An illustration of the pairwise epistasis

could be done by dividing the quantity by $I(X_i, X_j; Y)$. As a result, the gene epistasis ε_{ij} between gene i and gene j is defined as

$$\varepsilon_{ij} = \begin{cases} 1 - \dfrac{I(X_i; Y) + I(X_j; Y)}{I(X_i, X_j; Y)} & \text{if } I(X_i, X_j; Y) \neq 0 \\ 0 & \text{otherwise.} \end{cases} \quad (9)$$

Since the minimum and the maximum of the fraction in the Equation (9) are 0 and 2, respectively, the epistasis ranges from -1 to 1. It has a positive value if $I(X_i, X_j; Y) > I(X_i; Y) + I(X_j; Y)$ and it has a negative value otherwise. The former case means that the genes interact constructively with each other, and the latter case means that they interact destructively with each other. We call the epistasis of the former case *positive* gene epistasis, and that of the latter case *negative* gene epistasis. If the two genes are mutually independent, the gene epistasis is zero. Figure 2 shows an illustration of the above definition.

The mean absolute of the gene epistases of all gene pairs could be used as a measure of the epistasis of a problem, i.e., the problem epistasis η is defined as

$$\eta = \frac{1}{n(n-1)} \sum_{i=1}^{n} \sum_{j<i} |\varepsilon_{ij}|. \quad (10)$$

Since each ε_{ij} ranges from -1 to 1, η ranges from 0 to 1. The larger value η a problem has, the more epistatic the problem is.

3.3 Fitness Discretization

In general, the fitness function of a problem is defined on a continuous domain, while each gene has discrete allele values in many cases. So, the fitness value needs to be discretized to apply the measures described in Sect. 3.2. The most simple methods are *equal-width discretization* and *equal-frequency discretization* [12]. In the equal-width discretization, the whole range is divided into k intervals of equal widths, while the whole range is divided into k intervals that include the same number of samples in the equal-frequency discretization. We use equal-frequency discretization with $k = 10$ in Sects. 4.3 and 4.4. Ten is the most widely used number of intervals.

3.4 An Example

Table 2 shows two example fitness functions. The function f_{pos} has four feasible solutions, while f_{neg} has only three feasible solutions. We can compute the values of the measures, proposed in Sect. 3.2, of the example functions as follows. First, we make a joint pmf table for each function as Tables 3a–b. Then, we apply the equations in Sect. 2 to each of the tables to compute the entropies and mutual informations. Finally, we use the Equations (8), (9), and (10) to compute the gene significance, gene epistasis, and problem epistasis, respectively. Table 4 shows the intermediate values and the resultant measure values. We can see that the two genes of f_{pos} have positive gene epistasis, while the two genes of f_{neg} have negative gene epistasis. The table shows that the gene 1 of f_{neg} is more significant than gene 2, and the function f_{neg} is more epistatic than f_{pos}.

Table 2. Two example functions f_{pos} and f_{neg}

$x_1\ x_2$	f_{pos}	f_{neg}
0 0	0	0
0 1	0	undefined
1 0	0	1
1 1	1	1

Table 3. Joint pmf's of the example functions

(a) Joint pmf of f_{pos}.

$X_1\ X_2$	$Y=0$	$Y=1$	
0 0	1/4	0	1/4
0 1	1/4	0	1/4
1 0	1/4	0	1/4
1 1	0	1/4	1/4
	3/4	1/4	

(b) Joint pmf of f_{neg}.

$X_1\ X_2$	$Y=0$	$Y=1$	
0 0	1/3	0	1/3
0 1	0	0	0
1 0	0	1/3	1/3
1 1	0	1/3	1/3
	1/3	2/3	

Table 4. The calculation of the measures for the example functions f_{pos} and f_{neg}

	$H(Y)$	$I(X_1;Y)$	$I(X_2;Y)$	$I(X_1,X_2;Y)$	ξ_1	ξ_2	ε_{12}	η
f_{pos}	0.811	0.311	0.311	0.811	0.384	0.384	0.233	0.233
f_{neg}	0.918	0.918	0.252	0.918	1.000	0.274	−0.274	0.274

Table 5. Davidor's four example functions f_1, f_2, f_3, and f_4

$x_1\ x_2\ x_3$	f_1	f_2	f_3	f_4
0 0 0	0	0	0.0	7
0 0 1	1	0	0.5	5
0 1 0	2	0	1.0	5
0 1 1	3	0	1.5	0
1 0 0	4	0	2.0	3
1 0 1	5	0	2.5	0
1 1 0	6	0	3.0	0
1 1 1	7	28	17.5	8

Table 6. The gene significance ξ_i, gene epistasis ε_{ij}, and problem epistasis η of Davidor's four example functions f_1, f_2, f_3, and f_4

	ξ_1	ξ_2	ξ_3	ε_{12}	ε_{13}	ε_{23}	η	σ_ε^2
f_1	0.333	0.333	0.333	0.000	0.000	0.000	0.000	0
f_2	0.254	0.254	0.254	0.060	0.060	0.060	0.060	49
f_3	0.333	0.333	0.333	0.000	0.000	0.000	0.000	12.25
f_4	0.304	0.188	0.188	0.082	0.082	0.298	0.154	8.57

4 Experimental Results

4.1 Davidor's Examples

Davidor tested his epistasis variance on four example functions as shown in Table 5. They are a linear function (f_1), a delta function (f_2), a mixture of the linear function and delta function ($f_3 = \frac{f_1+f_2}{2}$), and a minimal deceptive function (f_4). Table 6 shows the gene significance ξ_i, gene epistasis ε_{ij}, and problem epistasis η of each example function. The epistasis variance σ_ε^2 in the final column was quoted from the Davidor's paper [5] for comparison. The results are somewhat different. The problem epistasis and the epistasis variance of f_1 are zero in common. But, the epistasis variance of f_3 is not zero, while the problem epistasis of the function is zero. At the same time, the most problem epistatic function among them is f_4, while the function with the largest epistasis variance is f_2. The difference comes mainly from the reasons in the following. The epistasis variance treats the fitness as a scalar quantity. But, the proposed measures treat the fitness as a categorical index, i.e., the proposed measures do not individually concern the magnitude of the fitness. The proposed measures only concern whether the fitness values of solutions are the same or not.

4.2 Royal Road Function

Royal Road function is a function proposed by Forrest and Mitchell [13] to investigate precisely and quantitatively how schema processing actually takes place during the typical evolution of a genetic algorithm. To do so, the function was designed to have obvious building blocks and an optimal solution. Royal Road function is defined as

$$f(x_1, x_2, \ldots, x_n) = \sum_i c_i \delta_i(x_i, x_2, \ldots, x_n) \qquad (11)$$

where c_i is a predefined coefficient corresponding to a schema s_i, and $\delta_i : \{0,1\}^n \to \{0,1\}$ is a function that returns 1 if the solution contains the schema s_i, and returns 0 otherwise. Generally, the coefficient c_i is defined as the order of schema s_i. Table 7 shows the two Royal Road functions used in our experiments. The function R_1 has four building blocks of order 2, while R_2 has the building blocks of R_1 and two more building blocks of order 4. Figures 3a–b shows the illustrations of the gene epistases of R_1 and R_2, respectively. We can see that the genes of the building blocks have relatively strong gene epistasis with each other. The figure shows that the gene epistases between the genes in order-2 building blocks are larger than those of order-4 building blocks. The problem epistasis η of R_1 and R_2 were 0.126 and 0.236, respectively. It means that R_2 has stronger problem epistasis than R_1.

4.3 NK-Landscape

The NK-landscape model is a model proposed by Kauffman [14] to define a family of fitness functions that have various dimensions of search space and degrees of epistasis (see also [15,16]). The functions are tuned by two parameters:

Table 7. Two Royal Road functions

(a) R_1

Schema	Coefficient
$s_1 = 1\,1\,*\,*\,*\,*\,*\,*$	$c_1 = 2$
$s_2 = *\,*\,1\,1\,*\,*\,*\,*$	$c_2 = 2$
$s_3 = *\,*\,*\,*\,1\,1\,*\,*$	$c_3 = 2$
$s_4 = *\,*\,*\,*\,*\,*\,1\,1$	$c_4 = 2$
$s_{opt} = 1\,1\,1\,1\,1\,1\,1\,1$	

(b) R_2

Schema	Coefficient
$s_1 = 1\,1\,*\,*\,*\,*\,*\,*$	$c_1 = 2$
$s_2 = *\,*\,1\,1\,*\,*\,*\,*$	$c_2 = 2$
$s_3 = *\,*\,*\,*\,1\,1\,*\,*$	$c_3 = 2$
$s_4 = *\,*\,*\,*\,*\,*\,1\,1$	$c_4 = 2$
$s_5 = 1\,1\,1\,1\,*\,*\,*\,*$	$c_5 = 4$
$s_6 = *\,*\,*\,*\,1\,1\,1\,1$	$c_6 = 4$
$s_{opt} = 1\,1\,1\,1\,1\,1\,1\,1$	

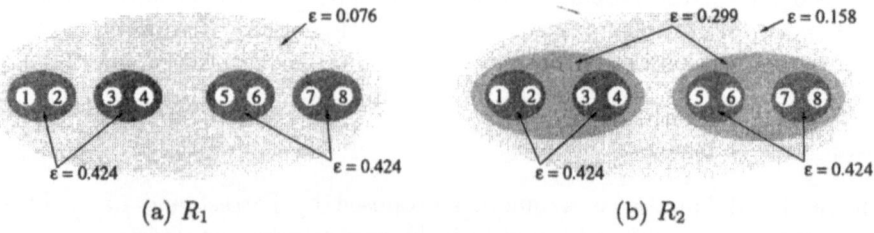

(a) R_1 (b) R_2

Fig. 3. Gene epistasis ε_{ij} of Royal Road functions

Table 8. Gene epistasis ε_{ij} of NK-landscape ($N = 12$)

K	\multicolumn{13}{c}{Co-Contribution Frequency}

K	0	1	2	3	4	5	6	7	8	9	10	11	12
2	0.072	0.206	0.249										
3	0.092	0.196	0.242	0.254									
4	0.112	0.186	0.228	0.263	0.262								
5	0.136	0.193	0.233	0.257	0.271	0.282	0.303						
6		0.231	0.249	0.269	0.289	0.305	0.307	0.310					
7			0.260	0.281	0.287	0.300	0.305	0.311	0.325				
8				0.295	0.305	0.309	0.311	0.317	0.321	0.327	0.317		
9					0.312	0.320	0.320	0.324	0.330	0.329	0.330		
10						0.315	0.337	0.328	0.334	0.338	0.339		
11													0.334

N and K. The parameters N and K determines the dimension of the problem space and the degree of epistasis between the genes constituting a chromosome, respectively. The fitness f of a solution $(x_1, x_2, \ldots, x_N)$ is defined as

$$f(x_1, x_2, \ldots, x_N) = \frac{1}{N} \sum_{i=1}^{N} f_i(x_i, x_{j_{i1}}, x_{j_{i2}}, \ldots, x_{j_{iK}}) \tag{12}$$

where the fitness contribution f_i depends on the value of gene i and the values of K other genes $j_{i1}, j_{i2}, \ldots, j_{iK}$. The function $f_i : \{0,1\}^{K+1} \to \mathbb{R}$ assigns a random number, distributed uniformly between 0 and 1, to each of its 2^{K+1} inputs. The values for $j_{i1}, j_{i2}, \ldots, j_{iK}$ are chosen from $\{1, 2, \ldots, N\}$ at random.

The gene values $x_i, x_{j_{i1}}, x_{j_{i2}}, \ldots, x_{j_{iK}}$ contribute together to the fitness contribution f_i. We define the *co-contribution frequency* of gene i and j as the number of cases that the gene values x_i and x_j contribute together to the fitness f. Intuitively, we can say that two genes are strongly correlated if the co-contribution frequency of the genes is high.

The gene epistases of gene pairs of NK-landscapes were listed along with their co-contribution frequencies in Table 8. We discretized the fitness into 10 intervals by the equal-frequency discretization method in computing the measures. For each K, 200 independently generated functions were used for statistical stability. In the table, the (i, j) entry represents the average gene epistasis of the gene pairs of the NK-landscapes of $K = i$, that co-contribute j times. We can see that the gene epistasis increases as the co-contribution frequency increases for small K's, but it tends to converge for larger K's. Figure 4 shows the problem epistases of the NK-landscapes for various K's. We can see that the problem epistasis increases as the K increases. Both of the results support our intuitive predictions.

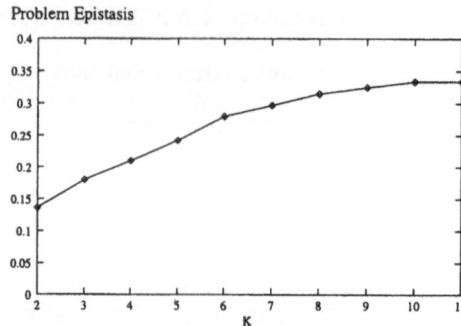

Fig. 4. Problem epistasis η of NK-landscapes ($N = 12$)

4.4 Traveling Salesman Problem

Given n cities, the traveling salesman problem (TSP) is the problem of finding a shortest Hamiltonian cycle visiting the cities. TSP is a well-known NP-hard problem [17]. It is one of the most popular optimization problems and has served as an initial proving ground for new problem solving techniques for decades.

We apply the locus-based encoding to the problem as in [18]; one gene is allocated for every city and the gene value represents the index of its next city in the Hamiltonian cycle. By the encoding, the fitness f of a solution $(x_1, x_2, \ldots, x_n)$ that represents a Hamiltonian cycle is written as

$$f(x_1, x_2, \ldots, x_n) = C_{max} - \sum_{i=1}^{n} d_{ix_i} \tag{13}$$

where d_{pq} is the distance from city p to city q and C_{max} is the cycle length of the worst solution. The subtraction in the equation forces the fitness to be nonnegative and the problem becomes a maximization problem. It is notable that the absolute value of C_{max} does not affect the epistasis measures when the equal-frequency discretization is used. We computed the problem epistasis η on TSP and compared it with a problem difficulty measure, *fitness distance correlation*.

The fitness distance correlation (FDC) is a measure of problem difficulty proposed by Jones and Forrest [19]. FDC is defined to be the correlation coefficient of the fitness and the distance to the nearest global optimum of sampled solutions. Thus, it ranges from -1 to 1. As the value approaches -1, a problem is believed to become easier.

When a genetic algorithm is hybridized with a local optimization algorithm, what the algorithm can see are only local optima. Thus, it is valuable to examine the space of local optima. For each problem instance, the solution set used for the computation of FDC and problem epistasis, was chosen as follows. First, we generate ten thousand solutions at random and apply a local optimization algorithm to them. Then, we discard the duplicated copies from the resultant solutions. We used 2-Opt [20] as the local optimization algorithm because it is

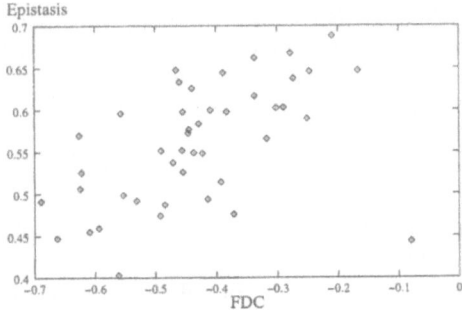

Fig. 5. Fitness-distance correlation (FDC) vs. problem epistasis of TSP

one of the most simple and basic heuristics. The fitness was discretized into 10 intervals by the equal-frequency discretization method as in the case of NK-landscape. As the distance measure, we used Hamming distance that is defined to be the number of genes with different values. Figure 5 shows the relationship between the problem epistasis and the FDC for 44 instances taken from TSPLIB [21]. They are all instances available whose numbers of cities lie inbetween 100 and 700. The figure shows that the two measures are strongly correlated. It means that the problem epistasis works well as the problem difficulty measure.

5 Conclusions

We provided a new framework to formulate and quantify the epistasis of a problem based on Shannon's entropy. With the framework, three measures were proposed: gene significance, gene epistasis, and problem epistasis. They are for choosing significant genes, detecting epistatic gene pairs, and quantifying the epistasis of a problem as a difficulty measure, respectively. They are different from Davidor's epistasis variance in the way of treating the fitness. They treat a fitness value as a categorical index, while the epistasis variance treats it as a scalar quantity. The experimental results on various well-known problems, such as Royal Road function, NK-landscape, and traveling salesman problem, support their usefulness and appropriateness. Future studies include extensions of the framework and applications of the measures.

Acknowledgments. This work was partly supported by Optus Inc. and Brain Korea 21 Project. The RIACT at Seoul National University provided research facilities for this study.

References

1. D. E. Goldberg. Simple genetic algorithms and the minimal deceptive problem. In *Genetic Algorithms and Simulated Annealing*, pages 74–88, 1987.

2. J. Horn and D.E. Goldberg. Genetic algorithm difficulty and the modality of fitness landscapes. In *Foundations of Genetic Algorithms 3*, pages 243–270. 1995.
3. H. Kargupta. Signal-to-noise, crosstalk, and long range problem difficulty in genetic algorithms. In *International Conference on Genetic Algorithms*, pages 193–200, 1995.
4. J. Holland. *Adaptation in Natural and Artificial Systems*. The University of Michigan Press, 1975.
5. Y. Davidor. Epistasis variance: Suitability of a representation to genetic algorithms. *Complex Systems*, 4:369–383, 1990.
6. C.R. Reeves and C.C. Wright. An experimental design perspective on genetic algorithms. In *Foundations of Genetic Algorithms 3*, pages 7–22. 1995.
7. P. Larrañaga and J.A. Lozano. *Estimation of Distribution Algorithms: A New Tool for Evolutionary Computation*. Kluwer Academic Publishers, 2002.
8. M. Pelikan, D.E. Goldberg, and F. Lobo. A survey of optimization by building and using probabilistic models. *Computational Optimization and Applications*, 21(1):5–20, 2002.
9. D.I. Seo and B.R. Moon. A survey on chromosomal structures and operators for exploiting topological linkages of genes. In *Genetic and Evolutionary Computation Conference*, 2003.
10. C.E. Shannon. A mathematical theory of communication. *Bell System Technical Journal*, 27:379–423, 623–656, 1948.
11. T. Cover and J. Thomas. *Elements of Information Theory*. John Wiley & Sons, 1991.
12. D. Chiu, A. Wong, and B. Cheung. Information discovery through hierarchical maximum entropy discretization and synthesis. In *Knowledge Discovery in Databases*. MIT Press, 1991.
13. S. Forrest and M. Mitchell. Relative building-block fitness and the building-block hypothesis. In *Foundations of Genetic Algorithms 2*, pages 109–126. 1993.
14. S.A. Kauffman. Adaptation on rugged fitness landscapes. In D. L. Stein, editor, *Lectures in the Sciences of Complexity*, pages 527–618. Addison-Wesley, 1989.
15. B. Manderick, M. de Weger, and P. Spiessens. The genetic algorithm and the structure of the fitness landscape. In *International Conference on Genetic Algorithms*, pages 143–157, 1991.
16. P. Merz and B. Freisleben. On the effectiveness of evolutionary search in high-dimensional NK-landscapes. In *IEEE Conference on Evolutionary Computation*, pages 741–745, 1998.
17. M.R. Garey and D.S. Johnson. *Computers and Intractability: A Guide to the Theory of NP-Completeness*. Freeman, 1979.
18. T.N. Bui and B.R. Moon. A new genetic approach for the traveling salesman problem. In *IEEE Conference on Evolutionary Computation*, pages 7–12, 1994.
19. T. Jones and S. Forrest. Fitness distance correlation as a measure of problem difficulty for genetic algorithms. In *International Conference on Genetic Algorithms*, pages 184–192, 1995.
20. G.A. Croes. A method for solving traveling salesman problems. *Operations Research*, 6:791–812, 1958.
21. TSPLIB.
http://www.iwr.uni-heidelberg.de/groups/comopt/software/TSPLIB95/.

A Survey on Chromosomal Structures and Operators for Exploiting Topological Linkages of Genes

Dong-Il Seo and Byung-Ro Moon

School of Computer Science & Engineering, Seoul National University
Sillim-dong, Kwanak-gu, Seoul, 151-742 Korea
{diseo,moon}@soar.snu.ac.kr
http://soar.snu.ac.kr/~{diseo,moon}/

Abstract. The building block hypothesis implies that the epistatic property of a given problem must be connected well to the linkage property of the employed representation and crossover operator in the design of genetic algorithms. A good handling of building blocks has much to do with topological linkages of genes in the chromosome. In this paper, we provide a taxonomy of the approaches that exploit topological linkages of genes. They are classified into three models: static linkage model, adaptive linkage model, and evolvable linkage model. We also provide an overview on the chromosomal structures, encodings, and operators supporting each of the models.

1 Introduction

A genetic algorithm maintains a population of solutions and repeatedly applies genetic operators to find optimal or near-optimal solutions. Through the genetic process, various patterns of genes, called *schemata*, are created, destroyed, and recombined in parallel. Holland named the phenomenon *implicit parallelism* [1]. The schemata of short defining length, low order, and high quality play an important role in the process. They are called *building blocks* and the genetic process is well explained as the juxtaposition of building blocks [2]. It means that the power of a genetic algorithm lies in its ability to find building blocks and to grow them efficiently to larger ones.

The solutions in the population have a structure of genes called *chromosomal structure* for the crossover operator. A typical chromosomal structure is one-dimensional array. But, for many problems, it is known that non-linear chromosomal structures are more advantageous than such a simple structure [3,4,5,6,7]. One of the reasons of adopting such non-linear structures is that we can minimize the loss of information contained in the given problem with more natural structures for the problem. Another reason is that many crossovers require special type of chromosomal structures. The reasons are ultimately related to the creation and growth of the building blocks. That is, the first one helps the chromosomal structure to reflect well the epistatic property of the problem and the second one allows us to be able to apply a crossover operator that creates and recombines well the building blocks.

There has been a significant amount of efforts for designing efficient chromosomal structures, encodings, and operators. In this paper, we provide a survey and bibliography of such studies.

The rest of this paper is organized as follows. The basic concepts are introduced in Sect. 2 and the chromosomal structures are explained in Sect. 3. The approaches concerned with the topic are classified and explained in Sect. 4. Finally, the summary of this paper is provided in Sect. 5.

2 Preliminaries

Building blocks appear in interactive gene groups. We can easily observe that the contribution of a gene to the chromosomal fitness depends on the values of other genes in the chromosome. This phenomenon is called *gene interaction* or *epistasis* [1,8,9]. For example, a bat must be able to hear high frequency ultrasonic waves if it generates high frequency ultrasonic squeaks, and it must be able to hear low frequency ultrasonic waves if it generates low frequency ultrasonic squeaks. So, for bats, the genes related to the organs that generate ultrasonic squeaks and those related to the organs that hear ultrasonic waves have strong interactions.

Building blocks are generated by the crossover and mutation. In this paper, we focus on the crossover. A crossover operator generates an offspring by recombining two parents. There are the number of genes minus one operators in one-point crossover with one-dimensional array representation. The creativity of a crossover is strongly related to the diversity of the operators. Generally, the more operators a crossover has, the more diverse new schemata it can generate [5]. However, existing schemata do not always survive through the crossover. A gene group is said to have a *strong linkage* if the genes are arranged so that the gene group has relatively high survival probability, and it is said to have a *weak linkage*, otherwise [1]. For example, in the case of one-point crossover on a chromosome of one-dimensional array, the strength of the linkage of a gene group is inversely proportional to the defining length of the corresponding schema. For the growth of building blocks to an optimal or near-optimal solution, they have to survive through the genetic process, specifically through the crossover. Consequently, we come to a proposition that mutually interactive genes need to have strong linkages [1,2,10]. In the case of one-point crossover, for example, it is helpful to make the strongly epistatic gene groups have short defining lengths.

The linkage of a gene group is, in many cases, dependent on the distribution of the genes in the chromosome and the crossover scheme. Particularly, there is a group of approaches where each gene is placed in an Euclidean or non-Euclidean space, called *chromosomal space*, to represent the linkages between genes. The linkage in this context is called *topological linkage*. There are many other problem-specific or analytic methods to exploit the linkages between genes. Estimation-of-distribution algorithms (EDAs) [11,12], also called probabilistic model-building genetic algorithms (PMBGAs), are examples. But, they are out of the scope of this paper.

3 Chromosomal Structures

In order to make the topological linkages reflect well the epistatic structure of a given problem, we need to choose an appropriate chromosomal structure. Generally, in representing a graph geometrically, it is known that considerably high dimensions of representation space are necessary to alleviate the degree of distortion [13]. This is a good reason that multi-dimensional representations are more advantageous than simple one-dimensional representations for highly epistatic problems. However, multi-dimensional representations require sophisticated crossover operators.

Figure 1 shows an illustration of several representative chromosomal structures. One-dimensional array (a) is a typical and the most popular chromosomal structure. By linking the two ends, a ringed array (b) is generated. This structure has a virtue of treating all the positions of genes symmetrically. We can also increase the dimension of chromosomes (c)–(d). Two-dimensional array (c) is frequently used. The distortion level in using an array may yet be lowered by adopting a real space. For example, a one-dimensional real space (line) (e) or a two-dimensional real space (plane) (f) may be used. A complete graph (g) is also possible. In the case of a complete graph, the genes are linked by weighted edges. The edge weight is sometimes called *genic distance*.

The chromosomal structures can be classified into two categories: *Euclidean chromosomal structures* and *non-Euclidean chromosomal structures*. The Euclidean chromosomal structures include arrays and real spaces and the non-Euclidean chromosomal structures include complete graphs. It should be noted that the chromosomal structure is not necessarily the same as the data structure used in the encoding. The term means only the conceptual structure of genes used in the crossover operator in this paper.

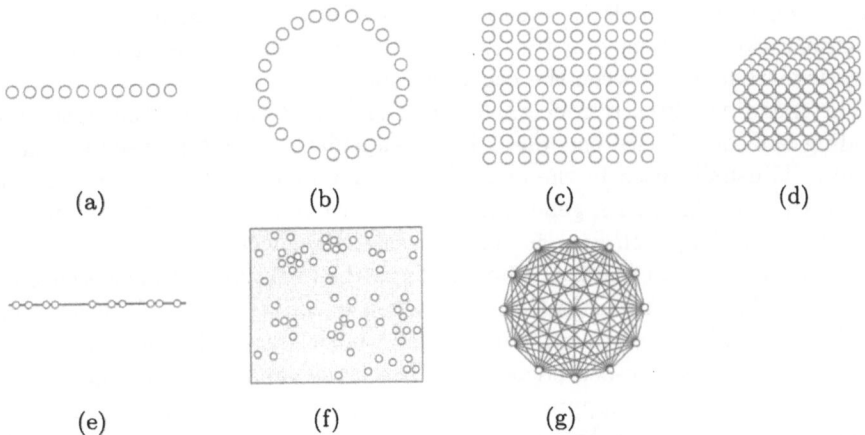

Fig. 1. Various chromosomal structures. (a) 1D array. (b) ringed array. (c) 2D array. (d) 3D array. (e) 1D real space. (f) 2D real space. (g) complete graph.

4 Linkage Models

In general, the interactions between genes are not simple. If they are, the genes can be divided into mutually independent subgroups. It means that we can divide the whole problem into a number of subproblems and can conquer them independently. In most cases, the epistatic structure is so complicated that it is extremely hard to find an optimal arrangement of the genes in the chromosome. For example, when a problem is encoded into n binary genes, the chromosome has totally $n!$ arrangements in an array representation. It means that the problem is more difficult than the original problem with a total of 2^n solutions. To make matters worse, we cannot exactly grasp the epistatic properties of the given problem unless we explore the whole solution space [8].

For decades, various approaches to the problem have been tried. They are divided into two classes: *static linkage model* and *dynamic linkage model*. The linkage is fixed statically during the genetic process in the static linkage model, while it changes dynamically in the dynamic linkage model. The dynamic linkage model is divided again into *adaptive linkage model* and *evolvable linkage model*. In the evolvable linkage model, the linkage itself evolves in parallel with the alleles of genes through the competition of individuals in the genetic process. In the adaptive linkage model, however, the linkage alters but does not evolve in the genetic process. Generally, the linkages are assigned analytically or heuristically using the prior information about the epistatic property of the given problem in the static linkage model. Thus, efficient methods for the assignment are needed. The evolvable linkage model is advantageous for the problems where prior information is not available. Such problems are sometimes called *black box optimization* [32]. The model, however, experienced many failures initially. The failures seem to be due to the loss in the race against allele selection because of the slow evolution speed. As mentioned before, the linkage assignment problem is at least as difficult as the original problem in many cases. So, the model requires efficient methods to speed up the evolution. The model also requires implementation overheads for the evolution of the linkage. The adaptive linkage model is a trade-off between the other two models.

The encodings directly concerned with the linkage issues are "encapsulated" encoding and locus-based encoding. The encapsulated encoding, used by Bagley [14] first, is usually used in the evolvable linkage model with array representation. In the encoding, each gene is an encapsulation of a position indicator and an allele value, i.e., each gene is encoded as (index, value) pair. The encoding allows the crossover operator to contribute to the evolutions of both alleles and linkages at the same time. In the locus-based encoding, the allele values have meaning only when associated with specific positions in the chromosome. In order to use the locus-based encoding in the array representation and the real space representation, a mapping table that maps each gene to a position in the chromosomal space is needed. In order to use the locus-based encoding in the complete graph representation, a table for the genic distances is needed.

The approaches that exploit topological linkages are summarized in Table 1. For each approach, chromosomal structure (CStr), linkage model (Mod), and

Table 1. A summary of the topological linkage-based approaches. For each approach, chromosomal structure (CStr), linkage model (LMod), and encoding (Enc) are given.

Approach	CStr	LMod	Enc	Remark
Bagley, 1967 [14]	1D A	E	E	inversion operator
Goldberg et al., 1989 [15]	1D A	E	E	messy GA
Goldberg et al., 1990 [16]	1D A	E	E	messy GA
Levenick, 1991 [17]	1D A	S	L	non-coding segment
Goldberg et al., 1993 [18]	1D A	E	E	fast messy GA
Bui & Moon, 1993 [19], 1996 [20]	1D A	S	L	DFS/BFS-reordering
Bui & Moon, 1994a [21]	1D A	S	L	DFS-reordering
Bui & Moon, 1994b [22]	1D A	S	L	local optimum reordering
Wu et al., 1994 [23]	1D A	S	L	non-coding segment
Wu & Lindsay, 1995 [24]	1D A	S	L	non-coding segment
Levenick, 1995 [25]	1D A	S	L	metabits
Kargupta, 1996 [26]	1D A	E	E	GEMGA
Bui & Moon, 1998 [27]	1D A	S	L	weighted-DFS reordering
Levenick, 1999 [28]	1D A	S	L	non-coding segment
Knjazew & Goldberg, 2000 [29]	1D A	E	E	OMEGA
Kwon & Moon, 2002 [30]	1D A	A	L	correlation coefficient
Harik & Goldberg, 1996 [31]	RA	E	E	LLGA
Harik, 1997 [32]	RA	E	E	LLGA
Lobo et al., 1998 [33]	RA	E	E	compressed intron LLGA
Harik & Goldberg, 2000 [34]	RA	E	E	prob. expression LLGA
Chen & Goldberg, 2002 [35]	RA	E	E	start expression gene LLGA
Cohoon & Paris, 1986 [3]	2D A	S	L	rectangle-style cross. (ψ_2)
Anderson et al., 1991 [4]	2D A	S	L	block-uniform crossover
Bui & Moon, 1995 [36]	2D A	S	L	Z3
Moon & Kim, 1997 [37]	2D A	S	L	geo. attraction-based embed.
Moon et al., 1998 [38]	2D A	S	L	geographic crossover, DFS-zigzag embedding
Moon & Kim, 1998 [39]	2D A	E	L	two-layered crossover
Kim & Moon, 2002 [40]	2D A	S	L	geographic crossover, neuron reordering
Kahng & Moon, 1995 [5]	2/3D A	S	L	geographic crossover, DFS-row-major embedding
Lee & Antonsson, 2001 [41]	1D RS	E	L	adaptive non-coding seg.
Jung & Moon, 2000 [42], 2002a [6]	2D RS	S	L	natural crossover
Jung & Moon, 2002b [43]	2D RS	S	L	natural crossover
Greene, 2002 [44]	CG	A	L	cutoff-value based crossover
Seo & Moon, 2002 [7], 2003 [45]	CG	S	L	Voronoi quantized crossover

encoding (Enc) are summarized in the table. The chromosomal structures are abbreviated as A (array), RA (ringed array), RS (real space), and CG (complete graph). The linkage models are abbreviated as S (static linkage model), A (adaptive linkage model), and E (evolvable linkage model). The encodings are abbreviated as E (encapsulated encoding) and L (locus-based encoding).

4.1 Static Linkage Models

Raising Dimensions. Cohoon and Paris's work [3] seems to be the first approach published employing multi-dimensional representation. They proposed a rectangle-style crossover (ψ_2) using two-dimensional array representation for VLSI circuit placement. In the crossover, the gene values in a $k \times k$ square section of one parent and the remaining gene values of the other parent are copied into the offspring, where k has a truncated normal distribution with mean 3 and variance 1.

Another approach based on two-dimensional array is the block-uniform crossover proposed by Anderson et al. [4]. The crossover divides the chromosome into $i \times j$ blocks where i and j are chosen at random. Then each block of one parent is interchanged randomly with the corresponding block of the second parent based on a preassigned probability. They applied it to Ising problem.

Bui and Moon [36], proposed an extension of the traditional k-point crossover to multi-dimensional array representation, called multi-dimensional k-point crossover (Z3) for grid-placement and graph partitioning. In the crossover, the chromosomal space is divided into hyper-subspaces by k random crossover points, i.e., k_i crossover points are chosen at random for each ith dimension such that $k_i \geq 0$ and $\sum_{i=1}^{n} k_i = k$, and, by them, the chromosomal space is divided into $\prod_{i=1}^{n}(k_i + 1)$ hyper-subspaces. Then the gene values in each adjacent subspaces are copied alternately from one of the two parents into the offspring.

Although multi-dimensional array representations can preserve more epistatic relations between genes than simple one-dimensional array representations, naive multi-dimensional crossovers have a potential weakness in the relatively small number of recombinations. In the case of one-point crossover for n genes, for example, one-dimensional array representation has $n-1$ operators, while the Z3 on two-dimensional and three-dimensional array representations have only $2(\sqrt{n} - 1)$ and $3(\sqrt[3]{n} - 1)$ operators, respectively. Kahng and Moon [5] proposed a flexible framework to deal with the problem. In their geographic crossover, the chromosomal space is divided into subspaces with monotonic hypersurfaces instead of hyperplanes. The crossover have been applied to graph partitioning [5], neural network optimization [40], and fixed channel assignment [46] recently.

Although multi-dimensional array representations and crossovers have the advantages mentioned above, array-based chromosomal structures have limits to represent non-regular epistatic relationships between genes. Jung and Moon [42,6] focused on the point and proposed 2D real space representation and the natural crossover. In the approach, the 2D image of the problem instance itself is used for crossover and arbitrary simple curves are used for chromosomal cutting. They are applied to 2D traveling salesman problem [6,42] and vehicle routing problem [43].

Figure 2(a)–(e) illustrates the crossovers described above.

Reordering. Bui and Moon [19] proposed the static linkage model that reorders the positions of genes in an array representation before starting the ge-

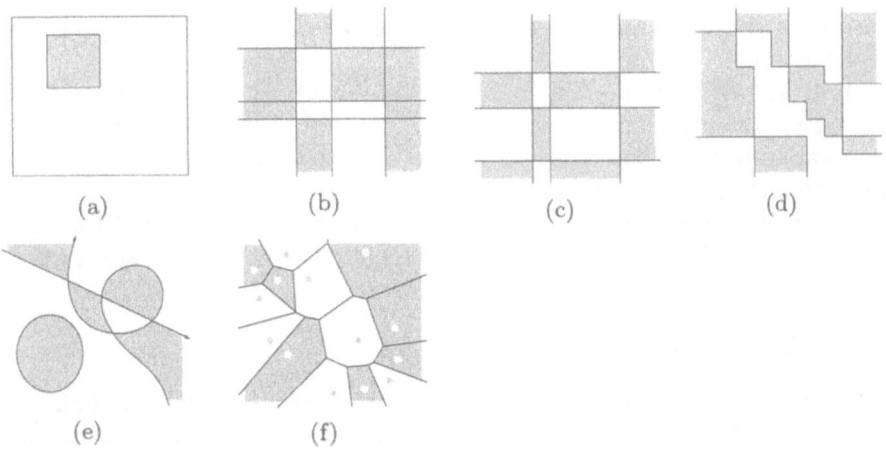

Fig. 2. Illustrations of the non-linear crossover operators. (a) Rectangle-style crossover. (b) Block-uniform crossover. (c) Multi-dimensional k-point crossover (Z3). (d) Geographic crossover. (e) Natural crossover. (f) Voronoi quantized crossover with genic distances defined on 2D Euclidean space.

netic algorithm. The approach is sometimes called *schema preprocessing* or *static reordering*.

Various reordering methods were proposed recently. Many of them are for graph problems and VLSI circuit design. They are breadth-first search (BFS) reordering [19,20], depth-first-search (DFS) reordering [19,20], and weighted-DFS reordering [27] for one-dimensional array representation, and DFS-row-major embedding [5], Euclidean embedding [5], DFS-zigzag embedding [38], and geographic attraction-based embedding [37] for multi-dimensional array representation. Using the order of cities in a local optimum solution for traveling salesman problem [22] and using the correlation coefficients of variables for function approximation [30] are other examples of reordering.

Other Approaches. Another remarkable approach is that of complete graph representation. In the representation, each gene pair is linked by a weighted edge that reflects explicitly the strength of the epistasis between the genes. The edge weights are also called *genic distances*. Seo and Moon [7,45] proposed heuristics for assigning genic distances and Voronoi quantized crossover (VQX) for traveling salesman problem and sequential ordering problem. The heuristics are based on the fact that the epistasis reflects the topological locality of the given cities. In VQX, the chromosomal space defined by the genic distances is divided into k Voronoi regions (nearest neighbor regions) determined by k randomly selected genes, then a sort of block-uniform crossover [4] is performed on the regions. This approach is discriminated from the others in this section in that the chromosomal structure is not necessarily based on a metric space. The crossover is illustrated visually in Fig. 2(f) with the assumption that genic distances are defined on a two-dimensional Euclidean space. The assumption is merely for the visualization.

It is known that the non-coding DNA, sometimes called *introns*, found in biological systems helps the evolution [47]. Several approaches motivated by the observation were proposed recently [17,23,24,25,28]. In the approaches, the chromosome includes non-coding segments that hold positions in the representation space but do not contribute to the evaluation of the solutions. LLGA [31,32] and its variants [33,34,35] described in Sect. 4.2 also use non-coding segments.

4.2 Dynamic Linkage Models

Adaptive Linkage Model. In the adaptive linkage model, the linkages are assigned similarly with the way in the static linkage model. But, it differs from the static linkage model in that the linkages are adjusted dynamically during the process of the genetic algorithm. An application that does not have prior information about the given problem is also possible by assigning the linkages based only on the analysis of the solutions in the population.

There are several approaches based on the adaptive linkage model. Kwon and Moon [30] applied the model to a function optimization problem. They adopted one-dimensional array representation and periodically rearranged the genes by a heuristic based on the correlation coefficients between the corresponding function variables.

Greene [44] proposed an adaptive linkage model employing complete graph representation. In the approach the genic distances are dynamically assigned based on the normalized entropy of the gene values. He proposed a crossover based on a cutoff value where an uncopied gene is repeatedly chosen at random and, at the same time, the values of the genes whose genic distances from the chosen gene are less than a preassigned cutoff value are copied together from one of the two parents alternately.

Evolvable Linkage Model. The beginning of the evolvable linkage model is Bagley's inversion operator [14]. The operator works by reversing the order of the genes lying inbetween a pair of randomly chosen points. To make the genes movable, the "encapsulated" encoding was used. The result of the approach was not successful. One of the reasons might be that the evolution of the linkage in the approach was so slow that it lost the race against allele selection. Messy genetic algorithm (mGA) [15,16] and fast messy genetic algorithm (fast mGA) [18,29] proposed by Goldberg *et al.* are the approaches dealing with the difficulty found in the Bagley's study. They use cut-and-splice operator for recombination. By the operator, genes in a solution migrate in group to other solutions in the population. The operator may cause over-specification and under-specification. The former occurs when a particular gene position is specified by two or more allele values and the other occurs, on the other hand, when a particular gene position is not specified by any allele value in the chromosome. They resolved the over-specification problem by scanning the chromosome in a left-to-right manner

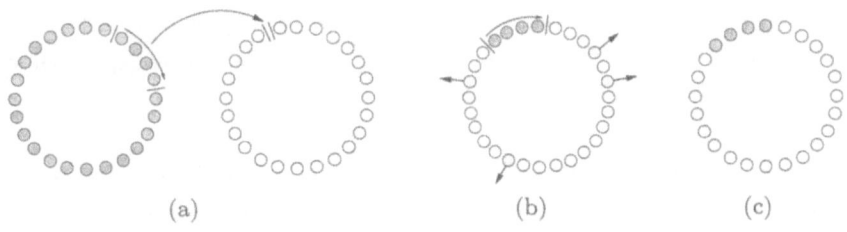

Fig. 3. An illustration of exchange crossover. (a) A randomly chosen segment of the donor is grafted to the recipient. (b) The duplicated genes are removed from the intermediate solution. (c) The final solution.

and choosing the allele that occurred first for each position, and resolved the under-specification problem by random assignment or a heuristic constructive filling in. The problems were avoided in Kargupta's gene expression messy genetic algorithm (GEMGA) [26], and Harik and Goldberg's linkage learning genetic algorithm (LLGA) [31,32]. In LLGA, the exchange crossover was used. It grafts a randomly chosen segment of one chromosome (called donor) on the other chromosome (called recipient) and deletes the genes in the recipient that have the same gene positions with the grafted genes to avoid duplication. Figure 3 shows an illustration of the crossover. There are a number of variants for the exchange crossover [33,34,35]. One-dimensional array representations were used in inversion, mGA, fast mGA, and GEMGA, and a ringed array representation was used in LLGA.

The adaptive non-coding segment representation proposed by Lee and Antonsson [41] is an approach that uses a one-dimensional real space. In the representation, genes are linked by edges with variable lengths. The edges represent the strengths of linkage between genes and perturbed by Gaussian random variable for adaptation. It is a variant of the approaches employing non-coding segments described in Sect. 4.1.

Moon and Kim [39] proposed a genetic algorithm that maintains two populations: one for the evolution of solutions and the other for the evolution of embeddings. In the approach, a solution is generated by a two-layered crossover; two embeddings are crossovered first and then the resultant embedding is used for the recombination of the solutions.

5 Summary

The results of previous theoretical and experimental studies suggest that the performance of a genetic algorithm highly depends on its ability to create and grow the building blocks. Various genetic algorithms motivated by the hypothesis have been proposed recently. A large part of them is based on the topological linkage of genes in the chromosome. In this paper, we summarized the approaches that exploit the topological linkages of genes and provided a taxonomy of them. The various chromosomal structures are byproducts of such studies. They include, for example, 1D array, ringed array, 2D array, 3D array, 1D real space (line), 2D real space (plane), and complete graph.

Acknowledgments. This work was partly supported by Optus Inc. and Brain Korea 21 Project. The RIACT at Seoul National University provided research facilities for this study.

References

1. J. Holland. *Adaptation in Natural and Artificial Systems*. The University of Michigan Press, 1975.
2. D.E. Goldberg. *Genetic Algorithms in Search, Optimization, Machine Learning*. Addison-Wesley, 1989.
3. J. Cohoon and D. Paris. Genetic placement. In *IEEE International Conference on Computer-Aided Design*, pages 422–425, 1986.
4. C. Anderson, K. Jones, and J. Ryan. A two-dimensional genetic algorithm for the Ising problem. *Complex Systems*, 5:327–333, 1991.
5. A.B. Kahng and B.R. Moon. Toward more powerful recombinations. In *International Conference on Genetic Algorithms*, pages 96–103, 1995.
6. S. Jung and B.R. Moon. Toward minimal restriction of genetic encoding and crossovers for the 2D Euclidean TSP. *IEEE Transactions on Evolutionary Computation*, 6(6):557–565, 2002.
7. D.I. Seo and B.R. Moon. Voronoi quantized crossover for traveling salesman problem. In *Genetic and Evolutionary Computation Conference*, pages 544–552, 2002.
8. Y. Davidor. Epistasis variance: Suitability of a representation to genetic algorithms. *Complex Systems*, 4:369–383, 1990.
9. D.I. Seo, Y.H. Kim, and B.R. Moon. New entropy-based measures of gene significance and epistasis. In *Genetic and Evolutionary Computation Conference*, 2003.
10. D. Thierens and D.E. Goldberg. Mixing in genetic algorithms. In *International Conference on Genetic Algorithms*, pages 38–45, 1993.
11. P. Larrañaga and J.A. Lozano. *Estimation of Distribution Algorithms: A New Tool for Evolutionary Computation*. Kluwer Academic Publishers, 2002.
12. M. Pelikan, D.E. Goldberg, and F. Lobo. A survey of optimization by building and using probabilistic models. *Computational Optimization and Applications*, 21(1):5–20, 2002.
13. N. Linial, E. London, and Y. Rabinovich. The geometry of graphs and some of its algorithmic applications. In *Foundations of Computer Science*, pages 577–591, 1994.
14. J.D. Bagley. *The Behavior of Adaptive Systems which Employ Genetic and Correlation Algorithms*. PhD thesis, University of Michigan, 1967.
15. D.E. Goldberg, B. Korb, and K. Deb. Messy genetic algorithms: Motivation, analysis, and first results. *Complex Systems*, 3(5):493–530, 1989.
16. D.E. Goldberg, K. Deb, and B. Korb. Messy genetic algorithms revisited: Studies in mixed size and scale. *Complex Systems*, 4:415–444, 1990.
17. J.R. Levenick. Inserting introns improves genetic algorithm success rate: Taking a cue from biology. In *International Conference on Genetic Algorithms*, pages 123–127, 1991.

18. D.E. Goldberg, K. Deb, H. Kargupta, and G. Harik. Rapid, accurate optimization of difficult problems using fast messy genetic algorithms. In *International Conference on Genetic Algorithms*, pages 56–64, 1993.
19. T.N. Bui and B.R. Moon. Hyperplane synthesis for genetic algorithms. In *International Conference on Genetic Algorithms*, pages 102–109, 1993.
20. T.N. Bui and B.R. Moon. Genetic algorithm and graph partitioning. *IEEE Transactions on Computers*, 45(7):841–855, 1996.
21. T.N. Bui and B.R. Moon. Analyzing hyperplane synthesis in genetic algorithms using clustered schemata. In *Parallel Problem Solving from Nature*, pages 108–118. 1994.
22. T.N. Bui and B.R. Moon. A new genetic approach for the traveling salesman problem. In *IEEE Conference on Evolutionary Computation*, pages 7–12, 1994.
23. A.S. Wu, R.K. Lindsay, and M.D. Smith. Studies on the effect of non-coding segments on the genetic algorithm. In *IEEE Conference on Tools with Artificial Intelligence*, pages 744–747, 1994.
24. A.S. Wu and R.K. Lindsay. Empirical studies of the genetic algorithm with non-coding segments. *Evolutionary Computation*, 3(2):121–147, 1995.
25. J.R. Levenick. Metabits: Generic endogenous crossover control. In *International Conference on Genetic Algorithms*, pages 88–95, 1995.
26. H. Kargupta. The gene expression messy genetic algorithm. In *IEEE Conference on Evolutionary Computation*, pages 814–819, 1996.
27. T.N. Bui and B.R. Moon. GRCA: A hybrid genetic algorithm for circuit ratio-cut partitioning. *IEEE Transactions on CAD*, 17(3):193–204, 1998.
28. J.R. Levenick. Swappers: Introns promote flexibility, diversity and invention. In *Genetic and Evolutionary Computation Conference*, pages 361–368, 1999.
29. D. Knjazew and D.E. Goldberg. OMEGA – Ordering messy GA: Solving permutation problems with the fast messy genetic algorithm and random keys. In *Genetic and Evolutionary Computation Conference*, pages 181–188, 2000.
30. Y.K. Kwon, S.D. Hong, and B.R. Moon. A genetic hybrid for critical heat flux function approximation. In *Genetic and Evolutionary Computation Conference*, pages 1119–1125, 2002.
31. G.R. Harik and D.E. Goldberg. Learning linkage. In *Foundations of Genetic Algorithms 4*, pages 247–262. 1996.
32. G.R. Harik. *Learning Gene Linkage to Efficiently Solve Problems of Bounded Difficulty Using Genetic Algorithms*. PhD thesis, University of Michigan, 1997.
33. F.G. Lobo, K. Deb, and D.E. Goldberg. Compressed introns in a linkage learning genetic algorithm. In *Third Annual Conference on Genetic Programming*, pages 551–558, 1998.
34. G.R. Harik and D.E. Goldberg. Learning linkage through probabilistic expression. *Computer Methods in Applied Mechanics and Engineering*, 186:295–310, 2000.
35. Y.P. Chen and D.E. Goldberg. Introducing start expression genes to the linkage learning genetic algorithm. Technical Report 2002007, University of Illinois at Urbana-Champaign, Illinois Genetic Algorithms Laboratory, 2002.
36. T.N. Bui and B.R. Moon. On multi-dimensional encoding/crossover. In *International Conference on Genetic Algorithms*, pages 49–56, 1995.
37. B.R. Moon and C.K. Kim. A two-dimensional embedding of graphs for genetic algorithms. In *International Conference on Genetic Algorithms*, pages 204–211, 1997.
38. B.R. Moon, Y.S. Lee, and C.K. Kim. GEORG: VLSI circuit partitioner with a new genetic algorithm framework. *Journal of Intelligent Manufacturing*, 9(5):401–412, 1998.

39. B.R. Moon and C.K. Kim. Dynamic embedding for genetic VLSI circuit partitioning. *Engineering Applications of Artificial Intelligence*, 11:67–76, 1998.
40. J.H. Kim and B.R. Moon. Neuron reordering for better neuro-genetic hybrids. In *Genetic and Evolutionary Computation Conference*, pages 407–414, 2002.
41. C.Y. Lee and E.K. Antonsson. Adaptive evolvability via non-coding segment induced linkage. In *Genetic and Evolutionary Computation Conference*, pages 448–453, 2001.
42. S. Jung and B.R. Moon. The natural crossover for the 2D Euclidean TSP. In *Genetic and Evolutionary Computation Conference*, pages 1003–1010, 2000.
43. S. Jung and B.R. Moon. A hybrid genetic algorithm for the vehicle routing problem with time windows. In *Genetic and Evolutionary Computation Conference*, pages 1309–1316, 2002.
44. W.A. Greene. A genetic algorithm with self-distancing bits but no overt linkage. In *Genetic and Evolutionary Computation Conference*, pages 367–374, 2002.
45. D.I. Seo and B.R. Moon. A hybrid genetic algorithm based on complete graph representation for the sequential ordering problem. In *Genetic and Evolutionary Computation Conference*, 2003.
46. E.J. Park, Y.H. Kim, and B.R. Moon. Genetic search for fixed channel assignment problem with limited bandwidth. In *Genetic and Evolutionary Computation Conference*, pages 1172–1179, 2002.
47. A.S. Wu and R.K. Lindsay. A survey of intron research in genetics. In *Parallel Problem Solving from Nature*, pages 101–110, 1996.

Cellular Programming and Symmetric Key Cryptography Systems

Franciszek Seredyński[1,2], Pascal Bouvry[3], and Albert Y. Zomaya[4]

[1] Polish-Japanese Institute of Information Technologies
Koszykowa 86, 02-008 Warsaw, Poland
[2] Institute of Computer Science of Polish Academy of Sciences
Ordona 21, 01-237 Warsaw, Poland
sered@ipipan.waw.pl
http://www.ipipan.waw.pl/~sered
[3] Luxembourg University of Applied Sciences
6, rue Coudenhove Kalergi, L-1359 Luxembourg-Kirchberg, Luxembourg
pascal.bouvry@ist.lu
http://www.ist.lu/users/pascal.bouvry
[4] School of Information Technologies, University of Sydney
Sydney, NSW 2006 Australia
zomaya@it.usyd.edu.au
http://www.cs.usyd.edu.au/~zomaya

Abstract. The problem of designing symmetric key cryptography algorithms based upon cellular automata (CAs) is considered. The reliability of the Vernam cipher used in the process of encryption highly depends on a quality of used random numbers. One dimensional, nonuniform CAs is considered as a generator of pseudorandom number sequences (PNSs). The quality of PNSs highly depends on a set of applied CA rules. To find such rules nonuniform CAs with two types of rules is considered. The search of rules is based on an evolutionary technique called cellular programming (CP). Resulting from the collective behavior of the discovered set of CA rules very high quality PNSs are generated. The quality of PNSs outperform the quality of known one dimensional CA-based PNS generators used in secret key cryptography. The extended set of CA rules which was found makes the cryptography system much more resistant on breaking a cryptography key.

1 Introduction

Confidentiality is mandatory for a majority of network applications including, for example, commercial uses of the Internet. Two classes of algorithms exist on the market for data encryption: secret key systems and public key systems. An extensive overview of currently known or emerging cryptography techniques used in both types of systems can be found in [14]. One of such a promising cryptography techniques are cellular automata (CAs). An overview on current state of research on CAs and their application can be found in [13].

CAs were proposed for public-key cryptosystems by Guan [1] and Kari [5]. In such systems two keys are required: one key is used for encryption and the other for decryption, and one of them is held in private, the other is published. However, the main concern of this paper are secret key cryptosystems. In such systems the same key is used for encryption and decryption. The encryption process is based on the generation of pseudorandom bit sequences, and CAs can be effectively used for this purpose. In the context of symmetric key systems, CAs were first studied by Wolfram [17], and later by Habutsu et al. [3], Nandi et al. [11] and Gutowitz [2]. Recently they were a subject of study by Tomassini and his colleagues (see, eg. [16]). This paper extends these recent studies and describes the application of one dimensional (1D) CAs for the secret key cryptography.

The paper is organized as follows. The following section presents the idea of an encryption process based on Vernam cipher and used in CA-based secret key cryptosystems. Section 3 outlines the main concepts of CAs, overviews current state of applications of CAs in secret key cryptography and states the problem considered in the paper. Section 4 outlines evolutionary technique called cellular programming and Sect. 5 shows how this technique is used to discover new CA rules suitable for encryption process. Section 6 contains the analysis of results and the last section concludes the paper.

2 Vernam Cipher and Secret Key Cryptography

Let P be a plain-text message consisting of m bits $p_1 p_2 ... p_m$, and $k_1 k_2 ... k_m$ be a bit stream of a key k. Let c_i be the $i-th$ bit of a cipher-text obtained by applying a XOR (exclusive-or) enciphering operation:

$$c_i = p_i \; XOR \; k_i.$$

The original bit p_i of a message can be recovered by applying the same operation XOR on c_i with use of the same bit stream key k:

$$p_i = c_i \; XOR \; k_i.$$

The enciphering algorithm called Vernam cipher is known to be [9,14] perfectly safe if the key stream is truly unpredictable and is used only one time. From practical point of view it means that one must find answers on the following questions: (a) how to provide a pure randomness of a key bit stream and unpredictability of random bits, (b) how to obtain such a key with a length enough to encrypt practical amounts of data, and (c) how to pass safely the key from the sender to receiver and protect the key.

In this paper we address questions (a) and (b). We will apply CAs to generate high quality pseudorandom number sequences (PNSs) and a safe secret key.

3 Cellular Automata and Cryptography

One dimensional CA is in a simpliest case a collection of two-state elementary automata arranged in a lattice of the length N, and locally interacted in a discrete time t. For each cell i called a central cell, a neighborhood of a radius r is defined, consisting of $n_i = 2r + 1$ cells, including the cell i. When considering a finite size of CAs a cyclic boundary condition is applied, resulting in a circle grid.

It is assumed that a state q_i^{t+1} of a cell i at the time $t+1$ depends only on states of its neighborhood at the time t, i.e. $q_i^{t+1} = f(q_i^t, q_{i1}^t, q_{i2}^t, ..., q_{ni}^t)$, and a transition function f, called a *rule*, which defines a rule of updating a cell i. A length L of a rule and a number of neighborhood states for a binary uniform CAs is $L = 2^n$, where $n = n_i$ is a number of cells of a given neighborhood, and a number of such rules can be expressed as 2^L. For CAs with e.g. $r = 2$ the length of a rule is equal to $L = 32$, and a number of such rules is 2^{32} and grows very fast with L. When the same rule is applied to update cells of CAs, such CAs are called uniform CAs, in contrast with nonuniform CAs when different rules are assigned to cells and used to update them.

S. Wolfram was the first to apply CAs to generate PNSs [17]. He used uniform, 1D CAs with $r = 1$, and rule 30. Hortensius *et al.* [4] and Nandi *et al.* [11] used nonuniform CAs with two rules 90 and 150, and it was found that the quality of generated PNSs was better than the quality of the Wolfram system. Recently Tomassini and Perrenoud [16] proposed to use nonuniform, 1D CAs with $r = 1$ and four rules 90, 105, 150 and 165, which provide high quality PNSs and a huge space of possible secret keys which is difficult for cryptanalysis. Instead to design rules for CAs they used evolutionary technique called cellular programming (CP) to search for them.

In this study we continue this line of research. We will use finite, 1D, nonuniform CAs. However, we extend the potential space of rules by consideration of two sizes of rule neighborhood, namely neighborhood of radius $r = 1$ and $r = 2$. To discover appropriate rules in this huge space of rules we will use CP.

4 Cellular Programming Environment

4.1 Cellular Programming

CP is an evolutionary computation technique similar to the diffusion model of parallel genetic algorithms and introduced [15] to discover rules for nonuniform CAs. Figure 1 shows a CP system implemented [10] to discover such rules. In contrast with the CP used in [16] the system has a possibility to evaluate nonuniform rules of two types. The system consists of a population of N rules (left) and each rule is assigned to a single cell of CAs (right). After initiating states of each cell, i.e. setting an initial configuration, the CAs start to evolve according to assigned rules during a predefined number of time steps. Each cell produces a stream of bits, creating this way a PNS.

After stopping evolving CAs all PNSs are evaluated. The entropy E_h is used to evaluate the statistical quality of each PNS. To calculate a value of the entropy each PNS is divided into subsequences of a size h. In all experiments the value $h = 4$ was used. Let k be the number of values which can take each element of a sequence (in our case of binary values of all elements $k = 2$) and k^h a number

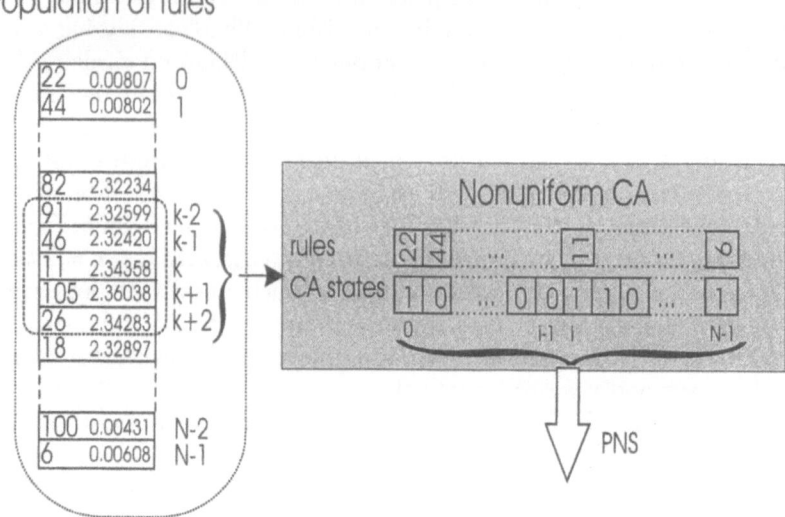

Fig. 1. CP environment for evolution of rules of nonuniform CAs.

of possible states of of each sequence ($k^h = 16$). E_h can be calculated in the following way:

$$E_h = -\sum_{j=1}^{k^h} p_{h_j} \log_2 p_{h_j},$$

where p_{h_j} is a measured probability of occurrence of a sequence h_j in a PNS. The entropy achieves its maximal value $E_h = h$ when the probabilities of the k_h possible sequences of the length h are equal to $1/k^h$. The entropy will be used as a fitness function of CP.

A single PNS is produced by a CA cell according to assigned rules and depends on a configuration c_i of states of CAs. To evaluate statistically reliable value of the entropy, CAs run with the same set of rules C times for different configurations c_i, and finally the average value of entropy is calculated and serves as a fitness function of each rule from the population of rules.

After evaluation of a fitness function of all rules of the population genetic operators of selection, crossover and mutation are locally performed on rules. The evolutionary algorithm stops after some predefined number of generations of CP. The algorithm can be summarized in the following way:

1: initiate randomly *population* of N rules of type 1 ($r = 1$) or type 2 ($r = 2$), or both types, and create CAs consisting of N cells

2: assign $k - th$ rule from the CP population to $k - th$ cell of CAs

3: **for** $i=1 \ldots C$ **do**
   { create randomly configuration c_i of CAs
     evolve CAs during M time steps
     evaluate *entropy* of each PNS }

4: Evaluate *fitness function* of each rule

5: Apply locally to rules in a specified sequence genetic operators of *selection, crossover* and *mutation*

6: if STOP condition is not satisfied return to 2.

4.2 Genetic Operators

In contrast with the standard genetic algorithm population, rules – individuals of CP population occupy specific place in the population and have strictly defined neighborhood. For example, the rule 11 (see Fig. 1) (also indexed by k) corresponds to $k-th$ cell of CAs, and rules 46 and 105 are its immediate neighbors. All rules shown in this figure belong to the first type of rules with $r = 1$, i.e. a transition function of the rule depends on 3 cells, a given cell and two cell-neighbors. However, in more general case considered in the paper, we assume that rules are either of type 1 ($r = 1$, short rules) or of type 2 ($r = 2$, long rules).

Additionally to a neighborhood associated with two types of rules we use also an evolutionary neighborhood, i.e. the neighborhood of rules which are considered for mating when genetic operators are locally applied to a given rule. The size and pattern of this neighborhood may differ from the neighborhood associated with types of rules. Figure 1 shows an example of the evolutionary neighborhood for the rule k which is created by rules $k-2, k-1, k, k+1, k+2$. It is assumed that the pattern of such a neighborhood is the same for all rules and is a predefined parameter of an experiment.

A sequence of genetic operators performed locally on a given rule depends on values of fitness function of rules (numbers on the right side of rule names, see Fig. 1) from the evolutionary neighborhood of this rule. Genetic operators are applied in the following way:

1. if the $k-th$ rule is the best (the highest value of the fitness function) in its evolutionary neighborhood then the rule survives (selection) and remains unchanged for the next generation; no other genetic operators are performed
2. if in the evolutionary neighborhood of the rule k only one rule exists which is better than considered rule then the rule k is replaced by the better rule (selection) only if both rules are of the same type, and next mutation on this rule is performed; the rule remains unchanged if the better rule is of the other type
3. if two rules better than the rule k exist in the neighborhood then a crossover on the pair of better rules is performed; a randomly selected child from a pair of children replaces rule k, and additionally a mutation is performed

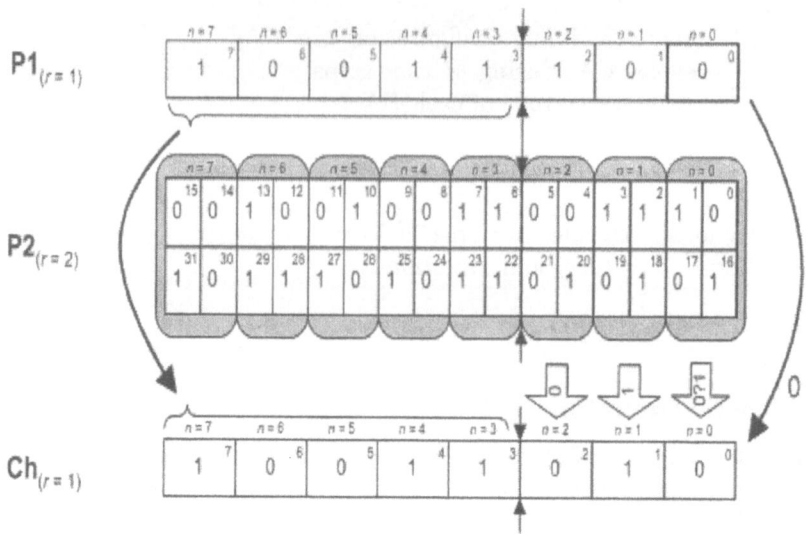

Fig. 2. Example of crossover resulting in a short child rule.

4. if more than two rules better than the rule k exist in the neighborhood then two randomly selected better rules create (crossover) a pair of children; on a randomly selected child a mutation is performed, and the child replaces the rule k.

Two types of rules existing in a CP population can be considered as two species of a coevolutionary algorithm. Therefore to perform a crossover between rules special regulations are required. It is assumed that two parental rules of the same species create a single child rule of the same species, which can replace either the first type of a rule or the second type of the rule. If rules of different types take part in the mating then a species of a child depends on species of a replaced rule, and is the same as a species of a rule to be replaced. Figure 2 shows a crossover between a short rule 156 ($r = 1$) and a long rule 617528021 ($r = 2$), and the result of crossover – a short rule 154.

The short rule $P1$ taking part in crossover consists of 8 genes ($n = 0, ..., 7$) which values correspond to values of transition function defined on 8 neighborhood states $\{000, 001, ..., 111\}$ existing for $r = 1$. The long rule $P2$ consists of 32 genes, each corresponding to values of transition function defined on 32 neighborhood states existing for $r = 2$. The long rule is folded because there is a strict relation between a state order number which corresponds to $j - th$ gene of $P1$ and states' order numbers corresponding to genes $2j, 2j+1$ and $2j+16, 2j+17$ of $P2$. These order numbers of states of $P2$ are just an extension of corresponding order number of a gene from $P1$. For example, the gene $n = 7$ of $P1$ corresponds to the neighborhood state $\{111\}$, and genes 15, 14 and 31, 30 of $P2$ correspond to states respectively $\{0\mathbf{111}1, 0\mathbf{111}0\}$ and $\{1\mathbf{111}1, 1\mathbf{111}0\}$ containing the state of $P1$ (marked in bold).

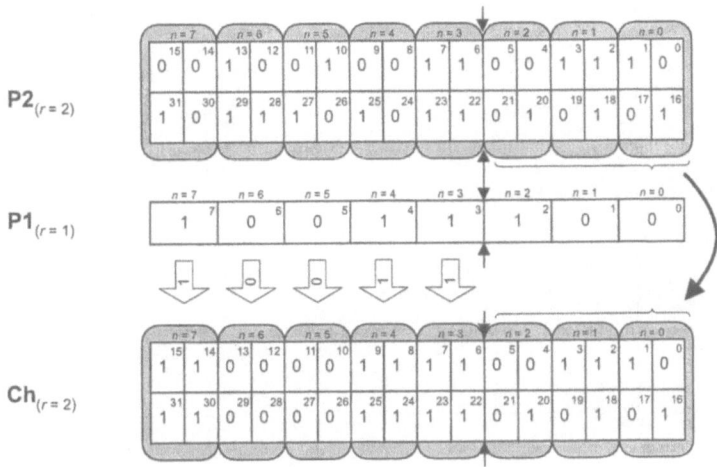

Fig. 3. Example of crossover resulting in a long child rule.

As Fig. 2 shows both rules $P1$ and $P2$ are crossed between genes 2 and 3 and a child Ch corresponding to a short rule ($r = 1$) is created. For this purpose the left part of the short rule is copied to the left part of the child. The right part of Ch is created according to the right part of $P2$ on the basis of majority of 0's or 1's in the corresponding genes. Figure 3 shows the crossover of two rules resulting in a long child rule. Last genetic operator is a flip-bit mutation performed with the probability $p_m - 0.001$.

5 Discovery of Rules in 1D, Nonuniform CAs

In all conducted experiments a population of CP and the size of nonuniform CAs were equal to 50 and the population was processing during 50 generations. The CAs with initial random configuration of states and a set of assigned rules evolved during $M = 4096$ time steps. Running CAs with a given set of rules was repeated for $C = 300$ initial configurations. Figure 4 shows an example of running CP for the evolutionary neighborhood $i-3, i-2, i, i+2, i+3$. One can see that whole CAs is able to produce very good PNSs after about 40 generations (see, the average value avg of the entropy close to 4).

A typical result of a single run of an evolutionary process starting with a random rules assigned to cells of CAs is discovering by CP a small set of good rules which divide the cellular space of CAs into domains – areas where the same rules, short ($r = 1$) or long ($r = 2$), live together. Evolutionary process is continued on borders of domains where different rules live. This process may result in increasing domains of rules which are only slightly better than neighboring rules, which domains will decrease and finally disappear. This happens in particular

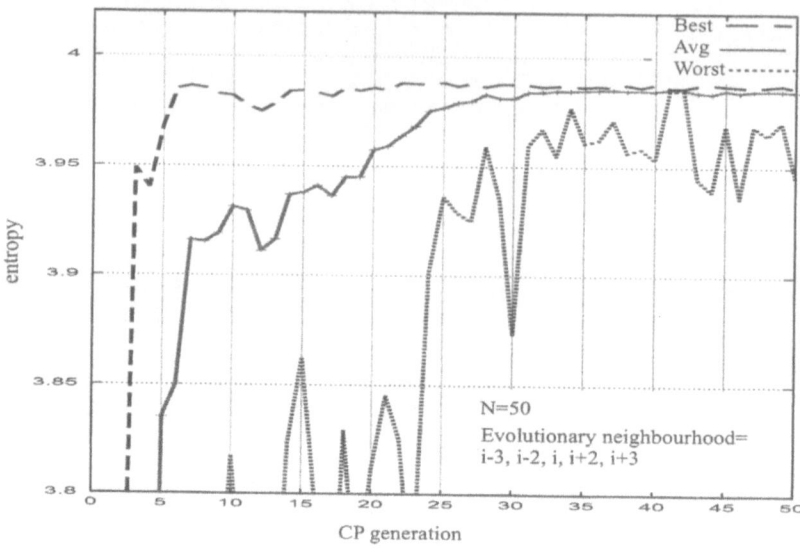

Fig. 4. A single run of CP evolutionary process.

when two neighboring domains are occupied respectively by the same short rules and the same long rules. The search space of short rules is much smaller than the search space of the long rules. Therefore better short rules are discovered faster than better long rules, and for this reason long rules are gradually replaced by short rules. To limit this premature convergence of short rules, the short and long rules are initially randomly assigned to cells in the proportion of 1:3 in all subsequent experiments.

The purpose of the experiments which followed was to discover an enlarged set of rules (to enlarge the key space of cryptography system) which working collectively would produce very high quality PNSs. It was noticed that in a single run of CP the evolutionary algorithm produces typically a small set of rules with a very high value of the entropy. In the result of evolutionary searching process a set of 8 short rules (including 5 rules found by [16]) and a set of 39 long rules was found.

6 Analysis and Comparison of Results

The entropy used as the fitness function for evolution CA rules producing high quality PNSs is only one of existing statistical tests of PNSs. None of them is enough strong to claim statistical randomness of a PNS in the case of passing a given test. For this purpose uniform CAs consisting of 50 cells evolved during 65536 time steps with each single discovered rule. Each PNS produced by CAs was divided into 4-bit words and tested on general statistical tests such as the entropy, χ^2 test, serial correlation test [6] (some weaker rules after this testing were removed), and next on a number of statistical tests required by the FIPS

140-2 standard [12], such as monobit test, poker test, runs test, and long runs test.

Figure 5 shows results of testing rules against the FIPS 140-2 standard. The best scores were achieved by rules 30, 86, 101, 153 and by 8 long rules. Rules 90, 105, 150 and 65 [16] working separately in uniform CAs obtained good results in test of entropy and long runs test, quite good results in serial correlation test and monobit test but were weak in χ^2 test, poker test and runs test. However this set of rules working collectively in nonuniform CAs achieves good results (see Fig. 6).

For this reason only 10 rules were removed from discovered set of rules which have passed the FIPS 140-2 standard testing. These rules were worse than Tomassini & Perrenoud rules. However passing all statistical tests does not exclude a possibility that the PNS is not suitable for cryptographic purposes. Before a PNS is accepted it should pass special cryptographic tests.

Therefore rules which passed tests were next submitted to a set of Marsaglia tests [7] – a set of 23 very strong tests of randomness implemented in the Diehard program. Only 11 rules passed all 23 Marsaglia tests. These are short rules 30, 86, 101, and long rules 869020563, 1047380370, 1436194405, 1436965290, 1705400746, 1815843780, 2084275140 and 2592765285.

The purpose of the last set of experiments was a selection of a small set of short and long rules for nonuniform CAs which working collectively would provide a generation of very high quality PNSs suitable for the secret key cryptography. Simple combination of different rules which passed all Marsaglia tests in nonuniform CAs have shown that resulting PNSs may have worse statistical characteristic than PNSs obtained using uniform CAs. On the other hand, experiments with Tomassini & Perrenoud rules show that rules that separately are working worse can provide better quality working collectively. For these reasons rules 153 and some long rules which obtained very good results in general tests but not passed all Marsaglia tests were also accepted for the set of rules to search a final set of rules.

In the result of combining rules into sets of rules and testing collective behavior of these sets working in nonuniform CAs the following set of rules has been selected: 86, 90, 101, 105, 150, 153, 165 ($r = 1$), and 1436194405 ($r = 2$). Among the rules are 4 rules discovered in [16]. The set of found rules have been tested again on statistical and cryptographic tests using nonuniform CAs with random assignment of rules to CA cells. Figure 6 shows results of testing this new set of rules and compares the results with ones obtained for Tomassini & Perrenoud rules. One can see that results of testing both sets on general tests and FIPS 140-2 tests are similar. However, the main difference between these results can be observed in passing Marsaglia tests: while the new discovered set of rules passes all 23 Marsaglia tests, the Tomassini & Perrenoud set of rules passes only 11 tests. Figure 7 shows a space-time diagram of both set of rules working collectively.

The secret key K which should be exchanged between two users of considered CA-based cryptosystem consists of a pair of randomly created vectors: the vector R_i informing about assigning 8 rules to N cells of CAs and the vector $C(0)$

No	Rules	Monobit Test	Poker Test	Run Test	Long Run Test
Rules found by Tomassini & Perrenoud					
1	30	50	50	50	50
2	90	46	0	23	50
3	105	41	0	4	50
4	150	45	0	12	50
5	165	46	0	14	50
Rules of radius r = 1 (No 6, 7, 8) and r = 2					
6	86	50	50	50	50
7	101	50	50	50	50
8	153	50	50	50	50
11	728094296	50	5	17	50
12	869020563	50	50	49	50
13	892695978	50	2	9	50
14	898995801	50	0	4	50
15	988725525	50	11	16	50
17	1042531548	38	0	12	50
18	1047380370	50	50	47	50
19	1367311530	50	5	5	50
20	1378666419	42	3	16	50
21	1386720346	50	20	32	50
22	1403823445	50	19	32	50
23	1427564201	46	1	27	50
24	1436194405	50	50	50	50
25	1436965290	50	50	50	50
27	1457138022	50	0	0	50
28	1470671190	50	50	49	50
29	1521202561	40	0	3	50
30	1537843542	50	48	37	50
31	1588175194	50	21	27	50
32	1704302169	50	50	50	50
33	1705400746	50	50	50	50
35	1721277285	49	1	4	50
37	1721325161	50	50	50	50
38	1746646725	49	6	2	50
39	1755030679	50	49	34	50
43	1815843780	50	50	50	50
45	2036803240	50	0	0	50
46	2084275140	50	50	50	50
47	2592765285	50	50	49	50

Fig. 5. Testing rules against the FIPS 140-2 standard

Test	Tomassini & Perrenoud Rules (90, 105, 150, 165)	Discovered Rules (86, 90, 101, 105, 150, 153, 165, 1436194405)
Min entropy	3,9988	3,9987
Max entropy	3,9998	3,9997
Min χ^2	5,0254	6,998
Max χ^2	26,396	30,805
Min correlation	0,00007	-0,00006
Max correlation	0,02553	0,01675
Monobit test	50	50
Poker test	50	50
Run test	50	50
Long run test	50	50
Number of passed Marsaglia tests	11	23

Fig. 6. Comparison of rules found by Tomassini & Perrenoud [16] and new set of discovered rules.

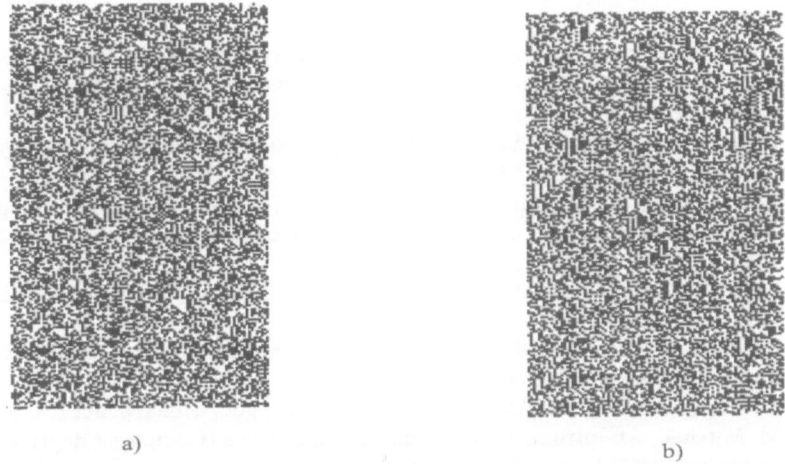

a) b)

Fig. 7. Space-time diagram of CAs with $N = 100$ and $M = 200$ time steps working collectively with (a) randomly assigned Tomassini & Perrenoud [16] rules, and (b) with new set of discovered rules.

describing an initial binary state of CA cells. The whole key space has therefore the size $8^N * 2^N$. The key space is much larger than the key space ($4^N * 2^N$) of 1D CA-based system [16]. Therefore the proposed system is much more resistant for cryptographic attacks.

7 Conclusions

In the paper we have reported results of the study on applying CAs to the secret key cryptography. The purpose of the study was to discover a set of CA rules which produce PNSs of a very high statistical quality for a CA-based cryptosystem which is resistant on breaking a cryptography key. The main assumption of our approach was to consider nonuniform 1D CAs operating with two types of rules. Evolutionary approach called CP was used to discover suitable rules. After discovery of a set of rules they were carefully selected using a number of strong statistical and cryptographic tests. Finally, the set consisting of 8 rules has been selected. Results of experiments have shown that discovered rules working collectively are able to produce PNSs of a very high quality outperforming the quality of known 1D CA-based secret key cryptosystems, which also are much more resistant for breaking cryptography keys that known systems.

References

1. Guan, P.: Cellular Automaton Public-Key Cryptosystem, *Complex Systems* 1, (1987) 51–56
2. Gutowitz, H.: Cryptography with Dynamical Systems, in E. Goles and N. Boccara (Eds.) *Cellular Automata and Cooperative Phenomena*, Kluwer Academic Press, (1993)
3. Habutsu, T., Nishio, Y., Sasae, I., Mori, S.: A Secret Key Cryptosystem by Iterating a Chaotic Map, *Proc. of Eurocrypt'91*, (1991) 127–140
4. Hortensius, P.D., McLeod, R.D., Card H.C.: Parallel random number generation for VLSI systems using cellular automata, *IEEE Trans. on Computers* 38, (1989) 1466–1473
5. Kari, J.:, Cryptosystems based on reversible cellular automata, personal communication, (1992)
6. Knuth, D.E.:, *The Art of Computer Programming*, vol. 1 & 2, *Seminumerical Algorithms*, Addison-Wesley, (1981)
7. Marsaglia, G.: Diehard http://stat.fsu.edu/~geo/diehard.html, (1998)
8. M. Mitchell, An Introduction to Genetic Algorithms (Complex Adaptive Systems), Publisher: *MIT Press*, ISBN: 0262133164
9. Menezes, A., van Oorschot, P., Vanstone, S.: *Handbook of Applied Cryptography*, CRC Press, (1996)
10. Mroczkowski,A.: Application of Cellular Automata in Cryptography, Master Thesis (in Polish), Warsaw University of Technology, (2002)
11. Nandi, S., Kar, B.K., Chaudhuri, P.P.: Theory and Applications of Cellular Automata in Cryptography, *IEEE Trans. on Computers*, v. 43, (1994) 1346–1357
12. National Institute of Standards and Technology, Federal Information Processing Standards Publication 140-2: *Security Requirements for Cryptographic Modules*, U.S. Government Printing Office, Washington (1999)
13. Sarkar, P.: A Brief History of Cellular Automata, *ACM Computing Surveys*, vol. 32, No. 1, (2000) 80–107
14. Schneier, B.: *Applied Cryptography*, Wiley, New York, (1996)
15. Sipper, M., Tomassini, M.: Generating parallel random number generators by cellular programming, *Int. Journal of Modern Physics C*, 7(2), (1996) 181–190

16. Tomassini, M., Perrenoud, M.: Stream Ciphers with One- and Two-Dimensional Cellular Automata, in M. Schoenauer at al. (Eds.) *Parallel Problem Solving from Nature – PPSN VI*, LNCS 1917, Springer, (2000) 722–731
17. Wolfram, S.: Cryptography with Cellular Automata, in *Advances in Cryptology: Crypto '85 Proceedings*, LNCS 218, Springer, (1986) 429–432

Mating Restriction and Niching Pressure: Results from Agents and Implications for General EC

R.E. Smith and Claudio Bonacina

The Intelligent Computer Systems Centre, Faculty of Computing
Engineering, and Mathematical Sciences, The University of The West of England, Bristol, UK
{robert.smith,c2-bonacina}@uwe.ac.uk
http://www.cems.uwe.ac.uk/icsc

Abstract. This paper presents results and observations from the authors' continuing explorations of EC systems where population members act as autonomous agents that conduct their own, independent evaluation of (and reproduction with) other agents. In particular, we consider diversity preservation in one such agent-based EC system, applied to the multi-peak functions often used to illustrate and evaluate the effects of fitness-sharing-like schemes in GAs. We show how (somewhat surprisingly) *mating restriction alone* yields stable niching in agent-based EC. This leads to a consideration of niching as a generalized phenomenon, and the introduction of *niching pressure* as a concept that parallels *selective pressure*, and which can yield insight. The utility of the niching pressure concept for general EC is explored, and directions for further research are discussed.

1 Introduction

The maintenance of diversity in an EC population is often the key to success in a given task. Diversity can be useful to prevent premature convergence. It is essential in multimodal function optimization, in multi-objective optimization, in dynamic function optimization, and in co-evolutionary machine learning (e.g., learning classifier systems). In many such systems, diversity is maintained through some sort of "niching" operator; an operator that encourages the population to break into separate sub-populations that occupy distinct "niches" of the search space. Mahfoud [8] defines niching methods as "... techniques that promote the formation and maintenance of stable subpopulations in the GA. Niching methods can be applied to the formation and maintenance of interim sub solutions on the way to a single, final solution. They are traditionally viewed, however, in the context of forming and maintaining multiple, final solutions". Despite the often-essential need for diversity preservation through niching, there is little understanding of the *dynamics* of niching operators.

The authors have recently investigated agent-based EC systems, where each population member acts as an autonomous agent that exchanges information with other such agents, evaluates them as potential mates, and conducts reproduction with them [7,13,14,15]. As has been previously pointed out by the authors, the dynamics of such systems can be non-trivially different from those of centralized EC, and may yield new insights into agent systems that are not specifically exploiting EC. Also, as will be shown in this paper, they can provide insight into EC in general.

In the following sections, we will discuss the problem of implementing diversity preservation (niching) mechanisms in an agent-based EC system, the testing of such mechanisms on the simple test problems extensively investigated by Deb and Goldberg [3], and results. The results will show that, although mating restrictions alone cannot yield stable niches in a centralized EC system [8], it can in an agent-based EC system. However, this result depends on a subtle balance of effects. Therefore, we introduce the notion of *niching pressure* as a phenomenon that parallels *selective pressure* in an EC system that has niching operators. The broader uses of the niching pressure concept are outlined, and future research directions are discussed.

2 Niching Operators in an Agent-Based EC System

One of the goals of the authors' current research program is to examine systems where EC operates as a means of interaction of individual, autonomous agents, with no centralized EC *per se*. Thus, in the system we consider here (which we will call the ECoMAS model, for evolutionary computation multi-agent system) we want to avoid any form of centralized operations. This complicates the implementation of some commonly used niching operators. We will briefly review common niching operators to clarify these issues.

One of the most successful methods to maintain diversity in EC is fitness sharing [6]. Given that this scheme is well known, it will only be briefly discussed here. Fitness sharing is a fitness-scaling scheme that is applied just before parent selection. Assuming application to a multimodal maximization problem, such schemes can be described as means of forcing similar individuals to share their payoff or fitness. This has the effect of limiting the number of individuals in any area of the fitness landscape, based on the relative height of the fitness peaks in that area. Theoretically, fitness sharing should distribute the number of individuals in various areas proportionally to the height of peaks in those areas. With a limited population size, only the "highest" regions will be covered. Fitness sharing works well with fitness proportional selection schemes. Tournament selection decreases the stability of the fitness sharing algorithm, but Oei et al. [12] proposed a solution to combine binary tournament selection and fitness sharing.

The adoption of fitness sharing in the ECoMAS model is problematic.

- Having a centralized perception of the fitness for all individuals is against the decentralized philosophy of our ECoMAS model. We could in principle design a distributed version of fitness sharing or apply a scheme similar to the one described by Yin and Gemany [17]. However, the asynchronous nature of the system would make the implementation of such schemes difficult.
- We do not intend to restrict our system to use non-negative fitness measures, and positive fitness values are an implicit assumption in most fitness sharing schemes. We could rescale fitness values to positive values, but a local imple-

mentation of fitness sharing (necessary in our scheme) would lead to different scaling in different subsets of individuals.
- In our system, individuals locally select their mating partners based on the limited number of *plumages* they receive (in an asynchronous fashion) from other individuals. This is most similar to a tournament selection in regular GAs. We could use the scheme suggested by Oei et al. [12] for combining tournament selection and fitness sharing, but this could also interfere with schemes we might adopt to solve the problems highlighted in the previous bullet points.

Another well-know method to maintain diversity is *crowding* [4]. Crowding works by assuring that new individuals replace similar individuals in the population. As in fitness sharing, similarity is defined by some distance measure between individuals. Unlike fitness sharing, crowding does not allocate solutions proportionally to the height of the peaks. The original crowding scheme proposed by De Jong [4] suffers from replacement errors, which prevent it from stably maintaining solutions close to desired peaks.

Mahfoud [9],[8] modified crowding in order to minimize replacement errors. In the new scheme, called *deterministic crowding*, the population is divided into $N/2$ *random* couples of parents. Two offspring are then generated from each couple. Two parent-offspring pairs are created, minimizing the sum of parent-to-offspring distance. For each pair, the most fit of the two individuals is copied into the new population.

Deterministic crowding is certainly more suitable to adoption in ECoMAS models than fitness sharing. Nevertheless, the following problems are still present:

- In order to implement deterministic crowding, we would have to enforce some level of centralization and synchronization in the system to ensure that each agent at each generation would father only one offspring. This would require centralized information management. We would also have to bind the two agents behaving as parents in our system, because (in principle) we could be willing to replace both of them with their offspring. In order to do so, an undesired level of synchronization would be introduced.
- Co-evolutionary scenarios are one of the aims of our agent-based EC research, and it is not clear what the effects of deterministic crowding would be in such settings.

Deb and Goldberg [3] introduced a mating restriction mechanism as an addition to fitness sharing, in order to prevent recombination between individuals in different niches that may result in low fitness offspring. Deb and Goldberg used a distance measure in the decoded parameter space (*phenotypic mating restriction*) and a threshold σ_{Mating}. Deb and Goldberg then set $\sigma_{Mating} = \sigma_{Share}$ for simplicity. Only individuals that are less than σ_{Mating} apart from each other are, in principle, allowed to mate. If no such individual is found, a mating partner is chosen at random. Deb and Goldberg found that the introduction of phenotypic mating restrictions enhanced the on-line performance of the fitness sharing algorithm.

Mahfoud [8] has proven that the mating restrictions introduced by Deb and Goldberg cannot maintain diversity in a centralized EC system on their own. However, this is not the case in agent-based EC, as we will later see. Moreover, mating restrictions have a straightforward, entirely local implementation in our agent-based scheme.

3 Preliminary Results on Mating Restriction in Agent-Based EC

This section describes a very basic test of our ECoMAS model with mating restrictions. In order to conduct a basic test of mating restrictions in agent-base EC, we use some of the multimodal functions described in [3]. Specifically, we consider the following two multimodal functions:

$$F1:\ f_1(x) = \sin^6(5\pi x), \qquad F2:\ f_2(x) = e^{-2\ln 2\left(\frac{x-0.1}{0.8}\right)^2} \sin^6(5\pi x)$$

F1 has five peaks of equal height and width, evenly spread in the range $0 \le x \le 1$. *F2* has five peaks of decreasing height and equal width, evenly spread in the range $0 \le x \le 1$.

In the ECoMAS system applied to these simple test functions, individuals send out plumages and collect plumages. A plumage is composed of features of the sending agent. This is typically its real-valued decoding (the "phenotype"). The fitness function value and genotype are also included for convenience, but we do not, in general, assume that a centralized fitness value is available, or that the genotype can be inferred from the plumage for purposes of reproduction, so that the decentralized intent of the ECoMAS model is not compromised. Each agent sends out *number_of_receivers* plumages initially, and each time it receives a plumage. Each agent gathers a list of plumages, which represents its list of possible partners. As soon as the list contains *max_partners* agents, the individual can behave as a "mother" and generate an offspring. An agent acting in the mother role sorts the list according to its own (fitness and niching) criteria. The agent tries to get the "best" individual in her list to father her offspring, by requesting his genetic material. If the father is no longer available in the system, the mother tries to get to the second best father in its list, and so on. If none of the individuals in the list are available, the mother produces the offspring asexually (via mutation only). If a father is successfully found, the offspring is the result of the recombination of the genotype of the mother and the father. After an agent produces *max_children* offspring in the mother role, it dies (eliminates itself from the system).

In the mating restriction scheme adopted here, each agent filters the individuals that it inserts into its list of partners. If an agent receives a plumage from an agent that is farther than a local threshold (*mating restriction threshold*) away (according to some distance metric), this plumage is not added to the partner list, and is simply discarded. We consider two distance metrics:

- Euclidian distance, based on the real-valued decoding of the binary string representing each individual. When using this metric we will call the mating restriction scheme *Euclidean mating restrictions*.

- Hamming distance, based directly on the binary genotype of each agent. When using this metric we will call the mating restriction scheme *Hamming mating restrictions*.

The terms *phenotypic* and *genotype* mating restriction (respectively) are used for these sorts of schemes in the literature. However, the authors feel that these terms aren't particularly accurate, given that the distinction in most cases, and particularly in our system, is only that of a neighborhood structure, not the *developmental* distinction of phenotype from genotype.

In the experiments presented in this section, 100 agents, each with a 30-bit genotype, are run for 200 generations, using two-point crossover with probability 0.9, and *number_of_recievers* = 10. Setting *max_children* to one insures the initial number of agents (population size) is constant. A mutation probability of zero is used for comparison of results to Deb and Goldberg's original experiments.

Results in Fig. 1 show that the mating restriction scheme yields stable niches on each of the peaks. Note that niches are formed in less than 30 generations, and after that time they are absolutely stable, by the nature of the system. This result is somewhat surprising, given that in regular GAs, mating restriction alone is not expected to yield stable niching. This effect is explored further in the following sections.

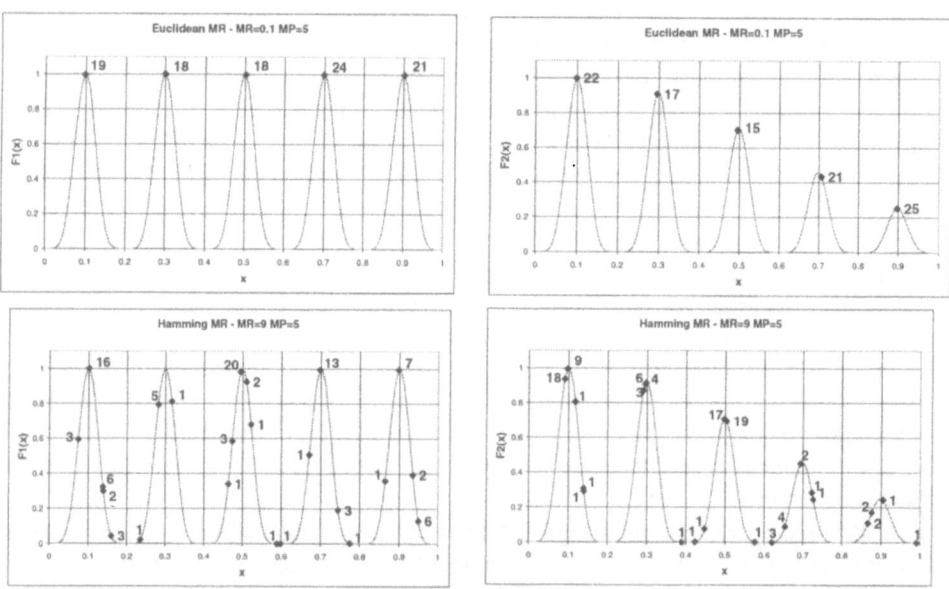

Fig. 1. Results of Mating Restriction in the ECoMAS model. Numbers in these figures are used to indicate the number of individuals on each peak. Settings for *max_partners* and *mating_restriction_threshold* are noted in sub-figure titles

4 Niching Pressure

In the ECoMAS model whose results are shown above, mating restrictions are applied *before* an individual selects its mate. If a mating restriction cannot be fulfilled, the individual in the "mother" role reproduces asexually. Effectively, individuals only interact reproductively (that is, in selection, recombination, and replacement) with individuals within *mating_restriction_threshold* distance of themselves. Over time, individuals that are within this distance from one another *separate* into subpopulations that are effectively isolated from one another. While it might be possible to simulate this effect in a non-agent-based system, it is a natural and desirable side effect of the agent perspective, which may have implications for other EC systems.

The behavior is in strict line with Mahfoud's [8] comments that "niching methods alter the selection algorithm to provide selective pressure within, but not across regions of the search space. The selective pressure within individual regions can be substantial, and still preserve niching properties."

Clearly (like with most aspects of EC) there is a subtle balance involved here. If the population divides into isolated subpopulations too quickly, selective pressure can only find peaks in a limited number of areas of the search space (each covered by a subpopulation), and some peaks may not be found. However, if the population does not divide quickly enough, selective pressure (or drift) will drive the population towards too few peaks. In either case, not all peaks will be maintained.

In an attempt to better understand the balances involved, this section will introduce the concept of *niching pressure* as a parallel to selective pressure. Thus, we first review the concept of selective pressure itself.

Selection typically emphasizes better solutions in the population in two different stages of the evolutionary algorithm:

- Selection of parent(s): the operator has to select the parent(s) from which the offspring will originate.
- Selection of survivor(s): after the generation of the offspring, the selection operator has to decide which members of the population are to be replaced by the new ones and which are going to survive.

Selective pressure is informally defined as the emphasis of selection of the best individuals. Selection in GAs typically yields an exponential form of growth in numbers of observed higher-fitness schemata. To provide a characteristic for this exponential growth, Goldberg and Deb [5] and Back [1] related selective pressure to *takeover time*, which Deb [2] defines as "... the speed at which the best solution in the initial population would occupy the complete population by repeated application of the selection operator alone." Clearly, takeover time, t_t, can vary between '1' and infinity. In Goldberg and Deb [5], selective pressure is inversely proportional to takeover time. The longer the takeover time, the lower the selective pressure; and vice versa. We will assume the selective pressure to vary in the range $[0, \infty)$. If $t_t = 1$ the selective pressure is infinite; if $t_t = \infty$, then the selective pressure is '0'.

As noted above, the effect of mating restrictions in the ECoMAS model is for isolated subpopulations to emerge and (in line with Mahfoud's comments) for selective pressure to then act only within those subpopulations. We will say that separation into subpopulations has occurred when a final, particular set of isolated subpopulations emerges. By isolated, we mean populations that do not influence selection or recombination in one another. Paraphrasing Deb's description of takeover time, separation time is the speed at which the final set of subpopulations emerges through repeated application of the selection and niching operators alone. In line with Mahfoud's comments, selection can still occur within these subpopulations (that is, particular individual types in a subpopulation may be eliminated) after separation, but we consider separation to be complete when these sets are stable, regardless of their contents.

Given the previous discussion, we will define niching pressure in a fashion similar to selective pressure. Specifically, we will identify niching pressure as inversely proportional to the time until separation into a final set of stable subpopulations (the separation time). Like takeover time and its inverse, selective pressure, separation time and niching pressure are simplified, single-number characterizations of a complex process, which in practical GAs is unlikely to be so simple. However, like selection pressure, niching pressure is a useful concept.

A rise in niching pressure (whose controls we have not yet discussed) should result in a reduction in separation time. However, note that niching pressure and selective pressure are not entirely separate. Raising selective pressure should reduce both takeover time and separation time. Selective pressure drives niching pressure, and niching (separation) ultimately defines the subpopulations in which selective pressure can act.

5 Balancing Selective and Niching Pressure

How is niching pressure controlled? First, consider how selective pressure is controlled in the previously discussed ECoMAS model (we will later generalize to other models). Assuming *max_children*=1, selective pressure is primarily controlled by *max_partners*, the size of the list upon which "mother" agents base their selective decisions, and (less directly) by *number_of_receivers*, the number of plumages an agent sends out each time in the father role. Increasing *max_partners* increases the number of individuals against which any potential father is evaluated, thus reducing the chances of mating for less-fit fathers. Increases in this parameter should lower takeover time, and thus increase the selective pressure. Note that this is similar to the effect of raising the tournament size in tournament selection.

As was discussed earlier, niching pressure is directly effected by controls on selective pressure. We accounted for selective pressure controls in the previous paragraph. However, we also want to account for controls on niching pressure that are separable from controls on selective pressure. Niching pressure is inversely proportional to separation time. Clearly, separation time (but not takeover time within any given subpopulation) is controlled by the *mating restriction threshold*. Thus, this is a control on niching pressure that does not directly affect selective pressure. Generally, the higher the *mating restriction threshold* the longer it will take for separation to occur. In the limiting cases, a threshold that nearly spans the space will result in a very long separation time, and a threshold of zero means the population is fully separated immediately. Thus, we theorize that the *mating restriction threshold* directly controls separation time, and *inversely* controls niching pressure. In the next paragraphs we will test the effects of the controls suggested here.

We consider Euclidean mating restrictions with three levels of the *mating_restriction_threshold* to control niching pressure: 0.05, 0.10, and 0.15. We consider four levels of *max_partners* to control selective pressure. These are 2, 5, 10, and 15. Similar investigations for Hamming mating restrictions have been performed, but those results cannot be presented here, for the sake of brevity, and will be presented in a later document. All results presented are averages over five runs. We adopted the Chi-squared-like measure introduced by Deb and Goldberg [3]. This measures the deviation from a desired distribution. In the case of F1, the desired distribution is equal numbers of individuals at the top of each peak. In the case of F2, the number of individuals on top of each peak in the desired distribution is proportional to the peak's height. This measure is not entirely informative, since the mating restriction scheme does not have the global perspective of fitness sharing, and does not explicitly attempt to "balance" the number of individuals on each peak. Since the mating restriction scheme works differently with regard to peak height (results of which we will later discuss), we evaluate the Chi-square measure against both an equal and a proportional distribution on the peaks for function F2. Like in Deb and Goldberg's study, we consider an individual to be "on" a peak if its fitness is more than 80% of the peak's value.

To more thoroughly consider cases where individuals are represented on every peak, but not necessarily distributed proportionally, we also consider the average number of peaks covered by more than 4 individuals (the expected number in a random population) and more than 10 individuals (the expected number if individuals were evenly placed in the "bins" defined in Deb and Goldberg's Chi-square-like measure).

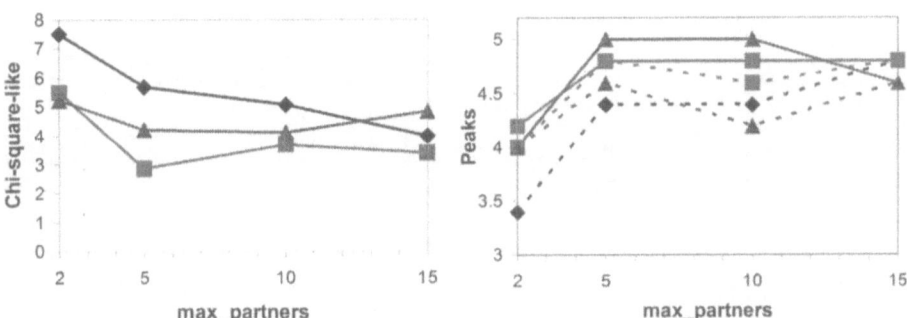

Fig. 2. Results of varying niching and selective pressure, function F1 (equal peaks), Euclidean mating restrictions. *Mating_restriction_threshold* values are 0.05=♦, 0.10=■, 0.15=▲. In the right-hand graph, solid lines are average number of peaks with over 4 individuals; dotted lines are peaks with over 10 individuals. Note coincident solid lines for ♦ and ■ in the right-hand graph

Fig. 2 shows the results for F1 and Euclidean mating restrictions. The graphs show a complex interplay of selective and niching pressure, which can be clarified by other data, which cannot be included here for the sake of brevity. Some explanation will suffice. Although error bars are not included in the graphs for the sake of visual clarity, the following statements are true with 90% confidence levels.

With the highest niching pressure (*mating_restriction_threshold* = 0.05), the population rapidly separates into many small niches. Some of these are off of the function's peaks. Therefore, increase in the selective pressure is necessary to drive individuals up the peaks before separation occurs. Thus, increase in selective pressure results in better Chi-square-like measure (left-hand graph) and in better distribution of individuals on peaks (right-hand graph, dotted line). For slightly lower niching pressure (*mating_restriction_threshold* = 0.1), niches are larger, and separation is slower. This allows for low values of selective pressure to generate better performance than in the previous case. For all but the lowest value of selective pressure, all peaks are represented in almost all runs. If the niching pressure is on its lowest value (*mating_restriction_threshold* = 0.15), coupled with moderate selective pressure, the best performance is obtained. While the "even" distribution indicated by the Chi-square-like measure is slightly worse, all peaks are found with great consistency. However, if the selective pressure is too high, drift occurs before separation, resulting in a worse distribution of individuals on peaks (right-hand graph, solid line). In fact, a few peaks are not even covered by four agents (which we interpret as peak loss). Also, the dotted line goes up, because the individuals that do not cover some peaks end up on over-represented peaks (which we interpret as drift). The drift phenomenon was also observed with Hamming mating restrictions. In that case, the effects were much more pronounced, and were particularly evident even for intermediate values of niching pressure. Drift had a greater effect, due to the more connected neighborhood structure of the Hamming space.

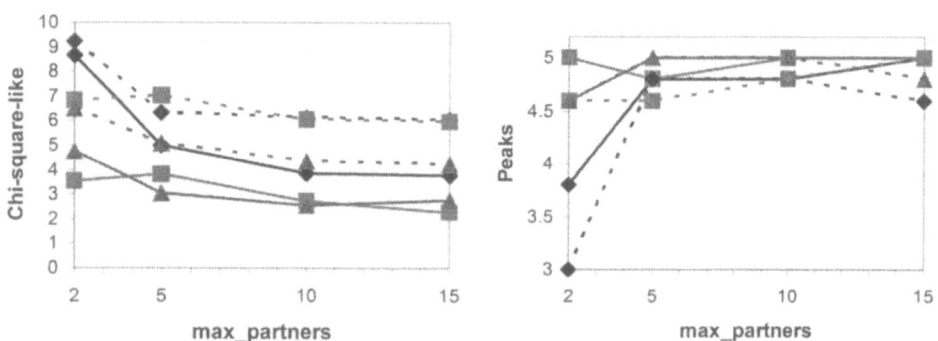

Fig. 3. Results of varying niching and selective pressure, function F2 (decreasing peaks), Euclidean mating restrictions. Symbols are generally as in the previous figure. In the left-hand figure, dotted lines represent deviations from the proportional distribution, and solid lines from the equal distribution

Fig. 3 shows results for F2 and the Euclidean mating restriction. In this case, increasing selective pressure generally enhances results for all levels of niching pressure. However, note that for the highest value of selective pressure ($max\_partners$=15) and the highest value of niching pressure ($mating\_restriction\_threshold$ = 0.05), peaks are beginning to be underrepresented (reflected in the right-hand graph) because separation occurs before all peaks represented adequately.

6 Implications for Other Niching Schemes

Given our consistent use of mating restrictions within the ECoMAS model as an illustrative example, it may at first seem that the introduction of niching pressure as a concept is particular to this system. The authors believe this is not the case, and that the concept can be usefully generalized.

We will call the group of individuals over which selection acts the *selection group*. In characterizing what we mean by the selection group, we must consider both roles of selection mentioned previously: selection to act as parents, and selection for replacement. Generally, we observe that *increase in the number of individuals in the selection group increases selective pressure*.

For instance, consider deterministic crowding, where selection to act as parents is random, and selection for replacement is within a parent-offspring couple. Thus, selective pressure could be controlled in a manner similar to tournament selection. One could enlarge the group of parents, or the number of children produced by parents.

We will call the maximum expected range of distances between any pair of individuals in a selection group that group's *span*. We further observe that *increase in the span of the selection group is likely to decrease niching pressure because it will increase separation time between subpopulations*.

In the ECoMAS model, the span of the selection group is directly controlled by the mating restriction threshold, whose effects were illustrated in the previous section. In deterministic crowding, the span is between a parent and the children with which it is paired (based on distance). The span of this group is an artifact of the representation, and the recombination operators. This complicates analysis. It is possible that full separation may never occur in deterministic crowding, since under given operators for a given representation, selection groups from a given subpopulation may always overlap other subpopulations. This may sometimes cause inability of deterministic crowding to maintain stable niches. Also note that controls on selective and niching pressure in deterministic crowding are more difficult to separate than in ECoMAS.

The previously introduced concepts can be generalized to other schemes, for instance, island model GAs [11] and fine-grained GAs [10]. Although these schemes are known to be unable to maintain stable niches, extensions of the concepts reveal the possibility of analogous controls on selective and niching pressures.

In the island model, subpopulations are defined by distinct islands. Selection groups are defined by those subpopulations and migrants from other islands. Note that in this case, the group for selection of parents could span the same space as the islands (depending on the migration policy), while the group for replacement remains local to each island. Thus, the control on niching pressure is on the span of migrants that may come from other islands. If this was dynamically adjusted towards zero over time, stable niches could be obtained. Note that the number of islands also influences niching pressure, independently of selective pressure. The authors believe that a careful theory of balancing niching pressure and selective pressure should yield insights into appropriate migration tuning policies for desired effects in island models.

In fine-grained GAs, an individual's local neighborhood defines both its subpopulation and its (parental and replacement) selection group. Once again, tuning the selective group size and span in a fashion that balances selective and niching pressure over time should yield desired niching.

Finally, consider the implications of the previously introduced concepts for learning classifier systems [6], particularly XCS [16]. In this system, both subpopulations and (parental) selection groups are defined by the match-and-act process, since this process forms the groups of individuals that participate in the non-panmictic GA. We believe an emerging theory of balance for selective and niching pressure (for which this paper is a start) should clarify niching in XCS.

Fitness sharing (perhaps the most popular niching scheme) [6] is something of an exception. Fitness sharing never separates the population into distinct subpopulations within which selection is isolated, and selective groups always span the population in its centralized scheme. Selection is always possible between all niches. While this sounds like a prescription for drift towards one peak over all others, fitness sharing corrects by globally renormalizing the fitness of peaks based on the number of individuals at each peak. While this is a powerful scheme when centralization is possible, it also explains the difficulty in transferring fitness sharing to more distributed GAs like ECoMAS.

7 Discussion and Final Comments

This paper has introduced results with an agent-based EC system where stable niches emerge through the effects of mating restrictions, and introduced the notion of niching pressure as a parallel to the well-known concept of selective pressure. Niching pressure is defined to be inversely related to separation time, which is in turn defined as the theoretical time until isolated subpopulations emerge. We experimentally considered controls on the balance of these two pressures, and how the concepts revealed can be extended to other systems, through the notion of the size and span of selection groups, in relationship to subpopulations.

Clearly, the results of this paper are preliminary. Thorough analytical and empirical treatments of the concept of niching pressure are now necessary. However, the authors strongly feel that the concepts introduced will provide much greater insight into an important, balancing effect in EC systems with niching.

References

[1] Back T. (1994) Selective pressure in evolutionary algorithms: A characterisation of selection mechanisms. In *Proceedings 1st IEEE Conference On Evolutionary Computation*. pp. 57–62. IEEE Press.

[2] Deb, K. (1997) Introduction to selection. In Back T., Fogel D. B. and Michalewicz Z. (eds.) *The Handbook of Evolutionary Computation*. pp. C2.1:1–4. Institute of Physics Publishing and Oxford University Press.

[3] Deb, K. and Goldberg, D. E. (1989). An investigation of niche and species formation in genetic function optimization. *Proceedings of the Third International Conference on Genetic Algorithms*. p. 42–50. Morgan Kaufmann.

[4] DeJong, K.A. (1975). *An analysis of the behavior of a class of genetic adaptive systems*. Unpublished Doctoral Dissertation. The University of Michigan.

[5] Goldberg D.E. and Deb K. (1991). A comparative analysis of selection schemes used in genetic algorithms. In Rawlins G. J. E. (ed.) *Foundations of Genetic Algorithms*. pp. 69–93. Morgan Kaufmann.

[6] Goldberg, D.E. (1989). *Genetic algorithms in search, optimization, and machine learning*. Addison-Wesley. Reading, MA.

[7] Kearney, P., Smith, R., Bonacina, C., and Eymann, T. (2000). Integration of Computational Models Inspired by Economics and Genetics. *BT Technology Journal*. v18n4. pp. 150–161.

[8] Mahfoud, S.W. (1995). *Niching method for genetic algorithms*. Unpublished Doctoral Dissertation. University of Illinois. IlliGAL Technical Report 95001.

[9] Mahfoud, S.W. (1992). Crowding and preselection revisited. In *The Proceedings of the Second Parallel Problem Solving From Nature Conference*. pp. 27–36. Springer.

[10] Manderick, M. and Spiessens, P. (1989). In *Proceedings of the Third International Conference on Genetic Algorithms*. pp. 428–422. Morgan Kaufmann.

[11] Martin, W.N., Lienig, J., and Cohoon, J.P. (1997) Island (migration) models: Evolutionary algorithms based on punctuated equilibria. In Back T., Fogel D. B. and Michalewicz Z. (eds.) *The Handbook of Evolutionary Computation*. pp. C6.3:1–16. Institute of Physics Publishing and Oxford University Press.

[12] Oei, C. K., Goldberg, D.E., and Chang, S. (1991). Tournament selection, niching, and the preservation of diversity. IlliGAL Technical Report 91011.

[13] Smith, R. E. and Bonacina, C. (2003). Evolutionary computation as a paradigm for engineering emergent behaviour in multi-agent systems. In Plekhanova, V. (ed.) *Intelligent Agent Software Engineering*. pp. 118–136. Idea Group Publishing.

[14] Smith, R.E. and Taylor, N. (1998). A framework for evolutionary computation in agent-based systems. In C. Looney and J. Castaing (eds.) *Proceedings of the 1998 International Conference on Intelligent Systems*. ISCA Press. p. 221–224.

[15] Smith, R.E., Bonacina C., Kearney P. and Merlat W. (2000). Embodiement of Evolutionary Computation in General Agents. *Evolutionary Computation*. 8:4. pp. 475–493.

[16] Wilson, S.W. (1995). Classifier fitness based on accuracy. *Evolutionary Computation*. 3(2), pp. 149–176.

[17] Yin X. and Germay N. (1993) A fast genetic algorithm with sharing scheme using cluster analysis methods in multimodal function optimization. In Albrecht R.F., Reeves C.R. and Steele N.C. (eds.) *Artificial Neural Nets and Genetic Algorithms: Proceedings of an International. Conference*. pp. 450–457. Springer.

EC Theory: A Unified Viewpoint

Christopher R. Stephens and Adolfo Zamora

Instituto de Ciencias Nucleares, UNAM, Circuito Exterior
A. Postal 70-543, México D.F. 04510
{stephens,zamora}@nuclecu.unam.mx

Abstract. In this paper we show how recent theoretical developments have led to a more unified framework for understanding different branches of Evolutionary Computation (EC), as well as distinct theoretical formulations, such as the Schema theorem and the Vose model. In particular, we show how transformations of the configuration space of the model – such as coarse-grainings/projections, embeddings and coordinate transformations – link these different, disparate elements and alter the standard taxonomic classification of EC and different EC theories. As an illustration we emphasize one particular coordinate transformation between the standard string basis and the "Building Block" basis.

1 Introduction

The development of Evolutionary Computation (EC) up to now has been principally empirical and phenomenological, theory only playing a relatively limited role. In its more "scientific" guise it is related to an older, more mature field – population genetics. Mathematically speaking it is the study of certain classes of heuristic algorithms based on populations of objects (for a rigorous exposition in the context of Genetic Algorithms (GAs) see [1]). From the "engineering" point of view it is an area where the analogical use of "natural selection" appears as a moulding force in the creation of more "competent" problem-solvers [2].

In distinction to other more mature areas of science there is not really a clear consensus on what should be the task of EC theory. Is it to provide recipes for practitioners, to provide exact computational models, to allow a deeper understanding of a complex system, all of these, none of these, or what?

Different approaches to EC theory have been proposed in the past, often with proponents who have been disparaging of the others' point of view. These include schema theories [3], the Vose model [4], the statistical mechanics approach [5] and more. Is there a model that is superior to all others? Often, models are judged by their clarity, simplicity, and ability to explain and predict. Is there a framework that does this best?

A theoretical model is also often judged by how well it unifies a range of phenomena. As there are many different flavours of Evolutionary Algorithm (EA) – GAs, Genetic Programming (GP), Evolution Strategies (ES) etc. – one may ask if there is a theoretical framework that encompasses them all? If not, then which is the framework with the broadest applicability?

In many sciences a large part of theory is associated with taxonomy – classification with respect to natural relationships. In EC, various high-level taxonomic labels are at our disposal, such as GP, GAs, ES etc. Whether these labels are optimal, or even useful other than in an historic sense, however, is a debatable point. Taxonomy allows us to understand commonality between different things. Subsequently we must understand why such commonality exists. For instance, the periodic table was initially an empirical and phenomenological construct until the atomic theory gave it a firm "microscopic" foundation. What is the "periodic table" for EC? Does such a construct exist? If nothing of this nature existed it would be deeply worrying as it would mean that a theoretical treatment of each and every EA and/or problem would be different. It is clear however that there is commonality. The question is more – can it be suitably formalized?

At the other extreme one could claim a type of "hyperuniversality", such as was present in the original version of the Building Block Hypothesis [6], which claimed that all GAs behaved in the same way in finding an optimum – via fit, short schemata. We now know that this, in its strict form, is wrong, being rather an engineering rule-of-thumb with only limited validity, and that such a degree of hyperuniversality does not exist. Nevertheless, a prime job of EC theory should be to tell us what EAs and problems, or classes of EAs and problems, are likely to lead to similar outcomes or behaviour. It does not need to be elaborated on that a deeper understanding of this would be of great use to practitioners.

In this article we wish to give a more unified presentation of EC theory showing how recent developments [7,8,9,10,11,12] have pointed to a new understanding of the relationships between the different branches of EC and the different theoretical formulations.

2 Some Fundamentals of EC Theory

We first describe, briefly and non-rigorously, some fundamentals of EC theory that could be applied to many different types of EA. We try to maintain as much generality as possible, in particular to show how a unified theoretical framework, encompassing most standard EAs, can be developed. Formally, an EA is a stochastic algorithm that takes as input a population of "objects" (strings, trees etc.) and a fitness function, at a given time, and gives as output the population at a later time. The objects live on a configuration space X, of dimensionality N_X, with elements $i \in X$. We denote a population by $\mathbf{P} = (n_1, n_2, ..., n_{N_X})$, where n_i represents the proportion of objects of type i in the population. Each object is assigned a quality or fitness, via a fitness function $f_X : X \to R^+$.

A dynamics is imposed via an evolution operator which in the infinite population limit leads to

$$\mathbf{P}(t+1) = \mathcal{S}\left(\mathbf{P}(t), \mathsf{f}\right) \circ \mathcal{T}\left(\mathbf{P}(t), \mathsf{p}\right), \qquad (1)$$

where $\circ$ denotes the Schur product of vectors. The transmission term $\mathcal{T}(\mathbf{P}(t), \mathsf{p})$ describes the probability of transforming one object into another one by mutation, crossover, or other genetic operators, the explicit transmission mechanism being encoded by the parameters p. The term $\mathcal{S}(\mathbf{P},\mathbf{f})$ describes the selection forces acting on P with the parameters f determining the fitness function.

For selection the number of parameters necessary depends on the type of fitness function and the amount of degeneracy of f_X. For instance, for one-max, only N fitness values are needed due to the degeneracy of the genotype-phenotype map. Mutation usually only depends on one parameter – the mutation probability. Two-parent recombination depends on the set of recombination distributions, $\{\lambda_{ijk}(m)\}$, that characterize the transferral of genetic material from parents to offspring, where $\lambda_{ijk}(m)$ is the probability to form an offspring object, i, given two parent objects, j and k, and a crossover "mode", m, i.e. a rule for redistributing genetic material between parent and offspring objects.

We mentioned previously that taxonomy is important without specifying what should be classified. One may think that it is EAs themselves. An EA alone however, is in some sense a "black box" which takes a "problem" (usually a fitness landscape and an initial population) as input and then gives an output (the population at a later time). A given EA, though, may have very different characteristics with respect to a given measure on one problem versus another, hence, one is led to study EA-problem pairs rather than EAs in isolation. We will call an EA/problem pair a "model", characterizing a particular model by a set of parameters p and f, and taking the models to live on a space, $\mathcal{E}$. In principle one could put a metric on $\mathcal{E}$ and talk about how "close" one model is to another. A less rigorous, but more pragmatic, approach is to think of two models as being "close" if they lead to "similar" behaviour. Of course, to do this one must define "similarity measures". A simple example might be the time taken for a particular correlation between two loci in a population to decrease by a given factor. Continuity on $\mathcal{E}$ would lead one to believe that models with similar parameter values should behave similarly.

Population flows take place on X. All the main branches of EC – GP, GAs, ES etc. – fall into this general framework. The chief differences lie more in what objects are being represented in X and what specific operators generate the dynamics. For instance, in GAs the i represent fixed length strings. In GP they are program trees and in machine code GP [13] or Grammatical evolution [14] they are variable length strings. "Coarse grained" representations, such as phenotypes or schemata also offer very useful representations.

3 Generic Genetic Dynamics

The space of models, $\mathcal{E}$, is very large if one thinks of all possible genetic operators. Selection, mutation and crossover, however, form a very generic set and we will now restrict attention to them. Formally at least, the following also applies to GP as well as GAs:

$$P_i(t+1) = \sum_j \mathcal{P}_{ij} P_j^c(t) \qquad (2)$$

where $P_i^c(t)$ is the probability to find objects of type i after selection and crossover. The matrix elements of the mutation matrix, $\mathcal{P}$, give the probability to mutate object j to object i. In the simple case of fixed length GAs for instance, $\mathcal{P}_{ij} = p^{d^H(i,j)}(1-p)^{N-d^H(i,j)}$, where d_{ij}^H is the Hamming distance between the two strings and N is the strings' length. For mutation Hamming distance is clearly a very natural metric. Explicitly $P_i^c(t)$ is given by

$$P_i^c(t) = (1 - p_c)P_i'(t) + \sum_m \sum_j \sum_k \lambda_{ijk}(m) P_j'(t) P_k'(t) \qquad (3)$$

where $P_i'(t)$ is the probability to select i. $P_i' = (f_i/\bar{f}(t))P_i$ for proportional selection, where $\bar{f}$ is the average population fitness. $\lambda_{ijk}(m)$ is an interaction term between objects, i.e. objects j and k are selected and crossed over ("interact") to potentially form an object i, depending not only on j and k but also on the particular recombination mode. In the case of homologous crossover these modes are just crossover masks with $\sum_m$ being the sum over all possible masks while for non-homologous crossover they are more general. Equations (2) and (3), as an exact representation of the dynamics, in the case of fixed-length GAs, where a crossover mode is simply a mask, are equivalent to the Vose model or, indeed, to earlier formulations in population biology. These equations however are also valid for objects other than fixed-length strings.

$\lambda_{ijk}(m) = 0$ unless the mode m creates object i from j and k. This is unlikely and hence the vast majority of interactions are zero. e.g. in GAs with binary alleles, for a given i and m, $\lambda_{ijk}(m)$ is a 2^N-dimensional square matrix. However, only of the order of 2^N matrix elements are non-zero. Thus, the microscopic representation is very inefficient. This also hold for more complicated types of object.

4 Coarse Graining and Coordinate Transformations

Rather than considering one type of representation as being more "general" or fundamental than another it is useful to think of transforming between different representations. We will consider three basic types of transformation: coarse grainings, "coordinate" transformations, and embeddings, concentrating mainly on the first two. Such transformations give flexibility in terms of what particular representation we may find most suitable for a problem and also give a more unified framework within which we may view different elements of EC, such as GP and GAs, in a more coherent light. More importantly they can facilitate an understanding of the dynamical equations associated with the true effective degrees of freedom of the model. These effective degrees of freedom will more often than not be aggregations of the underlying "microscopic" degrees of freedom and may be made more manifest via a coordinate transformation, embedding or coarse-graining/projection. Additionally, it may be the case that effective degrees of freedom most naturally emerge in an approximation to the dynamics rather than the exact dynamics.

4.1 Coarse Graining

The generic dynamics discussed in the previous section is described by an exponentially large number of coupled, non-linear difference equations representing the microscopic degrees of freedom, i.e. the completely specified objects themselves. In the absence of recombination, the equations are essentially linear and the problem reduces down to finding the eigenvalues and eigenvectors of the selection-mutation matrix. However, save in very simple problems, even this simpler problem is formidable. Crossover adds yet another layer of complexity. Naturally, in such problems one wishes to find the correct effective degrees of freedom so as to be able to affect an effective reduction in the dimensionality of the problem. Such a reduction can be affected by an appropriate coarse graining.

We can formalize these considerations by introducing a general coarse graining operator, $\mathcal{R}(\eta, \eta')$, which coarse grains from the variable $\eta \in X_\eta$ to the variable $\eta' \in X_{\eta'} \subset X_\eta$. Thus, the action of $\mathcal{R}$ is a projection. Given two such coarse grainings we have

$$\mathcal{R}(\eta, \eta') P_\eta(t) = P_{\eta'}(t) \qquad \mathcal{R}(\eta, \eta'') P_\eta(t) = P_{\eta''}(t) \qquad (4)$$

However, given that $\mathcal{R}(\eta', \eta'') P_{\eta'}(t) = P_{\eta''}(t)$ we deduce that

$$\mathcal{R}(\eta, \eta'') = \mathcal{R}(\eta, \eta') \mathcal{R}(\eta', \eta'') \qquad (5)$$

i.e., the space of coarse grainings has a semi-group structure. This type of structure is known, by an abuse of language, as the Renormalization Group. The naturalness of a particular coarse graining transformation will be to a large extent determined by how the transformed dynamics looks.

Considering (1), then given that $\mathcal{R}(\eta, \eta') P_\eta(t) = P_{\eta'}(t)$ the dynamics under a coarse graining is governed by $\mathcal{R}(\eta, \eta') \mathcal{S}(\mathbf{P}_\eta(t), \mathsf{f}) \circ \mathcal{T}(\mathbf{P}_\eta(t), \mathsf{p})$, where $\mathcal{S}(\mathbf{P}_\eta(t), \mathsf{f})$ and $\mathcal{T}(\mathbf{P}_\eta(t), \mathsf{p})$ are the dynamical operators associated with the variables η. If this can be written in the form $\mathcal{S}(\mathbf{P}_{\eta'}(t), \mathsf{f}') \circ \mathcal{T}(\mathbf{P}_{\eta'}(t), \mathsf{p}')$ with suitable "renormalizations", $\mathsf{f} \longrightarrow \mathsf{f}'$ and $\mathsf{p} \longrightarrow \mathsf{p}'$ of the model's parameters, then the dynamics is form covariant or invariant under this coarse graining. Note that we are here considering a more general notion of invariance than the idea of "compatibility" [1] (see [15] for a discussion of the relationship between the two). In the case of selection only, the coarse graining transforms the fitness

$$f_\eta \longrightarrow f_{\eta'} = \mathcal{R}(\eta, \eta') f_\eta = \sum_{\eta \in \eta'} f_\eta P_\eta(t) / \sum_{\eta \in \eta'} P_\eta(t). \qquad (6)$$

An important point to note here is that, generically, a coarse graining gives rise to a *time dependent* coarse-grained fitness.

Of course, there are many types of coarse graining procedure all of which lead to a dimensional reduction. Such reductions can sometimes come about in a relatively "trivial" fashion, such as in the case of the genotype-phenotype map, where the dynamics is invariant in the case of selection and in the absence of mixing operators. In fact, it is strictly invariant not just form invariant, as there is no "renormalization" necessary of any parameter or variable and we have $f_{\eta'} =$

$\mathcal{R}(\eta,\eta')f_{\eta} = f_{\eta}$, where, here, η' represents the phenotype and η the genotype. A concrete example is that of the "needle-in-a-haystack" landscape where the fitness landscape is degenerate for all genotypes except one, the "needle". For selection only, as there are only two phenotypes, there is a reduction in the size of the configuration space from 2^N to 2, i.e. a reduction in the number of degrees of freedom from N to 1. However, if we include in the effect of mutation we see that there is an induced breaking of the genotype-phenotype symmetry due to the fact that sequences close to the master sequence in Hamming distance have more offspring than the equally fit genotypes that are further away. In this case $\mathcal{R}(\eta,\eta')\mathcal{S}\left(\mathbf{P}_{\eta}(t),\mathbf{f}\right) \circ \mathcal{T}\left(\mathbf{P}_{\eta}(t),\mathbf{p}\right) \neq \mathcal{S}\left(\mathbf{P}_{\eta'}(t),\mathbf{f'}\right) \circ \mathcal{T}\left(\mathbf{P}_{\eta'}(t),\mathbf{p'}\right)$ and the natural effective degrees of freedom are Hamming classes rather than phenotypes.

Another important class of coarse grainings is that of "schemata", which we will denote by α with $P_{\alpha}(t)$ its relative frequency at time t. In this case the action of the coarse graining operator is: $\mathcal{R}(x,\alpha)P_x(t) = P_{\alpha}(t) = \sum_{x \in \alpha} P_x(t)$. Schemata have a simple geometric interpretation in the binary case, a particular schema being represented by an $(N-N_2)$-dimensional hyperplane in X which passes through the 2^{N-N_2} vertices that represent the loci that have been coarse grained. A schema partition then consists of 2^{N_2} of such 2^{N-N_2}-dimensional hyperplanes. Thus, there is an effective dimensional reduction from a 2^N-dimensional configuration space X to a 2^{N_2}-dimensional one, X_{α}, and a corresponding reduction in the number of degrees of freedom from N to N_2.

Unlike the simple case of the coarse graining to phenotype, here the coarse-grained fitness is time-dependent, with the "renormalized" fitness now being a highly non-trivial function of the original microscopic fitness, being defined as $f_{\alpha} = \mathcal{R}(x,\alpha)f_x = \sum_{x \in \alpha} f_x P_x(t) / \sum_{x \in \alpha} P_x(t)$. As there exist 3^N possible schemata a full schema basis is over-complete and non-orthonormal. However, the space of schemata is not the natural one for recombination, as we shall see.

4.2 Embeddings

In the case of embeddings one passes from a lower to a higher dimensional configuration space. An example would be that of passing from a representation where objects are represented by variable-length strings, of up to maximum size N_m with binary alleles, to a fixed-length representation of size N_m, by including a third allele value that specifies that there was no corresponding bit in the variable length case. The original configuration space is of dimension $2(2^{N_m}-1)$. However, due to the addition of a third allele the dimension of the embedding space is 3^{N_m}. Of course, for these more general transformations development of the operators and the theory necessary to maintain syntactic correctness of the offspring is a open issue. In this case, one might be better off using the theory for variable length structures already developed in GP. Lest the reader think that this type of transformation is somewhat exotic we may mention the case of protein structure where the protein can be specified at the level of a linear sequence of amino acids but which later forms a two-dimensional secondary structure and finally a three-dimensional tertiary structure.

4.3 Coordinate Transformations

Coordinate transformations are a standard tool in the mathematical analysis of physical models where one seeks a set of variables that is well-adapted to the internal structure of the model and hence simplifies the structure of the equations that have to be analyzed. For our purposes we will only need *linear* transformations which can be described in terms of matrices. We restrict our discussion of explicit examples to the case of binary strings. In this case the standard string basis is: $x = (x_1, \ldots, x_N)$ with $x_i = 0, 1$ and the configuration space the N-dimensional Boolean hypercube.

The three alternative bases we will consider: the Walsh basis, the Taylor basis and the Building Block Basis are all related to the standard basis via linear transformations. In (14), (15) and (16) we show the explicit transformation matrices for the case of three loci.

Walsh (Fourier) Basis. Probably the most important alternative basis is the Walsh basis ψ, consisting of *Walsh functions*, $\psi_I(x) = 1/\sqrt{|X|} \prod_{j \in I} x_j$, where I is a subset of $\{1, \ldots, N\}$ and $x_j = \pm 1$. The Walsh functions are normed and orthogonal and of *order* $|I|$, the number of loci that are multiplied. The Walsh-transform, $\hat{f}$, of a function f is defined implicitly by $f(i) = \sum_I \hat{f}(I)\psi_I(i)$ Multiplying with $\psi_K(x)$ and summing over all $i \in X$ we obtain

$$\sum_{i \in X} f(i)\psi_K(i) = \sum_I \hat{f}(I) \sum_{i \in X} \psi_I(i)\psi_K(i) = \sum_I \hat{f}(I)\delta_{IK} = \hat{f}(K) \qquad (7)$$

In matrix form $\hat{f} = \Psi f$, where the matrix Ψ has the Walsh functions ψ_K as its rows. One of the most important properties of the Walsh functions is that they are eigenfunctions of the mutation operator $\mathcal{P}$ that satisfies $\mathcal{P}\psi_I = (1 - 2|I|/N)\psi_I$ The mutation operator is therefore diagonal in the Walsh basis. Equation (2) reads in these coordinates

$$\hat{P}_I(t+1) = \sum_i (\Psi \mathcal{P})_{Ii} P_i^c = \left(1 - 2\frac{|I|}{N}\right) \hat{P}_I^c \qquad (8)$$

The Walsh basis will be particularly useful if the transformed selection-crossover term $\hat{P}_I^c$ also has a simple form.

Taylor Series. While the standard basis and the Walsh basis ψ are orthonormal, this is not necessarily the case in general. In [16], for instance, the *Taylor series* of a landscape on the Boolean hypercube is introduced in terms of the polynomials $\tau_I(i) = \prod_{i \in I} \tilde{x}_i$, such that $f(i) = \sum_I \tilde{f}(I)\tau_I(i)$. Let us write $i \subset I$ if $\tilde{x}_i = x_i = 1$ for all elements of I. We define the matrix Υ by $\Upsilon_{Ii} = 1$ if $i \subset I$ and $\Upsilon_{Ii} = 0$ otherwise, i.e., $\Upsilon_{Ii} = \tau_I(i)$. Thus, we can write the Taylor series expansion in the form $f = \Upsilon \tilde{f}$, i.e., $\tilde{f} = \Upsilon^{-1} f$. The matrix Υ is invertible [16] but is neither normalized nor orthogonal. i.e. the basis functions $\tau_I(i)$ do not form an orthonormal basis. In fact, we have $\sum_{i \in X} \tau_I(i)\tau_k(i) = 2^{N-|I \cup K|}$ since $\tau_I(i)\tau_k(i) = 1$ whenever $x_i = 1$ for all elements of $I \cup K$, and 0 otherwise.

The Building Block Basis. Coordinate transformations such as the Walsh transform and the Taylor series are general. Here, however, we wish to consider a particular coordinate transformation that arises as an almost inevitable consequence of the action of recombination. We saw that representing recombination in terms of the fundamental, microscopic objects is very inefficient due to the sparsity of the interaction matrix. This is an indication that individual objects are not the natural effective degrees of freedom for recombination. So what are? To form the string 111 with a mask 100 one can join strings 111, 110, 101, and 100 with either 111 or 011. In other words, for the first parent the second and third bit values are unimportant and for the second the first bit value is unimportant. Thus, it is natural to coarse grain over those strings that give rise to the desired target for a given mask.

If one picks, arbitrarily, a vertex in X, associated with a string i, one may perform a linear coordinate transformation $\Lambda : X \to \tilde{X}$ to a basis consisting of all schemata that contain i. For instance, for two bits $X = \{11, 10, 01, 00\}$, while $\tilde{X} = \{11, 1*, *1, **\}$. The invertible matrix Λ is such that $\Lambda_{\alpha i} = 1 \iff i \in \alpha$.

$$\Lambda = \begin{array}{c|cccccccc} & 111 & 110 & 101 & 011 & 100 & 010 & 001 & 000 \\ \hline 111 & 1 & 0 & 0 & 0 & 0 & 0 & 0 & 0 \\ 11* & 1 & 1 & 0 & 0 & 0 & 0 & 0 & 0 \\ 1*1 & 1 & 0 & 1 & 0 & 0 & 0 & 0 & 0 \\ *11 & 1 & 0 & 0 & 1 & 0 & 0 & 0 & 0 \\ 1** & 1 & 1 & 1 & 0 & 1 & 0 & 0 & 0 \\ *1* & 1 & 1 & 0 & 1 & 0 & 1 & 0 & 0 \\ **1 & 1 & 0 & 1 & 1 & 0 & 0 & 1 & 0 \\ *** & 1 & 1 & 1 & 1 & 1 & 1 & 1 & 1 \end{array} \quad (9)$$

$$\Upsilon = \begin{array}{c|cccccccc} & 000 & 100 & 010 & 001 & 110 & 101 & 011 & 111 \\ \hline 111 & 1 & 1 & 1 & 1 & 1 & 1 & 1 & 1 \\ 110 & 1 & 1 & 1 & 0 & 1 & 0 & 0 & 0 \\ 101 & 1 & 1 & 0 & 1 & 0 & 1 & 0 & 0 \\ 011 & 1 & 0 & 1 & 1 & 0 & 0 & 1 & 0 \\ 100 & 1 & 1 & 0 & 0 & 0 & 0 & 0 & 0 \\ 010 & 1 & 0 & 1 & 0 & 0 & 0 & 0 & 0 \\ 001 & 1 & 0 & 0 & 1 & 0 & 0 & 0 & 0 \\ 000 & 1 & 0 & 0 & 0 & 0 & 0 & 0 & 0 \end{array} \quad (10)$$

$$\Psi = \frac{1}{\sqrt{8}} \begin{array}{c|cccccccc} & 111 & 110 & 101 & 011 & 100 & 010 & 001 & 000 \\ \hline 111 & -1 & +1 & +1 & +1 & -1 & -1 & -1 & +1 \\ 110 & +1 & -1 & -1 & -1 & -1 & -1 & +1 & +1 \\ 101 & +1 & -1 & +1 & -1 & -1 & +1 & -1 & +1 \\ 011 & +1 & -1 & -1 & +1 & +1 & -1 & -1 & +1 \\ 100 & -1 & -1 & -1 & +1 & -1 & +1 & +1 & +1 \\ 010 & -1 & -1 & +1 & -1 & +1 & -1 & +1 & +1 \\ 001 & -1 & +1 & -1 & -1 & +1 & +1 & -1 & +1 \\ 000 & +1 & +1 & +1 & +1 & +1 & +1 & +1 & +1 \end{array} \quad (11)$$

We denote the associated coordinate basis the Building Block basis (BBB)[1] in that one may think of the elements of this basis as the BBs[2] that are joined together by crossover to form i.[3] The BBB is complete but not orthonormal. Note that the vertex i by construction is a fixed point of this transformation. Apart from the vertex i, the points in $\tilde{X}$, being schemata, correspond to higher dimensional objects in X. For instance, 1* and *1 are one-planes in X while ** is the whole space. In the BBB one may transform (3) to find

$$\tilde{P}_\alpha^c(t+1) = (1-p_c)\tilde{P}_\alpha'(t) + \sum_{m=1}^{2^N} \sum_{\beta,\gamma} \tilde{\lambda}_{\alpha\beta\gamma}(m)\tilde{P}_\beta'(t)\tilde{P}_\gamma'(t) \qquad (12)$$

where $\tilde{\lambda}_{\alpha\beta\gamma}(m) = \Lambda_{\alpha i}\lambda_{ijk}\Lambda_{\beta j}^{-1}\Lambda_{\gamma k}^{-1}$.

The advantage of this new representation is that the properties and symmetries of crossover are much more transparent. For instance, $\tilde{\lambda}_{\alpha\beta\gamma}(m)$ is such that for a given mask only interactions between BBs that construct the target schema are non-zero, i.e., $\tilde{\lambda}_{\alpha\beta\gamma}(m) = 0$, unless γ corresponds to a schema which is the complement of β with respect to α. Also, $\tilde{\lambda}_{\alpha\beta\gamma}(m) = 0$ unless β is equivalent to m, whereby equivalent means that for any 1 in the mask we have a 1 at the corresponding locus in β and for any 0 we have a *. These two important properties mean that the summations over β and γ in (12) disappear to leave only the sum over masks with an "interaction" constant $p_c(m)$ which depends only on the mask. For example, for two bits, if we choose as vertex 11, then 11 may interact only with **, while 1* may interact only with *1.

In X this has the interesting interpretation that for a target schema α of dimensionality $(N-d)$ only geometric objects "dual" in the d-dimensional subspace of X that corresponds to α may interact. In other words, a k-dimensional object recombines only with a $(N-d-k)$-dimensional object. Additionally, a $(N-d)$-dimensional object may only be formed by the interaction of higher dimensional objects. In this sense interaction is via the geometric intersection of higher dimensional objects. For example, the point 11 can be formed by the intersection of the two lines 1* and *1. Similarly, 1111 can be formed via intersection of the three-plane 1 * ** with the line *111 or via the intersection of the two two-planes 11 * * and * * 11.

As mentioned, one of the primary advantages of the BBB representation is that the sums over j and k disappear thus obtaining

$$P_i^c(t) = (1-p_c)P_i'(t) + \sum_{m=1}^{2^N} p_c(m) P_{i_m}'(t) P_{i_{\bar{m}}}'(t) \qquad (13)$$

---

[1] This basis is implicit in the work [7,8] but has only been considered in more detail recently [17].
[2] Note that these BBs are not the same as their well known counterparts in GA theory [6] being dynamic, not static, objects. Neither are they necessarily short or fit.
[3] Given the arbitrariness of the choice of vertex there are in fact 2^N equivalent BBBs each transformable to any other by a permutation.

where $P'_{i_m}(t)$ is the probability to select the BB i_m (note that the mask uniquely specifies which element, i_m, of the BBB to choose) and $P'_{i_{\bar{m}}}(t)$ is the probability to select the BB $i_{\bar{m}}$, which is uniquely specified as the complement of i_m in i. Both i_m and $i_{\bar{m}}$ are elements of the BBB associated with i. The above equation clearly shows that recombination is most naturally considered in terms of the BBB. In the standard basis there were of the order of 2^{2N} elements of λ_{ijk} to be taken into account for a fixed i. In the BBB there is only one term. Of course, the coarse grained averages of i_m and $i_{\bar{m}}$ contain 2^N terms, still, the reduction in complication is enormous. Thus, crossover naturally introduces the idea of a coarse graining, the associated effective degrees of freedom being the BBs we have defined. This is an important point as it shows that evolution is acting in the presence of crossover most naturally at the level of populations, the BBs representing populations with a certain degree of "kinship" to the target object.

Inserting (13) in (2) we can try to solve for the dynamics. However, in order to do that we must know the time dependence of x_m and $x_{\bar{m}}$. Although the number of BB basis elements is 2^N we may generalize and consider the evolution of an arbitrary schema, α. To do this we need to sum with $\sum_{x \in \alpha}$ on both sides of the equation (2). This can simply be done to obtain [7,8] again the form (2), where this time the index α runs only over the 2^{N_2} elements of the schema partition and where again $\mathbf{M}_{\alpha\beta} = p^{d^H(\alpha,\beta)}(1-p)^{N-d^H(\alpha,\beta)}$. In this case however $d^H(\alpha,\beta)$ is the Hamming distance between the two schemata. For instance, for strings with three loci the schemata partition associated with the first and third loci is $\{1*1, 1*0, 0*1, 0*0\}$. In this case $d^H(1,2) = 1$ and $d^H(1,4) = 2$. $P_\alpha^c(t) = \sum_{x \in \alpha} P_x^c(t)$ is the probability of finding the schema α after selection and recombination. Note the form invariance of the equation after coarse graining. To complete the transformation to schema dynamics we need the schema analog of (13). This also can be obtained by acting with $\sum_{x \in \alpha}$ on both sides of the equation. One obtains

$$P_\alpha^c(t) = (1 - p_c N_\alpha) P'_\alpha(t) + \sum_{m \in \mathcal{M}_r} p_c(m) P'_{\alpha_m}(t) P'_{\alpha_{\bar{m}}}(t) \qquad (14)$$

where α_m represents the part of the schema α inherited from the first parent and $\alpha_{\bar{m}}$ that part inherited from the second. $N(\alpha)$ is the number of crossover masks that affect α, relative to the total number of masks with $p_c(m) \neq 0$, the set of such masks being denoted by $\mathcal{M}_r$. Obviously, these quantities depend on the type of crossover implemented and on properties of the schema such as defining length. Note that the BBB naturally coarse grains here to the BBB appropriate for the schema α as opposed to the string x.

Thus, we see that the evolution equation for schemata is form invariant, there being only a simple multiplicative renormalization of the recombination probability p_c. This form invariance, shown in [7], demonstrates that BB schemata in general are a preferred set of coarse grained variables and, more particularly, the BBB is a preferred basis in the presence of recombination. It has also been shown [1] that schemata, more generally, are the only coarse graining that leads to invariance in the presence of mutation and recombination.

Considering again the structure of (13) and (14) we see that variables associated with a certain degree of coarse graining are related to BB "precursors" at an earlier time, which in their turn ... etc. This hierarchical structure terminates at order-one BBs as these are unaffected by crossover. Thus, for example, the level one BB combinations of 111, i.e., BBs that lead directly upon recombination to 111 are: 11* : **1, 1*1 : *1* and 1** : *11. The level two BBs are 1**, *1* and **1. Thus, a typical construction process is that BBs 1** and *1* recombine at $t = t_1$ to form the BB 11* which at some later time t_2 recombines with the BB **1 to form the sequence 111.

In the case of recombination note also that the coarse graining operator associated with the BBs satisfies

$$\mathcal{R}(\eta, \eta') = \mathcal{R}(\eta^m, \eta'^m)\mathcal{R}(\eta^{\bar{m}}, \eta'^{\bar{m}}) \tag{15}$$

where $\mathcal{R}(\eta^m, \eta'^m)$ represents the action of the coarse graining on the BB $\mathcal{S}$ while $\mathcal{R}(\eta^{\bar{m}}, \eta'^{\bar{m}})$ represents the action on the BB $\bar{m}$.

5 Conclusions

In this paper, based on a coarse-grained, or schema-based, formulation of genetic dynamics we have seen how various branches of EC, in particular GAs and GP, can be understood in a more unified framework. Additionally, we have explicitly demonstrated how to pass between different formulations of genetic dynamics, such as the Vose model and schema-based models, via coordinate transformations on the configuration space, showing how the most natural basis for crossover – the BBB – can be obtained from the standard string basis. The emphasis here has been on how transformations of the configuration space of the model – such as coarse-grainings/projections, embeddings and coordinate transformations – link these different, disparate elements and alter the standard taxonomic classification of EAs and theories of EAs. We firmly believe that a more profound understanding of EC in general will result from a deeper understanding of these, and other, transformations. Finally, it is important to note that such a unified viewpoint has already led to several notable advances, such as proofs of the well known Geiringer's theorem to the case of variable-length strings in the case of subtree crossover [18] and homologous crossover [19], as well as an extension, in the case of fixed length strings, to the case of a non-flat landscape with weak selection [9].

Acknowledgements. CRS would like to thank Riccardo Poli and Peter Stadler for very fruitful collaborations and acknowledges support from DGAPA project ES100201 and Conacyt project 30422-E. AZ would like to thank the Instituto de Ciencias Nucleares for their hospitality.

References

1. Michael D. Vose. *The simple genetic algorithm: Foundations and theory.* MIT Press, Cambridge, MA, 1999.
2. David E. Goldberg. *The Design of Innovation.* Kluwer Academic Publishers, Boston, 2002.
3. J. Holland. *Adaptation in Natural and Artificial Systems.* University of Michigan Press, Ann Arbor, USA, 1975.
4. A.E. Nix and M.D. Vose. Modeling genetic algorithms with Markov chains. *Annals of Mathematics and Artificial Intelligence*, 5(1):79–88, April 1992.
5. Adam Prügel-Bennett and Jonathan L. Shapiro. An analysis of genetic algorithms using statistical mechanics. *Physical Review Letters*, 72:1305–1309, 1994.
6. David E. Goldberg. *Genetic Algorithms in Search, Optimization and Machine Learning.* Addison Wesley, Reading MA, 1989.
7. C.R. Stephens and H. Waelbroeck. Effective degrees of freedom of genetic algorithms and the block hypothesis. In T. Bäck, editor, *Proceedings of ICGA97*, pages 31–41, San Francisco, CA, 1997. Morgan Kaufmann.
8. C.R. Stephens and H. Waelbroeck. Schemata evolution and building blocks. *Evol. Comp.*, 7:109–124, 1999.
9. C.R. Stephens. Some exact results from a coarse grained formulation of genetic dynamics. In *Proceedings of the Genetic and Evolutionary Computation Conference (GECCO-2001)*, pages 631–638, San Francisco, California, USA, 7-11 July 2001. Morgan Kaufmann.
10. Riccardo Poli. Exact schema theorem and effective fitness for GP with one-point crossover. In *Proceedings of the Genetic and Evolutionary Computation Conference*, pages 469–476, Las Vegas, July 2000. Morgan Kaufmann.
11. Riccardo Poli. Exact schema theory for genetic programming and variable-length genetic algorithms with one-point crossover. *Genetic Programming and Evolvable Machines*, 2(2):123–163, 2001.
12. W.B. Langdon and R. Poli. *Foundations of Genetic Programming.* Springer Verlag, Berlin, New York, 2002.
13. Peter Nordin. *Evolutionary Program Induction of Binary Machine Code and its Applications.* PhD thesis, der Universitat Dortmund am Fachereich Informatik, 1997.
14. Michael O'Neill and Conor Ryan. Grammatical evolution. *IEEE Transaction on Evolutionary Compuation*, 5(4), August 2001.
15. A. Aguilar, J. Rowe, and C.R. Stephens. Coarse graining in genetic algorithms: Some issues and examples. In E. Cantú-Paz, editor, *Proceedings of GECCO 2003*, Chicago, 9-13 July 2003. Springer-Verlag.
16. Edward D. Weinberger. Fourier and Taylor series on fitness landscapes. *Biological Cybernetics*, 65:321–330, 1991.
17. C.R. Stephens. The renormalization group and the dynamics of genetic systems. *Acta Phys. Slov.*, 52:515–524, 2002.
18. R. Poli, J.E. Rowe, C.R. Stephens, and A.H. Wright. Allele diffusion in linear genetic programming and variable-length genetic algorithms with subtree crossover. In *Proceedings of EuroGP 2002, LNCS, Vol. 2278*, pages 211–227. Springer-Verlag, 2002.
19. R. Poli, C.R. Stephens, J.E. Rowe, and A.H. Wright. On the search biases of homologous crossover in linear genetic programming and variable-length genetic algorithms. In *Proceedings of the Genetic and Evolutionary Computation Conference (GECCO-2002)*, pages 211–227. Morgan Kaufmann, 2002.

Real Royal Road Functions for Constant Population Size

Tobias Storch and Ingo Wegener

Department of Computer Science, University of Dortmund
44221 Dortmund, Germany
tobias.storch@uni-dortmund.de
wegener@ls2.cs.uni-dortmund.de

Abstract. Evolutionary and genetic algorithms (EAs and GAs) are quite successful randomized function optimizers. This success is mainly based on the interaction of different operators like selection, mutation, and crossover. Since this interaction is still not well understood, one is interested in the analysis of the single operators. Jansen and Wegener (2001a) have described so-called real royal road functions where simple steady-state GAs have a polynomial expected optimization time while the success probability of mutation-based EAs is exponentially small even after an exponential number of steps. This success of the GA is based on the crossover operator *and* a population whose size is moderately increasing with the dimension of the search space. Here new real royal road functions are presented where crossover leads to a small optimization time, although the GA works with the smallest possible population size – namely 2.

1 Introduction and History

Genetic algorithms (GAs) and evolution strategies (ESs) have many areas of application. Here we consider the maximization of pseudo-boolean functions $f_n : \{0,1\}^n \to \mathbb{R}_0^+$. The success of GAs and ESs depends on the interaction of the different operators, among them the so-called search (or genetic) operators which create new individuals from existing ones. A search operator working on one individual is called mutation and a search operator working on two (or more) individuals is called crossover. We only investigate the best-known crossover operators namely one-point crossover and uniform crossover. There have been long debates whether mutation or crossover is "more important". This paper does not contribute to this debate. Our purpose is to investigate when and why crossover is essential.

The problem of premature convergence and the problem to maintain diversity in the population are well known. There are many ideas how to cope with these problems: multi-starts, fitness sharing, niching, distributed GAs, and many more. They all have shown their usefulness in experiments. It has also been also possible to analyse highly specialized GAs on some functions. However, the rigorous analysis of GAs is still in its infancy. This motivates the investigation of special properties of single operators (like the takeover time).

The success of GAs is based on the use of populations *and* the use of crossover operators. Holland (1975) has discussed why crossover is a good search operator. This has led to the building-block hypothesis and the schema theory (see also Goldberg (1989)). We are interested in a rigorous analysis and we concentrate on the following parameters: $T_{A,f}$ describes the random number of fitness evaluations until the algorithm A evaluates an optimal search point for f. The *expected optimization time* $E(T_{A,f})$ is the expected value of $T_{A,f}$ and the *success probability function* $t \to \text{Prob}(T_{A,f} \leq t)$ describes the probability of a successful search within a given number of steps. An algorithm is called *efficient* if the expected optimization time is polynomially bounded (with respect to the problem dimension n) or if at least the success probability within a polynomial number of steps converges to 1 (with respect to n). Crossover is essential for a sequence $f = (f_n)$ of fitness functions if a simple GA (without specialized modules) is efficient while all mutation-based evolutionary algorithms (EAs) are not efficient. Mitchell, Forrest, and Holland (1992) have looked for such functions and presented the so-called royal road functions $\text{RR}_{n,k} : \{0,1\}^n \to \mathbb{R}_0^+$. The input string a is partitioned to m blocks of length k each where $m = n/k$ is an integer. Then $\text{RR}_{n,k}(a)$ equals the number of blocks of a containing ones only. This is a nice example function since it seems to be a "royal road" for the building-block hypothesis and the application of one-point crossover. However, Mitchell, Holland, and Forrest (1994) (see also Mitchell (1996)) have shown that this intuition is wrong. Crossover is not essential when optimizing the royal road functions. Other "GA-friendly" functions like H-IFF (see, e.g., Watson and Pollack (1999)) have been presented but their analysis is not rigorous. Some rigorous analysis has been performed for highly specialized GAs (see, e.g., Dietzfelbinger, Naudts, van Hoyweghen, and Wegener (2002)). Jansen and Wegener (2001a) have presented so-called real royal road functions for uniform crossover and for one-point crossover and proved rigorously that the expected optimization time of a simple steady-state GA is polynomially bounded while each mutation-based EA needs exponential time until the success probability is not exponentially small. The results hold for populations of size n but not for populations whose size is independent of n.

Hence, the steady-state GA needs a population of moderate size *and* the appropriate crossover operator to be efficient. This raises the question of whether populations without crossover can be essential and the question of whether crossover needs populations whose size grows with n in order to be essential. Jansen and Wegener (2001b) have presented functions where mutation-based EAs working with large populations are efficient while all mutation-based EAs with population size 1 are not efficient. Here we answer the second question by presenting *real royal road functions for populations of size* 2.

More precisely, we describe in Sect. 2 a steady-state GA (called (2+1)GA) working with the smallest possible population size allowing crossover, namely population size 2. This GA is not specialized. It only guarantees that the population contains two *different* individuals. In Sect. 3, the real royal road function for uniform crossover and constant population size is presented. It is proven that the (2+1)GA is efficient in the sense, that the success probability after a polynomial number of steps is $1 - o(1)$, i.e., converging to 1 as $n \to \infty$ (Theorem 4). Each mutation-based EA needs exponential time until the success probability is not exponentially small (Proposition 1). In Sect. 4, the function is changed into a real royal road function for one-point crossover and population size 2. The (2+1)GA needs only a polynomial number of steps to reach a success probability of $1 - o(1)$ (Theorem 6), but the expected optimization time grows exponentially. The reason is that with small probability some bad event happens. This event implies a very large optimization time leading to the large expected optimization time. Therefore, we present in Sect. 5 a strong real royal road function for one-point crossover and population size 2. For this function the (2+1)GA even has a polynomially bounded expected optimization time (Theorem 11) while mutation-based EAs still need exponential time until the success probability is not exponentially small (Proposition 7). We finish with some conclusions.

2 The Steady-State (2+1)GA

We describe a simple steady-state GA working on the smallest possible population size allowing crossover namely population size 2.

(2+1)GA

Initialization: Randomly choose two different individuals $x, y \in \{0,1\}^n$.
Search: Produce an individual z, more precisely,
 - with probability 1/3, z is created by mutate(x),
 - with probability 1/3, z is created by mutate(y),
 - with probability 1/3, z is created by mutate(crossover(x,y)).
Selection: Create the new population $\mathcal{P}$.
 - If $z = x$ or $z = y$, then $\mathcal{P} := \{x, y\}$.
 - Otherwise, let $a \in \{x, y, z\}$ be randomly chosen among those individuals with the worst f-value. Then $\mathcal{P} := \{x, y, z\} - \{a\}$.

The reader may wonder why all three possibilities of the search step have probability 1/3. This choice is not essential. Our results hold for all probabilities p_1, p_2, p_3, even if they are based on the fitness of x, y, and z as long as they are bounded below by a positive constant $\epsilon > 0$.

We apply the usual mutation operator flipping each bit independently with probability $1/n$. In Sects. 4 and 5, we apply the usual one-point crossover but create only one child, i.e., choose $i \in \{1, \ldots, n-1\}$ randomly and set crossover(x, y) := $(x_1, \ldots, x_i, y_{i+1}, \ldots, y_n)$. Here the order of x and y is chosen randomly. In Sect. 3, we apply uniform crossover where each bit of the child is chosen from each of the parents with probability 1/2.

3 Real Royal Roads for the (2+1)GA and Uniform Crossover

Before presenting our new function we emphasize that our purpose is to prove rigorously that the (2+1)GA can outperform all mutation-based EAs. As in many other cases, the first example functions with certain properties are artificial fitness functions designed only to prove the results under considerations.

Our example function has one area of global optima and two local optima of different fitness. It should be difficult to create a globally optimal point by mutation of one of the local optima, but it should be easy to do so by uniform crossover of the two local optima. Hence, the GA has to realize a population consisting of the local optima. The fitness function gives hints to reach the better local optimum first. Since we have always two different individuals, one of them is only close to the local optimum and gets hints to look for the second local optimum. These ideas are now made precise.

To simplify the notation we assume that $m := n/6$ is an even integer. Let $|x|$ be the length of x, $\|x\| := \text{ONEMAX}(x) := x_1 + \cdots + x_n$ denote the number of ones of x, and 0^k a string of k zeros. The Hamming distance $H(x, y)$ equals the number of indices i where $x_i \neq y_i$. A path is a sequence $a_1, \ldots, a_p$ such that $H(a_i, a_{i+1}) = 1$ and the points a_i are pairwise distinct. The definition of the new real royal road function R_n^u (u indicates that we use uniform crossover) is based on a path P and a target region T. The path $P = (a_0, \ldots, a_{7m})$ contains $7m + 1$ search points: For $i \leq 6m$ let $a_i := 0^{n-i} 1^i$ and for $i = 6m + j$ let $a_i := 1^{n-j} 0^j$. Let $\text{R}_n^u(a_i) := n + i$ for all $i \neq 5m$ and $\text{R}_n^u(a_{5m}) := n + 8m$. This implies that we have two local optima on P, namely $a^* := a_{5m}$ and $a^{**} := a_{7m}$. If the (2+1)GA first finds a^*, the second individual can search for a^{**}. Hence, we like to have a good chance of creating an optimal search point by uniform crossover from a^* and a^{**}. Let T contain all points $b1^{4m}c$ where $|b| = |c| = m$ and $\|b\| = \|c\| = m/2$. Uniform crossover between $a^* = 0^m 1^{5m}$ and $a^{**} = 1^{5m} 0^m$ preserves the $4m$ ones in the middle part. The probability of creating $m/2$ ones in the prefix equals $\binom{m}{m/2} 2^{-m} = \Theta(1/m^{1/2})$ (by Stirling's formula). The same holds independently for the suffix. Hence, the probability that uniform crossover on $\{a^*, a^{**}\}$ creates a target point and mutation does not destroy this property equals $\Theta(1/m) = \Theta(1/n)$. Now we give a complete definition of R_n^u which is illustrated in Fig. 1. Let $P_1 := (a_0, \ldots, a_{5m-1})$ and $P_2 := (a_{5m+1}, \ldots, a_{7m})$.

$$\text{R}_n^u(x) := \begin{cases} 15m & \text{if } x \in T \\ 14m & \text{if } x = a^* \\ 6m + i & \text{if } x = a_i \in P_1 \cup P_2 \\ 6m - \|x\| & \text{if } x \in R := \{0, 1\}^n - P - T \end{cases}$$

Proposition 1. *Evolution strategies (without crossover) need with probability $1 - o(1)$ exponentially many steps w.r.t. n to optimize R_n^u.*

We omit the proof of this proposition. The probability to create a target point by mutation from a path point is exponentially small. Hence, one has to search within R for a small target where the fitness function only gives the advice to decrease the number of zeros. This makes it exponentially unlikely to hit T. A complete proof follows the lines of the proof of Proposition 6 of Jansen and Wegener (2001a).

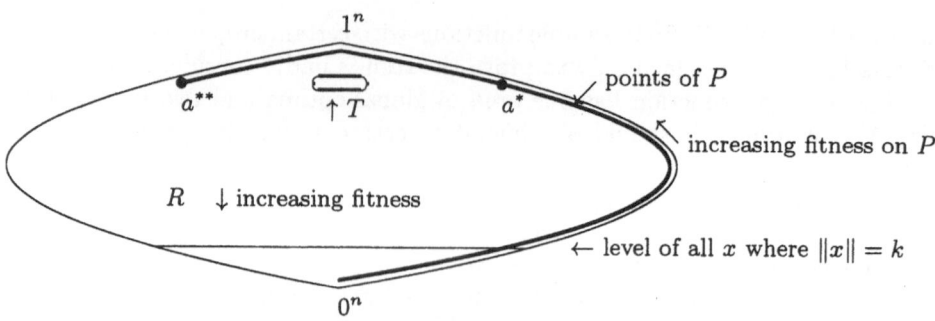

Fig. 1. An illustration of R_n^u

Lemma 2. *The probability that the (2+1)GA does not find a search point from $P_2 \cup T \cup \{a^*\}$ within $c_1 n^2$ steps is $2^{-\Omega(n)}$ (c_1 an appropriate constant).*

Proof. We consider the fitness levels L_i, $1 \leq i < 11m$, containing all search points x where $R_n^u(x) = i$. These fitness levels contain all search points outside $P_2 \cup T \cup \{a^*\}$. Each of these search points has a better Hamming neighbor. A population $\mathcal{P} = \{x, y\}$ belongs to L_i if $\max\{f(x), f(y)\} = i$. The probability to leave L_i is at least the probability of choosing the right individual for the right 1-bit-mutation and, therefore, at least $p = 1/(3en)$. The steps where crossover is chosen can be ignored. They only may increase the success probability. Hence, we have to wait for at most $11m$ successes in an experiment where each trial has a success probability of at least p. Therefore, the result follows by an application of Chernoff bounds (see, e.g., Motwani and Raghavan (1995)). □

Lemma 3. *If the population contains a^*, the probability that the (2+1)GA does not find an optimal search point, namely a search point from T, within $c_2 n^2$ steps is $2^{-\Omega(n)}$ (c_2 an appropriate constant).*

Proof. By the selection procedure, the population will contain a^* until a search point of T is created. The probability to create some $a \in P_2$ from a^* is bounded below by $p = 1/(3en)$. Afterwards, $\mathcal{P} = \{a^*, a_i\}$, $i > 5m$. We now consider the level defined by the index i until $\mathcal{P} \subseteq \{a^*, a^{**}\} \cup T$. The probability to increase i is bounded below by p in each step. Finally, the probability to create some target point from the population $\{a^*, a^{**}\}$ equals by our arguments above $\Theta(1/n)$. Moreover, only target points are accepted if $\mathcal{P} = \{a^*, a^{**}\}$. Hence, we have to wait for $2m + 1$ successes in experiments where each trial has a success probability of $\Theta(1/n)$. Therefore, Lemma 3 follows by the same arguments as Lemma 2. □

Theorem 4. *The success probability that the (2+1)GA with uniform crossover optimizes R_n^u within cn^2 steps, c an appropriate constant, is $1 - O(1/n)$.*

Proof. Applying Lemmas 2 and 3 we are left with the problem whether the (2+1)GA creates a search point from P_2 before a search point from $T \cup \{a^*\}$

(called bad event). It is sufficient to bound the probability of the bad event by $O(1/n)$. Here we have to cope with the "undesired" effects of uniform crossover. Remember that all search points of $P_2 \cup T \cup \{a^*\}$ contain $4m$ ones in their "middle part".

By Chernoff bounds, the probability that the initial population contains a search point with more than $(7/2)m$ ones is exponentially small. As long as no point from $P \cup T$ is created, we only have search points with at most $(7/2)m$ ones if we start with such strings. Hence, each of the search points has at least $m/2$ wrong bits in the middle part. The probability to correct a wrong bit by uniform crossover is at most $1/2$ and the probability to correct it by mutation is $1/n$. Hence, it is very likely to create a search point a_i, $i \leq (15/4)m$, before creating a point from $P_2 \cup T \cup \{a^*\} \cup \{a_i \mid (15/4)m < i < 5m\}$ (the failure probability is exponentially small). We can repeat these arguments to prove that at some point of time the population contains a_i and a_j, $0 \leq i < j < 4m$. Uniform crossover applied to a_i and a_j creates some a_k, $i \leq k \leq j$. Therefore, better points are created by mutation from some a_k, $k < 4m$. The probability of creating some point of $P_2 \cup \{a^*\}$ is exponentially small. Thus, we will obtain in $O(n^2)$ steps a_i and a_j, $4m \leq i < j < 5m$, with overwhelming probability. Then better points are created from some a_k, $4m \leq k < 5m$. In this case, there is exactly one $(5m-k)$-bit mutation to create a^* and exactly one $(5m-k+l)$-bit mutation, $1 \leq l \leq 2m$, to create the point a_{5m+l} from P_2. The probability of the $(5m-k)$-bit mutation equals $q_1 := (1/n)^{5m-k}(1-1/n)^{n-5m+k}$ and the probability of all the $(5m-k+l)$-bit mutations, $1 \leq l \leq 2m$, altogether is the sum q_2 of all $(1/n)^{5m-k+l}(1-1/n)^{n-5m+k-l}$. Since, $q_2/q_1 = O(1/n)$, the probability to create a point from P_2 before creating a point from $T \cup \{a^*\}$ is altogether bounded by $O(1/n)$. □

Having this essential result we can play with the definition of R_n^u. Let $\mathrm{R}_{n,k}^u$ be the variant of R_n^u where $a_{5m+1}, \ldots, a_{5m+k}$ belong to the region R. Proposition 1 and Lemma 2 hold also for $\mathrm{R}_{n,k}^u$. In Lemma 3, we now have to wait for the event to flip the right $k+1$ bits of a^* to obtain a_{5m+k+1}. The expected time for this is bounded above by en^{k+1}. After $c_2 n^{k+1} \log n$ steps the probability of not creating a point of P_2 is bounded above by

$$\left(1 - \frac{1}{en^{k+1}}\right)^{c_2 n^{k+1} \log n} \leq e^{-(c_2/e)\log n}.$$

This can be made smaller then $1/n^k$ by choosing c_2 appropriately. However, in the proof of Theorem 4 now $q_2/q_1 = O(1/n^k)$, since we need at least $(k+1)$-bit mutations to create points from P_2. This leads to the following result.

Theorem 5. *The success probability that the (2+1)GA with uniform crossover optimize $\mathrm{R}_{n,k}^u$ within $cn^{k+1} \log n$ steps, c an appropriate constant, is $1 - O(1/n^k)$.*

4 A Variant for One-Point Crossover

In order to obtain a real royal road function for the (2+1)GA and one-point crossover we can consider an appropriate variant of R_n^u. The probability of creating 1^n by one-point crossover from $a^{**} = 1^{5m}0^m$ and $a^* = 0^m1^{5m}$ is bounded below by a positive constant. The idea is to make 1^n the only target point which cannot be reached easily by mutation from a point on P_2. For this reason we replace the path between a^* and a^{**} by a path which is far away from 1^n. The function R_n^1 is defined in the same way as R_n^u with two exceptions. The target set T is replaced by $T := \{1^n\}$. The points $(a_{5m+1}, \ldots, a_{7m})$ are now defined by $a_{5m+2i-1} := 1^i 0^{m-i} 1^{5m-i+1} 0^{i-1}$ and $a_{5m+2i} := 1^i 0^{m-i} 1^{5m-i} 0^i$, $1 \le i \le m$. Then the points of P_2 have $5m$ or $5m+1$ ones and are far from 1^n and far from the typical points created in the initialization step. Thus Proposition 1 also holds with the same arguments for R_n^1.

Theorem 6. *The success probability that the (2+1)GA with one-point crossover optimizes R_n^1 within cn^2 steps, c an appropriate constant, is $1 - O(1/n)$.*

Proof. It is easy to prove the claims of Lemmas 2 and 3 also for R_n^1. The old proofs can be used without changes. If the population consists of a_i and a_j, $0 \le i < j < 4m$, we can use the arguments of the proof of Theorem 4 to conclude that the probability of creating a point from P_2 before creating a point from $T \cup \{a^*\}$ is bounded by $O(1/n)$. Let x and y be the search points of the initial population. All search points from P_2 have $4m$ ones in their middle part. By Chernoff bounds, the probability that x and y have not at least $m/2$ common zeros in this middle part is exponentially small. In order to obtain a search point from P_2 it is necessary to flip each of these positions at least once. In a phase of cn steps the probability of not flipping a bit at a special position equals $(1-1/n)^{cn}$ and the probability of flipping it is therefore $1 - (1-1/n)^{cn}$ and thus bounded above by a constant $\epsilon < 1$. Since the bit positions are treated independently by the mutation operator, the probability of flipping $m/2$ given positions is exponentially small. (These calculations are related to the coupon collector's theorem (see Motwani and Raghavan (1995)).) If this event does not happen we are in the situation of the fitness function $6m - \|x\| = n-\text{ONEMAX}(x)$. The standard analysis of ONEMAX leads to the result that we can expect after cn steps a population of two search points of at most m ones. Each step has a constant probability of decreasing the number of ones in the population. By Chernoff bounds, cn steps for an appropriate c are enough to decrease the number of ones from at most $11m$ ones to at most $2m$ ones. The failure probability again is exponentially small. If both search points have at most m ones, crossover can only create a seach point with $2m$ ones and even then the probability of creating $4m$ ones by mutation is exponentially small. Hence, we create some a_i and a_j, $0 \le i < j < 4m$, before some point from P_2 with a probability exponentially close to 1. □

With the same notations and arguments leading to Theorem 5 we can prove a similar theorem for $R_{n,k}^1$.

5 Real Royal Roads for the (2+1)GA and One-Point Crossover

The aim is to define a real royal road function R_n^{1*} for the (2+1)GA and one-point crossover which even has the property that the expected optimization time of the (2+1)GA is bounded by some polynomial $p(n)$. This implies by Markoff's inequality a success probability of at least $1/2$ within $2p(n)$ steps. Since the bound on the expected optimization time holds for *all* initial populations, the success probability within $O(p(n) \log n)$ steps can be bounded by $1 - O(n^{1/k})$ for each constant k.

The definition of R_n^{1*} is influenced by the function R_n^1. We modify R_n^1 in the following way. All a_i, where $i = 5m + 2j - 1$ and $1 \leq j \leq m$, $i = 5m + 2$ or $i = 5m + 4$, now belong to the bad region R. Finally, all other individuals a_i, $i > 5m$, have the same fitness $13m$ and define a plateau of constant fitness. The full definition of R_n^{1*} is:

$$R_n^{1*}(x) := \begin{cases} 15m & \text{if } x \in T := \{1^n\} \\ 14m & \text{if } x = 0^m 1^{5m} =: a_0 \\ 13m & \text{if } x \in \{a_i := 1^i 0^{m-i} 1^{4m} 1^{m-i} 0^i \mid 3 \leq i \leq m\} =: P \\ 6m + i & \text{if } x = 0^{n-i} 1^i, \, 0 \leq i < 5m \\ 6m - \|x\| & \text{otherwise} \end{cases}$$

The following result can be proved like Proposition 1.

Proposition 7. *Evolution strategies (without crossover) need with a probability exponentially close to 1 exponentially many steps w.r.t. n to optimize R_n^{1*}.*

Finally, we analyse the (2+1)GA with one-point crossover on R_n^{1*}.

Lemma 8. *The expected time until the population of the (2+1)GA contains a search point from $Q := P \cup T \cup \{a_0\}$ is bounded by $O(n^2)$. With a probability of $1 - O(1/n^6)$ this search point is contained in $T \cup \{a_0\}$.*

Proof. As in the proof of Theorem 6 we get an expected time of $O(n^2)$ for creating a search point from Q. Again we have to create with overwhelming probability the search point from Q by mutation of some $0^{n-i} 1^i$, $0 \leq i < 5m$. The probability that this happens for some $i \leq 4m$ is exponentially small. Otherwise, $H(0^{n-i} 1^i, a_3) = H(0^{n-i} 1^i, a_0) + 6$ and we have to compare k-bit mutations with $(k+6)$-bit mutations. Hence, we can apply the same arguments as in the proof of Theorem 5. □

Let us first consider what happens if the population contains only individuals of P. Remember that all these individuals have the same fitness.

Lemma 9. *If the population contains only search points from P, the expected time until the population of the (2+1)GA contains an element from $T \cup \{a_0\}$ is bounded by $O(n^9)$.*

	I	II	III	IV	V
a_i	$1\cdots 1$	$0\cdots 0$	$0\cdots 01\cdots 1$	$1\cdots 1$	$0\cdots 0$
a_j	$1\cdots 1$	$1\cdots 1$	$0\cdots 01\cdots 1$	$0\cdots 0$	$0\cdots 0$
a_i	$1\cdots 1$	$0\cdots 0$	$0\cdots 01\cdots 1$	$1\cdots 1$	$0\cdots 0$
length	i	$j-i$	$n-2j$	$j-i$	i

Fig. 2. One-point crossover between a_i and a_j resp. a_j and a_i

Proof. We only have to consider the situation where the population consists of a_i and a_j, $3 \leq i < j \leq m$. A step is called *essential* if a search point is created, which has a chance of being accepted for the next population, namely a search point from $Q - \{a_i, a_j\}$. The probability of producing $a_{i \pm d}$, $d \geq 1$, $3 \leq i \pm d \leq m$, from a_i by mutation equals $\Theta(1/n^{2d})$, since we consider a special $2d$-bit mutation. In order to investigate mutate(crossover(a_i, a_j)) resp. mutate(crossover(a_j, a_i)) we describe the search points in Fig. 2.

If the cut position falls into Region I or Region V (including the borders), we obtain a_i or a_j, altogether each of them with the same probability. If the cut position falls into Region III (including the borders) we obtain $1^i 0^{m-i} 1^{4m} 1^{m-j} 0^j$ or $1^j 0^{m-j} 1^{4m} 1^{m-i} 0^i$, for Region II we obtain some $1^i 0^{s-i} 1^{j-s} 0^{m-j} 1^{4m} 1^{m-j} 0^j$ or $1^s 0^{m-s} 1^{4m} 1^{m-i} 0^i$, $i < s < j$, and for Region IV some $1^i 0^{m-i} 1^{4m} 1^{m-s} 0^s$, or $1^j 0^{m-j} 1^{4m} 1^{m-j} 0^{j-s} 1^{s-i} 0^i$, $i < s < j$. The situation is almost symmetric if $3 \leq i < j \leq m$. More precisely, the following individuals have the same chance to be created by crossover and a following mutation from the pair (a_i, a_j):
- a_{i-d} and a_{j+d} (as long as $i - d \geq 3$ and $j + d \leq m$) and
- a_{i+d} and a_{j-d} (as long as $i + d \leq j - d$).

And at least one bit has to flip to obtain an element from $Q - \{a_i, a_j\}$ from crossover(a_i, a_j) resp. crossover(a_j, a_i). Hence, the total probability of an essential step is bounded by $O(1/n)$ and $\Omega(1/n^2)$. In order to prove the lemma it is sufficient to show a bound of $O(n^7)$ on the number of essential steps until a_0 or 1^n is produced.

If $i = 3$, the probability that the next essential step produces a_0 by mutation from a_i is bounded by $\Omega(1/n^5)$. Hence, the expected number of essential steps with a population containing a_3 until a_0 is included in the population is bounded by $O(n^5)$. We are done by proving a bound of $O(n^2)$ on the number of essential steps until the population contains a_3, a_0, or 1^n if we start with a_i and a_j, $3 \leq i < j \leq m$.

For this purpose, it is sufficient to prove that for a phase of cn^2 essential steps, c an appropriate constant, there is a probability of at least a constant $\epsilon > 0$ that we produce a_3, a_0, or 1^n. We ignore the chance of producing a_0 or 1^n. Let $\{a_{i'}, a_{j'}\}$ be the population created from $\{a_i, a_j\}$ in an essential step. The *gain* of this step is defined by $(i' + j') - (i + j)$. To produce a_{i-d} or a_{j+d}, $d \geq 2$, mutation alone has to flip $2d$ bits and the mutation following a crossover at least $2d$ bits. Hence, this happens with a probability of $\Theta(1/n^2)$ in an essential step.

There is a probability of $(1-\Theta(1/n^2))^{cn^2} \leq \epsilon'$ where $\epsilon' < 1$ is a constant that the phase does not contain such a step. This implies that $\{a_i, a_j\}$ can create only a_k, $i-1 \leq k \leq j+1$. Further, the probability of good steps of length 1 ($a_i \to a_{i-1}$, $a_j \to a_{j-1}$) is the same as the probability of bad steps of length 1 ($a_i \to a_{i+1}$, $a_j \to a_{j+1}$). This even holds for steps of length d, $d \geq 2$ ($a_j \to a_{j-d}, a_i \to a_{i+d}$). However, if we produce a_j from a_i, we cannot accept the copy of a_j. This does not disturb the symmetry since with the same probability we produce a_i from a_j. If $j = m$, a_{j+1} does not exist. This disturbs the symmetry but supports a positive gain.

The steps of length 1 can be considered as independent coin tosses with success probability $1/2$. Applying Chernoff bounds, $\Theta(n^2)$ coin tosses are enough to have a surplus of n wins with a probability of at least a constant $\epsilon'' > 0$. For each fixed point of time the probability that the steps of length d, $d \geq 2$, lead to a non-negative gain is by symmetry at least $1/2$ (it is not exactly $1/2$, since the total gain of these steps may be 0). This implies that we create a_3 with a probability of at least ϵ in one phase if c is chosen large enough. (Since the search points $a_3, \ldots, a_m$ are a plateau of constant fitness of R_n^{1*}, we have used some ideas contained in the paper of Jansen and Wegener (2001c).) □

Lemma 10. *If the population contains a_0 and some a_j, $3 \leq j \leq m$, the probability for the (2+1)GA that the population contains 1^n or a_0 and a_m within the next cn^4 steps, c an appropriate constant, is bounded below by some constant $\epsilon > 0$.*

Proof. We can apply many ideas of the proof of Lemma 9. The probability of creating some a_k, $3 \leq k \leq m$, by mutation of a_0 is bounded by $O(1/n^6)$. Further, the probability of producing $a_{j \pm d}$, $d \geq 2$, by mutation of a_j is bounded by $O(1/n^4)$. Let us investigate the effect of crossover, namely mutate(crossover(a_0, a_j)) and mutate(crossover(a_j, a_0)). We are in the same situation as in Fig. 2 but now $i = 0$. Thus, Region I and Region V are empty. If the cut position falls into Region III (including the borders) we obtain $0^m 1^{4m} 1^{m-j} 0^j$ or $1^j 0^{m-j} 1^{4m} 1^m$. To produce $a_{j \pm d}$, $d \geq 1$, the first (or the last) $3 \leq j \pm d \leq m$ bits of the string and d of the last (or the first) m bits must flip by the mutation following crossover. For each of these cut positions, this leads to a probability of $O(1/n^4)$ for a successful crossover. If the cut position falls into Region II, we obtain some $0^s 1^{j-s} 0^{m-j} 1^{4m} 1^{m-j} 0^j$ (II.a) or $1^s 0^{m-s} 1^{4m} 1^m$ (II.b), $0 < s < j$, and for Region IV we obtain some $0^m 1^{4m} 1^{m-s} 0^s$ (IV.a) or $1^j 0^{m-j} 1^{4m} 1^{m-j} 0^{j-s} 1^s$ (IV.b), $0 < s < j$. For (II.a) and (IV.b) we distinguish the cases $s \geq 2$ and $s = 1$.

Case 1. $s \geq 2$. To obtain $a_{j \pm d}$ at least four bits have to flip in the mutation following crossover:
  – the first 2 resp. last 2 ones, since $s \geq 2$,
  – at least one other *special* bit of the first resp. last m positions, since $d \geq 1$,
  – at least $d \geq 1$ *special* bits of the last resp. first m positions.

This again leads for each of these cut positions to a probability of $O(1/n^4)$ for a successful mutate-crossover step.

Case 2. $s = 1$. Here we can guarante only three flipping bits at selected positions. We can use the same arguments but have to take into accout that $s = 1$. However, the case $s = 1$ refers to only one cut position of crossover. This leads to a probability of $O(1/n^3)$ for a successful mutate-crossover step.

For (II.b) and (IV.a) the mutation following a crossover has to flip $k + |s - k|$ bits to generate a_k. This leads for $s \neq 3$ to a probability of $O(1/n^4)$ and for $s = 3$ to a probability of $O(1/n^3)$ for a successful mutate-crossover step. Therefore, since each cut position has the same probability of $1/(n-1)$, altogether we get a probability of

$$\frac{(n-3) \cdot O(1/n^4) + 2 \cdot O(1/n^3)}{n-1} = O(1/n^4)$$

for creating some a_k. Hence, the probability of a successful crossover or a mutation flipping 4 or even more bits is bounded by $O(1/n^4)$. The probability of including a_m in the population can be analysed in the same way as the probability of including a_3 in the proof of Lemma 9. □

Theorem 11. *The expected time until the (2+1)GA has optimized* R_n^{1*} *is bounded above by* $O(n^6)$.

Proof. By Lemmas 8 and 9, an expected number of

$$(1 - O(1/n^6))O(n^2) + O(1/n^6) \max(O(n^2), O(n^9)) = O(n^3)$$

steps are enough to obtain a population containing a_0 or 1^n. In the second case we are done. Otherwise, the population will contain a_0 until it has optimized R_n^{1*}. If the second indidividual of the population does not equal some a_j, $3 \leq j \leq m$, a 6-bit mutation will create one from a_0, what needs an expected time of $O(n^6)$. By Lemma 10 the expected time until the population then contains $a_0 = 0^m 1^{5m}$ and $a_m = 1^{5m} 0^m$ is bounded by $O(n^4)$. Afterwards, the probability of creating 1^n is at least $1/(10e)$ (choose crossover, its cut position s in the middle, namely $s \in \{m, \ldots, 5m\}$ and do not mutate any bit). Hence, with a probability of $1-o(1)$ we produce 1^n before a_m is replaced by some a_j, $3 \leq j \leq m - 1$. In this case, we can repeat the arguments. The expected number of these phases is $1 + o(1)$. □

We also have defined a real royal road function for uniform crossover where also the expected time of the (2+1)GA is polynomially bounded. The construction of the function and the analysis of the (2+1)GA are more complicated than the results presented here. The reason is the following. If uniform crossover is applied to two search points a and b with a small Hamming distance, the same result can be obtained with a not too small probability by mutating a or b. If the Hamming distance is large, each point which can be created by uniform crossover from a and b has an exponentially small probability of being created. This differs from one-point crossover where each possible search point has a probability of at least $1/(n-1)$ to be created. This implies that uniform crossover can be useful only if there are many good search points "between a and b". This again makes it more difficult to control the "undesired effects" of uniform crossover – in particular, in small populations.

Conclusions

The question whether crossover without special methods to ensure the diversity of the population and without a population whose size grows with the dimension of the search space can improve a mutation-based EA significantly has been solved. Fitness functions have been presented where a simple GA with population size 2 is efficient while mutation-based EAs need with overwhelming probability exponentially many steps. Efficiency is defined as a success probability of $1-o(1)$ within a polynomial number of steps or even as a polynomial expected optimization time. The most important types of crossover, namely uniform crossover and one-point crossover, have been investigated. These are the first results to prove for some examples that crossover can be essential even for populations of size 2. Nevertheless, in most cases of application, it is useful to have larger populations and some method to preserve the diversity in the population.

References

Dietzfelbinger, M., Naudts, B., van Hoyweghen, C., and Wegener, I. (2002). The analysis of a recombinative hill-climber on H-IFF. Submitted for publication in IEEE Trans. on Evolutionary Computation.

Goldberg, D.E. (1989). *Genetic Algorithms in Search, Optimization and Machine Learning.* Addison–Wesley, Reading, Mass.

Holland, J.H. (1975). *Adaptation in Natural and Artificial Systems.* The Univ. of Michigan Press, Ann. Arbor, Mich.

Jansen, T. and Wegener, I. (2001a). Real royal road functions – where crossover provably is essential. Proc. of GECCO'2001, 375–382. Morgan Kaufmann, San Francisco, Calif.

Jansen, T. and Wegener, I. (2001b). On the utility of populations in evolutionary algorithms. Proc. of GECCO'2001, 1034–1041. Morgan Kaufmann, San Francisco, Calif.

Jansen, T. and Wegener, I. (2001c). Evolutionary algorithms – how to cope with plateaus of constant fitness and when to reject strings with the same fitness. IEEE Trans. on Evolutionary Computation 5, 589–599.

Mitchell, M. (1996). *An Introduction to Genetic Algorithms.* Chapter 4.2. MIT Press, Cambridge, Mass.

Mitchell, M., Forrest, S., and Holland, J.H. (1992). The royal road for genetic algorithms: Fitness landscapes and GA performance. Proc. of First European Conf. on Artificial Life. MIT Press, Cambridge, Mass.

Mitchell, M., Holland, J.H., and Forrest, S. (1994). When will a genetic algorithm outperform hill-climbing? Advances in NIPS 6, Morgan Kaufmann, San Mateo, Calif.

Motwani, R. and Raghavan, P. (1995). *Randomized Algorithms.* Cambridge Univ. Press.

Watson, R.A. and Pollack, J.B. (1999). Hierarchically-consistent test problems for genetic algorithms. Proc. of the Congress on Evolutionary Computation (CEC), 1406–1413. IEEE Press.

Two Broad Classes of Functions for Which a No Free Lunch Result Does Not Hold

Matthew J. Streeter

Genetic Programming, Inc.
Mountain View, California
mjs@tmolp.com

Abstract. We identify classes of functions for which a No Free Lunch result does and does not hold, with particular emphasis on the relationship between No Free Lunch and problem description length. We show that a NFL result does not apply to a set of functions when the description length of the functions is sufficiently bounded. We consider sets of functions with non-uniform associated probability distributions, and show that a NFL result does not hold if the probabilities are assigned according either to description length or to a Solomonoff-Levin distribution. We close with a discussion of the conditions under which NFL can apply to sets containing an infinite number of functions.

1 Introduction

The No Free Lunch theorems [3,8,12] loosely state that all search algorithms have the same performance when averaged over all possible functions. Given this fact, an important question for anyone interested in the design of black-box search algorithms is whether a No Free Lunch result holds for the subset of all possible problems that represents actual problems of interest. In this paper we approach this question from two angles: Bayesian learning and description length. In Sect. 2 we show that a No Free Lunch result applies only when a certain form of Bayesian learning is impossible, and suggest that the possibility of this form of Bayesian learning may be characteristic of interesting problems. Sections 3 and 4 focus on the relationship between No Free Lunch and problem description length. Section 5 discusses circumstances under which No Free Lunch applies to infinite sets. Section 6 discusses limitations of this work. Section 7 is the conclusion. The remainder of this section introduces terminology that will be used throughout this paper.

1.1 Terminology

Search Algorithm Framework. Our framework is essentially that given in [10]. Let X be a finite set of points in a search space, and let Y be a finite set of cost values assigned to the points in X by a cost function $f:X \rightarrow Y$. For simplicity, we assume in this paper that the elements of X and Y are integers. Define a trace of size m to be a sequence of pairs $T_m \equiv \langle (x_0,y_0), (x_1, y_1), ..., (x_{m-1},y_{m-1}) \rangle$ where for $0 \leq i \leq m-1$, $x_i \in X$ and $y_i \in Y$. Adopt the notation: $T_m^x \equiv \langle x_0, x_1, ..., x_{m-1} \rangle$; $T_m^y \equiv \langle y_0, y_1, ..., y_{m-1} \rangle$.

Define a search algorithm $A:T,X \rightarrow X$ to be a function that takes as input a trace T and a domain X, and returns as output a point x_i where $x_i \in X$ and $x_i \notin T^x$. Under this formulation, we consider only non-retracing, deterministic black-box search algorithms (hereafter, "search algorithms"). However, the conclusions we will draw about such search algorithms can be extended to stochastic, potentially retracing algorithms using the arguments made by Wolpert and Macready [12] to show that the No Free Lunch theorem applies to such algorithms.

A search algorithm A operating on cost function $f:X \to Y$ is evaluated according to the following steps:

1. Let $T=T_0$, where $T_0 \equiv \langle \rangle$ is the empty trace.
2. Evaluate $A(T,X)$ to obtain a point $x_i \notin T^x$.
3. Evaluate $f(x_i)$, and append the pair $(x_i, f(x_i))$ to T.
4. As long as there exists some x_j such that $x_j \in X$ and $x_j \notin T^x$, return to step 2.

Let $T_m(A,f)$ denote the length m trace generated by algorithm A and function f (i.e. the trace T obtained after executing steps 2-3 m times). We define $V_m(A,f) \equiv (T_m(A,f))^y$ to be the length m *performance vector* generated by A and f. When the subscript is omitted, the performance vector will be assumed to have length $|X|$ (i.e., $V(A,f) \equiv V_{|X|}(A,f)$). We let $P(v,A)$ denote the probability of obtaining performance vector v when running algorithm A against a cost function chosen at random under P.

No Free Lunch Results. Define a *performance measure* $M:V \to \Re$ to be a function that takes a performance vector as input and produces a non-negative real number as output. Let F be a set of functions, and let P be a probability distribution over F. We define the overall performance, $M_O(A)$, of a search algorithm A as:

$$M_O(A) \equiv \sum_{f \in F} P(f) M(V(A,f)).$$

We say that a *No Free Lunch result* applies to the pair (F,P) iff., for any performance measure M and any pair of search algorithms A and B, $M_O(A) = M_O(B)$. In other words, a No Free Lunch result applies to (F,P) iff. the overall performance with respect to any performance measure M is independent of the chosen search algorithm. This is essentially the definition of a No Free Lunch result used by Wolpert and Macready [12]. Schumacher [10] has proposed a related definition of a No Free Lunch result that applies when P is uniform (and is equivalent to our definition in this case). In some cases we shall omit P and simply say that a No Free Lunch result applies to a set of functions F. By this we shall mean that a No Free Lunch result applies to (F,P), where P is a uniform probability distribution over F.

Set-Theoretic Terminology. A multiset is a set that may contain multiple copies of a given element (e.g. $\{0,0,1,0\}$). A *generalized permutation* is an ordered listing of the elements in a multiset. We define $G(Y_{mult})$ to be the number of generalized permutations of some multiset Y_{mult}. Where $f:X \to Y$ is a function, we let $Y_{mult}(f)$ denote the multiset containing one copy of $f(x_i)$ for each $x_i \in X$. $Y_{mult}(f)$ differs from the range of f in that $Y_{mult}(f)$ may contain more than one copy of a given $f(x_i) \in Y$.

Let $f:X \to Y$ be a function and let $\sigma:X \to X$ be a permutation. We define the permutation σf of f to be the function $\sigma f(x) = f(\sigma^{-1}(x))$. We say that a set of functions F is closed under permutation iff., for any $f \in F$ and any σ, $\sigma f \in F$. By the words "set of functions", we will mean a finite set of functions having a common, finite domain, unless otherwise specified.

1.2 The No Free Lunch Theorems

The following is essentially the No Free Lunch theorem proved by Wolpert and Macready [12].

NFL: Let X and Y be finite sets, and let $F \equiv Y^X$ be the set of all functions having domain X and codomain Y. Let P be a uniform distribution over F. A No Free Lunch result applies to (F,P).

Schumacher [10] has provided a sharpened version of this theorem that gives both necessary and sufficient conditions when P is uniform. We refer to this as NFLP.

NFLP: Let F be a set of functions, and let P be a uniform probability distribution. A No Free Lunch result applies to (F,P) iff. F is closed under permutation.

Though Schumacher uses a different definition of No Free Lunch result than the one used in this paper, it can easily be shown that the two definitions are equivalent when P is uniform. The relationship between No Free Lunch and permutations of functions has also been studied by Whitley [11].

2 No Free Lunch and Bayesian Learning

In this section we investigate the relationship between No Free Lunch and Bayesian learning. We show that a No Free Lunch result applies to a set of functions if and only if a certain form of Bayesian learning is not possible. We then argue that only very weak assumptions about the set of "interesting" functions are necessary in order to guarantee that this type of Bayesian learning is possible for interesting functions.

Let F be a set of functions, and let f be a function chosen at random from F under some probability distribution P. Let S be a set of points in the search space that have already been visited (i.e., that are known to be part of f). Let x_i be a point in the search space that has not yet been visited. Let $P(f(x_i)=y \mid S)$ denote the conditional probability, given our knowledge that f contains the points in S, that $f(x_i)=y$. The task of the search algorithm is to choose the value of i (subject to the constraint that x_i has not yet been visited). Clearly if $P(f(x_i)=y \mid S)$ is independent of x_i for all y, the decision made by the search algorithm has no influence on the next y-value that is obtained. Thus, intuitively, one might expect that all search algorithms would perform identically if the value of $P(f(x_i)=y \mid S)$ is independent of i for all S and y. What we establish in this section is that this independence is both a necessary and sufficient condition for a No Free Lunch result to apply to (F,P). We first prove the following Lemma.

Lemma 2.1. Let F be a set of functions, and let P be a probability distribution over F. A No Free Lunch result holds for the pair (F,P) iff., for any performance vector v and any search algorithms A and B, $P(v,A) = P(v,B)$.
Proof: (IF) Trivial. (ONLY IF) Suppose that $P(v,A) \neq P(v,B)$ for some v, A, and B. In this case we can define M to be a performance measure that assigns a value of 1 to v while assigning a value of 0 to all other performance vectors. We will thus have $M_O(A)=P(v,A)$ whereas $M_O(B)=P(v,B)$, so $M_O(A) \neq M_O(B)$. □

Theorem 2.2. Let F be a set of functions, and let P be a probability distribution over F. Let f be a function selected at random from F under P, and let $S=\{(x_{f,0},y_{f,0}), (x_{f,1},y_{f,1}), ..., (x_{f,n-1},y_{f,n-1})\}$ denote a (possibly empty) set of n pairs that are known to belong to f. Let X_S denote the domain of S, and let $P(f(x_i)=y \mid S)$ denote the conditional probability that $f(x_i)=y$, given our knowledge of S. Let $x_i, x_j \notin X_S$ be two points in the search space whose cost values are not yet known. Then a No Free Lunch result applies to (F,P) iff., for any S and y, the equation

$$P(f(x_i)=y \mid S) = P(f(x_j)=y \mid S) \tag{2.1}$$

holds for all x_i, x_j.

Proof: (ONLY IF) By way of contradiction, suppose that a No Free Lunch result applies to (F,P), but that $P(f(x_i)=y \mid S) \neq P(f(x_j)=y \mid S)$ for some x_i, x_j. Let F_S denote the subset of F containing functions that are consistent with S. Let $X_{LEX} = \langle x_0, x_1, ..., x_{|X|} \rangle$ denote some fixed (perhaps lexicographic) ordering of the elements of X.

Let A and B be two search algorithms that each unconditionally evaluate the n points $x_{f,0}, x_{f,1}, ..., x_{f,n-1}$ as the first n steps of their execution. If the cost values obtained by evaluating these n points are not exactly the cost values in S, then both A and B continue to evaluate the remaining points in the order specified by X_{LEX}. If the observed cost values *are* those in S, then A chooses x_i as the next point whereas B chooses x_j. From this point onward, A and B both evaluate the remaining points in the order specified by X_{LEX}.

Let V_{PRE} denote the set of performance vectors that begin with the prefix $\langle y_{f,0}, y_{f,1}, ..., y_{f,n-1}, y \rangle$, and let $P_{VPRE}(a)$ denote the probability of obtaining a performance vector that is a member of V_{PRE} using search algorithm a (run against a function drawn at random from F under P). Let $P_S(a)$ denote the probability of obtaining the n points in S as the result of the first n evaluations of a search algorithm a. We have:

$$P_{VPRE}(A) = P(f(x_i)=y \mid S)*P_S(A) \text{ and } P_{VPRE}(B) = P(f(x_j)=y \mid S)*P_S(B).$$

Because A and B behave identically for the first n steps of their execution, it is clear that $P_S(A)=P_S(B)$. It cannot be the case that $P_S(A)=0$, because A obtains the points in S when running against f. Thus the fact that $P_S(A)=P_S(B)$, in combination with the assumption that $P(f(x_i)=y \mid S) \neq P(f(x_j)=y \mid S)$, establishes that $P_{VPRE}(A) \neq P_{VPRE}(B)$. There must therefore be some $v \in V_{PRE}$ that satisfies the equation $P(v,A) \neq P(v,B)$. By Lemma 2.1, this establishes that a No Free Lunch result does not apply to (F,P), which is a contradiction.

(IF) If S is the empty set, then equation 2.1 becomes $P(f(x_i)=y) = P(f(x_j)=y)$, so the initial choice made by the search algorithm cannot effect the probability of obtaining any particular performance vector v_1 (of length 1). Assume that the search algorithm cannot affect the probability of obtaining a performance vector v_n (of length n), and let S denote the first n pairs observed by some search algorithm. The equation $P(f(x_i)=y \mid S) = P(f(x_j)=y \mid S)$ guarantees that the choice of i made by the search algorithm cannot affect the probability of obtaining any particular value y on the next evaluation. Therefore the performance vector v_{n+1} of length $n+1$ obtained on the next evaluation will also be independent of the search algorithm. Thus, by induction, the probability of obtaining any particular performance vector is independent of the search algorithm, which by Lemma 2.1 establishes that a No Free Lunch result holds for (F,P). Note that this second half of our proof is similar to the derivation of the No Free Lunch theorem itself [12]. □

2.1 Discussion

Equation 2.1 provides a necessary and sufficient condition for a No Free Lunch result to hold for (F,P). We now examine some of the consequences this equation. Suppose equation 2.1 holds for some (F,P), where F is a set of functions and P is a probability distribution. Letting S be the empty set, we have $P(f(x_i)=y) = P(f(x_j)=y)$ for all x_i, x_j. Since the probability that $f(x_i)$ is equal to any particular value y is independent of i, the expected fitness of x_i is also independent of i, so $E[f(x_i)]=E[f(x_j)]$ for all x_i, x_j. Thus, by linearity of expectations,

$$E[|f(x_i)-f(x_j)|] = E[|f(x_k)-f(x_l)|] \qquad (2.2)$$

for all x_i, x_j, x_k, x_l.

Equation 2.2 is of particular relevance for genetic algorithms. Suppose the x_i are chromosomes, and the cost values $f(x_i)$ are their fitness values. Equation 2.2 tells us that even if x_i and x_j have 98% of their genetic material in common, while x_k and x_l have only 2% of their genetic material in common, we are to make no assumption that the fitness of x_i and x_j is likely to be closer than that of x_k and x_l. As another illustration of the consequences of equation 2.2, suppose that $k=i$, that x_j is an individual obtained by randomly mutating one gene of x_i, and that x_l is an individual obtained by randomly mutating all the genes of x_i. Equation 2.2 tells us that the expected effect of this point mutation on fitness is the same as the expected affect on fitness of replacing x_i with a chromosome generated at random. In short, equation 2.2 expresses the assumption that there is no correlation between genotypic similarity and similarity of fitness.

Under (a form of) the assumption that nearby points in the search space do tend to have similar cost values, Christensen and Oppacher [3] have shown that a simple algorithm called SUBMEDIAN-SEEKER outperforms random search.

As a further consequence of equation 2.1, note that if the probability of obtaining any particular value y is independent of the choice of i, then the probability of obtaining any range of y-values is also independent of i, so that equation 2.1 implies:

$$P(y_{min} \le f(x_i) \le y_{max} \mid S) = P(y_{min} \le f(x_j) \le y_{max} \mid S) \qquad (2.3)$$

Equation 2.3 is particularly relevant to the analysis of genetic algorithms by Holland involving schema [5]. As an illustration, suppose the search space X consists of all 32-bit chromosomes, and the set of cost values Y are interpreted as rational numbers between 0 and 1. Let $s_1 \equiv$ ab0c???? and $s_2 \equiv$ f18a???? denote two schemata, where the chromosomes are specified in hexadecimal and where the ? characters are 4-bit wildcards. Suppose each of these schemata are sampled at random (without replacement) 1000 times, and the observations are placed into a set S containing 2000 pairs. Let the schema s_1 have an observed fitness distribution with mean 0.7 and variance 0.1, while that of s_2 is observed to have mean 0.3 and variance 0.01. Now suppose that two additional (unique) samplings are made of s_1 and s_2, and that one is asked to bet on which sampling returns the higher value. If one enters this scenario with equation 2.3 as an assumption, one shall regard the fact that the points in s_1 and s_2 are distributed so differently as an extreme coincidence, but one will not make any extrapolations about unseen points in s_1 or s_2. Yet clearly there are a wide variety of interesting problems for which such statistical inferences are possible.

Given that equation 2.1 holds if and only if a No Free Lunch result applies to the pair (F,P), both equations 2.2 and 2.3 can be regarded as consequences of the No Free Lunch theorem. However, because equation 2.1 is both a necessary and sufficient condition for a No Free Lunch result to hold for (F,P), equations 2.2 and 2.3 can also be regarded as *assumptions* required by the No Free Lunch theorem. Thus, if one does not accept these assumptions, one has grounds for ignoring No Free Lunch.

Other restrictions on the circumstances under which a No Free Lunch result can apply have been provided by Igel and Toussaint [6], who show that a No Free Lunch does not apply to F when F includes only those functions having less than the maximum number of local optima or less than the maximum steepness (as measured w.r.t. some neighborhood structure defined on the search space). Arguments similar to the ones above have been made by Droste, Jansen, and Wegener [4].

3 Functions of Bounded Description Length

In the previous section, we defined various statistical properties that the set of functions F and probability distribution P must satisfy in order for a No Free Lunch result to apply to (F,P). We now focus on properties that the cost functions in F must have in order for a No Free Lunch result to apply, assuming that the cost functions are implemented as programs. In this section, we assume P is a uniform distribution, and that the restrictions on the admissible cost functions are couched in terms of description length. In the next section, we will consider the case where P is defined as a function of description length.

The (minimum) description length, $K_U(f)$, of a function $f:X \to Y$ with respect to some universal Turing machine U is the length of the shortest program that runs on U and implements f. By a program that implements f, we mean a program that produces output $f(x)$ for input $x \in X$, and that produces an error code for all inputs $x \notin X$. Description length is also known as Kolmogorov complexity [7]. The Compiler theorem [7] shows that for two universal Turing machines U and V, the difference $|K_U(f) - K_V(f)|$ is bound by some constant that depends on U and V but not on f. For this reason, it is customary to omit the subscript and simply write the description length as $K(f)$.

3.1 Relevance of Description Length

Consider running a genetic algorithm on a problem involving a 500 byte chromosome and 4 byte fitness values. A fitness function $f:X \to Y$ defines a mapping from the set of chromosomes X to the set of fitness values Y. An explicit representation of f in memory (i.e., one that simply listed all pairs $(x_i, y_i) \in f$) would require $|X|*(\lg|X|+\lg|Y|) \approx 2^{4012}$ bits of storage. Even if we allow the domain of f to be implicit (i.e., we simply list the y-values in $Y_{mult}(f)$ in some fixed order), the amount of storage required is still $|X|*\lg|Y|=2^{4005}$ bits. No amount of memory in any way approaching this quantity is available on current computer systems. Rather, evaluation of a fitness function f (assuming that f is implemented in software) would typically involve a call to compiled code occupying perhaps tens of megabytes of memory. In such a case, implementing f as a program rather than an explicit list provides a compression ratio of over 10^{1000}:1. Thus, we say that real-world fitness functions are highly compressible.

Given this property of real-world fitness functions, the question arises as to whether a No Free Lunch result applies to sets of functions whose definition reflects this property. Droste, Jansen, and Wegener have shown that a No Free Lunch result does not hold for certain classes of functions having highly restricted description length (e.g., functions representable by a boolean circuit of size at most 3) [4]. Schumacher has shown that a No Free Lunch result can hold for *some* sets of functions that are compressible [10] (e.g., needle-in-the-haystack functions), but this does not mean that a NFL result will apply to a subset of $F \equiv Y^X$ defined by placing a bound on description length.

The remaining theorems in this section will make use of the following Lemma.

Lemma 3.1. Let F be the permutation closure of some function $f:X \to Y$ (i.e., $F \equiv \{\sigma f, \sigma:X \to X$ is a permutation$\}$). Then $|F|=G(Y_{mult}(f))$, for any $f \in F$.

Proof: Because all functions in F are permutations of one another, $Y_f \equiv Y_{mult}(f)$ denotes the same multiset for any $f \in F$. The functions in F also have the same domain X. If we establish some ordering for the elements of X, then any function $f \in F$ can be uniquely specified by listing in order the y-values (among those in Y_F) that f assigns to the elements of X. Thus, there is a one-to-one correspondence between elements of F and generalized permutations of Y_F, so $|F|=G(Y_F)$. □

The following theorem defines bounds on description length that are sufficient to ensure that a No Free Lunch result does not hold for a set of functions (under a uniform probability distribution).

Theorem 3.2 ("Hashing theorem"). Let h be a hashing function, and let $F \equiv Y^X$ be a set of functions. Let F_k be the subset of F containing only those functions of description length k bits or less, where $K(h) \leq k$. Then a No Free Lunch result does not apply to F_k so long as:

$$k < \lg(G(Y_{mult}(h))+1)-1.$$

Proof: The number of halting programs of description length k bits is at most 2^k. Therefore the number of halting programs of description length k or less is at most $2^{k+1}-1$, so $|F_k| \leq 2^{k+1}-1$. Let H be the set of functions containing all permutations of h. By Lemma 3.1, $|H|=G(Y_{mult}(h))$. The inequality $k<\lg(G(Y_{mult}(h))+1)-1$ can be rewritten as $2^{k+1}-1<G(Y_{mult}(h))$, so that we have $2^{k+1}-1<|H|$. This implies $|F_k|<|H|$, which means that H is not a subset of F_k. Let $h_k \in H$ denote a function not in F_k. By the definition of H, h_k must be a permutation of h. But h must be a member of F_k since we have assumed $K(h) \leq k$. Thus F_k is not closed under permutation, so by NFLP a No Free Lunch result does not apply to F_k. □

We refer to h as a "hashing function" because hashing functions (used to assign a key to a random position in a hash table) typically have small description length $K(h)$ and are designed to generate many different y-values, which maximizes $G(Y_{mult}(h))$. Thus, the theorem will tend to provide sharper upper and lower bounds when h is a hashing function. Of course, the theorem applies equally to any function h.

We now prove a special case of Theorem 3.2 that will allow us to illustrate some of its implications.

Theorem 3.3. Let X be a set containing the first $|X|$ non-negative integers, and let Y be a set containing the first $|Y|$ non-negative integers. Let h be the hashing function $h(x) = x \bmod |Y|$, and let the description length of this function be denoted by $k_{mod} \equiv K(h)$. Let F_k be the subset of Y^X containing only those functions of description length k bits or less, where $k_{mod} \leq k$. Then a No Free Lunch result does not apply to F_k so long as $k<\lg(G(Y_{mult}(h))+1)-1$, where the value of $G(Y_{mult}(h))$ is given by:

$$G(Y_{mult}(h)) = \frac{|X|!}{\left(\left\lceil \frac{|X|}{|Y|} \right\rceil!\right)^{(|X| \bmod |Y|)} \left(\left\lfloor \frac{|X|}{|Y|} \right\rfloor!\right)^{(|Y|-(|X| \bmod |Y|))}} \tag{3.1}$$

Proof: The Hashing theorem establishes that a No Free Lunch result does not apply to F_k so long as $k<\lg(G(Y_{mult}(h))+1)-1$. It remains only to establish the value of $G(Y_{mult}(h))$. Let n denote the number of distinct elements in the multiset $Y_{mult}(h)$, and let $a_0, a_1, \ldots, a_{n-1}$ be counts of the number of occurrences of each of these n elements in $Y_{mult}(h)$. $G(Y_{mult}(h))$ is given by the standard formula for counting the number of generalized permutations of a multiset [1]:

$$G(Y_{mult}(h)) = \frac{|Y_{mult}(h)|!}{(a_0!)(a_1!)\ldots(a_{n-1}!)} \tag{3.2}$$

We now establish that the right hand sides of equations 3.1 and 3.2 are equal. Because there is one element in $Y_{mult}(h)$ for every element in X, clearly $|X|=|Y_{mult}(h)|$, so that the numerators in the two expressions are equal. To see that the denominators are also equal, note that the first (|X| mod |Y|) integers will appear the most times in $Y_{mult}(h)$. Specifically, these integers will each appear ceiling(|X|/|Y|) times in $Y_{mult}(h)$. The remaining (|Y|-(|X| mod |Y|)) integers will appear only floor(|X|/|Y|) times in $Y_{mult}(h)$. Letting L=(|X| mod |Y|), c=ceiling(|X|/|Y|), and f=floor(|X|/|Y|), we see that the equation $a_i=c$ holds for L distinct values of i, while the equation $a_i=f$ for holds for |Y|-L values of i. The value of the denominator is thus $((c)!)^L*(f!)^{|Y|-L}$, so that:

$$G(Y_{mult}(h)) = \frac{|X|!}{(c!)^L (f!)^{(|Y|-L)}} \quad (3.3)$$

Expanding equation (3.3) in terms of c, f, and L establishes the theorem. □

As an illustration of Theorem 3.3, let X contain all n-bit integers ("chromosomes"), and let Y contain all m-bit integers ("fitness values"). Let F_k be the subset of Y^X containing only those functions of description length k bits or less. Theorem 3.3 ensures that a No Free Lunch result does not apply to F_k so long as $k_{mod}<k<\lg(G(Y_{mult}(h))+1)-1$, where the value of $G(Y_{mult}(h))$ is given by equation 3.1. Table 1 presents the values of these lower and upper bounds on k (which is itself an upper bound on the description length of functions in F_k) as a function of n and m. The values in the last column of the table were computed using Stirling's approximation for factorials. All table entries are in bits. Note that the values in the rightmost column of the table are the upper bounds $\lg(G(Y_h)+1)-1$, and not the numbers of generalized permuations $G(Y_h)$.

Table 1. Upper bounds on description length necessary to ensure that a No Free Lunch result does not hold

Chromosome length n	Fitness value length m	Least upper bound	Greatest upper bound $\lg(G(Y_{mult}(h))+1)-1$
16	1	k_{mod}	$6.55*10^4$
32*8	8	k_{mod}	$9.26*10^{77}$
500*8	32	k_{mod}	$4.22*10^{1205}$

The second row of the table tells us that for a No Free Lunch result to apply to F_k where n=32 bytes and m=1 byte, we must be prepared to allow F_k to contain functions of description length $9.26*10^{77}$ bits. To put this number into perspective, the number of atoms in the universe is estimated to be approximately 10^{80}. For any reasonable machine architecture, the lower bound k_{mod} will be measured in tens of bytes. Thus, if we assume that P is uniform and define F_k as above, only very weak assumptions about k are required in order to ensure that a No Free Lunch result does not apply to (F_k,P).

4 Distributions Defined by Description Length

The previous section concerned sets of functions that are defined by placing a bound on description length. In this section, we will be concerned with probability distributions that are defined in ways that are closely related to description length. We begin by introducing some complexity-theoretic terminology that is relevant to our discussion.

Let U be a special type of Turing machine that runs only self-delimiting programs (i.e., no halting program can be the prefix of another). The Solomonoff universal prior (or Solomonoff-Levin distribution), P_U, is defined as:

$$P_U(f) = \sum_{p:\, p \text{ implements } f \text{ on } U} 2^{-|p|}$$

The Solomonoff universal prior has been well-studied in theoretical computer science [7]. This distribution has the characteristic that for any computable probability distribution P_C, $P_C(f)/P_U(f)$ is bound by a constant dependent on P_C but not on f, which is the origin of the name "universal prior". It has been argued that the Solomonoff universal prior is a formalization of Occam's Razor [9]. In the context of genetic algorithms, the Solomonoff universal prior has the desirable property that it assigns high probabilities to problems that have actually been studied in the GA literature (e.g. TSP, maximum clique), while assigning low probability to most of the functions that could only be described as random.

Theorem 4.1 lays the groundwork both for the remaining theorem proved in this section and for our discussion of infinite sets in Sect. 5. Theorem 4.2 shows that a No Free Lunch result does not apply to (F,P) if P is a Solomonoff universal prior.

Theorem 4.1. Let F be a set of functions, where each $f \in F$ has a finite (but not necessarily common) domain X_f and range Y_f, and let P be a probability distribution over F. A No Free Lunch result holds for (F,P) iff., for any $f \in F$ and any permutation g of f, $P(g)=P(f)$, where we define $P(g \notin F) \equiv 0$.

Proof: (IF) Suppose that for any $f \in F$ and any permutation g of f, $P(g)=P(f)$. The set F can thus be partitioned into subsets which are each closed under permutation and assigned uniform probability by P. By NFLP, a No Free Lunch result applies to each of these subsets. Thus, because the overall performance is simply the sum of the performance on each of these subsets, a No Free Lunch result applies to (F,P).

(ONLY IF) Suppose $P(g) \neq P(f)$ for some g that is a permutation of f. Let H denote the set of all functions that are permutations of f. Schumacher has shown that, when run against all functions in H, any search algorithm will obtain each possible performance vector exactly once [10]. Let A be a search algorithm, and let $V(A,f)$ denote the performance vector that A obtains against f. Let B be a search algorithm that, for all domains $X \neq X_f$, behaves identically to A (B can recognize such cases because the domain X is an input to the search algorithm). However, for $X=X_f$, let B be the search algorithm such that $V(B,g)=V(A,f)$. Such a B must exist because f and g are permutations of one another (and B can simply visit points in whatever order is necessary to obtain $V(A,f)$ when running against g). Furthermore, for all $h \in H$, $h \neq g$ we have $V(B,h) \neq V(A,f)$ because (as just mentioned), when running over all functions in H, B must obtain each performance vector exactly once.

Now consider the performance measure M that assigns a value of 1 to $V(A,f)$ and a value of 0 to all other performance vectors. Let X_i represent the domain of some arbitrary function i, and suppose we run both A and B on i. For $X_i \neq X_f$, A and B will be assigned the same performance because they are defined to perform identically in this case. For $Y_{mult}(i) \neq Y_{mult}(f)$, A and B will both be assigned 0 performance, because the performance vector $V(A,f)$ can only be obtained when the available multiset of y-values are those in $V(A,f)$. Thus the difference $M_O(f)-M_O(g)$ must depend only on those i for which $X_i = X_f$ and $Y_{mult}(i) = Y_{mult}(f)$ (i.e., only on those $i \in H$). When run over all $i \in H$, A will obtain $V(A,f)$ only on f while B obtains $V(A,f)$ only on g. Thus, over all $i \in H$, A will have performance $P(f)$ while B has performance $P(g)$. Thus the overall performance difference $M_O(f)-M_O(g)$ is precisely equal to $P(f)-P(g)$. Since we have hypothesized that $P(f) \neq P(g)$, it follows that $M_O(f) \neq M_O(g)$, so a No Free Lunch result does not hold for (F,P). This establishes the theorem. □

Theorem 4.1 can be seen as a generalization of NFLP simultaneously to non-uniform probability distributions, to sets of functions that do not have a common domain, and (as will be seen in Sect. 5) to infinite sets of functions. The only significant restriction we have placed on each $f \in F$ is that it have a finite domain. Thus, aside from this one restriction, Theorem 4.1 is the most general possible form of the No Free Lunch theorem.

Theorem 4.2. Let F and P be defined as in Theorem 4.1, let h be a hashing function, and let P_U be a Solomonoff universal prior. Then a No Free Lunch result does not hold for (F,P_U) so long as $K_U(h) < \lg(G(Y_{mult}(h)))$.

Proof: Let $k \equiv K_U(h)$. Because h has description length k, there must be at least one program of length k that computes h. Thus h must be assigned a probability of at least 2^{-k} under P_U. Since the sum of the probabilities P_U assigns to all functions in F must not exceed 1, there can be at most 2^k functions assigned probability $P_U(h)$. If a No Free Lunch Result holds for (F,P_U), then by Theorem 4.1 all functions that are permutations of h must be assigned probability $P_U(h)$. The number of such functions is $G(Y_{mult}(h))$. Thus a No Free Lunch result cannot hold so long as $2^k < G(Y_{mult}(h))$. □

A similar proof can be used to establish the same result for P that assign probability in a manner that is strictly decreasing w.r.t. description length.

5 No Free Lunch and Infinite Sets

The No Free Lunch theorems are commonly said to apply to "all possible functions". However, the original No Free Lunch theorems applied only to any finite set of functions $F \equiv Y^X$ having a specified finite domain X and specified finite codomain Y. In this section we discuss circumstances under which a No Free Lunch result will apply when F is infinite.

When F may be infinite, our terminology from Sect. 1 essentially still works, with two caveats. First, we must now distinguish between search functions and search algorithms. We define a search function $A:T,X\rightarrow X$ in the same way we defined a search algorithm in Sect. 1.1, and we now define a search algorithm as a computable search function. Second, we must require that the values returned by a performance measure M not be unbounded (i.e., there must always exists some C_M such that $M(v)<C_M$ for all v), so that the summation required to compute the overall performance M_O will always be guaranteed to converge.

The output of the performance measure M used in second half of the proof of Theorem 4.1 is certainly bounded, since it can only take on the values of 0 and 1. The only objection one might have to the assertion that Theorem 4.1 applies to infinite sets is that the proof of this theorem describes A and B as search algorithms, whereas using our above terminology they may only be search functions. However, it turns out that both A and B can be search algorithms. Because the choice of A in the proof of the theorem was arbitrary, A may be a search algorithm by hypothesis, whereas B is a search algorithm because it emulates A in all but a finite number of cases. Thus, Theorem 4.1 applies to infinite sets of functions.

When considering a finite set of functions $F \equiv Y^X$, a No Free Lunch result applies to (F,P) so long as P is uniform. Though we would argue against such a choice of P even in this case, such a choice of P can at least be justified under the grounds of "making no assumptions". However, when F is infinite, a uniform distribution is not possible, and a No Free Lunch result will apply to (F,P) only if P satisfies the partitioning criteria given in the statement of Theorem 4.1. It can be shown that these partitioning criteria are satisfied iff., for any $f \in F$, P can be fully specified as a function of the domain X_f and the multiset $Y_{mult}(f)$ (i.e., $P(f)=P_{X,Y}(X_f,Y_{mult}(f))$) for all $f \in F$, for some $P_{X,Y}$). What can be the argument for assuming the real world's P is of this form?

One could of course argue that memory limitations restrict X and Y to be finite in practice, but these limitations also restrict description length. And, as we have seen in Sect. 3, only very moderate bounds on description length are required to ensure that a No Free Lunch result does not apply to Y^X.

6 Limitations

The purpose of the proofs in this paper has been identify pairs (F,P) for which a No Free Lunch result does and does not hold. We must acknowledge, however, that the mere fact that a No Free Lunch result does not apply to some pair (F,P) does not guarantee that the development of black-box algorithms for (F,P) is worthwhile. All that is guaranteed is that, if a No Free Lunch result does not hold for (F,P), some search algorithms will perform better than *some* others in terms of *some* performance measure. We have not shown that any algorithms outperform either random or exhaustive search. Also, we have not shown that the performance measures are of the kind that typically would be used in optimization. We hope to address both of these limitations in future work.

7 Summary and Conclusions

We have presented two broad classes of functions for which a No Free Lunch result does not hold: functions of bounded description length, and functions whose probability is defined according to description length. We have argued that such probability distributions are more relevant to search and optimization than are the ones required by NFL. Finally, we have identified the circumstances under which a No Free Lunch result may apply to infinite sets of functions.

References

1. V.K. Balakrishnan. *Introductory Discrete Mathematics*. New York: Dover Publications; 1991.
2. S. Christensen and F. Oppacher. What can we learn from no free lunch? A first attempt to characterize the concept of a searchable function. In *Proc. 2001 Genetic and Evolutionary Computation Conf.*, 2001, pp. 1219–1226.
3. J. Culberson. On the futility of blind search. *Evolutionary Computation*, 6(2):109–127, 1999.
4. S. Droste, T. Jansen, and I. Wegener. Perhaps Not a Free Lunch But At Least a Free Appetizer. In *Proc. 1999 Genetic and Evolutionary Computation Conf.*, 1999, pp. 833–839.
5. J. Holland. *Adaptation in Natural and Artificial Systems*. Cambridge, MA: The MIT Press; 1992.
6. C. Igel and M. Toussaint. On classes of functions for which no free lunch results hold. Los Alamos e-Print Archive cs.NE/0108011. 2001.
7. M. Li and P. Vitányi. *An Introduction to Kolmogorov Complexity and Its Applications*. New York: Springer-Verlag; 1993.
8. N.J. Radcliffe and P.D. Surry. Fundamental limits on search algorithms: Evolutionary computing in perspective. In J. van Leeuwen, ed., *Lecture Notes in Computer Science 1000*. Springer-Verlag, 1996.
9. J. Schmidhuber. Discovering solutions with low Kolmogorov complexity and high generalization capability. In *Proc. 12th Intern. Conf. on Machine Learning*, 1995, pp. 488–496.
10. C. Schumacher, M.D. Vose, and L.D. Whitley. The no free lunch and problem description length. In *Proc. 2001 Genetic and Evolutionary Computation Conf.*, 2001, pp. 565–570.
11. D. Whitley. Functions as permutations: regarding no free lunch, walsh analysis and summary statistics. In Schoenauer et al., eds., *Parallel Problem Solving from Nature 6*, 2000, pp. 169–178.
12. D.H. Wolpert and W.G. Macready. No free lunch theorems for optimization. *IEEE Transactions on Evolutionary Computation*, 4:67–82, 1997.

Dimensionality Reduction via Genetic Value Clustering

Alexander Topchy and William Punch

Computer Science Department, Michigan State University
East Lansing, MI, 48824, USA
{topchyal,punch}@cse.msu.edu

Abstract. Feature extraction based on evolutionary search offers new possibilities for improving classification accuracy and reducing measurement complexity in many data mining and machine learning applications. We present a family of genetic algorithms for feature synthesis through clustering of discrete attribute values. The approach uses new compact graph-based encoding for cluster representation, where size of GA search space is reduced exponentially with respect to the number of items in partitioning, as compared to original idea of Park and Song. We apply developed algorithms and study their effectiveness for DNA fingerprinting in population genetics and text categorization.

1 Introduction

Abundance of features and their measurement complexity is the central problem in many real-world applications of statistical pattern recognition and machine learning. Typical data models, which are the part of the classification and prediction systems encountered in data mining, may include thousands of variables and databases with millions of entries. Feature selection and extraction algorithms seek to provide a lower dimensional data representation that preserves most of the available information [1]. Many application domains often require that knowledge extracted from the data be comprehensible and compact [2,3]. Finding the best subset of variables relevant to the computational model is the goal of feature *selection*, while feature *extraction* synthesizes new features from the original variables, where features (variables) refer to the components of the pattern measurement vector. Potential benefits of reducing the data dimensions include: better modeling (classification/prediction) accuracy, simplification of the developed model, faster learning with fewer parameters, lower measurement costs, and improved reliability of parameter estimation.

Numerous dimensionality reduction algorithms are known and actively used in image recognition, text processing, computational biology and other domains [2]. Most of them fall into two broad categories. The first category, filter methods, estimates the relevance of a candidate feature transformation by analyzing the data distribution, often heuristically, *without* running the actual model. The second category, wrapper methods, optimizes the actual classification accuracy achieved by the selected (extracted) features [4]. In both cases the evaluation function is usually nonlinear and highly multimodal. Furthermore, the search space tends to be astronomically large resulting in a difficult optimization problem.

Evolutionary algorithms (EAs), in particular genetic algorithms (GAs), are a unique kind of optimization method as applied to feature selection/extraction. Unlike conventional sequential search of feature subsets [1], GA is certainly not sensitive to the non-monotonicity of an evaluation function and therefore not as prone to the difficulties involved in finding hidden, nested sets of features [2]. Siedlecki and Sklansky pioneered a GA approach to large-scale feature selection in [5], where classical, direct representation of feature sets was introduced. More sophisticated schemes were proposed in [6,7,8]. Building on this seminal work [5], GAs were also applied to feature extraction coupled with k nearest neighbor classifiers in [9,10] and several other algorithms [11,12]. Typically new features are created through the learning of real-value weights applied to the original features [13] or through a sequence of primitive operators encoded in the chromosome [14]. Recent attribute construction algorithms successfully utilize GA for data mining tasks [15]. A variety of relevant algorithms is reviewed in [3,16,17].

In this study we introduce a different GA approach to dimensionality reduction that does not explicitly operate with whole features, but instead operates on individual values assumed by the feature variables. Features are extracted by clustering several "old" values into a new meta-value which substitutes for the old values in the feature vector. Therefore, new features are created by clustering the values of variables. A GA is used as a search engine for this value clustering. If features assume nominal values then clustering could be viewed as a grouping problem and there are a number of known genetic algorithms that can be used [18]. Grouping as well as clustering GAs is an important research area in itself [18]. Scalability and redundancy are two characteristic problems of grouping algorithms. Here we present a grouping GA with an improved graph-based encoding, that extends work by Park and Song [19]. The proposed cluster representation has lower redundancy while preserving the simplicity of decoding and evaluation. This encoding is not limited to value clustering and therefore applicable to other grouping tasks.

There is an additional motivation behind value clustering. One outcome of dimensionality reduction could be more important than simplified data and knowledge representation, namely such reduction can also help to diminish the effect of the "curse of dimensionality". The curse of dimensionality phenomenon manifests itself in a decrease in classification accuracy when a large number of features are included in the model. Estimation errors will inevitably lead to accuracy degradation for the model induced from a fixed training sample [20]. It is important to note that dimensionality increase occurs not only in the number of features but also in the number of values assumed by the individual variables. In general, appropriate input transformations may eliminate either features or feature values or both. Effectively, such a transformation reduces measurement complexity, which is proportional to the total number of cells in the input value space. Overcoming the curse of dimensionality by lowering the measurement complexity is essential in many real-world tasks.

We analyze the performance of the value clustering genetic algorithms in two application domains with discrete variables:

1. Text categorization. Classification of text using naïve Bayes classifier and a "bag of words" data model where words correspond to different feature values. The number of features is determined by the text's length.
2. Population genetics. Genotype based assignment of individuals to their putative populations of origin. The data model is an extension of a multinomial model with multiple features and certain value constraints within features. The allelic values of multilocus individuals are subject to grouping (merging).

In the remainder of the paper we consider existing value clustering methods. A feature extraction genetic algorithm is then developed. We briefly review grouping GAs (GGAs) and describe an enhanced graph-based chromosome encoding of clusters. Experimental problems and implementation details of empirical studies are presented. Finally we discuss the results and interesting future work.

2 Background on Value Clustering

Data mining applications typically deal with instances represented as a set of input attributes (features), where each attribute assumes one of several possible values. We are primarily interested in data represented by nominal values. Throughout the paper, the term "value clustering" refers to a method that replaces several original values of an attribute by a new meta-value. This meta-value is completely artificial and simply a substitute for all the values used in the cluster. Such a clustering affects the data model and consequently the classification rule. This clustering of values represents a kind of generalization or value abstraction which is used to improve classification. We consider important related works below. However, it is instructive to start with a simple example illustrating the idea of value clustering.

2.1 Clustering in Naïve Bayes Model

Consider a naïve Bayes classifier. The classifier learns the conditional probability $P(A_i|C)$ of each attribute A_i given the class label C. Classification is then done by applying Bayes rule to compute the probability of C given the particular instance of $A_1,...,A_n$ and then predicting the class with the highest posterior probability (MAP hypothesis):

$$MAP \equiv \arg\max_i P(c_i | a_1,...,a_n), \qquad (1)$$

where c_i is an instance of C (a class) and a_j is an instance of A_j. There is one major assumption behind this: all the attributes are conditionally independent given the value of the class C. Therefore, by using Bayes rule:

$$P(c_i | a_1,...,a_n) = \frac{P(a_1,...,a_n,c_i)}{P(a_1,...,a_n)} \propto P(c_i) \prod_j P(a_j | c_i) \qquad (2)$$

Let us assume a multinomial probability distribution for the values of each attribute. Using the maximum likelihood principle one can estimate the class-conditional probabilities $\{P(a_j|c_i)\}$ from training samples as a ratio of number of instances with a specific value of a_j within class c_i to the total number of instances in the same class. For example, if there are only three possible values of the first attribute $a_1 \in \{\alpha,\beta,\gamma\}$, one would estimate the three terms $P(a_1=\alpha|c_i)$, $P(a_1=\beta|c_i)$ and $P(a_1=\gamma|c_i)$. However, if values α and β are merged into one cluster, the meta-value x is constructed $x= (\alpha \text{ or } \beta)$, and only two model parameters remain to be estimated: $P(a_1=x|c_i)$ and $P(a_1=\gamma|c_i)$. In the multinomial model, the probability of the meta-value x obeys:

$$P(a_1=x|c_i) = P(a_1=\alpha|c_i) + P(a_1=\beta|c_i). \tag{3}$$

As a result of this clustering all the occurrences of α or β are replaced by x. Of course, the estimation of parameters is classifier specific.

2.2 Related Work

In traditional database research there are several studies for finding an approximate answer for a database query with no exact answer [21]. Automatic query expansion systems offer approximations from a cluster of similar values, obtained by term clustering [22]. Term clustering creates groups of related words by analyzing their co-occurrence in a document collection. A hierarchy of values can either be garnered from experts or obtained automatically. In [23] the attribute values are clustered automatically in an attribute abstraction hierarchy. This hierarchy is discovered using rules derived from database instances. For rules with the same consequence, values found in the premise are clustered. This process is called pattern based knowledge induction. However, information retrieval approaches cannot be easily combined with arbitrary classifiers.

In machine learning studies we find a number of flexible value clustering methods. Perhaps the most interesting method is the information bottleneck by Tishby et al [24]. The information bottleneck method replaces the original random variable X by a compact representation $\tilde{X}$, which tries to keep as much information as possible about the random variable Y. In particular, variable X could stand for feature values and variable Y is a class label. The information bottleneck method maximizes mutual information $I(\tilde{X};Y)$ between $\tilde{X}$ and Y, conditioned on the information content $I(\tilde{X};X)$ resulting from the clustering $\tilde{X}$ with respect to X. Most significantly, the optimal solution of this information theoretic problem can be found in terms of probability distributions $p(\tilde{x}|x)$, $p(y|\tilde{x})$, and $p(\tilde{x})$ with an arbitrary joint distribution $p(x,y)$. The exact solution is given by a set of nonlinear equations [24]. Iterative solving requires the user-defined cardinality of $\tilde{X}$, and $\tilde{X}$ is usually initialized to random values that potentially may lead to sub-optimal local solutions. The resulting clustering is not "hard", where each value of X belongs to a single cluster in $\tilde{X}$, but rather "soft" with membership probabilities $p(\tilde{x}|x)$. However, hard clustering can be achieved in the agglomerative information bottleneck method [25]. The agglomerative information bottleneck method is a hierarchical process that uses distance measures between distributions and greedy search by merging clusters pair-wise. The multivariate information bottleneck method [26, 27] further extends the approach for multivariate distributions, maximizing mutual information between Y and several "bottlenecked" variables simultaneously.

Recently, the value abstraction approach (clustering in our terms) led to considerable progress in linkage analysis for genetic mapping [28], and was expanded for more general likelihood computations and faster Bayesian network inference in [29]. All these works can be characterized as filter-type feature extraction, since classifier accuracy or induction step feedback is not used for value clustering.

3 Feature Extraction by Value Clustering

Genetic search for better clusters is the core of the proposed dimensionality reduction methodology. As such it is the part of a complete feature extraction system, which also involves classifier induction and performance evaluation. The overall view of the system is shown on Fig. 1.

The fidelity of the computational model can be judged by its classification accuracy. Cross-validation or holdout estimators of true accuracy are typically used during the search [4]. A candidate solution receives a fitness value proportional to the estimated classification accuracy of a corresponding classifier. However, our ultimate goal is to improve the predictive accuracy on a previously unseen sample(s). That is why final performance estimation is necessary for the best solution once the search is completed. Our induction step is classifier specific. For example, induction of a naïve Bayes classifier consists of estimating the probabilities of the feature values and is relatively efficient. For other models, training may be quite computationally consuming, e.g. for neural networks, or especially in GA since it operates with an entire population of solutions. We provide further details on the classifiers used in the experimental results section.

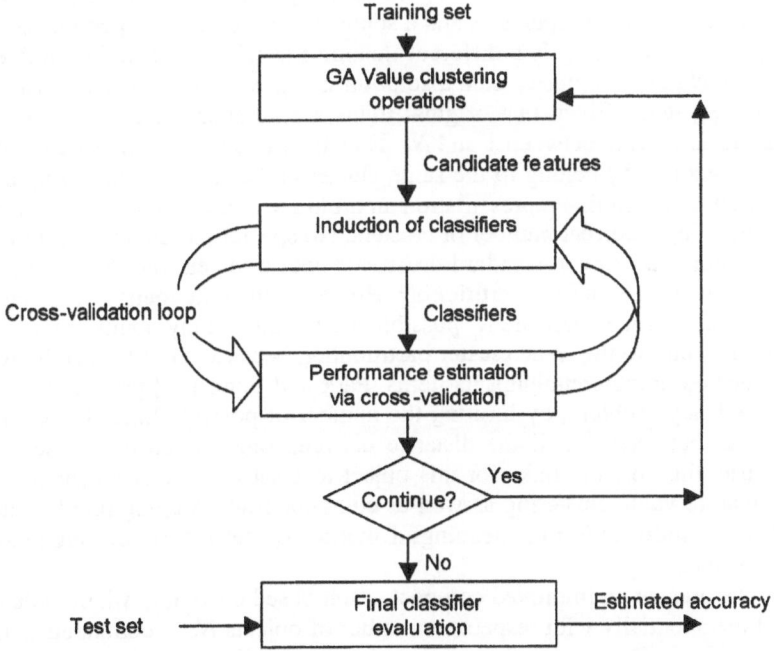

Fig. 1. Feature extraction wrapper-type approach with GA search engine

3.1 Cluster Representation and Genetic Operators

Value clustering can be regarded as a grouping problem, where a number of objects (items) must be placed in several groups. Examples of grouping problems include bin packing, graph coloring and line balancing, all of which are NP-hard. The search space size for partitioning N objects in M groups grows exponentially with N:

$$\frac{1}{M!}\sum_{i=1}^{M}(-1)^{M-i}\binom{M}{i}i^{N} \qquad (4)$$

The major difficulty of typical GA approaches is in designing a representation of a grouping as a chromosome. A good encoding must ensure the complete coverage of the space of possible partitions with minimal redundancy. This encoding should also allow meaningful inheritance of information from parents to children. Finally, it is important that validity checks and chromosomes repair have insignificant computational overhead, if in fact such a repair is necessary.

Several genetic algorithms are specifically known to solve grouping problems. The most straightforward approach uses a standard GA with group-numbers, one gene per object, where each gene contains a group-number for the object. Unfortunately, standard n-point crossover fails to transmit grouping information between chromosomes, since group-numbers do not carry meaning by themselves and could be arbitrarily permuted for any given grouping. This representation also requires knowing the correct number of clusters in advance. Similar difficulties are encountered with permutation encodings. In contrast, GGA algorithm [18] guarantees meaningful inheritance using permutation with separators and a sophisticated crossover operator with repairs.

Graph-based encoding is a different method for solving clustering problems [19]. If there are N items to cluster then a solution is represented by a string with N genes, one gene per item. Given that original items are numbered from 1 to N, then each gene assumes a value between 1 and N. Thus if i-th gene contains value j, then items with numbers i and j belong to the same cluster and can be connected by a link in a graph of all items. In this representation an arbitrary partitioning can be reached without pre-defining a correct number of clusters. No special crossover is required for this representation, since even standard crossover both transfers the item's link and preserves information about the partitioning. However, the redundancy of the encoding is extremely large, since there are N^N possible individuals. Many points in the GA search space correspond to the same cluster partitioning, because a cluster can be formed by any connected graph containing its items. Park and Song [19] proposed a remedy to this redundancy problem, by limiting the number of possible links, based on domain knowledge. For example, if the distance between objects can be defined, then one may restrict the possible links for this object to a set of nearest neighbors. Unfortunately, feature value clustering as well as combinatorial grouping problem in general do not allow introduction of meaningful metrics or definitions of neighborhood for nominal values.

Here we present an improved compact graph-based encoding, where redundancy is reduced exponentially with respect to number of objects N, as compared to the original idea of [19]. The proposed representation keeps all the good qualities of the raw graph-base encoding, without sacrificing its representation power, as we strictly prove below.

First of all, we note that the regular graph-based encoding picks N links out of a total N^2 available, because the chromosome has N genes with N possible values per gene. However, even a complete graph on N vertices has only $N(N-1)/2$ edges. This results in a major combinatorial explosion in the number of ways to encode the same cluster. Furthermore, to represent any partition of items on a graph we need at most

N-1 edges. Thus an arbitrary tree connecting all the objects within a cluster would be sufficient. Indeed, the largest possible cluster contains N items that could be represented by a tree with N-1 edges.

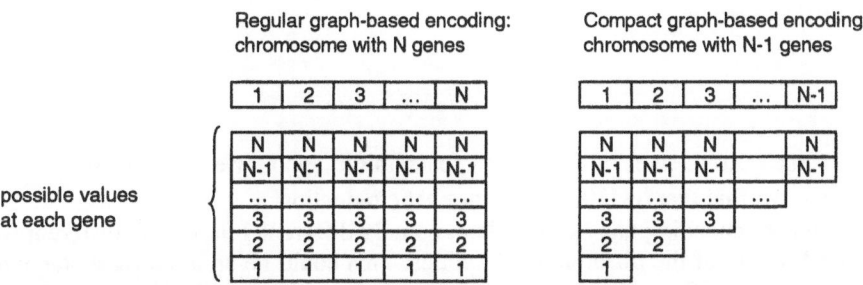

Fig. 2. Regular graph-based clustering GA selects N genes from the total of N^2 edges, while compact encoding needs only N-1 from $N(N-1)/2$ edges. GA search space size is reduced to $N!$ from N^N

We propose an enhanced encoding that draws gene values only from $N(N-1)/2$ edges and needs N-1 genes. The main idea is to remove duplicated edges from a pool of possible gene values. Retaining only values from N to i, for the i-th gene, does this exactly. The first gene assumes values in $\{1,...,N\}$, the second gene in $\{2,...,N\}$, etc. The N-th gene is unnecessary, since it always is set to N, and can be omitted. If the gene value happens to be equal to the gene's number, it means that no edge is placed on the graph. We cannot have less than $N(N-1)/2$ edges of a complete graph, because clusters with only two objects must be obtained by a unique link between them. Fig. 2 compares the regular and compact graph-based encoding. Fig.3 shows graph representations of the sample chromosomes.

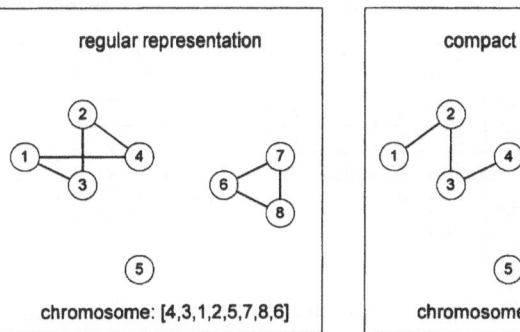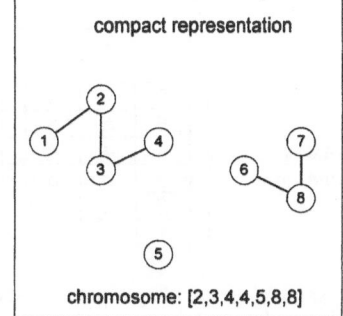

Fig. 3. Sample chromosomes and corresponding clusterings are shown by connected subgraphs for regular and compact encodings

In the compact encoding, each gene obtains its values from an alphabet of a different cardinality. This raises a question as to how required clusters are formed. It is easy to prove that any possible clustering is still attainable in new encoding. For this purpose consider a cluster containing objects with arbitrary numbers. One can sort objects in a cluster by their number. Any such cluster can be formed, by connecting an object with next higher numbered object from the sorted list, because the required edges are always available: any object has access to edges incident to any object with greater number. Therefore, the proposed encoding has the same representational power at the original encoding but has an exponentially lower space, as we approximately estimate from:

$$\delta = \frac{N^N}{N!} \approx \frac{e^N}{\sqrt{2\pi N}}, \quad (5)$$

where δ is the ratio of sizes of GA search space in old and new encodings. For example, if $N=20$ we find that $\delta > 43000000$, a dramatic reduction.

The compact encoding has one problem, namely different genes carry different information because of the different cardinalities. This could be an issue for a standard crossover (e.g. 1- or 2-point) since lower numbered positions on the chromosome are more important than the others. As an alternative, we offer a modification of the compact encoding that equalizes the cardinality of alphabets for different genes. The genes with the most available edges can transfer some of their edges to other genes, so we still have the very same pool of $N(N-1)/2$ edges at our disposal. For example, we can delete the value $(N-1)$ in the alphabet of the first gene, and insert value 1 in the alphabet of $(N-1)$th gene, leaving us with the same edges. Equalization creates cardinality $1+N/2$ for every gene, and can be done exactly with even N, with minor deviation at some positions when N is an odd number. In this equalized compact representation, the reduction in size of search space is still greater than before, namely:

$$\delta = \frac{N^N}{(1+N/2)^{N-1}} = \left(1+\frac{N}{2}\right)\left(\frac{2}{1+2/N}\right)^N \quad (6)$$

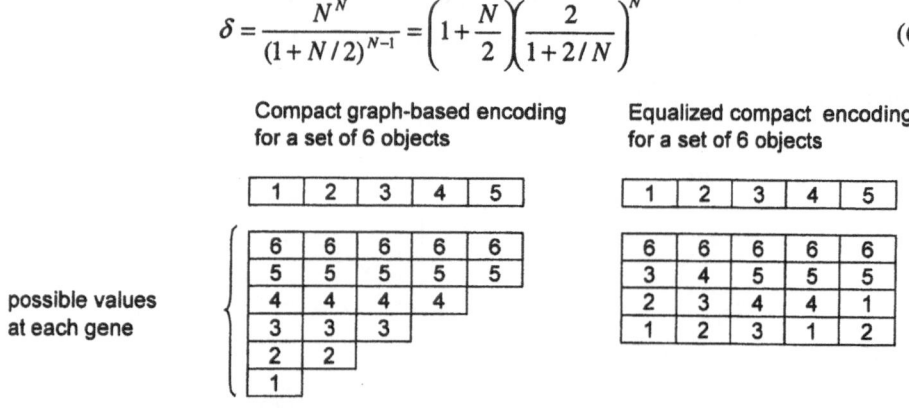

Fig. 4. An example of encoding alphabets at each chromosome position in compact and equalized compact graph-based representations of a set of 6 items

Let us prove that it is possible to realize any clustering in the equalized compact representation. We will show that a cluster with arbitrary objects can be created. The alphabet of the genes (corresponding to any chosen objects) contains all possible links between these objects, because equalization preserves the edges by simple transfer. Now an auxiliary construction is necessary: imagine a complete graph on chosen objects. In this complete graph we can assign direction to the edge from object I to J if the alphabet of gene I contains value J, otherwise the edge goes in the opposite direction. This directed graph is called a tournament since it is a complete and directed graph. Graph theory provides us with the important fact: every tournament has a Hamiltonian path [30]. Hamiltonian path connects all the objects following the arcs and visits each and every object only once. Therefore, arcs from such a directed path are equivalent to valid assignment of edges to genes. Since arbitrary chosen objects are connected, we can obtain any partitioning.

For our experiments we used equalized compact encoding and 2-point crossover. The next section presents the details of our empirical study of compact encodings for value clustering in some applications.

4 Empirical Study

The main goal of the empirical study is to compare the performance of select classifiers with and without GA value clustering. Even though an overall accuracy improvement is very valuable, we are also interested in reduction of measurement complexity.

4.1 Implementation Details

The driver GA program included the following major steps:
1. Initialization. The population is initialized with random individuals using the equalized compact graph-based encoding as described before. Each individual is created by selecting its genes from the alphabets of respective chromosome positions. For the i-th position, the probability to select the value i is set to $(N-1)/N$, to prevent formation of a single big cluster covering all the objects. If the value i, is not generated, we initialize this gene to one of the remaining available values (edges) for this position using a uniform probability. Prevention of excessive clustering is motivated by a result from graph theory on critical probability of connectedness of random graphs [31]. Finally, we always seeded a chromosome $[1,2,...,N]$, corresponding to a N non-merged objects (feature values).
2. Fitness evaluation of each individual in the population. Chromosomes are easily decoded to the actual cluster partitioning of feature values as described above. Model parameters are estimated according to the extracted clustering. Classifier accuracy is evaluated using hold-out samples or 10-fold cross-validation (in one experiment).
3. Termination criteria check. The number of generations was the measure of computational effort. The search was stopped after about 200 generations in each run
4. Tournament selection (tournament size = 2) of parents. Pairs are selected at random with replacement and their number is equal to the population size. The better of the two individuals becomes a parent at the next step.
5. Crossover and reproduction. Standard 2-point crossover is used. Each pair of parents produces two offspring. Mutation with small probability $p_m=0.01$ is applied to each position. In addition, elitism was always used and the best individual of the population survives unchanged.
6. Continue to step 2.

4.2 Test Problems and Results

Two different applications were chosen as benchmarks: text categorization and assignment of individuals to their putative population based on their DNA markers. Feature extraction is very important here, because computations in all these domains suffer from excessive number of features. Certainly, our goal is not necessarily to outperform existing classification algorithms in the area, but rather to demonstrate that GA value clustering effectively reduces feature dimensions as well as improving accuracy in comparison to direct classification.

Text Categorization. Text categorization attempts to automatically place documents into one of a set of predefined classes. Training samples for the experiments were taken from the Reuters-21578 documents collection. We use a classical "bag of words" data model using a naïve Bayes classifier. The number of features is equal to the number of words in a document and each feature can assume any value from the observed dictionary of size N. The dictionary is created from all the training data. The classifier is induced by estimating the probabilities of words in the entire training documents collection. If previously unseen word appears during the inference phase, its probability is set to a small value $0.1/N$. Note that the number of features changes from document to document, while the number of feature values is constant for a fixed training set. It is our goal to cluster the values within features. The probability distribution for each feature is the same due to the independence assumption used in naïve Bayes. Therefore, the clustering of feature values is the same for all features. Thus each GA chromosome represents a clustering of words in the dictionary. The full dictionary of the Reuters data set contains tens of thousands of words (values to be clustered). It is commonly acknowledged that preprocessing is necessary to keep reduce computational costs to a reasonable level. We performed preprocessing by retaining features having the highest mutual information within the category label. Mutual information was assessed independently for each feature. In two different setups we selected 10 and 50 words respectively, corresponding to chromosome lengths with 9 and 49 genes in the equalized compact encoding.

The experiments were designed to distinguish between Reuters articles on "acquisitions" or "earnings" with 2308 and 3785 non-overlapping documents in each category respectively. The known expected accuracy is extremely high, so we made the task more challenging by considering only the first two lines in the body of every document. 1500 documents from each class were split between training and holdout sets (for classifier fitness evaluation), the rest of documents were used to estimate the true accuracy of the best found solution. After the search phase, the best classifier was induced from a combination of training and holdout sets before the final testing. The comparison between the three methods has been made, namely, a naïve non-clustered approach, our proposed clustering GA and a pure best-first hill climbing with the same number of fitness evaluations. We performed 100 runs of each algorithm with different sizes of splits between training and holdout sets. Our GA approach ran for 200 generations with a population of size 100. The results for the estimated true accuracy and percentage of runs with improvement are reported in Table 1.

Table 1. Comparison of test accuracy for the best solutions found by GA value clustering and best-first hillclimbing with "naive" non-clustered solution. First two columns, are training and holdout set sizes. Last column shows percentage of runs when GA found better than naive solution. Each row is the average over 100 runs, with training and holdout sets sampled at random in each run from original data

	Tr. set	Holdout	"Naïve" model	Hill-climbing	Clustering GA	Best runs, %
10 words	400	1100	79.9	79.9	82.3	87
	600	900	79.8	80	81.6	81
The	800	700	79.7	79.9	81.5	78
results	1000	500	79.9	79.9	81.1	74
50 words	600	900	88.2	88.7	90.8	73
	800	700	88.4	88.8	90.5	60

The results show that GA typically improves the predictive accuracy by 1-2% and most runs are successful. Perhaps the main advantage of GA was in dimensionality reduction. The average number of clusters for the 10-words problem was 7.9 and 38.6 for 50-words problem.

DNA Fingerprinting. Assignment of individuals to their putative populations of origin is a practical problem in population genetics. In a classical framework [32], to assign an individual it is necessary to compute likelihood functions for each source population. The population having the highest likelihood value is deemed to be the most probable population of origin. Genetic markers, taken from DNA samples, serve as features and their values (allelic configurations) are assumed to be multinomially distributed [33]. Classification is then performed using a naïve Bayes approach. The large number of possible alleles (e.g.10-20) for each feature seriously complicates the assignment in many practical situations, because the reliability of the estimated parameters is low, since a typical baseline sample consists of only 50-100 individuals. Value clustering is quite promising for binning the alleles together to improve the assignments.

The experiments were done with data set of two lake trout hatchery strains. The data has been collected from trout populations in Lake Seneca and Lake Superior, USA and kindly made available to us by Dr. Kim Scribner. We used 100 individuals per sampled population. Each individual is given as a set of two alleles per locus (feature) with 8 loci in total. While most of the features can have only 2-4 different alleles, there are 10 possible alleles for one of the loci. We ran GA value clustering only for the allelic values at this particular locus. The classification step and GA search are very similar to what was done for text categorization. However, the probabilities of individuals are computed differently, because for trout genotypes there are two alleles specified at each locus (feature). Namely, the probability of allelic configuration (i,j) at each locus is estimated as $2p_ip_j$, if alleles i and j are different, or as p_i^2, if $i=j$, where p_i is an estimate of i-th allele probability at a locus. This modification of the regular naïve Bayes classifier is trivial and appears only in computations of fitness for candidate solutions. Also the fitness was estimated by 10-fold cross-validation due to the limited number of individuals in the original data set. As a main result, the value clustering GA was able to improve classification accuracy (cross-validated) from 77.5% to 81.0%. In most runs, the best solution for a locus was consistently represented by only 6 meta-values instead of the 10 original values.

5 Future Work and Conclusion

The novelty of the approach is in adopting a genetic algorithm for clustering the values of variables. In contrast, traditional genetic algorithm approaches for data mining operate on the features as a whole by including/excluding them from a subset or adjusting the appropriate weights. We make a step further and organize the search in the entire input space. The enhanced graph-based encoding has a much less redundant chromosome while maintaining complete coverage of possible partitions, as we strictly prove. Performed experiments demonstrate reduction in measurement complexity and improvement in classification accuracy due to the better reliability of meta-values.

In our future work we want to generalize and apply genetic clustering to a difficult problem of parametric learning in the context of Bayesian networks. Bayesian network classifiers, even augmented to the trees over the feature nodes, requires estimation of a much greater number of parameters than a typical naïve Bayes classifier. We optimistically expect that the conditional probability distribution in a Bayesian network can be considerably simplified and learned from data with the help of GA value clustering.

Acknowledgements. The authors are grateful to Dr. Kim Scribner for providing DNA data samples. The work of Alexander Topchy has been supported by graduate research award from the Research Excellence Fund Center for Biological Modeling at Michigan State University.

References

1. Devijver, P.A. and Kittler, J.: Pattern Recognition: A Statistical Approach. Prentice-Hall International, (1982)
2. Jain A.K., Duin R.P. and Mao J, Statistical pattern recognition: a review. IEEE Transactions on Pattern Analysis and Machine Intelligence 22, (2000) 4–37
3. Freitas A.A.: Data Mining and Knowledge Discovery with Evolutionary Algorithms. Springer-Verlag, (2002)
4. Kohavi R. and John G.: Wrappers for Feature Subset Selection. Artificial Intelligence Journal 97 (1–2), (1997) 273–324
5. Siedlecki W. and Sklansky J.: On automatic feature selection. International Journal of Pattern Recognition and Artificial Intelligence, 2, (1988) 197–220
6. Vafaie H., and De Jong K.: Robust feature selection algorithms. In Proc. of the 5th IEEE International Conference on Tools for Artificial Intelligence, Boston, MA, (1993) 356–363
7. Whitley D., Beveridge R., Guerra C. and Graves C.: Messy Genetic Algorithms for Subset Feature Selection. International Conference on Genetic Algorithms. T. Baeck, ed. Morgan Kaufmann, (1997)
8. Yang J. and Honavar V.: Feature Subset Selection Using a Genetic Algorithm. In: Feature Extraction, Construction, and Subset Selection: A Data Mining Perspective. Motoda, H. and Liu, H. (Eds.) New York, Kluwer, (1998)
9. Punch W.F., Goodman E.D., Pei M., Chia-Shun L., Hovland P. and Enbody R.: Further Research on Feature Selection and Classification Using Genetic Algorithms. In Proc. 5th International Conference on Genetic Algorithms, Urbana-Champaign IL, (1993) 557–562
10. Raymer M., Punch W., Goodman E., Sanschagrin P., and Kuhn L., Simultaneous Feature Extraction and Selection using a Masking Genetic Algorithm. In Proc. of 7th International Conference on Genetic Algorithms (ICGA), San Francisco CA, (1997) 561–567

11. Vafaie H. and DeJong K.: Feature Space Transformation Using Genetic Algorithms. IEEE Intelligent Systems 13(2), (1998) 57–65
12. Lin C. and Wu J.: Automatic facial feature extraction by genetic algorithms. IEEE Trans. on Image Processing, vol. 8(6), (1999) 834–845
13. Raymer M.L., Punch W.F., Goodman E.D., Kuhn L.A. and Jain A.K.: Dimensionality Reduction Using Genetic Algorithms. IEEE Trans. on Evolutionary Computations 4(2), (2000) 164–171
14. Brumby S.P., Theiler J., Perkins S.J., Harvey N.R., Szymanski J.J., Bloch J.J., and Mitchell M.: Investigation of Feature Extraction by a Genetic Algorithm. Proc. SPIE 3812, (1999) 24–31
15. Larsen O., Freitas A.A. and Nievola J.C.: Constructing X-of-N attributes with a genetic algorithm. In Proc. 4th Int. Conf. on Recent Advances in Soft Computing, (2002) 326–331
16. Pudil P. and Novovicová J.: Feature Subset Selection Using a Genetic Algorithm in Feature Extraction. In: Huan Liu, Hiroshi Motoda (eds.): Construction and Selection: A Data Mining Perspective, Kluwer (1998)
17. Martin-Bautista M. and Vila M.-A.: A survey of genetic feature selection in mining issues. In Proceedings of the Congress on Evolutionary Computation (CEC 99), (1999) 13–23
18. Falkenauer E., Genetic Algorithms and Grouping Problems. John Wiley & Son Ltd., (1998)
19. Park Y-J. and Song M-S.: A genetic algorithm for clustering problems. In Proc. 3rd Annual Conf. on Genetic Programming, (1998) 568–575.
20. Trunk, G.V.: A problem of dimensionality: a simple example. IEEE Trans. Patt. Anal. Mach. Intell. 1, (1979) 306–307
21. Minker, J., Wilson, G.A., Zimmerman, B.H., An evaluation of query expansion by the addition of clustered terms for a document retrieval system. Information Storage and Retrieval 8(6), (1972) 329–348
22. Spark-Jones K. and Jackson D.M.: The use of automatically-obtained keyword classifications for information retrieval. Information Processing and Management 5, (1970) 175–201
23. Merzbacher M. and Chu W. W.: Pattern-based clustering for database attribute values. In Proc. of AAAI Workshop on Knowledge Discovery in Databases, Wash., D.C., (1993)
24. Tishby N., Pereira F.C., and Bialek W.: The information bottleneck method. In Proc. of the 37-th Annual Allerton Conference on Communication, Control and Computing, (1999) 368–377
25. Slonim N. and Tishby N.: Agglomerative Information Bottleneck. In Advances in Neural Information Processing Systems (NIPS–12), MIT Press, (1999) 617–623
26. Friedman N., Mosenzon O., Slonim N., and Tishby N.: Multivariate Information Bottleneck. In Proc. of the Seventeenth Conference on Uncertainty in Artificial Intelligence (UAI), (2001)
27. Slonim N., Friedman N., and Tishby N.: Agglomerative Multivariate Information Bottleneck. In Advances in Neural Information Processing Systems (NIPS–14), (2001)
28. O'Connell J.R. and Weeks D.E.: The VITESSE algorithm for rapid exact multilocus linkage analysis via genotype set-recoding and fuzzy inheritance. Nature Genetics 11, (1995) 402–408
29. Friedman N., Geiger D., and Lotner N.: Likelihood Computation with Value Abstraction. In Proc. Sixteenth Conf. on Uncertainty in Artificial Intelligence (UAI), (2000)
30. Chartrand, G. and Oellermann O.R.: Applied and Algorithmic Graph Theory. McGraw-Hill, Inc., New York (1993)
31. Bollob'as B.: Random Graphs. Academic Press, London, (1985)
32. Waser P.M. and Strobeck C.: Genetic signatures of interpopulation dispersal. Trends Ecol Evol 13, (1998) 43–44
33. Guinand, B., Topchy A., Page K.S., Burnham-Curtis M.K., Punch W.F., and Scribner K. T.: Comparisons of likelihood and machine learning methods of individual classification. Journal of Heredity 93(4), (2002) 260–269

The Structure of Evolutionary Exploration: On Crossover, Buildings Blocks, and Estimation-Of-Distribution Algorithms

Marc Toussaint

Institut für Neuroinformatik, Chair of Theoretical Biology
Ruhr-Universität Bochum ND-04, 44780 Bochum, Germany
toussaint@neuroinformatik.rub.de

Abstract. Correlations between alleles after selection are an important source of information. Such correlations should be exploited for further search and thereby constitute the building blocks of evolutionary exploration. With this background we analyze the structure of the offspring probability distribution, or *exploration distribution*, for a simple GA with mutation only and a crossover GA and compare them to Estimation-Of-Distribution Algorithms (EDAs). This will allow a precise characterization of the structure of exploration w.r.t. correlations in the search distribution for these algorithms. We find that crossover transforms, depending on the crossover mask, mutual information between loci into entropy. In total, it can only decrease such mutual information. In contrast, the objective of EDAs is to estimate the correlations between loci and exploit this information during exploration. This may lead to an effective *increase* of mutual information in the exploration distribution, what we define *correlated exploration*.

1 Introduction

In the realm of evolutionary computation the notion of building blocks has been developed in Holland's original works [5,6] to describe the effect of crossover. In that respect, building blocks are composed of genes with more or less linkage between them. This is one to one with the notion of schemata and eventually lead to the schema theories (also first developed in these papers) which describe the evolution of these building blocks.

Since crossover is a biologically inspired concept, Holland's notion of building blocks is also relevant in understanding natural evolution. In the biology literature though, there exists a second notion of building blocks which has quite a different connotation. As a paradigm we choose the following phenomenon. In their experiments, Halder, Callaerts, & Gehring [4] forced the mutation of a single gene, called *eyeless gene*, in early ontogenesis of a Drosophila Melanogaster fly. This rather subtle genotypic variation results in a severe phenotypic variation: An additional functionally complete eye grows at some place it was not supposed to. Here, the notion of a building block refers to the eye as a functional module which can be grown phenotypically by triggering a single gene.

In other words, a single mutation of a gene leads to a highly complex, in terms of cell properties highly correlated phenotypic variation. Such properties of the genotype-phenotype mapping are considered as the basis of complex adaptation [12]. Recently, a theory on the evolution of complex phenotypic variability was proposed [10].

Besides the discussion of crossover in GAs and that of functional modularity in natural evolution, there is a third field of research that relates to the discussion of building blocks: Estimation-of-Distribution Algorithms (EDAs, [8]). These algorithms are a direct implementation of the idea of correlated exploration in the framework of heuristic search algorithms. They explicitly encode the search distribution (i.e., offspring probability distribution) by means of some chosen distribution model, e.g., a product of marginals (PBIL, [1]), dependency trees [2], or a Bayesian network (BOA, [7]). To our point of view, the key of these algorithms is that they are capable to induce this *second* notion of building blocks. For instance, consider a dependency tree where the leaves encode the phenotypic variables. Offspring are generated by *sampling* this probabilistic model, i.e., by first sampling the root variable of the tree, then, according to the dependencies encoded on the links, sampling the root's successor nodes, etc. Now, if we assume that the dependencies are very strong, say, deterministic, it follows that a single variation at the root leads to a completely correlated variation of all leaves. Hence, we may define a set of leaves which, due to their dependencies, always vary in high correlation as a functional phenotypic module in the same sense as for the eyeless paradigm.

What is the principle difference in the exploration induced by crossover in a simple GA and the one we exemplified in the context of biology and EDAs? We will propose a criterion to distinguish these two kinds of exploration depending on whether the exploration distribution can comprise *more* mutual information than the parent population had. We show that this can never be the case for crossover and mutation but give an example, similar to the one just mentioned, where this is the case for an EDA.

After we setup our formalism in the next section, Sects. 3 and 4 will present some theorems on the structure of the search distribution after mutation and crossover. With structure we mean the correlational structure that we measure by means of mutual information. Many of our arguments will be based on the increase and decrease of mutual information in relation to increase or decrease of entropy in the search distribution. Section 5 finally defines the notion of correlated exploration and thereby pinpoints the difference between linkage correlations in crossover GAs and correlated variability in EDAs.

2 Formalism

The Simple GA [11]. We represent a population as a distribution $p \in \Lambda^\Omega$ over genotype space Ω. In this paper we assume that a genotype is composed of a fixed number of genes, $\Omega = \Omega^1 \times \cdots \times \Omega^N$, where the space Ω^i of alleles of the ith gene may be arbitrary. We represent also finite populations as a distribution

$p \in \Lambda^\Omega$ over Ω, namely, if the population is given as a multiset $A = \{x_1, .., x_\mu\}$, we (bijectively) represent it as the *finite distribution* given by $p = \frac{1}{\mu}\sum_{i=1}^{\mu} \delta_{x_i}$ where δ_x is the delta distribution at x, i.e., $p(x) = \frac{|A \cap \{x\}|}{|A|} = \frac{\text{multiplicity of } x \text{ in } A}{|A|}$. Crossover and mutation are represented as operators $\Lambda^\Omega \to \Lambda^\Omega$ that map a parental (finite or infinite) population to an offspring distribution. Given some operator $\mathcal{U} : \Lambda^\Omega \to \Lambda^\Omega$ we will use the notation $\Delta_{\mathcal{U}} B = B(\mathcal{U}p) - B(p)$ to denote the difference of a quantity $B : \Lambda^\Omega \to \mathbb{R}$ under transition, e.g., the quantity may be the entropy $H(p)$ of a distribution.

In that framework we may write the evolution equation of a crossover GA as

$$p^{(t+1)} = \mathcal{S}^\mu \mathcal{F}^{(t)} \mathcal{S}^\lambda \mathcal{M} \mathcal{C} p^{(t)} , \qquad (1)$$

with crossover $\mathcal{C}$, mutation $\mathcal{M}$, offspring sampling $\mathcal{S}^\lambda$, fitness $\mathcal{F}$, and parent sampling $\mathcal{S}^\mu$. A sampling operator $\mathcal{S}^n : \Lambda^\Omega \to \Lambda^\Omega$ draws n independent samples from a distribution and maps this multiset of samples to the respective finite distribution; note that $\lim_{n \to \infty} \mathcal{S}^n = \text{id}$. The sampling operators are the only stochastic operators in this equation. Fitness $\mathcal{F}^{(t)} : \Lambda^\Omega \to \Lambda^\Omega$ re-weights a distribution proportional to some functional $f^{(t)}$ that gives the selection probability, $(\mathcal{F}^{(t)} p)(x) = \frac{f^{(t)}(x)\, p(x)}{\sum_{x'} f^{(t)}(x')\, p(x')}$. (This presumes either "fitness-proportional" selection or that $f^{(t)}$ may arbitrarily depend on the current offspring population.) The concatenation $\mathcal{S}^\mu \mathcal{F}^{(t)}$ is also called selection. We define mutation and crossover more precisely as follows:

Definition 1 (Mutation). *We define mutation as an operator $\mathcal{M} : \Lambda^\Omega \to \Lambda^\Omega$ defined by the conditional probability $\mathcal{M}(y|x)$ of mutating from $x \in \Omega$ to $y \in \Omega$:*

$$(\mathcal{M}p)(y) = \sum_x \mathcal{M}(y|x)\, p(x) .$$

A typical mutation operator fulfills the constraints of symmetry $\mathcal{M}(y|x) = \mathcal{M}(x|y)$ and component-wise independence $\mathcal{M}(x|y) = \prod_{i=1}^{N} \mathcal{M}^i(x^i|y^i)$. In the following we will refer to the simple mutation operator *for which all component-wise mutation operators are such that the probability of mutating from x to y is constant for $x \neq y$:*

$$\forall i : \mathcal{M}^i = \mathcal{M}^*, \quad \forall x \neq y \in \Omega^* : \mathcal{M}^*(x|y) = \frac{\alpha}{n}, \quad \mathcal{M}^*(x|x) = 1 - \frac{\alpha(n-1)}{n},$$

where $n = |\Omega^*|$ and $0 < \alpha \leq 1$ denotes the mutation rate parameter.

Definition 2 (Crossover). *We define crossover as an operator $\Lambda^\Omega \to \Lambda^\Omega$ parameterized by a crossover mask distribution $c \in \Lambda^{\{0,1\}^N}$ over the space $\{0,1\}^N$ of bit-masks, where N is the number of loci (or genes) of a genome in Ω:*

$$\mathcal{C} : \Lambda^\Omega \to \Lambda^\Omega , \quad (\mathcal{C}p)(x) = \sum_{x_0, x_1 \in \Omega} \mathcal{C}(x|x_0, x_1)\, p(x_0)\, p(x_1) ,$$

$$\mathcal{C}(x|x_0, x_1) = \sum_{m \in \{0,1\}^N} c(m)\, [x = x_0 \otimes_m x_1] ,$$

where the ith allele of the m-crossover-product $x_0 \otimes_m x_1$ is the ith allele of the parent x_{m_i}, i.e., $(x_0 \otimes_m x_1)^i = (x_{m_i})^i$. The bracket expression $[A = B]$ equals 1 for $A = B$ and 0 for $A \neq B$. We only consider symmetric crossover, where $c(m) = c(\bar{m})$ and $\bar{m}$ is the conjugate of the bit-string m.

It is important to realize that, in our formalism, crossover and mutation are deterministic operators over the space of distributions. The stochasticity is solely captured by the offspring sampling operator $\mathcal{S}^\lambda$. Hence, when we will derive statements about $\mathcal{M}$ and $\mathcal{C}$ in the following, they will not account for the stochasticity of offspring sampling.

Estimation-Of-Distribution Algorithms. Concerning EDAs, we write their dynamics as
$$y^{(t+1)} = \mathcal{H}(\mathcal{F}\tilde{q}^{(t)}, \tilde{q}^{(t)}, y^{(t)}) \quad \text{where } \tilde{q}^{(t)} = \mathcal{S}^\lambda \Phi y^{(t)} \;,$$
where, instead of a parent population, some other parameters $y^{(t)}$ (e.g., a Bayesian graph or dependency tree) determine the offspring distribution $\Phi y^{(t)}$, which is sampled to the offspring population $\tilde{q}^{(t)}$, evaluated, and, instead of a simple parent sampling, mapped back on new parameters $y^{(t+1)}$ by some update operator $\mathcal{H}$. The operator $\mathcal{H}$ is called *heuristic rule* and, in the case of Estimation-of-Distribution Algorithms, is such that the new search distribution $\Phi y^{(t+1)}$ *estimates* the experienced fitness distribution $\mathcal{F}^{(t)} \mathcal{S}^\lambda \Phi y^{(t)}$. The generic implementation of this idea is
$$y^{(t+1)} = y^* = \mathcal{E}(\mathcal{F}^{(t)} \mathcal{S}^\lambda \Phi y^{(t)}) \;, \quad \text{where } \mathcal{E}(p) = \operatorname*{argmin}_{y \in Y} D(p \,\|\, \Phi y) \;. \quad (2)$$
We call $\mathcal{E}$ estimation, Y is the space of feasible parameters y, and $D(\,\cdot\,\|\,\cdot\,)$ denotes the Kullback-Leibler distance. In fact, the MIMIC algorithm [3], which uses a dependency chain to parameterize the search distribution, realizes exactly this scheme. Other algorithms [7,2,1] differ in some details, e.g., they use distance measures other than the Kullback-Leibler divergence or realize a gradual adaptation of continuous parameters y of the style "$y^{(t+1)} = \alpha\, y^* + (1-\alpha)\, y^{(t)}$". See [10] for a survey on the relation between EDAs and the evolution of genetic representations (σ-*evolution*) in the context of non-trivial genotype-phenotype mappings.

3 The Structure of the Mutation Distribution

This section derives a theorem that simply states that mutation increases entropy and decreases mutual information. (It is surprising how non-trivial it is to prove this intuitively trivial statement.)

Lemma 1 (Component-wise mutation). *Consider the component-wise simple mutation operator $\mathcal{M}^*$ as given in Definition 1. It follows that*

a)
$$\mathcal{M}^* p(x) = (1-\alpha)\, p(x) + \alpha\, \frac{1}{n} \;,$$

which is a linear mixture between p and the uniform distribution ("$\frac{1}{n}$") with mixture parameter α.

b) *For every non-uniform population p, the entropy of $\mathcal{M}^*p$ is greater than the entropy of p,*
$$H(\mathcal{M}^*p) > H(p).$$

Proof. a)
$$\mathcal{M}^*p(x) = \left[\sum \frac{\alpha}{n} p(y)\right] - \frac{\alpha}{n} p(x) + \left(1 - \frac{\alpha(n-1)}{n}\right) p(x) = \frac{\alpha}{n} + (1-\alpha) p(x).$$
b) We generally show that the entropy increases if you mix a distribution with the uniform distribution. We prove this by considering the first two derivatives of the entropy functional with respect to the mixture parameter α. Let
$$q(x) = (1-\alpha) p(x) + \frac{\alpha}{n},$$
and recall $H(q) = -\sum_x q(x) \ln q(x)$ and $(X \ln X)' = X'((\ln X) + 1)$. It follows
$$\frac{\partial}{\partial \alpha} H(q) = -\sum_x \left[-p(x) + \frac{1}{n}\right] (\ln q(x) + 1) = \sum_x \left[p(x) - \frac{1}{n}\right] \ln q(x),$$
$$\frac{\partial}{\partial \alpha} H(q)\big|_{\alpha=1} = \sum_x \left[p(x) - \frac{1}{n}\right] \ln \frac{1}{n} = 0,$$
$$\frac{\partial^2}{\partial \alpha^2} H(q) = -\sum_x \frac{(p(x) - \frac{1}{n})^2}{q(x)} < 0 \quad \text{if } p \text{ is non-uniform.}$$

What we found is that (i) the entropy is maximal for the extreme case $\alpha = 1$ since its derivative w.r.t. α at this point vanishes (of course, this corresponds to the case where q becomes the uniform distribution) and (ii) the second derivative is always negative if p is non-uniform. Hence, the plot of H versus α is comparable to an upside-down parabola with maximum at $\alpha = 1$. It follows that for all $\alpha < 1$ (to the left of the maximum) the derivative $\frac{\partial}{\partial \alpha} H(q)$ is positive. Entropy continuously increases with α. And hence, for every $0 < \alpha \leq 1$ and every non-uniform population p, $H(\mathcal{M}^*p) > H(p)$. □

Theorem 1. *Consider the simple mutation operator $\mathcal{M}(x|y) = \prod_i \mathcal{M}^*(x^i|y^i)$ as given in Definition 1. If $p \in \Lambda^\Omega$ is non-uniform it follows that entropy increases, $H(\mathcal{M}p) > H(p)$, and mutual information decreases, $I(\mathcal{M}p) < I(p)$.*

Proof. We first prove that the cross entropy decreases. Assuming only two genes, the compound mutation distributions reads
$$\mathcal{M}p(x,y) = (1-\alpha)^2 p(x,y) + (1-\alpha)\alpha p(x) \frac{1}{n} + (1-\alpha)\alpha \frac{1}{n} p(y) + \alpha^2 \frac{1}{n}\frac{1}{n}$$
$$= (1-\alpha)\left[(1-\alpha) p(x,y) + \alpha \frac{1}{n} p(x)\right] + \alpha \frac{1}{n}\left[(1-\alpha) p(y) + \alpha \frac{1}{n}\right]$$
$$= (1-\alpha) q(x,y) + \alpha \frac{1}{n} q(y),$$
where $q(x,y) = (1-\alpha) p(x,y) + \frac{\alpha}{n} p(x)$, $q(x) = p(x)$, $q(y) = (1-\alpha) p(y) + \frac{\alpha}{n}$

We call q a one-component α-mixture since only in one component the uniform distribution was mixed to p. This shows that the compound distribution $\mathcal{M}p$ for two genes is a one-component α-mixture of a distribution q, which is itself a one-component α-mixture. For compound distributions with more than two genes this will be recursively the case and generally the mutation operator can be expressed as concatenation of one-component α-mixtures. Hence, it suffices when we prove that the mutual information decreases for one such step of one-component α-mixing.

We use the same technique of calculating derivatives with respect to the mixture parameter to prove decreasing cross entropy. To simplify the notation we use the abbreviations:

$$A=q(x,y)\,,\ A\big|_{\alpha=1}=\frac{\alpha\,p(x)}{n}\,,\ A'=\frac{\partial}{\partial\alpha}A=-p(x,y)+\frac{p(x)}{n}\,,\ A''=0\,,\ B''=0\,,$$

$$B=q(x)\,q(y)=p(x)\Big[(1-\alpha)\,p(y)+\frac{\alpha}{n}\Big]\,,\ B\big|_{\alpha=1}=A\big|_{\alpha=1}\,,\ B'=p(x)\,\Big(-p(y)+\frac{1}{n}\Big)\,.$$

With these abbreviations (keeping the dependencies on x, y, and α in mind) we can write:

$$I(q)=\sum_{x,y}A\ln\frac{A}{B}\,,\quad \frac{\partial}{\partial\alpha}I(q)=\sum_{x,y}\Big[A'\ln\frac{A}{B}+A'-\frac{A\,B'}{B}\Big]$$

$$\frac{\partial}{\partial\alpha}I(q)\big|_{\alpha=1}=\sum_{x,y}\Big[A'\big|_{\alpha=1}\ln 1+\Big[-p(x,y)+\frac{p(x)}{n}\Big]-\Big[p(x)\,(-p(y)+\frac{1}{n})\Big]\Big]=0$$

$$\frac{\partial^2}{\partial\alpha^2}I(q)=\sum_{x,y}\Big[A'\frac{B}{A}\Big[\frac{A'}{B}-\frac{A\,B'}{B^2}\Big]+0-\frac{A'B'}{B}+\frac{A\,(B')^2}{B^2}\Big]$$

$$=\sum_{x,y}\Big[\frac{(A')^2}{A}-2\frac{A'B'}{B}+\frac{A\,(B')^2}{B^2}\Big]=\sum_{x,y}\Big[\frac{(B\,A'-A\,B')^2}{A\,B^2}\Big]\geq 0$$

So, what we found is that (i) for $\alpha=1$ the cross entropy is minimal since its derivative w.r.t. α at this point vanishes (of course, this corresponds to the case where $q(x,y)=p(x)\frac{1}{n}$) and (ii) for all other points the second derivative is positive. The plot of I versus α is comparable to an upwards parabola with minimum at $\alpha=1$. It follows that for $\alpha<1$ (to the left of the minimum) the derivative $\frac{\partial}{\partial\alpha}I(q)$ is negative and thus the cross entropy continuously decreases with increasing α.

Concerning increasing entropy, it is obvious that the marginals of the mutation distribution $\mathcal{M}p$ are simply $(\mathcal{M}p)^i=\mathcal{M}^*p^i$. For the component-wise mutation operators we proved that entropy increases (for non-zero α and non-uniform p) and thus $\Delta_\mathcal{M}H^i>0$. Consequently, $\Delta_\mathcal{M}H=\sum_i\Delta_\mathcal{M}H^i-\Delta_\mathcal{M}I>0$. □

4 The Structure of the Crossover Distribution

What is the structure of the crossover search distribution $\mathcal{C}p$, given $p\in\Lambda^\Omega$ and $c\in\Lambda^{\{0,1\}^N}$? The first theorem can directly be derived from our definition of the crossover operator. It captures the most basic properties of the crossover operator with respect to the correlations it *destroys* in the search distribution:

Theorem 2. *Let $H(p)$, p^i, $H^i(p) = H(p^i)$, and $I(p) = \sum_i H^i(p) - H(p)$ denote the entropy, the ith marginal distribution, the marginal entropies, and the mutual information of a distribution p. For any crossover operator C and any population p it holds*
a) $\forall i : (Cp)^i = p^i$, $\Delta_C H^i = 0$, *i.e., the marginals and hence their entropies do not change,*
b) $\Delta_C I = -\Delta_C H \le 0$, *i.e., the increase of entropy is equal to the decrease of mutual information.*

Proof. Let us first calculate the marginals after crossover. Let a be an allele of the ith gene.

$$(Cp)^i(a) = \sum_{x_0, x_1} \sum_m c(m) \, [a = (x_{m_i})^i] \, p(x_0) \, p(x_1) ,$$

$$= \sum_{x_0, x_1} \left[\sum_{m : m_i = 0} c(m) \, [a = (x_0)^i] + \sum_{m : m_i = 1} c(m) \, [a = (x_1)^i] \right] p(x_0) \, p(x_1) ,$$

$$= p^i(a) \left[\sum_{m : m_i = 0} c(m) \right] + p^i(a) \left[\sum_{m : m_i = 1} c(m) \right] = p^i(a) .$$

Since the marginals are not changed by crossover, the marginal entropies do not change either. Statement *b)* follows from the definition of the mutual information:

$$\Delta_C H + \Delta_C I = H(Cp) - H(p) + I(Cp) - I(p)$$
$$= H(Cp) - H(p) + \sum_i H^i(Cp) - H(Cp) - \left[\sum_i H^i(p) - H(p) \right]$$
$$= \sum_i H^i(Cp) - \sum_i H^i(p) = 0 . \qquad \square$$

The following theorem makes this more concrete when focusing on two specific genes (generally, two arbitrary subparts of arbitrary length) of a genome. We calculate the mutual information between these two genes in the search distribution Cp—which is a measure for the *linkage* between them. Let it be the ith and jth gene. We use a and b as alleles; $p^{ij}(a,b) = \sum_{x \in \Omega} [x^i = a] \, [x^j = b] \, p(x)$ denotes the probability that the ith gene has allele a and the jth gene allele b. Analogously, let c^{ij} be the marginal of the crossover mask distribution with respect to the two genes, i.e., $c^{ij}_{01} = \sum_{m \in \{0,1\}^N} [m^i = 0] \, [m^j = 1] \, c(m)$.

Theorem 3. *For any crossover operator C and any population p it holds:*
a) *The compound distribution of two genes after crossover is given by*
$$(Cp)^{ij}(a,b) = 2 c^{ij}_{00} \, p^{ij}(a,b) + 2 c^{ij}_{01} \, p^i(a) \, p^j(b) ,$$
i.e., a linear combination of the original compound distribution $p^{ij}(a,b)$ and the decorrelated product distribution $p^i(a) \, p^j(b)$.
b) *The mutual information $I(Cp)^{ij}$ in the compound distribution of two specific genes is*
$$I(Cp)^{ij} = \sum_{a,b} \left(2 c^{ij}_{00} \, p^{ij}(a,b) + 2 c^{ij}_{01} \, p^i(a) p^j(b) \right) \ln \left(2 c^{ij}_{00} \frac{p^{ij}(a,b)}{p^i(a) p^j(b)} + 2 c^{ij}_{01} \right) ,$$

c) and we have
$$0 \leq 2c_{00}^{ij}\left(I(p)^{ij} + \ln(2c_{00}^{ij})\right) \leq I(Cp)^{ij} \leq I(p)^{ij}.$$

The two left $\leq$ are exact for complete crossover, $c_{00}^{ij} = 0$, $c_{01}^{ij} = \frac{1}{2}$, the right $\leq$ is exact for no crossover, $c_{00}^{ij} = \frac{1}{2}$, $c_{01}^{ij} = 0$.

Proof. a)
$$Cp^{ij}(a,b) = \sum_{x_0,x_1}\sum_m c(m)\,[(x_{m_0})^0 = a]\,[(x_{m_1})^1 = b]\,p(x_0)\,p(x_1)$$
$$= \sum_{x_0,x_1}\left(c_{00}^{ij}\,[(x_0)^0 = a][(x_0)^1 = b] + c_{01}^{ij}\,[(x_0)^0 = a][(x_1)^1 = b] + \right.$$
$$\left. c_{10}^{ij}\,[(x_1)^0 = a][(x_0)^1 = b] + c_{11}^{ij}\,[(x_1)^0 = a][(x_1)^1 = b]\right)p(x_0)\,p(x_1)$$
$$= 2\sum_{x_0} c_{00}^{ij}\,[(x_0)^0 = a][(x_0)^1 = b]\,p(x_0)$$
$$+ 2\sum_{x_0,x_1} c_{01}^{ij}\,[(x_0)^0 = a][(x_1)^1 = b]\,p(x_0)\,p(x_1)$$
$$= 2c_{00}^{ij}\,p^{ij}(a,b) + 2c_{01}^{ij}\,p^i(a)\,p^j(b).$$

b&c)
$$I(Cp)^{ij} = H(Cp^i) + H(Cp^j) - H(Cp) = H(p^i) + H(p^j) - H(Cp)$$
$$\leq H(p^i) + H(p^j) - H(p) = I(p)^{ij}$$
$$H(Cp) = -\sum_{a,b}(Cp)^{ij}(a,b)\left[\ln\left(2c_{00}^{ij}\frac{p^{ij}(a,b)}{p^i(a)p^j(b)} + 2c_{01}^{ij}\right) - \ln p^i(a) - \ln p^j(b)\right]$$
$$= -\sum_{a,b}(Cp)^{ij}(a,b)\left[\ln\left(2c_{00}^{ij}\frac{p^{ij}(a,b)}{p^i(a)p^j(b)} + 2c_{01}^{ij}\right)\right] + H(p^i) + H(p^j)$$
$$I(Cp)^{ij} = \sum_{a,b}\left(2c_{00}^{ij}\,p^{ij}(a,b) + 2c_{01}^{ij}\,p^i(a)\,p^j(b)\right)\ln\left(2c_{00}^{ij}\frac{p^{ij}(a,b)}{p^i(a)p^j(b)} + 2c_{01}^{ij}\right)$$
$$\geq \sum_{a,b}\left(2c_{00}^{ij}\,p^{ij}(a,b)\right)\ln\left(2c_{00}^{ij}\frac{p^{ij}(a,b)}{p^i(a)p^j(b)}\right) = 2c_{00}^{ij}\left(I(p)^{ij} + \ln(2c_{00}^{ij})\right)\quad \Box$$

Let us summarize what we actually found in the above theorems:
- The marginal distributions do not change at all. There is no exploration w.r.t. the alleles of single genes.
- The more entropy crossover introduces in a population, the more the mutual dependencies between alleles are destroyed. Actually, crossover destroys mutual information in the parent population by *transforming* it into entropy in the crossed population. In particular, if there is no mutual information in the parent population, crossover will not generate any more entropy. That's linkage equilibrium.

- The last theorem shows how the crossover mask distribution c determines *which* correlations are destroyed and transformed into entropy.

The purpose of these theorems is to propose a probably non-standard point of view on what crossover actually does: Actually, a *non*-crossover GA comprises the strongest and most natural building blocks; individuals as such are the building blocks that carry the mutual information between their alleles. Crossover is a means to break these maximal building blocks apart into smaller pieces by converting mutual dependencies into entropy. As a result it induces smaller, more fine-grained building blocks with, in total, less mutual information in the crossed population. Hence, the correlational structure in the crossed population is not more complex—it is simpler since it carries less information. In the limit of linkage equilibrium (or uniform c), all correlations have been destroyed and the crossed population becomes a product distribution.

5 Correlated Exploration and EDAs

Exploration essentially means to add entropy to the search distribution, $\Delta H > 0$. For instance, mutation typically adds entropy to the search distribution by adding independent noise to each marginal. However, adding independent noise reduces the mutual information between alleles, $\Delta I < 0$, see Lemma 1. Using crossover to add entropy, Theorem 2b tells us that all the entropy is added at the expense of mutual information, $\Delta H = -\Delta I > 0$. Generally, it seems difficult to add entropy to a distribution without destroying mutual information. But, instead of only preserving the mutual information that exists in the parent population, we could go even further and ask: How could one *extrapolate* this mutual information from the parent population to the new explorations, i.e., how could one ensure that the exploration, measured by $\Delta H > 0$, comprises the same structural correlations that have been present in the parent population such that in total $\Delta I > 0$? A possibility is to first estimate the structure of the mutual information in the parent population and then to add entropy while respecting that same structure. In our view, this is the core of EDAs (except for those that do not estimate correlations, like the PBIL [1]).

Let us consider a simple example that shows how an EDA, similar to the MIMIC [3], can in principle realize this latter kind of correlated exploration:

Example 1. Consider a two gene scenario with $\Omega = \Omega^* \times \Omega^*$, $\Omega^* = \{0, 1, 2, 3, 4\}$. As a distribution model consider the dependency chain $p(x^0, x^1) = p(x^0) p(x^1|x^0) \in \Lambda^\Omega$ with two parameters $\alpha \in \Omega^*$, $\beta \in [0, 1]$ and

$$p(x^0) = \begin{cases} 1/2 & x^0 = \alpha \\ 1/8 & \text{otherwise} \end{cases}, \quad p(x^1|x^0) = \begin{cases} 1 - 4\beta & x^1 = x^0 \\ \beta & \text{otherwise} \end{cases}.$$

Let the parent population comprise the three individuals $\{(0,0), (1,1), (3,3)\}$. An EDA would estimate a deterministic dependence $\beta = 0$ and $\alpha = 0, 1,$ or 3, which lead to the same minimal Kullback-Leibler divergence within the distribution model. The mutual information in the parent population is $I = H^0 + H^1 - H = \log_2 3 + \log_2 3 - \log_2 3 \approx 1.6$. The EDA search distribution p has mutual

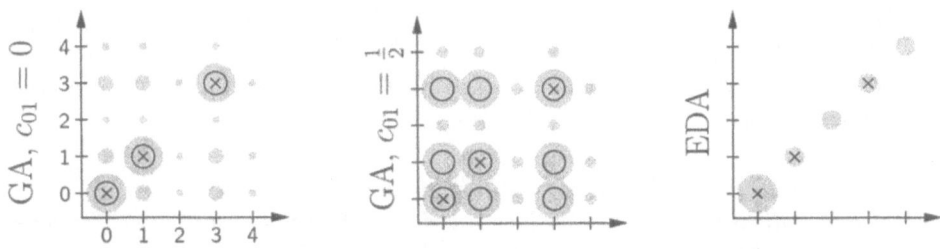

× individuals in the finite parent population p
○ individuals in the crossed population $\mathcal{C}p$
● exploration distribution $\mathcal{MC}p$ respectively $\Phi(\mathcal{E}(p))$, radius ≈ probability

Fig. 1. Illustration of the types of correlations in GAs with and without crossover in comparison to correlated exploration in EDAs. The search space $\Omega = \{0, 1, 2, 3, 4\}^2$ is composed of two genes (in the case of bit-strings these loci would refer to several bits and c_{01} denoted the probability for 1-point crossover between these groups of bits). The radius of the gray shade circles indicates the probabilities in the exploration distribution. The degree to which the gray shading is aligned with the bisecting line indicates correlatedness. The crossover GA in the middle destroys correlations whereas EDAs may induce high correlations, see Example 1

information $I = H^0 + H^1 - H = 2 + 2 - 2 = 2$. Hence the mutual information as well as entropy is increased, $\Delta I > 0$, $\Delta H > 0$.

The example is illustrated and compared to crossover and mutation in Fig. 1. In a finite population of 3 individuals, marked by crosses, the values at the two loci are correlated, here illustrated by plotting them on the bisecting line. The crossed population $\mathcal{C}p$ comprises at most 9 different individuals; in the special cases $c_{01}^{ij} = 0$ and $c_{01}^{ij} = \frac{1}{2}$ the population is even finite and comprises 3 respectively 9 equally weighted individuals marked by circles. Mutation adds independent noise, illustrated by the gray shading, to the alleles of each individual. The two illustrations for the GA demonstrate that crossover destroys correlations between the alleles in the initial population: the gray shading is not focused on the bisecting line. Instead, an EDA can first estimate the distribution of the individuals in p. Depending on what probabilistic model is used, this model can capture the correlations between the alleles; in the illustration the model could correspond to Example 1 and the estimation of the correlations in p leads to the highly structured search distribution which comprises more mutual information than the parent population.

We capture this difference in the following definition:

Definition 3 (Correlated exploration). *Let $\mathcal{U} : \Lambda^\Omega \to \Lambda^\Omega$ be an operator. The following conditions need to hold for almost all $p \in \Omega$ which means: for all the space Ω except for a subspace of measure zero. We define*
- *$\mathcal{U}$ is explorative $\iff \Delta_\mathcal{U} H > 0$ for almost all $p \in \Omega$,*

- $\mathcal{U}$ is marginally explorative $\iff$ $\mathcal{U}$ is explorative and $\exists i:\ \Delta_{\mathcal{U}} H^i > 0$ for almost all $p \in \Omega$,
- $\mathcal{U}$ is correlated explorative $\iff$ $\mathcal{U}$ is explorative and $\Delta_{\mathcal{U}} I > 0$, or equivalently $0 < \Delta_{\mathcal{U}} H < \sum_i \Delta_{\mathcal{U}} H^i$, for almost all $p \in \Omega$.

Corollary 1. *From this definition it follows that*
a) If and only if there exist two loci i and j such that the marginal crossover mask distribution c_{01}^{ij} for these two loci is non-vanishing, $c_{01}^{ij} = c_{10}^{ij} > 0$, then crossover $\mathcal{C}$ is explorative. For every mask distribution $c \in \Lambda^{\{0,1\}^N}$, crossover $\mathcal{C}$ is neither marginally nor correlated explorative.
b) Simple mutation $\mathcal{M}$ is marginally but not correlated explorative.
c) $\mathcal{M} \circ \mathcal{C}$ is marginally but not correlated explorative.
d) EDAs can be correlated explorative.

Proof. a) That $\mathcal{C}$ is neither marginally nor correlated explorative follows directly from Theorem 2a, which says that for every $c \in \Lambda^{\{0,1\}^N}$ and any population $p \in \Lambda^{\Omega}$ the marginals of the population do not change under crossover, $\Delta_{\mathcal{C}} H^i = 0$. But under which conditions is $\mathcal{C}$ explorative?

If, for two loci i and j, c_{01}^{ij} is non-vanishing, it follows that $\mathcal{C}$ reduces the mutual information between these two loci (Theorem 3c). The subspace of populations p that do not have any mutual information I^{ij} between these two loci is of measure zero. Hence, for almost all p, $\Delta_{\mathcal{C}} I^{ij} < 0$ and, following Theorem 2b this automatically leads to an increase of entropy $\Delta_{\mathcal{C}} H^{ij} > 0$ in the compound distribution of the two loci and, since $\Delta_{\mathcal{C}} H \geq \Delta_{\mathcal{C}} H^{ij}$, also of the total entropy.

The other way around, if, for every two loci i and j, c_{01}^{ij} vanishes it follows that there is no crossover, i.e., only the all-0s and all-1s crossover masks have non-vanishing probability. Hence, $\mathcal{C} = \mathrm{id}$ and is not explorative.

b) In Lemma 1 we prove that for every non-uniform population p $\Delta_{\mathcal{M}} H > 0$, $\Delta_{\mathcal{M}} H^i > 0$, and $\Delta_{\mathcal{M}} I < 0$.

c) Since both mutation and crossover are not correlated explorative, it follows that their composition is also not correlated explorative:
$$\Delta_{\mathcal{C}} I \leq 0,\ \Delta_{\mathcal{M}} I \leq 0 \quad \Rightarrow \quad \Delta_{\mathcal{M}\mathcal{C}} I \leq 0.$$

d) Example 1 demonstrates this possibility. □

Finally, if crossover and mutation cannot be correlated exploration, how can biology realize correlated exploration as we mentioned it in the introduction? In nature there exists a *non-trivial* genotype-phenotype mapping (see [10] for the concept of non-trivial genotype-phenotype mappings). The assumptions we made about the mutation operator (component-wise independence) refer to the genotype space, not to the phenotype space: On the genotype space mutation kernels are product distributions and mutative exploration is marginally explorative but not correlated; projected on phenotype space, the mutation kernels are in general not anymore product distributions and hence phenotypic mutative exploration can be correlated. The same arguments hold for crossover. In the language of [10], the definition of mutation and of crossover do not commute with phenotype equivalence. Thus, mutation as well as crossover can be *phenotypically* correlated explorative. See [9] for a demonstration of evolution of complex phenotypic exploration distributions.

6 Conclusions

The evolutionary process, as given in Eq. (1) is a succession of increase and decrease of entropy in the population. The fitness operator adds information to the process by decreasing the entropy (it typically maps a uniform finite distribution on a non-uniform with same support). And crossover and mutation add entropy in order to allow for further exploration.

If the crossover mask distribution is well adapted to the problem at hand, crossover can be understood as a tool to freely regulate where mutual information between loci is preserved and where it is decreased. However, we proved that crossover can never *increase* the mutual information between loci in the search distribution compared to what has been present in the parent population.

Why should one intent to increase the mutual information? The idea is that the mutual information in the parent population, which is an important source of information about the problem at hand, can be exploited for further exploration. One way of exploitation is to extrapolate this mutual information from the parent populations to search distribution. This means, that the exploration, measured by $\Delta H > 0$, exhibits the same correlational structure as the parent population such that in total the mutual information in the search distribution will be greater than the one in the parent population.

Our definition of *correlated exploration* distinguishes algorithms depending on whether they can or cannot increase the mutual information. We proved that crossover and mutation cannot be correlated explorative while EDAs can.

There is another (well-known) difference between EDAs and (crossover) GAs with respect to the self-adaptation of the exploration distribution. EDAs always adapt their search distribution (including correlations) according to the distribution of previously selected solutions. In contrast, the crossover mask distribution, that determines where correlations are destroyed or not destroyed, is usually not self-adaptive.

Finally, we mentioned how correlated exploration by means of mutation and crossover is possible (e.g., in natural evolution) when accounting for non-trivial genotype-phenotype mappings. In [10,9] we present a theory and a demonstration of the self-adaptation of complex phenotypic exploration distributions.

References

1. S. Baluja. Population-based incremental learning: A method for integrating genetic search based function optimization and competitive learning. Technical Report CMU-CS-94-163, Comp. Sci. Dep., Carnegie Mellon U., 1994.
2. S. Baluja and S. Davies. Using optimal dependency-trees for combinatorial optimization: Learning the structure of the search space. In *Proceedings of the Fourteenth Int. Conf. on Machine Learning (ICML 1997)*, pages 30–38, 1997.

3. J.S. de Bonet, C.L. Isbell, Jr., and P. Viola. MIMIC: Finding optima by estimating probability densities. In M. C. Mozer, M. I. Jordan, and T. Petsche, editors, *Advances in Neural Information Processing Systems*, volume 9, page 424. The MIT Press, 1997.
4. G. Halder, P. Callaerts, and W. Gehring. Induction of ectopic eyes by targeted expression of the eyeless gene in Drosophila. *Science*, 267:1788–1792, 1995.
5. J. Holland. *Adaptation in Natural and Artificial Systems*. University of Michigan Press, Ann Arbor, USA, 1975.
6. J.H. Holland. Building blocks, cohort genetic algorithms, and hyperplane-defined functions. *Evolutionary Computation*, 8:373–391, 2000.
7. M. Pelikan, D.E. Goldberg, and E. Cantú-Paz. Linkage problem, distribution estimation, and Bayesian networks. *Evolutionary Computation*, 9:311–340, 2000.
8. M. Pelikan, D.E. Goldberg, and F. Lobo. A survey of optimization by building and using probabilistic models. Technical Report IlliGAL-99018, Illinois Genetic Algorithms Laboratory, 1999.
9. M. Toussaint. Demonstrating the evolution of complex genetic representations: An evolution of artificial plants. In *2003 Genetic and Evolutionary Computation Conference (GECCO 2003)*, 2003. In this volume.
10. M. Toussaint. On the evolution of phenotypic exploration distributions. In C. Cotta, K. De Jong, R. Poli, and J. Rowe, editors, *Foundations of Genetic Algorithms 7 (FOGA VII)*. Morgan Kaufmann, 2003. In press.
11. M.D. Vose. *The Simple Genetic Algorithm*. MIT Press, Cambridge, 1999.
12. G.P. Wagner and L. Altenberg. Complex adaptations and the evolution of evolvability. *Evolution*, 50:967–976, 1996.

The Virtual Gene Genetic Algorithm

Manuel Valenzuela-Rendón

ITESM, Monterrey
Centro de Sistemas Inteligentes
C.P. 64849, Monterrey, N.L., México
valenzuela@itesm.mx
http://www-csi.mty.itesm.mx/~mvalenzu

Abstract. This paper presents the *virtual gene genetic algorithm* (vgGA) which is a generalization of traditional genetic algorithms that use binary linear chromosomes. In the vgGA, traditional one point crossover and mutation are implemented as arithmetic functions over the integers or reals that the chromosome represents. This implementation allows the generalization to virtual chromosomes of alphabets of any cardinality. Also, the sites where crossover and mutation fall can be generalized in the vgGA to values that do not necessarily correspond to positions between bits or digits of another base, thus implementing *generalized digits*. Preliminary results that indicate that the vgGA outperforms a GA with binary linear chromosomes on integer and real valued problems where the underlying structure is not binary are presented.

1 Introduction

Traditional genetic algorithms evolve populations of individuals that are represented by linear chromosomes defined in a small cardinality alphabet, usually binary [1,2]. Traditional crossover and mutation create new individuals by manipulating these bit strings. This paper shows that traditional one point crossover and mutation can be simulated as arithmetic functions over the integers represented by the binary chromosomes. In this way, a genetic algorithm for integer individuals where traditional operation are performed, not over the genotype, but rather simulated over the phenotype, can be implemented.

Binary chromosomes can be used to represent real values in many ways; one of the simplest is the use of a linear mapping [1, page 82]. If a linear mapping is used, traditional one point crossover and mutation can also be simulated as arithmetic functions over the reals represented by the binary chromosomes. Even though there is a large body of work which explores the use of real valued individuals in an evolutionary algorithm (see for example [3,4,5,6]), most of this work is oriented at creating new operators with effects that cannot be easily seen as manipulations of the bit representations of the individuals, and therefore is not directly related to the work here presented.

The basic idea of the paper is then generalized. Points where crossover or mutation can occur can be visualized not only as bit positions (or digit positions in the case of non-binary chromosomes) but rather as the value these bits or digits represent, and therefore crossover and mutation can occur at generalized values, what we will call, at *generalized digits*.

In the rest of this paper, we will see the mathematical basis of the *virtual gene genetic algorithm* (vgGA), an algorithm that implements crossover and mutation as arithmetic functions of the phenotype of the individuals, and preliminary experiments that show that the vgGA can outperform a traditional genetic algorithm in problems with an underlying that is not binary.

2 Traditional Crossover and Mutation

Let p be an integer represented by a binary chromosome of length N, where the string of all zeros, $000\cdots 0_2$, represents the integer zero, and the string of all ones, $111\cdots 1_2$, represents the integer $2^N - 1$. The *lower part* of p below and including bit m can be obtained as

$$L_m(p) = p \bmod 2^m, \tag{1}$$

where

$$x \bmod y = \begin{cases} x - y \lfloor x/y \rfloor, & \text{if } y \neq 0; \\ x, & \text{if } y = 0; \end{cases} \tag{2}$$

The *higher part* of p above bit m can be obtained as

$$H_m(p) = p - L_m(p) = p - p \bmod 2^m = 2^m \lfloor p/2^m \rfloor. \tag{3}$$

By substituting an arbitrary base B for 2, the above formulas can be generalized to chromosomes in an alphabet of cardinality B.

$$L_m(p) = p \bmod B^m. \tag{4}$$

$$H_m(p) = p - p \bmod B^m = B^m \lfloor p/B^m \rfloor. \tag{5}$$

Using the lower and higher parts of an integer individual, it is possible to extract parts of a chromosome. The value represented by the digits of the higher part, which we will call the *higher part value* of p above bit m can be obtained as

$$\hat{H}_m(p) = \frac{H_m(p)}{B^m} = \lfloor p/B^m \rfloor. \tag{6}$$

The m-th digit (where the least significant digit is numbered 1, and the most significant digit is numbered N) can be obtained as

$$\text{digit}_m(p) = L_1(\hat{H}_{m-1}(p)) = \hat{H}_{m-1}(L_m(p)). \tag{7}$$

The *segment* of digits $m_1 + 1, m_1 + 2, \ldots, m_2$ can be obtained as

$$\text{segment}_{m_1,m_2}(p) = L_{\Delta m}(\hat{H}_{m_1}(p)), \tag{8}$$

where $\Delta m = m_2 - m_1$.

With the definitions of lower part and higher part, we can now express the crossover of two chromosomes over an alphabet of cardinality B as an arithmetic operation over the integers these chromosomes represent. Let p_1 and p_2 be two integers over base B, one point crossover produces two offspring h_1 and h_2 that can be expressed in the following way:

$$h_1 = \text{crossover}_m(p_1, p_2) = L_m(p_1) + H_m(p_2); \tag{9}$$

$$h_2 = \text{crossover}_m(p_2, p_1) = L_m(p_2) + H_m(p_1). \tag{10}$$

Therefore, one point crossover is simply the exchange between two integers of their lower and higher parts at a given crossover point. A simplified expression for crossover can be obtained by substituting the expressions for lower and higher part, obtaining the following:

$$h_1 = \text{crossover}_m(p_1, p_2) = p_2 + \chi_m(p_1, p_2); \tag{11}$$

$$h_2 = \text{crossover}_m(p_2, p_1) = p_1 - \chi_m(p_1, p_2); \tag{12}$$

where $\chi_m(p_1, p_2) = p_1 \bmod B^m - p_2 \bmod B^m = -\chi_m(p_2, p_1)$.

In traditional mutation for binary chromosomes, mutation of a bit is the same as complementing its value, in other words, flipping the bit from 1 to 0, or from 0 to 1. For alphabets of higher cardinality, the most natural definition of mutation of a digit is to replace it with a random value that is *not* the original value in that position. To facilitate its application when non-binary chromosomes are used, we define mutation in a slightly different manner. We will define mutation as the operation that given an integer p, removes a segment of consecutive digits, and replaces it with a random segment of the same number of digits. The mutation of the segment of digits $m_1 + 1, m_1 + 2, \ldots, m_2$ of an integer p can be expressed as

$$\text{mutation}_{m_1,m_2}(p) = L_{m_1}(p) + H_{m_2}(p) + B^{m_1} \lfloor B^{\Delta m} \text{rand}() \rfloor, \tag{13}$$

where $\Delta m = m_2 - m_1$ and rand() is a function that generates a random number in $[0, 1)$ with uniform distribution.

3 Generalized Crossover and Mutation

The concepts of lower part and higher part presented above were defined in terms of the m-th digit. The formulas include the term B^m, which is the *weight* of bit $m + 1$. A generalization of these formulas can be produced by substituting B^m with n, an integer that is not necessarily an integer power of B. Let us define *generalized lower part* and *generalized higher part* as follows:

$$L(p, n) = p \bmod n; \tag{14}$$

$$H(p, n) = p - p \bmod n = n \lfloor p/n \rfloor. \tag{15}$$

Notice that $L(p, n)$ and $H(p, n)$ refer to the generalized lower and higher parts, and that $L_m(p)$ and $H_m(p)$ refer to the lower and higher parts.

We can also find an expression for the *generalized higher part value* in the following way:
$$\hat{H}(p, n) = \frac{H(p, n)}{n} = \lfloor p/n \rfloor. \tag{16}$$

Note what n means: digit m has a weight of B^{m-1} in the value of p, i.e., if digit m has a value of d_m, it will contribute with $d_m B^{m-1}$ to the value of p. We will call *generalized digit n* of base B what is obtained when the following operation is performed:
$$\text{digit}(p, n, B) = L(\hat{H}(p, n/B), B). \tag{17}$$

This generalized digit has a weight of n/B in the value of p. To avoid the use of traditional digits, we define *generalized segment* in terms of an initial value and a segment width. The *generalized segment* of width δ starting at value n is given by the following expression:
$$\text{segment}(p, n, \delta) = L(\hat{H}(p, n), \delta), \tag{18}$$

where δ is an integer greater or equal than B. These definitions modify the meaning of parts of a chromosome to the point where it is more useful to think about chromosomes, not as strings of characters, but rather as integer values.

We can now express crossover and mutation in terms of the generalized operations defined above. The *generalized crossover* of integers p_1 and p_2 at value n results in two offspring that can be expressed as
$$h_1 = \text{crossover}(p_1, p_2, n) = L(p_1, n) + H(p_2, n); \tag{19}$$
$$h_2 = \text{crossover}(p_2, p_1, n) = L(p_2, n) + H(p_1, n). \tag{20}$$

This can also be written as the following:
$$h_1 = \text{crossover}(p_1, p_2, n) = p_2 + \chi(p_1, p_2, n); \tag{21}$$
$$h_2 = \text{crossover}(p_2, p_1, n) = p_1 - \chi(p_1, p_2, n); \tag{22}$$

where $\chi(p_1, p_2, n) = p_1 \bmod n - p_2 \bmod n = -\chi(p_2, p_1, n)$.

The *generalized mutation* for a segment of width δ starting at value n is defined as the following:
$$\text{mutation}(p, n, \delta) = L(p, n) + H(p, n\delta) + n\lfloor \delta\,\text{rand}() \rfloor. \tag{23}$$

It can be shown that traditional operators are a special case of generalized operators by substituting B^m for n. For crossover and mutation we have that
$$\text{crossover}_m(p_1, p_2) = \text{crossover}(p_1, p_2, B^m); \tag{24}$$
$$\chi_m(p_1, p_2) = \chi(p_1, p_2, B^m); \tag{25}$$
$$\text{mutation}_{m_1, m_2}(p) = \text{mutation}(p, B^m, B^{\Delta m+1}). \tag{26}$$

In the rest of this paper we will only be using the generalized expressions and therefore we will drop the word generalized.

4 Real Valued Individuals

We now proceed to adapt the formulas developed before for real valued individuals. Let r be a real number in the interval $[r_{\min}, r_{\max})$. Using a linear mapping r can be represented by a chromosome of N digits. We can see this chromosome as an integer p that can take a value in the set $\{0, 1, 2, 3, \ldots, B^N - 1\}$. The transformation between p and r is given by the following formula:

$$r = p\Delta\bar{r} + r_{\min}, \qquad (27)$$

where $\Delta\bar{r} = (r_{\max} - r_{\min})/B^N$. We define *lower part*, *higher part*, and *segment* of a real individual r in terms of those same operations over the corresponding integer p. In this way, the lower part of r is given by

$$L(r, k, r_{\min}) = L(p, n)\Delta\bar{r} + r_{\min}; \qquad (28)$$
$$= (r - r_{\min}) \bmod (k - r_{\min}) + r_{\min}. \qquad (29)$$

The higher part of a real number is given by

$$H(r, k, r_{\min}) = H(p, n)\Delta\bar{r} + r_{\min}; \qquad (30)$$
$$= r - (r - r_{\min}) \bmod (k - r_{\min}). \qquad (31)$$

We define the crossover of two real valued individuals r_1 and r_2 as

$$h_1 = \text{crossover}(r_1, r_2, k, r_{\min}) = (L(p_1, n) + H(p_2, n))\Delta\bar{r} + r_{\min}; \qquad (32)$$
$$h_2 = \text{crossover}(r_1, r_2, k, r_{\min}) = (L(p_2, n) + H(p_1, n))\Delta\bar{r} + r_{\min}; \qquad (33)$$

where h_1 and h_2 are the offspring produced. Simplifying, the above can also be written in the following way:

$$h_1 = \text{crossover}(r_1, r_2, k, r_{\min}) = r_2 + \chi(r_1, r_2, k, r_{\min}); \qquad (34)$$
$$h_2 = \text{crossover}(r_2, r_1, k, r_{\min}) = r_1 - \chi(r_1, r_2, k, r_{\min}); \qquad (35)$$

where

$$\chi(r_1, r_2, k, r_{\min})$$
$$= (r_1 - r_{\min}) \bmod (k - r_{\min}) - (r_2 - r_{\min}) \bmod (k - r_{\min}) \qquad (36)$$
$$= -\chi(r_2, r_1, k, r_{\min}). \qquad (37)$$

The mutation of a real valued individual r, at value k, with a mutation width of δ is given as

$$\text{mutation}(r, k, \delta, r_{\min}) = (\text{mutation}(p, n, \delta))\Delta\bar{r} + r_{\min}. \qquad (38)$$

Simplifying we arrive at the following:

$$\text{mutation}(r, k, \delta, r_{\min}) = L(r, k, r_{\min}) + H(r, \delta[k - r_{\min}] + r_{\min}, r_{\min})$$
$$+ (k - r_{\min})\lfloor \delta\,\text{rand}()\rfloor - r_{\min}. \qquad (39)$$

We can treat integer individuals as a special case of real valued individuals. and thus, the formulas presented above can be also applied to integers.

Not all values of δ produce valid results. If we want mutation to produce only individuals in the interval $[r_{min}, r_{max})$, the following condition must be met:

$$L(r, k, r_{min}) + H(r, (k - r_{min})\delta + r_{min}, r_{min}) \leq r + r_{min}, \quad (40)$$

for $\delta > 1$. Substituting the expression for lower part and higher part, and simplifying the following is arrived at:

$$\delta \left\lfloor \frac{r - r_{min}}{(k - r_{min})\delta} \right\rfloor \leq \left\lfloor \frac{r - r_{min}}{k - r_{min}} \right\rfloor. \quad (41)$$

This inequality is satisfied in the general case only if δ is an integer.

5 Generating Crossover Points

In the formulas developed to perform crossover, n for integers and k for reals is a random number with a given probability function. In a traditional genetic algorithm, crossover falls between traditional digits, i.e., at integer powers of B. Crossover sites that have this distribution can be produced as $n = B^{\lfloor N \text{ rand}() \rfloor}$. A probability function that has the same form but uses generalized digits can be obtained if the crossover sites are generated by $n = \lfloor B^{N \text{ rand}()} \rfloor$.

For real valued individuals, we can find similar formulas to those developed for integers, but additionally, we have the option of having continuous distributions by dropping the use of the floor function. Table 1 summarizes possible ways to generate crossover sites. Figures 1 and 2 show the cumulative distribution function for the crossover point distributions for integer and real valued individuals mentioned above.

If traditional digits are being used, crossover cannot produce invalid results, but for generalized digits it is possible that the result is greater than r_{max}. Given that for integers the sum of the offspring is equal to the sum of the parents, $p_1 + p_2 = h_1 + h_2$, we know that a condition that insures that crossover will

Table 1. Different ways to produce crossover sites

crossover points distribution	(integers) n	(reals) k
traditional	$B^{\lfloor N \text{ rand}() \rfloor}$	$B^{\lfloor N \text{ rand}() \rfloor} \Delta \bar{r} + r_{min}$
generalized	$\lfloor B^{N \text{ rand}()} \rfloor$	$\lfloor B^{N \text{ rand}()} \rfloor \Delta \bar{r} + r_{min}$
continuous		$B^{N \text{ rand}()} \Delta \bar{r} + r_{min}$

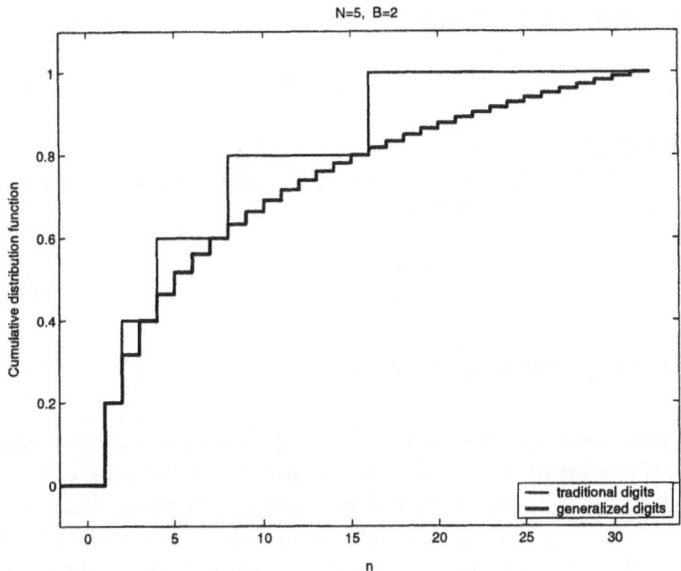

Fig. 1. Cumulative distribution functions for crossover sites for a binary chromosome of length 5 representing an integer

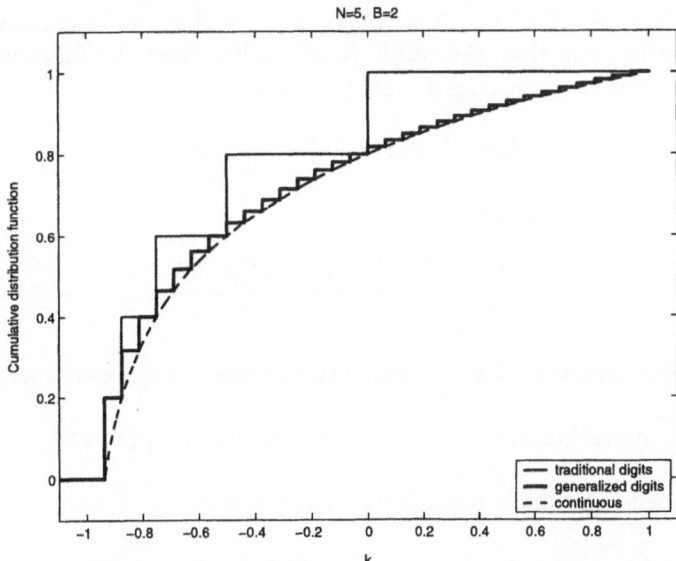

Fig. 2. Cumulative distribution functions for crossover sites for a chromosome of base 3 and length 5 that represents a real individual in $[-1, -1)$

produce valid individuals for any crossover site is the following:

$$p_1 + p_2 < B^N. \tag{42}$$

If the above condition is not met, we apply crossover with traditional digits to insure that the result will be valid. We call this the *crossover overflow correction*. The condition for reals is as follows:

$$r_1 + r_2 - r_{\min} < r_{\max}. \tag{43}$$

6 Generating Mutation Sites

It is known that the implementation efficiency of a binary genetic algorithm can be increased if mutation is controlled by means of a *mutation clock* [7]. According to the mutation clock idea, instead of generating a random number for each bit in the population to decided if it should be mutated, a random number with the proper probability distribution is generated so that it tells in how many bits the next mutation should occur. A mutation clock for the vgGA for traditional and generalized digits, and for integer and real valued individuals, was developed.

As in crossover, mutation at generalized digits could produce invalid results. Mutation removes a segment of the individual and substitutes it with a random segment in the set of all possible values. If we call γ the maximum value the segment can have so that the result is still valid, then the following equation expresses the relation between an integer p and its γ:

$$L(p, n) + H(p, \delta n) + n\gamma = B^N. \tag{44}$$

From the above, we can obtain γ as

$$\gamma = \frac{B^N - L(p, n) - H(p, n\delta)}{n}. \tag{45}$$

Now, we define mutation for integers with the *gamma correction* as

$$\text{mutation}(p, n, \delta) = L(p, n) + H(p, n\delta) + n \lfloor \gamma \, \text{rand}() \rfloor. \tag{46}$$

For real valued individuals the value of γ is given by

$$\gamma = \frac{r_{\max} + r_{\min} - L(r, k, r_{\min}) - H(r, [k - r_{\min}]\delta + r_{\min}, r_{\min})}{k - r_{\min}}. \tag{47}$$

Mutation of reals with the *gamma correction* is defined as the following:

$$\text{mutation}(k, r, \delta, r_{\min}) = L(r, k, r_{\min})$$
$$+ H(r, [k - r_{\min}]\delta + r_{\min}, r_{\min}) + (k - r_{\min}) \lfloor \gamma \, \text{rand}() \rfloor - r_{\min}. \tag{48}$$

7 Experiments

One could expect the vgGA that implements generalized digits to outperform a traditional GA on problems where the underlying representation is not binary. To test this assumption, a generalization of the well known one-max problem was defined. In the consecutive one-max problem, or *c-one-max problem,* the evaluation depends on the lengths groups of consecutive digits that are equal to one when the individual is expressed in given base. Each group contributes to the fitness function with its length to a power α. For example, for $\alpha = 2$, an individual with phenotype of 412281_{10}, which can be expressed as 101143111_5, has an evaluation $1^2 + 2^2 + 3^2 = 14$ in the c-one-max problem of base 5 as shown in Fig. 3. For binary GAs, the c-one-max problem in any base that is not a multiple of 2 should be a relatively hard problem (at least harder than the problem in base 2). On the other hand, since the vgGA is not tied to a given base, its performance on this problem should be higher.

A vgGA, where individuals are vectors of integers or reals, was implemented in MATLAB, and tested with the c-one-max problem of base 2 and base 5 (this is the base of the problem and not of the individuals in the vgGA) and $\alpha = 2$. Table 2 summarizes the parameters of the vgGA used for these tests. Binary tournament selection [8] was used. Figures 4 and 5 shows the results for the c-one-max problem of base 2 and base 5, respectively. These plots are the average of the best-found-so-far of 100 runs. For the base 2 problem, traditional digits, i.e. a traditional genetic algorithm, outperform generalized digits. For the base 5 problem the results are the opposite, as expected.

A real valued version of the c-one-max problem can be obtained if the evaluation depends on the number of digits that are equal to those of a given irrational constant, expressed on a given base. The *c-pi-max problem* will be defined as the problem of finding the digits of π where the evaluation depends on the number

Table 2. Parameters of the vgGA used in all experiments

runs	100
generations	150
population size	20
N	40
p_c	1.0
p_m	0.1
B	2
δ	2

$$\underbrace{1}_{1^2} \; 0 \; \underbrace{1\;1}_{2^2} \; 4 \; 3 \; \underbrace{1\;1\;1}_{3^2}$$

Fig. 3. Example of c-one-max base 5 problem. The evaluation of $412281_{10} = 101143111_5$ is $1^2 + 2^2 + 3^2 = 14$

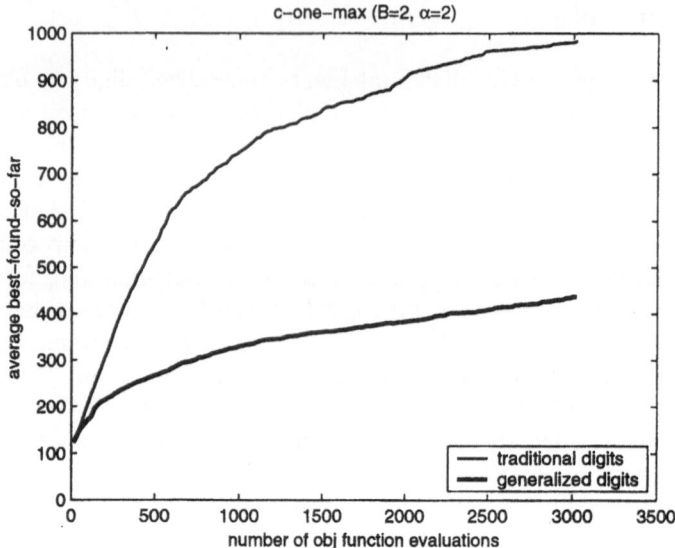

Fig. 4. Average of 100 runs of the best-found-so-far for the c-one-max problem of base 2 with $\alpha = 2$

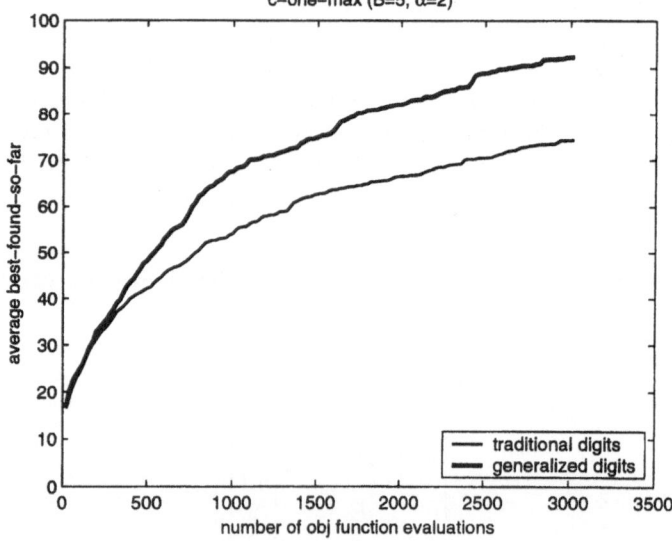

Fig. 5. Average of 100 runs of the best-found-so-far for the c-one-max problem of base 5 with $\alpha = 2$

$$\underbrace{0\ 3}_{2^2}\ .\ 0\ \underbrace{1\ 5}_{2^2}\ 0\ \underbrace{9\ 2\ 6}_{3^2}\ 0$$

Fig. 6. Example of c-pi-max base 10 problem. The evaluation of 03.01509260_{10} is $2^2 + 2^2 + 3^2 = 17$

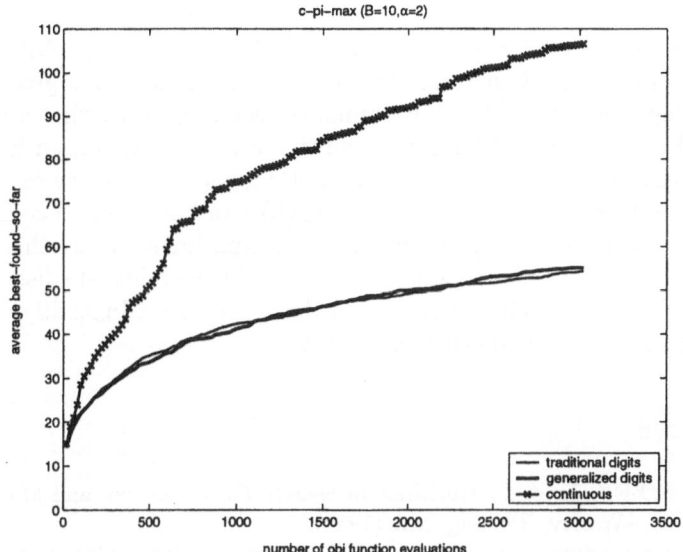

Fig. 7. Average of 100 runs of the best-found-so-far for the c-pi problem of base 5

of consecutive digits on a given base as described for the c-one-max problem. For example, an individual with phenotype of 03.01509260_{10} has an evaluation of $2^2 + 2^2 + 3^2 = 17$ for the c-pi-max problem with $\alpha = 2$, base 10, and considering two digits to the left and eight digits to the right of the decimal point as shown in Fig. 6. Figure 7 shows the results for the c-pi-max problem of base 10 with 2 integer positions and 49 decimal positions. The vgGA implements real valued individual in $[0,5)$. As the figure shows, the vgGA that uses a continuous distribution of crossover points and mutation sites has the best performance on this problem.

8 Conclusions

This paper shows that traditional one point crossover and mutation can be mapped to arithmetic operations over integers, the formulas found can be generalized to chromosomes of any base, and also, to real valued individuals represented by a linear mapping. A *virtual gene genetic algorithm* (vgGA) which works on the phenotype and can produce the same results as a traditional binary GA, has been implemented in MATLAB.

The vgGA is a generalization of a traditional GA with binary linear chromosomes. It is a generalization because by mapping traditional crossover and mutation to operations over the phenotype, it can simulate linear chromosomes of any integer base, not necessarily binary. Additionally, the vgGA extends where crossover and mutation sites may fall, allowing the simulation of generalized digits.

The sites where crossover and mutation fall in a traditional GA can be generalized to values that do not correspond to integer powers of a given base, thus implementing *generalized digits*. Preliminary results indicate that a vgGA using generalized digits can outperform a traditional binary GA on an integer problem where the underlying structure does not depend on a binary representation. When solving a real valued problem, the vgGA can implement a continuous distribution of crossover and mutation sites. An experiment where this continuous distribution produces better results than the traditional discrete distribution was presented. The experiments presented in this paper are admittedly very limited and should be extended to other problems.

References

1. Goldberg, D.E.: Genetic Algorithms in Search, Optimization, and Machine Learning. Addison-Wesley, Reading, MA (1989)
2. Holland, J.H.: Adaptation in Natural and Artificial Systems. University of Michigan Press, Ann Arbor, MI (1975)
3. Wright, A.H.: Genetic algorithms for real parameter optimization. In Rawlins, G.J.E., ed.: Foundations of Genetic Algorithms. Morgan Kaufmann, San Mateo, CA (1991) 205–218
4. Eschelman, L.J., Schaffer, J.D.: Real-coded genetic algorithms and interval-schemata. In Whitley, L.D., ed.: Foundations of Genetic Algorithms 2. Morgan Kaufmann, San Mateo, CA (1993) 187–201
5. Michalewicz, Z.: Genetic Algorithms + Data Structures = Evolution Programs. 2nd. edn. Springer-Verlag, Berlin (1994)
6. Surry, P.D., Radcliffe, N.J.: Real representations. In Belew, R.K., Vose, M.D., eds.: Foundations of Genetic Algorithms 4. Morgan Kaufmann, San Mateo, CA (1997) 187–201
7. Golbderg, D.E.: Personal communication. (1990)
8. Goldberg, D.E., Deb, K.: A comparative analysis of selection schemes used in genetic algorithms. In Rawlins, G.J.E., ed.: Foundations of Genetic Algorithms. Morgan Kaufmann, San Mateo, CA (1991) 69–93

Quad Search and Hybrid Genetic Algorithms

Darrell Whitley, Deon Garrett, and Jean-Paul Watson

Department of Computer Science, Colorado State University
Fort Collins, Colorado 80523, USA
{whitley,garrett,watsonj}@cs.colostate.edu

Abstract. A bit climber using a Gray encoding is guaranteed to converge to a global optimum in fewer than $2(L^2)$ evaluations on unimodal 1-D functions and on multi-dimensional sphere functions, where L bits are used to encode the function domain. Exploiting these ideas, we have constructed an algorithm we call **Quad Search**. Quad Search converges to a local optimum on unimodal 1-D functions in not more than $2L + 2$ function evaluations. For unimodal 1-D and separable multi-dimensional functions, the result is the global optimum. We empirically assess the performance of steepest ascent local search, next ascent local search, and Quad Search. These algorithms are also compared with Evolutionary Strategies. Because of its rapid convergence time, we also use Quad Search to construct a hybrid genetic algorithm. The resulting algorithm is more effective than hybrid genetic algorithms using steepest ascent local search or the RBC next ascent local search algorithm.

1 Background and Motivation

There are several advantages to using a reflected Gray code binary representation in conjunction with genetic algorithms and local search bit climbers. For both unimodal 1-dimensional functions and separable multi-dimensional unimodal functions, proofs show that steepest ascent local search is guaranteed to converge to a global optimum after $2L$ steps when executed on an L-bit local search neighborhood. The proofs assume that the functions are bijections, so that search cannot become stuck on plateaus of equally good solutions. Furthermore, the proofs demonstrate that there are only 4 neighbors per step (or move) that are critical to global convergence [1].

Using a reflected Gray code representation, it can be proven that a reduction from L to $L - 1$ dimensions is always possible after at most 2 moves (even if the search repeatedly "steps across" the optimum). Hence the global optimum of any 1-dimensional unimodal real-valued function can be reached in at most $2L$ steps. Under steepest ascent local search, each step requires $\mathcal{O}(L)$ evaluations. Thus, it follows that steepest ascent local search actually requires up to $\mathcal{O}(L^2)$ evaluations to converge to the global optimum. In practice, next ascent local search methods are often much faster than steepest ascent. However, the worst case order of complexity of next ascent algorithms is exponential. This is due to the fact that next ascent algorithms in the worst case can take very small steps at each iteration.

The current paper exploits the ideas behind these convergence proofs to construct a new algorithm which we call **Quad Search**. Quad Search is a specialized form of steepest ascent that operates on a reduced neighborhood. The algorithm cuts the search space into four quadrant and then systematically eliminates quadrants from further consideration. On unimodal functions Quad Search converges to the global optimum in at most $\mathcal{O}(L)$ evaluations, as opposed to $\mathcal{O}(L^2)$ evaluations for regular steepest ascent. For multi-dimensional functions, Quad Search converges to a point that is locally optimal in each dimension.

The new algorithm is tested on different types of unimodal functions. Quad Search is compared to steepest ascent and Davis' Random Bit Climber, RBC, which is a next ascent local search method [2]. Quad Search using a Gray code representation converges after at most $2L + 2$ evaluations on classes of functions such as sphere functions where convergence proofs have also been developed for Evolution Strategies and Evolutionary Programming [1]. Given the convergence results on unimodal functions, we compare Quad Search with Evolution Strategies. The representations used by Evolution Strategies are continuous while Quad Search uses a discrete representation, but in practice, both encodings can be high-precision.

Finally, one of the most exciting uses of Quad Search is in combination with genetic algorithms. We combine Quad Search with the "Genitor" steady-state genetic algorithm. We also look at hybrid forms of Genitor that use steepest ascent and RBC. Genitor in combination with Quad Search proved to be the most effective hybrid genetic algorithm.

2 The Gray Code Representation

Gray codes have two important properties. First, for any 1-D function and in each dimension of any multi-dimensional problem, adjacent neighbors in the real space are also adjacent neighbors in the Gray code hypercube graph [3]. Second, the standard Binary reflected Gray code folds the search space in each dimension.

For each reference point R, there is exactly one neighbor in the opposite half of the search space. For example, in a 1-D search space of points indexed from 0 to $2^L - 1$, the points i and $(2^L - 1 - i)$ are neighbors. In effect these neighbors are found by folding or reflecting the 1-D search space about the mid-point between 2^{L-1} and $2^{L-1} + 1$.

There are also reflections between each quartile of the search space. Thus, starting at some point i in the first quartile of the search space we can define a set of four points that will be critical to Quad Search. These points are found at locations given by the integer indices $i, (2^{L-1} - 1 - i), (2^{L-1} + i)$ and $(2^L - 1 - i)$.

An inspection of these integer values shows that each of the points is in a different quadrant of the search space. This follows from the following observations. Point i is in the first quadrant and $i < 2^{L-2}$. Furthermore, $2^{L-1} - 1$ is the last point in the second quadrant and therefore $(2^{L-1} - 1 - i)$ must also be in the second quadrant. Similarly 2^{L-1} is the first point in the third quadrant and therefore $(2^{L-1} + i)$ must be in the third quadrant. Finally $2^L - 1$ is the last point in the fourth quadrant and therefore $(2^L - 1 - i)$ is in the fourth quadrant.

If we interpret each of these integers as Gray encoded bit strings, we can find each of these points using exclusive-or (denoted by $\oplus$) as follows:

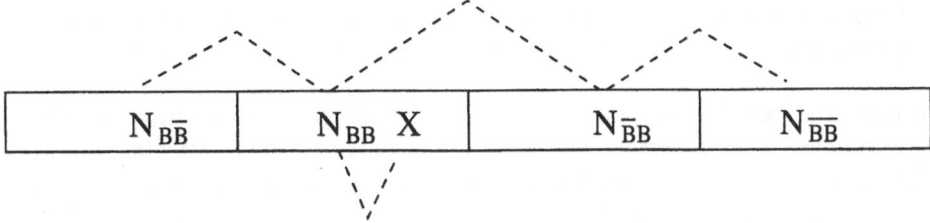

Fig. 1. The neighbor structure across and within quadrants

$$\begin{aligned} i &= i \\ (2^{L-1} - 1) - i &= i \oplus (2^{L-1} - 1) \\ (2^L - 1) - i) &= i \oplus (2^L - 1) \\ (2^{L-1} + i) &= i \oplus (2^L - 1) \oplus (2^{L-1} - 1) \end{aligned} \quad (1)$$

Note that under a Gray code $2^L - 1$ is a string of all zeros except for a single 1 bit in the first position and $2^{L-1} - 1$ is a string of all zeros except for a single 1 bit in the second position. Therefore, moving from one of these points to the other involves flipping the first bit, the second bit, or both. (Also note that $i \oplus (2^L - 1) \oplus (2^{L-1} - 1)$ is a Hamming distance 1 neighbor of i under a Binary encoding [1].)

Also, since the exclusive-or operator really just applies a mask that flips the first or second bit, it works regardless of the quadrant in which the point i actually appears.

On average, flipping these bits moves the search the greatest distance. Recently published proofs show that these are the only moves that are critical in order to achieve convergence in a linear number of steps under a steepest local search algorithm algorithm [1]. We next construct the Quad Search algorithm to exploit these same moves.

3 Quad Search

Consider a unimodal function encoded using an L bit Gray code. We can break the space into four quadrants, defined by the two high order bits of the current individual. Note that any given point has two neighbors in quadrants outside of its own quadrant; they are the points generated by flipping the two high order bits. The highest order unfixed bit is called the *major* bit. The second highest order unfixed bit is called the *minor* bit. Flipping any other bit must generate a neighbor inside the current quadrant.

Consider a point N_{BB} located inside Quadrant 2, as illustrated in Fig. 1. In general, BB denotes the first two bits that map to an arbitrary quadrant. We can flip the major bit to generate the point at $N_{\overline{B}B}$ and the minor bit to generate the point at $N_{B\overline{B}}$. We flip both bits to reach the point at $N_{\overline{BB}}$. We will also refer to the quadrants as Quadrant BB, each with its respective index.

In Fig. 1 the point located at X is a neighbor of point N_{BB} that is located in Quadrant BB. X can be left or right of N_{BB}. We will express the problem in terms of minimization. One of the 4 quadrant neighbors is the current point from which the search is being carried out.

Theorem: *Quad Search converges to the global optimum on bijective unimodal 1-D functions after at most $2L + 2$ evaluations from an arbitrary starting point.*

Proof: The proof is recursive in nature. Assume the search space contains at least 8 points. means there are at least 2 points in each quadrant. After initializing the search, we show the search space can be cut in half after at most 2 new evaluations.

Sample a point N_{BB} and its reflected neighbors $N_{\overline{B}B}, N_{B\overline{B}}$ and $N_{\overline{BB}}$.

Let Z represent a string of zeros when interpreted as a string and the number of zeros when interpreted as an integer. Without loss of generality, we can relabel the points $\{N_{BB}, N_{\overline{B}B}, N_{B\overline{B}}, N_{\overline{BB}}\}$ using exclusive-or so that the minimal neighbor is point N_{000Z}. Then $\{000Z, 100Z, 010Z, 110Z\}$ denotes the same strings where the first 3 bits are as shown, and the remaining bits are all zero. We add one additional sample at point $001Z$.

Because the function is unimodal, the global minimum cannot be in Quadrant 11. Note that samples $000Z$ and $001Z$ are from Quadrant 00 which contains the current best known solution. Without loss of generality, assume that $100Z + c < 000Z + c < 001Z + c < 010Z + c$ when the strings are interpreted as integers and the integers are shifted by constant c using addition mod 2^L. (The function can be shifted by a constant, reversed, or relabeled using exclusive-or without changing the neighborhood structure: all that is really required for the proof is that there is 1 out of quadrant point to the left and the right and 2 samples in the middle quadrant.)

1. If $f(000Z) > f(001Z) < f(010Z)$, then the global optimum must reside in Quadrant 00 or 01. The major bit is fixed to 0. The points $000Z, 001Z, 010Z$, becomes the points $00Z, 01Z, 10Z$ in the reduced space. Only 2 additional points are needed to continue the search. First we evaluate $11Z$. Then the space is remapped again so that the minimum of these 4 points is at $00Z$. Then we evaluate $001(Z-1)$.
2. If $f(100Z) > f(000Z) < f(001Z)$, then the global optimum must reside in Quadrant 00 or 10. The minor bit is fixed to 0. The points $000Z, 001Z, 100Z$, becomes the points $00Z, 01Z, 10Z$ in the reduced space. Only 2 additional points are needed to continue the search. First we evaluate $11Z$. Then the space is remapped again so that the minimum of these 4 points is at $00Z$. Then we evaluate $001(Z-1)$.

After the first 5 evaluations, the search space contains $2^{(L-1)}$ points. At each iteration, the search space is cut in half after at most 2 new evaluations. After (L-3) iterations the search space is reduced to at most 4 points, since $2^{(L-1)}/2^{(L-3)} = 4$. However, at this point, we have already evaluated $00, 01$ and 10 (remapped so that 00 is the best solution so far). Only 11 remains to be evaluated. The total number of evaluation is $5 + 2(L-3) + 1 = 2(L) + 2$. □

We assume the function is a bijection. Obviously, this is not true for many unimodal functions. While Quad Search cannot make progress over functions that are largely flat, it can step across small to moderate size flat regions. There are a number of ways ties might be handled. The current implementation does not always guarantee global convergence if there are flat regions, but its behavior is similar to that of next ascent. In general, convergence times that are less than exponential cannot be guaranteed if the space is largely flat (even if the space is unimodal).

At the beginning of the algorithm, the ordering of the quadrants is completely determined by the two high order bits. However, as bits are fixed, the high order bits used in the computation refer to the two highest order bits *which remain unfixed*. After a "1" bit has been fixed in a Gray code, this ordering is destroyed. The major bit is always the leftmost bit that has not been fixed. The minor bit is always the second bit from the left that has not been fixed. The major and minor bits need not be adjacent, but all bits to the right of the minor bit are unfixed. The only bit to the left of the minor bit that is unfixed in the major bit. Thus, "quadrants" are not always physically adjacent. However, the search space can always be remapped to a new physical space such that the major and minor bits can be used to move between adjacent quadrants in the wrapped space.

4 Performance of Quad Search versus Local Search

We examined the following variants of local search in this study: next-ascent local search, steepest-ascent local search, and Quad Search. We used Davis' Random Bit Climbing algorithm (RBC) [2] as our implementation of next-ascent. RBC is a simple next-ascent hill climber that randomizes the order in which bits are flipped. A random permutation is generated and used to determine the order in which to test bits. All L bits flips are tested before any bit flips are retested; after all bits have been flipped, a new permutation is generated and the process repeated. RBC also accepts bit flips resulting in equal fitness. The steepest-ascent algorithm, which we call steepest-ascent bit climbing, or SABC, differs in that it tests all bit flips, and accepts only the one that results in the largest improvement in fitness; ties are broken randomly.

Our implementation of Quad Search is based on an earlier proof that showed that not more than 8L evaluations were required [1]. It evaluates all of the potential Hamming 1 neighbors before evaluating any Hamming distance 2 neighbors. Also, the current implementation does not "reuse" samples inside of the middle quadrant. Obviously, this results in unnecessary computation; empirically, the current implementation uses just under $3L$ evaluations. This will be corrected in a future implementation.

We executed each local search algorithm on the set of three common test functions shown in Table 1. Each parameter was encoded using a 32-bit reflected Gray representation, and we considered 10, 20, and 30 dimensional variations of each test function. Each trial was executed until the global optimum was located.

We compare local search against Evolution Strategies because Evolution Strategies are proven to converge on unimodal functions and outperform other genetic algorithms in unimodal spaces. We did not use correlated mutation for 20 and 30 dimensional test functions due to the explosion of the number of correlation-strategy variables (for k object variables there are $\mathcal{O}(k^2)$ correlation-strategy variables). In addition, a restart mechanism was included in the ES to allow them to escape the condition noted by Mathias et. al., [4] whereby the strategy variables become so small as to halt the progress of the search. The specific restart mechanism employed is drawn from the CHC genetic algorithm [5]. Upon *detection of convergence* (defined to be all strategy variables in the population falling below some critical threshold) the best individual is copied into the new population.

Table 1. Test functions used in this study. We consider $n = 10$, $n = 20$, and $n = 30$.

Function	Definition	Domain
Sphere	$\sum_{i=1}^{n} (x_i + 5)^2$	$-20 \leq x_i \leq 20$
Parabolic Ridge	$-10x_1 - \sum_{x=2}^{n} (x_i)^2 - 1000$	$-100 \leq x_i \leq 100$
Sharp Ridge	$-10x_1 - \sqrt{\sum_{x=2}^{n} x_i^2} - 1000$	$-100 \leq x_i \leq 100$

The remainder of the population is filled with mutated copies of the best individual, and all strategy variables are reset to new values in the range $[-1.0, 1.0]$. In this study, the threshold for restarting was chosen as 1.0×10^{-6}, and the mutations were performed on 35% of the variables in each individual. In our implementation, restarting does not modify the object variables in the population, it simply resets the strategy variables. To facilitate comparisons with local search, we used a 32-bit floating-point representation to encode each parameter. All trials were terminated after 1,000,000 fitness evaluations or when the fitness of the best individual reached or exceeded -1.0×10^{-10}.

Table 2 shows the performance of each algorithm on the Sphere and Parabolic Ridge functions; the results for the Sharpe Ridge function are similar to the Parabolic Ridge function and are not shown. All local search methods are guaranteed to find the optimum. RBC and Quad Search empirically display linear time search behavior. The number of evaluations needed for the SABC steepest ascent algorithm to find the optimum is, as expected, $\mathcal{O}(L^2)$. The performance of the Evolution Strategies is roughly comparable to SABC.

The reader should be aware that this is probably not the most efficient form of Evolution Strategies for these test problems. Evolution Strategies using a 1/5 rule to control the mutation rate should lead to linear time convergence on the sphere function.

One problem with the Evolution Strategies we have implemented is that they are often unable to make the fine adjustments necessary to reach optimality to the desired accuracy. Often all but one (or a small number) of the parameters are set to the optimal values, while the remaining parameters are near-optimal. Sometimes the ES is not able to optimize the near-optimal parameters without disturbing the values of the optimal parameters; the values of the strategy variables become too small to be effective and the search stagnates.

Restarts rarely help on these particular functions – restarting resets several strategy variables, causing future mutations to disturb the parameters that are already optimal. It may also be the case that the Evolution Strategies used here would have displayed much better performance if we had used a less conservative convergence criteria.

5 A Quad Search/Genetic Algorithm Hybrid

We now evaluate the performance of Quad Search on a suite of multimodal test functions which are relatively difficult for most optimization algorithms [6]. The specific functions we consider are shown in Table 3. Powell's function is 4-dimensional, and we encode each parameter using 20 bits under a reflected Gray representation. For the other three functions, we use 10-dimensional variants, encoding each parameter using 10 bits under

Table 2. Number of evaluations to find the optimum of the Sphere and Parabolic Ridge functions. The Evolution Strategies used in this study used a restart mechanism. **# Solved** refers to the number of times the optimal solution was found out of 30 independent trials. **# Evaluations** refers to the number of evaluations that were required.

Algorithm	Sphere 10-D		Sphere 20-D		Sphere 30-D	
	# Solved	# Evaluations	# Solved	# Evaluations	# Solved	# Evaluations
Quad Search	30	932.7	30	1801.0	30	2659.7
RBC	30	5875.3	30	12115.9	30	18420.2
SABC	30	51478.7	30	208198.0	30	458078.0
(1+1) ES	5	563967.0	0	-	0	-
(1+5) ES	8	412074.3	0	-	0	-
(10+10) ES	30	13206.2	30	151866.0	6	574046.0
(10+50) ES	30	17957.7	30	146173.0	17	477820.0
(50+50) ES	30	30707.3	30	130571.0	28	500807.0
(50+250) ES	30	50106.3	30	153628.0	28	418223.0
(100+100) ES	30	50868.3	30	191504.1	28	595862.3
(100+500) ES	30	89413.7	30	242475.6	28	544634.4

Algorithm	Parabolic Ridge 10-D		Parabolic Ridge 20-D		Parabolic Ridge 30-D	
	# Solved	# Evaluations	# Solved	# Evaluations	# Solved	# Evaluations
Quad Search	30	951.9	30	1825.6	30	2670.3
RBC	30	5918.1	30	13013.3	30	19605.4
SABC	30	50151.9	30	201809.0	30	446225.5
(1+1) ES	9	457005.1	0	-	0	-
(1+5) ES	9	298370.4	0	-	0	-
(10+10) ES	30	15438.8	30	228059.1	4	531212.2
(10+50) ES	30	21234.3	30	122101.0	11	568849.0
(50+50) ES	30	36132.1	30	120264.7	28	539147.3
(50+250) ES	30	60439.7	30	169586.2	29	413286.7
(100+100) ES	30	61381.3	30	236389.5	28	665666.7
(100+500) ES	30	108480.2	30	284724.0	23	564163.3

Table 3. Test functions: Griewangk, EF101, and EF102 are 10-Dimensional and Powell is 4-Dimensional. EF101 and EF102 were also tested in 20-Dimensional variants. Powell's function was encoded with 20 bits per parameter. All others used 10-bit encodings.

Name	Function	Global Optimum
Griewangk	$1 + \sum_{i=1}^{N} \frac{x_i^2}{4000} - \prod_{i=1}^{N}(cos(\frac{x_i}{\sqrt{i}}))$	0.0
Powell	$(x_1 + 10x_2)^2 + (\sqrt{5}(x_3 - x_4))^2 +$ $((x_2 - 2x_3)^2)^2 + (\sqrt{10}(x_1 - x_4)^2)^2$	0.0
EF101	$-xsin(\sqrt{|x - \frac{y+47}{2}|}) - (y + 47)sin(\sqrt{|y + 47 + \frac{x}{2}|})$	-939.8
EF102	$xsin(\sqrt{|y + 1 - x|})cos(\sqrt{|x + y + 1|}) +$ $(y + 1)cos(\sqrt{|y + 1 - x|})sin(\sqrt{|x + y + 1|})$	-511.7

Table 4. Performance of local search and genetic algorithms on multimodal test functions. **Mean** refers to the mean value of the best solution averaged over 30 runs. **# Evaluations** refers to the mean number of evaluations averaged over 30 runs. **# Solved** refers to the number of trials (out of 30) in which the optimum was located. If **# Solved**=0, then 500,000 evaluations were consumed in all trials. σ denotes the standard deviation of the quantity in the adjacent left column.

Function	Algorithm	Mean	σ	# Solved	# Evaluations	σ
Griewangk	QuadSearch	0.073	0.032	2	233222.0	92175.1
	RBC	0.009	0.014	21	181605.3	129114.3
	SABC	0.0	0.0	30	85053.8	82216.0
	Genitor	0.108	0.058	2	13761.5	570.3
	CHC	0.001	0.005	29	62179.0	80013.5
Powell	QuadSearch	1624.330	1519.975	0	—	—
	RBC	2.62e-4	5.56e-5	0	—	—
	SABC	2.06e-4	9.94e-5	0	—	—
	Genitor	2.92e-4	7.24e-5	0	—	—
	CHC	0	0.0	30	200998.1	94196.4
EF101	QuadSearch	-724.880	25.322	0	—	—
	RBC	-759.683	28.087	0	—	—
	SABC	-756.576	34.588	0	—	—
	Genitor	-817.61	69.70	5	83151.2	117792.3
	CHC	-939.82	0.02	30	17171.3	11660.1
EF102	QuadSearch	-434.873	8.744	0	—	—
	RBC	-446.421	9.905	0	—	—
	SABC	-442.754	11.301	0	—	—
	Genitor	-443.45	17.87	0	—	—
	CHC	-495.56	5.53	0	—	—

a reflected Gray representation. EF101 and EF102 are both based on the 2-dimensional functions shown in Table 3; the 10-dimensional versions are constructed using a wrapped weighting scheme [6].

For each local search algorithm (QuadSearch, RBC, and SABC), we execute 30 independent trials on each test function for a maximum of 500,000 evaluations. Search is terminated if a global optimum is encountered, and is re-initiated from a randomly generated starting point when descent terminates at local optima. For comparative purposes, we also consider two standard GAs: Genitor and CHC. For Genitor, we used a population size of 300. CHC is a state-of-the-art GA with its own form of diversity management [5]. We execute each GA for 500,000 evaluations or until a globally optimal solution is located. Such a methodology favors algorithms that produce the best results possible given a large number of evaluations; an algorithm might give good quick approximate results but not perform well in the current competition.

We show the results in Table 4. The data indicate that Quad Search performs poorly on multimodal, multi-dimensional test functions. This is not surprising. Quad Search behaves much like a greedy search algorithm: after finding the best move among 4 neighbors, it eliminates half the space from further consideration at each step. In contrast,

RBC and SABC continue to sample the entire search space. But this also means that Quad Search often does much less work than RBC or SABC.

The advantage of Quad Search seems to be that it quickly locates the first local optimum it can find while evaluating only a few points in each dimension of the search space. It is interesting to note that RBC is usually the best of the local search methods when used in isolation, except on Griewangk, where SABC finds the optimal solution every time. Additionally, Quad Search can work within the normal bit representations often used by genetic algorithms. It is therefore natural to combine Quad Search with a genetic algorithm. Such algorithms are called hybrid genetic algorithms [7]. There has also been a recent trend to refer to such algorithms as memetic algorithms [8].

We consider hybrid genetic algorithms based on the Genitor algorithm [9]. Genitor is an elitist steady state GA which utilizes rank-based selection. It has proven to be amenable to hybridization, due to a number of factors. Selection is based solely on rank, which helps to maintain diversity; this can be especially important for hybrid genetic algorithms since local search often produces highly fit individuals. Under proportional selection, such individuals can quickly come to dominate the population.

All of the hybrid genetic algorithms we examine use uniform crossover and bitwise mutation with probability 0.001 per bit. The population size was 100. In addition, two-point reduced surrogate [10] and HUX crossover (a variant of uniform crossover in which exactly half of the bits are exchanged) [5] were tested and yielded negligible differences in performance. We investigate hybrid genetic algorithms based on the following local search algorithms: Quad Search, RBC, and SABC. Each algorithm can be executed in the following modes:

1. **Full Local Search:** Use a local search method to improve each member of the population until a local optimum is reached.

2. **Stochastic Local Search:** Use local search to improve each member of the population with probability 0.05; again local search is run until a local optimum is reached.

3. **Restricted Local Search:** Apply local search to every member of the population until a single improving move is found.

For the Restricted and Stochastic forms of local search, only the best results are given for SABC and RBC. The local search methods were not hybridized with CHC because we were not able to develop a hybrid form of CHC that improved on the original form of CHC.

Overall, the hybrid genetic algorithm combining Genitor and Quad Search is the best hybrid genetic algorithm; it was best on Powell, EF101 and Rana's function (EF102). The RBC and SABC algorithms were best on Griewangk, but Griewangk is also solved by using SABC alone.

Both CHC and the Genitor/Quad Search Hybrid solved EF101 every time, but CHC did so faster. CHC was somewhat better than the Genitor/Quad Search Hybrid on Powell's functions, but the Genitor/Quad Search algorithm was dramatically better than CHC on Rana's function (EF102). To the best of our knowledge, this is the first time any search

Table 5. Performance of Hybrid GAs on 4 multimodal test functions. For RBC and SABC, results for the Stochastic and Restricted forms of local search are reported, depending on which worked best. **Mean** is the mean value of the best solution averaged over 30 runs. **# Evaluations** refers to the mean number of evaluations averaged over 30 runs. **# Solved** is the number of trials (out of 30) that found the optimum. If **# Solved**=0, then 500,000 evaluations were consumed in all trials. σ denotes the standard deviation of the quantity in the adjacent left column.

Function	Local Search	Mode	Mean	σ	# Solved	# Evaluations	σ
Griewangk	Quad Search	Full	0.016	0.014	13	100090.3	45138.8
Griewangk	RBC	Full	0	0	30	144024.0	45507.5
Griewangk	SABC	Full	0	0	30	418697.8	14614.9
Griewangk	RBC	Restricted	0.011	0.016	20	123474.5	41413.3
Griewangk	SABC	Stochastic	0.013	0.018	19	76160.2	37452.1
Powell	Quad Search	Full	3.46e-9	9.07e-9	22	262931.2	57478.2
Powell	RBC	Full	2.14e-4	8.44e-5	0	—	—
Powell	SABC	Full	9.87e-5	8.34e-5	0	—	—
Powell	RBC	Stochastic	1.94e-7	7.83e-7	5	351407.2	44262.3
Powell	SABC	Restricted	5.55e-5	6.28e-5	0	—	—
EF101	Quad Search	Full	-939.84	0.01	30	216299.3	38496.7
EF101	RBC	Full	-924.63	34.95	25	395994.2	52569.5
EF101	SABC	Full	-772.59	45.55	1	448305.0	—
EF101	RBC	Stochastic	-906.79	55.21	22	237970.1	87683.9
EF101	SABC	Stochastic	-892.53	70.75	20	283078.8	80614.8
EF102	Quad Search	Full	-510.29	2.53	26	267792.7	42256.3
EF102	RBC	Full	-471.45	7.09	0	—	—
EF102	SABC	Full	-447.55	13.14	0	—	—
EF102	RBC	Stochastic	-484.08	7.66	0	—	—
EF102	SABC	Stochastic	-463.64	9.56	0	—	—

algorithm has solved a 10-D version of Rana's function to optimality. Overall, the Genitor/Quad Search Hybrid is better than any of the other hybrid genetic algorithms we tested and is competitive with and sometimes superior to CHC.

What make the Genitor/Quad Search combination so effective? We explored two hypotheses in an attempt to explain its behavior.

1. **Hypothesis 1**: Quad search improves strings quickly with minimal effort.
2. **Hypothesis 2**: Quad search works because it sustains genetic diversity.

The two hypotheses are not mutually exclusive, but experiments were performed to try to determine if one of these factors might be more important.

Figure 5 shows two graphs. The test function in this case was Rana's EF102. On the left, we show the speed with which improving moves are found using Genitor in combination with Quad Search (QS), RBC and SABC. On the right, we measured the entropy in the population.

$$Entropy(Population) = -B1 \log B1 - B0 \log B0$$

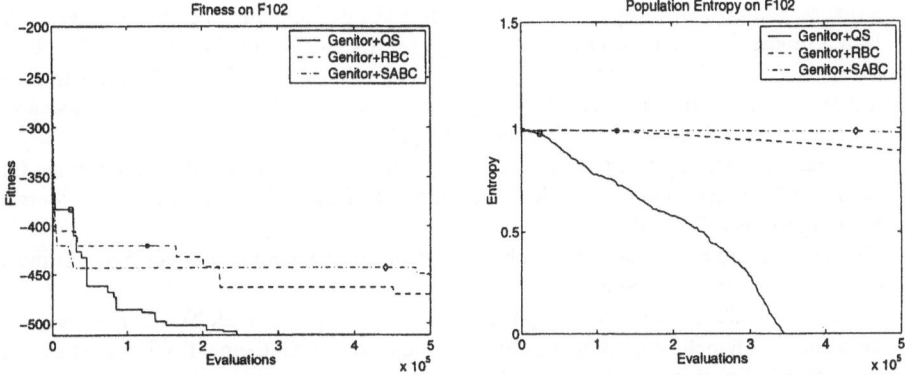

Fig. 2. Performance of a Hybrid Genetic Algorithm utilizing Quad Search, RBC, and SABC as local optimizers. Fitness of the best individual versus number of evaluations is shown in the left figure. Entropy of the population versus number of evaluations is shown in the right figure. The test function was Rana's EF102

where $B1$ counts the number of 1 bits in the population and $B0$ counts the number of 0 bits in the population. The figure shows that the Hybrid GA combining Genitor and Quad Search does not preserve diversity. Instead, it drives diversity out of the population faster than the other two local search methods when used as part of a hybrid genetic algorithm. It also shows that Genitor using Quad Search finds improvements more rapidly that Genitor used in combination with RBC or SABC.

6 Conclusions

The theoretical analysis of Quad Search presented in the paper, along with the empirical data, shows that Quad Search can be both extremely fast and effective on a special subclass of problems.

Because Quad Search is so fast, it may also have important practical uses in the construction of hybrid genetic and evolutionary algorithms. Empirical studies show that a hybrid genetic algorithm combining Genitor and Quad Search is generally superior to a hybrid genetic algorithm based on RBC or SABC.

Acknowledgments. This work was supported by National Science Foundation grant number IIS-0117209.

References

1. Whitley, D., Barbulescu, L., Watson, J.: Local Search and High Precision Gray Codes. In: Foundations of Genetic Algorithms FOGA-6, Morgan Kaufmann (2001)
2. Davis, L.: Bit-Climbing, Representational Bias, and Test Suite Design. In Booker, L., Belew, R., eds.: Proc. of the 4th Int'l. Conf. on GAs, Morgan Kaufmann (1991) 18–23

3. Rana, S., Whitley, D.: Representations, Search and Local Optima. In: Proceedings of the 14th National Conference on Artificial Intelligence AAAI-97, MIT Press (1997) 497–502
4. Mathias, K., Schaffer, J., Eshelman, L., Mani, M.: The effects of control parameters and restarts on search stagnation in evolutionary programming. In Eiben, G., Bäck, T., Schoenauer, M., Schwefel, H.P., eds.: ppsn5, Springer-Verlag (1998) 398–407
5. Eshelman, L.: The CHC Adaptive Search Algorithm: How to Have Safe Search When Engaging in Nontraditional Genetic Recombination. In Rawlins, G., ed.: FOGA -1, Morgan Kaufmann (1991) 265–283
6. Whitley, D., Mathias, K., Rana, S., Dzubera, J.: Evaluating Evolutionary Algorithms. Artificial Intelligence Journal **85** (1996) 1–32
7. Davis, L.: Handbook of Genetic Algorithms. Van Nostrand Reinhold, New York (1991)
8. Moscato, P.: Memetic Algorithms: A Short Introduction. In D. Corne, F. Glover, M.D., ed.: New Ideas in Optimization, McGraw-Hill (1999)
9. Whitley, L.D.: The GENITOR Algorithm and Selective Pressure: Why Rank Based Allocation of Reproductive Trials is Best. In Schaffer, J.D., ed.: Proc. of the 3rd Int'l. Conf. on GAs, Morgan Kaufmann (1989) 116–121
10. Booker, L.: Improving Search in Genetic Algorithms. In Davis, L., ed.: Genetic Algorithms and Simulated Annealing. Morgan Kaufmann (1987) 61–73

Distance between Populations

Mark Wineberg[1] and Franz Oppacher[2]

[1]Computing and Information Science
University of Guelph
Guelph, Canada
wineberg@cis.uoguelph.ca
[2]School of Computer Science
Carleton University
Ottawa, Canada
oppacher@scs.carleton.ca

Abstract. Gene space, as it is currently formulated, cannot provide a solid basis for investigating the behavior of the GA. We instead propose an approach that takes population effects into account. Starting from a discussion of diversity, we develop a distance measure between populations and thereby a population metric space. We finally argue that one specific parameterization of this measure is particularly appropriate for use with GAs.

1   Introduction: The Need for a Population Metric

All previous attempts to characterize gene space have focused exclusively on the Hamming distance and the hypercube. However, this 'chromosome space' cannot fully account for the behavior of the GA.

An analysis of the GA using chromosome space implicitly assumes that the fitness function alone determines where the GA will search next. This is not correct. The effect that the population has on the selection operation can easily be seen in the following (obvious) examples: In fitness proportional selection (fps) the fitness values associated with a chromosome cannot be derived from the fitness function acting on the chromosome alone, but also takes into account the fitness of all other members in the population. This is because the probability of selection in fps is based on the ratio of the 'fitness' of the individual to that of the total population. This dependence on population for the probability of selection is true not just for fitness proportional selection, but also for rank selection as the ranking structure depends on which chromosomes are in the population, and tournament selection since that can be reduced to a subset of all polynomial rank selections. Finally, and most glaringly, the probability of selecting a chromosome that is not in the population is zero; this is true no matter the fitness of the chromosome! Consequently the fitness value associated with the chromosome is meaningless when taken independently of the population.

As the above examples demonstrate, any metric that is used to analyze the behavior of the GA must include population effects. These effects are not made evident if only the chromosome space is examined. Therefore the metric used must include more information than just the distance between chromosomes; we must look to the population as a whole for our unit of measure. In other words, we need a distance between populations.

There are four sections in this paper. The first section examines the well-known population measure 'diversity' since the definitions and methodologies developed for it will form the basis of the distance measures. In the two sections after, two different approaches are introduced that attempt to determine the distance between populations. The first approach, the all-possible-pairs approach, is a natural extension of the traditional diversity definition. The second approach describes the mutation-change distance between populations. In the final section, a synthesis of these two distance concepts is developed eventually leading to a single definition of the distance between populations

2 Diversity

Before attempting to find a relevant distance between populations, it will be instructive to first discuss the related concept of 'diversity'.

There are three reasons for this. First, diversity is a known measure of the population that is independent of the fitness function. Since the distance between populations should likewise be independent of the fitness, useful insights may be derived from a study of diversity. Second, several techniques shall be introduced in this section that will become important later when discussing the distance between populations. Finally, the concept of diversity itself will be used in the analysis of the distance between populations.

3 All-Possible-Pairs Diversity

The simplest definition of diversity comes from the answer to the question "how different is everybody from everybody else?" If every chromosome is identical, there is no difference between any two chromosomes and hence there is no diversity in the population. If each chromosome is completely different from any other, then those differences add, and the population should be maximally diverse. So the diversity of a population can be seen as the difference between all possible pairs of chromosomes within that population.

While the above definition makes intuitive sense, there is one aspect not covered: what do we mean by different? If a pair of chromosomes is only different by one locus, it only seems reasonable that this pair should not add as much to the diversity of the population as a pair of chromosomes with every locus different. Consequently the difference between chromosomes can be seen as the Hamming distance or chromosome distance, and hence the population diversity becomes the sum of the Hamming distances between all possible pairs of chromosomes [1]. In cluster theory this is called the *statistic scatter* [2].

Now, since the Hamming distance is symmetric, and is equal to 0 if the strings are the same, only the lower (or, by symmetry, only the upper) triangle in a chromosome-pairs-table need be considered when computing the diversity. Consequently the all-possible-pairs diversity can be formalized as

$$\text{Div}(P) = \sum_{i=1}^{n}\sum_{j=1}^{i-1} \text{hd}(c_i, c_j) \qquad (1)$$

where P is the population, n is the population size, chromosome $c_i \in P$, l is the length of a chromosome and $\text{hd}(c_i, c_j)$ is the Hamming distance between chromosomes.

The Reformulation of the All-Possible-Pairs Diversity: A Linear Time Algorithm

A problem with formula (1) is its time complexity. Since the Hamming distance between any two pairs takes $O(l)$ time and there are n^2 possible pairs (actually $\frac{1}{2}n(n-1)$ pairs when symmetry is taken into account), then the time complexity when using (1) is $O(l\,n^2)$. Since the time complexity of the GA is $O(l \cdot n)$ calculating the diversity every generation would be expensive.

Surprisingly, a reformulation of definition (1) can be converted into an $O(l \cdot n)$ algorithm to compute the all-possible-pairs diversity.

Gene Counts and the Gene Frequencies

We will now introduce two terms that not only will be used to reformulate the definition of the all-possible-pairs diversity, but also will become ubiquitous throughout this paper. They are the *gene count* across a population, and the *gene frequency* of a population.

The gene count $c_k(\alpha)$ is the count across the population of all genes at locus k that equal the symbol α. This means that

$$c_k(\alpha) = \sum_{i=1}^{n} \delta_{i,k}(\alpha) \qquad (2)$$

where $\delta_{i,k}(\alpha)$ is a Kronecker δ that becomes 1 when the gene at locus k in chromosome i equals the symbol α, and otherwise is 0. Later in the paper we will frequently write $c_k(\alpha)$ as $c_{\alpha,k}$, or just as c_α if the locus k is understood in the context.

The array of the gene counts of each locus will be called the *gene count matrix*.

The gene frequency $f_k(\alpha)$ is the ratio of the gene count to the size of the population. In other words,

$$f_k(\alpha) = \frac{c_k(\alpha)}{n} \qquad (3)$$

Again, later in the paper we will frequently write $f_k(\alpha)$ as $f_{\alpha|k}$, or just as f_α if the locus k is understood in the context.

The array of the gene frequencies of each locus will be called the *gene frequency matrix*.

The Reformulation

With the notation in place we can present the alternate form of writing the all-possible-pairs diversity:

Theorem 1: The all-possible-pairs diversity can be rewritten as

$$\text{Div}(P) = \frac{n^2}{2l} \sum_{k=1}^{l} \sum_{\forall \alpha \in A} f_k(\alpha)(1 - f_k(\alpha)) \quad (4)$$

Proof: Let us first examine a chromosome c_i that at locus k has gene α. When computing all of the possible comparison pairs, 0 is obtained when compared to all of the other chromosomes that also have gene α at locus k. There are $n f_k(\alpha)$ of those. Consequently there are $n - n f_k(\alpha)$ comparisons that will return the value 1. So the component of the distance attributable to c_i is $n - n f_k(\alpha)$. Since there are $n f(\alpha)$ chromosomes that have the same distance component, the total distance contributed by chromosomes with gene α at locus k is $n f_k(\alpha)(n - n f_k(\alpha))$, which simplifies to $n^2 f_k(\alpha)(1 - f_k(\alpha))$. Summing over all α will give us double the comparison count (since we are adding to the count both $hd_k(c_i, c_j)$ and $hd_k(c_j, c_i)$). So the true comparison count is $\frac{n^2}{2} \sum_{\forall \alpha \in A} f_k(\alpha)(1 - f_k(\alpha))$. Averaging over all loci gives us the result we want. ∎

Normalizing (4) assuming that the alphabet size $a < n$ and that a divides into n evenly, we get

$$\overline{\text{Div}(P)} = \frac{a}{l(a-1)} \sum_{k=1}^{l} \sum_{\forall \alpha \in A} f_k(\alpha)(1 - f_k(\alpha)) \quad . \quad (5)$$

Since in the majority of GA implementations a binary alphabet is used with an even population size (because crossover children fill the new population in pairs), the above equation becomes

$$\overline{\text{Div}(P)} = \frac{2}{l} \sum_{k=1}^{l} \sum_{\forall \alpha \in A} f_k(\alpha)(1 - f_k(\alpha)) \quad . \quad (6)$$

The gene frequencies can be pre-computed for a population in $O(l \cdot n)$ time. Consequently, the formula above can be computed in $O(a \cdot l \cdot n)$, which reduces to $O(l \cdot n)$ since a is a constant of the system (usually equal to 2). Thus we show that the all-possible-pairs diversity can be computed in $O(l \cdot n)$ time, which is much faster than the $O(l \cdot n^2)$ time that the original naïve all-possible-pairs algorithm would take.

4 An All-Possible-Pairs "Distance" between Populations

The obvious extension of the all-possible-pairs diversity of a single population would be an all-possible-pairs distance between populations. Here we would take the Cartesian product between the two populations producing all possible pairs of chromosomes, take the Hamming distance between each of those pairs of chromosomes, and sum the results. Since there are $O(n \cdot m)$ such pairs (where n and m are the two population sizes) then assuming $m \propto n$, there would be $O(n^2)$ distances being combined. Consequently the resulting summation, if it turns out to be a distance, would be a squared distance. So formally we have:

$$\text{Dist}'(P_1, P_2) = \sqrt{\sum_{i=1}^{n_1} \sum_{j=1}^{n_2} \text{hd}(chr1_i, chr2_j)} \qquad (7)$$

where P_1 and P_2 are populations with population sizes of n_1 and n_2 respectively, $chr1_i \in P_1$ and $chr2_j \in P_2$, and i and j are indices into their respective population. The reason we are using the function name Dist' instead of Dist shall be explained in the next subsection. This 'distance' between populations is used in some pattern recognition algorithms and is called the *average proximity function*[1].

Following the same argument as with diversity presented when reformulating the diversity to become a linear algorithm, a frequency-based version of the same formula can be produced:

$$\text{Dist}'(P_1, P_2) = \sqrt{\frac{nm}{l} \sum_{k=1}^{l} \sum_{\forall \alpha \in A} f_{1,k}(\alpha)(1 - f_{2,k}(\alpha))} \qquad (8)$$

where $f_{1,k}(\alpha)$ is the gene frequency of the gene α at locus k across population P_1, and $f_{2,k}(\alpha)$ is the corresponding gene frequency for population P_2.

Problems

While initially attractive for its simple intuitiveness, the all-possible-pairs "distance" is unfortunately not a distance. While it is symmetric and non-negative, thus obeying distance properties M_2 and M_3, it fails on properties M_1 and M_4[2].

The failure of property M_1 is readily seen. M_1 states that the distance must be 0 iff the populations are identical; consequently the all-possible-pairs "distance" of a population to itself should be equal to 0. Instead it is actually the all-possible-pairs diversity measure, which is typically greater than 0. In fact, the diversity only equals 0 when all of the chromosomes in the population are identical!

Furthermore the all-possible-pairs "distance" also fails to satisfy M_4, the triangle inequality. This can be seen from the following example. Let A be a binary alphabet {0, 1} from which the chromosomes in all three populations that form the triangle

---

[1] [3] pg. 378.
[2] See Appendix A.

will be drawn. Let populations P_1 and P_3 both have a population size of 2 and P_2 have in it only a single chromosome. To make the situation even simpler, in all populations let each chromosome consist of only 1 locus. Now look at an example where the population make-up is as follows:

$$P_1 = \{<chr_{1,0},0>,<chr_{1,1},0>\},$$
$$P_2 = \{<chr_{2,0},0>\}$$
$$P_3 = \{<chr_{3,0},1>,<chr_{3,1},1>\}.$$

The corresponding gene frequencies are $f_1(0) = 1$, $f_1(1) = 0$, $f_2(0) = 1$, $f_2(1) = 0$, $f_3(0) = 0$ and $f_3(1) = 1$. Using the all-possible-pairs "distance" definition (8) we can calculate that $\text{Dist}(P_1,P_2) + \text{Dist}(P_2,P_3) = \sqrt{0} + \sqrt{2} = \sqrt{2},$ and that $\text{Dist}(P_1,P_3) = \sqrt{4} = 2$. Consequently $\text{Dist}(P_1,P_2) + \text{Dist}(P_2,P_3) < \text{Dist}(P_1,P_3)$ and so the triangle inequality does not hold.

Thus the all-possible-pairs "distance" cannot be considered a metric[3]. It is for this reason that we put the prime after the 'distance function' that has been developed so far.

Correcting the All-Possible-Pairs Population Distance

We will now modify the formula to turn it into a true distance.

We shall first deal with the failure to meet the triangle inequality. Definition (8) was written to be as general as possible. Consequently, it allows for the comparison of two populations of unequal size. In the counter-example showing the inapplicability of the triangle inequality, unequal sized populations were used. When populations of equal size are examined no counter-example presents itself. This holds even when the largest distance between P_1 and P_3 is constructed and with a P_2 specially chosen to produce the smallest distance to both P_1 and P_3. Generalizing this, we could redefine the definition (8) such that small populations are inflated in size while still keeping the equivalent population make-up. The same effect can be produced by dividing definition (8) by the population sizes, or in other words through normalization.

Now let us address the problem of non-zero self-distances. As noted in the previous subsection, this property fails because the self-distance, when comparing all possible pairs, is the all-possible-pairs diversity, which need not be zero. To rectify the situation we could simply subtract out the self-distances of the two populations from the all-possible-pairs distance equation[4]. Again we are removing the problems through normalization.

To summarize the above, looking first only at a single locus and normalizing the squared distance (which simplifies the calculation) we get:

---

[3] It is not even a measure. See [3] pg. 378 under the properties of the *average proximity function*.

[4] The $\frac{a-1}{2a}$ term in front of the two normalized diversities in the resulting distance equation is a re-normalization factor. It is needed to ensure that the resulting distance cannot go below zero, i.e. the distance stays normalized as required.

$$\text{Dist}_k^2(P_1,P_2) = \left(\overline{\text{Dist}'_k(P_1,P_2)}\right)^2 - \frac{a-1}{2a}\overline{\text{Div}_k(P_1)} - \frac{a-1}{2a}\overline{\text{Div}_k(P_2)} \qquad (9)$$

Now, let us substitute (8), the definition of $\text{Dist}'(P_1,P_2)$, into the above equation. Also let $\overline{\text{Div}_k(P)} = \frac{a}{(a-1)}\sum_{\forall \alpha \in A} f_k(\alpha)\cdot(1-f_k(\alpha))$, the normalized diversity from the diversity reformulation section modified for a single locus. Then (9) becomes

$$\text{Dist}_{L_2,k}(P_1,P_2) = \sqrt{\frac{1}{2}\sum_{\forall \alpha \in A}\left(f_{1,k}(\alpha)-f_{2,k}(\alpha)\right)^2} \qquad (10)$$

(the use of the L_2 subscript will become apparent in the next section).

Notice that the above distance is properly normalized[5]. Furthermore, this process has actually produced a distance (or rather a pseudo-distance):

Theorem 2: The function $\text{Dist}_{L_2,k}(P_1,P_2)$ is a pseudo-distance at a locus k.

Proof: First notice that $f_{1,k}(\alpha) - f_{2,k}(\alpha)$ forms a set of vector spaces (with k being the index of the set). Now $\sqrt{\sum_{\forall \alpha \in A}\left(f_{1,k}(\alpha)-f_{2,k}(\alpha)\right)^2}$ is the L_2-norm on those vector spaces. As noted in Appendix B, we know that the norm of a difference between two vectors $\|v-w\|$ obeys all distance properties. Consequently, the equation $\sqrt{\sum_{\forall \alpha \in A}\left(f_{1,k}(\alpha)-f_{2,k}(\alpha)\right)^2}$ is a distance. Any distance multiplied by a constant (in this case $\frac{1}{\sqrt{2}}$) remains a distance. However, $\text{Dist}_{l_2,k}(P_1,P_2)$ is a distance between gene frequency matrices, and of course there is a many-to-one relationship between populations and a gene frequency matrix. For example, you can crossover members of a population thus producing a new population with different members in it but with the same gene frequency matrix. Hence you can have two distinct populations with a distance of 0 between them. Consequently, $\text{Dist}_{L_2,k}(P_1,P_2)$ is a distance for gene frequency matrices, but only a pseudo-distance for populations. ∎

Using the L_2-norm, we can combine the distances for the various loci into a single pseudo-distance:

$$\text{Dist}_{L_2}(P_1,P_2) = \frac{1}{\sqrt{2l}}\sqrt{\sum_{k=1}^{l}\sum_{\forall \alpha \in A}\left(f_{1,k}(\alpha)-f_{2,k}(\alpha)\right)^2} \qquad (11)$$

While it would be nice to have an actual distance instead of a pseudo-distance between populations, most properties of metrics are true of pseudo-metrics as well. Furthermore, since the distances between gene frequency matrices are actual dis-

---

[5] The maximum occurs when $f_{1|k}(\alpha_1) = f_{2|k}(\alpha_2) = 1$ and $f_{1|k}(\alpha \neq \alpha_1) = f_{2|k}(\alpha \neq \alpha_2) = 0$.

tances, their use connotes a positioning in gene space, albeit with some loss of information.

5 The Mutational-Change Distance between Populations

While, in the section above, we were able to derive a population distance using an all-possible-pairs approach, it is a bit disappointing that to do so we needed to perform ad-hoc modifications. In this section we will approach the matter from a different perspective. We will define the distance between populations as the minimal number of mutations it would take to transform one population into the other.

The above definition of population distance is the generalization of the Hamming distance between chromosomes. With the distance between chromosomes we are looking at the number of mutations it takes to transform one chromosome into the other; with the distance between populations we directly substitute into the entire population each mutational change to create an entirely new population.

There are, of course, many different ways to change one population into another. We could change the first chromosome of the first population into the first chromosome of the other population; or we could change it into the other population's fifth chromosome. However, if we just examine one locus, it must be true that the gene counts of the first population must be transformed into those of the second by the end of the process. The number of mutations that must have occurred is just the absolute difference in the gene counts (divided by 2 to remove double counting).

There is one slight problem with the above definition. It only makes sense if the two populations are the same size. If they are of different size, no amount of mutations will transform one into the other. To correct for that, we transform the size of one population to equal that of the other.

To give the intuition behind the process that will be used, imagine two populations, one double the size of the other. If we want to enlarge the second population to the size of the first population, the most obvious approach is to duplicate each chromosome. The effect that this has is the matching of the size of the second population to the first while still maintaining all of its original *gene frequencies*. Since a population will not always be a multiple of the other, we duplicate each population n times, where n is the other population's size. Now both populations will have the same population size. So the duplication factor in front of the first population is n_2, the duplication factor in front of the second population is n_1, and the common population size is $n_1 n_2$. So we can now define the mutational-change distance between two populations at a locus as

$$\text{Dist}_{L_1,k}(P_1, P_2) = \sum_{\forall \alpha, \alpha \in A} \left| n_2 c_{1,k}(\alpha) - n_1 c_{2,k}(\alpha) \right|$$

$$= n_1 n_2 \sum_{\forall \alpha, \alpha \in A} \left| \frac{c_{1,k}(\alpha)}{n_1} - \frac{c_{2,k}(\alpha)}{n_2} \right|$$

$$= n_1 n_2 \sum_{\forall \alpha, \alpha \in A} \left| f_{1,k}(\alpha) - f_{2,k}(\alpha) \right|$$

which, when normalized, becomes

$$\text{Dist}_{L_1,k}(P_1,P_2) = \frac{1}{2} \sum_{\forall \alpha, \alpha \in A} |f_{1,k}(\alpha) - f_{2,k}(\alpha)| \qquad (12)$$

Notice the similarity between the above and the all-possible-pairs distance at a locus (10). We basically have the same structure except that the L_2-norm is replaced by the L_1-norm (hence the use of the L_1 and L_2 subscripts). Therefore, the argument that was used to prove Theorem 2 applies here as well. Consequently the mutational-change distance between populations at a locus is also a pseudo-distance.

Finally, averaging across the loci produces the mutational-change pseudo-distance between populations:

$$\text{Dist}_{L_1}(P_1,P_2) = \frac{1}{2l} \sum_{k=1}^{l} \sum_{\forall \alpha, \alpha \in A} |f_{1,k}(\alpha) - f_{2,k}(\alpha)| \qquad (13)$$

6 The L_k-Norms and the Distance between Populations

In the previous two sections we have seen two different distances (actually pseudo-distances) between populations derived through two very different approaches. Yet there seems to be the same underlying structure in each: the norm of the differences between gene frequencies. In one case the norm was the L_1-norm, in the other the L_2-norm, otherwise the two results were identical. Generalizing this, we can define an L_k-distance on the population:

$$\text{Dist}_{L_k}(P_a,P_b) = \sqrt[k]{\frac{1}{2l} \sum_{i=1}^{l} \sum_{\forall \alpha, \alpha \in A} |f_{a,i}(\alpha) - f_{b,i}(\alpha)|^k} \qquad (14)$$

and

$$\text{Dist}_{L_\infty}(P_a,P_b) = \max_{\substack{\forall \alpha, \alpha \in A \\ \forall i, i \in [1,l]}} \left(|f_{a,i}(\alpha) - f_{b,i}(\alpha)|\right) . \qquad (15)$$

Interestingly, the L_∞-distance restricted to a single locus can be recognized as the Kolmogorov-Smirnov test. The K-S test is the standard non-parametric test to determine whether there is a difference between two probability distributions.

Realizing that there are an infinite number of possible distance measures between populations, the question naturally arises: is one of the distance measures preferable or will any one do?

Of course, to a great degree the choice of distance measure depends on matching its properties to the purpose behind creating that distance measure in the first place; i.e. different distances may or may not be applicable in different situations.

That being said, there is a property possessed by the distance based on the L_1-norm which none of the others possess, making it the preferable distance. This property becomes evident in the following example. Let us examine 4 populations; the chromosomes in each population are composed of a single gene drawn from the quaternary alphabet {a, t, c, g}. The 4 populations are:

$$P_{1a} = \{<chr_1, a>, <chr_2, a>, <chr_3, a>, <chr_4, a>\}$$
$$P_{1b} = \{<chr_1, c>, <chr_2, c>, <chr_3, c>, <chr_4, c>\}$$
$$P_{2a} = \{<chr_1, a>, <chr_2, a>, <chr_3, t>, <chr_4, t>\}$$
$$P_{2b} = \{<chr_1, c>, <chr_2, c>, <chr_3, g>, <chr_4, g>\}$$

and so

$$f_{1a}(a) = 1, \quad f_{1a}(t) = 0, \quad f_{1a}(c) = 0, \quad f_{1a}(g) = 0,$$
$$f_{1b}(a) = 0, \quad f_{1b}(t) = 0, \quad f_{1b}(c) = 1, \quad f_{1b}(g) = 0,$$
$$f_{2a}(a) = \tfrac{1}{2}, \quad f_{2a}(t) = 0, \quad f_{2a}(c) = \tfrac{1}{2}, \quad f_{2a}(g) = 0,$$
$$f_{2b}(a) = 0, \quad f_{2b}(t) = \tfrac{1}{2}, \quad f_{2b}(c) = 0, \quad f_{2b}(g) = \tfrac{1}{2}.$$

Now, lets look at the two distances $Dist_{L_k}(P_{1a}, P_{1b})$ and $Dist_{L_k}(P_{2a}, P_{2b})$. In both cases the populations have no genes in common. We should therefore expect the distance between both pairs of populations to be the maximum distance that can be produced. It is true that the diversity within each of the first two populations is 0, while the diversity within each of the second two is greater than 0; however that should have nothing to do with the distances between the populations. One expects both distances to be equally maximal. Working out the distances from the equation

$$Dist_{L_k}(P_a, P_b) = \sqrt{\frac{1}{2} \sum_{\forall \alpha, \alpha \in A} |f_a(\alpha) - f_b(\alpha)|^k}$$

we get

$$Dist_{L_k}(P_{1a}, P_{1b}) = \left(\frac{1}{2} \cdot 2 \cdot (1)^k + \frac{1}{2} \cdot 2 \cdot (0)^k\right)^{\frac{1}{k}} = 1 \quad \text{and}$$

$$Dist_{L_k}(P_{2a}, P_{2b}) = \left(\frac{1}{2} \cdot 4 \cdot \left(\frac{1}{2}\right)^k\right)^{\frac{1}{k}} = 2^{\frac{k-1}{k}}$$

The only value of k for which the two distances will be equal (and since 1 is the maximum, they will be both maximal) is when $k = 1$. For the L_∞-norm, $Dist_{L_\infty}(P_{1a}, P_{1b}) = 1$ and $Dist_{L_\infty}(P_{2a}, P_{2b}) = \tfrac{1}{2}$, so it is only under the L_1-norm that the two distances are equal and maximal. The above property of the L_1-norm holds for any alphabet and population sizes.

7 Conclusion

The purpose of this paper is to develop a distance measure between populations. To do so we first investigated population diversity. Using our analysis of diversity as a template, we defined two notions of population distance, which we then generalized into the L_k- distance set. We picked the L_1-distance as the most appropriate measure for GAs because it is the only measure that consistently gives maximum distance for populations without shared chromosomes. This distance forms a metric space on populations that supersedes the chromosome-based gene space. We feel that this enhancement to the formulation of gene space is important for the further understanding of the GA.

References

1. Louis, S. J. and Rawlins, G. J. E.: Syntactic Analysis of Convergence in Genetic Algorithms. In: Whitley, L. D (ed.): Foundations of Genetic Algorithms 2. Morgan Kaufmann, San Mateo, California, (1993) 141–151
2. Duran, B. S. and Odell, P. L.: Cluster Analysis: A Survey. Springer-Verlag, Berlin (1974)
3. Theodoridis, S. and Koutroumbas, K.: Pattern Recognition, Academic Press, San Diego (1999)
4. Lipschutz, S. (1965). Schaum's Outline of Theory and Problems of General Topology. McGraw-Hill, Inc., New York (1965)

Appendix A: Distances and Metrics

While the concept of 'distance' and 'metric space' is very well known, there are many equivalent but differing definitions found in textbooks. A metric space is a set of points with an associated "distance function" or "metric" on the set. A distance function d acting on a set of points X is such that $d: X \times X \to R$, and that for any pair of points $x, y \in X$, the following properties hold:

M_1 $\qquad d(x, y) = 0$ iff $x = y$

M_2 $\qquad d(x, y) = d(y, x)$ $\qquad$ (Symmetry)

M_3 $\qquad d(x, y) \geq 0$

as well as a fourth property: for any 3 points $x, y, z \in X$,

M_4 $\qquad d(x, y) + d(y, z) \geq d(x, z)$ $\qquad$ (Triangle Inequality)

If for $x \neq y$, $d(x, y) = 0$, which is a violation of M_1, then d is called a pseudo-distance or pseudo-metric. If M_2 does not hold, i.e. the 'distance' is not symmetric, than d is called a quasi-metric. If M_4 (the triangle inequality) does not hold, d is called a semi-metric. Finally note that if d is a proper metric then M_3 is redundant, since it can be derived from the three other properties when z is set equal to x in M_4.

Appendix B: Norms

Norms are also a commonly known set of functions. Since we make use of norms so extensively, we felt that a brief summary of the various properties of norms would be helpful. A norm is a function applied to a vector in a vector space that has specific properties. From the Schaum's Outline on Topology[6] the following definition is given: "Let **V** be a real linear vector space ...[then a] function which assigns to each vector $v \in \mathbf{V}$ the real number $\|v\|$ is a *norm* on **V** iff it satisfies, for all $v, w \in \mathbf{V}$ and $k \in \mathbf{R}$, the following axioms:

[N_1] $\|v\| \geq 0$ and $\|v\| = 0$ iff $v = 0$.
[N_2] $\|v + w\| \leq \|v\| + \|w\|$.
[N_3] $\|kv\| = |k| \|v\|$.

The norm properties hold for each of the following well-known (indexed) functions:

$$L_k - norm = \| < a_1, \ldots, a_m > \| = \sqrt[k]{\sum |a_i|^k}.$$

Taking the limit as $k \to \infty$ of the L_k-norm produces the L_∞-norm:

$$L_\infty - norm = \max(|a_1|, |a_2|, \ldots, |a_m|).$$

The norm combines the values from the various dimensions of the vector into a single number, which can be thought of as the magnitude of the vector. This value is closely related to the distance measure. In fact, it is well known that the norm of the difference between any two vectors is a metric.

---

[6] [4] pg. 118.

The Underlying Similarity of Diversity Measures Used in Evolutionary Computation

Mark Wineberg[1] and Franz Oppacher[2]

[1]Computing and Information Science
University of Guelph
Guelph, Canada
wineberg@cis.uoguelph.ca
[2]School of Computer Science
Carleton University
Ottawa, Canada
oppacher@scs.carleton.ca

Abstract. In this paper we compare and analyze the various diversity measures used in the Evolutionary Computation field. While each measure looks quite different from the others in form, we surprisingly found that the same basic method underlies all of them: the distance between all possible pairs of chromosomes/organisms in the population. This is true even of the Shannon entropy of gene frequencies. We then associate the different varieties of EC diversity measures to different diversity measures used in Biology. Finally we give an $O(n)$ implementation for each of the diversity measures (where n is the population size), despite their basis in an $O(n^2)$ number of comparisons.

1 Introduction

In recent years there has been a growing interest in genetic diversity in the Evolutionary Computation field [1,2]. Diversity maintenance procedures are beginning to be emphasized, especially in the areas of multi-objective optimization [3], and dynamic environments in evolutionary systems [4,5].

There are many different diversity measures that can be found in the literature. The standard diversity measure is the sum of the Hamming distances between all possible pairs of chromosomes. Another popular measure is the use of the Shannon Entropy from Information Theory on gene frequencies. Even variance and standard deviation can be viewed as a diversity measure, especially when using real valued genes. This naturally leads to the question: which of these varied diversity measures is best?

Surprisingly, there appear to be a deep similarity between all of the above different measures. In this paper we analyze all of the above diversity measures and expose this similarity. We also present efficient procedures for computing these diversity measures. For example the time complexity for the "all possible pairs" diversity measure would naively be thought to be $O(n^2)$ because there are $O(n^2)$ chromosome pairings in a population of size n. The algorithm we present computes the value in $O(n)$ time.

2 All-Possible-Pairs Diversity

The simplest definition of diversity comes from the answer to the question "how different is everybody from everybody else?" If every chromosome is identical, there is no difference between any two chromosomes and hence there is no diversity in the population. If each chromosome is completely different from one another, then those differences add, and the population should be maximally diverse. So the diversity of a population can be seen as the difference between all possible pairs of chromosomes within that population.

While the above definition makes intuitive sense, there is one aspect not covered: what do we mean by different. If a pair of chromosomes is only different by one locus, it only seems reasonable that this pair should not add as much to the diversity of the population as a pair of chromosomes with every locus different. Consequently the difference between chromosomes can be seen as the Hamming distance or chromosome distance, where the Hamming distance is the sum of all loci where the two chromosomes have differing genes. Hence the population diversity becomes the sum of the Hamming distances between all possible pairs of chromosomes; see [6]. In cluster theory this is called the *statistic scatter*, see [7].

Now, since the Hamming distance is symmetric, and is equal to 0 if the strings are the same, only the lower triangle in a chromosome-pairs table need be considered when computing the diversity. Consequently the all-possible-pairs diversity can be formalized as

$$\text{Div}(P) = \sum_{i=1}^{n} \sum_{j=1}^{i-1} \text{hd}(c_i, c_j) \qquad (1)$$

where P is the population and chromosome $c_i \in P$ and n is the population size and $\text{hd}(c_i, c_j)$ is the Hamming distance between two chromosomes. This notation, as well as l being the length of a chromosome, shall be used throughout the paper.

3 The Reformulation of the All-Possible-Pairs Diversity: A Linear Time Algorithm

A problem with formula (1) is its time complexity. Since the Hamming distance between any two pairs takes $O(l)$ time, and there are n^2 possible pairs (actually $\frac{1}{2}n(n-1)$ pairs when symmetry is taken into account), then the time complexity when using (1) is $O(l\,n^2)$. Since the time complexity of the GA is $O(l \cdot n)$ calculating the diversity every generation would be expensive.

Fortunately a reformulation of definition (1) can be converted into an $O(l \cdot n)$ algorithm to compute the all-possible-pairs diversity. This will be developed directly.

It seems unlikely such an efficient algorithm would not already be present somewhere in the literature. Unfortunately, most mathematical textbooks that deal with sets of binary strings seem to be interested in the maximum of the distances between all possible pairs and not the average. Unlike the sum or average, the time complexity of the maximum Hamming Distance between all possible pairs cannot be reduced from $O(l\,n^2)$ to $O(l \cdot n)$ time, so the issue is ignored. While the maximum distance between all possible pairs of chromosomes can also be considered as a diversity measure on a GA population, it is not a particularly good one since it is too sensitive to outlier chromosomes, and does not give fine grain information about changes in the population.

Gene Counts and the Gene Frequencies

Before we can give the reformulation, we shall introduce two terms that will be extensively used throughout the paper: the **gene count** across a population, and the **gene frequency** of a population.

The gene count $c_k(\alpha)$ is the count across the population of all genes at locus k that equals the symbol α. This means that

$$c_k(\alpha) = \sum_{i=1}^{n} \delta_{i,k}(\alpha) \qquad (2)$$

where $\delta_{i,k}(\alpha)$ is a Kronecker δ that becomes 1 when the gene at locus k in chromosome i equals the symbol α, and otherwise is 0. The array of the gene counts of each locus will be called the **gene count matrix**[1].

The gene frequency $f_k(\alpha)$ is the ratio of the gene count to the size of the population. In other words,

$$f_k(\alpha) = \frac{c_k(\alpha)}{n} \qquad (3)$$

The array of the gene frequencies of each locus will be called the **gene frequency matrix**[2].

The Reformulation

With the notation in place we can present the alternate form of writing the all-possible-pairs diversity:

Theorem 1a: The all-possible-pairs diversity can be rewritten as

---

[1] For non-evolutionary systems, such as those used for cluster analysis or symbolic machine learning, the gene count matrix could be called the *symbol count matrix*.
[2] Just as with the gene count matrix, the gene frequency matrix could be called the *symbol frequency matrix* when dealing with non-evolutionary systems.

$$\text{Div}(P) = \frac{n^2}{2l} \sum_{k=1}^{l} \sum_{\forall \alpha \in A} f_k(\alpha)(1 - f_k(\alpha)) \qquad (4)$$

Proof: Let us first examine a chromosome ch_i that at locus k has gene α. When computing all of the possible comparison pairs, the value 0 is obtained from comparisons of ch_i with any other chromosomes that also have the same gene α at locus k. There are $n\,f_k(\alpha)$ chromosomes with gene α at locus k (including ch_i). Consequently there are $n - n\,f_k(\alpha)$ comparisons with chromosomes that do not have gene α at locus k, and hence will return the value 1. So the component of the distance attributable to ch_i is $n - n\,f_k(\alpha)$. Since there are $n\,f(\alpha)$ chromosomes that have the same gene at locus k, the total distance contributed by chromosomes with gene α at locus k is $n\,f_k(\alpha)(n - n\,f_k(\alpha))$, which simplifies to $n^2 f_k(\alpha)(1 - f_k(\alpha))$. Summing over all α will give us double the comparison count (since we are adding to the count both $hd_k(ch_i, ch_j)$ and $hd_k(ch_j, ch_i)$). So the true comparison count is $\frac{n^2}{2} \sum_{\forall \alpha \in A} f_k(\alpha)(1 - f_k(\alpha))$. Averaging over all loci gives us the result we want. ∎

Theorem 1b: The normalized all-possible-pairs diversity can be rewritten as

$$\overline{\text{Div}(P)} = \begin{cases} \dfrac{a}{l \cdot \left((a-1) - \dfrac{r(a-r)}{n^2}\right)} \sum_{k=1}^{l} \sum_{\forall \alpha \in A} f_k(\alpha) \cdot (1 - f_k(\alpha)) & a < n \\[2ex] \dfrac{n}{l \cdot (n-1)} \sum_{k=1}^{l} \sum_{\forall \alpha \in A} f_k(\alpha) \cdot (1 - f_k(\alpha)) & a \geq n \end{cases}$$

where $r = n \bmod a$.

Proof. To normalize this formula we must first find the maximum diversity value that can be obtained under any make-up of the population. Once this is found, the normalized diversity is just the regular diversity divided by this value.

Case 1) The alphabet size is strictly less than the size of the population ($a < n$)
In this case the diversity becomes maximal when all gene frequencies are as equal as possible[3], i.e. when $f_k(\alpha) = \frac{1}{a}$. Substituting this into the diversity equation (4) when modified slightly to handle the case when a does not evenly divide into n produces

$$\max(\text{Div}(P)) = \frac{n^2}{2a}\left((a-1) - \frac{r(a-r)}{n^2}\right) \qquad (5)$$

which gives us the result we want when divided into equation (4).

---

[3] This can be proven easily by solving for each of the frequencies from the set of equations produced by $\nabla \text{Div}(P) = 0$ after taking into account the constraint $\sum_{\forall \alpha} f_k(\alpha) = 1$.

Case 2) When the alphabet size is greater then or equal to the size of the population ($a \geq n$) each frequency becomes $\frac{1}{n}$. This is because there can only be the n different symbols in the population when maximally diverse. Also since $n \mid n$, $r = 0$. So, by setting $a = n$ and $r = 0$ in equation (5) and simplifying, then dividing this maximum into equation (4), the result required is obtained. ∎

In most cases $a < n$ and $a \mid n$, so r is 0 and the normalized all-possible-pairs diversity can be written as

$$\overline{Div(P)} = \frac{a}{l(a-1)} \sum_{k=1}^{l} \sum_{\forall \alpha \in A} f_k(\alpha)(1 - f_k(\alpha)) \qquad (6)$$

Since, in the majority of GA implementations a binary alphabet is used with an even population size (because crossover children fill the new population in pairs), the above equation becomes

$$\overline{Div(P)} = \frac{2}{l} \sum_{k=1}^{l} \sum_{\forall \alpha \in A} f_k(\alpha)(1 - f_k(\alpha)) \qquad (7)$$

An $O(l \cdot n)$ All-Possible-Pairs Diversity Algorithm

The normalized all-possible-pairs diversity measure (6) can be translated into an algorithm that computes this diversity in $O(l \cdot n)$ time.

To keep its time complexity down, each alphabet symbol is replaced in the chromosome with its corresponding index in the alphabet set. For example, if the alphabet is $A = \{a, t, c, g\}$, then the corresponding indices are a=1, t=2, c=3 and g=4, and the chromosome *aatccgctatag* becomes 112334321214. This is done to allow constant time indexing into the gene frequency array, based on the gene values.

The calculation is done in two parts. First the gene frequencies are found in 'findGeneFrequency' then the diversity is computed in 'APP_Diversity':

```
function findGeneFrequencies(p, lgth, a)
```

args: **population** p - *a population of chromosome*
          int lgth - *the length of the chromosome*
          int a - *the size of the alphabet*

vars: **float**[lgth,a] geneFreq; **int**[lgth,a] geneCount
          int[lgth] chr; **int** gene, i, j, k

code: geneFreq := makeArray(**float**, lgth , a)   geneCount := makeArray(**int**, lgth , a)   initAllValues(geneCount, 0)
              for each chr **in** p
                for k := 1 **to** lgth
                  gene := chr[k]
                  geneCount[k,gene] := geneCount[k,gene] + 1

```
                for i := 1 to lgth
                  for j := 1 to a
                    geneFreq[i,j] := geneCount[i,j] / size(p)
                return geneFreq

function APP_Diversity(p, lgth, a)
```

args: *same as in findGeneFrequencies( )*

vars: **float**[lgth,a] geneFreq
         int max; **float** diversity = 0

code: geneFreq := findGeneFrequencies(p, lngth, a)
         max := lgth * (a - 1) / a   *assumes a | lgth;*
         if a doesn't divide lgth, use max from Theorem1b
         for k := 1 to lgth
           for α := 1 to a
             diversity := diversity +
                          geneFreq[k,α] * (1-geneFreq[k,α])
         return diversity / max

Finding the gene frequencies costs $O(l \cdot n)$ time while the actual diversity is calculated in $O(l \cdot a)$ time. Since the alphabet size is considered to be a constant of the system (this is definitely true with the standard binary alphabet), the overall time complexity is $O(l \cdot n) + O(l) = O(l \cdot n)$. This is optimal time for this problem (each chromosome must at least be looked at once to affect the diversity value), so the all-possible-pairs diversity algorithm can be computed in $\Theta(l \cdot n)$ time. This is much faster than the $O(l \cdot n^2)$ time that the original naïve all-possible-pairs algorithm would take.

4   Diversity as Informational Entropy

In information theory, entropy is defined as [8]

$$H(X) = \sum_{x \in A} p(x) \log \frac{1}{p(x)} \tag{8}$$

where X is a discrete random variable with values taken from alphabet A and a probability mass function $p(x) = \Pr\{X = x\}$, $x \in A$. Equating the population at a locus with X and the gene frequencies at that locus $f_k(\alpha)$ with the probabilities $p(x)$, the entropic diversity at a locus can be written as:

$$\text{Div}_k(P) = \sum_{\alpha \in A} f_k(\alpha) \log \frac{1}{f_k(\alpha)} \tag{9}$$

Averaging over all loci gives the actual entropic diversity of the population:

$$\text{Div}(P) = \frac{1}{l}\sum_{k=1}^{l}\sum_{\alpha \in A} f_k(\alpha) \log \frac{1}{f_k(\alpha)} \tag{10}$$

To distinguish between the two different types of diversity, the all-possible-pairs diversity shall be symbolized by $\text{Div}_X(P)$ and the entropic diversity by $\text{Div}_H(P)$ (the X represents the complete Cartesian cross of the all possible pairs, and the H represents the entropy).

The entropic diversity is closely tied to the all-possible-pairs diversity in behavior. Preliminary experiments done by the authors show that the correlation between the two is very close, although not identical to 1. If we compare the definition (4) of all-possible-pairs diversity with the entropic diversity definition above, we see that aside from the constants in front, the two forms are remarkably similar. The only difference is the use of $\log \frac{1}{f_k(\alpha)}$ term in the entropic diversity instead of the $(1 - f_k(\alpha))$ term as used in the all-possible-pairs diversity. Furthermore both diversities can be seen as just the expected value of those terms. However, if we perform a Taylor expansion of $\log \frac{1}{f_k(\alpha)}$ around $\alpha = 1$, we get

$$\log \frac{1}{f_k(\alpha)} = (1 - f_k(\alpha)) + \frac{(1 - f_k(\alpha))^2}{2} + \frac{(1 - f_k(\alpha))^3}{3} + \cdots + \frac{(1 - f_k(\alpha))^i}{i} \tag{11}$$

Notice that the first term in the Taylor series is the same as the one used in the all-possible-pairs diversity definition. Also notice that the other terms are all less than 1 and rapidly approach 0 and consequently the early terms will dominate. So we can now see that $\text{Div}_X(P)$ is just the first term in the Taylor expansion about 1 of $\text{Div}_H(P)$, which accounts for the similarity in their behavior.

Now $(1 - f_k(\alpha))$ can be thought of as the "probability" that, if selected at random, the gene at this location won't be α. The other terms then can be seen as the probability under random selection that gene α won't be selected after i selections. Therefore, the diversity can be regarded as the expectation that the selected gene will be "some other gene". Since in each generation the GA selects from the population multiple times, and since the Taylor series above rapidly converges, the entropic diversity is used as a more accurate measure of diversity.

Finally, it is well known that the maximum of the Shannon entropy occurs under a uniform probability distribution. This, as we would expect, is the same as we found with the all-possible-pairs diversity. Consequently, the normalized form for the entropic diversity in the usual case when $a < n$ and $n \mid a$ is

$$\overline{\text{Div}(P)} = \frac{1}{l \log a}\sum_{k=1}^{l}\sum_{\alpha \in A} f_k(\alpha) \log \frac{1}{f_k(\alpha)} \tag{12}$$

(notice that if we choose the base of a for the logarithm, the original definition of entropic diversity (10) is already normalized). The corresponding formulae for the cases when $a < n$ but $n \nmid a$, and when $a \geq n$) are analogous to those developed for the all-possible-pairs diversity.

5 The Diversity Measures as Used in Biology and EC

Both of the all-possible-pairs formulations as well as the entropy diversity measure have been used in various biological fields of study such as genetics and ecology.

This diversity definition is actually used in the field of molecular genetics, although modified slightly to reflect the fact that, in practice, one only has a sampling of DNA sequences from a population. The modified formula is proportional to the all-possible-pairs definition of diversity,

$$\text{Div}(P) = \frac{2}{l \cdot n \cdot (n-1)} \sum_{i=1}^{P} \sum_{j=1}^{i-1} \text{hd}(c_i, c_j) \qquad (13)$$

and is called *nucleotide diversity*. For more details, see *Molecular Evolution* [9] pp. 237–238.

The reformulation of the all-possible-pairs diversity also has a biological interpretation; this time it is not in the recent area of molecular genetics, but in the older field of population genetics. Here the diversity is used to measure the variation of alleles in a population and is known as either the *gene diversity* or the *expected heterozygocity*. It is normally only defined for a single locus and is usually given in the form $1 - \sum_{\forall a \in A} f_a^2$, where A is the set of alleles at that locus. Remember, an allele at a locus may comprise an entire sequence, or even many sequences of nucleotides and so is working at a much higher level than the nucleotide diversity, even though the underlying mathematics is identical. For details, see [8] pp. 52–53.

This version of the (normalized) all-possible-pairs diversity has appeared in the GA literature [10] with reference made to the biological definition of heterozygocity, although the formula given had been modified to deal with binary values only. In the paper, no attempts were made to connect this diversity definition with the standard all-possible-pairs formulation.

The diversity measure based on the Shannon entropy also is commonly used in biology in the field of ecology, where it is used to compute the diversity of species, see [11] pp. 7–8. While less common, entropic diversity has also been used for the genetic diversity of populations in the EC field [12].

6 Diversity in Populations with Real Valued Genomes

Until now we have concentrated our attention on populations with binary and symbolic gene valued chromosomes. There we found that many measures that are naively thought to be distinct are in fact connected by the underlying concept of an all-possible-pairs comparison. It would be reasonable to expect that the situation would be very different when considering real valued genes. However, this is not the case.

At first one might think that the all-possible-pairs diversity for a locus k would be

$$\frac{1}{2} \sum_{j=1}^{n} \sum_{i=1}^{n} D(x_{i,k}, x_{j,k}) \qquad (14)$$

where D is the Euclidean distance between chromosomes. However, a simple summation of the distances is equivalent to taking the L_1-norm of the all-possible-pairs 'vector'. Furthermore, since D is based on the L_2-norm, it only makes sense to match the method of combination of all of the possible pairs with that used for the distance itself. There is nothing special about the averaging done by the L_1-norm. Rather the L_2-norm can be thought of as an 'average' that emphasizes the effects of larger differences. Consequently, using the L_2-norm, the all-possible-pairs diversity at a locus k becomes

$$Dv_k^2(P) = \frac{1}{2}\sum_{j=1}^{n}\sum_{i=1}^{n} D^2(x_{i,k}, x_{j,k}) \tag{15}$$

To obtain the actual all-possible-pairs diversity by combining all the diversities at the various loci, we again use the L_2-norm producing

$$Dv^2(P) = \frac{1}{2}\sum_{k=1}^{l}\sum_{j=1}^{n}\sum_{i=1}^{n} D^2(x_{i,k}, x_{j,k}) = \frac{1}{2}\sum_{k=1}^{l}\sum_{j=1}^{n}\sum_{i=1}^{n}(x_{i,k} - x_{j,k})^2 \tag{16}$$

Reformulating the Real Valued All-Possible-Pairs Diversity for a Linear Time Algorithm

The above formula, coded as is, would produce an $O(l \cdot n^2)$ algorithm. However, just as with systems that use symbolic genes, the formula can be rearranged to produce a program that has $O(l \cdot n)$ time complexity. For the symbolic gene systems, this was accomplished by introducing the frequency of a gene at a locus. While this cannot be used directly for systems using numeric genes, there is an analogous measure: the average gene value at a locus. This can be used to produce a reformulation of the all-possible-pairs diversity:

Theorem 2: Let $\overline{x_k} = \frac{1}{n}\sum_{i=1}^{n} x_{i,k}$ and $\overline{x_k^2} = \frac{1}{n}\sum_{i=1}^{n} x_{i,k}^2$.

Then $Dv^2(P) = n^2 \sum_{k=1}^{l}(\overline{x_k^2} - (\overline{x_k})^2)$

Proof.
$$Dv^2(P) = \frac{1}{2}\sum_{k=1}^{l}\sum_{j=1}^{n}\sum_{i=1}^{n}(x_{i,k} - x_{j,k})^2 \quad \text{from (15)}$$

$$= \frac{1}{2}\sum_{k=1}^{l}\sum_{j=1}^{n}\sum_{i=1}^{n}x_{i,k}^2 + \frac{1}{2}\sum_{k=1}^{l}\sum_{j=1}^{n}\sum_{i=1}^{n}x_{j,k}^2 - \sum_{k=1}^{l}\sum_{j=1}^{n}\sum_{i=1}^{n}x_{i,k}x_{j,k}$$

$$= \frac{n}{2}\sum_{k=1}^{l}\sum_{i=1}^{n}x_{i,k}^2 + \frac{n}{2}\sum_{k=1}^{l}\sum_{j=1}^{n}x_{j,k}^2 - \sum_{k=1}^{l}\left(\sum_{i=1}^{n}x_{i,k}\right)\left(\sum_{j=1}^{n}x_{j,k}\right)$$

$$= \frac{n^2}{2}\sum_{k=1}^{l}\overline{x_k^2} + \frac{n^2}{2}\sum_{k=1}^{l}\overline{x_k^2} - n^2\sum_{k=1}^{l}(\overline{x_k})^2$$

$$= n^2\sum_{k=1}^{l}(\overline{x_k^2} - (\overline{x_k})^2) \quad \blacksquare$$

Looking at the diversity equation in the theorem, we see that the all-possible-pairs diversity is dependent directly on population size and implicitly on the square of the chromosome length. Since intuitively the diversity of a population should not increase by simply duplicating the exact population or by exactly duplicating genes, we will define the true diversity as

$$Div(P) = \frac{1}{l}\sqrt{\sum_{k=1}^{l}(\overline{x_k^2} - (\overline{x_k})^2)} \tag{17}$$

which is simply the all-possible-pairs diversity divided by the population size and chromosome length.

We will now turn to the time complexity of an algorithm that implements the above definition of diversity. Since the average gene value, $\overline{x_k}$, and the average of the gene value squared, $\overline{x_k^2}$, can be computed in $O(n)$ time for a single locus, and since there are l loci, all of the average gene values and gene values squared can be computed in $O(l \cdot n)$ time. Furthermore, once the average gene values are obtained, the diversity squared can be computed in $O(l)$ time. Consequently the total time complexity of an algorithm to find the squared diversity is $O(l \cdot n)$.

All-Possible-Pairs Diversity and Statistical Variance

The common statistical measure for scatter around the mean is the variance (the square root of which is the standard deviation). The definition of the variance is the sum of the squared difference between each value and the overall mean: $\sigma^2(Y) = E[(Y - \mu)^2]$, where $\mu = E(Y)$ is the expected value of the random variable Y also called the mean. If the probability of each value in Y is unknown then if Y

is tested by taking n samples, then $E(Y) = \frac{1}{n}\sum_{i=1}^{n} Y_i$, which is called the sample mean and is frequently written as $\overline{Y}$. The sample variance would therefore be

$$Var(Y) = \frac{1}{n}\sum_{i=1}^{n}(Y_i - \overline{Y})^2 = \overline{(Y^2)} - \overline{Y}^2 \tag{18}$$

This is a well-known result and can be found in any statistical textbook.

Now compare the above sample variance with (17) the linear time representation of the all-possible-pairs diversity formula. What we have is the square root variance in the population of each gene averaged across all loci. In other words the all-possible-pairs diversity corresponds to the 'average' standard deviation of each gene in a chromosome.

7 Conclusion

Many diversity measures exist in both the biological and EC literature: gene diversity or expected heterozygocity, nucleotide diversity, entropy and variance and standard deviation. In this paper we have shown that all are restatements or slight variants of the basic sum of the distances between all possible pairs of the elements in a system. Consequently, experiments need not be done to distinguish between the various measures, trying to find which of them provides a better measure of diversity, as they are really all the same measure.

By recognizing how all of the diversity measures have been formed, different diversity measures of this family can easily be created: merely change the distance measure used between the pairs. In the diversity measured examined in this paper, the Hamming Distance was used for binary and symbolic chromosomes while the Euclidean Distance was used for chromosomes with real valued genes.

However, we have pointed out that care must be taken to match the way the distances are to be combined. If the distance is formed by the combination of differences of component parts, the method used to combine the distances should match the method used to combine the parts. For example, the Hamming distance uses the L1-norm to combine the differences between genes and so the L1-norm was used to combine the distances between chromosomes; the L2-norm is used to combine the differences between genes in the Euclidean distances used for real valued genomes and so the L2-norm was used to combine the corresponding distances between pairs of chromosomes with real valued genes.

Finally we have shown that care must be taken when implementing all-possible-pair style diversity measures, as it is frequently easy to take the $O(n^2)$ comparisons and manipulate them so that it takes only $O(n)$ time to compute. We have given the associated $O(n)$ algorithms for all the diversity measures studied in this paper.

References

1. Hutter M.: Fitness Uniform Selection to Preserve Genetic Diversity. In: Fogel D.B. (ed.): CEC'02. IEEE Press (2002) 783–788
2. Motoki, T.: Calculating the Expected Loss of Diversity of Selection Schemes. In Evolutionary Computation 10–4. MIT Press, Cambridge MA (2002) 397–423
3. Ang, K.H., Chong, G. and Li, Y.: Preliminary Statement on the Current Progress of Multi-Objective Evolutionary Algorithm Performance Measurement. In: Fogel D.B. (ed.): CEC'02. IEEE Press (2002) 1139–1144
4. Oppacher F. and M. Wineberg (1999). The Shifting Balance Genetic Algorithm: Improving the GA in a Dynamic Environment. In Banzhaf W. et al. (Eds.): GECCO'99. Morgan Kaufmann, San Francisco (1999) 504–510
5. Garrett, S.M., Walker, J.H.: Genetic Algorithms: Combining Evolutionary and 'Non'-Evolutionary Methods in Tracking Dynamic Global Optima. In Langdon W. B. et. al. (Eds.): GECCO-2002. Morgan Kaufmann, San Francisco (2002) 359–366
6. Louis, S.J. and Rawlins, G.J.E.: Syntactic Analysis of Convergence in Genetic Algorithms. In Whitley, L.D. (ed.); Foundations of Genetic Algorithms 2. Morgan Kaufmann, San Mateo, California. (1993) 141–151
7. Duran, B.S. and Odell, P.L.: Cluster Analysis: A Survey. Springer-Verlag, Berlin (1974)
8. Cover, T.M. and Thomas J.A.: Elements of Information Theory. John Wiley & Sons, New York (1991)
9. Li, W.-H.: Molecular Evolution. Sinauer Associates, Sunderland MA (1997)
10. Collins, R.J., and Jefferson, D.R.: Selection in Massively Parallel Genetic Algorithms. In: Belew, R.K., and Booker L.B. (eds.): 4th ICGA. Morgan Kaufmann, San Mateo, California (1991) 249–256
11. Pielou, E.C.: Ecological Diversity. John Wiley & Sons, New York (1975)
12. Rosca, J.: Entropy-driven adaptive representation. In: J. Rosca (ed.): Proceedings of the Workshop on Genetic Programming: From Theory to Real-World Applications. Tahoe City, California (1995) 23–32

Implicit Parallelism

Alden H. Wright[1], Michael D. Vose[2], and Jonathan E. Rowe[3]

[1] Dept. of Computer Science, University of Montana, Missoula, Montana 59812, USA
wright@cs.umt.edu
[2] Computer Science Department, University of Tennessee, Knoxville, TN 37996, USA
vose@cs.utk.edu
[3] School of Computer Science, University of Birmingham, Birmingham B15 2TT, Great Britain
J.E.Rowe@cs.bham.ac.uk

Abstract. This paper assumes a search space of fixed-length strings, where the size of the alphabet can vary from position to position. Structural crossover is mask-based crossover, and thus includes n-point and uniform crossover. Structural mutation is mutation that commutes with a group operation on the search space. This paper shows that structural crossover and mutation project naturally onto competing families of schemata. In other words, the effect of crossover and mutation on a set of string positions can be specified by considering only what happens at those positions and ignoring other positions. However, it is not possible to do this for proportional selection except when fitness is constant on each schema of the family. One can write down an equation which includes selection which generalizes the Holland Schema theorem. However, like the Schema theorem, this equation cannot be applied over multiple time steps without keeping track of the frequency of every string in the search space.

1 Introduction

This paper describes the remarkable properties of structural crossover and mutation with respect to families of competing schemata. Recall that a schema defines certain components as taking on fixed values, while the remaining components are free to vary. Two schemata are in the same family if they specify fixed values for the same set of components. They are competing if they specify different values at these components. Given a set of components, the set of all possible fixed values that can be assigned to them gives us a whole competing family of schemata.

Vose and Wright, 2001 showed that each schema corresponds to a vector in population space. These schema vectors make up the components of a generalized (and redundant) coordinate system for population space. As the genetic algorithm moves from point to point in population space, it might be said to "process" schemata in the sense that each point in population space determines the frequency of each schema. This processing of many schemata has traditionally been called "implicit parallelism". However, the components corresponding to nontrivial schemata families are not independently processed and thus any implication that the genetic algorithm gains processing leverage due to the processing of schemata is misguided.

The set of components corresponding to a competing family of schemata may be picked out by a binary mask, and we can view such a mask as being a projection onto a smaller search space (in which we ignore what happens at the other positions). The remarkable result that we will prove is that structural crossover and mutation project naturally onto these families. That is, we can specify the effect of crossover and mutation just on the set of positions under consideration, ignoring what happens at other positions. Because this result applies to all schemata families simultaneously, we dub it the "Implicit Parallelism Theorem". In doing so, we hope to eradicate the previous usage of this phrase (which did not apply to a theorem at all, but to a fundamental mistake) and rescue it for a more suitable usage.

The result that operators project naturally onto schemata families applies only to crossover and mutation. It would be nice if it also applied to selection, and this possibility is investigated. The conclusion, however, is that it cannot (except in the trivial case where fitness is a constant for each schema in a schema family). One can, of course, write down an equation which involves selection, as well as crossover and mutation. This equation is, in fact, more elegant and meaningful than Holland's original Schema Theorem, but it suffers from the same fundamental flaw. In order to compute the average fitness of a schema at a given generation, one needs to know all the details of the entire population at that generation. It is therefore impossible to project onto a schemata family and ignore what happens at the other components. One implication of this is that one cannot iterate the equation: like the Schema Theorem, it can be applied over more than one time-step only by keeping track of the frequency of every string in the search space.

2 Structural Search Spaces and Operators

We generalize Vose, 1999 by introducing a class of genetic operators associated with a certain subgroup structure (for more details, see Rowe et al., 2002). Suppose the search space forms a group $(\Omega, \oplus)$ which has nontrivial subgroups $A_0, \ldots, A_{\ell-1}$ such that for all i, j, x,

1. $\Omega = A_0 \oplus \ldots \oplus A_{\ell-1}$
2. $i \neq j \implies A_i \cap A_j = \{0\}$
3. $x \oplus A_i = A_i \oplus x$

Then Ω is the *internal direct sum* of the A_i (which are *normal* subgroups of Ω) and each element $x \in \Omega$ has a unique representation $x = x_0 \oplus \ldots \oplus x_{\ell-1}$ where $x_i \in A_i$. The map

$$x \longmapsto \langle x_0, \ldots, x_{\ell-1} \rangle$$

is an isomorphism between Ω and the product group $A_0 \times \ldots \times A_{\ell-1}$ (see Lang, 1993), where

$$\langle x_0, \ldots, x_{\ell-1} \rangle \oplus \langle y_0, \ldots, y_{\ell-1} \rangle = \langle x_0 \oplus y_0, \ldots, x_{\ell-1} \oplus y_{\ell-1} \rangle$$

and

$$\ominus \langle x_0, \ldots, x_{\ell-1} \rangle = \langle \ominus x_0, \ldots, \ominus x_{\ell-1} \rangle$$

(where $\ominus$ indicates the inverse of an element). In this case the search space is called *structural*. Assume for the remainder of this paper that Ω is structural, and identify $x \in \Omega$ with $\langle x_0, \ldots, x_{\ell-1} \rangle$.

As an example, consider a length ℓ string representation where the cardinality of the alphabet at string position j is c_j. The alphabet at position j can be identified with $\mathcal{Z}_{c_j}$ (the integers modulo c_j), and the $\oplus$ operator can be componentwise addition (modulo c_j at position j). Then Ω is isomorphic to the direct product $\mathcal{Z}_{c_0} \times \ldots \times \mathcal{Z}_{c_{\ell-1}}$. This example extends the situation considered in Koehler et al., 1997 where $c_0 = c_1 = \cdots = c_{\ell-1}$. The standard example of fixed-length binary strings is a special case in which $c_j = 2$ for all positions j.

A concrete example of the above is: $\ell = 2$, $c_0 = 3$, $c_1 = 2$, so Ω is isomorphic to $\mathcal{Z}_3 \times \mathcal{Z}_2$. When we write elements of Ω as strings, the standard practice of putting the least significant bit to the right is followed. Thus,

$$\Omega = \{00, 01, 10, 11, 20, 21\} = \{0, 1, 2, 3, 4, 5\}$$

The group operator works by applying addition modulo 3 to the left bit, and addition modulo 2 to the right bit. For example

$$21 \oplus 11 = 00$$

The element 00 is the identity.

The set $\mathcal{B}$ of *binary masks* corresponding to Ω is

$$\mathcal{B} = \{\langle b_0, \ldots, b_{\ell-1}\rangle : b_i \in \mathcal{Z}_2\}$$

where $\mathcal{Z}_2$ is the set of integers modulo 2. Note that $\mathcal{B}$ is an Abelian group under component-wise addition modulo 2. It is notationally convenient to let $\oplus$ also denote the group operation on $\mathcal{B}$; hence $\oplus$ is *polymorphic*.[1] Let $\otimes$ denote component-wise multiplication on $\mathcal{B}$, and let $1 \in \mathcal{B}$ be the identity element for $\otimes$. For $b \in \mathcal{B}$, define $\bar{b}$ by

$$\bar{b} = 1 \oplus b$$

It is notationally convenient to extend $\otimes$ to a commutative operator acting also between elements $b \in \mathcal{B}$ and elements $x \in \Omega$ by

$$\langle b_0, \ldots, b_{\ell-1}\rangle \otimes \langle x_0, \ldots, x_{\ell-1}\rangle = \langle b_0 x_0, \ldots, b_{\ell-1} x_{\ell-1}\rangle$$

where $0 x_i = 0 \in \Omega$ and $1 x_i = x_i \in \Omega$. Here the right hand sides are elements of Ω; hence $\otimes$ is polymorphic.

It is easy to check that for all $x, y \in \Omega$ and $u, v \in \mathcal{B}$

$$x = x \otimes 1$$
$$1 = u \oplus \bar{u}$$
$$0 = u \oplus u$$
$$0 = u \otimes \bar{u}$$
$$(x \oplus y) \otimes u = (x \otimes u) \oplus (y \otimes u)$$
$$(x \otimes u) \oplus (y \otimes \bar{u}) = (y \otimes \bar{u}) \oplus (x \otimes u)$$
$$(x \otimes u) \otimes v = x \otimes (u \otimes v)$$
$$\ominus(u \otimes x) = u \otimes (\ominus x)$$

---

[1] An operator is polymorphic when its definition depends upon the type of its arguments.

To simplify notation, $\otimes$ takes precedence over $\oplus$ by convention. If $b \in \mathcal{B}$ is a mask then $\#b$ denotes the number of ones it contains.

Let $\mathcal{X}$ be a probability distribution over the set of binary masks,

$$\mathcal{X}_b = \text{the probability of mask } b$$

Structural crossover with distribution $\mathcal{X}$ applied to parents u and v corresponds to choosing binary mask b with probability $\mathcal{X}_b$ and then producing the offspring $u \otimes b \oplus \bar{b} \otimes v$. The probability that parents $u, v \in \Omega$ have child k is therefore

$$r(u, v, k) = \sum_{b \in \mathcal{B}} \mathcal{X}_b [u \otimes b \oplus \bar{b} \otimes v = k]$$

The corresponding crossover scheme $\mathcal{C}$ is also called structural and satisfies

$$\mathcal{C}(p)_k = \sum_{u,v} p_u p_v \sum_{b \in \mathcal{B}} \frac{\mathcal{X}_b + \mathcal{X}_{\bar{b}}}{2} [u \otimes b \oplus \bar{b} \otimes v = k]$$

where $p \in \mathbb{R}^{|\Omega|}$ is a distribution over the search space Ω. That is, p is a *population vector* in which p_k is the proportion of the population made up of copies of element k. $\mathcal{C}(p)$ gives the expected distribution after the application of crossover. For example, for uniform crossover with crossover rate u, the probability distribution $\mathcal{X}$ is given by $\mathcal{X}_0 = 1 - u + u/2^\ell$ and $\mathcal{X}_b = u/2^\ell$ for $b \neq 0$.

Let μ be a probability distribution over Ω,

$$\mu_k = \text{the probability of } k$$

Structural mutation with distribution μ applied to $v \in \Omega$ corresponds to choosing k with probability μ_k and then producing the result $v \oplus k$. The probability that v mutates to u is therefore

$$U_{u,v} = \mu_{\ominus v \oplus u}$$

The corresponding mutation scheme $\mathcal{U}$ is also called structural. $\mathcal{U}(p) = Up$ gives the effect of applying mutation to population vector $p \in \mathbb{R}^{|\Omega|}$.

3 Masks as Projections

This section generalizes Vose and Wright, 2001. Assume Ω is structural, crossover is structural, and mutation is structural. Each binary mask b has associated subgroup

$$\Omega_b = b \otimes \Omega$$

The map

$$x \longmapsto b \otimes x$$

is a homomorphism from Ω to Ω_b since $b \otimes (x \oplus y) = b \otimes x \oplus b \otimes y$. The kernel is the normal subgroup $\Omega_{\bar{b}}$, and, therefore, the following map from the image Ω_b to the quotient group $\Omega/\Omega_{\bar{b}}$ is an isomorphism Lang, 1993,

$$z = z \otimes b \longmapsto \Omega_{\bar{b}} \oplus z$$

The quotient group $\Omega/\Omega_{\bar{b}} = \{\Omega_{\bar{b}} \oplus z : z \in \Omega_b\}$, being comprised of disjoint schemata, is referred to as the *schema family corresponding to b*, and schema $\Omega_{\bar{b}} \oplus z$ is referred to as the *schema corresponding to $z \in \Omega_b$*.

For $b \in \mathcal{B}$, define Λ_b as

$$\Lambda_b = \left\{ p \in \mathbb{R}^{|\Omega_b|} : p_k \geq 0, \sum p_k = 1 \right\}$$

The linear operator $\Xi_b : \mathbb{R}^{|\Omega|} \longrightarrow \mathbb{R}^{|\Omega_b|}$ with matrix

$$(\Xi_b)_{i,j} = [j \otimes b = i]$$

is called the *operator associated with the schema family corresponding to b*; it has rows indexed by elements of Ω_b and columns indexed by Ω. Notice that $\Xi_b(\Lambda) \subseteq \Lambda_b$. To simplify notation, we will refer simply to Ξ when the binary mask b is understood.

For the example of the fixed length string representation where Ω is isomorphic to $\mathcal{Z}_3 \times \mathcal{Z}_2$, for $b = 10$,

$$\Xi_{10} = \begin{bmatrix} 1 & 1 & 0 & 0 & 0 & 0 \\ 0 & 0 & 1 & 1 & 0 & 0 \\ 0 & 0 & 0 & 0 & 1 & 1 \end{bmatrix}$$

and for $b = 01$,

$$\Xi_{01} = \begin{bmatrix} 1 & 0 & 1 & 0 & 1 & 0 \\ 0 & 1 & 0 & 1 & 0 & 1 \end{bmatrix}$$

Note that

$$\sum_i (\Xi p)_i = \sum_{i \in \Omega_b} \sum_j [j \otimes b = i] p_j$$
$$= \sum_j p_j \sum_{i \in \Omega_b} [j \otimes b = i]$$
$$= \sum_j p_j$$

Hence if $p \in \Lambda$ is a probability vector, then $\Xi p \in \Lambda_b$ is a probability vector. As the following computation shows, the ith component of Ξp is simply the proportion of the population p which is contained in the schema $\Omega_{\bar{b}} \oplus i$ which corresponds to $i \in \Omega_b$,

$$(\Xi p)_i = \sum_j [j \otimes b = i] p_j$$
$$= \sum_j [j \otimes \bar{b} \oplus j \otimes b = j \otimes \bar{b} \oplus i] p_j$$
$$= \sum_j [j = j \otimes \bar{b} \oplus i] p_j$$
$$\leq \sum_j [j \in \Omega_{\bar{b}} \oplus i] p_j$$

Conversely, given $i \in \Omega_b$,

$$\sum_j [j \in \Omega_{\bar{b}} \oplus i] p_j \leq \sum_j [b \otimes j \in b \otimes \Omega_{\bar{b}} \oplus b \otimes i] p_j$$
$$= \sum_j [j \otimes b = i] p_j$$

The matrix Ξ therefore projects from the distribution over all possible strings to a distribution over a family of competing schemata. For example, using traditional schema notation, we could write:

$$\Xi_{10}(p_{00}, p_{01}, p_{10}, p_{11}, p_{20}, p_{21}) = (p_{0*}, p_{1*}, p_{2*})$$

and

$$\Xi_{01}(p_{00}, p_{01}, p_{10}, p_{11}, p_{20}, p_{21}) = (p_{*0}, p_{*1})$$

Let $\mathcal{B}_b = b \otimes \mathcal{B}$. It is notationally convenient to make Ξ_b polymorphic by extending it to also represent the linear map $\Xi_b : \mathbb{R}^{|\mathcal{B}|} \longrightarrow \mathbb{R}^{|\mathcal{B}_b|}$ with matrix

$$(\Xi_b)_{i,j} = [j \otimes b = i]$$

Here the rows are indexed by elements of $\mathcal{B}_b$ and columns are indexed by $\mathcal{B}$. Again, we will drop the subscript and refer simply to Ξ when the mask is understood.

For the example of the fixed length string representation where Ω is isomorphic to $\mathcal{Z}_3 \times \mathcal{Z}_2$, the set of masks is $\mathcal{B} = \{00, 01, 10, 11\}$. For $b = 10$,

$$\Xi_{10} = \begin{bmatrix} 1 & 1 & 0 & 0 \\ 0 & 0 & 1 & 1 \end{bmatrix},$$

and for $b = 01$,

$$\Xi_{01} = \begin{bmatrix} 1 & 0 & 1 & 0 \\ 0 & 1 & 0 & 1 \end{bmatrix}.$$

Note that

$$\sum_i (\Xi x)_i = \sum_{i \in \mathcal{B}_b} \sum_j [j \otimes b = i] x_j$$
$$= \sum_j x_j \sum_{i \in \mathcal{B}_b} [j \otimes b = i]$$
$$= \sum_j x_j$$

Hence if $x \in \mathbb{R}^{|\mathcal{B}|}$ is a probability vector, then $\Xi x \in \mathbb{R}^{|\mathcal{B}_b|}$ is a probability vector.

4 The Implicit Parallelism Theorem

Given that $\Omega = A_0 \oplus \cdots \oplus A_{\ell-1}$ is structural, Ω_b is also structural,

$$\Omega_b = A_{k_0} \oplus \cdots \oplus A_{k_{\#b-1}}$$

where $\{k_0, \ldots, k_{\#b-1}\} = \{i : b_i = 1\}$. Moreover, Λ_b is precisely the Λ previously defined as corresponding to the search space, if the search space is chosen to be Ω_b. Likewise, $\mathcal{B}_b$ is precisely the $\mathcal{B}$ previously defined as corresponding to the search space, if the search space is chosen to be Ω_b. Therefore, since $\Xi\chi$ and $\Xi\mu$ are probability vectors indexed by $\mathcal{B}_b$ and Ω_b (respectively), they have corresponding structural crossover and mutation schemes $\mathcal{C}_b$ and $\mathcal{U}_b$ which represent crossover and mutation on the state space Λ_b.

Theorem 1. *If* $\mathcal{M} = \mathcal{U} \circ \mathcal{C}$, *then* $\Xi\mathcal{M}(x) = \mathcal{U}_b \circ \mathcal{C}_b(\Xi x)$

Proof Let $\mathcal{M}_b = \mathcal{U}_b \circ \mathcal{C}_b$. The k th component of $\Xi\mathcal{M}(x)$ is

$$\sum_{k' \in \Omega_{\overline{b}} \oplus k} \mathcal{M}(x)_{k'} = \sum_{k' \in \Omega_{\overline{b}}} \mathcal{M}(x)_{k \oplus k'}$$

$$= \sum_{u,v \in \Omega_b} \sum_{u',v' \in \Omega_{\overline{b}}} x_{u \oplus u'} x_{v \oplus v'} \sum_{k' \in \Omega_{\overline{b}}} \mathcal{M}_{u \oplus u' \ominus k \ominus k', v \oplus v' \ominus k \ominus k'}$$

$$= \sum_{u,v \in \Omega_b} \sum_{u' \in \Omega_{\overline{b}}} x_{u \oplus k \oplus u'} \sum_{v' \in \Omega_{\overline{b}}} x_{v \oplus k \oplus v'} \sum_{k' \in \Omega_{\overline{b}}} \mathcal{M}_{u \ominus k', v \oplus v' \ominus u' \ominus k'}$$

The innermost sum above is

$$\sum_{k' \in \Omega_{\overline{b}}} \sum_{i \in \Omega_b} \sum_{i' \in \Omega_{\overline{b}}} \sum_{j \in \mathcal{B}_b} \sum_{j' \in \otimes \mathcal{B}_{\overline{b}}} \mu_{i \oplus i'} \frac{\chi_{j \oplus j'} + \chi_{\overline{j \oplus j'}}}{2}$$

$$[((i \oplus i') \oplus (u \ominus k')) \otimes (j \oplus j') \oplus (\overline{j \oplus j'}) \otimes (v \oplus v' \ominus u' \ominus k')) = 0]$$

Note that the indicator function above is equivalent to

$$[i' \ominus k' \otimes j' \oplus (\overline{b} \oplus j') \otimes (v' \ominus u' \ominus k') = 0][i \oplus u \otimes j \oplus (b \oplus j) \otimes v = 0]$$

The first factor above is equivalent to $[i' \oplus (\overline{b} \oplus j') \otimes (v' \ominus u') = k']$ which determines k'. This is most easily seen by choosing the search space to be $\Omega_{\overline{b}}$, in which case the first factor is an expression over the search space and its binary masks, and $\overline{b}$ is the identity element for $\otimes$; in that context $(\overline{b} \oplus j')$ is $\overline{j}'$ and the first factor becomes

$$[i' \ominus k' \otimes j' \oplus \overline{j}' \otimes (v' \ominus u' \ominus k') = 0]$$
$$= [i' \ominus k' \otimes j' \oplus \overline{j}' \otimes (v' \ominus u') \ominus \overline{j}' \otimes k' = 0]$$
$$= [i' \oplus \overline{j}' \otimes (v' \ominus u') \ominus k' \otimes j' \ominus \overline{j}' \otimes k' = 0]$$
$$= [i' \oplus \overline{j}' \otimes (v' \ominus u') \ominus k' = 0]$$
$$= [i' \oplus \overline{j}' \otimes (v' \ominus u') = k']$$

It follows that the sum above is

$$\sum_{i \in \Omega_b} \sum_{j \in B_b} [i \oplus u \otimes j \oplus (b \oplus j) \otimes v = 0] \sum_{i' \in \Omega_{\bar{b}}} \mu_{i \oplus i'} \sum_{j' \in \otimes B_{\bar{b}}} \frac{\chi_{j \oplus j'} + \chi_{\overline{j \oplus j'}}}{2}$$

$$= \sum_{i \in \Omega_b} \sum_{j \in B_b} [i \oplus u \otimes j \oplus (b \oplus j) \otimes v = 0] (\Xi\mu)_i \frac{(\Xi\chi)_j + (\Xi\chi)_{\bar{j}}}{2}$$

$$= (M_b)_{u,v}$$

Where M_b is the mixing matrix for $\mathcal{M}_b$. Therefore, the The kth component of $\Xi\mathcal{M}(x)$ is

$$\sum_{u,v \in \Omega_b} (M_b)_{u,v} \sum_{u' \in \Omega_{\bar{b}}} x_{u \oplus k \oplus u'} \sum_{v' \in \Omega_{\bar{b}}} x_{v \oplus k \oplus v'}$$

$$= \sum_{u,v \in \Omega_b} (M_b)_{u,v} (\Xi x)_{u \oplus k} (\Xi x)_{v \oplus k}$$

$$= \mathcal{M}_b(\Xi x)_k$$

□

Corollary 1 (Implicit Parallelism).

$$\mathcal{M} = \mathcal{U} \circ \mathcal{C} \implies \Xi\mathcal{M} = \mathcal{M}_b \circ \Xi \text{ where } \mathcal{M}_b = \mathcal{U}_b \circ \mathcal{C}_b$$
$$\mathcal{M} = \mathcal{C} \circ \mathcal{U} \implies \Xi\mathcal{M} = \mathcal{M}_b \circ \Xi \text{ where } \mathcal{M}_b = \mathcal{C}_b \circ \mathcal{U}_b$$

In particular, $\Xi\mathcal{U} = \mathcal{U}_b \circ \Xi$ and $\Xi\mathcal{C} = \mathcal{C}_b \circ \Xi$.

Proof The first implication is theorem 1. A special case is $\mathcal{C} = \mathcal{I}$, in which case the conclusion is

$$\Xi\mathcal{U}(x) = \mathcal{U}_b(\Xi x)$$

Another special case is $\mathcal{U} = \mathcal{I}$, in which case the conclusion is

$$\Xi\mathcal{C}(x) = \mathcal{C}_b(\Xi x)$$

Consequently,

$$\Xi\mathcal{C} \circ \mathcal{U} = \mathcal{C}_b \circ \Xi \circ \mathcal{U} = \mathcal{C}_b \circ \mathcal{U}_b \circ \Xi$$

□

Corollary 1 speaks to schemata through the isomorphism $\Omega_b \cong \Omega/\Omega_{\bar{b}}$ given by

$$x \longmapsto \Omega_{\bar{b}} \oplus x$$

Therefore, $\mathcal{M}_b$ represents mixing (i.e., crossover and mutation) on a search space of schemata (i.e., the schema family $\Omega/\Omega_{\bar{b}}$).

A consequence of corollary 1 is that, independent of the order of crossover and mutation, *the following commutative diagram holds, in parallel, for every choice of schema family, simultaneously*

$$\begin{array}{ccc} x & \longrightarrow & \mathcal{M}(x) \\ \Xi \downarrow & & \downarrow \Xi \\ \Xi x & \longrightarrow & \mathcal{M}_b(\Xi x) \end{array}$$

Because this result does speak to parallelism and schemata—subjects which implicit parallelism has classically dealt with—Vose (1999) has redefined the phrase "implicit parallelism" to refer to it.[2] This use of the term conflicts with that employed by GA practitioners (for example in Holland, 1975,Goldberg, 1989). To the extent "implicit parallelism" has traditionally indicated that some kind of "processing leverage" is enjoyed by GAs, traditional usage has been misguided; genetic algorithms exhibit no such behaviour, nor can the theorems of Holland establish such a result. Because corollary 1 *does* address exactly what happens within all schema families, in parallel, simultaneously, it is proposed as an appropriate alternative to take over the "Implicit Parallelism" label, in the hope that the misguided and incorrect traditional notion be eradicated.

Example

Let us consider the implicit parallelism of mutation on the example $\Omega = \mathcal{Z}_3 \times \mathcal{Z}_2$. Firstly, let us define our mutation operator by the probability distribution:

$$\mu_j = \begin{cases} 0.9 & \text{if } j = 00 \\ 0.02 & \text{otherwise} \end{cases}$$

That is, there is a probability of 0.9 that no mutation will take place. Otherwise, we pick an element $j \in \Omega$ at random (uniformly) and apply it to our current individual. Now suppose that we are interested in what happens in the first component. That is, we are concerned with the effect of mutation on the family of schemata $0*, 1*, 2*$. One way to calculate this would be to work out the effect of mutation on the whole population and then sum up the results for each schema in the family. The implicit parallelism theorem tells us that we don't need to do this. Instead, we can find a mutation operator that acts on the family of schemata itself, and has the exact equivalent effect.

For a concrete example, consider the population vector

$$p = (p_{00}, p_{01}, p_{10}, p_{11}, p_{20}, p_{21}) = (0.1, 0.2, 0.1, 0.2, 0.25, 0.15)$$

Our family of schemata corresponds to the mask $b = 10$. We have already seen that this gives us a matrix

$$\Xi_{10} = \begin{bmatrix} 1 & 1 & 0 & 0 & 0 & 0 \\ 0 & 0 & 1 & 1 & 0 & 0 \\ 0 & 0 & 0 & 0 & 1 & 1 \end{bmatrix}$$

---

[2] ... in the binary case. This paper establishes the result more generally.

Multiplying p by this matrix gives us the distribution of the population over the family of schemata:

$$\Xi_{10} p = (p_{0*}, p_{1*}, p_{2*}) = (0.3, 0.3, 0.4)$$

We now have to define a mutation operator for this reduced search space. This is given by

$$\Xi_{10} \mu = (0.92, 0.04, 0.04)$$

So our mutation operator acting on our family of schemata consists of picking an element of $\{0*, 1*, 2*\}$ according to the above probability distribution and applying it to the element to be mutated. Notice that in this quotient group the element $0*$ is the identity. Constructing the mutation operator that acts on Λ_b from this distribution gives us

$$\mathcal{U}_{10}(x) = \begin{bmatrix} 0.92 & 0.04 & 0.04 \\ 0.04 & 0.92 & 0.04 \\ 0.04 & 0.04 & 0.92 \end{bmatrix} x$$

So in our example, we calculate the effect of mutation on the family of schemata as being

$$\begin{bmatrix} 0.92 & 0.04 & 0.04 \\ 0.04 & 0.92 & 0.04 \\ 0.04 & 0.04 & 0.92 \end{bmatrix} \begin{bmatrix} 0.3 \\ 0.3 \\ 0.4 \end{bmatrix} = \begin{bmatrix} 0.304 \\ 0.304 \\ 0.392 \end{bmatrix}$$

Notice that to make this calculation we did not need to know the details of the population p. We only needed to know how many elements were in each schema (given by Ξp). We can check this result by working out the effect of mutation on the whole population and then summing over the schemata. The implicit parallelism theorem tells us that we will get exactly the same result.

5 Implicit Parallelism and Fitness-Based Selection

It would be especially useful if, in the commutative diagram above, $\mathcal{M}$ could be generalized to $\mathcal{G}$ so the effects of selection could be incorporated. For proportional selection at least, Vose has pointed out the difficulties involved and concluded that such commutativity is in general not possible Vose, 1999. In an attempt to *force* commutativity, a selection scheme $\mathcal{F}_b$ might be defined on the quotient by

$$\mathcal{F}_b(\Xi x) = \Xi \mathcal{F}(x)$$

The problem here is that $\mathcal{F}_b$ is not well defined; the right hand side might depend on the particular x involved even though the left hand side does not (i.e., even if Ξx does not). In an attempt to ignore this complication, one might define a "fitness vector" f_b (over Ω_b) for which

$$\mathcal{F}_b(\Xi x) = \frac{\text{diag}(f_b) \Xi x}{f_b^T \Xi x}$$

Since the complication *cannot* be ignored, the vector f_b *must depend on x*. If f_b is defined as
$$f_b = \mathrm{diag}(\Xi x)^{-1} \, \Xi \, \mathrm{diag}(f) \, x$$
then
$$\begin{aligned} f_b^T \Xi x &= \sum_i (\Xi x)_i^{-1} (\Xi \, \mathrm{diag}(f) \, x)_i (\Xi x)_i \\ &= \sum_i (\Xi \, \mathrm{diag}(f) \, x)_i \\ &= \sum_j (\mathrm{diag}(f) \, x)_j \\ &= f^T x \end{aligned}$$

Therefore, by way of notational sleight of hand,
$$\begin{aligned} \mathcal{F}_b(\Xi x) &= \frac{\mathrm{diag}(f_b) \, \Xi x}{f^T x} \\ &= \frac{\mathrm{diag}\left(\mathrm{diag}(\Xi x)^{-1} \Xi \, \mathrm{diag}(f) \, x\right) \Xi x}{f^T x} \\ &= \frac{\Xi \, \mathrm{diag}(f) \, x}{f^T x} \\ &= \Xi \mathcal{F}(x) \end{aligned}$$

Of course, this definition for f_b is precisely the one given in the "schema theorem" Holland, 1975. Using this definition, one could define
$$\mathcal{G}_b = \mathcal{M}_b \circ \mathcal{F}_b$$
and appeal to implicit parallelism to conclude
$$\Xi \mathcal{G}(x) = \Xi \mathcal{M} \circ \mathcal{F}(x) = \mathcal{M}_b \circ \Xi \circ \mathcal{F}(x) = \mathcal{M}_b \circ \mathcal{F}_b(\Xi x) = \mathcal{G}_b(\Xi x)$$
thereby "extending" implicit parallelism from $\mathcal{M}$ to $\mathcal{G}$. Unlike Holland's result, the relation $\Xi \mathcal{G}(x) = \mathcal{G}_b(\Xi x)$

- is an equality which in every case provides nonvacuous information,
- says something nontrivial about new elements produced by mixing,
- makes explicit the relationships between the genetic operators and the underlying group structure of the search space.

However, it should be noted that because f_b depends on x, the "extension"
$$\Xi \mathcal{G}(x) = \mathcal{G}_b(\Xi x)$$
speaks only to what happens over a single time step (like Holland's result) and the information provided is insufficient to characterize the next generation (even in Λ_b). In particular, it cannot be used to map out population trajectories, and it certainly cannot

be used to justify talk of "above average fitness building blocks being selected exponentially". The "fitness" of schemata cannot be well-defined (it is not an attribute of schemata Ξx, but is determined instead by x). The various claims about GAs that are traditionally made under the name of the *building block hypothesis* have, to date, no basis in theory, and, in some cases, are simply incoherent. One exception is when the fitness function is a constant for each schema of a schema family (in which case the remaining constituents are redundant). Otherwise, one must take account of the fact that schemata "fitnesses" are *dynamic* quantities that change from population to population (see Stephens and Waelbroeck, 1999 for such a view of the building block hypothesis). Moreover, the fitness of such a "building block" at a given time depends on the entire microscopic structure of the population at that time.

Example

Consider the family of schemata $0*, 1*, 2*$ on the search space $\Omega = \mathcal{Z}_3 \times \mathcal{Z}_2$. Let the fitness vector be $f = (f_{00}, f_{01}, f_{10}, f_{11}, f_{20}, f_{21})$. Then f_b can be calculated as:

$$f_b = \begin{bmatrix} \frac{f_{00}x_{00}+f_{01}x_{01}}{x_{00}+x_{01}} \\ \frac{f_{10}x_{10}+f_{11}x_{11}}{x_{10}+x_{11}} \\ \frac{f_{20}x_{20}+f_{21}x_{21}}{x_{20}+x_{21}} \end{bmatrix}$$

We can verify that $\mathcal{F}_b(\Xi x) = \Xi \mathcal{F}(x)$.

$$\mathcal{F}_b(\Xi x) = \begin{bmatrix} \frac{f_{00}x_{00}+f_{01}x_{01}}{x_{00}+x_{01}} & 0 & 0 \\ 0 & \frac{f_{10}x_{10}+f_{11}x_{11}}{x_{10}+x_{11}} & 0 \\ 0 & 0 & \frac{f_{20}x_{20}+f_{21}x_{21}}{x_{20}+x_{21}} \end{bmatrix} \begin{bmatrix} x_{00}+x_{01} \\ x_{10}+x_{11} \\ x_{20}+x_{21} \end{bmatrix} = \Xi \mathcal{F}(x)$$

Notice that the fitness of a schema depends on the details of the whole population x and not just on the corresponding schemata family.

6 Conclusions

This paper has developed a framework for the theory of genetic algorithms that use a fixed-length string representation where the cardinality of the alphabet at each string position is arbitrary. Structural crossover and mutation represent the natural ways to define crossover and mutation in this framework. An implicit paralllelism is proved. This theorem states that structural crossover and mutation project naturally onto all competing families of schemata. This kind of projection does not work for proportional selection except when fitness is constant on each schema of the family. An exact equation which generalizes the Holland Schema theorem can be proved, but like the Holland Schema theorem, it cannot be applied in realistic situations for more than one time step.

References

Goldberg, D.E. (1989). *Genetic Algorithms in Search, Optimization & Machine Learning*. Addison Wesley, Reading, MA.

Holland, J. (1975). *Adapdation in Natural and Artificial Systems*. University of Michigan Press, Ann Arbor, Michigan.

Koehler, Bhattacharyya, and Vose (1997). General cardinality genetic algorithms. *Evolutionary Computation*, 5(4):439–459.

Lang, S. (1993). *Algebra*. Addison-Wesley, Reading, MA, third edition.

Rowe, J.E., Vose, M.D., and Wright, A.H. (2002). Group properties of crossover and mutation. *Evolutionary Computation*, 10(2):151–184.

Stephens, C. and Waelbroeck, H. (1999). Schemata evolution and building blocks. *Evolutionary Computation*, 7(2).

Vose, M.D. (1999). *The Simple Genetic Algorithm: Foundations and Theory*. MIT Press, Cambridge, MA.

Vose, M.D. and Wright, A.H. (2001). Form invariance and implicit parallelism. *Evolutionary Computation*, 9(3):355–370.

Finding Building Blocks through Eigenstructure Adaptation

Danica Wyatt[1] and Hod Lipson[2]

[1] Physics Department
[2] Mechanical & Aerospace Engineering, Computing & Information Science
Cornell University, Ithaca, NY 14853, USA

Abstract. A fundamental aspect of many evolutionary approaches to synthesis of complex systems is the need to compose atomic elements into useful higher-level building blocks. However, the ability of genetic algorithms to promote useful building blocks is based critically on genetic linkage - the assumption that functionally related alleles are also arranged compactly on the genome. In many practical problems, linkage is not known *a priori* or may change dynamically. Here we propose that a problem's Hessian matrix reveals this linkage, and that an eigenstructure analysis of the Hessian provides a transformation of the problem to a space where first-order genetic linkage is optimal. Genetic algorithms that dynamically transforms the problem space can operate much more efficiently. We demonstrate the proposed approach on a real-valued adaptation of Kaufmann's NK landscapes and discuss methods for extending it to higher-order linkage.

1 Introduction

A fundamental aspect of many evolutionary algorithms is the need to compose atomic elements into higher-level building blocks. This compositional process should continue recursively to generate increasingly complex modules from lower level components, until the desired solution is attained. The importance of discovery of partial building blocks was initially stressed by Holland in "The building block hypothesis" [5] that described how Genetic Algorithms (GAs) work. GAs promote useful building blocks represented as schemata and compose them through the process of crossover. As the Schema Theorem shows, however, the ability of GAs to promote useful building blocks is based critically on genetic linkage - the assumption that functionally related alleles are also arranged compactly on the genome. If genetic linkage is poor (i.e., there is little correlation between functional dependency and genetic proximity) then the crossover operator is more likely to break useful building blocks than it is likely to compose them.

The effect of poor genetic linkage can be dramatic. Before proceeding to describe previous work and our proposed solution, we demonstrate the grave effect of poor linkage in Fig. 1. The graph shows the best fitness of a GA running on a hard test problem. The test problem consists of 16 real-valued dimensions,

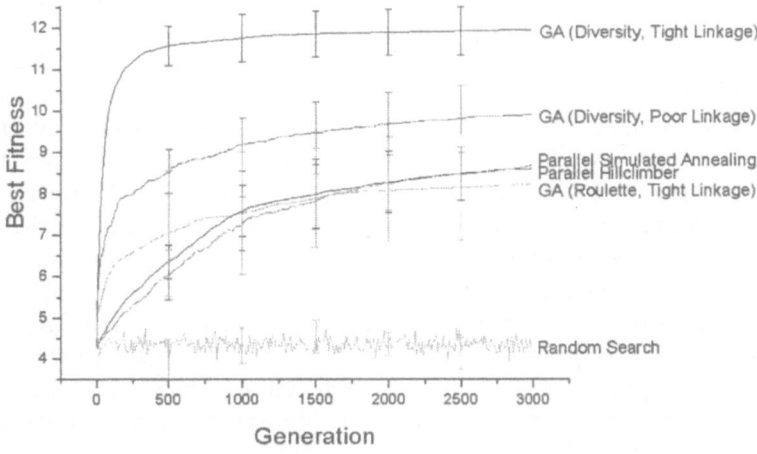

Fig. 1. Performance of a GA on a 16 dimensional problem with poor and tight linkage, with and without diversity maintenance. A parallel Simulated Annealing optimizer, a parallel hillclimber, and a random search are provided for comparison. All methods perform equal number of evaluations per generation. Error bars of one standard deviation are based on 10 runs on the same problem

each of which is deceptive (gradients lead in the wrong direction) and contains significant coupling between the variables. The test problem will be described in detail later; for now, it will suffice to notice the difference in performance of the GA between the top curve, where the genome is ordered so that coupled variables are close to each other (tight linkage) versus the lower curve, where the genome is shuffled so that coupled variables are far from each other (poor linkage.) In both of these cases, diversity maintenance techniques were also used. Without diversity, the crossover operator has little or no effect, and the GA's performance is inferior even to standard optimizers such as a well-tuned parallel simulated-annealing optimizer and a parallel random mutation hillclimber (a basic gradient optimizer.) A random search process is shown for reference. All methods perform an equal number of samplings per generation.

2 Prior Work

Since genetic linkage is not necessarily known in advance and may change dynamically as solution progresses or as the problem varies, new evolutionary algorithms have been proposed that either change linkage dynamically, or that do *not rely on* genetic linkage at all (at an additional computational cost.) Originally, Holland [5] proposed an inversion operator that would reorder alleles on the genome, with the expectation that genomes with better orderings (tighter genetic linkage) would gradually take over the population. However, this operator turned out to be too slow in practice. A variety of algorithms were proposed that do not assume linkage and build up the genome from scratch starting with the founding building blocks. Goldberg et al [4], and later Watson and Pollack [10], developed GAs that co-evolve partial solutions in which linkage can be adapted through new kinds of combination and split operators. Kargupta introduced new operators which search explicitly for relationships between genes [12]. Genetic programming [7] combines building blocks in tree hierarchies, allowing arbitrary branches of solutions to be swapped thereby permitting promotion of useful subcomponents without relying on linkage. More recently, Harik and Goldberg [11] proposed a linkage learning genetic algorithm (LLGA) that intersperses alleles in the partial solutions with variable-sized gaps (introns) allowing the GA more control over linkage tightening and exchange of complete building blocks. A more recent version of this algorithm uses evolved start expressions to assist in the process of nucleation of tightly linked groups [1]. All of these methods increase computational cost through the exploration of many compositional permutations.

3 Identifying Linkage through the Hessian

Here we propose an alternative way of identifying building blocks through dynamic eigenstructure analysis of the problem landscape's Hessian matrix. The Hessian matrix H is defined by measuring the cross-correlation effect of the variation of each of the n variables x_i on the variation effect of each other locus of the genome on the fitness function $F(X)$. Essentially, the Hessian matrix determines the first-order functional dependencies between gene locations. The Hessian is defined as

$$H_{ij} = \frac{\partial^2 F(X)}{\partial x_i \partial x_j} \qquad (1)$$

By definition, it is a symmetric matrix. For example, if

$$X = (x_1, x_2, x_3, x_4) \qquad (2)$$

and the fitness function to be optimized is

$$F(X) = \sin(2x_1 \, x_3) + \sin(3x_2 \, x_4) \qquad (3)$$

then computing [1] H and evaluating at $X = 0$ would yield

$$H = \begin{bmatrix} 0 & 0 & 2 & 0 \\ 0 & 0 & 0 & 3 \\ 2 & 0 & 0 & 0 \\ 0 & 3 & 0 & 0 \end{bmatrix} \qquad (4)$$

---

[1] Note that the Hessian can easily be computed numerically with the fitness given as a black box.

Highly coupled variable pairs are represented by large magnitude elements in H. Large off-diagonal elements imply coupling between variables that are not adjacent on the genome; to improve linkage, variables can be rearranged so that coupled pairs are proximate on the genome, bringing their corresponding Hessian coefficient closer to the diagonal. Rearranging the order of parameters on the genome is equivalent to swapping rows and columns of H, effectively transforming it by a permutation transformation. To bring the elements of H in the example above closer to the diagonal (effectively tightening the genetic linkage) we can use the permutation matrix T, where

$$T = \begin{bmatrix} 0 & 1 & 0 & 0 \\ 0 & 0 & 1 & 0 \\ 1 & 0 & 0 & 0 \\ 0 & 0 & 0 & 1 \end{bmatrix} \qquad (5)$$

to yield a new Hessian H'

$$H' = T^T H T = \begin{bmatrix} 0 & 2 & 0 & 0 \\ 2 & 0 & 0 & 0 \\ 0 & 0 & 0 & 3 \\ 0 & 0 & 3 & 0 \end{bmatrix} \qquad (6)$$

and the improved genome ordering X'

$$X' = XT = (x_3, x_1, x_2, x_4) \qquad (7)$$

H' above is as close as we can bring the non-zero elements to the diagonal by reordering, and the resulting genome ordering of Eq. (7) has indeed improved its linkage compared to the original ordering of Eq. (2). But is there a way to bring them even closer to the diagonal?

3.1 Generalized Genome Reordering

The optimal linkage ordering is not necessarily sequential. The permutation matrix simply reorders elements on the genome. However, there might not be a perfect sequential ordering if variables are coupled in any way but a linear serial chain. For example, if three variables are coupled equally, there is no way to arrange them in linear progression so that all are equally close to each other (the first will be less proximate to the last than it is to the middle variable.) Similarly, in the previous example of Eq. (3), x_1 is coupled to x_3, but not to x_2 so the genome ordering provided in Eq. (7) is still not optimally tight, and that is why H' is not exactly diagonal.

However, since the permutation matrix is nothing but an orthogonal linear transformation, we can think of any reordering process as merely a linear transformation. In this form, by allowing the transformation to have non-integer elements, we can find even more compact orderings. The optimal transformation is essentially the one that will bring all elements of the Hessian matrix *exactly* to the diagonal. We thus seek the optimal transformation T_0 to a diagonal matrix λ:

$$T_0^T H T_0 = \lambda \tag{8}$$

Solving for T_0 yields the Hessian's eigenvectors. Because the Hessian matrix is always symmetric, the eigenvectors have no imaginary component. In our example,

$$T_0 = \frac{\sqrt{2}}{2} \begin{bmatrix} 0 & -1 & -1 & 0 \\ 1 & 0 & 0 & 1 \\ 0 & 1 & -1 & 0 \\ -1 & 0 & 0 & 1 \end{bmatrix} \tag{9}$$

and

$$H_0 = T^T H T = \begin{bmatrix} -3 & 0 & 0 & 0 \\ 0 & -2 & 0 & 0 \\ 0 & 0 & 2 & 0 \\ 0 & 0 & 0 & 3 \end{bmatrix} \tag{10}$$

and the optimal genome ordering X_0 is given by

$$X_0 = X T_0 = \frac{\sqrt{2}}{2}(<x_2, -x_4>, <x_3, -x_1>, <-x_1, -x_3>, <x_2, -x_4>) \tag{11}$$

To understand why the linkage-optimal vector X_0 above has even less coupling than the compactly ordered vector $X' = (x_3, x_1, x_2, x_4)$ of Eq. (7), consider the following. A small positive mutation δ applied to x_1, a gene of X', will result in either an increase or decrease of the fitness function. Whether it will be an increase or a decrease depends on the sign of another gene, x_3. Thus there is still coupling between alleles on the genome. On the other hand, a small positive mutation in δ applied to (multiplied by) $<x_1, x_3>$ will always result in an increase in fitness, regardless of the states of other genes. Therefore, $<x_1, x_3>$ is more suited to be a gene than any single variable. Similarly, a small positive mutation δ applied to the complementary vector $<x_1, -x_3>$, the second gene of X_0, will do nothing to the fitness function, independent of the value of the other variables. These two genes thus span the search space much more effectively because they are uncoupled.

3.2 What Can We Learn from the Eigenvalues?

The eigenvalues λ hold the scaling factors of the new space, by which any variation operators can be calibrated; in our example, these are 2 and 3. Dividing the mutation operators by these factors would allow all blocks to be explored at the same resolution. Similarly, subspaces with large eigenvalues are more dominant and should be explored first if gradual complexification is employed [15].

Degenerate eigenvalues(two or more equal eigenvalues) indicate a subspace of the landscape space, spanned by the corresponding eigenvectors, where variables are uniformly coupled (or decoupled.) Negative eigenvalues may indicate collapsed subspaces.

3.3 Using Transformations in a GA

Based on the analysis above, we conclude that

- The genome reordering is equivalent to a linear transformation of the fitness landscape, therefore,
- There exist non-discrete genome orderings, and
- The ordering that yields optimal genetic linkage at a point in the landscape is given by the eigenvectors of the Hessian at that point.
- Eigenvalues of the Hessian reveal properties of the search space.

Once a linkage-optimal ordering transformation has been determined by computing the eigenvectors of the Hessian of $F(X)$, a GA can proceed regularly by evolving individuals h_i but evaluating $F(T_0 h_i)$ instead of directly $F(h_i)$ Many varieties of GAs, such as those incorporating partial solutions, learning, and sophisticated composition operators can easily be modified to use this formulation.

4 Test Function

We tested the eigenvector reordering process on evolution of a solution to a multidimensional function of real variables, $Z(X)$, that composes n base functions $\Psi(x)$:

$$Z(X) = \frac{1}{2} \sum_{i=1}^{n} \Psi(b_i) \qquad (12)$$

where

$$\Psi(u) = \frac{\cos(u)+1}{|u|+1} \qquad (13)$$

and

$$b_i = \sum_{j=1}^{k} a_{ij}\, x_{(i+j) \bmod n} - c_i \qquad (14)$$

This function is a real-valued adaptation of Kauffman's NK landscapes [6]. Kauffman defined a function with n bits, in which each bit's fitness contribution depends arbitrarily on its k neighbors. NK landscapes thus have "tunable

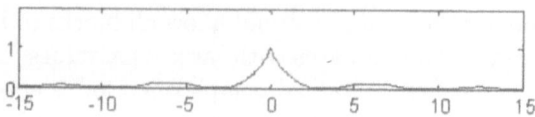

Fig. 2. The base function $\Psi(u)$ used to construct the test function. The full test function is a shuffled kth-order mixing of n base functions

ruggedness" and are often used to test GAs. Here we defined a similar problem that uses real values instead of bits. First, we define a multimodal base function $\Psi(u)$ with one global maximum at $u = 0$ and several smaller maxima, shown in Fig. 2. This base function is deceptive if the search space is larger than $\pm 2\pi$ around $u = 0$ because then in most of the space the gradients lead to local optima. The function Z is a sum of the n single dimensional base functions; however the blocks are not separable: Each base function is evaluated at position b_i, which is a mixing function of k elements of the argument vector X. The mixing is obtained through an array of coefficients a_{ij} and an offset c_i, each set arbitrarily. Because of the mixing of order k, optimizing one dimension may (and usually does) lead to adverse effect in the other $k - 1$ dimensions. Finally, the genome elements $x_{i..n}$ are shuffled so that variables that contribute to the same base function are maximally apart on the genome, so as to create the worst linkage. The function is defined so that it always has a single global optimum with the value n, and it occurs at $X = A^{-1}c$ (where A is a k-diagonal matrix with the elements of a on its diagonals, and X is the unshuffled vector.) In our study, we generate the coefficients a_{ij} randomly with a uniform distribution in the range of ± 1, and the coefficients c_i with a uniform distribution in the range of ± 8. These values were selected arbitrarily, and were not tuned.

5 Experimental Results

We carried out a series of tests, with different setting of n and k. All problems have bad shuffling (poor linkage.) In each experiment, the problem coefficients were generated randomly and were not tuned. We then carried out several runs to collect average performance and standard deviation. We also ran a parallel hillclimber as a baseline control. Individuals were initialized randomly in $X = \pm 10$ with uniform distribution. Single-point crossover and mutations of maximum size ± 0.1 were used for all offspring. In all experiments we used a GA with a diversity maintenance technique known as "Deterministic Crowding" [9] in which a more successful offspring replaces the parent it most closely resembles after pairwise matching. Diversity maintenance is important to avoid premature convergence, in which case crossover operators would do little and the effect we wish to study would vanish.

First, let us look closely at the effect of using a Hessian Eigenvector transformation. Figure 3 shows average performance of an Eigenvector GA on a problem with $n = 16$ and $k = 4$. Population size is 100, and average and standard devia-

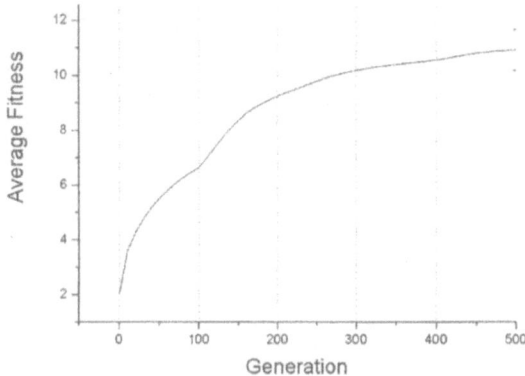

Fig. 3. Performance of a GA on a shuffled real-valued NK landscape with $n = 16$ and $k = 4$. Reordering transformation is recomputed every 100 generations (at vertical lines,) and performance boosts are visible at those instances. Statistics are of 20 runs. Eigenvector GA uses 5% more evaluations than regular GA

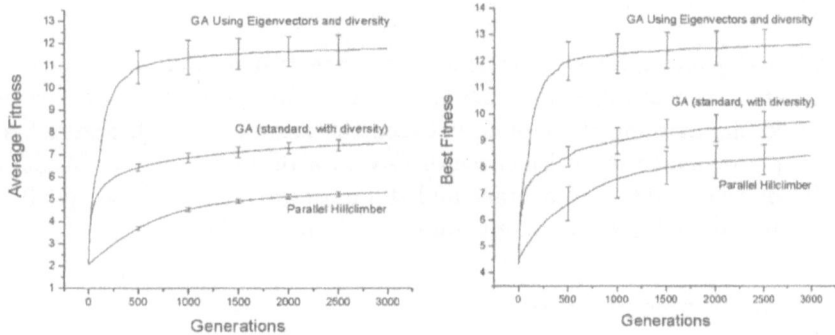

Fig. 4. Performance comparison on a shuffled real-valued NK landscape with n=16 and k=4. Reordering transformation is recomputed every 100 generations. Statistics are of 20 runs

tion is based on 20 runs. The transformation is re-computed numerically every 100 generations, around the currently best solution, adding 5% evaluations. We will discuss the cost of this computation in the next section. The points of re-computation are marked with a vertical solid line; note how the curve "boosts" its optimization rate at those intervals. Figure 4 shows the long-term behavior of the solver over 3000 generations using the same parameter settings. Both average fitness of the population and best fitness in the population are shown. We see that the eigenvector GA has significantly outperformed the regular GA. The contribution of the eigentransformation has been exhausted after 500 generations or so. After that period, both algorithms progress at roughly the same rate. We hypothesize this point is where contribution of first-order (linear) linkage has been exhausted.

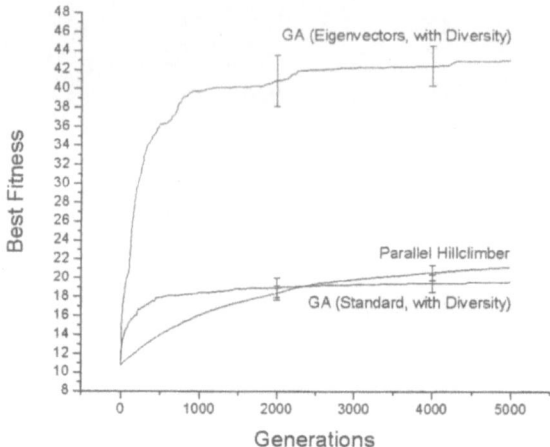

Fig. 5. Performance comparison on a shuffled real-valued NK landscape with $n = 64$ and $k = 8$. Reordering transformation is recomputed every 100 generations. Statistics are of 10 runs

The performance boost provided by the eigentransformation becomes more significant as the problem becomes harder both in the number of parameters (n) and in the amount of coupling between the parameters (k). Figure 5 shows average performance of an Eigenvector GA on a problem with $n=64$ and $k=8$. Population size is 100, and average and standard deviation is based on 10 runs. At $n=256$ and $k=16$, we observed similar performance boosts.

6 Computational Cost

Additional computational cost is incurred by the linear transformation and eigenstructure calculation, as well as by additional evaluations. The transformation cost adds $\mathcal{O}(n^2)$ arithmetic operations per evaluation and $\mathcal{O}(n^2)$ arithmetic operations per Hessian calculation for computing derivatives from the samples, and computing the eigenvectors. Both of these are typically negligible compared to the cost of a single evaluation of a hard practical problem.

6.1 Evaluation Cost

The evaluation cost is especially critical because direct numerical calculation of the Hessian matrix involves approximately $2n^2$ samplings of the search space. If each individual is at a different location of the landscape, this may amount to $2n^2p$ extra evaluations per generation, where p is the population size. This cost is prohibitive, and so more efficient schemes must be found. If linkage properties of the landscape are consistent over time, significantly fewer than n^2p samplings are necessary. For example, the results shown in Fig. 4 required only one Hessian calculation per 100 generations. Since for that problem $n=16$, this amounts to only 5% additional evaluations, and the advantage gained is well worth it. The number of evaluations grows as $\mathcal{O}(n^2)$ but the interval at which Hessian-recalculation is required increases too since propagation of building blocks is slower. If the linkage properties of the landscape are consistent over space, the Hessian does not need to be tracked for every individual. For the results shown in Fig. 4 we recalculated the Hessian only around the currently best individual, but use it for all individuals in the diverse population. A harder class of problems exists where linkage properties of the landscape change over time and over space (nonlinear linkage.) Prior work dealing with various forms of linkage learning and composition typically assumed that the linkage properties are fixed for the given problem. For such linkage-variable problems we might either spend more evaluations to track separate Hessian matrices for clusters of individuals, and also update them more frequently. Alternatively, we could extract Hessian information indirectly from the samplings already being done by the GA anyway.

6.2 Indirect Computation of the Hessian

The Hessian can be computed indirectly from the GA samples using cross-correlation statistics. As described by the "Two-armed Bandit" problem [3], there is a trade off between exploration and exploitation. However, with a linkage-learning algorithm as proposed here, it is possible to use all samples to probe linkage properties and thus enhance exploration without incurring additional function evaluations. The key to indirect extraction of linkage information is the understanding that linkage can be learned just as efficiently even from individuals with low fitness. Hessian approximation techniques used in quasi-Newton methods [13][14] could be adapted to make use of this information. Another way to gather linkage properties from arbitrary, unstructured samplings of the search space is through least-squares fitting to a linkage modeling function. For example, assume a two dimensional landscape $F(x_1, x_2)$ can be described very roughly by the conic section equation

$$F(x_1, x_2) = ax_1^2 + 2bx_1x_2 + cx_2^2 + dx_1 + ex_2 + f \qquad (15)$$

The linkage coefficient we are interested in is the magnitude of parameter b with respect to parameters a and c. These parameters reveal how the landscape is influenced by the combination (x_1x_2). Note that the function is not used to model the landscape for direct optimization; that is, we do not proceed to compute the optimum of F, because there is no guarantee whatsoever that the landscape is of degree 2 (if it was then much more efficient optimization techniques could be used.) Instead, we only use it to probe how variables are dependent on each other, by fitting a conic surface locally to a small patch of the landscape. A landscape with n parameters will have $(n+1)(n+2)/2$ coefficients, and so $\mathcal{O}(n^2)$ samplings will be needed to determine the coefficients through least squares modeling. Direct computation of the Hessian also requires $\mathcal{O}(n^2)$ samplings, and so the new formulation is not a improvement in terms of the number of evaluations. However, the direct computation method required structured samplings on a grid, whereas the new formulation can use an unstructured pattern of samples and can therefore use samples performed anyway by the GA in course of its normal operation, provided they are close enough. The quality of the linkage approximation degrades as the sampling radius δ increases, just like direct numerical computation of the Hessian degrades with $\mathcal{O}(\delta^2)$ according to Taylor expansion around the center point ($\delta \simeq 0.1$ for the test function described in this paper.) It is therefore necessary to use samples that are within a small region with respect to nonlinearities in the function. In a diverse set of samples this can be done by clustering samples into groups, at the cost of cluster management.

7 Higher-Level Compositions

Direct eigenstructure transformation resolves first-order linkage, but nonlinear transformations may resolve even higher-order linkage. We say a linkage is first order when the linkage between two variables does not depend on any other, third variable or on time. If it does, we would have a second-order linkage. Higher-order linkage effects can be identified using Kronecker tensor products of partial derivatives of arbitrary orders. Large eigenvalues of these tensors indicates a strong high-order dependency that can be resolved through a nonlinear transformation. These higher-order transformations are again provided by (nonlinear) eigenstructure analysis of tensors [8]. The approach proposed here is akin to support vector machine (SVM) methods [2] that transform the problem space so that data classes that were originally convoluted are now linearly separable. In Eq. (15) we used second-order polynomial kernels that transform the search space so that first-order linkage is optimal; other kernels, including higher-order polynomials, could be used to probe and compensate for higher-order linkage.

8 Conclusions

Performance of genetic algorithms is critically based on both diversity maintenance and genetic linkage. Here we propose that transformations can effectively be used to reorder the genome, and that the first-order linkage-optimal transformation can be found through Eigenstructure analysis of the landscape's Hessian matrix. In a series of experiments using a highly coupled, nonlinear, deceptive and shuffled function, we show how the presented algorithm produces significantly superior performance, at relatively low additional computational cost. We also propose that when high-order linkage is systematic (unlike random or "needle-in-a-haystack" landscapes) it can be resolved dynamically using high-order statistical probes that use existing evaluations. We further suggest that kernel methods that are traditionally used in machine learning to transform convoluted problems into linearly separable problems can be brought to bear on evolutionary computation to decompose high-order linkage problems into linkage optimal space, as a principled form of adaptation of the genomic representation.

Acknowledgments. This work has been supported in part by the US Department of Energy (DOE), grant DE-FG02-01ER45902, and by the US National Science Foundation Integrative Graduate Education and Research Traineeship (IGERT) program at Cornell University.

References

[1] Chen, Y.-P., Goldberg, D.E. *Introducing start expression genes to the linkage learning genetic algorithm.* Proceedings of the Parallel Problem Solving from Nature Conference (PPSN VII). Berlin, Germany:Springer, 351–360, 2002.
[2] Cristianini N., Shawe-Taylor J. *An Introduction to Support Vector Machines and Other Kernel-based Learning Methods.* Cambridge University Press, 2000.
[3] De Jong, K. *An analysis of the behaviour of a class of genetic adaptive systems.* PhD thesis, University of Michigan, 1975.
[4] Goldberg, D., Korb, B., and Deb, K. *Messy Genetic Algorithms: Motivation, Analysis, and First Results.* Complex Systems, 4:415–444, 1989.
[5] Holland, J.H. *Adaptation in Natural and Artificial Systems* University of Michigan Press, Ann Arbor, 1975.
[6] Kauffman, S. *The Origins of Order: Self-Organization and Selection in Evolution.* Oxford University Press, 1993.
[7] Koza J. *Hierarchical genetic algorithms operating on populations of computer programs.* 11th Int. joint conference on genetic algorithms, 768–774, 1989.
[8] Lipson H., Siegelmann H.T. *High Order Eigentensors as Symbolic Rules in Competitive Learning.* in Hybrid Neural Systems, S. Wermter, R. Sun (Eds.) Springer, LNCS 1778, 286–297, 2002.
[9] Mahfoud S. *Niching Methods for Genetic Algorithms.* IlliGAL Report No. 95001, University of Illinois at Urbana-Champaign, 77–80, 1995.
[10] Watson, R.A., Pollack, J.B. *A Computational Model of Symbiotic Composition in Evolutionary Transitions Biosystems.* to appear in 2003.
[11] Harik, G., and Goldberg, D.E. *Learning linkage.* Foundations of Genetic Algorithms 4:247–262, 1997.
[12] Kargupta, H. *The gene expression messy genetic algorithm.* Proceedings of the 1996 IEEE International Conference on Evolutionary Computation. Nagoya University, Japan, 631–636, 1996.
[13] Bertsekas, D.P. *Non-linear Programming.* Athena Scientific, Belmont, MA, 134–141, 1995.
[14] Shanno, D.F. *Conditioning of quasi-Newton methods for function minimization.* Mathematics of Computation 24:647–656, 1970.
[15] Stanley, K.O., Miikkulainen, R. *Achieving High-Level Functionality Through Complexification.* Proceedings of the AAAI 2003 Spring Symposium on Computational Synthesis. Stanford, CA: AAAI Press, 2003.

A Specialized Island Model and Its Application in Multiobjective Optimization

Ningchuan Xiao[1] and Marc P. Armstrong[1,2]

[1] Department of Geography, The University of Iowa, Iowa City, IA 52242, USA
{ningchuan-xiao,marc-armstrong}@uiowa.edu
[2] Program in Applied Mathematical and Computational Sciences
The University of Iowa, Iowa City, IA 52242, USA

Abstract. This paper discusses a new model of parallel evolutionary algorithms (EAs) called the specialized island model (SIM) that can be used to generate a set of diverse non-dominated solutions to multiobjective optimization problems. This model is derived from the island model, in which an EA is divided into several subEAs that exchange individuals among them. In SIM, each subEA is responsible (i.e., specialized) for optimizing a subset of the objective functions in the original problem. The efficacy of SIM is demonstrated using a three-objective optimization problem. Seven scenarios of the model with a different number of subEAs, communication topology, and specialization are tested, and their results are compared. The results suggest that SIM effectively finds non-dominated solutions to multiobjective optimization problems.

1 Introduction

Though the parallel characteristics of evolutionary algorithms (EAs) have been recognized for decades [14], intensive research on parallelism in EAs did not take place until the 1980s [4,12]. Since that time, two major types of parallel models have been designed for EAs [1]: cellular and distributed models. For a cellular model, the population of an EA is subdivided into a large number of subpopulations (each often containing one individual) and communication is only permitted among neighbor subpopulations. For a distributed model (also called an island or coarse grained model), a population is partitioned into a small number of subpopulations and each performs as a complete EA; a sparse exchange of individuals among subpopulations is conducted.

The goal of parallelizing an EA is to: 1) relieve the computational burden mainly imposed by evaluation (computing objective functions) and evolutionary processes (e.g., selection and recombination), and 2) improve EA results by increasing population diversity. While it has been reported that parallelization can be used to help with both goals [13,17], the use of parallel models in multiobjective optimization has mainly focused on the first [16,18,26], though some progress has recently been made on the second goal [15].

The purpose of this paper is to describe a new approach to modeling parallel EAs with a specific emphasis placed on multiobjective optimization. This approach, called the specialized island model (SIM), is loosely based on the general concept used by the island model, and can be easily implemented on different kinds of parallel computers. In SIM, an EA is divided into a number of subEAs, and each is specialized to solve a modified version of a multiobjective optimization problem using a subset of the original objective functions. The use of SIM is demonstrated using a three-objective optimization problem; seven scenarios of the model with a different number of subEAs, communication topology, and specialization are tested, and their results are compared.

2 Background

The central idea in multiobjective optimization is Pareto optimality, which is derived from Vilfredo Pareto's original work [21]. To illustrate this concept, consider an optimization problem with k objectives:

$$\begin{array}{ll} \text{Minimize} & \boldsymbol{f}(\boldsymbol{x}) = [f_1(\boldsymbol{x}), f_2(\boldsymbol{x}), \ldots, f_k(\boldsymbol{x})]^T \\ \text{subject to} & g_i(\boldsymbol{x}) \geq 0 \quad i \in [1, \ldots, m] \\ & h_i(\boldsymbol{x}) = 0 \quad i \in [1, \ldots, p] \, , \end{array} \quad (1)$$

where $\boldsymbol{x}$ is a vector of decision variables, g_i is an inequality constraint, and h_i is an equality constraint.

A solution $\boldsymbol{x}^*$ is said to dominate solution $\boldsymbol{x}$ (or $\boldsymbol{x}^* \prec \boldsymbol{x}$) if and only if

$$\forall i \ f_i(\boldsymbol{x}^*) \leq f_i(\boldsymbol{x}) \wedge \exists i \ f_i(\boldsymbol{x}^*) < f_i(\boldsymbol{x}), i \in \{1, \ldots, k\} \ . \quad (2)$$

In practice, especially when objectives conflict, it is often impossible to find a single optimum that dominates all other solutions. In contrast, it is common to encounter many non-dominated solutions. A solution $\boldsymbol{x}$ is said to be non-dominated if no other solution dominates it. Obviously, the ultimate goal of multiobjective optimization is to find all non-dominated solutions to a problem.

Numerous evolutionary methods have been developed to solve multiobjective optimization problems [5,7,8,10,26,27,28]. Most of these approaches use the global population structure of an EA and encourage the evolution of a set of diverse non-dominated solutions using a variety of techniques such as sharing, niching, and elitism. The method proposed in this paper is based on the island model of parallel EAs that allows each subpopulation to evolve in different "local" directions (in terms of the combination of objective functions) such that a diverse population will emerge. More importantly, the solutions found by each subpopulation can also be improved, through exchange of individual solutions among subpopulations.

3 Specialized Island Model

SIM is derived from concepts used in the island model of EAs. There are many variations of island models [2,3,12,19,23,25]. In general, in an island model, there are N subEAs (or subpopulations), each running in parallel as a complete EA.

A migration matrix $\mathbf{M} = \{0,1\}^{N \times N}$ is used to determine the communication topology among subpopulations. After a certain number of iterations (defined as migration frequency, m_f), a subpopulation P_i will send a number of individuals (defined as migration rate, m_r) to subpopulation P_j if and only if $m_{ij} = 1$, where $m_{ij} \in \mathbf{M}$. The total number of individuals that subpopulation i can receive is $m_r \cdot \sum_j^N m_{ij}$. After receiving individuals, an assimilation strategy is applied on each subpopulation to incorporate a portion of the "alien" individuals (defined as assimilation rate, m_a) into its own population. The formal procedure of the island model is described below, where PAR and END PAR blocks indicate the beginning and end of parallel sections.

Island Model
1   $t := 0$
2   PAR
3   Initialize each subpopulation $P_i(t), 1 \le i \le N$
4   Evaluate individuals in $P_i(t)$
5   END PAR
6   REPEAT
7   PAR {regular EA }
8   Evaluate individuals in $P_i(t)$
9   Apply recombination and mutation on $P_i(t)$
10   END PAR
11   PAR
12   Migration
13   END PAR
14   PAR
15   Assimilation in $P_i(t)$
16   Evaluate individuals in each subpopulation $P_i(t)$
18   END PAR
19      $t := t+1$
20 UNTIL termination criterion is satisfied

The difference between the SIM and island model is that, when SIM is used for multiobjective optimization, a subpopulation is not required to search for non-dominated solutions with respect to all objectives. Instead, some subpopulations are used only to search for solutions with respect to a subset of original objectives. This idea can be regarded as a generalized version of the vector evaluated genetic algorithm or VEGA [22], where individuals in each subpopulation are evaluated and selected using only one objective function. But in VEGA, crossover and mutation operate on the entire population, while in SIM these operations are used on each subpopulation. Figure 1 shows an example of a SIM with seven subpopulations for a problem with three objectives (f_1, f_2, f_3). Each node in the figure represents a subpopulation and the number(s) inside each node indicate the objective(s) for which the subpopulation is specialized. For instance, the node marked as "2, 3" is a subEA that is designed to search for optimal solutions to a problem with two objectives (f_2, f_3). Arrows in Fig. 1 indicate the source and destination of migration.

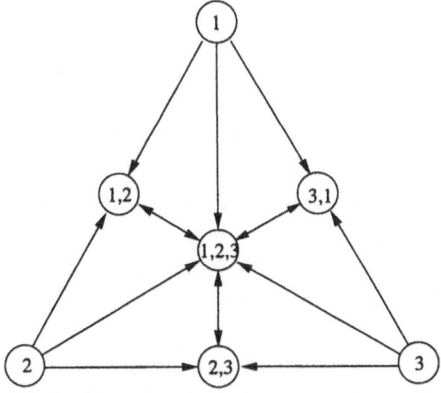

Fig. 1. An example of specialized island model for a 3-objective optimization problem

The specialization of a subEA can be depicted using a binary string $\{0,1\}^k$, where k is the number of objectives in the original problem, and there must be at least one "1" in the string. The i-th bit (starting from the left) is 1 if and only if the i-th objective in the original problem is included in the set of objectives for the subEA. Consequently, in Fig. 1, the specialization for the subEA marked as "2, 3" is "011".

4 Application of SIM

To demonstrate the use of SIM, the following optimization problem was designed:

$$\begin{aligned}\text{Minimize} \quad & \boldsymbol{f} = [f_1(\boldsymbol{x}), f_2(\boldsymbol{x}), f_3(\boldsymbol{x})]^T \\ \text{subject to} \quad & \boldsymbol{x} = (x_1, \ldots, x_m) \\ & 0 \le x_i \le 1, i \in \{1, \ldots m\} \\ & m = 30 \ .\end{aligned} \quad (3)$$

The objective functions are defined as:

$$\begin{cases} f_1 = \sqrt{x_1^2 + x_2^2} \\ f_2 = h_2/g_2 \\ f_3 = g_3 \cdot h_3 \\ g_2 = 1 + 10 \cdot (m-1) - \sum_{i=2}^{m}[\log(x_i^2 + 1) - 10 \cdot \cos(4\pi x_i)] \\ h_2 = 1 - m \cdot \sqrt{x_1/g_2} \\ g_3 = 1 - \sqrt{x_1/g_3} - (x_1/g_3) \cdot \sin(10\pi x_i) \\ h_3 = (1 - \frac{9}{m-1} \cdot \sum_{i=2}^{m} x_i) \ , \end{cases} \quad (4)$$

Note that f_2 and f_3 are modified versions of functions originally suggested in [6,27]. More specifically, they are modified from functions f_2's in test problems T_4 and T_3, respectively, in [27]. The reason for the modification is that the original f_2 functions in T_4 and T_3 were designed for problems with two objectives and did not necessarily conflict with each other.

4.1 Settings

Seven scenarios of SIM were designed to test the use of SIM on the problem described in Equations 3 and 4. Each scenario will be called a model in the remainder of this paper. Differences between the models are based on the number of subEAs used, as well as the migration strategy and specialization for each subEA. A description of each model is given in Table 1 and the settings of each model are provided in Table 2.

The parameters used by each subpopulation in the seven models are listed in Table 3. Two selection techniques are used in each subEA: roulette and tournament. In Table 3, a positive tournament size indicates that only tournament selection is used and the tournament size is specified as such; a negative tournament size in Table 3 means that a selection method will be randomly chosen (chance for each method is 50%), and when a tournament approach is used, the tournament size is the absolute value of the negative number. For subEAs specialized for 2 or 3 objectives, a non-dominated sorting approach is used [10,24]; for single-objective subpopulations, a sharing technique is applied [11]. Elitism is used for single-objective subpopulations.

Migrations among the subpopulations are conducted by selecting m_r individuals from subpopulations that are specified as migration sources. This selection process uses a tournament approach with the tournament size equal to two. The selected individuals are sent to a destination according to the migration matrix (Table 2). The destination subEA receives individuals from all source subEAs and puts them into a pool. After a subEA has accepted all alien individuals migrated from other subEAs, m_a individuals are randomly selected from the pool and copied into its own population. When elitism is applicable (for single objective subEAs in this research), the best solution in the population must not be overwritten.

Table 1. Descriptions of the seven models

Model	Description
A	Seven connected subpopulations
A1	An isolated version of A, no connections among subpopulations
B	Four connected subpopulations, one for three objectives, three for two objectives
B1	An isolated version of B, no connections among subpopulations
C	Four connected subpopulations, one for three objectives, three for one objective
C1	An isolated version of C, no connections among subpopulations
D	One subpopulation, specialized for three objectives

Table 2. Settings for the seven models

Model	Subpopulation (island)			Illustration
	Specialization	Minimizing	Migration matrix	
A	111	f_1, f_2, f_3	0 1 1 1 1 1 1	
	110	f_1, f_2	1 0 0 0 0 0 0	
	011	f_2, f_3	1 0 0 0 0 0 0	
	101	f_1, f_3	1 0 0 0 0 0 0	
	100	f_1	1 1 0 1 0 0 0	
	010	f_2	1 1 1 0 0 0 0	
	001	f_3	1 0 1 1 0 0 0	
A1	111	f_1, f_2, f_3	0 0 0 0 0 0 0	
	110	f_1, f_2	0 0 0 0 0 0 0	
	011	f_2, f_3	0 0 0 0 0 0 0	
	101	f_1, f_3	0 0 0 0 0 0 0	
	100	f_1	0 0 0 0 0 0 0	
	010	f_2	0 0 0 0 0 0 0	
	001	f_3	0 0 0 0 0 0 0	
B	111	f_1, f_2, f_3	0 1 1 1	
	110	f_1, f_2	1 0 0 0	
	011	f_2, f_3	1 0 0 0	
	101	f_1, f_3	1 0 0 0	
B1	111	f_1, f_2, f_3	0 0 0 0	
	110	f_1, f_2	0 0 0 0	
	011	f_2, f_3	0 0 0 0	
	101	f_1, f_3	0 0 0 0	
C	111	f_1, f_2, f_3	0 1 1 1	
	100	f_1	1 0 0 0	
	010	f_2	1 0 0 0	
	001	f_3	1 0 0 0	
C1	111	f_1, f_2, f_3	0 0 0 0	
	100	f_1	0 0 0 0	
	010	f_2	0 0 0 0	
	001	f_3	0 0 0 0	
D	111	f_1, f_2, f_3		

Table 3. Settings for each subEA

Parameter	Specialization of subEA		
	3-obj	2-obj	1-obj
Size of population	30	30	30
Tournament size	5	3	-3
Crossover probability	0.95	0.95	0.95
Mutation probability	0.15	0.1	0.1
Migration frequency	2	2	2
Migration rate	5	5	5
Assimilation rate	5	5	5
σ_{share} in the sharing function	0.3	0.6	0.6
α in the sharing function	1	1	1

4.2 Performance Metrics

In this research, each model was run 10 times, all solutions generated in the 10 runs were gathered, and the non-dominated ones were selected. Let O_i denote the non-dominated solutions generated by model i ($i \in$ {A, A1, B, B1, C, C1, D}). Assume operation $|\cdot|$ returns the size of a set and $||\cdot||$ is a distance measurement. Two metrics (C and M_2^*) designed in [27] are used to compare the performance of the seven models.

$$C(O_i, O_j) = \frac{|\{b \in O_j \text{ and } \exists a \in O_i : a \prec b \text{ or } a = b\}|}{|O_j|}. \tag{5}$$

Function $C(O_i, O_j)$ measures the fraction of set O_j that is covered by (i.e., dominated by or equal to) solutions in O_i, and $0 \leq C(O_i, O_j) \leq 1$. $C(O_i, O_j) = 1$ if all solutions in O_j are covered by solutions in O_i; $C(O_i, O_j) = 0$ if no solutions in O_j are covered by those in O_i. Note that $C(O_i, O_j)$ is not necessarily equal to $1 - C(O_j, O_i)$. Generally speaking, if $C(O_i, O_j) > C(O_j, O_i)$, then model i is considered to generate more non-dominated solutions than model j. However, two models will be considered to have similar performance characteristics if $C(O_i, O_j) \approx C(O_j, O_i)$.

$$M_2^*(O_i) = \frac{1}{|O_i| - 1} \sum_{p \in O_i} |\{q \in O_i \text{ and } ||f(p) - f(q)|| > \sigma\}|. \tag{6}$$

Function M_2^* is essentially a measure of the extent of the objective function values in a multi-dimensional space formed by the objectives. A high value of M_2^* indicates the ability of the corresponding model to generate optimal solutions spanning a wide extent.

Two additional metrics were designed to compare the overall non-domination of each model. To calculate these metrics, the non-dominated solutions generated by each model are gathered into a single set, from which the overall non-dominated solutions are picked to form a set $\mathcal{O}$. Let $\mathcal{O}_i$ be the subset of $\mathcal{O}$ that is formed by solutions generated by model i.

$$X_i = \frac{|\mathcal{O}_i|}{\sum_j^N |\mathcal{O}_j|}.$$ (7)

$$E_i = \frac{|\mathcal{O}_i|}{|O_i|}.$$ (8)

Here, X_i indicates the overall domination of model i when it is considered with all other models. E_i measures how many non-dominated solutions generated by model i are still non-dominated when the results from all models are considered. E_i can be regarded as the efficiency of model i in generating overall non-dominated solutions. $0 \leq E_i \leq 1$ because $0 \leq |\mathcal{O}_i| \leq |O_i|$.

4.3 Results

Figure 2 shows the non-dominated solutions generated by all seven models. Table 4 shows the comparative results using function C, M_2^*, X_i, and E_i. The portion of this table relating to function C is a 7×7 matrix that can be examined along rows and columns. For row $i \in \{A, A1, B, B1, C, C1, D\}$, the numbers in that row (denoted as $C(\mathcal{O}_i, \cdot)$) indicate the tendency for solutions generated by model i to cover solutions from other models. For column j, the numbers in that column mean (denoted as $C(\cdot, \mathcal{O}_j)$) indicate the tendency for solutions generated by model j to be covered by solutions from other models.

It was found that model A gives the best performance since $C(\mathcal{O}_A, \cdot) > C(\cdot, \mathcal{O}_A)$, and $M_2^*(\mathcal{O}_A)$, X_A and E_A are all the highest values for each metric. It can also be noted that model B is relatively competitive with respect to function C. On the other hand, it was found that models C and C1 performed poorly, except that model C has a relatively large range; this is reasonable because model C is specialized to find extreme values for each objective. Model D gives the lowest value in M_2^*, indicating the disadvantage of a single population in finding extreme values for all objective functions. However, model D demonstrates a relatively high value in E_i, which suggests that including a subpopulation specialized in

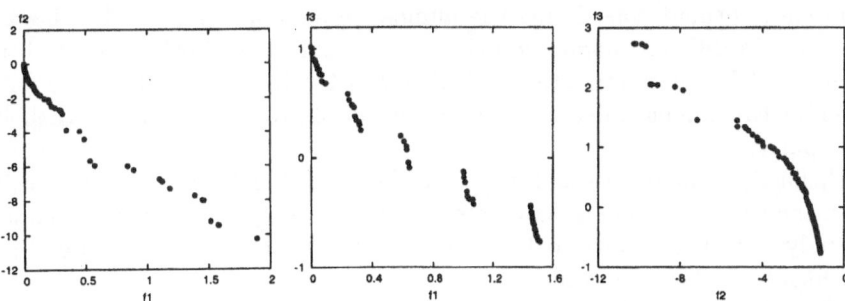

Fig. 2. Scattered plots showing non-dominated solutions for each pair of objective functions

Table 4. Comparison among seven models using metrics C, M_2^*, X_i, and E_i

Model	\multicolumn{7}{c}{C}	M_2^*	X_i	E_i
	A	A1	B	B1	C	C1	D			
A	-	0.44	0.29	0.39	0.82	0.44	0.26	12.077	0.320	0.63
A1	0.14	-	0.16	0.29	0.50	0.47	0.19	8.802	0.145	0.33
B	0.24	0.45	-	0.44	0.81	0.54	0.30	8.587	0.265	0.55
B1	0.15	0.30	0.15	-	0.52	0.45	0.19	9.059	0.146	0.36
C	0.02	0.12	0.02	0.13	-	0.18	0.09	9.458	0.013	0.07
C1	0.07	0.11	0.07	0.13	0.37	-	0.10	7.862	0.030	0.16
D	0.13	0.23	0.12	0.24	0.47	0.55	-	6.937	0.081	0.55

all objective functions can be very helpful in a SIM model. For models A, A1, B, B1, the connected models always outperform their isolated counterparts. This observation may be used to justify the use of migration among subpopulation in a SIM model.

5 Discussion and Conclusions

SIM is a model of parallel evolutionary algorithms for multiobjective optimization. It is relatively straightforward to implement SIM using tools such as the parallel virtual machine (PVM) [9] or message passing interface (MPI) [20] on different parallel computer architectures. The experimental results in this study suggest that, when connections are properly designed, SIM is effective in helping diversify the population of an EA, and ensures that a large number of non-dominated solutions are found. More generally, SIM can also be regarded as a hybrid approach in which each subEA can be designed to use a set of distinctive settings. Consequently, the entire EA is a hybridization of different evolutionary approaches that can be useful in areas outside multiobjective optimization.

The results generated by the SIM approach appear to be quite promising. However, additional research is needed on the following issues:

- Population size. This study used identical population sizes for all subpopulations, without considering the number of subpopulations. This, however, results in different population sizes among scenarios. Further research may be needed to determine the equivalency of the total population size for the entire EA among different scenarios according to the settings of their subpopulations.
- Niching parameters (α and σ_{share} in this study). EA outcomes are sensitive to the choice of niching parameters. It is necessary to conduct a sensitivity analysis of the results to these parameters so that each subpopulation is tuned to its best performance.
- Migration rate and frequency. These parameters play a critical role in diversifying the receiving subpopulation. A sensitivity analysis is needed to determine the optimal migration strategy for different settings.

- Assimilation rate. A higher assimilation rate can help diversify a population, but can also destroy the original population.

In addition, it is necessary to test SIM using a larger set of benchmark problems that include more than two objectives. Some guidelines for the design of such problems were suggested in [7, pp 346-348]. However, the problems obtained using these guidelines could not guarantee conflicts among objective functions from f_1 to f_{M-1}, where M is the number of objectives. To reflect problems encountered in the real-world, additional test problems with known optimal solutions are needed.

References

1. E. Alba and M. Tomassini. Parallelism and evolutionary algorithms. *IEEE Transactions on Evolutionary Computation*, 6(5):443–462, 2002.
2. T.C. Belding. The distributed genetic algorithm revisited. In L. J. Eshelman, editor, *Proceedings of the 6th International Conference on Genetic Algorithms*, pages 114–121. 1995.
3. E. Cantú-Paz. *A Survey of Parallel Genetic Algorithms*. University of Illinois at Urbana-Champaign, 1997.
4. E. Cantú-Paz. *Efficient and Accurate Parallel Genetic Algorithms*. Kluwer Academic Publishers, Boston, 2000.
5. C.A.C. Coello. An updated survey of GA-based multiobjective optimization techniques. *ACM Computing Surveys*, 32(2):109–143, 2000.
6. K. Deb. Multi-objective genetic algorithms: Problem difficulties and construction of test problems. *Evolutionary Computation*, 7(3):205–230, 1999.
7. K. Deb. *Multi-Objective Optimization Using Evolutionary Algorithms*. John Wiley & Sons, Chichester, 2001.
8. C.M. Fonseca and P.J. Fleming. An overview of evolutionary algorithms in multi-objective optimization. *Evolutionary Computation*, 3(1):1–16, 1995.
9. A. Geist, A. Beguelin, J. Dongarra, W. Jiang, R. Manchek, and V. Sunderam. *PVM: Parallel Virtual Machine, A Users' Guide and Tutorial for Networked Parallel Computing*. MIT Press, Cambridge, MA, 1994.
10. D.E. Goldberg. *Genetic Algorithms in Search, Optimization and Machine Learning*. Addison-Wesley, Reading, MA, 1989.
11. D.E. Goldberg and J. Richardson. Genetic algorithms with sharing for multimodal function optimization. In J.J. Grefenstette, editor, *Genetic Algorithms and Their Applications: Proceedings of the Second International Conference on Genetic Algorithms*, pages 41–49. Lawrence Erlbaum Associates, Hillsdale, NJ, 1987.
12. P.B. Grosso. *Computer Simulations of Genetic Adaptation: Parallel Subcomponent Interaction in a Multilocus Model*. PhD thesis, University of Michigan, Ann Arbor, MI, 1985.
13. F. Herrera and M. Lozano. Gradual distributed real-coded genetic algorithms. *IEEE Transactions on Evolutionary Computation*, 4(1):43–63, 2000.
14. J.H. Holland. *Adaptations in Natural and Artificial Systems*. University of Michigan Press, Ann Arbor, MI, 1975.
15. J. Knowles, editor. *The Second Workshop on Multiobjective Problem Solving from Nature (In association with The Seventh International Conference on Parallel Problem Solving from Nature)*, Granada, Spain, September 7-11 2002. http://iridia.ulb.ac.be/~jknowles/MPSN-II.

16. R.A.E. Mäkinen, J.Päriaux, M. Sefrioui, and J. Toivanen. Parallel genetic solution for multiobjective MDO. In A. Schiano, A. Ecer, J. Periaux, and N. Satofuka, editors, *Parallel CFD '96 Conference*, pages 352–359, Capri, 1996. Elsevier.
17. N. Marco and S. Lanteri. A two-level parallelization strategy for genetic algorithms applied to optimum shape design. *Parallel Computing*, 26:377–397, 2000.
18. N. Marco, S. Lanteri, J.-A. Desideri, and J.Périaux. A parallel genetic algorithm for multi-objective optimization in computational fluid dynamics. In K. Miettinen, M.M. Mäkelä, P. Neittaanmäki, and J.Périaux, editors, *Evolutionary Algorithms in Engineering and Computer Science*, pages 445–456. Wiley & Sons, Chichester, UK, 1999.
19. W.N. Martin, J. Lienig, and J.P. Cohoon. Island (migration) models: evolutionary algorithms based on punctuated equilibria. In T. Bäck, D.B. Fogel, and Z. Michalewicz, editors, *Handbook of Evolutionary Computation*, pages C6.3:1–16. Oxford University Press/IOP, New York, 1997.
20. MPI Forum. *MPI-2: Extensions to the Message-Passing Interface*. University of Tennessee, 1997.
21. V. Pareto. *Manual of Political Economy*. Augustus M. Kelley, New York, 1971.
22. J.D. Schaffer. Multiple objective optimization with vector evaluated genetic algorithms. In J.J. Grefenstette, editor, *Proceedings of the First International Conference on Genetic Algorithms and Their Applications*, pages 93–100. Lawrence Erlbaum, Hillsdale, NJ, 1985.
23. F. Seredynski. New trends in parallel and distributed evolutionary computing. *Fundamenta Informaticae*, 35:211–230, 1998.
24. N. Srinivas and K. Deb. Multiobjective optimization using nondominated sorting in genetic algorithms. *Evolutionary Computation*, 2(3):221–248, 1995.
25. R. Tanese. Distributed genetic algorithms. In J.D. Schaffer, editor, *Proceedings of the 3rd International Conference on Genetic Algorithms*, pages 434–439. 1989.
26. D.A. Van Veldhuizen and G.B. Lamont. Multiobjective evolutionary algorithms: analyzing the state-of-the-art. *Evolutionary Computation*, 8(2):125–147, 2000.
27. E. Zitzler, K. Deb, and L. Thiele. Comparison of multiobjective evolutionary algorithms: empirical results. *Evolutionary Computation*, 8(2):173–195, 2000.
28. E. Zitzler, K. Deb, L. Thiele, C.A.C. Coello, and D. Corne, editors. *Evolutionary Multi-Criterion Optimization: Proceedings of the First International Conference*, Berlin, 2001. Springer.

Adaptation of Length in a Nonstationary Environment

Han Yu[1], Annie S. Wu[1], Kuo-Chi Lin[2], and Guy Schiavone[2]

[1] School of EECS
University of Central Florida
P.O. Box 162362
Orlando, FL 32816-2362
{hyu,aswu}@cs.ucf.edu

[2] Institute for Simulation and Training
University of Central Florida
Orlando, FL 32826-0544
{klin,guy}@pegasus.cc.ucf.edu

Abstract. In this paper, we examine the behavior of a variable length GA in a nonstationary problem environment. Results indicate that a variable length GA is better able to adapt to changes than a fixed length GA. Closer examination of the evolutionary dynamics reveals that a variable length GA can in fact take advantage of its variable length representation to exploit good quality building blocks after a change in the problem environment.

1 Introduction

A typical genetic algorithm (GA) tends to use problem representations that are orderly, fixed, and somewhat arbitrary. Individuals are of a fixed length with information encoded at fixed, programmer-defined locations on the individuals. These representations are very efficient, well organized, and encoded in ways that are very logical or easy for humans to interpret. Extending a GA to use a variable length problem representation brings about a host of new issues that must be addressed, including how to encode information and modifications to traditional genetic operators. Nevertheless, the advantages of a more adaptable and evolvable problem representation are thought to outweigh the additional effort.

In this paper, we explore the adaptability of a variable length representation in a nonstationary environment. Previous work has suggested that in periods of heavy search, e.g. those periods immediately following a change in the environment or target solution, a variable length GA will tend to favor longer individuals because longer individuals provide more resources to the search process [1,2]. We test this theory on a variable length GA applied to the problem of multiprocessor task scheduling [3]. Although our initial results are somewhat surprising, a detailed analysis of the evolutionary dynamics provide an interesting and positive explanation.

2 Related Work

Early work on variable length representations includes Smith's LS-1 learning system [4], Goldberg's messy GA [5], Koza's genetic programming (GP) [6], and Harvey's SAGA [7]. These studies laid much of the groundwork in terms of defining the issues that need to be addressed with variable length representations and exploring potential solutions.

Since then, much research has been conducted on the variation of individual length during evolution. Langdon and Poli [2] explore the cause of bloat in the variable length genetic programming (GP) representation. They conclude that longer individuals are favored in the selection process because they have more ways to encode solutions than shorter individuals. Soule et al. [8,9] perform a detailed investigation on code growth in GP and conclude that code growth is dominated by non-functional code. As a result, parsimony pressure can affect the search quality in GP. The relationship between size and fitness in a population is useful in predicting the GP's search performance in a long run. Burke et al. [1] study the adaptation of length in a variable length GA. They found that, without parsimony pressure, a GA tends to generate individuals of arbitrary length. If parsimony pressure is applied, the average individual length increases quickly in early evolution, followed by a gradual decrease until stabilization. The early increase was thought to be a period of growth of resources: increasing individual length increases the probability of finding building blocks. All of the above studies examine variable length representation in a stable problem environment.

A variety of studies have looked at GA behavior in changing environments [10]. Some of these approaches focus on maintaining the population diversity during GA search, such as the use of random immigrants [11], hypermutation [12], adaptive GA operators [13], and the TDGA [14,15]. Other strategies attempt to improve the search by storing duplicate information with redundant representation schemes [16,17,18], using alternative memory systems [19], or encouraging the maintenance of multiple "species" within a GA population [20,21, 22].

3 Problem Description

We perform our experiments on a variable length GA applied to the problem of multiprocessor task scheduling [23]. The task scheduling problem begins with a task dependency graph which specifies the dependencies among a number of tasks that together compose a larger complete task. Each task has a fixed execution time that is given in the problem. Figure 1 shows two example task dependency graphs. The goal of the GA is to assign tasks to four parallel processors such that all tasks can be completed and total execution time is minimized. The same task may be assigned to multiple processors but not the same processor twice. The data dependencies that exist among tasks place restrictions on the order in which tasks can be assigned to processors. Dependent tasks that are assigned to different processors may incur additional communication delays. A

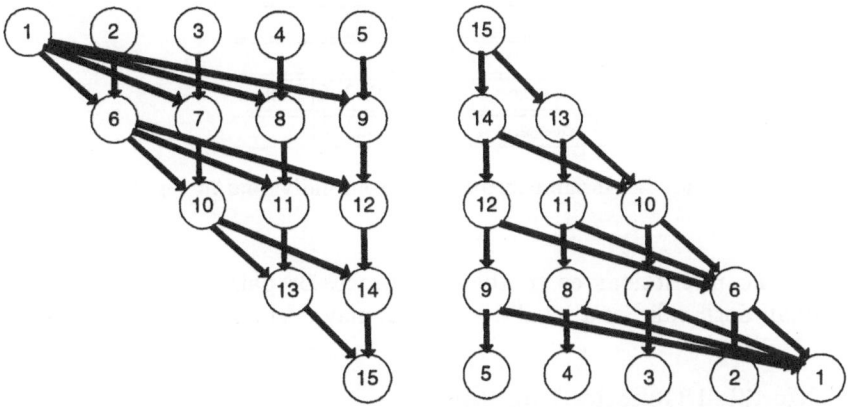

Fig. 1. Task dependency graph for problems G1 (left) and G2 (right)

valid solution is one in which there is at least one copy of every task and all task orderings are valid.

4 Implementation Details

We highlight some of the key features of the GA used in this study. A full description of this system is available in [3].

4.1 Variable Length Problem Representation

Each individual in the population consists of a vector of cells or genes. A *cell* is defined as a task-processor pair, (t, p), which indicates that a task t is assigned to processor p. The number of cells in an individual may be fixed or may vary during evolution. The order of the cells of an individual determines the order in which tasks are assigned to processors: cells are read from left to right and tasks are assigned to corresponding processors. If the same task-processor pair appears multiple times in an individual, only the first cell contributes to the fitness evaluation. This cell is called a coding gene. Any additional identical cells are ignored by the fitness function but are still subject to action by genetic operations. The *coding length* refers to the total number of distinct cells in an individual. In variable length evolution, the number of cells in an individual is limited to ten times the number of tasks in the problem. Figure 2 shows an example individual. Figure 3 shows the corresponding task assignment. Because

$(4,1)(2,4)(3,3)(2,3)(4,1)(5,4)(6,3)(1,1)(3,2)$

Fig. 2. An example individual

Processor 1	Task 4	Task 1	
Processor 2	Task 3		
Processor 3	Task 3	Task 2	Task 6
Processor 4	Task 2	Task 5	

Fig. 3. Assignment of tasks from individual in Fig. 2

there are no restrictions as to the task processor pairs that may exist in an individual, both valid and invalid solutions may occur in the population.

4.2 Modified Genetic Operators

Recombination is performed on the cell level and crossover points are restricted to falling only in between cells. In variable length runs, we use random one-point crossover which randomly selects a crossover point independently on each parent. As a result, the length of offspring may be different from their parents. In fixed length runs, we use simple one point crossover.

Each cell has equal probability to take part in mutation. If a cell is selected to be mutated, then either the task number or the processor number of that cell will be randomly changed.

4.3 Fitness Function

The fitness function is a weighted sum of two components, the *task_fitness* and the *processor_fitness*:

$$fitness = (1-b) * task\_fitness + b * processor\_fitness.$$

The value of b ranges from 0.0 to 1.0.

The *task_fitness* evaluates whether tasks have been scheduled in valid orders and whether all tasks have been included in a solution. Calculation of *task_fitness* consists of three steps.

1. Because we do not limit our GA to only generate valid solutions, we use an incremental fitness function in order to give partial credit to invalid solutions that may contain some valid subsequences of tasks. This fitness function starts out rewarding for simpler goals and gradually increases the criteria of goals until a complete valid solution is found. Partial credit is determined using the *era* counter which indicates the length of the subsequences to be checked for validity, $era = 0, 1, 2, \ldots, T$. Initially, *era* is set to zero. For all tasks assigned to the same processor, we check the sequence of every pair of adjacent tasks. The *era* counter increases when the average population fitness exceeds a user defined threshold value and a fixed percentage of the population consists of valid individuals. Each time *era* is increased, we increase the number of tasks in the sequences checked by one. Thus, the length

of the sequences checked always equals $era + 2$. We calculate $raw\_fitness$ using the following equation:

$$raw\_fitness = \frac{\text{number of valid task groups in an assignment}}{\text{total number of task groups in an assignment}} \quad (1)$$

2. We next calculate $task\_ratio$ using the following equation:

$$task\_ratio = \frac{\text{number of distinct tasks specified on an individual}}{\text{total number of tasks in the problem}}. \quad (2)$$

3. Finally, $task\_fitness = raw\_fitness \times task\_ratio$.

The *processor_fitness* evaluates the execution time of a valid task schedule, favoring schedules that minimize execution time.

$$processor\_fitness = \frac{P * serial\_len - t}{P * serial\_len}.$$

where P is the number of processors in the problem, t is the execution time of a solution, and *serial_len* is the execution time of all the tasks if they are assigned serially to a single processor.

Additional details regarding the fitness function are available in [3].

4.4 System Parameter Settings

Table 1 gives the default parameter settings used in our experiments. In a variable length GA, the initial population of individuals are initialized to length fifteen (the total number of tasks in the problem), and the maximum allowed length is 150. Experimental tests indicate that initial population length does not significantly affect GA performance. In the fixed length GA, we use individuals of length 150.

Table 1. Parameter settings used in our experiments

Parameter	Value
Population size	200
Number of generations	1500
Crossover type	random one-point
Crossover rate	1.0
Mutation rate	0.005
Selection scheme	Tournament (2)
b	0.2
Fitness threshold	0.75

5 Experimental Results

We compare the behavior of fixed and variable length GAs in a nonstationary environment. In these experiments, the target solution oscillates between two different problems, G1 and G2, shown in Fig. 1 every 100 generations. To emphasize the differences between the two target solutions and increase the difficulty of finding and maintaining solutions for both targets, we designed G2 to have the completely opposite set of task dependencies as G1. The results presented are from individual runs that are representative of overall GA behavior.

We compare four experimental scenarios:

1. A fixed length GA in which *era* is not reset.
2. A fixed length GA in which *era* is reset after each target change.
3. A variable length GA in which *era* is not reset.
4. A variable length GA in which *era* is reset after each target change.

In the runs where *era* is not reset, the *era* counter increases normally throughout a run, independent of target changes. In the runs where *era* is reset, the value of *era* is reset to zero after each target change; *era* increases normally in between two consecutive target changes.

5.1 Fixed versus Variable Length

Figure 4 shows the typical variation in average population fitness and average coding length during a fixed length GA run where *era* is not reset. A sharp drop in fitness occurs after each target change indicating that the GA has difficulty retaining both solutions in its population. The alternating high and low peaks suggest that the GA population has converged primarily to one of the two target solutions. When the converged solution is the target solution, the average population fitness reaches above 0.8. When the non-converged solution is the target solution, the average population fitness is unable to exceed 0.6.

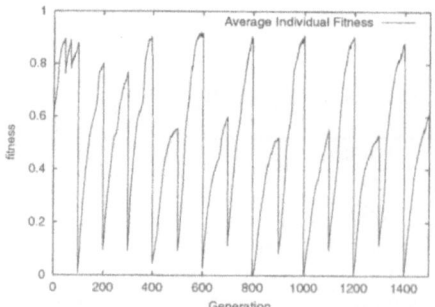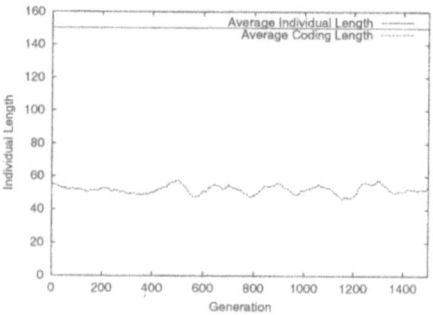

Fig. 4. Typical plots of average population fitness (left) and average coding length (right) for a fixed length (150) GA with no resetting *era* in a nonstationary environment

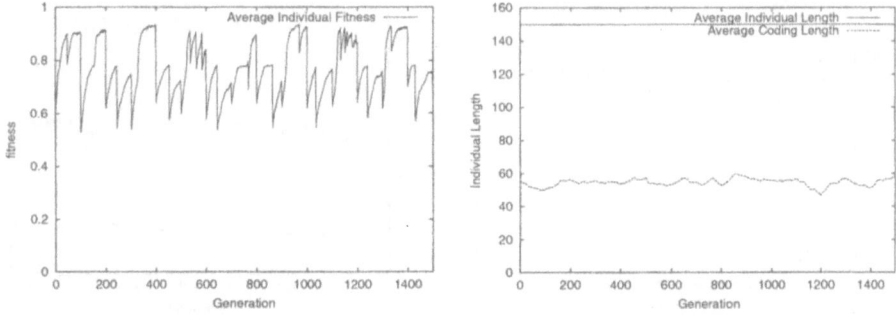

Fig. 5. Typical plots of average population fitness (left) and average coding length (right) for a fixed length (150) GA with resetting *era* in a nonstationary environment

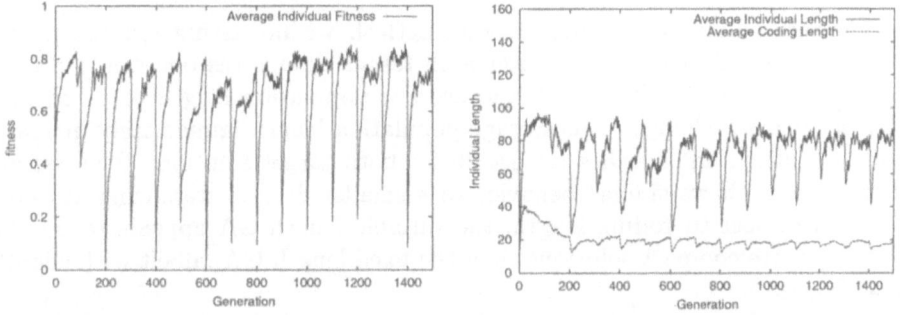

Fig. 6. Typical plots of average population fitness (left) and average length (right) for a variable length GA with no resetting *era* in a nonstationary environment

Figure 5 shows the typical variation in average population fitness and average coding length during a fixed length GA run where *era* is reset after each target change. The result is very similar to the run shown in Fig. 4. We see sharp drops in fitness after every target change. The high and low peaks in fitness indicate that GA is evolving solutions toward only one target. The primary difference is the existence of smaller oscillations in fitness within the larger peaks due to increases in the *era* counter.

Figure 6 shows typical plots of the average population fitness and average population length of a variable length GA where *era* is not reset. We again see sharp drops in fitness after each target change indicating difficulty in retaining multiple solutions in the population. These data exhibit two notable differences from the corresponding fixed length results in Fig. 4. First, there is little evidence of alternating peaks; the GA appears to be able to find equally strong solutions for both targets. Average fitness values are slightly lower, but comparable, to those achieved by the fixed length GA. Second, the rise in fitness following each drop is much faster and sharper. These differences suggest that a variable length GA is better able to adapt to changes in the target solutions.

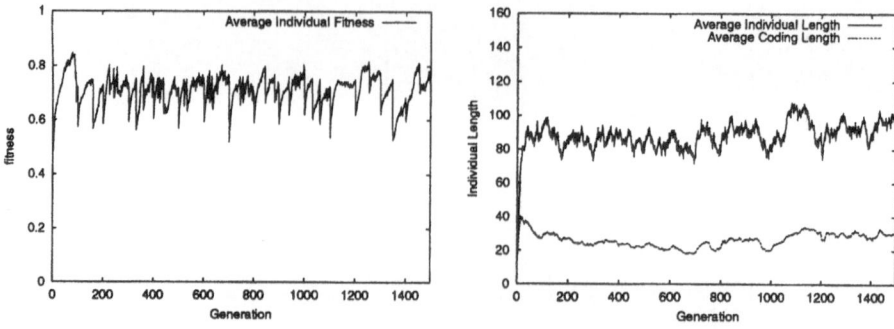

Fig. 7. Typical plots of average population fitness (left) and average length (right) for a variable length GA with resetting *era* in a nonstationary environment

The variation in average population length shows interesting and unexpected behavior. Instead of increasing after each target change (as one would predict based on previous studies [1,2]), the average population length drops sharply after each target change. The average population length immediately increases after each drop, and reaches almost 90 for both target solutions. The average coding length shows similar behavior to a smaller degree, stabilizing at about 20. With respect to coding length, the variable length GA appears to be able to evolve more compact solutions than the fixed length GA, albeit with slightly lower fitness.

Figure 7 shows typical plots of the average population fitness and average population length of a variable length GA where *era* is reset after each target change. We observe much smaller drops in individual fitness after target changes and variations in individual length that are less clearly correlated with target changes. Average fitness remains comparable to the fitness achieved in the other three experiments.

We further evaluate the adaptability of our GAs to changing environments by examining the probability of finding valid solutions within the interval of two consecutive target changes. The results, given in Table 2, are collected from twenty runs in each test case. Results show that variable length GAs, as expected, are more adaptable than fixed length GAs. Surprisingly, not resetting *era* actually results in much better adaptation for both the variable length and the fixed length GAs, showing more chances of finding valid solutions after tar-

Table 2. The probability of finding a valid solution between target changes

Test Cases	The Probability of Finding a Valid Solution Between Target Changes
Variable length GA (*era* not reset)	97.5%
Variable length GA (*era* reset)	87.1%
Fixed length GA (*era* not reset)	72.9%
Fixed length GA (*era* reset)	59.3%

get changes. This result was unexpected as we expected the *era* counter to guide the GA in finding a solution by rewarding for partial solutions.

5.2 Variable Length Dynamics

Our original hypothesis was that average length would increase following each target change to provide the GA search process with more resources in the form of longer individuals. Our variable length GA runs with no resetting *era*, however, exhibit completely opposite behavior. Closer examination reveals that the GA is, in fact, selecting for *better* resources instead of *more* resources.

We examine the distribution of fitness with respect to length in the generations immediately preceding and following a target change. Figure 8 shows this data for a sample run. Within a GA run, we take all individuals that are in populations immediately preceding a target change (in our case, generations 99, 199, 299, etc.), group those individuals by length, and plot the average fitness of each group. Similarly, we plot the fitness distribution of all individuals that are in generations immediately following a target change (generations 100, 200, 300, etc.). Immediately before a target change (generation "X99"), longer individuals have higher average fitness than shorter individuals. Immediately after a target change (generation "X00"), shorter individuals appear to be fitter than longer individuals. This same dynamic is seen to a stronger degree in similar plots of the coding length of a population. Figure 9 shows the distribution of fitness with respect to coding length.

Examination of individual members of the population provides an explanation. Both target problems used here consist of fifteen tasks. Thus, individuals must have coding lengths of at least fifteen to encode a complete solution. As expected, the data in Fig. 9a indicates that the average fitness peaks at coding lengths that are just slightly longer than fifteen and levels off with increasing length. For lengths shorter than fifteen, there is a steady linear decrease in fitness. Immediately before a target change, the GA has had time (in the runs

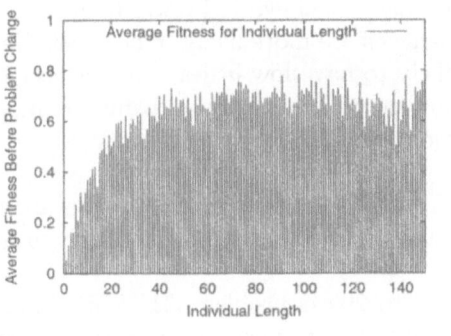
a
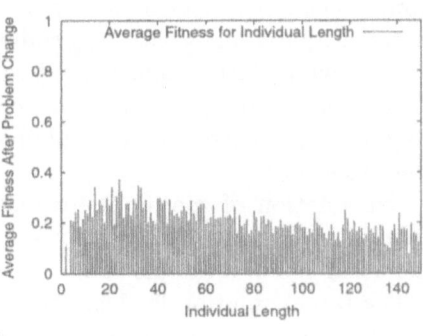
b

Fig. 8. Distribution of average fitness of individuals of equal length before (a) and after (b) a target change

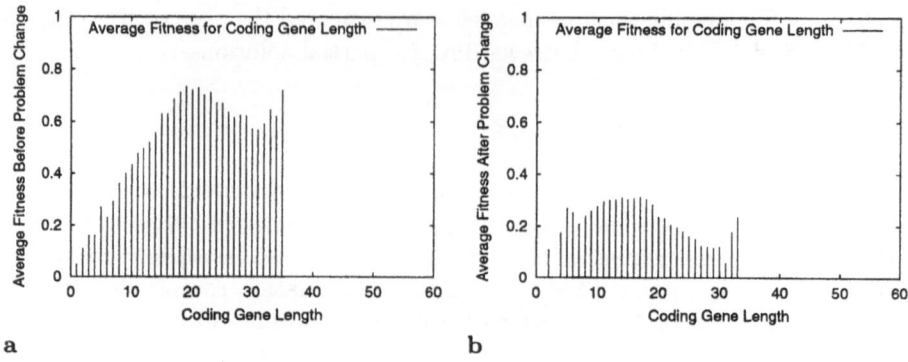

Fig. 9. Distribution of average fitness of individuals of equal coding length before (a) and after (b) a target change

Proc 1	5	8
Proc 2	3	5
Proc 3	7	11
Proc 4	4	

(a)

Proc 1	5	14	13	15					
Proc 2	1	3	2	6	4	5			
Proc 3	7	6	9	8	12	11	10	13	
Proc 4	4	9	12						

(b)

Fig. 10. Small (a) and large (b) solutions from generation 299 of sample run

here, 99 generations) to evolve towards the current target. Most individuals in the population will have some characteristics of the current target. Longer individuals are more likely to contain complete solutions, and therefore, are also more likely to be specialized towards the current target.

Figure 9b shows the fitness distribution in the generations immediately following a target change. Individuals with coding lengths less than fifteen tend to have higher relative fitness while individuals with longer coding lengths exhibit steadily decreasing fitness values. Immediately after a target change, much of the population still consists of individuals that have been evolved towards the previous target. Individuals with long coding lengths are more likely to be specialized to the previous target and, thus, more likely to have low fitness with respect to a new target. Individuals with short coding lengths, on the other hand, are more likely to contain less specific building blocks that may be applicable to more target solutions.

Examination of specific individuals from a population supports the above explanation. Figures 10a,b show a small and a large solution, respectively, from generation 299 of a sample run.

The target solution for generations 200 to 299 is problem G1. Accordingly, all task sequences in both solutions are valid with respect to G1. With respect to problem G2, all task sequences in the smaller solution are valid, but only the following subset of task sequences (18 out of 52 total) from the longer solution are valid:

[14-13] [1-3] [3-2] [1-3-2] [6-4] [4-5] [6-4-5] [7-6] [6-9] [9-8] [7-6-9] [6-9-8] [7-6-9-8]
[8-12] [12-11] [11-10] [12-11-10] [4-9]

The longer solution is a complete solution to problem G1. It is very specialized for G1, and consequently, noticeably less fit for G2. The shorter solution does not specify a complete solution for either G1 or G2, but is equally fit for both. With respect to the percent of valid sequences, the shorter solution actually scores higher than the longer solution and consequently appears fitter immediately after a target change.

Thus, the sharp drops in average population length seen in Fig. 6 appear to be due to selection for more general building blocks following a target change. In this particular problem encoding, shorter individuals tend to consist of more general building blocks, longer individuals tend to consist of problem specific building blocks. By selecting for shorter individuals immediately after a target change, the GA increases its population resources by increasing the number of more general building blocks.

The above study also explains why there are no evident drops in individual length due to a problem change in Fig. 7. In GAs where *era* is reset, the fitness function is again checking the order of every two adjacent tasks after each target change. Long individuals, though more specialized to the previous target, are still likely to contain short task sequences that are valid. As a result, long individuals drop less in fitness than they do in GAs where *era* is not reset and remain competitive with shorter, more general individuals. Surprisingly, resetting *era* is not beneficial in this problem as it actually retards a GA's adaptation to a new target.

6 Conclusions

In this paper, we investigate the behavior of a variable length GA in a nonstationary problem environment. We examine how variations in length can help a GA adapt to new environments.

We perform our experiments on a task scheduling problem under an oscillating environment. Experimental results indicate that a variable length GA has better and quicker adaptation to a new environment than a fixed length GA.

An interesting and somewhat surprising result shows the variable length GA undergoing sharp drops in length after each target change. This behavior is the opposite of what was expected based on previous work. Closer analysis reveals that the fitness function favors short individuals after a target change because short individuals contain more general building blocks. Long individuals, on the other hand, are more likely to contain very specific solutions adapted to the previous target. Our GA successfully exploits the flexibility of its variable length representation to better recognize and retain good building blocks.

Our study indicates that variable length representation provides a flexible way for GA to reorganize building blocks after problem changes. A good fitness function, where building blocks are properly defined, is able to reinforce this flexibility and improve a GA's adaptation to changing environments.

Acknowledgments. This research is supported in part by the Air Force Research Laboratory.

References

1. Burke, D.S., De Jong, K.A., Grefenstette, J.J., Ramsey, C.L., Wu, A.S.: Putting more genetics into genetic algorithms. Evolutionary Computation **6** (1998) 387–410
2. Langdon, W.B., Poli, R.: Fitness causes bloat. Soft Computing in Engineering Design and Manufacturing (1997) 13–22
3. Wu, A.S., Jin, S., Schiavone, G., Lin, K.C.: An incremental fitness function for partitioning parallel tasks. In: Proc. Genetic and Evolutionary Computation Conference. (2001)
4. Smith, S.F.: A learning system based on genetic adaptive algorithms. In: PhD thesis, Dept. Computer Science, University of Pittsburgh. (1980)
5. Goldberg, D.E., Korb, B., Deb, K.: Messy genetic algorithms: Motivation, analysis, and first results. Complex Systems **3** (1989) 493–530
6. Koza, J.R.: Genetic programming. MIT Press (1992)
7. Harvey, I.: Species adaptation genetic algorithms: A basis for a continuing saga. In: Proceedings of the First European Conference on Artificial Life. (1992) 346–354
8. Soule, T., Foster, J.A., Dickinson, J.: Code growth in genetic programming. In: Proc. Genetic Programming(GP). (1996) 400–405
9. Soule, T., Foster, J.A.: Effects of code growth and parsimony pressure on populations in genetic programming. Evolutionary Computation **6** (1998) 293–309
10. Branke, J.: Evolutionary approaches to dynamic optimization problems - updated survey. In: Genetic and Evolutionary Computation Conference Workshop Program. (2001) 27–30
11. Grefenstette, J.J.: Genetic algorihtms for changing environments. Parallel Problem Solving from Nature (PPSN) **2** (1992) 137–144
12. Cobb, H.G.: An investigation into the use of hypermutation as an adaptive operator in genetic algorithms having continuous, time-dependent nonstationary environments. Technical Report AIC-90-001, Naval Research Laboratory (1990)
13. Grefenstette, J.J.: Evolvability in dynamic fitness lanscapes: a genetic algorithm approach. In: Congress on Evolutionary Computation. (1999) 2031–2038
14. Mori, N., Kita, H., Nishikawa, Y.: Adaptation to a changing environment by means of the thermodynamical genetic algorithm. In: PPSN. (1996) 513–522
15. Kita, H., Yabumoto, Y., Mori, N., Nishikawa, Y.: Multi-objective optimization by means of the thermodynamical genetic algorithm. In: PPSN IV. (1996) 504–512
16. Goldberg, D.E., Smith, R.E.: Nonstationary function optimization using genetic algorithms with dominance and diploidy. In: Intl. Conf. on Genetic Algorithms (ICGA). (1987) 59–68
17. Ng, K.P., Wang, K.C.: A new diploid scheme and dominance change mechanism for non-stationary function optimization. In: Proc. 6th ICGA. (1995) 159–166
18. Smith, R.E.: Diploidy genetic algorithms for search in time varying environments. In: Annual Southeast Regional Conference of the ACM. (1987) 175–179
19. Branke, J.: Memory enhanced evolutionary algorithms for changing optimization problems. In: CEC. (1999) 1875–1882
20. De Jong, K.A.: An analysis of the behavior of a class of genetic adaptive systems. PhD thesis, University of Michigan (1975)
21. Deb, K., Goldberg, D.E.: An investigation of niche and species formation in genetic function optimization. In: Proc. ICGA. (1989) 42–50

22. Liles, W., De Jong, K.: The usefulness of tag bits in changing envrionments. In: CEC. (1999) 2054–2060
23. El-Rewini, H., Lewis, T.G., Ali, H.H.: Task scheduling in parallel and distributed systems. Prentice Hall (1994)

Optimal Sampling and Speed-Up for Genetic Algorithms on the Sampled OneMax Problem

Tian-Li Yu[1], David E. Goldberg[2], and Kumara Sastry[3]

Illinois Genetic Algorithms Laboratory (IlliGAL)
Department of General Engineering
University of Illinois at Urbana-Champaign
104 S. Mathews Ave, Urbana, IL 61801
{tianliyu,deg,kumara}@illigal.ge.uiuc.edu

Abstract. This paper investigates the optimal sampling and the speed-up obtained through sampling for the sampled OneMax problem. Theoretical and experimental analyses are given for three different population-sizing models: the decision-making model, the gambler's ruin model, and the fixed population-sizing model. The results suggest that, when the desired solution quality is fixed to a high value, the decision-making model prefers a large sampling size, the fixed population-sizing model prefers a small sampling size, and the gambler's ruin model has no preference for large or small sizes. Among the three population-sizing models, sampling yields speed-up only when the fixed population-sizing model is valid. The results indicate that when the population is sized appropriately, sampling does not yield speed-up for problems with subsolutions of uniform salience.

1 Introduction

Over the last few decades, significant progress has been made in the theory, design and application of genetic and evolutionary algorithms. A decomposition design theory has been proposed and *competent* genetic algorithms (GAs) have been developed (Goldberg, 1999). By competent GAs, we mean GAs that can solve hard problems quickly, reliably, and accurately. Competent GAs render problems that were intractable by first generation GAs tractable requiring only a polynomial (usually subquadratic) number of function evaluations.

However, in real-world problems, even the time required for a subquadratic number of function evaluations can be very high, especially if the function evaluation is a complex model, simulation, or computation. Therefore, GA practitioners have used a variety of *efficiency enhancement* techniques to alleviate the computational burden. One such technique is *evaluation relaxation* (Sastry and Goldberg, 2002), in which the accurate, but costly fitness evaluation is replaced by a cheap, but less accurate evaluation. Partial evaluation through sampling is an example of evaluation relaxation, and sampling has empirically shown to yield significant speed-up (Grefenstette & Fitzpatrick, 1985). Evaluation relaxation through sampling has also been analyzed by developing facetwise and dimensional models (Miller & Goldberg 1996a; Miller, 1997; Giguère & Goldberg,

1998). This study extends the theoretical analyses and investigates the utility of sampling as an evaluation-relaxation technique.

The objective of this paper is to extend the work of Giguère and Goldberg (1998) and incorporate the effect of sampling on both convergence time and population sizing of GAs. Specifically, we concentrate on problems with substructures of uniform salience. This paper is composed of four primary parts: (1) background knowledge including previous work and an introduction to the sampled OneMax problem, (2) the derivation of the optimal sampling sizes for the decision-making model, the gambler's ruin model, and the fixed population-sizing model, (3) empirical results, and (4) extensions and conclusions of this work.

2 Past Work

Grefenstette and Fitzpatrick (1985) achieved a great success in applying sampling techniques to the image registration problem. Their success motivated Miller and Goldberg (1996b), which gave a theoretical analysis of a related problem. They derived and empirically verified the theoretical optimal sampling size when the fitness function is clouded by an additive Gaussian noise. The sampling methods in the above two papers have a subtle but significant difference. In Grefenstette and Fitzpatrick (1985), the variance becomes zero if full sampling is used. However, in the additive Gaussian noisy OneMax problem in Miller and Goldberg (1996b), the variance of fitness function varies at the sampling size. The variance can be very small when the sampling size is large, but it will never be zero. More recently, Giguère and Goldberg (1998) investigated a sampled OneMax problem (SOM) where the fitness value is calculated by sampling. Their results showed that sampling is not really useful when the gambler's ruin population-sizing model (Harik, Cant'u-Paz, Goldberg & Miller, 1997; Miller, 1997) is adopted. Even though Giguère and Goldberg (1998) have investigated all population-sizing models discussed in this paper, they did not consider convergence time. In addition, detailed analytical models are needed for a better understanding of the sampling schemes as a technique of evaluation relaxation.

3 Sampled OneMax Problem (SOM)

Before starting the derivation of our models of computational requirement, let us first define the sampled OneMax problem (SOM). The SOM is basically the OneMax or the counting ones problem, except the fitness value is computed by sampling without replacement.

$$F_n(\bar{x}) = \frac{l}{n}(\Sigma_{i \in S} x_i), \qquad (1)$$

where $\bar{x}$ is a chromosome (a binary string in SOM), x_i is the value of the i-th gene in the chromosome (0 or 1), n is the sampling size ($0 < n \leq l$), S is a subset of $\{1, 2, \cdots, l\}$ with a restriction that $|S| = n$, and l is the chromosome length. The term $\frac{l}{n}$ is just for normalization so that the expectation of the fully sampled fitness $F(\bar{x}) = \Sigma_{i=1}^{l} x_i$ is the same as the expectation of F_n for all n. The variance of the noise introduced by sampling can be expressed as follows (Giguère & Goldberg, 1998):

$$\sigma_n^2 = \frac{l^2 p(t)(1 - p(t))}{n} \frac{l - n}{l - 1}, \quad (2)$$

where $p(t)$ is the performance model defined in Thierens and Goldberg (1994). Initially, $p(0) = 0.5$. The number of function evaluations is defined as how many bits the GA has sampled before convergence. We choose the SOM because (1) it is linear and easy to analyze, and (2) the SOM is considered as a GA-easy problem, the speed-up obtained for this problem should give an idea about how sampling acts on other problems.

4 Optimal Sampling for the Sampled OneMax Problem (SOM)

This section gives theoretical and empirical analyses of the optimal sampling sizes for three different population-sizing models: the decision-making model, the gambler's ruin model, and the fixed population-sizing model for the SOM.

This section starts by deriving the number of function evaluations required for the SOM as a function of the sampling size. Then with the number of function evaluation model derived, the optimal sampling size is derived for each of the three population sizing models. Finally, experimental results are shown to verify the optimal sampling sizes derived.

In this section, we fix the solution quality (the average fitness of the population) to $(l - 1)$, where l is the chromosome length, and try to minimize the number of function evaluations through sampling. The desired solution quality is set so high that the convergence time model is valid. Nearly full convergence is one of the presumptions of the derivations of the convergence time models in Sastry and Goldberg (2002).

4.1 Model for Number of Function Evaluations for the SOM

Now let us derive the model for number of function evaluations for the SOM. Sastry and Goldberg (2002) indicated the time to convergence for those problems with uniformly-scaled building blocks (BBs) corresponds to the squared root of the variance of the fitness function:

$$t_{conv} = \frac{\ln(l)}{I} \sqrt{\sigma_f^2 + \sigma_n^2}, \quad (3)$$

where I is selection intensity (Mühlenbein & Schlierkamp-Voosen, 1993). For binary tournament selection, $I = 1/\sqrt{\pi} \sim 0.5642$ (Bäck, 1994). By Eq. (2) and approximating σ_f^2 by the initial variance $lp(0)(1-p(0))$, t_{conv} can be rewritten as

$$t_{conv} = C_1 \sqrt{\frac{l^2}{n} - 1}, \qquad (4)$$

where $C_1 = \frac{\ln(l)}{I} \cdot \sqrt{\frac{l}{l-1}p(0)(1-p(0))}$.

The number of function evaluations is given by $n_{fe} = nGN$, where n is the sampling size, G is the number of generations, and N is the population size. By substituting G with t_{conv} obtained in Eq. (4), the number of function evaluations can be expressed as

$$n_{fe} = C_1 \cdot nN \sqrt{\frac{l^2}{n} - 1}. \qquad (5)$$

Now that we have the model for the number of function evaluations for the SOM, we are ready to derive the optimal sampling size for different population-sizing models.

4.2 Optimal Sampling for the Decision-Making Model

The decision-making model (Goldberg, Deb, & Clark, 1992) is given by

$$N = \Gamma(\sigma_f^2 + \sigma_n^2), \qquad (6)$$

where N is the population size, Γ is the population coefficient defined in Miller & Goldberg (1996b), σ_f^2 is the initial population fitness variance, and σ_n^2 is the variance of the fitness noise.

By Eq. (2) and substituting N into Eq. (5), n_{fe} for the decision-making model can be written as

$$n_{fe} = C_2 \cdot n \left(\frac{l^2}{n} - 1\right)^{\frac{3}{2}}, \qquad (7)$$

where $C_2 = \frac{\ln(l)}{I} \Gamma \left(\frac{l}{l-1}p(0)(1-p(0))\right)^{\frac{3}{2}}$.

Equation (7) has a minimum at $n = l$, which means no sampling at all (Fig. 1). This is not surprising because the population sizing model is known as an overestimation than needed for the desired solution quality. When sampling is adopted, the noise introduced makes the population size even larger than needed according to the decision-making model. As a result, the larger population size results in a larger number of function evaluations.

4.3 Optimal Sampling for the Gambler's Ruin Model

The gambler's ruin model (Harik, Cant'u-Paz, Goldberg & Miller, 1997; Miller, 1997) is given by

$$N = \Gamma'(\sqrt{\sigma_f^2 + \sigma_n^2}), \qquad (8)$$

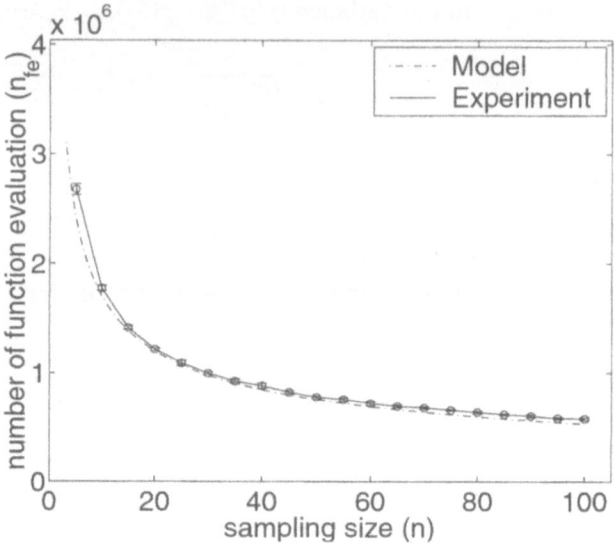

Fig. 1. The relationship between n_{fe} and n for the decision-making model

where N, σ_f^2, and σ_n^2 are defined the same as those in Eq. (6), and Γ' is another coefficient defined as

$$\Gamma' = -\frac{2^{k_{max}-1}\sqrt{\pi}\ln(\psi)}{d_{min}}. \tag{9}$$

k_{max} is an estimate of the maximal length of BBs. ψ is the failure rate, defined as the probability that a particular partition in the chromosome that fails to converge to the correct BBs. In other words, $(1-\psi)$ is the expected proportion of the correct BBs in an individual. d_{min} is an estimate of the minimal signal difference between the best and the second best BB. In other words, d_{min} is the smallest BB signal that GAs can detect. In OneMax domain, both k_{max} and d_{min} are 1, which yields a simpler form for Γ':

$$\Gamma' = -\sqrt{\pi}\ln(\psi) \tag{10}$$

By a similar algebraic process as in the previous subsection, the number of function evaluations is expressed as

$$n_{fe} = C_3 \cdot (l^2 - n), \tag{11}$$

where $C_3 = \frac{\ln(l)}{l}\Gamma'\left(\frac{l}{l-1}p(0)(1-p(0))\right)$.

The minimum of Eq. (11) occurs at $n = l$, which means, again, no sampling at all (Fig. 2). The gambler's ruin model still prefers a larger sampling size, but

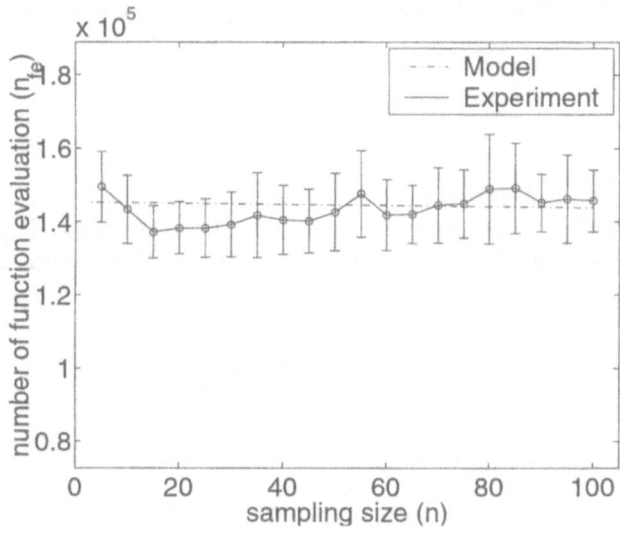

Fig. 2. The relationship between n_{fe} and n for the gambler's ruin model

only with a slight preference. This can be shown by comparing the case of $n = 1$ and $n = l$.

$$\frac{n_{fe}(n=1)}{n_{fe}(n=l)} = \frac{l^2 - 1}{l^2 - l} = \frac{l+1}{l} \qquad (12)$$

For a 100-bit SOM problem, the difference of n_{fe} is only 1%, which is so small that can be neglected compared with the approximations in the derivations. As a result, for the gambler's ruin model, sampling does not make much difference. The conclusion agrees with Giguère and Goldberg (1998).

4.4 Fixed Population-Sizing Model

The fixed population-sizing model is not an accurate model of GAs since it does not account for the effects of problem size and noise. However, it is widely adopted in real-world applications, such as in Grefenstette and Fitzpatrick (1985). Therefore, it is worthwhile to investigate this model as well. In this section, the assumption is that the fixed population size is larger than needed so that the GA can converge to an $(l-1)$ solution quality. This assumption is needed for applying the convergence time model. Since the population size is larger than needed, sampling should obtain speed-up. In addition, we still fix the desired solution quality to be $(l-1)$ because we stop the GA when a $(l-1)$ solution quality is reached.

The number of function evaluations for a fixed population size is given by

$$n_{fe} = C_4 \cdot n \left(\frac{l^2}{n} - 1\right)^{\frac{1}{2}}, \qquad (13)$$

where $C_4 = \frac{\ln(l)}{I} N \left(\frac{l}{l-1} p(0)(1-p(0))\right)^{\frac{1}{2}}$.

Equation (13) has a minimum at $n = 1$. If the overhead (α) is taken into account, and the cost of sampling one bit is β, the total run duration is expressed as

$$T = C_4 \cdot (\alpha + n\beta) \left(\frac{l^2}{n} - 1\right)^{\frac{1}{2}} \qquad (14)$$

For large l, $\frac{l^2}{n} \gg 1$, Eq. (14) can be approximated as

$$T = C_4 \cdot l(\alpha n^{-1/2} + \beta n^{1/2}) \qquad (15)$$

By differentiating Eq. (15) by n, and then setting it to be zero, the minimum is found at

$$n_{op} = \frac{\alpha}{\beta} \qquad (16)$$

It is interesting to compare Eq. (16) with what Miller and Goldberg (1996b) got ($\sqrt{\frac{\alpha}{\beta}} \sqrt{\frac{\sigma_n^2}{\sigma_f^2}}$). The term σ_n in Miller and Goldberg's result vanishes because now σ_n is controlled by the sampling size n. If the constant term is ignored, the result here is the square of Miller and Goldberg's optimal sampling size.

4.5 Experimental Results

Experimental results are obtained for the SOM problem with chromosome length $l = 100$. Binary tournament selection and uniform crossover are adopted. For all experiments, the crossover probability (p_c) is one and the mutation probability (p_m) is zero, which means, crossover always takes place and there is no mutation. All experimental results are averaged over 50 independent runs. Sastry and Goldberg (2002) used the variance of fitness of the initial population to estimate σ_f^2. This is, of course, an overestimation. In fact, the variance of fitness becomes smaller and smaller during the runs of a GA. For the OneMax problem, it almost becomes zero when the average fitness converges to $l-1$ bits ($p(t) \simeq 1$). Therefore, for a tighter estimation, $p(t) = 0.75$ (half convergence) is used in the calculation of σ_f^2 and σ_n^2.

Figures 1, 2, and 3 show the relationship of n_{fe} versus n for the decision-making model, the gambler's ruin model, and the fixed population-sizing model, respectively. In Fig. 1, the population size is obtained from $N = 8(\sigma_f^2 + \sigma_n^2)$ (Goldberg, Deb, & Clark, 1992). Figure 2 uses $N = 9.4\sqrt{\sigma_f^2 + \sigma_n^2}$ (Miller &

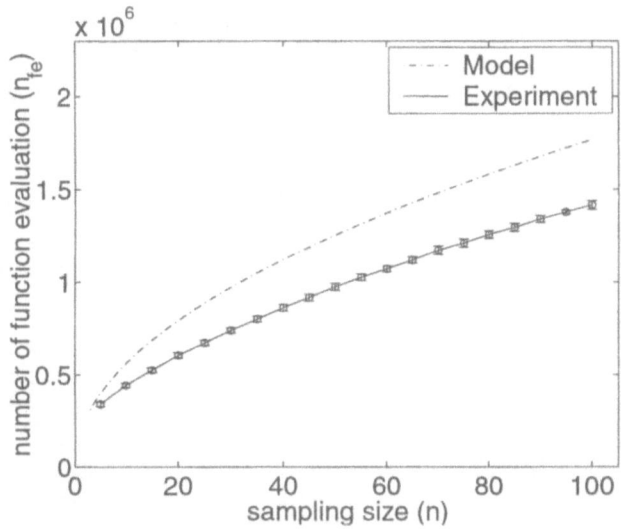

Fig. 3. The relationship between n_{fe} and n for fixed population $N = 500$. $\frac{\alpha}{\beta} = 0$

Goldberg, 1996b). In Fig. 2, the experimental results show slight minima in-between $N = 10$ and $N = 30$. It agrees with the observation in Giguère and Goldberg, 1998, which has not been explained by mathematical models so far. The fixed population size is set to be $N = 500$ and the results are shown in Fig. 3. The fixed population size is set so large to prevent failure of convergence. Finally, the total run duration for $\alpha/\beta = 20$ is shown in Fig. 4.

The experimental results agree remarkably well with the models derived in the previous section. The model of number of function evaluations for the decision-making model especially matches experimental results. The model for fixed population size overestimates somewhat the number of function evaluations required. Nevertheless, as Fig. 4 indicates, our model accurately predicts the optimal sampling size.

5 Apparent Speed-Up

From Sect. 4, the following key observations can be made:

1. Sampling does not yield speedup when the population is sized appropriately by either using the decision-making model or the gambler's ruin model. Furthermore, when the population is sized according to the decision-making model, the optimal solution is to sample all the bits, that is, $n = l$. On the other hand, when the population is sized according to the gambler's ruin model, there is no preference for a specific sampling size. That is, the same *number of function evaluations* is required when any valid sampling size is used.

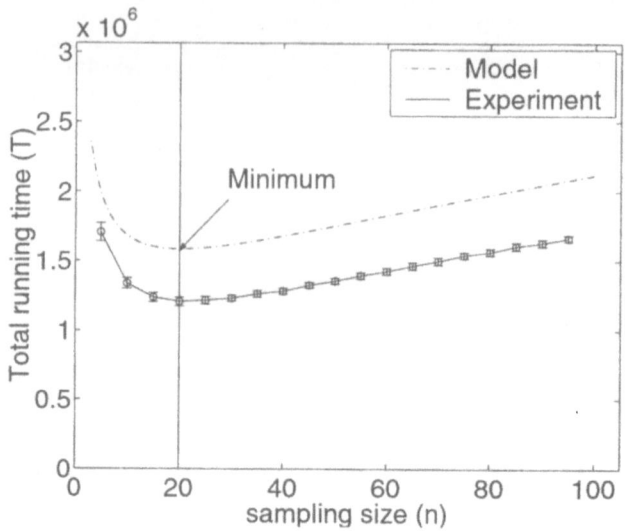

Fig. 4. The relationship between T and n for fixed population $N = 500$. $\frac{\alpha}{\beta} = 20$

2. When the population size is fixed arbitrarily, and usually to a large number (fixed population-sizing model), then speed-up can be obtained through sampling. The optimal sampling size in this case is given by Eq. (16).

Therefore, this section focuses on the fixed population-sizing model and investigates the speed-up obtained through sampling. Since the speed-up is gained only when the population is not sized appropriately, it is called *apparent speed-up*.

Equation (15) can be rewritten as a function of n as following:

$$T(n) = C_4 \cdot l(\alpha n^{-1/2} + \beta n^{1/2}). \tag{17}$$

The speed-up gained is

$$SP = \frac{T(n=l)}{T(n=n_{op})}, \tag{18}$$

where n_{op} is given by Eq. (16). With some algebraic simplifications, the speed-up can be expressed as a function of $\frac{\alpha}{\beta}$:

$$SP(\frac{\alpha}{\beta}) = \frac{1}{2}\left[l^{-\frac{1}{2}}\left(\frac{\alpha}{\beta}\right)^{\frac{1}{2}} + l^{\frac{1}{2}}\left(\frac{\alpha}{\beta}\right)^{-\frac{1}{2}}\right]. \tag{19}$$

Note that Eq. (19) is only valid when $1 \leq \frac{\alpha}{\beta} \leq l$, because $1 \leq n_{op} \leq l$. For $\frac{\alpha}{\beta} < 1$, the speed-up is $SP(1)$, and for $\frac{\alpha}{\beta} > l$, the speed-up is $SP(l) = 1$, which means no real speed-up is gained.

Figure 5 shows the relationship between the speed-up gained and $\frac{\alpha}{\beta}$ for a 100-bit SOM. The experiments were done using the same parameter settings

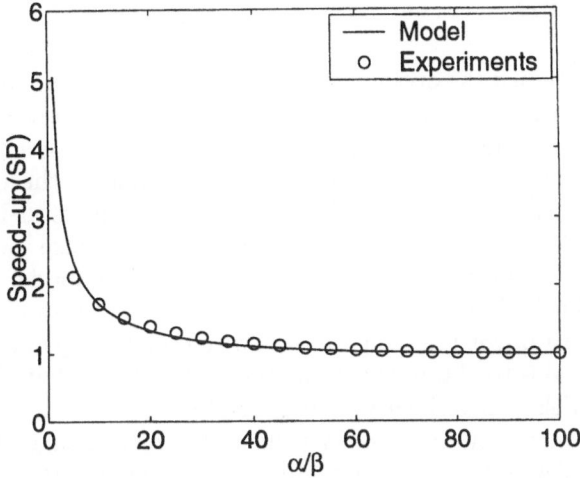

Fig. 5. The speed-up gained through sampling for a 100-bit SOM using a fixed population size 500

in the previous section. Again, all experimental results are averaged over 50 independent runs. The experimental results agree with the model derived. When the overhead (α) is relatively small, a higher speed-up is gained. When the overhead is relatively large, as one can expect, the speed-up becomes smaller. As an extreme case, when the overhead is so large that $\frac{\alpha}{\beta} > l$, sampling will not speed up the GA at all.

6 Future Work

This paper has analyzed sampling as an evaluation-relaxation technique on problems with substructures of uniform-salience. There are a number of avenues for future work which are both interesting as well as practically significant. Some of them are listed below.

- Further investigation needs to be done to bound the effectiveness of sampling schemes by considering an array of adversarial problems that contain one or more facets of problem difficulty such as deception, noise, and scaling.
- In this paper, the assumption that the GA converges to a high solution quality is needed for applying the convergence time model. Additional modeling and analyses are required to understand the time-critical condition where the number of function evaluations is limited.
- The long term goal of this work is to better understand the limit of sampling schemes applied to GAs: where and when sampling gives speed-up (or not).

We believe that the analysis and methodology presented this paper should carry over or can be extended in a straightforward manner on most of the above issues.

7 Conclusions

This paper has studied the optimal sampling and the speed-up obtained through sampling for the sampled OneMax problem (SOM). Based on Sastry and Goldberg, 2002, facetwise models for solution quality as a function of sampling size were derived for three population-sizing models, namely, the decision-making model (Goldberg, Deb, & Clark, 1992), the gambler's ruin model (Harik, Cantú-Paz, Goldberg & Miller, 1997; Miller, 1997),and the fixed population-sizing model. Then the speed-up for the fixed population-sizing model is analyzed and empirically verified. The optimal sampling size and speed-up obtained by sampling are analyzed under the scenario: we fix the solution quality to a very high value and our goal is to obtain it with minimum number of function evaluations. Each of the models was verified with empirical results.

When the desired solution quality is fixed, the results suggest that the decision-making model prefers a larger sampling size, and that the fixed population sizing model prefers a smaller sampling size. Sampling does not make much difference for the gambler's ruin model.

The results show that sampling does not give speedup for problems with subsolutions of uniform salience, if the population is sized appropriately to handle both the stochastic decision making and the noise introduced by sampling. On the other hand, if the population size is fixed without accounting for either decision-making or sampling, then the results presented in this paper show that sampling does indeed yield speed-up and an optimal sampling size exists.

Acknowledgement. This work was sponsored by the Air Force Office of Scientific Research, Air Force Materiel Command, USAF, under grant F49620-00-0163, the National Science Foundation under grant DMI-9908252. The U.S. Government is authorized to reproduce and distribute reprints for government purposes notwithstanding any copyright notation thereon.

The views and conclusions contained herein are those of the authors and should not be interpreted as necessarily representing the official policies or endorsements, either expressed or implied, of the Air Force Office of Scientific Research, the National Science Foundation, or the U.S. Government.

References

Bäck, T. (1994). Selective pressure in evolutionary algorithms: A characterization of selection mechanisms. *Proceedings of the First IEEE Conference on Evolutionary Computation*, 57–62.

Giguère, P., & Goldberg, D.E. (1998). Population sizing for optimum sampling with genetic algorithms: A case study of the Onemax problem. In *Genetic Programming 98* (pp. 496–503).

Goldberg, D.E. (1999). The race, the hurdle, and the sweet spot: Lessons from genetic algorithms for the automation of design innovation and creativity. *Evolutionary Design by Computers*, 105–118.

Goldberg, D.E., Deb, K., & Clark, J.H. (1992). Genetic algorithms, noise, and the sizing of populations. *Complex Systems, 6*, 333–362.

Grefenstette, J.J., & Fitzpatrick, J.M. (1985). Genetic search with approximate function evaluations. In *Proceedings of an International Conference on Genetic Algorithms and Their Applications* (pp. 112–120).

Harik, G., Cantú-Paz, E., Goldberg, D.E., & Miller, B.L. (1997). The gambler's ruin problem, genetic algorithms, and the sizing of populations. In *Proceedings of 1997 IEEE International Conference on Evolutionary Computation* (pp. 7–12). Piscataway, NJ: IEEE.

Miller, B.L. (1997). *Noise, sampling, and efficient genetic algorithms.* doctoral dissertation, University of Illinois at Urbana-Champaign, Urbana. Also IlliGAL Report No. 97001.

Miller, B.L., & Goldberg, D.E. (1996a). Genetic algorithms, selection schemes, and the varying effects of noise. *Evolutionary Computation, 4*(2), 113–131.

Miller, B.L., & Goldberg, D.E. (1996b). Optimal sampling for genetic algorithms. *Proceedings of the Artificial Neural Networks in Engineering (ANNIE '96) conference, 6*, 291–297.

Mühlenbein, H., & Schlierkamp-Voosen, D. (1993). Predictive models for the breeder genetic algorithm: I. Continuous parameter optimization. *Evolutionary Computation, 1*(1), 25–49.

Sastry, K., & Goldberg, D.E. (2002). Genetic algorithms, efficiency enhancement, and deciding well with differing fitness variances. *Proceedings of the Genetic and Evolutionary Computation Conference*, 528–535. (Also IlliGAL Report No. 2002002).

Thierens, D., & Goldberg, D.E. (1994). Convergence models of genetic algorithm selection schemes. In *Parallel Problem Solving fron Nature, PPSN III* (pp. 119–129).

Building-Block Identification by Simultaneity Matrix

Chatchawit Aporntewan and Prabhas Chongstitvatana

Intelligent System Laboratory, Department of Computer Engineering
Chulalongkorn University, Bangkok, 10330, Thailand
43718043@student.chula.ac.th, Prabhas.C@chula.ac.th

We propose a BB identification by simultaneity matrix (BISM) algorithm. The input is a set of ℓ-bit solutions denoted by S. The number of solutions is denoted by $n = |S|$. The output is a partition of bit positions $\{0, \ldots, \ell - 1\}$. The BISM is composed of Simultaneity-Matrix-Construction and Fine-Valid-Partition algorithms. Algorithm SMC is outlined as follows (a_{ij} denotes the matrix element at row i and column j, $\text{Count}_S^{ab}(i,j) = |\{x \in \{0, \ldots, n-1\} : s_x[i] = a \text{ and } s_x[j] = b\}|$ for all $(i,j) \in \{0, \ldots, \ell-1\}^2$, $(a,b) \in \{0,1\}^2$, $s_x[i]$ denotes the ith bit of xth solution, Random(0,1) gives a real random number between 0 and 1).

Algorithm SMC(S)
1. $a_{ij} \leftarrow 0$ for all $(i,j) \in \{0, \ldots, \ell-1\}^2$; // initialize the matrix
   for $(i,j) \in \{0, \ldots, \ell-1\}^2$, $i < j$ do
   $a_{ij} \leftarrow \text{Count}_S^{00}(i,j) \cdot \text{Count}_S^{11}(i,j) + \text{Count}_S^{01}(i,j) \cdot \text{Count}_S^{10}(i,j)$;
2. for $(i,j) \in \{0, \ldots, \ell-1\}^2$, $i < j$ do
   $a_{ij} \leftarrow a_{ij} + \text{Random}(0,1)$; // perturb the matrix
3. for $(i,j) \in \{0, \ldots, \ell-1\}^2$, $i < j$ do
   $a_{ji} \leftarrow a_{ij}$; // copy upper-triangle matrix to lower-triangle matrix
4. return $A = (a_{ij})$; // return the matrix

Algorithm FVP is outlined in the next page. It searches for a *valid* partition for the $\ell \times \ell$ simultaneity matrix M. The main idea is to recursively break the largest valid partition subset $\{0, \ldots, \ell - 1\}$ into many smaller valid partition subsets. If the size of partition subset is smaller than a constant c, the recursion will be terminated.

In the experiment, a simple GA is tailored as follows. Every generation, the SM is constructed. The FVP algorithm is executed to find a valid partition. The solutions are reproduced by a restricted crossover: bits governed by the same partition subset must be passed together. Two parents are chosen by roulette-wheel method. The mutation is turned off. The diversity is maintained.

A single run on a sum of five 3-bit trap functions is shown in the next page. The population size is set at 50. The bit positions are randomly shuffled. It is seen that the partition is improved along with the evolving population. The correct partition is found in the sixth generation. The optimal solution is found in the next generation. The partitioning approach is limited to separable functions. We are looking for a more general approach which will enable the SM to nonseparable functions.

Algorithm FVP(M)
return FVP-CORE($\{0,\ldots,\ell-1\}, M$)

Algorithm FVP-CORE(B, M)
1. **if** ($|B| \leq c$) **then return** $\{B\}$;
2. array $L \leftarrow \{$members of B sorted in ascending order$\}$;
3. **for** $i = 0$ **to** $|B| - 1$ **do**
    array $T \leftarrow \{$matrix elements in row i sorted in descending order$\}$;
    for $j = 0$ **to** $|B| - 1$ **do**
        $R[i][j] \leftarrow x$ where $a_{ix} = T[j]$;
    endfor
  endfor
4. $P \leftarrow \emptyset$;
  for $i = 0$ **to** $|B| - 1$ **do**
    if ($i \notin B_x$ for all $B_x \in P$) **then**
        $B_1 \leftarrow \{i\}$; $B_2 \leftarrow \{i\}$;
        for $j = 0$ **to** $|B| - 3$ **do**
            $B_1 \leftarrow B_1 \cup \{R[i][j]\}$;
            if (VALID(B_1, M)) **then**
                $B_2 \leftarrow B_1$;
            endif
        endfor
        $P \leftarrow P \cup \{B_2\}$;
    endif
  endfor
5. $P' \leftarrow \emptyset$;
  for $B_x \in P$ **do**
    $M' \leftarrow M$ shrunk by removing row i and column i, $i \notin B_x$;
    for $b \in B_x$ **do** replace b with $L[b]$;
    $P' \leftarrow P' \cup$ FVP-CORE(B_x, M');
  endfor
6. **return** P';

Algorithm VALID(B, M)
return (for all $b \in B$ the largest $|B| - 1$ elements of M's row b
      are found in columns of $B \setminus \{b\}$);

Gen.	Partition subsets
1	$\{0\}, \{1\}, \{2\}, \{3\}, \{4, 8\}, \{5\}, \{6\}, \{7\}, \{9\}, \{10\}, \{11\}, \{12, 14\}, \{13\}$
2	$\{0\}, \{1\}, \{2\}, \{3\}, \{4\}, \{5\}, \{6\}, \{7\}, \{8, 11\}, \{9\}, \{10\}, \{12\}, \{13, 14\}$
3	$\{0, 3\}, \{1\}, \{2\}, \{4\}, \{5\}, \{6\}, \{7\}, \{8, 11\}, \{9\}, \{10\}, \{12, 14\}, \{13\}$
4	$\{0, 3\}, \{1, 7, 13\}, \{2, 5\}, \{4\}, \{6\}, \{8, 11\}, \{9\}, \{10, 14\}, \{12\}$
5	$\{0, 3, 9\}, \{1, 7, 13\}, \{2, 4, 5\}, \{6\}, \{8, 11\}, \{10, 12, 14\}$
6	$\{0, 3, 9\}, \{1, 7, 13\}, \{2, 4, 5\}, \{6, 8, 11\}, \{10, 12, 14\}$

A Unified Framework for Metaheuristics

Jürgen Branke, Michael Stein, and Hartmut Schmeck

Institute AIFB, University of Karlsruhe, 76128 Karlsruhe, Germany
{branke,stein,schmeck}@aifb.uni-karlsruhe.de

1 The Framework

Over the past decades, a multitude of new search heuristics, often called "metaheuristics" have been proposed, many of them inspired by principles observed in nature. What distinguishes them from random search is primarily that they maintain some sort of memory of the information gathered during the search so far, and that they use this information to select the location where the search space should be tested next. Based on this observation, we propose a general unified framework which is depicted in Fig. 1: A memory is used to construct one or more new solutions which are then evaluated and used to update the memory, after which the cycle repeats.

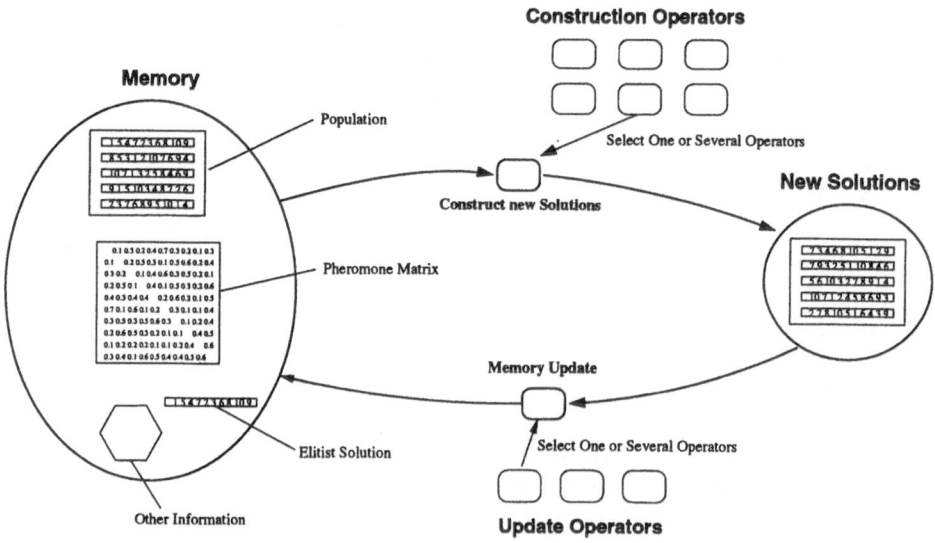

Fig. 1. Unified framework for iterative search algorithms

In the following, we will show in more detail how this general framework can be used to describe some of the aforementioned metaheuristics.

2 All Metaheuristics are Just One

EAs store information about the previous search in form of a set of solutions (population). New solutions are constructed by selecting two solutions (parents), combining them in some way (crossover) and performing some local modifications (mutation). Then, the memory is updated by inserting the new solutions into the population. Although there exists a variety of EA variants, they all fit into this general framework. For example, evolution strategies with self-adaptive mutation can be specified by extending the memory to also maintain information about the strategy parameters. Steady state genetic algorithms update the population after every newly generated solution, while genetic algorithms with generational reproduction generate a whole new population of individuals before updating the memory.

Simulated annealing only maintains a single solution in the memory. In addition to that, it keeps track of time by a temperature variable. New solutions are created by local modifications more or less equivalent to the mutation operator in EAs. The new solution replaces the current memory solution depending on the quality difference and the temperature.

Tabu search, just as simulated annealing, also maintains a single solution from which new solutions are constructed, but additionally maintains a tabu list of recently visited solutions that should not be re-visited. New solutions are created by local modifications, but taking into account the tabu list.

Ant colony optimization has a completely different way to store information about the search conducted so far. Instead of storing complete solutions, information about which partial decisions have been successful when constructing solutions from scratch is accumulated in a (pheromone) matrix. This matrix, is used to construct new solutions, giving a higher probability to decisions which have yielded successful solutions in the past. Usually, several new solutions are generated that way, and then the best solution found is used to update the matrix such that the decisions made to construct that particular solution become more preferable. An elitist ant (best solution found so far) can be modeled by an additional (complete) solution stored in the memory.

3 Benefits

Given a description of the different metaheuristics in this general form has many benefits. First of all, it creates a common language, which allows researchers from different fields to understand each other's approaches easily. Second, it moves the focus from the complete algorithms to the components. And third, it provides the interfaces for the different components to work together. It suggests a natural way for hybridization, basically turning the variety of metaheuristics into one large toolbox from which an algorithm designer can choose those parts that seem most appropriate for the application at hand.

The Hitting Set Problem and Evolutionary Algorithmic Techniques with ad-hoc Viruses (HEAT-V)

Vincenzo Cutello[1] and Francesco Pappalardo[1]

Dept. of Mathematics and Computer Science, University of Catania
V. le A. Doria, 6-I, 95125 Catania, Italy
{cutello,francesco}dmi.unict.it

Introduction. The *Weighted Minimum Hitting Set Problem (WMHSP)* and the standard *Minimum Hitting Set Problem (MHSP)*, are combinatorial problems of great interest for many applications. Although, these problems lend themselves quite naturally to an evolutionary approach. To our knowledge, there are no significant results of evolutionary algorithms applied to either the WMHSP or the MHSP, except for the results contained in Cutello et al., (2002). We will now formally introduce the optimization problem and recall that the corresponding decision problem is $\mathcal{NP}$-complete

- **Instance**: A finite set U, with $|U| = m$; a collection of sets $\mathcal{C} = \{S_1, \ldots, S_n\}$ such that $S_i \subseteq U$ $\forall i = \{1, \ldots, n\}$. A weight function $w : U \to \Re^+$.
- **Solution**: A hitting set for $\mathcal{C}$, that is to say $H \subseteq U$ such that $H \cap S_i \neq \emptyset$, $\forall i = 1, \ldots, n$.
- **Optimal Solution**: a hitting set H such that $w(H) = \sum_{s \in H} w(s)$ is minimal.

The above definition is very general and, by simply putting $w(s) = 1$, $\forall s \in U$, we obtain the standard definition of the Minimum Hitting Set problem.

Theoretical results show (Feige, 1998) that optimal solutions cannot be approximated within $(1 - \epsilon) \ln m$ $\forall$ $\epsilon > 0$, unless $\mathcal{NP} \subset \mathcal{DTIME}(m^{\log \log m})$.

The Description of Our Algorithm. Our evolutionary approach and the resulting genetic algorithm, denoted by HEAT-V, is based on the idea of a *mutant virus*, which somehow acts as a non-purely random mutation operator. Each chromosome in the population is a binary string of fixed length (see below for details). The selection operator is *tournament selection* and the selected individuals mate with probability $p = 1$. Reproduction uses uniform crossover (however this does not involve the virus part as we will describe later). Elitism is used on three specific elements (not necessarily distinct) of the population: best fitness element; hitting set of smaller cardinality; and, hitting set of smaller weight.

Chromosomes contain some extra genes, specifically $2 + \lceil \log |U| \rceil$. These genes represent the genetic patrimony of the *virus*. As a consequence, the total length of a chromosome is $|U| + 2 + \lceil \log |U| \rceil$. The extra $\lceil \log |U| \rceil$ bits uniquely identify one of the first $|U|$ *loci* of the chromosome. Viruses will hit an individual if the two extra control bits have both value 1.

We used three different fitness functions to test our algorithm. The best results were obtained using the function $f_3 : \mathcal{P} \to \mathcal{N} \setminus \{0\}$, that HEAT-V tries to <u>minimize</u>, and which is defined as follows:

$$f_3(c) = w(c) + w(\mathcal{L}_{c,m})$$

where

$$\mathcal{L}_{c,m} = \{e : (\exists K \subseteq U) \text{ s.t. } K \cap c = \emptyset \wedge e \in K \wedge w(e) = \min\{w(e') : e' \in K\}\}.$$

Intuitively, f_3 is computed by adding to the weight of a chromosome, the minimum weight of elements of sets which c does not hit. Thus, f_3 acts as a strict upper-bound to the fitness function of any chromosome that could become a hitting set by including c.

Computational Results. We compared HEAT-V to a greedy algorithm which approximates the optimal solution to a factor of $O(\ln m)$. Basically, the procedure greedy chooses at every step the element that maximizes the ratio between the number of hit sets (among the remaining ones) and its weight. The hit sets are eliminated. Many tests were performed. For each test HEAT-V was tested three times. The population contained 200 individuals and each test ran for 500 generations. We also checked HEAT-V against *vertex cover*, which can be easily reduced to MHS. In particular, we used the regular graphs proposed by Papadimitriou and Steiglitz (PS-rg) (Papadimitriou and Steiglitz, 1982) built so that the classical greedy strategy fails. We ran HEAT-V with PS-rg's of degrees $k = 32, k = 66, k = 100$. HEAT-V always finds the optimal solution.

Conclusions and Future Work. We are now testing HEAT-V on a dynamic version of the WMHSP. More formally, we are dealing with cases in which with a given probability p_d a subset of the given family $\mathcal{C}$ disappears, and appears, instead, with probability p_a. Such a framework can be applied to many real scenarios, such as for instance, a computer network where each of the machine in the network offers specific services. Computers have a certain probability to go down, and for all of them the probability of going down is different, since it depends on factors such as the quality of the component, maintenance service, number of access, etc. To guarantee that the network is always able to provide a set of basic services, it is necessary to know which is the minimum number of computers that should be running in order to overcome some probabilistic failures. First experiments, conducted using Gaussian probability distributions, show that the population not only contains a good hitting set, but also good hitting sets in case some subsets disappear or appear.

References

1. Cutello, V., Mastriani, E., and Pappalardo, F. An evolutionary algorithm for the T-constrained variation of the Minimum Hitting Set Problem. In *Proceedings of 2002 IEEE Congress on Evolutionary Computation (CEC2002)* Vol. 1, pp. 366–371.
2. Feige, U. A threshold of $\log n$ for approximating set cover. *Journal of ACM 45* (1998), pp. 634–652.
3. Papadimitriou, C.H., and Steiglitz, K. *Combinatorial Optimization*. Prentice Hall (1982), p. 407.

The Spatially-Dispersed Genetic Algorithm

Grant Dick

Department of Information Science
University of Otago, PO Box 56, Dunedin, New Zealand
gdick@infoscience.otago.ac.nz

Abstract. Spatially structured population models improve the performance of genetic algorithms by assisting the selection scheme in maintaining diversity. A significant concern with these systems is that they need to be carefully configured in order to operate at their optimum. Failure to do so can often result in performance that is significantly under that of an equivalent non-spatial implementation. This paper introduces a GA that uses a population structure that requires no additional configuration. Early experimentation with this paradigm indicates that it is able to improve the searching abilities of the genetic algorithm on some problem domains.

1 The Spatially-Dispersed Genetic Algorithm

The spatially-dispersed genetic algorithm (sdGA) is an alternative method of incorporating population genetics models into genetic algorithms by using a two dimensional Euclidean space to hold the members of the population [1]. This space is infinite and continuous. The placing of individuals on a plane restricts them so that they can breed only with individuals that are "visible" to them.

In the sdGA, the first parent is selected from the entire population via whichever selection method the system is currently using. A second parent is then selected from the deme that is visible to the initial parent. This concept of visibility essentially creates subpopulations based on the location of individuals on the surface.

When offspring are created, they are placed in a location that maintains the relationship with their parents' coordinates. To introduce new offspring into the population, one parent is chosen randomly and the new individual is placed randomly at coordinates which fall within that parent's visibility radius.

The sdGA does not require significant tuning of parameters in order to achieve its maximum potential, since the spatial behaviour of the population rapidly becomes independent of the selected visibility and initial scale space [2]. This differs from other spatially-structured population strategies (for example, an Island GA), which often require extensive parameter tuning [3].

2 Experimental Results

The sdGA was tested on the Travelling Salesman Problem (TSP) and Holland's Royal Road function. Each problem was tested in a generational and steady-state implementation. The results were compared with a genetic algorithm that used a non-spatially structured population.

The generational sdGA shows a significant improvement in performance over that of the generational standard GA for the TSP. Not only is the quality of the discovered solution better, the time taken to find an optimal solution is significantly reduced. In contrast, the steady-state sdGA appears to outperform the steady-state standard GA in the early iterations, but the advantage tapers out in the later stages of execution.

The Royal Road function appeared to behave in a completely opposite manner to the TSP. For the Royal Road problem, the steady-state sdGA performs better than the steady-state standard GA. The difference is discernible from as early as 2000 fitness evaluations. The generational sdGA appears to offer equal performance to that of the generational standard GA.

3 Conclusion and Future Work

The spatially-dispersed GA is a new technique for incorporating population genetics concepts into genetic algorithms. A significant advantage that the sdGA has over its predecessors is that it requires no user intervention or parameter setting to operate closer to its optimum level. Current results show that it is able to improve the performance of a GA on some problem domains.

3.1 Future Work

The sdGA is still in its infancy. More work needs to be done to properly understand the effect that its population structure has on GAs. Of particular interest would be an investigation into the population structure's effect on the schema theorem.

The ultimate goal of the sdGA is to use it in a distributed implementation. This would result in a parameterless distributed GA. Such a system should be of great interest to the GA community and the implications of a distributed implementation, in terms of computational overhead and scalability, is something that future work should investigate.

References

1. Dick, G., Whigham, P.: Spatially constrained selection in evolutionary computation. In: Australisia-Japan Joint Workshop on Intelligent and Evolving Systems. (2002)
2. Whigham, P., Dick, G.: A study of spatial distribution and evolution. In: The 14th Annual Colloquium of the Spatial Information Research Centre, Wellington, New Zealand. (2002) 157–166
3. Cantú-Paz, E.: A survey of parallel genetic algorithms. Technical Report 97003, Illinois Genetic Algorithms Laboratory(IlliGAL) (1997)

Non-universal Suffrage Selection Operators Favor Population Diversity in Genetic Algorithms

Federico Divina, Maarten Keijzer, and Elena Marchiori

Department of Computer Science
Vrije Universiteit
De Boelelaan 1081a, 1081 HV Amsterdam
The Netherlands
{divina,mkeijzer,elena}@cs.vu.nl

State-of-the-art concept learning systems based on genetic algorithms evolve a redundant population of individuals, where an individual is a partial solution that covers some instances of the learning set. In this context, it is fundamental that the population be diverse and that as many instances as possible be covered. The universal suffrage selection (US) operator is a powerful selection mechanism that addresses these two requirements. In this paper we compare experimentally the US operator with two variants, called Weighted US (WUS) and Exponentially Weighted US (EWUS), of this operator in the system ECL [1]. The US selection operator operates in two steps:

1. randomly select n examples from the positive examples set;
2. for each selected positive example e_i, $1 \leq i \leq n$, let $Cov(e_i)$ be the set of individuals covering e_i. If $Cov(e_i) = \emptyset$ then call a seed procedure for creating an individual that covers e_i. If $Cov(e_i) \neq \emptyset$, choose one individual from $Cov(e_i)$ with a roulette wheel mechanism, where the sector associated to an individual $x \in Cov(e_i)$ is proportional to the ratio between the fitness of x and the sum of the fitness of all individuals occurring in $Cov(e_i)$.

The basic idea behind this operator is that individuals are candidates to be elected, and positive examples are the voters. In this way, each positive example has the same voting power, i.e. has the same probability of being selected in the first step of the US selection operator. The WUS and EWUS operators modify the first step of the US operator. Within these operators, each example is assigned a weight, and then in the first step of the selection examples are chosen with a roulette wheel mechanism, where the probability of choosing an example depends on the weight of the example. The weights used by the WUS and the EWUS operators are respectively: $w_i = \frac{|Cov(e_i)|}{|Pop|}$ and $w_i = \frac{e^{-|Cov(e_i)|}}{\sum_{j=1}^{P} e^{-|Cov(e_j)|}}$.

Weights are adjusted at each iteration of the GA. In this way examples harder to cover will be selected with higher probabilities. The validity of each selection operator is tested on three datasets. The first is an artificially generated dataset, the other two are well known datasets: the mutagenesis and the vote datasets. The first dataset consists of five hundred positive examples and five hundred negative examples. Each example can be described by three attributes q, p and

r, that can assume respectively the values $\{a,b\}$, $\{c,d,e\}$ and $\{f,g\}$. Results are shown in Tables 1 and 2, where Unc stands for the average number of positive examples uncovered at the end of the evolution, Cov for the average number of individuals covering a positive example, Div for the average number of different clauses in the final population and Acc for the average accuracy.

Table 1. Table 1.1 shows the diversity in the final populations obtained by using the three selection operators. The first column shows the kind of clause, e.g. f means a clause of the form $z(X) \leftarrow r(X,f)$. For each kind of clause the percentage of individuals in the final population representing it is given in the table for each variant of the selection operator

Clause	US	WUS	EWUS
f	0.35 (0.16)	0.26 (0.16)	0.09 (0.02)
a	0.24 (0.05)	0.28 (0.02)	0.08 (0.02)
c	0.30 (0.10)	0.22 (0.16)	0.13 (0.03)
a,d	0.03 (0.05)	0.03 (0.00)	0.02 (0.02)
a,f	0.05 (0.02)	0.03 (0.00)	0.01 (0.02)
a,c	0.03 (0.05)	0.15 (0.07)	0.06 (0.02)
d	0	0	0.1 (0.07)
a,f,d	0	0	0.01 (0.02)
f,c	0	0	0.01 (0.02)
others	0	0.03 (0.04)	0.49 (0.02)

Table 1.1

Clause	US	WUS	EWUS
Unc	13.5 (3.53)	13 (0.01)	0.67 (0.58)
Cov	19.80 (0.24)	17.20 (3.11)	9.00 (0.23)

Table 1.2

Table 2. Results for the mutagenesis and the vote dataset. Standard deviation is given between brackets

	Mutagenesis				Vote		
	US	WUS	EWUS		US	WUS	EWUS
Div	8 (0.82)	9.34 (3.2)	21 (1.63)	Div	8.67 (1.47)	13.5 (1.50)	13 (1.12)
Unc	10 (2.45)	7.67 (3.09)	0.33 (0.47)	Unc	2.33 (0.47)	0.80 (0.75)	0 (0)
Cov	17.36 (7.85)	15.33 (7.35)	6.97 (2.81)	Cov	21.13 (7.42)	20.50 (8.38)	22.87 (5.16)
Acc	0.85 (0.08)	0.86 (0.08)	0.89 (0.06)	Acc	0.92 (0.05)	0.93 (0.06)	0.94 (0.04)

The results suggest that 'less' universal selection schemes are more effective for promoting diversity while maintaining the key property of the US selection operator of covering many positive examples.

References

1. F. Divina and E. Marchiori, *Evolutionary concept learning*, in GECCO 2002: Proceedings of the Genetic and Evolutionary Computation Conference, New York, 9–13 July 2002, Morgan Kaufmann Publishers, pp. 343–350.

Uniform Crossover Revisited: Maximum Disruption in Real-Coded GAs

Stephen Drake

COGS, University of Sussex
Falmer, Brighton, BN1 9QH, UK

Abstract. A detailed comparison is presented of five well-known crossover operators on a range of continuous optimization tasks to test longheld beliefs concerning disruption, with respect to real-coded GAs. It is suggested that, contrary to traditional assumptions, disruption is useful.

1 Introduction

Five crossover operators were compared on a wide variety of continuous optimization tasks. The operators compared were traditional one-point (1PX), two-point (2PX), and uniform (UX, $P_0 = 0.5$)[6] crossover operators, together with two operators specifically designed for use with real-coded GAs: linear (LX) [7] and blend (BLX-0.5) [2].

2 Experiments

The comparison testbed included 'traditional' test functions, two versions of a real-world problem (aircraft wing-box optimization [5]) and an approximation to a real-world problem [4]; see [1] for details.

A geographically distributed GA as described in [1] was used. Experiments were repeated incorporating a number of parameter variations. Fifty runs of 300 generations were carried out on each function for each operator.

3 Results

Results showed that UX appears to be the operator which invariably achieves a good (if not the best) quality solution quickly. While BLX achieves the best solution on six problems, it achieves the worst on five others. LX's performance similarly vacillates between being the best and the worst. (Fitness increase is slowest when using BLX.) Significantly, it was on the real-world problems which UX performed the best while LX and BLX performed the worst.

Unsurprsingly, fitness increase corresponded to a proportion of crossovers resulting in children fitter than their parents; however, while some construction occurred, fitness increased regardless of the proportion of destructive crossovers. BLX exhibited a significantly higher proportion of destructive crossovers than constructive (and by far the highest destructive proportion of all the operators) at all times – even on those functions where it was the best-performing operator.

3.1 Operators' Effects on Fitness Landscapes

The concept of a fitness landscape was introduced in as a framework for biological evolution; in EC, GAs' recombination operators determine the structure of a fitness landscape. To assess how the operators affected the ruggedness of each test function's corresponding fitness landscape, correlation lengths were measured for each case using the method described in [3].

1PX and 2PX had the longest correlation lengths (low 30s on average); the remaining operators exhibited considerably shorter lengths, with UX and BLX averaging just 5. This suggests that 1PX and 2PX flatten landscapes considerably; the more disruptive operators magnify landscapes' ruggedness by retaining local structure. Clearly, this retention of local structure owes itself to the higher level of exploration effected by the more disruptive operators: many more points on a landscape are reachable via a single application.

4 Conclusion

The empirical evidence presented here appears to question conventional GA wisdom: the results suggest that the assumption that disruption (of schemata) is to be avoided may be flawed (at least in the case of real-coded GAs). Exploration, or more specifically, ergodicity is key to a crossover operator's success. During a run, each variable of a solution is likely to be subjected to fewer changes of value under a less disruptive operator than it would under a highly disruptive one. Moreover, by combining a disruptive operator with a sensible selection strategy the best individuals will automatically be retained while a vigorous exploration of other parts of the search space is carried out [2].

References

1. Drake, S.: Crossover Operators in Real-Coded GAs: An Empirical Study. Cognitive Science Research Paper No. 560. University of Sussex (2003).
2. Eshelman, L. J., Schaffer, J. D.: Real-Coded Genetic Algorithms and Interval Schemata. In: Whitley, D. (ed.): FOGA 2. Morgan Kaufmann (1992).
3. Hordijk, W.: A Measure of Landscapes. Evolutionary Computation Vol. 4 (1996) 335–360.
4. Keane, A.: Experiences with Optimizers in Structural Design. In Parmee, I. C. (ed.): Proc. of the 1st Int. Con. on Adaptive Computing in Engineering Design and Control. University of Plymouth (1994).
5. McIlhagga, M., Husbands, P., Ives, R.: A Comparison of Search Techniques on a Wing-Box Optimisation Problem. In Voigt H.-M., Ebeling, W., Rechenberg, I., Schwefel H.-P. (eds.): PPSN 4. Springer (1996).
6. Syswerda, G.: Uniform Crossover in Genetic Algorithms. In Schaffer, J. D. (ed.): ICGA 3. Morgan Kaufmann (1989).
7. Wright, A. H.: Genetic Algorithms for Real Parameter Optimization. In Rawlins, G.J.E. (ed.): FOGA. Morgan Kaufmann (1991).

The Master-Slave Architecture for Evolutionary Computations Revisited

Christian Gagné, Marc Parizeau, and Marc Dubreuil

Laboratoire de Vision et Systèmes Numériques (LVSN)
Département de Génie Électrique et de Génie Informatique
Université Laval, Québec (QC), Canada, G1K 7P4
{cgagne,parizeau,dubreuil}@gel.ulaval.ca

The recent availability of cheap Beowulf clusters has generated much interest for Parallel and Distributed Evolutionary Computations (PDEC) [1]. Another often neglected source of CPU power for PDEC are networks of PCs, in many case very powerful workstations, that run idle each day for long periods of time. To exploit efficiently both Beowulfs and networks of heterogeneous workstations we argue that the classic master-slave distribution model is superior to the currently more popular island-model [1].

The key features of a good PDEC capable of exploiting networks of heterogeneous workstations are *transparency* for the user, *robustness*, and *adaptivity*. Transparency is essential to both the user of the PDEC and the user of the workstation, as none want to deal with the other. One way to implement such a PDEC is as a screen-saver. Robustness is very important because evolutionary computations may execute over long periods of time during which different types of failures are expected: *hard failures* caused by network problems, system crashes or reboots, and *soft failures* that stem from the use of the workstation for other tasks (e.g. when the user deactivates the screen-saver). Finally, adaptivity refers to the capability of the PDEC to exploit new or compensate for lost computing resources (dynamical network configuration). The classical island-model is not designed to deal with these features, essentially because populations (demes) are tightly coupled with processing nodes.

In contrast, the master-slave model has all required features. One issue that needs to be addressed, however, is its ability to scale with a large number of slave nodes, knowing that there is a communication bottleneck with the master node. In the rest of this short paper, we build a mathematical model of the master-slave and show that, given current Local Area Network (LAN) technologies, a quite large PDEC can be built before reaching this bottleneck.

For real world applications, assuming that the time needed for fitness evaluation is the dominant time factor for evolutionary algorithms, the speedup of a master-slave system over that of a single processor can be modeled by NT_f/T_p, where N is the population size, T_f is the time needed to evaluate the fitness of a single individual, and T_p is the time needed to evaluate all individuals using P processors. Possible distribution policies range from separating the population into P sets and sending each of them to a different slave, or sending the individuals one-by-one to available slaves until all are evaluated. Let S designate the average size of the sets that are sent to processing nodes, and $C = \lceil N/PS \rceil$ the

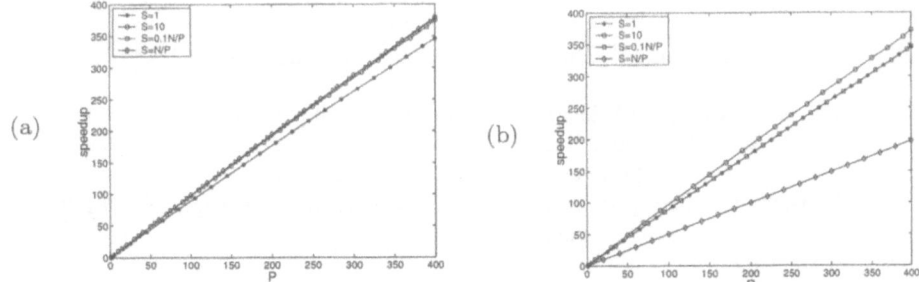

Fig. 1. Speedup for $S = \{1, 10, \frac{0.1N}{P}, \frac{N}{P}\}$ as a function of number of processors $P \in [1, 400]$, (a) with no failure ($K = 0$); and (b) with five failures ($K = 5$). Other parameters are $N = 500000$, $T_f = 1$ s, $T_c = 1.4 \times 10^{-4}$ s, and $T_l = 0.1$ s.

number of communication cycles needed to complete a generation. Then:

$$T_p = \underbrace{CST_f}_{\text{computation}} + \underbrace{CPST_c}_{\text{communication}} + \underbrace{CT_l}_{\text{latency}} + \underbrace{T_k}_{\text{failure}} \quad (1)$$

where T_c is the transmission time for one individual, T_l the average connection latency, and T_k the delay associated to the observation of $K \in [0, P]$ failures:

$$T_k = \underbrace{(1 - 0.5^K)ST_f}_{\text{synchronization}} + \underbrace{KST_c}_{\text{communication}} + \underbrace{[K/P]T_l}_{\text{latency}} \quad (2)$$

The synchronization term assumes that each failure follows a Poisson process and occurs on average at half-way time during an evaluation cycle. Figure 1 presents the speedup curves of different S values in a plausible scenario, with (a) no failure, and (b) exactly five failures per generation. With no failure, the figure shows linear speedup close to optimal for all S except when $S = 1$, where performance starts to degrade given the relatively large latency T_l. However, when failures occur, the figure shows that a value $S = \frac{N}{P}$ no longer achieves linear speedup, and that the intermediary value of $S = 10$ (for this scenario) makes a good compromise between efficiency and robustness. These curves thus globally show that parameter S should be adjusted dynamically in order to optimize performance.

For the above model, using the conservative parameters of Fig. 1 (that assumes 7 MB/sec peek throughput for 100Mb/sec LANs), it can be shown that a master-slave system of 7K processing nodes could be build, at least in theory, with a speedup of about 50% of the optimal value (i.e. 3.5K). Even if this result may be overly optimistic in practice, it goes to show that the scalability limitation of the master-slave model is a relative matter that needs to be put in perspective with its advantages. Moreover, if one needs to experiment with an island-model, it can always be simulated using a multidemic framework (e.g. Open BEAGLE; http://www.gel.ulaval.ca/~beagle) over a master-slave.

References

1. E. Cantú-Paz. *Efficient and accurate parallel genetic algorithms.* Kluwer Academic Publishers, Boston, MA, USA, 2000.

Using Adaptive Operators in Genetic Search

Jonatan Gómez[1], Dipankar Dasgupta[2], and Fabio González[1]

[1] Division of Computer Science, The University of Memphis, Memphis TN 38152
and Universidad Nacional de Colombia, Bogotá, Colombia
{jgomez,fgonzalz}@memphis.edu
[2] Division of Computer Science, The University of Memphis, Memphis TN 38152
dasgupta@memphis.edu

Abstract. In this paper, we provided an extension of our previous work on adaptive genetic algorithm [1]. Each individual encodes the probability (rate) of its genetic operators. In every generation, each individual is modified by only one operator. This operator is selected according to its encoded rates. The rates are updated according to the performance achieved by the offspring (compared to its parents) and a random learning rate. The proposed approach is augmented with a simple transposition operator and tested on a number of benchmark functions.

1 Simplified Hybrid Adaptive Evolutionary Algorithm (SHA-EA)

Parameter Adaptation (**PA**) eliminates the parameter setting of evolutionary algorithms (**EAs**) by adapting those through the execution of the EA [2,3]. Several PA techniques have been developed [2,4,5]. We introduced an hybrid technique for adapting genetic operator probabilities in our previous work [1]. The operator probabilities along with a learning rule rate, which determines the reward/punishment factor on the operator rate, were encoded in each individual. In this paper, we removed the encoding of the learning rate and simulated it with a uniform random rate [0,1]. Also, we used standard operators that only modify the solution part. When a non-unary operator is selected, the additional parents are chosen with a local selection strategy. The operator rates are evolved according to the performance achieved by the offspring (compared to its parent) and the random learning rate generated, according to the Algorithm 1. We tested several functions using different sets of operators. We used elitist selection as replacement strategy, with each EA run for 1000 iterations, and reported the results averaging 50 run. We used a population of 100 individual for binary functions and 200 for real valued functions, using 32 bits for encoding each real value. Table 1, compares the results obtained by SHA-EA with some reported results in the literature.

2 Conclusions

We presented a simplified version of the adaptive evolutionary algorithm proposed by Gomez and Dasgupta [1]. Experimental results showed that the proposed EA was able to determine the usefulness of some operators finding the solution. The performance of the proposed EA is, in general, better than several EAs with well tuned parameters or with the parameter adaptation strategies found in the literature [2,6,7].

Algorithm 1 Simplified Hybrid Adaptive Evolutionary Algorithm (SHA-EA)

SHA-EA(λ, termination_condition)
P_0 = initPopulation(λ),
$t_0 = 0$
while(termination_condition(t, P_t) is false)
P_{t+1} = GENERATEPOPULATION(P_*)
t = t+1

GENERATEPOPULATION(P, operators)
P' = {}
for each ind $\in$ P
    rates = extract_rates(ind)
    δ = random(0,1)   // learning rate
    oper = select(operators, rates)
    parents = select(arity(oper)-1, P, ind) $\cup$ {ind}
    ind' = apply(oper, parents)
    if(fitness(ind') > fitness(ind)) then
        rates[oper] = (1.0 + δ)*rates[oper] //reward
    else rates[oper] = (1.0 - δ)*rates[oper] //punish
    normalize_rates(ind')
    P' = P' + select{ ind, ind' }
return P'

Table 1. Best solutions found with different strategies. Third row shows the SHA-EA using mutation (M) and crossover (C) operations, but without using simple transposition (T) operation

EA	RoyalRoad	Deceptive-3	Rosenbrock	Schwefel	Rastrigin	Griewangk
M-SHA-EA	19.04	292.64	0.00012756	366.005	8.351	0.0480
MC-SHA-EA	53.44	298.16	0.01060971	584.553	2.640	0.3212
MT-SHA-EA	20.00	292.08	0.00001939	120.645	0.589	0.1155
MCT-SHA-EA	64.00	300.00	0.00314993	29.741	0.000	0.0008
Tuson [2]	40.64	289.68	-	-	-	-
Digalakis [6]	-	-	0.40000000	-	10.000	0.7000
Patton [7]	-	-	-	-	4.897	0.0043

References

1. J. Gomez and D. Dasgupta, "Using competitive operators and a local selection scheme in genetic search," in *Late-breaking papers GECCO 2002*, 2002.
2. A. Tuson and P. Ross, "Adapting operator settings in genetic algorithms," *Evolutionary Computation*, 1998.
3. F. Lobo, *The parameter-less genetic algorithm: rational and automated parameter selection for simplified genetic algorithm operation.* PhD thesis, Nova University of Lisboa, 2000.
4. L. Davis, "Adapting operator probabilities in genetic algorithms," in *Third International Conference on Genetic Algorithms and their Applications*, pp. 61–69, 1989.
5. B. Julstrom, "What have you done for me lately? adapting operator probabilities in a steady-state genetic algorithm," in *Sixth International Conference on Genetic Algorithms*, pp. 81–87, 1995.
6. J. Digalakis and K. Margaritis, "An experimental study of benchmarking functions for genetic algorithms," in *IEEE Conferences Transactions, Systems, Man and Cybernetics*, vol. 5, pp. 3810–3815, 2000.
7. A. Patton, T. Dexter, E. Goodman, and W. Punch, "On the application of cohort-driven operators to continuous optimization problems using evolutionary computation," *Evolutionary Programming*, no. 98, 1998.

A Kernighan-Lin Local Improvement Heuristic That Solves Some Hard Problems in Genetic Algorithms

William A. Greene

Computer Science Department, University of New Orleans
New Orleans, LA 70148, USA
bill@cs.uno.edu

Abstract. We present a Kernighan-Lin style local improvement heuristic for genetic algorithms. We analyze the run-time cost of the heuristic. We demonstrate through experiments that the heuristic provides very quick solutions to several problems which have been touted in the literature as especially hard ones for genetic algorithms, such as hierarchical deceptive problems. We suggest why the heuristic works well.

In this research, population members (chromosomes) are bit strings, all of the same length, which we denote by N. We will refer to population members as individuals.

A *local improvement heuristic* is a procedure which is applied to an individual, with the intention of modifying it into an related individual of higher fitness. Typically it is applied to a child chromosome, after its manufacture by some crossover operator but before the child is entered into the population. One form of local improvement heuristic is *hill-climbing*. One step of hill-climbing consists of flipping that bit in the individual which results in a maximal increase in fitness. To improve an individual, we might perform one step of hill-climbing, or several, or until the morphed individual has reached a local optimum and no further bit flip will increase fitness.

At a far extreme, we might argue that we do not need either a population or genetic forces at all. Any single individual differs from an optimal individual by having the wrong bit values on some subset of its bits. If we loop through all subsets of bits, each time flipping those bits in the individual, ultimately we will identify an optimal individual. But of course this approach has exponential cost, which is intolerable.

Our Kernighan-Lin local improvement heuristic falls in between hill-climbing and the extreme just described. It is inspired by the graph bisection work of Kernighan and Lin [1]. Let an individual be given. We build a tower $S_1, S_2, ..., S_{MAX}$ of subsets of its bits. Set S_{k+1} is built from S_k by including one more bit. The extra bit is that one which is optimal with respect to change of fitness when flipped, but we allow the change to be a negative change (one no worse than the change from any other bit). We allow negative changes (hill descent), in the hope that perhaps later bit flips, in concert with previous flips, will result in yet more improvement to fitness. At most how many bits will we try flipping? Half of them ($MAX = N/2$) is a reasonable first guess. Then our heuristic actually flips those bits of whichever subset $S_1, S_2, ..., S_{MAX}$ produces the highest fitness. A complexity analysis of our local improvement heuristic shows it to have cost $Q(N^2 g(N))$, where function g is the cost to recalculate an individual's fitness after flipping one of its bits. Function g may or may not be cheaper than the cost $F(N)$ to fully calculate the fitness of an individual that is N bits wide.

For our experiments we use a genetic algorithm that holds few surprises. It is generational, meaning that the population $P(t+1)$ at time $t+1$ is produced by creating enough children from the population $P(t)$ at time t, through crossover. It uses elitism: the best two individuals of $P(t)$ always survive into $P(t+1)$. The crossover operator is 2-point crossover. The only surprise is that fairly heavy mutation is practiced, and it is graduated, in that less fit individuals undergo more mutation. It is also stochastic, in that we make mutation attempts, but each actually results in a bit flip with only 50% chance.

We apply our local improvement heuristic first to a child after it has been created by crossover. Then we apply the heuristic again after the child has undergone mutation. Then the child is entered into the next generation $P(t+1)$. The heuristic is also applied once to the members of the initial population $P(0)$, which began as random strings.

We tried our heuristic on four problems from the literature, which have been touted as especially hard problems for genetic algorithms. The problems are: one-dimensional Ising [2], k-fold, 3-deceptive [3], hierarchical if-and-only-if (HIFF) [4], and hierarchical trap [5]. In all cases, we used a population of 40 individuals (which is smaller than in the referenced researches), and conducted 20 trials, each of which was allowed if need be to run to 500 generations. In all cases we used individuals of N bits where N was at least as wide as in the referenced researches. We experimented with various choices for MAX, the maximum number of bits which could be flipped on a local improvement step, starting at $N/2$ and decreasing. For these four problems, our approach was consistently able to find optimal individuals, in a small number of generations. Moreover, the time cost was just several seconds (using a 1.7 GHz Pentium processor), when we used optimized versions of the function g. Space allows us only to report on one problem here. It will be the first hierarchical trap function of reference [5]. As in [5], individuals are 243 bits wide. In [5], costs are given in terms of fitness calculations (225,000 on average); we show our costs in terms of clocked runtimes.

Table 1. Hierarchical-Trap-#1. Size of individual = 243 bits; maximum fitness = 1215

max flippable bits	avg. final gen. #	avg. trial time (sec)	avg. best fitness
121	1.0	3.94	1215
81	1.0	2.74	1215
27	3.6	3.45	1215
9	13.55	6.29	1215
3	484.5	153.17	832.3

References

1. Kernighan, B., and Lin, S.: An Efficient Heuristic Procedure for Partitioning Graphs. Bell Systems Technical Journal, Vol. 49 (1970), pp. 291–307
2. Van Hoyweghen, C., Goldberg, D.E., and Naudts, B.: Building Block Superiority, Multimodality, and Synchronization Problems. In: Proceedings of the Genetic and Evolutionary Computation Conference (GECCO 2001), pp. 694-701. Morgan Kaufmann Publishers, San Francisco
3. Pelikan, M., Goldberg, D.E., and Sastry, K.: Bayesian Optimization Algorithm, Decision Graphs, and Occam's Razor. In: Proceedings of the Genetic and Evolutionary Computation Conference (GECCO 2001), pp. 519-526. Morgan Kaufmann Publishers, San Francisco
4. Watson, R. A., Hornby, G.S., and Pollack, J.B.: Modeling Building Block Interdependency. In: Parallel Problem Solving from Nature 1998 (PPSN V), pp. 97-106. Springer-Verlag, Berlin
5. Pelikan, M., and Goldberg, D. E.: Escaping Hierarchical Traps with Competent Genetic Algorithms. In: Proceedings of the Genetic and Evolutionary Computation Conference (GECCO 2001), pp. 511–518. Morgan Kaufmann Publishers, San Francisco

GA-Hardness Revisited

Haipeng Guo and William H. Hsu

Department of Computing and Information Sciences
Kansas State University, Manhattan, KS 66502
{hpguo,bhsu}@cis.ksu.edu

1 Introduction

Ever since the invention of Genetic Algorithms (GAs), researchers have put a lot of efforts into understanding what makes a function or problem instance hard for GAs to optimize. Many measures have been proposed to distinguish so-called *GA-hard* from *GA-easy* problems. None of these, however, has yet achieved the goal of being a reliable predictive GA-hardness measure. In this paper, we first present a general, abstract theoretical framework of instance hardness and algorithm performance based on Kolmogorov complexity. We then list several major misconceptions of GA-hardness research in the context of this theory. Finally, we propose some future directions.

2 Instance Hardness and Algorithm Performance

Intuitively, an algorithm a performs better if it compiles more information about the input instance. An instance f is hard if it does not have much structure information for any algorithm to exploit to solve it. Since both algorithm and problem instance can be encoded as finite strings, we can build an abstract framework of problem instance hardness and algorithm performance base on Kolmogorov complexity as follows to capture these intuitions. We measure the hardness of f by $K(f)$, the Kolmogorov complexity of f, which is defined as the size of the smallest program that can produce f. It can be seen as the absolute information, or the "randomness", of f. The information in a about f is defined by $I(a:f) = K(f) - K(f|a)$. It is a natural indicator of the performance of a on f. Similarly, $K(a)$ measures the randomness of the algorithm a. Random instances are the "hardest" ones because they are incompressible and contain no internal structures for any algorithm to exploit. Random search algorithms are the least efficient algorithms because they convey no information about the problem and just visit the search space randomly. Given any f, can we deliberately design an a that performs *worse* than random search? If f happens to be a random one, we cannot design an algorithm to perform either better or worse than a random search. If it is a structured problem but we do not know any information about its structure, it is still hard for us to do so since we do not know what we should deceive. The only case we can do that is when f contains structural information one and we know what it is. The resulting algorithm can be called a *deceptively-solving algorithm* for its "purpose" is to seek deceptiveness and perform worse than random search. It is a structured algorithm since it contains structural information about the problem. But it uses this information in a "pathological" way. In this sense we can say a contains "negative information" about f. In the other hand, there are algorithms that are structured and perform better than random search. These algorithms can be called *straightforwardly-solving algorithms*. Therefore, there are three factors that cause an instance to be hard for a particular algorithm: *randomness*, *mismatch* and *deception*.

3 Misconceptions in GA-Hardness Research

There are three major misconceptions in the *problem space*. The first misconception is blurring the differences among the three above-mentioned instance hardness factors and feeding GAs with problems that are too hard to be meaningful. The second misconception is using a few, specific problem instances to support a general result. The third one is applying GAs on problems that are too general to be realistic. In the *algorithm space*, the most common misconception is considering GAs as a single algorithm and seeking a universal GA dynamics and general separation of GA-hard and GA-easy problems. For a given instance, if we change the parameters of GAs we can get totally different convergence results. So instead of taking GAs as a whole, researches into GA hardness should be done separately on different subclasses of GAs. The main misconception in the *performance space* is taking for granted the existence of a general *a priori* GA-hardness measure that can be used to predict a GA's performance on the given instance. Ideally, we want to have a program (a Turing machine) that can take as inputs a GA and an instance f of optimization problem and return how hard f is for the GA. Rice's theorem states that any non-trivial property of recursive enumerable languages is undecidable. This means that there are no nontrivial aspects of the behavior of a program which are algorithmically determinable given only the text of the program. It implies the nonexistence of general *a priori* predictive measure for problem hardness and algorithm performance based only on the description of the problem instance and the algorithm. This limits the general analysis, but it does not rule out the possibility of inducing an algorithm performance predictive model from some elaborately designed experimental results.

4 Future Directions: Experimental Approaches

We propose that future researches should focus more on real world NP-complete optimization problems rather than man-made functions; should study the classification of various GAs rather than considering them as a whole; and should give up analytically seeking a priori GA-hardness measures based *only* on the descriptive information of the problem and the GA, in favor of experimental, inductive approaches to learn the predictive model from the posterior results of running various GAs on problem instances with designed parameter settings.

Barrier Trees For Search Analysis

Jonathan Hallam and Adam Prügel-Bennett

Southampton University, Southampton SO17 1BJ, UK
jbh02r@ecs.soton.ac.uk, apb@ecs.soton.ac.uk

The development of genetic algorithms has been hampered by the lack of theory describing their behaviour, particularly on complex fitness landscapes. Here we propose a method for visualising and analysing the progress of a search algorithm on some hard optimisation problems using barrier trees. A barrier tree is a representation of the search space as a tree where the leaves of the tree represent local optima and nodes of the tree show the fittest saddle points between subtrees. The depth of the leaves and nodes corresponds to the fitness of the local optima and saddle points. Each configuration in search space can be mapped to a position on the tree depending on which optima are accessible without ever decreasing the fitness below its starting value. Having computed a barrier tree we can easily visualise how any search algorithm explores the search space.

Barrier trees have been used in the study of many physical and chemical systems, such as disordered spin systems and bio-polymer folding. A generalisation of these trees, and a formal definition, is given by Flamm et al.[1]. Conventionally barrier trees have their leaves (fittest configurations) at the bottom of the tree. In building a barrier tree we need to specify the neighbourhood structure that we are interested in. Here, we have used a Hamming neighbourhood. The novelty of our approach is to consider every configuration in the problem space as a point on the barrier tree. This allows us to see how the search space is being explored. For a genetic algorithm each member of the population can be represented as a point on the barrier tree. At each generation these points move around (usually down) the tree.

Figure 1 shows two example barrier trees. The trees have been truncated above the root – all configurations above this fitness are mapped to a single point on the tree. The two barrier trees are for two classic NP-hard problems, the binary or Ising perceptron and MAX-3-SAT. As can be seen, there are clear differences visible between different types of problems. Various explanations for poor or good performance can be thought of after watching a genetic algorithm solving a problem. The majority of the fitness landscape has a worse fitness than the root of the tree, and so only a fairly small proportion is mapped to anywhere else on the tree. In the examples we have looked at, which admittedly are all fairly small problems, genetic algorithms very quickly (approximately ten iterations) descend into the tree, and then take a lot longer to reach the optimal solution.

The barrier tree allows us to see the influence of changing parameters of the search algorithm. For example, we can see what happens as we change the mu-

---

[1] Christoph Flamm, Ivo L. Hofacker, Petr F. Stadler, and Michael T. Wolfinger. Barrier trees of degenerate landscapes. *Z. Phys. Chem.*, 216:155-173, 2002.

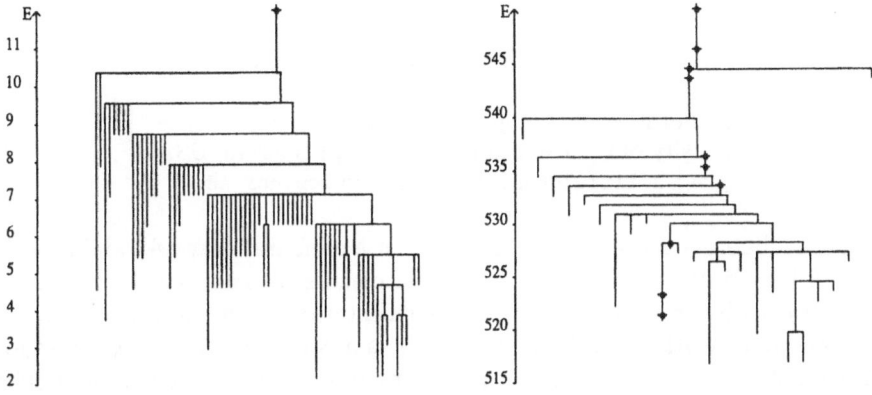

Fig. 1. Some example barrier trees, on minimisation problems. L.h.s a 17 variable binary perceptron training problem, with 1000 training patterns. The entire genetic algorithm population is plotted at the root of the tree, as all the members are less fit than the least fit saddle point, and can descend into any basin. R.h.s a seventeen variable MAX-3-SAT problem, with 4675 clauses. After seventy iterations, the genetic algorithm has descended into the tree, and has become trapped in a basin. It takes another seventy iterations to find a better solution

tation rate or use a different type of crossover. In addition, we can examine how the evolution changes if we use a deme GA (do the demes examine different local optima) or some other population structure. As well as visualising the landscape we can study the evolution systematically. For example, we can determine the probability of reaching the different local optima using a hill-climber or we can study the probability of jumping between branches of the tree if we crossover two parents from the same subtree.

One limitation of using barrier trees is that we need to enumerate all solutions in the search space. This limits the problems we can investigate to around twenty-five binary variables (around 35 million total configurations). However, for some problems it is possible to efficiently enumerate all the configuration above a certain fitness. We can then construct the barrier tree (or barrier forest) for these fit configurations. This should allow us to investigate much harder problems, although we have still to fully implement this enhancement.

Barrier trees provide a new insight into how search algorithms explore complex fitness landscapes. We hope that these investigations will encourage the development of a theoretical framework which is applicable to hard optimisation problems – that is, those problems of most interest for practitioners. Our preliminary investigations suggest that this approach yields many insights about how genetic algorithm explore complex landscapes.

A Genetic Algorithm as a Learning Method Based on Geometric Representations

Gregory A. Holifield and Annie S. Wu

School of Electrical Engineering and Computer Science
University of Central Florida, Orlando, FL 32816, USA
greg.holifield@us.army.mil,aswu@cs.ucf.edu

A number of different methods combining the use of neural networks and genetic algorithms have been described [1]. This paper discusses an approach for training neural networks based on the geometric representation of the network. In doing so, the genetic algorithm becomes applicable as a common training method for a number of machine learning algorithms that can be similarly represented. The experiments described here were specifically derived to construct claim regions for Fuzzy ARTMAP Neural Networks [2,3].

The adaptation of several principles to guarantee the success in traditional training of neural networks provided a similar level of success for the GA. Especially exciting is that this method provides the rare case of a GA guaranteeing a perfect solution with an increasing probability with each generation.

One neural network that has demonstrated good success, specifically for classification problems, is Fuzzy Adaptive Resonance Theory-MAP Neural Networks [2]. Fuzzy-ARTMAP Neural Networks (FAM-NN) provide numerous benefits, the most notable being online learning and notification of inputs that cannot be classified with the current learning [2,3]. FAM-NN constructs claim regions which can be viewed as geometric shapes, i.e. rectangles, cubes or hyper-cubes, in n-dimensional space. Each region has an assigned a category. Based simply on the classification rules, points within these regions are classified as being part of that claim region's category. Georgiopoulos provides a more detailed discussion on this point as it applies to the rules within FAM-NN [3].

The genetic algorithm for this application used a variable length representation. Each individual consists of m-pairs of points in n-dimensional space. Each pair constitutes the vertices of a rectangle, cube or hyper-cube, depending on the dimensionality of the representational space. Just as in the claim regions in FAM-NN, each pair is also assigned a category according to the classification categories in the training data.

The operators selected for this work concentrate on achieving success comparable to the traditional FAM-NN training. By adapting operators that emphasize the best traditional performance, the GA performance is improved and new characteristics emerge further improving the training. The genetic algorithm uses single-point crossover with random selection of parents. Mutation consists of several possibilities, the latter of which are novel.

Typical mutation of a randomly selected dimension of a particular vertex occurs at a given probability. Additionally, pairs of points are randomly added, deleted or swapped. Identical pairs of points based on particular input patterns

are added. By doing this, the new point is guaranteed to correctly classify the
particular input pattern. The input pattern is randomly selected from a queue
that is filled with patterns that were misclassified in the previous fitness evaluations. To counter the increased number of claim regions and achieve highest
possible compression, a mutation operator is utilized which combines vertices of
like category regions.

We used three test patterns in this study. They include random points generated within a grid, a circle in a square and two spirals – Fig. 1.

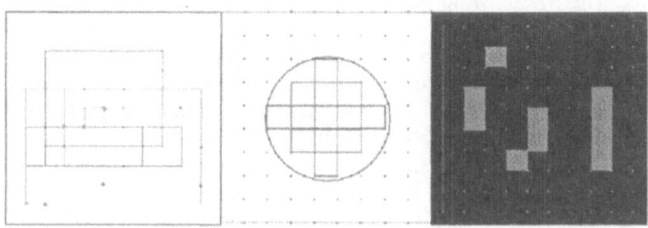

Fig. 1. Final Solution (Left) Random Points. Larger dots and lighter gray indicate category 1. The smaller dots and darker gray represent category 2.(Center) Circle in a Square (Right)Two spirals

The solution for the classification of a data set of random points was optimized with 6 regions for 25 training patterns resulting in a compression rate of 76%. The solution for the circle in a square problem was found at 1700 generations. It is optimized with 4 regions for 100 points resulting in a compression rate of 96%. The solution for the two spirals problem was found around generation 1400 with a compression of 9 regions of 100 or a 91% compression rate.

The combination of the techniques used in traditional FAM-NN provides an excellent springboard for an effective solution to training a good online classifying system. The method leverages the geometric representation allowing for an expansion to other machine learning algorithms on the basis of their geometric representation. The ability of a GA to provide a guarantee of convergence with an increasing probability with each generation provides a novel complement to remaining question of GA convergence. The extension of this work to hybrid algorithms that utilize a wider variety of geometric representation could provide a classification method beyond those currently developed.

References

1. J.T. Alander. An indexed bibliography of genetic algorithms and neural networks.
2. G.A. Carpenter et al. Fuzzy ARTMAP: A neural network architecture for incremental supervised learning of multidimensional maps. *IEEE Transactions on Neural Networks*, 3(5):698–713., 1992.
3. Michael Georgiopoulos and Christos Christodoulou. *Applications of Neural Networks in Electromagnetics*. Artech House, 2001.

Solving Mastermind Using Genetic Algorithms

Tom Kalisker and Doug Camens

Software Engineering Masters Program
Engineering and Information Science Division
Penn State Great Valley School of Graduate Professional Studies
30 East Swedesford Road
Malvern, PA 19355-1443
{trk139,dmc123@psu.edu}@psu.edu

Abstract. The MasterMind game involves decoding a secret code. The classic game is a code of six possible colors in four slots. The game has been analyzed and optimal strategies have been posed by computer scientists and mathematicians. In this paper we will survey previous work done on solving MasterMind, including several approaches using Genetic Algorithms. We will also analyze the solution sets and compare our results using a novel scoring system inside a GA against previous work using Genetic and Heuristic algorithms. Our GA is performing closer to optimal then previously published work. The GA we present is a Steady State GA using Fitness Proportional Reproduction (FPR), where the fitness function incorporates a simple heuristic algorithm. We also present a scoring method that is simpler then those used by other researchers. In larger games such as 10 colors and 8 slots our GA clearly outperform the heuristic algorithm. In fact if one wishes to tradeoff a higher average number of guesses to a faster running time, extremely large games such as 12 x10 can be solved in a reasonable time (i.e. minutes) of run time.

MasterMind is a game where a secret code is discovered through decoding successive guesses with hints as to the quality of each guess.

There have been many papers published about strategies to solve MasterMind, several papers including one by Knuth are summarized here.

Other researchers have looked at rule based or heuristic approaches to solving 6x4 MasterMind. The best to date is that by Radu Rosu.

The solutions using GAs by Bento et al and Team Geneura are reviewed.

The score for any size game of MasterMind can be given by the following formula:

Let N = total number of pegs awarded i.e.(B+W)
B = number of black pegs
W= number of white pegs

$$\text{Score} = (2*B + W) + \sum_{i=1}^{N-1} i$$

A genome represents a potential guess in the game of MasterMind. It is encoded using an array of integer genes. The alleles possible for each gene are the set of integers from 0 to c-1 where c is the number of colors in the game. The GA consists of a custom fitness function based on the three components listed below.

1. The heuristic algorithm introduced by Radu Rosu.
2. A scoring system based on the number of black and white pegs awarded in previous guesses of the game.
3. Any individual that has already been played in a previous guess automatically gets a fitness of zero. We had hoped that this alone would eliminate the play of multiple guesses.

To calculate fitness for an individual, it is evaluated against all previous guesses as if the previous guess is a secret code and the individual is a guess to that code. Simple single-point crossover was used exclusively Mutation is done by randomly switching a gene (i.e. peg) to a different color chosen at random from the colors available in the gene color pool.

We ran our GA with the same size populations as those reported by Bento et al. and show the following results for averages of 500 trials.

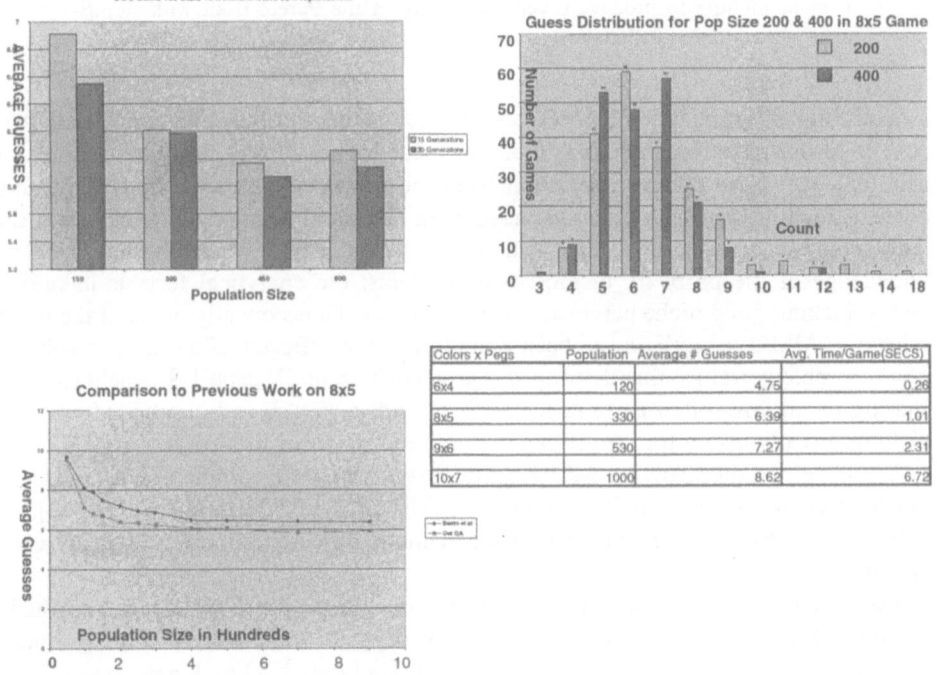

Acknowledgments. The authors would like to thank Dr. Walter Cedeño and Penn State Great Valley for providing a course in Genetic Algorithms.

Evolutionary Multimodal Optimization Revisited*

Rajeev Kumar[1] and Peter Rockett[2]

[1] Department of Computer Science & Engineering
Indian Institute of Technology, Kharagpur 721 302, India
rkumar@cse.iitkgp.ernet.in

[2] Department of Electronic & Electrical Engineering
Mappin St, University of Sheffield, Sheffield S1 3JD, England
p.rockett@sheffield.ac.uk

Abstract. We revisit a class of multimodal function optimizations using evolutionary algorithms reformulated into a multiobjective framework where previous implementations have needed niching/sharing to ensure diversity. In this paper, we use a steady-state multiobjective algorithm which preserves diversity *without* niching to produce diverse sampling of the Pareto-front with significantly lower computational effort.

Multimodal optimization (MMO) and multiobjective optimization (MOO) are two classes of optimizations requiring multiple (near-)optimal solutions: having found a solution set, a user makes a selection from the (hopefully) diverse options. In this context, niching/sharing techniques have been commonly employed to ensure a diverse solution set although such techniques work the best when one has *a priori* knowledge of the problem. In most real-problems, the analytical form is unknown and so picking good niche parameters is problematic. Consequently, most of the work related to MMO using EAs has been done to *test* the efficacy of the EAs in solving *known* problems rather than *solving* the problem *per se*. Watson [1] concluded that sharing-based GAs often perform *worse* than random search and questioned whether niching is really useful for identifying multiple fitness peaks in MMOs.

We have revisited solving MMO using EAs without any problem-dependent parameters using the same reformulation of MMO into a MOO framework as [2], to obtain good diversity in objective space without any *explicit diversity-preserving* operator.

Deb [2] has recast of a number of single-objective MMO problems into dual-objective MOO problems and empirically investigated the effects of sharing. Many have studied the promotion of diversity using sharing for MOO problems – see [3] for a review. We have used a MOO algorithm [3] which, to the best of our knowledge, is the only implementation which does not need *any* explicit sharing mechanism; we demonstrate its efficacy in achieving diversity for two sample MMO problems, *F1* (Sect. 4.1 of [2]) and *F2* (Sect. 5.3 of [2]) which were considered by earlier researchers using multiobjective methods. We have used the same formulation, as far as is

* Partially supported by the Ministry of Human Resource Development, Government of India

known, for fair comparison. We repeated the experiments many hundreds of times, each with a different initial population to check the consistency of the results. Typical results selected on the basis of their *average* performance are presented below.

The multi-modal *F2* function can be recast as a multiobjective problem requiring the simultaneous minimization of $f_{21}(x_1)$ and f_{22}; see Sect. 5.3 of reference [2] for more details. Figure 1a shows the initial population (size = 100) where we obtain individuals near the optima by random chance. Figure 1b shows the population after 100 epochs which can be compared with the results for 500 generations in [2] and are superior to those in [2] both in terms of proximity to the Pareto-optimal front and diversity. Significantly this has been achieved at reduced computational cost. We have also studied the *F1* function (Sect. 4.1 of [2]) and find a result entirely consistent with that we have observed with function *F2* – see [3].

Explicit diversity preserving methods need prior knowledge and their efficacy depends on parameter fine-tuning; without proper values they cannot be beneficial. Claims of the superiority of variable- *vs.* objective space sharing are unfounded, problem dependent and nothing general can be said on the selection of proper values for niching/sharing.

In conclusion, we have shown that we can solve multimodal problems by recasting them as multiobjective ones *without an explicit niching/sharing*. Comparing our results with previous work [2], the algorithm employed here provided superior diversity and proximity to the true Pareto-front at reduced computational cost.

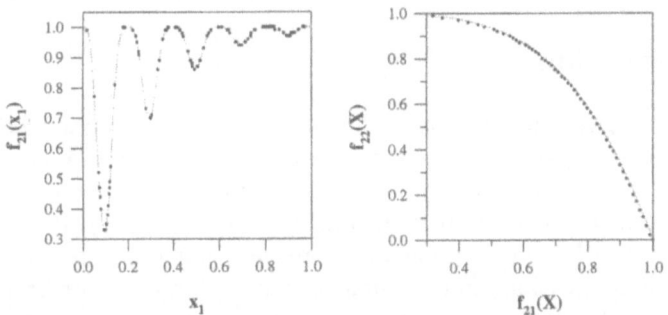

Fig. 1. Function *F2* – Initial population shown in (a) x_1 vs. f_{21} and (b) f_{21} vs. f_{22} plots.

References

1. Watson, J.P.: A Performance Assessment of Modern Niching Methods for Parameter Optimization Problems. Gecco-99 (1999)
2. Deb, K.: Multiobjective Genetic Algorithms: Problem Difficulties and Construction of Test Problems. Evolutionary Computation 7 (1999) 1–26
3. Kumar, R., Rockett, P.I.: Improved Sampling of the Pareto-Front in Multiobjective Genetic Optimizations by Steady-State Evolution: A Pareto Converging Genetic Algorithm. Evolutionary Computation 10(3): 282–314, 2002

Integrated Genetic Algorithm with Hill Climbing for Bandwidth Minimization Problem

Andrew Lim[1], Brian Rodrigues[2], and Fei Xiao[3]

[1] Department of Industrial Engineering and Engineering Management
Hong Kong University of Science and Technology, Clear Water Bay, Hong Kong
iealim@ust.hk
[2] School of Business, Singapore Management University, 469 Bukit Timah Road
Singapore 259756
br@smu.edu.sg
[3] Department of Computer Science, National University of Singapore
3 Science Drive 2, Singapore 117543
xiaofei@comp.nus.edu.sg

Abstract. In this paper, we propose an integrated Genetic Algorithm with Hill Climbing to solve the matrix bandwidth minimization problem, which is to reduce bandwidth by permuting rows and columns resulting in the nonzero elements residing in a band as close as possible to the diagonal. Experiments show that this approach achieves the best solution quality when compared with the GPS [1] algorithm, Tabu Search [3], and the GRASP with Path Relinking methods [4], while being faster than the latter two newly-developed heuristics.

1 Introduction

For $A = \{a_{ij}\}$ a symmetric matrix, the matrix bandwidth minimization problem is to find a permutation of rows and columns of the matrix A so as to bring all the non-zero elements of A to reside in a band that is as close as possible to the main diagonal, that is to $Minimize\{max\{|i-j| : a_{ij} \neq 0\}\}$. In

The bandwidth minimization problem has been found to be relevant to a wide range of applications. For example, in solving large linear systems, Gaussian elimination can be performed in $O(nb^2)$ time on matrices of bandwidth b, which is much faster than the normal $O(n^3)$ algorithm if $b << n$. In this work, we propose a Genetic Algorithm (GA) integrated with Hill Climbing to solve the bandwidth minimization problem. Genetic Algorithms [2], have been widely used in solving combinatorial optimization problems. The Genetic Algorithm, while having been shown to be especially good in global search, is not well suited for tuned search. In developing a solution for the bandwidth minimization problem, we have therefore combined a Genetic Algorithm with a Hill Climbing Algorithm. Computational Results show that this integrated algorithm performs well. Compared with the classical GPS algorithm, and the newly-developed techniques of Tabu Search and GRASP with Path Relinking, it provides the best solution quality while being faster than the latter two heuristics.

2 Computational Experiments

we compare our GA_HC with best known heuristic methods on two sets of instances from Harwell-Boeing Matrix Collection
(http://math.nist.gov/MatrixMarket/data/Harwell-Boeing).

Table 1. Performance comparison according to problem size

	GPS	TS	GRASP_PR	GA_HC
33 instances with n=30,..,199				
$B_f(G)$	31.42	23.33	22.52	22.48
Deviation	39.77%	3.78%	0.18%	0.00%
CPU seconds	0.003	2.36	4.21	2.54
80 instances with n=200,...,1000				
$B_f(G)$	156.38	100.78	99.43	97.02
Deviation	61.18%	3.88%	2.49%	0.00%
CPU seconds	0.11	121.66	323.19	85.22

We can find in Table 1 that the best solution in quality is obtained for the first test set by the GA_HC which is about 40 percent better than the classical GPS algorithm. On the second test set, the GA_HC also obtains the best solution in quality and is faster than the TS and GRASP_PR. Overall, our GA_HC obtains the best solution in quality comparing with the best known algorithms. And it is faster than the newly developed TS and GRASP_PR algorithm.

3 Conclusions

We have proposed a combined GA with Hill Climbing for solving the bandwidth minimization problem. Extensive experiments have shown that our new GA with Hill Climbing obtains best solutions for the bandwidth minimization problem in reasonable time.

References

1. Gibbs, N.E., Poole, W.G., Stockmeyer, P.K., 1976, An algorithm for reducing the bandwidth and profile of sparse matrix. SIAM Journal on Numerical Analysis 13 (2), 236–250.
2. Holland, J.H., 1975, Adaption in Natural and Artificial Systems, University of Michigan Press, Ann Arbor.
3. Marti, R., Laguna, M., Glover, F. and Campos, V., 2001, Reducing the Bandwidth of a Sparse Matrix with Tabu Search, European Journal of Operational Research, 135(2), pp. 211–220.
4. Pinana, E., Plana, I., Campos, V. Marti, R., 2002, in print, European Journal of Operational Research. http://www.uv.es/~rmarti/.

A Fixed-Length Subset Genetic Algorithm for the p-Median Problem

Andrew Lim[1,2] and Zhou Xu[2]

[1] Dept Industrial Engineering and Engineering Management, Hong Kong University of Science and Technology, Clear Water Bay, Kowloon, Hong Kong
[2] Dept of Computer Science, National University of Singapore, 3 Science Drive 2, Singapore 117543
{ialim@ust.hk,xuzhou@comp.nus.edu.sg}

Abstract. In this paper, we review some classical recombination operations and devise new heuristic recombinations for the fixed-length subset. Our experimental results on the classical p-median problem indicate that our method is superior and very close to the optimal solution.

1 Fixed-Length Subset Recombinations

We study the Fixed Length Subset Genetic Algorithm (*FLS-GA*), whose candidate solutions are represented by the fixed-length subset (*FLS*), which can be defined as any subset with a fixed size for a given set. In FLS-GA, we adopt a *subset encoding* [CHWS97], which uses a list of elements to represent the candidate FLS. [Rad93] studies two *pure recombinations* for FLS, which are Random Respectful Recombination (*RRR*) and Random Assorting Recombination (*RAR*). We extend them to *heuristic recombinations*.

1. Construct candidate set S', and inherited pattern s_0, from FLSs A and B;
2. Choose sub-optimal FLS from S' using the heuristic procedure $H(S', s_0)$. Let the result of $H(S', s_0)$ be the child of recombinations;

For FLS-GA, we could use RRR and RAR respectively to construct candidate set S', and inherited pattern s_0, leading the following two heuristic recombinations.

- Heuristic RRR (*H-RRR*)
  The inherited pattern $s_0 = A \cap B$ and the candidate set $S' = A \cup B - A \cap B$. Thus the size of the inherited pattern $|s_0|$ is equal to $|A \cap B|$, while the size of the candidate set $|S'|$ is $|A \cup B| - |A \cap B|$;
- Heuristic RAR (*H-RAR*)
  Each element of the inherited pattern s_0, is chosen from s_0 with probability q. And each element of the candidate set S', is chosen from $A \cap B - s_0$ with probability p_0, or from $A \cup B - A \cap B$ with probability p_1, or from $S - A \cup B$ with probability p_2.

Moreover, to balance the diversity and pattern, an *adaptive heuristic recombinations* can be designed as follows.

- Threshold H-RAR (*T-H-RAR*)
  Different heuristic recombinations are used in different situations of diversity.
  1. If the diversity of the current population is less than a threshold H, we adopt H-RAR with a smaller q and bigger p_2 to increase the diversity;
  2. If the diversity of the current population is larger than a threshold H, we adopt H-RAR with a bigger q and smaller p_2 to decrease the diversity;

2 FLS-GA Application to P-Median Problem (PMP)

PMP [ODE03] can be formulated as a FLS optimization problem.

- Instance:
  1. A set $S = \{v_1, .., v_m\}$ with m vertices;
  2. A positive number $p \leq m$;
  3. A distance matrix $D_{m \times m}$, where D_{ij} represents the distance between vertex v_i and vertex v_j;
  4. A function $F(s) = \sum_{\forall v_i \in S} (\min_{\forall v_j \in s} D_{ij}), \forall s \subseteq S$;
- Constraint: $|s| = m$
- Output: A fixed length subset s^*, where $s^* \subseteq S$ and $|s^*| = m$.
- Objective: To minimize the value of $F(s^*)$;

Two heuristic procedures are adoptted for when we applying FLS-GA to PMP.

- Decreasing Heuristics (*DH*)
  For a given candidate set S' and an inherited pattern s_0, firstly add all elements of S to the result FLS. Then select the rest $(p - |s_0|)$ elements from $S' - s_0$ one by one. In each round, the element, which decreases the objective function most, is added to the result FLS.
- Increasing Heuristics (*IH*) [ODE03]
  For a given candidate set S' and an inherited pattern s_0, delete $(|S'| - p)$ elements from S' one by one. In each round, the element, which is excluded from s_0 and increases the objective function the least, is deleted from S'.

The experimental results have shown that the FLS-GAs with heuristic recombinations over-perform the GA with pure genetic operations, and that the hybrid heuristic genetic recombinations exhibit the best results over all. For most problems, FLS-GAs with T-H-RAR can obtain solutions that are very close to the optimal solution (the gap $< 0.1\%$).

References

[CHWS97] Kelly D. Crawford, Cory J. Hoelting, Roger L. Wainwright, and Dale A. Schoenefeld. A study of fixed-length subset recombination. In Richard K. Belew and Michael D. Vose, editors, *Foundations of Genetic Algorithms 4*, pages 365–378. Morgan Kaufmann, San Francisco, CA, 1997.

[ODE03] Alp O., Z. Drezner, and E. Erkut. An efficient genetic algorithm for the p-median problem. *Annals of Operations Research*, 2003. forthcoming.

[Rad93] Nicholas J. Radcliffe. Genetic set recombination. In *Foundations of Genetic Algorithms 2*, San Mateo, CA, 1993. Morgan Kaufmann Publishers.

Performance Evaluation of a Parameter-Free Genetic Algorithm for Job-Shop Scheduling Problems

Shouichi Matsui, Isamu Watanabe, and Ken-ichi Tokoro

Communication & Information Research Laboratory (CIRL)
Central Research Institute of Electric Power Industry (CRIEPI)
2-11-1 Iwado-kita, Komae-shi, Tokyo 201-8511, Japan
{matsui,isamu,tokoro}@criepi.denken.or.jp

1 Introduction

The job-shop scheduling problem (JSSP) is well known as one of the most difficult NP-hard combinatorial optimization problems. Genetic Algorithms (GAs) for solving the JSSP have been proposed, and they perform well compared with other approaches [1].

However, the tuning of genetic parameters has to be performed by trial and error, making optimization by GA *ad hoc*. To address this problem, Sawai et al. have proposed the Parameter-free Genetic Algorithm (PfGA), for which no control parameters for genetic operation need to be set in advance [3].

We have proposed an extended parameter-free GA for JSSP [2], and reported that the GA performed well without tedious parameter-tuning. This paper reports the result of empirical evaluation of the GA to a wider range of problem instances. The simulation results show that the GA performs well for many problem instances, and the performance can be improved greatly by increasing the number of subpopulations in the parallel distributed version.

2 Computational Results

The GA is tested by the benchmark problems from ORLib. We tested a wider range of instances, namely 10 tough problems, ORB01–ORB10, SWV01–SWV20, and TA01–TA30, but only the results for TA01–TA30 are shown in Table 1. The GA was run for each problem instance using 50 different random seeds. The maximum number of fitness evaluation was set to 1,000,000 for all cases.

- The GA can find the makespan that is equal to the optimal or the best upper bound in TA02, TA03, TA05, and TA07 when we increase the number of subpopulations.
- As we increase the number of subpopulations, the average makespan is reduced, but the best makespan does not always shorten.

Table 1. Simulation results for TA problems

Prob.	$n \times m$	Best bounds[1]	$N=1$ Best	μ	$N=16$ Best	μ	$N=32$ Best	μ	$N=64$ Best	μ
TA01	15×15	1231	1265	1288.2	1249	1263.1	1248	1259.6	1241	1254.3
TA02	15×15	1244	1252	1280.9	**1244**	1264.3	**1244**	1261.0	**1244**	1255.8
TA03	15×15	1218(1206)	1228	1256.8	1219	1232.4	<u>1218</u>	1229.3	<u>1218</u>	1226.4
TA04	15×15	1175(1170)	1191	1222.6	1181	1195.0	1181	1191.3	1181	1188.4
TA05	15×15	1228(1210)	1253	1272.4	1236	1252.6	1236	1246.4	<u>1228</u>	1244.9
TA06	15×15	1240(1210)	1249	1286.8	1248	1263.7	1248	1258.6	1246	1255.4
TA07	15×15	1228(1223)	1250	1264.0	1236	1247.3	1233	1242.6	<u>1228</u>	1240.6
TA08	15×15	1217(1187)	1233	1266.8	1221	1242.0	1224	1238.2	1223	1233.1
TA09	15×15	1274(1247)	1314	1342.9	1296	1316.4	1292	1310.2	1284	1305.6
TA10	15×15	1241	1244	1299.0	1244	1265.1	1244	1256.3	1244	1246.5
TA11	20×15	1364(1321)	1416	1464.0	1395	1430.0	1408	1426.8	1398	1419.9
TA12	20×15	1367(1321)	1428	1471.6	1406	1435.4	1408	1430.8	1396	1421.8
TA13	20×15	1350(1271)	1407	1443.1	1378	1408.1	1361	1399.6	1371	1393.5
TA14	20×15	1345	1374	1405.5	1352	1378.0	1355	1373.6	1351	1365.8
TA15	20×15	1342(1293)	1412	1461.3	1388	1416.1	1379	1412.6	1384	1402.1
TA16	20×15	1368(1300)	1417	1462.8	1397	1426.9	1400	1420.2	1380	1413.8
TA17	20×15	1478(1458)	1498	1546.5	1496	1519.9	1494	1514.1	1490	1508.3
TA18	20×15	1396(1369)	1468	1519.1	1461	1489.1	1449	1477.6	1443	1468.9
TA19	20×15	1341(1276)	1419	1454.2	1398	1419.9	1390	1415.2	1382	1405.8
TA20	20×15	1353(1316)	1414	1453.3	1399	1420.0	1389	1414.1	1380	1406.8
TA21	20×20	1647(1539)	1729	1773.6	1687	1728.7	1688	1723.1	1685	1716.1
TA22	20×20	1603(1511)	1699	1736.2	1677	1697.3	1665	1689.7	1646	1683.3
TA23	20×20	1558(1472)	1635	1683.9	1595	1641.9	1598	1636.9	1589	1625.1
TA24	20×20	1651(1594)	1712	1767.8	1695	1730.3	1698	1722.0	1693	1716.2
TA25	20×20	1598(1496)	1665	1715.6	1651	1682.8	1644	1671.3	1650	1668.3
TA26	20×20	1655(1539)	1728	1766.3	1700	1728.7	1704	1724.7	1699	1719.7
TA27	20×20	1689(1616)	1787	1827.3	1737	1777.9	1725	1761.3	1741	1760.1
TA28	20×20	1615(1591)	1698	1739.4	1658	1691.5	1653	1685.9	1654	1676.4
TA29	20×20	1625(1514)	1689	1730.5	1655	1693.3	1656	1685.9	1653	1678.1
TA30	20×20	1596(1468)	1656	1717.8	1639	1679.0	1638	1667.7	1638	1661.5

- The relative error to the best upper bound is small enough in TA01–TA10, below 1%, when $N \geq 64$. But in TA11–TA30 it varies from 0.4% to 3.4% even in $N = 64$.

References

1. Jain, A.S., and Meeran, S.: Deterministic job-shop scheduling: past, present and future, *European Journal of Operational Research*, vol.113, pp. 390–434, 1999.
2. Matsui, S., Watanabe, I., and Tokoro, K.: Real-coded parameter-free genetic algorithm for job-shop scheduling problems, *Proc. Seventh Parallel Problem Solving from Nature – PPSN VII*, pp. 800–810, 2002.
3. Sawai, H., Kizu, S.: Parameter-free genetic algorithm inspired by "disparity theory of evolution", *Proc. Seventh Parallel Problem Solving from Nature – PPSN V*, pp. 702–711, 1998.

SEPA: Structure Evolution and Parameter Adaptation in Feed-Forward Neural Networks

Paulito P. Palmes, Taichi Hayasaka, and Shiro Usui

Department of Information and Computer Sciences
Toyohashi University of Technology, Toyohashi, 441-8580 Japan
{ppalmes,hayasaka,usui}@bpel.ics.tut.ac.jp
http://www.bpel.ics.tut.ac.jp/

Abstract. In developing algorithms that dynamically changes the structure and weights of ANN (Artificial Neural Networks), there must be a proper balance between network complexity and its generalization capability. SEPA addresses these issues using an encoding scheme where network weights and connections are encoded in matrices of real numbers. Network parameters are locally encoded and locally adapted with fitness evaluation consisting mainly of fast feed-forward operations. Experimental results in some well-known classification problems demonstrate SEPA's high consistency performance in classification, fast convergence, and good optimality of structure.

1 SEPA Strategy

Until now, ANN architecture design remains to be one of the most important areas of research due to the lack of general criteria in finding an optimal network topology for a particular class of problems. To address this issue, SEPA uses a GA-based model of ANN where weight adaptation is not gradient-based but stochastic (Fig. 1). To achieve a proper balance in ANN's network complexity and generalization capability, SEPA's fitness function considers three important parameters: training error; hidden nodes; and connections.

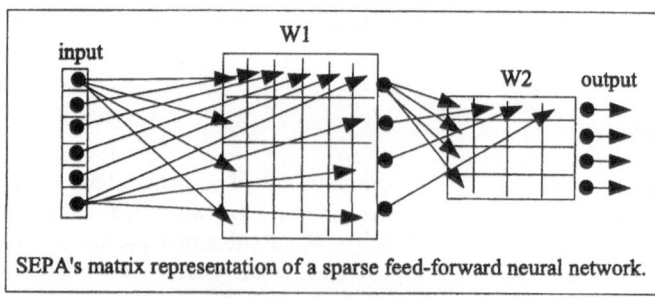

Fig. 1. Encoding scheme

Equation (1) is SEPA's fitness function where: α, β are complexity constants; n, N are no. of active hidden nodes and its total; and c, C are no. of active connections and its total. The degree of influence in network complexity is controlled by setting α and β parameters appropriately. SEPA uses $\alpha = 0.70$ and $\beta = 0.30$.

$$Q = \frac{1}{N}\sum_{i=1}^{N}(T_i - O_i)^2 + \alpha(\frac{n}{N}) + \beta(\frac{c}{C}) \qquad (1)$$

SEPA's structure evolution is carried out by mutation using (2) and (3) with mutation probability set to 0.01. Structure exploration is carried out by adapting the mutation operator using Gaussian perturbation:

$$m' = m + \mathbf{N}_0(0, \sigma) \qquad (2)$$

$$w'_{ij} = w_{ij} + \mathbf{N}_0(0, m') \qquad (3)$$

where σ is the step size parameter (SSP) influencing the strength of mutation. SEPA uses step sizes of 5 (SSP5) and 100 (SSP100) in the experiment. SEPA's selection criterion uses an elitist roulette-wheel method where only the two best parents are retained.

2 Summary of Results

SEPA exhibits very low classification error variability (maximum=0.02) and small network size in all problems (Table 1). The experiments suggest that SSP can serve as a local optimizer in SEPA. For well-behaved classification problems, larger SSP produces smaller networks, faster convergence, and good classification performance. On the other hand, smaller SSP produces larger network structure with slower convergence rate but with good adaptability in noisy data.

Table 1. Experiments

HIDDEN NODES							CLASSIFICATION ERROR						
a) cancer		p=0.00	b) iris		p=0.00		a) cancer		p=0.64	b) iris		p=0.01	
SSP	ave	var	SSP	ave	var		SSP	ave	var	SSP	ave	var	
5	5.84	2.71	5	4.28	3.39		5	3.8%	0.00	5	10.4%	0.00	
100	3.84	1.65	100	3.04	1.30		100	3.6%	0.00	100	7.2%	0.00	
c) wine		p=0.00	d) glass		p=0.00		c) wine		p=0.35	d) glass		p=0.01	
SSP	ave	var	SSP	ave	var		SSP	ave	var	SSP	ave	var	
5	4.78	3.60	5	5	0.00		5	8.6%	0.00	5	40.0%	0.02	
100	3.24	1.45	100	5.9	0.10		100	7.8%	0.00	100	57.0%	0.00	

ACTIVE CONNECTIONS							MAXIMUM GENERATION						
a) cancer		p=0.00	b) iris		p=0.00		a) cancer		p=0.00	b) iris		p=0.00	
SSP	ave	var	SSP	ave	var		SSP	ave	var	SSP	ave	var	
5	60.72	235.39	5	23.68	41.86		5	1093.0	2779005.10	5	452.00	148269.4	
100	32.44	235.88	100	19.20	24.33		100	77.2	4816.49	100	139.20	18346.3	
c) wine		p=0.00	d) glass		p=0.00		c) wine		p=0.00	d) glass		p=inf	
SSP	ave	var	SSP	ave	var		SSP	ave	var	SSP	ave	var	
5	102.02	268.02	5	58.20	14.40		5	1311.4	2460293.92	5	5000	0	
100	79.62	376.36	100	74.90	0.10		100	233.8	74693.43	100	5000	0	

Real-Coded Genetic Algorithm to Reveal Biological Significant Sites of Remotely Homologous Proteins

Sung-Joon Park and Masayuki Yamamura

Interdisciplinary Graduate School of Science and Engineering
Tokyo Institute of Technology
R1-418, 4259 Nagatsuta, Midori, Yokohama, 226-8502, Japan
{park,my}@es.dis.titech.ac.jp
http://www.es.dis.titech.ac.jp/

1 Introduction

Discovering biological importance from protein structures needs to utilize heuristic approaches. Since three-dimensional(3D) protein structures play a crucial role in biological reactions, comparing new proteins with well-studied proteins is vital for understanding the native functions. A few residues forming in geometrically close positions activate such biological functions.

Our goal is to find biological significant sites from protein pairs. Since a pair belonging different protein families has lower global similarity but similar function (so called *remotely homologous*), we develop a GA-based alignment tool to emphasize small regions of protein pairs geometrically similar.

2 Method

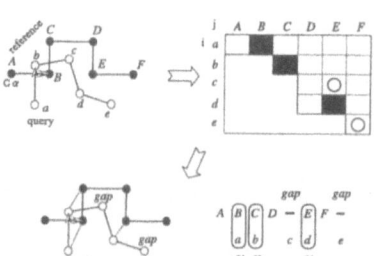

Fig. 1. Estimation of equivalent $C\alpha$ atom pairs in fitness function.

The proposed Real-coded GA, GSA (Genetic Structural Alignment), optimizes an isometric transformation consisting of Euler's angle $\mathcal{R}$ for rotation and a translation vector $\vec{T}$ for superposition. $\mathcal{R}$ and $\vec{T}$ code individuals in GSA, i.e. six-dimensional function optimization. When the pair of a query protein (Stc_1) and a reference (Stc_2) are given ($length(Stc_1) \leq length(Stc_2)$), the 1st $C\alpha$[1] atom of Stc_1 connects to the last $C\alpha$ of Stc_2. The structures move to the absolute origin. A 3D rectangle containing Stc_1 and Stc_2 is defined, and $\vec{T}$'s of the initial population are plotted in the rectangle. A $\vec{T}$ defines the position of the 1st $C\alpha$ of Stc_1, and a $\mathcal{R}$ of an individual rotates Stc_1; the 1st $C\alpha$ is the origin of rotation.

---

[1] α-carbon. The side chain of an amino acid links to a $C\alpha$. The backbone of a protein is consecutive $C\alpha$ atoms.

UNDX [1] generates two sets of six-dimensional real numbers for two children at a cross time by using the positional relationship of three parents. The best offspring and a randomly selected member replace the parents.

Estimating equivalent $C\alpha$ atoms is prerequisite to evaluate individuals. If d_{ij} between the ith $C\alpha$ of Stc_1 transformed by a individual and the jth of Stc_2 is the nearest pair and less than δ, it is added to equivalent set C. Otherwise, the ith atom corresponds to a gap (Fig. 1). Once $C = \{c_1, c_2, \ldots, c_z\}$ and the number of gaps g are defined, the fitness function f evaluates the individual;

$$s = \sum_{i=1}^{z} e^{\varepsilon \times d_{c_i}} \quad (1) \qquad f = \frac{s + 1.0}{g + 1.0} \quad (2)$$

3 Results and Conclusion

We adjusted parameters for GSA and finally set to generation=3000, population=50, cross-time=100, UNDX α=0.5, β=0.3, ε=-0.8, δ=2.24. Fig. 2 shows distribution of $P_1 = \frac{z}{length(Stc_1)} \times 100$ and $P_2 = \frac{z'}{z} \times 100$, where z' is the number of equivalent pairs being less than 0.5 in distance. The protein pairs in Fig. 2 have less than 30% sequence identity.

GA_FIT [2], a SGA using dynamic programming(DP), has found more z than GSA. It is conspicuous, however, that GSA possesses a great number of z' in the equivalent set. Such geometrically conserved $C\alpha$ atoms in the protein pair have a significant possibility to present similar biological function. DP-based methods can find topological similarity from global protein structures. On the other hand, a few $C\alpha$ atoms are lost for superimposing global structures.

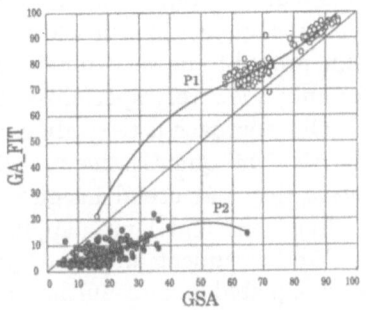

Fig. 2. Distribution of equivalences.

The backbone of a protein is rigorously changed by the amino acid side chains, but the backbone is conservative for keeping its function [3]. For this paradox, comparing proteins has to consider both 3D coordinates of the backbone and chemical properties. We intend to add extra information to the fitness function for finding biologically reliable alignments.

References

1. Ono, I., Kobayashi, S.: A Real-coded Genetic Algorithm for function optimization using unimodal normal distribution crossover. Proc. the 7th ICGA (1997) 246–253
2. May, A.C.W., Johnson, M.S.: Improved genetic algorithm-based protein structure comparisons: pairwise and multiple superpositions. Protein Engng. 8 (1995) 873–882
3. Chothia, C., Lesk, A.M.: The relation between the divergence of sequence and structure in proteins. EMBO J. 5 (1986) 823–826

Understanding EA Dynamics via Population Fitness Distributions

Elena Popovici and Kenneth De Jong

Department of Computer Science
George Mason University
Fairfax, VA 22030

1 Motivation and Methodology

This paper introduces a new tool to be used in conjunction with existing ones for a more comprehensive understanding of the behavior of evolutionary algorithms. Several research groups including [1,3,4] have shown how deeper insights into EA behavior can be obtained by focusing on the changes to the entire population fitness distribution rather than just "best-so-far" curves. But characterizing how repeated applications of selection and reproduction modify this distribution over time proved to be very difficult to achieve analytically and was done successfully for only a few very specialized EAs and/or very simple fitness landscapes.

Our approach is to study empirically derived fitness distributions, both qualitatively and quantitatively, believing that they have the potential for providing interesting and useful insights into the behavior of EAs. The methodology is quite general and can be applied to any EA and to any fitness landscape.

We instrument an EA to provide snapshots of the population fitness data at designated points during the evolutionary process, and display them in a histogram-like manner. While visualizing population fitness distributions can be quite insightful, we go a step further and perform a quantitative analysis of the observed fitness distributions, by estimating how likely the empirically generated distributions reflect an underlying standard distribution (e.g. normal, exponential, etc.). The statistical technique used was the generation of "Q-Q" plots from which we computed an R^2 value representing the likelihood that an empirically generated distribution is due to a certain theoretical distribution [2].

2 Experiments and Results

Our initial experiments were designed to provide some insight into the following questions: 1) how dependent are these observations on the particular type of EA being used, 2) how dependent are the observed population fitness distributions on a particular fitness landscape, and 3) can useful insights be gained by taking snapshots within a generation to study the effects of selection and reproduction on population fitness distributions?

As earlier work had focused on the population fitness distributions of standard generational GAs, for contrast we focused on an EA using a binary representation, standard crossover and mutation, but $\mu + \lambda$ population dynamics.

We performed experiments on two problems: 1) the F1 function of the De Jong test suite, used with dimensionality 4 on $[-5.12, 5.12]$ and 2) evolving rules for two-dimensional cellular automata that generate some predefined pattern.

Our experiments showed that the $\mu + \lambda$ EA, due to its strong truncation selection pressure, dramatically distorts the shape of the fitness distributions and steadily decreases their mean and variance. By contrast, a GA on the same problems reaches a steady state rather quickly.

The various operators used have different effects on the fitness distribution as well. There is little difference in the distortions produced by crossover and mutation in the early generations. But, as the population converges, the difference in distortions becomes more apparent (crossover fails to match any of the theoretical distributions while mutation still fits some of them quite well).

Experiments with the CA domain show the dependency of fitness distributions on the landscape. However, the interesting thing is that although the population fitness distributions for the CA problem are quite different in shape, the distortion effects due to crossover and mutation are much the same as we saw on F1. In the early generations they have pretty much the same effect, but increasingly differ in their effects as evolution proceeds.

3 Conclusions

In summary we are optimistic that the methodology presented here will prove to be a useful addition to the current set of tools for analyzing the behavior of EAs. Even the simple experiments presented here have yielded useful insights, including the fact that population fitness distributions are *seldom* observed to be normally distributed, that the shapes of these distributions are heavily dependent on both the fitness landscape and the EA selection pressure, and that the differences between the population fitness distribution distortions due to crossover and mutation are only significant in the later stages of the evolutionary process. In addition, using this methodology, one can actually verify the validity of various assumptions about fitness distributions that theoretical work in this area makes in order to keep the mathematics tractable.

References

1. Lee Altenberg. The Schema Theorem and Price's Theorem. In L. Darrell Whitley and Michael D. Vose, editors, *Foundations of Genetic Algorithms 3*, pages 23-49, Estes Park, Colorado, USA, 1995. Morgan Kaufmann.
2. R. Jain. *The Art of Computer Systems Performance Analysis*. John Wiley and Sons, Inc., New York, 1991.
3. H. Mühlenbein and D. Schlierkamp-Voosen. Predictive models for the breeder genetic algorithm. *Evolutionary Computation*, 1(1):25-49, 1993.
4. J. Shapiro, A. Prügel-Bennett, and M. Rattray. A statistical mechanical formulation of the dynamics of genetic algorithms. In Terence C. Fogarty, editor, *Evolutionary Computing, AISB Workshop*, volume 993 of *Lecture Notes in Computer Science*. Springer, 1994.

Evolutionary Feature Space Transformation Using Type-Restricted Generators

Oliver Ritthoff and Ralf Klinkenberg

Chair of Artificial Intelligence, Department of Computer Science
University of Dortmund, 44221 Dortmund, Germany
{ritthoff,klinkenberg}@ls8.cs.uni-dortmund.de
http://www-ai.cs.uni-dortmund.de/

Abstract. Data preprocessing, especially in terms of feature selection and generation, is an important issue in data mining and knowledge discovery tasks. Genetic algorithms proved to work well on feature selection problems where the search space produced by the initial feature set already contains the target hypothesis. In cases where this precondition is not fulfilled, one needs to construct new features to adequately extend the search space. As a solution to this representation problem, we introduce a framework combining feature selection and type-restricted feature generation in a wrapper-based approach using a modified canonical genetic algorithm for the feature space transformation and an inductive learner for the evaluation of the constructed feature set.

A crucial aspect for successfully solving a learning task at hand is the language in which the hypotheses, i.e. possible solutions, are represented. Two learning tasks that handle the representation problem by properly transforming an inadequate feature space are *feature selection* and *feature generation* [2]. Models of *feature selection* assume that the description language contains a superset of the features that are sufficient to describe the target hypothesis. Thus, learning comprises the selection of a feature subset that maximizes the learning performance in a classification or regression task. There are a number of heuristic feature selection strategies, incrementally choosing feature subsets that lead to the highest performance increase in one iteration. Yet, in contrast to genetic algorithms (GAs) [3], a major shortcoming of such methods is their lacking ability to cope with complex, multimodal search spaces. The main goal of *feature generation* is to reveal feature dependencies that need to be recognized to find the target concept. This is done by transforming the original feature set into a feature set more suitable for the learning task at hand.

The overall approach is structured as follows. The modified GA produces individuals by varying and recombining given feature sets and conducts the search for a good feature set using the learning algorithm for its evaluation. The training data set the learning algorithm is run on is partitioned into internal training and hold out sets, with the feature sets removed from and added to the data that were acquired in the GA search step. The process of creating feature sets, using the modified GA and evaluating these sets is repeated until a given

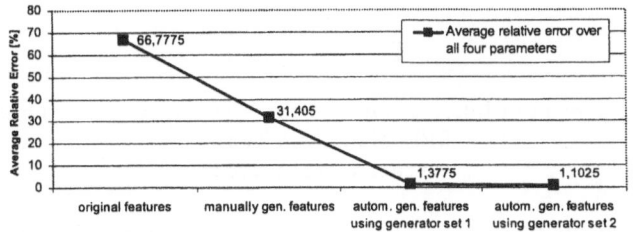

Fig. 1. Average relative error over all four parameters on a two-substance mixture

termination criterion is fulfilled. The resulting feature set is chosen as the final set on which to run the learning algorithm. The final evaluation of the resulting classifier is done using an independent test set not used during the learning step. The feature generation process is based on two classes of feature types, namely the *value type* and the *block type*, together with their ontologies, each describing a hierarchical is-superset-of relation. The *value type* specifies the data type (e.g. nominal or real) of an individual feature, whereas the *block type* contains some meta-data about the feature, e.g. if it is just an individual feature or a part of a time series. The idea is to restrict the constructible features to those matching the required types of the feature generator at hand, e.g. the "plus" generator should only be applied to numeric features, discarding e.g. all nominal features.

The experiment shown in Fig. 1, which has been conducted with our flexible learning environment YALE [1], investigates our approach focusing on the aspect of type-restrictions for feature generation in the field of chromatography. The learning task was to predict four characteristic coefficients of a two component mixture given the corresponding chromatogram time series, representing the original feature set. We compared the performance of the presented approach, automatically generating an optimized feature set using different generator sets, with a manually created feature set and the original feature set as baselines for the prediction performance. The first generator set contained the arithmetic generators *plus, minus, multiply*, and *divide* (generator set 1), the second set additionally comprised the generator *time series*, producing several function characteristics (generator set 2). Not adapting the feature space, and even manual feature generation proved to be far less effective than our automatic transformation approach in terms of predictive error. Furthermore, including domain knowledge using explicit ontology-based type-restrictions significantly limits the feature space and thus accelerates the search process.

References

1. S. Fischer, R. Klinkenberg, I. Mierswa, and O. Ritthoff. Tutorial for YALE: Yet Another Learning Environment. Technical Report CI 136/02, SFB 531, University of Dortmund, June 2002. http://yale.uni-dortmund.de/.
2. H. Liu and H. Motoda. *Feature Extraction, Construction, and Selection: A Data Mining Perspective*. Kluwer, Dordrecht, NL, 1998.
3. M. Mitchell. *An Introduction to Genetic Algorithms*. MIT Press, 1996.

On the Locality of Representations

Franz Rothlauf

Department of Information Systems 1, University of Mannheim
68131 Mannheim/Germany
franz@rothlauf.com

Abstract. It is well known that using high-locality representations is important for efficient evolutionary search. This paper discusses how the locality of a representation influences the difficulty of a problem when using mutation-based search approaches. The results show that high-locality representations do not change problem difficulty. In contrast, low-locality representations randomize the search process and make problems that are phenotypically easy for mutation-based search more difficult and phenotypically difficult problems more easy.

1 Metrics and Locality

When considering representations it must be distinguished between phenotypes x_p and genotypes x_g. Thus, an optimization problem can be decomposed into two parts. The first maps the genotypic space Φ_g to the phenotypic space Φ_p, and the second maps Φ_p to the fitness space $\mathbb{R}$:

$$f_g(x_g) : \Phi_g \to \Phi_p,$$
$$f_p(x_p) : \Phi_p \to \mathbb{R},$$

where the overall optimization problem is defined as $f = f_p \circ f_g = f_p(f_g(x_g))$. The genotype-phenotype mapping f_g is the used representation and f_p is the used fitness function.

When using search algorithms, a metric has to be defined on the search space Φ. Based on the metric the distance d_{x_a,x_b} between two individuals $x_a \in \Phi$ and $x_b \in \Phi$ describes how similar the two individuals are. The larger the distance, the more different two individuals are. Two individuals are neighbors if the distance between two individuals is minimal. For example, when using the Hamming metric for binary strings the minimal distance between two individuals is $d = 1$.

If we use a representation f_g there are two different search spaces, Φ_g and Φ_p. Therefore, different metrics can be used for Φ_g and Φ_p. In general, the metric used on the phenotypic search space Φ_p is determined by the specific problem that should be solved and describes which problem solutions are similar to each other. In contrast, the metric defined on Φ_g is not given a priori but depends on the used genotypes. As different genotypes can be used to represent the same phenotypes, different metrics can be defined on Φ_g. Therefore, in general, different metrics are used for Φ_p and Φ_g which imply a different neighborhood structure in Φ_g and Φ_p.

The locality of a representation describes how well neighboring genotypes correspond to neighboring phenotypes. The locality of a representation is high if all neighboring genotypes correspond to neighboring phenotypes. In contrast, the locality of a representation is low if some neighboring genotypes do not correspond to neighboring phenotypes.

2 Influence on Problem Difficulty

The phenotypic difficulty of an optimization problem depends on the metric that is defined on the phenotypes and the function f_p which assigns a fitness value to every phenotype. Based on the phenotypic metric a local search operator can be defined (for the phenotypes). By the use of a representation which assigns a genotype to every phenotype a new genotypic metric is introduced that depends on the used search operator and that can differ from the phenotypic metric. Therefore, the character of the search operator can be different for genotypes and phenotypes. If the locality of a representation is high, then the mutation operator has the same effect on the phenotypes as on the genotypes. As a result, genotypic and phenotypic problem difficulty is the same and the difficulty of a problem remains unchanged by the use of an additional representation f_g. Phenotypically easy problems remain genotypically easy and phenotypically difficult problems remain genotypically difficult. Figure 1 (left) illustrates the effect of mutation for high-locality representations. The search operator (mutation) has the same effect on the phenotypes as on the genotypes.

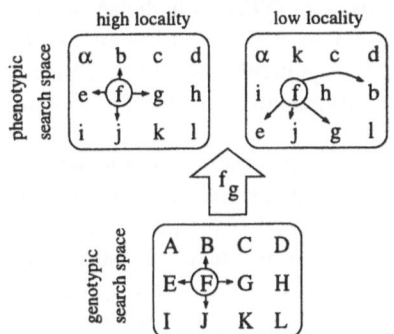

Fig. 1. The effect of mutation for high versus low-locality representations

The situation is different when focusing on low-locality representations. Then the influence of the representation on the difficulty of a problem depends on the considered optimization problem. If the considered problem f_p is easy and the structure of the search space guides the mutation-based search method to the optimal solution, a low-locality representation f_g randomizes the problem and makes the overall problem f more difficult. When using low-locality representations a small change of a genotype does not correspond to a small change of the phenotype but larger changes of the phenotype are possible (compare Fig. 1 (right)).

Finally, we have to consider deceptive and misleading problems. On average, the use of low-locality representations transform such problems into easier problems as the problems become more randomized. Therefore, mutation-based search is less misled by the fitness landscape and the problem difficulty for mutation-based search is reduced. On average, low-locality representations "destroy" the deceptiveness of the problem and make it easier.

3 Conclusions

This work discusses that representations can easily change the difficulty of problems. Only when using high-locality representations problem difficulty does not change. Phenotypically easy problems remain overall easy and phenotypically difficult problems remain overall difficult when using mutation-based search approaches. In contrast, low-locality representations randomize the search process. Therefore, problem difficulty changes and phenotypically easy problems which guide mutation-based search to good solutions become more difficult and phenotypically difficult problems that mislead mutation become more easy.

New Subtour-Based Crossover Operator for the TSP

Sang-Moon Soak and Byung-Ha Ahn

Department of Mechatronics, Kwang-ju Institute of Science and Technology
Oryong-dong, Buk-gu, Gwangju 500-712, Republic of Korea
{soakbong,bayhay}@kjist.ac.kr

Genetic algorithm (GA) is a very useful method for the global search of large search space and has been applied to various problems. It has two kinds of important search mechanisms, crossover and mutation. Because the performance of GA depends on these operators, a large number of operators have been developed for improving the performance of GA. Especially many researchers have more interested in crossover operator than mutation operator because crossover operator has charge of the responsibility of local search. We only deal with crossover operator.

In this paper we propose subtour preservation crossover (SPX), which uses a similar subtour enumeration method to other subtour-based crossovers but has an amount of difference in method that generates a valid tour. SPX generates offsprings as many as we wish by using genetic information of parents propagated over a lot of generations.

And the most severe drawback of subtour-based crossovers is they cannot generate a different offspring when two identical parents are selected for crossover. At our experiments, in case of over 200 generations, the average number of times which two identical parents are selected is over 20. That is, it shows the identical parents are selected about over 30% for total crossovers (Pc = 0.6). So if we do not consider a supplementary method for avoiding this problem, the improvement of solutions will be more and more difficult because crossover operator cannot generate new offspring.

So our method for generating new offspring is as follows. If identical parents are selected in SPX, it first generates two solutions randomly and then applies a local search method to each solution for competing with good solution existing already in population.

Figure 1 indicates the procedure of SPX. The first step of SPX is to enumerate common subtours and then reconnect each subtour using reconnection rule. Reconnection rule is as follow.

(1) Select shorter one between two alternative edges shown in parents.
(2) If shorter edge was already selected, another edge is selected.
(3) If the selected edge has the length of subtour more than 1, all elements included in the subtour are connected in turn.
(4) If both edges were already selected at previous step, random selection is performed among endpoints not selected at previous step.
(5) If all subtours are connected, reconnection procedure is terminated.

Procedure of Subtour Preservation Crossover
Begin
  Set the number of offspring (*Num_Of_Offspring*)
  Enumerate all common subtours;
  Counter_Offspring = 0;
  while *Counter_Offspring* != *Num_Of_Offspring* **do**
    Counter_Offspring = *Counter_Offspring* - 1;
    Choose an arbitrary starting city (C_i) among endpoints of subtours or cities, which do
      not construct subtour;
    if the number of subtour = 1 **then**
      Generate arbitrarily two new offspring;
      Apply a local search method to new offspring;
    else
      while termination condition (complete tour) != yes **do**
        Find the adjacent cities (C_j......C_k) of the city (C_i) in two parents;
        if common subtour (S_{it}) exists = yes **then**
          $C_i := C_t$;
        else
          Compare the distances (d_{ij}......d_{ik}) and find the shortest edge,
          $d_{it} = \min(d_{ij},......,d_{ik})$; $C_i := C_t$;
      end
    end
  end
end

Fig. 1. Subtour Preservation Crossover (SPX)

Finally we compared the proposed operator to several crossover operators for traveling salesman problem (TSP) for showing the performance of proposed crossover and introduced an efficient simple hybrid genetic algorithm using proposed operator.

From an experiment, we showed SPX to appear the features corresponding to "schema theorem" and "building block hypothesis" and pointed out the necessity of a supplementary method in subtour-based crossovers. By means of several computational experiments, it has been confirmed that SPX can get much better results than permutation based crossovers and other crossover operators using parental information. And a hybrid GA was also implemented by combining 2-opt, nearest neighbor algorithm and SPX. Though HGA was not fine-tuned, it indicated very encouraging results below 3% over the optimum in all test cases (the number of cities is from 100 to 1002).

Is a Self-Adaptive Pareto Approach Beneficial for Controlling Embodied Virtual Robots?

Jason Teo and Hussein A. Abbass

Artificial Life and Adaptive Robotics (A.L.A.R.) Lab
School of Computer Science, University of New South Wales
Australian Defence Force Academy Campus, Canberra, Australia
{j.teo,h.abbass}@adfa.edu.au

Abstract. A self-adaptive Pareto Evolutionary Multi-objective Optimization (EMO) algorithm is proposed for evolving controllers for a virtually embodied robot. The main contribution of the self-adaptive Pareto approach is its ability to produce controllers with different locomotion capabilities in a single run, therefore reducing the evolutionary computational cost significantly. The aim of this paper is to verify this hypothesis.

1 Methods

Creature Morphology: The creature is a basic quadruped with 4 short legs, with dimensions 4 x 1 x 2cm for the torso and 1 x 1 x 1cm for a limb. A leg has two limbs connected through a hinge joint and the upper limb is connected to the torso via a similar joint. Hinges rotate between 0 to 1.57 radians. Each hinge is actuated with a motor controlled using an artificial neural network (ANN).

Genotype Representation: The genotype encodes both the ANN weights and active hidden units using a real-valued matrix and binary vector respectively.

Experiment 1: SPANN – A Self-Adaptive Pareto EMO Algorithm: A modified version of the Self-adaptive Pareto Differential Evolution algorithm (SPDE) [1] is used. SPDE uses elitism through breeding children only from the current Pareto set. It uses a *differential evolution* (DE) crossover operator with the fixed step in the original DE algorithm replaced with a Gaussian step in the SPDE algorithm. The evolutionary parameters are as follows: 1000 generations, 30 individuals, maximum of 15 hidden units, 500 timesteps and 10 repeated runs.

Experiment 2: A Hand-Tuned EMO Algorithm: In this set of experiments, we used an EMO algorithm with user-defined crossover and mutation rates rather than self-adapting parameters in SPANN. Apart from the non-self-adapting crossover and mutation rates, the hand-tuned EMO algorithm is otherwise similar to SPANN in all other respects. Three different crossover and mutation rates were used: 10%, 50% and 90% in both cases giving a total of 9 different combinations. All other evolutionary and simulation parameters remain the same.

Experiment 3: A Weighted Sum EMO Algorithm: Here, we used an EMO algorithm with a single-objective that combined the two objectives using a weighted sum. Apart from the change to the manner in which the objectives are evaluated, the weighted sum EMO algorithm is otherwise similar to SPANN in all other respects. 10 different values were used for the relative weights. A $(\lambda + \mu)$ strategy is used where the 15 best individuals of the population are carried over to the next generation intact. All other parameters remain the same.

Experiment 4: A Single-Objective Evolutionary Algorithm (EA): Finally, we used a conventional EA which optimizes only one objective of maximizing the locomotion distance achieved by the ANN controller while keeping the hidden layer size fixed. As in the weighted sum EMO algorithm, the $(\lambda + \mu)$ strategy is used in this single-objective EA. Sixteen separate sets of evolutionary runs were conducted corresponding to each one of the different number of hidden units ranging from 0 to 15, which is the range allowed in the multi-objective runs.

2 Results and Discussions

A summary of the results are presented in Table 1. The total computational cost is estimated using the total number of hidden unit activations registered during the search process for each algorithm since most of the computational time is spent on the evaluation of evolved genotypes representing different ANN controllers within the physics-based simulator. The total computational cost (C) will differ between different algorithms as a function of the number of hidden unit activations required to evaluate the fitness of each newly generated genotype (A), the number of new genotypes generated per evolutionary run (G) and the number of evolutionary runs per algorithm (R), as given by $C = A \times G \times R$.

Table 1. Comparison of best locomotion distance obtained and corresponding total computational cost using SPANN against all other algorithms

Algorithm	Locomotion Distance	$+/-\%$ of SPANN	Computational Cost	$+/-\%$ of SPANN
SPANN	17.6994	-	909,520,500	-
Hand-Tuned EMO	19.5051	+15.9%	7,529,814,400	+727.9%
Weighted Sum EMO	21.8228	+23.3%	3,073,867,500	+238.0%
Single-Objective EA	22.4069	+26.5%	1,441,441,000,000	+15748.4%

In summary, although controllers evolved using the hand-tuned, weighted sum and single-objective algorithms achieved higher locomotion distances, the trade-off in terms of computational cost was staggeringly high compared to SPANN. Hence, the self-adaptive Pareto approach adopted in SPANN is computationally efficient while producing sufficiently good locomotion controllers.

References

1. Hussein A. Abbass. The self-adaptive Pareto differential evolution algorithm. In *Proceedings of the 2002 Congress on Evolutionary Computation (CEC2002)*, volume 1, pages 831–836, Piscataway, NJ, 2002. IEEE Press.

A Genetic Algorithm for Energy Efficient Device Scheduling in Real-Time Systems

Lirong Tian[1,2] and Tughrul Arslan[1]

[1] Aeronautical Computing Technology Research Institute, Xi'an, China
[2] School of Engineering and Electronics, Edinburgh University, UK
{Li-rong.Tian,Tughrul.Arslan}@ee.ed.ac.uk

1 Introduction

Power is becoming a critical constraint for designing embedded applications,because the amount of power available to these systems is limited due to limitation of battery life [1]. For this reason energy consumption is an important parameter in evaluating performance of embedded systems.DPM (dynamic power management) has gained considerable attention over the last few years as a way to save energy in device that can be turned on and off by operating system control [2].Some successful designs have been made using this technique [3,4].

Genetic algorithms are an artificial intelligence technique based on the principles of evolutionary theory. it extensively used in solving many real-world problem. In this paper, we investigate a way to implement real-time I/O-centric DPM scheduling by using GA(Genetic Algorithm), and compare the efficiency with other exsiting schedulers.

2 Problem Description

In order to save energy, when a task is running, the unused devices can stay in the sleeping mode. But frequently switching a device between the active and sleep states also results in energy consumption. A schedule for these tasks must reduce the transitions as much as possible. The scheduler we present here takes as input a predetermined set of tasks and a device usage list for each task. The output is a task execution sequence such that the power consumption of the devices is near minimized.

3 Genetic Algorithm Implementation

3.1 The Chromosome

Each individual parameter is defined as a sub-chromosome, representing the delay between start time and submit time for each job. The chromosome simply concatenate many individual parameters.

3.2 Fitness Function and the Reduction of Device Power Consumption

What we are trying to do is to reduce the device transitions between the working and the sleeping states through intelligent scheduling. From this point of view, if two or more jobs which are using the same device can be executed successively, it will reduce the transitions of the device and extend the sleep time for other devices. To find a schedule with many appearances of such a condition is our objective. The fitness is defined as the power saved by avoided transitions through reasonable schedule. It is calculated through the following eqution:

$$Fitness = 2 * P_t * T_t * Score \qquad (1)$$

$$Score = T_{max} - T_s. \qquad (2)$$

Where P_t is defined as the power consumed by wake up or shut down, T_t is defined as the time for transaction. T_{max} is defined as the maximum number of transitions without reasonable schedule, and T_s represent the transitions needed by the evaluated schedule.

4 Results and Conclusions

The shortcomings in memory and time requirements become obvious for traditional algorithms which deal with the scheduling problem as the problem size is increased. Experiment shows that our GA-based scheduler is not only attractive in less memory and time requirments, but also the well scaled feature for complex problems.

References

1. Mike Tien-Chien Lee, Vivek Tiwari, Sharad Malik, and Masahiro: Power analysis and minimization techniques for embedded DSP software. IEEE Transactions on Very Large Scale Intergration (VLSI) Systems, Vol 5, No. 1, March 1997.
2. Sandy Irani, Sandeep Shukla, and Rajesh Gupta: Competitive analysis of dynamic power management strategies for systems with multiple power saving states. Design Automation and Test in Europe Conference and Exhibition,2002. Proceedings, pp. 117–123, 2002.
3. Jinwoo Suh, Dong-In Kang,and Craga, S.P: Dynamic power management of Multiprocessor systems. Parallel and Distributed Processing Symposium, Proceedings International, IPDPS pp. 97–104, 2002.
4. J.Lorch and A.Smith: Software strategies for portable computer energy management. IEEE Personal Communication Mag, vol. 3, No.5, pp. 60–73, June 1998.

Metropolitan Area Network Design Using GA Based on Hierarchical Linkage Identification

Miwako Tsuji, Masaharu Munetomo, and Kiyoshi Akama

Hokkaido University, North 11, West 5, Sapporo, 060-0811, Japan
{m_tsuji,munetomo,akama}@cims.hokudai.ac.jp

Network design is a difficult combinatorial optimization problem which searches an optimal network configuration that satisfies geographical constraints, user traffic constraints etc. from a huge number of candidates. In solving the problem with GAs, building block (BB) destructions occur frequently which misleads the algorithms to unfavorable local optima, because appropriate encoding only with prior knowledge is not usually an easy task. To overcome this difficulty, linkage identification techniques that identify linkage groups – sets of loci tightly linked to form BBs – must be employed. In addition, some real-world problems, especially large and complex problems, seem to have interactions not only among loci but among linkage groups in a hierarchical manner. Therefore we propose a hierarchical version the $LIEM^2$ [1] (Linkage Identification with Epistasis Measure considering Monotonicity) and apply it to the network design problem.

Previous studies [3, 2] showed that to solve hierarchical problems efficiently it is essential to encode strings appropriately and to maintain diversity over schemata. For the first condition, we decompose a problem into sub-problems hierarchically as the following manner: in the initial step, our method identifies sets of loci tightly linked to make linkage groups with the $LIEM^2$. An *epistasis* measure – tightness of linkage between a pair of loci – is defined by the amount of violation of the monotonicity condition along perturbations of their variables ($0 \to 1$ or $1 \to 0$). In the following steps, it checks interdependence of the merged groups and this process continues recursively. The epistasis between linkage groups is calculated by BB-wise perturbations like $BB_1 \to BB_2$ or $BB_2 \to BB_1$, where BB_1 and BB_2 are schemata that with the largest and the second largest frequencies in the population. For instance, if linkage groups $\{0,1\}$ and $\{3,4\}$ have BB candidates 11***,00*** and ***01,***11 respectively and fitness changes along the perturbations 00*01 $\to$ 11*01$\to$ 11*11 or 00*01 $\to$ 00*11$\to$ 11*11 cause larger epistasis than other pair of linkage groups, then $\{0,1\}$ and $\{3,4\}$ should be merged. A specified number of strings are selected from the original population and the amounts of violation are calculated, and then epistasis is obtained as the maximum value of the amounts. To obtain optimal solutions in higher levels, multiple trials of the combination of the good schemata from lower levels are necessary. For this second condition, niching is introduced to maintain diversity over schemata.

We employ this hierarchical linkage identification technique to design Metropolitan Area Networks (MANs). The purpose of the network design problem is to construct a network with the lowest cost under some constraints. In our problem, geographical constraints that limits where we can lay down links (we

Table 1. cost (unit : JAY 10,000)

	average	standard deviation	minimum
uniform crossover	147753	911	146137
single layer LIEM^2	147290	1304	145292
hierarchical LIEM^2	145540	682	144517

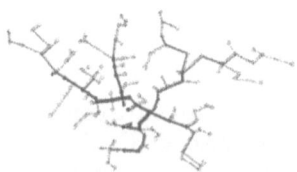
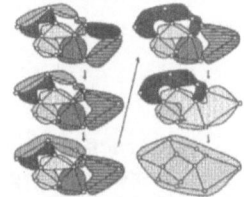

Fig. 1. test network   Fig. 2. best solution   Fig. 3. linkage groups

assume that links can be laid down only along major roads in Fig.1) and user traffic constraints must be satisfied. The i-th allele s_i of string s define whether the link is constructed or not in the i-th candidate location. Table. 1 shows results of the experiment in 20 runs and the GA with hierarchical linkage identification performed the best. The solution obtained from the proposed method is shown in Fig.2. Fig.3 shows the process of merging linkage groups in a part of the test network. Variables (candidate locations for laying down a link) merged with neighborhoods at first levels gradually become large groups that cover large areas. Although GAs only with single layer linkage identification can improve some quality of solution, GAs with hierarchical linkage identification can solve MAN problems with large solution space and complex natures more effectively.

The employed test problem is simple and real problems may have many additional matters such as future expansion, server allocations, more detailed geographical constraints, routing policies, and so on. At this point if one or more of these conditions are changed, GAs tailored to a specific network problem need to change their encoding and genetic operators. On the other hand, the proposed GA should be applicable to various kinds of network design problems because it employs information only on strings and their fitness values – general information of the problems – and doesn't depend on problem domains.

References

1. Munetomo, M. : Linkage Identification with Epistasis Measure Considering Monotonicity Conditions. In Proc. of SEAL 02 (2002), 550–554
2. Pelikan, M., Goldberg, D.E. : Escaping Hierarchical Traps with Competent Genetic Algorithms. In Proc. of GECCO-2001 (2001), 511–518
3. Watson, R.A., Hornby, G.S., Pollack, J.B. : Modeling Building-Block Interdependency. In Proc. of PPSN V (1998), 97–106

Statistics-Based Adaptive Non-uniform Mutation for Genetic Algorithms

Shengxiang Yang

Department of Mathematics and Computer Science
University of Leicester
University Road, Leicester LE1 7RH, UK
s.yang@mcs.le.ac.uk

Abstract. A statistics-based adaptive non-uniform mutation (SANUM) is presented for genetic algorithms (GAs), within which the probability that each gene will subject to mutation is learnt adaptively over time and over the loci. SANUM uses the statistics of the allele distribution in each locus to adaptively adjust the mutation probability of that locus. The experiment results demonstrate that SANUM performs persistently well over a range of typical test problems while the performance of traditional mutation operators with fixed rates greatly depends on the problems. SANUM represents a robust adaptive mutation that needs no advanced knowledge about the problem landscape.

1 Introduction

Holland's *schema theorem* states that building blocks receive an exponentially increasing trials in the subsequent generations. Usually with the progress of the GA, the frequency of 1's in the alleles of these loci where building blocks reside will eventually converge to 1 (or 0). SANUM makes use of this convergence information as feedback information to control the mutation by adjusting the mutation probability for each locus. Let $f_1(i,t)$ ($i = 1\ldots L$ where L is the string length) denote the frequency of 1's in the alleles in locus i over the population at time (generation) t and $p_m(i,t)$ ($i = 1\ldots L$) denote the mutation probability of locus i at time t. Then, $p_m(i,t)$ can be calculated from $f_1(i,t)$ as follows:

$$p_m(i,t) = P_{max} - 2 * |f_1(i,t) - 0.5| * (P_{max} - P_{min}) \qquad (1)$$

where $|.|$ is an absolute function, P_{max} and P_{min} are the maximum and minimum allowable mutation probabilities for a locus respectively. Now during the evolution of the GA, after a new population t has been generated, we first calculate $f_1(i,t)$ for each locus i over the population and from this obtain $p_m(i,t)$ for gene locus i. Then we can perform SANUM operations similarly as traditional bit mutation except that SANUM uses $p_m(i,t)$ for each locus i instead of a global mutation probability p_m for all the loci.

The motivation behind SANUM lies in the fact that with the progress of the genetic search SANUM helps protecting building blocks found so far while still exploiting new building blocks. With the progress of the GA, when building blocks are partially found, SANUM decreases the mutation probabilities of those loci where these building blocks reside according to Eq. (1). In this way, SANUM can protect building blocks found so far. While on the other hand, for those unconverged loci the mutation probabilities remain high within SANUM. This is useful because there may be building blocks not expressed on these loci yet. SANUM strikes to balance the construction of new building blocks and protection of found building blocks with time adaptively.

As the population converges, with traditional bit mutation fewer and fewer offsprings generated by mutating converged loci survive the next generation. That is, many mutation operations are wasted on converged loci. SANUM can save these wasted mutations and hence wasted fitness evaluations through adaptively decreasing $p_m(i,t)$ from P_{max} to P_{min} for those converged loci.

2 Experimental Results

In order to test SANUM, it is compared with traditional bit mutation with a series of recommended fixed probabilities: $1/L$, 0.01, $1.75/(N*L^{1/2})$ where N is the population size, and 0.001 over a range of typical test problems. Within SANUM, $p_m(i,t)$ varied adaptively between $P_{max} = 1/L$ and $P_{min} = 10^{-4}$ according to Eq. (1). In the experiment all GAs are generational and use the fitness proportionate selection with the stochastic universal sampling and elitist model, 2-point crossover with crossover probability 0.6, and the population size of 100. For each run, the best-so-far fitness was recorded every 100 evaluations. SANUM performs persistently well on the test problems. SANUM performs much better than traditional mutation operators on royal road functions R_1 and R_2 (see Fig. 1) due to the strong building blocks built in them. After certain evaluations, when the GA has built up some useful schemas, SANUM efficiently avoids mutating those converged loci, i.e., found building blocks.

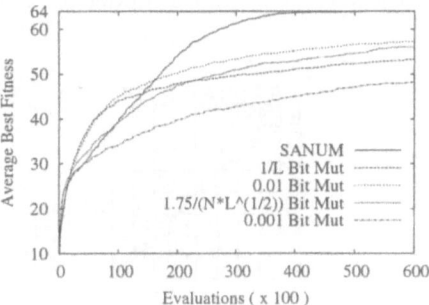 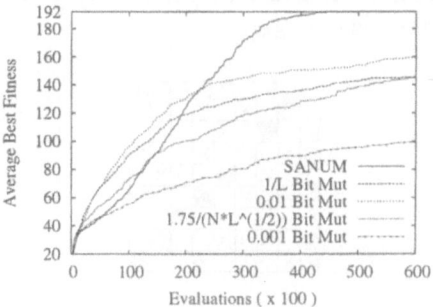

Fig. 1. Average best-so-far fitness against evaluations on Royal Road Functions (*Left*) R_1 and (*Right*) R_2. Experiment results were averaged over 100 independent runs

Genetic Algorithm Design Inspired by Organizational Theory: Pilot Study of a Dependency Structure Matrix Driven Genetic Algorithm

Tian-Li Yu, David E. Goldberg, Ali Yassine, and Ying-Ping Chen

University of Illinois at Urbana-Champaign, Urbana, IL 61801
{tianliyu,deg,yassine,ychen21}@uiuc.edu

Abstract. This study proposes a dependency structure matrix driven genetic algorithm (DSMDGA) which utilizes the dependency structure matrix (DSM) clustering to extract building block (BB) information and use the information to accomplish BB-wise crossover. Three cases: tight, loose, and random linkage, are tested on both a DSMDGA and a simple genetic algorithm (SGA). Experiments showed that the DSMDGA is able to correctly identify BBs and outperforms a SGA.

MDL-Based DSM Clustering. A dependency structure matrix (DSM) is a matrix where each entry d_{ij} represents the dependency between node i and node j. This study concentrates on the 0-1 domain, where $d_{ij} = 0$ means that there is no dependency between node i and node j, and 1 means that node i and node j are dependent. The goal of DSM clustering is to find subsets of DSM elements (i.e., clusters) so that nodes within a cluster are maximally dependent and clusters are minimally interacting.

The minimum description length principle (MDL) provides as a metric to cluster DSMs. Our MDL clustering metric is defined as follows:

$$f_{DSM}(M) = n_c \log(n_c) + \log(n_n) \Sigma_{i=1}^{n_c} cl_i + |S|(2\log(n_n) + 1),$$

where n_c is the number of clusters, n_n is the number of nodes, and cl_i is the number of nodes in the i-th cluster. The objective is to find a model M that minimizes f_{DSM}.

Fig. 1. A slightly complicated DSM. Clustering is not so obviously at first glance.

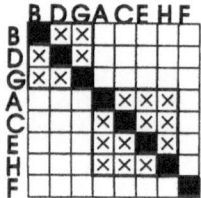
Fig. 2. After being reordered, the same DSM can be cleanly clustered as ((B,D,G)(A,C,E,H),(F)).

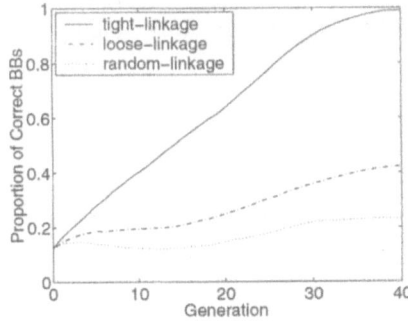
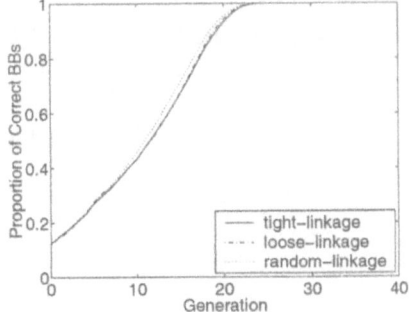

Fig. 3. The performance of the SGA with two-point crossover.

Fig. 4. The performance of the DSMDGA with BB crossover.

The DSMDGA. There are two levels of evolutionary algorithms in the DSMDGA: one is to solve the given problem, called the *primary GA*, and other one is to solve the building block (BB) identification problem, called the *auxiliary evolutionary strategy* (ES). The basic idea of the DSMDGA is to use the auxiliary ES to identify BBs, then the primary GA solves the problem by utilizing the BB information.

Empirical Results. The test function is a 30-bit Maxtrap problem composed of 10, 3-bit trap functions. Three linkage cases were tested: tight linkage, loose linkage, and random linkage. Figures 3 shows the performance of a simple GA (SGA) using two-point crossover. The SGA worked only for the tight linkage case. For loose and random linkage cases, SGA did not work because of BB disruption. Correspondingly, figure 4 illustrates the performance of the DSMDGA using BB-wise two-point crossover. The DSMDGA converged for all three tests. Even in the tight linkage test, the DSMDGA (converged at the 22th generation) outperformed the SGA (converged at the 40th generation) because the DSMDGA disrupted fewer BBs.

For more details of this study, please refer to Yu, Goldberg, Yassine, and Chen (2003).

Acknowledgment. This work was sponsored by the Air Force Office of Scientific Research, Air Force Materiel Command, USAF, under grant F49620-00-0163, the National Science Foundation under grant DMI-9908252. The U.S. Government is authorized to reproduce and distribute reprints for government purposes notwithstanding any copyright notation thereon.

Reference

Yu, T.-L., Goldberg, D.E., Yassine, A., & Chen, Y.-P., (2003), *A genetic algorithm design inspired by organization theory: a pilot study of a dependency structure matrix driven genetic algorithm.* (IlliGAL Technical Report No. 2003007). Urbana, IL: University of Illinois at Urbana-Champaign.

Are the "Best" Solutions to a Real Optimization Problem Always Found in the Noninferior Set? Evolutionary Algorithm for Generating Alternatives (EAGA)

Emily M. Zechman and S. Ranji Ranjithan

North Carolina State University, CB 7908, Raleigh, NC 27695, USA
{emzechma,ranji}@eos.ncsu.edu

1 Introduction

Evolutionary algorithms (EAs) continue to offer an effective, powerful, and sometimes exclusive way to search for solutions to real optimization problems. While these algorithms can help solve a complex optimization problem, whether the results represent the "best" choices for making decisions about a solution to a real problem is questionable. In decision-making problems that are ill posed, all objectives may not be defined clearly and therefore not quantitatively captured in the optimization model [1]. The noninferior set of solutions to the optimization model being solved may not necessarily contain the best solution to the actual problem.

The search for "good" solutions to a real optimization problem with unmodeled objectives should not be focused only on the noninferior set. Exploring the inferior region is important to make better decisions. As described in [2], the Modeling to Generate Alternatives (MGA) approach implements a systematic exploration to generate a small number of alternative solutions that are good within the modeled objective space while being maximally different in the decision space. A target constraint in the objective value is specified to allow search in a small region of the non-inferior space. Resulting alternative solutions are likely to provide truly different choices, all performing similarly with respect to the modeled objectives but differently with respect to unmodeled objectives, enabling exploration of the decision space for good solutions while considering unmodeled objectives when making decisions. The focus of this paper is to present a new EA-based approach for generating good alternative solutions for real problems with unmodeled objectives.

2 Evolutionary Algorithm for Generating Alternatives (EAGA)

EAGA is designed to generate a small number of good but maximally different alternatives, where subpopulations collectively and simultaneously search for different solutions. Each solution is represented by one subpopulation that undergoes an evolutionary search procedure. The survival of solutions in each subpopulation depends upon the performance of a solution with respect to the modeled objectives as well as upon how far that solution is from the others in decision space. The main steps of the algorithm are described below.

Step 1. Initialization – create an initial population with P subpopulations (each with a population size of K), where P is the number of alternative solutions being sought. Let SP_p (p=1, ...,P) represent the index for subpopulation p. First subpopulation (SP_1) is dedicated to the search for the optimal solution to the modeled problem, and this solution will serve as the benchmark for setting the relaxation constraint.
Step 2. In SP_1, evaluate and identify the best solution with respect to the modeled objective.
Step 3. In SP_p, p=2, ..., P, evaluate all solutions with respect to the modeled objective. Solutions that meet the target constraint are assigned a "feasible" flag, and the others an "infeasible" flag.
Step 4. Apply elitism operator to all subpopulations SP_p.
Step 5. Check for termination criteria. Stop the algorithm if termination criteria (e.g., a maximum number of iterations) are met. Otherwise, go to Step 6.
Step 6. For each SP_p, identify the centroid in decision space.
Step 7. For each solution k in subpopulation q (q≠1), calculate a distance measure $D^{k,q}$ in the decision space between that solution and other subpopulations. This distance represents the minimum distance between solution k in subpopulation q and the centroids of all other subpopulations.
Step 8. In each subpopulation SP_p, apply binary tournament selection. In SP_1, the selection is based on how good the solution is with respect to the modeled objectives. In SP_p, p≠1, the selection is based on the goodness of the solution with respect to the modeled objectives (feasibility) as well as its distance from other subpopulations ($D^{k,p}$).
Step 9. In each subpopulation, apply crossover and mutation operators to the solutions selected in Step 8, and repeat Step 2.

3 Final Remarks

The new method EAGA extends the powerful alternatives generation notions that are established in the mathematical programming and operations research literature to evolutionary search. By enabling the EA to systematically search in slightly inferior regions for maximally different solutions in the decision space, their value in offering good solutions to real problems is enhanced.

References

1. Liebman, Jon C.: Some simple-minded observations on the role of optimization in public systems decision–making. Interfaces, Vol. 6 (4), 1976, pp. 102–108
2. Brill, E. Downey, Jr.: Use of Optimization Models in Public-Sector Planning. Management Science, Vol. 25 (5). 1979, pp. 413–422

Population Sizing Based on Landscape Feature

Jian Zhang, Xiaohui Yuan, and Bill P. Buckles

Department of Electrical Engineering and Computer Science
Tulane University, New Orleans, LA 70001

Abstract. Population size for Evolutionary Algorithms is usually an empirical parameter. We study the population size from aspects of fitness landscapes' ruggedness and Probably Approximately Correct (PAC) learning theory.

1 Introduction

Evolutionary Algorithms (EAs) have been applied to various kinds of problems [1,2]. One important aspect that affects the global convergence in EA is deciding population size, which has become a focus of research in recent years [3,4]. In order to estimate population size in real-coded EAs, we employ PAC learning theory and propose concepts on characterizing the fitness landscapes.

2 Population Size of Evolutionary Algorithms

The intuitive notion of *ruggedness* is related to the difficulty of optimizing over a given landscape [5]. Several distinctive approaches have been proposed to quantify ruggedness [6,7,8,9].

Intuitively, if the initial population of real-coded EAs is large enough to include points that are in the neighborhood of the global optimum, EAs have a great chance of locating the target point in subsequent generations. It is therefore necessary to define a metric inferring closeness to optima points.

Definition 1. *Granularity of Fitness Function*
A fitness function $f(x)$ can be represented by the linear combination of a set of orthonormal basis functions $\{\varphi_k(x)\} = \{\varphi(k_1 x), \varphi(k_2 x), \ldots, \varphi(k_n x)\}$, that is $f(x) = \sum_k a_k \varphi_k(x)$. The granularity of fitness function f is defined as $\tau = 1/max(k_i), i = 1, \ldots, n$, where k_i characterizes the frequency information of the basis function $\varphi(k_i x)$.

In order to infer the nearness of organisms in real value domain, we adopt the definition of ϵ-cover provided by Vidyasagar [10].

Theorem 1. *Given the fitness function $f(x)$ with granularity τ, where $f(x)$ is defined on $S \subset R^n$, the PAC population size m is bounded by $\tilde{m}$, that is, $m \geq \tilde{m}$*

$$\tilde{m} = \lceil \frac{1}{\phi}(ln\lceil \frac{1}{\phi}\rceil + ln\frac{1}{\delta})\rceil \tag{1}$$

where $\phi = g(\tau)/S, g(\tau) = \tau$ when $n = 1$; $g(\tau) = \pi\tau^2/4$ when $n = 2$. $\lceil 1/\phi \rceil$ defines the size of hypothesis space, such that with confidence δ, $0 < \delta < 1$, the initial population forms an ϵ-cover of S with probability greater than $1 - \delta$ and $\epsilon = \tau$.

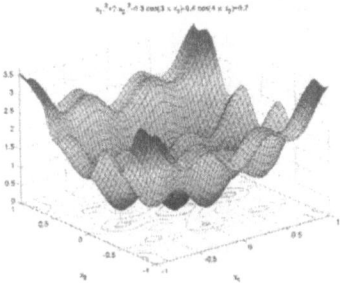 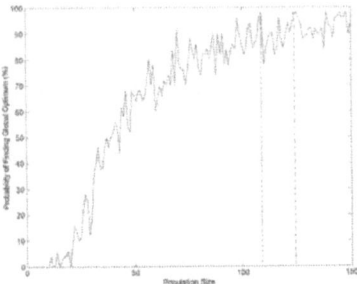

Fig. 1. Convergence graphs for Bohachevsky function: $f(x_1, x_2) = x_1^2 + 2x_2^2 - 0.3\cos(3\pi x_1) - 0.4\cos(4\pi x_2) + 0.7$, $x_1, x_2 \in [-1, 1]$.

Assume the initial population is drawn according to a uniform distribution. If the initial population forms an ϵ-cover of the solution space, then no point in that space is more than ϵ-away from a member in ϵ-cover. Several approaches for measuring the ruggedness have been considered, such as the number of local minima and the correlation length. We adopted a granularity measure τ from the decomposition of fitness landscapes.

Figure 1 shows that enlarging the population size increases the probability of finding the global optimum. However, after it reaches a certain point, the improvement is no longer dramatic. The PAC population size is at the threshold that gives high convergence rate and minimizes the computational expense.

References

1. Dasgupta, D., Michalewicz, Z., eds.: Evolutionary Algorithms in Enigneering Applications. Springer, Berlin, Germany (1997)
2. Zhang, J., Yuan, X., Buckles, B.P.: An evolution strategies based approach to image registration. In: Proceeding of Genetic Evolutionary Computation Conference, New York (2002)
3. Goldberg, D.E., Rudnick, M.: Genetic algorithms and the variance of fitness. Technical report no. 91001, Illinois GAs laboratory, University of Illinois, Champaign-Urbana (1991)
4. Hernández-Aguirre, A., Buckles, B.P., Martinez-Alcantara, A.: The PAC population size of a genetic algorithm. In: the 12th Int'l Conf. on Tools with Artificial Intelligence. (2000) 199–202
5. Jones, T., Forrest, S.: Fitness distance correlation as a measure of problem difficulty for genetic algorithms. In Eshelman, L., ed.: Proceedings of the Sixth International Conference on Genetic Algorithms, San Francisco, CA, Morgan Kaufmann (1995) 184–192
6. Sorkin, G.: Combinatorial optimization, simulated annealing and fractals. No. 61253, IBM Research Report RC13674 (1988)
7. Weinberger, E.D.: Correlated and uncorrelated fitness landscapes and how to tell the difference. Biological Cybernetics **63** (1990) 325–336
8. Kauffman, S.A., Levin, S.: Towards a general theory of adaptive walks on rugged landscapes. Journal of Theoretical Biology **128** (1987) 11–45
9. Palmer, R.: Optimization on rugged landscapes. Molecular Evolution on Rugged Landscapes: Proteins, RNA, and the Immune System (1991) 3–25
10. Vidyasagar, M.: A Theory of Learning and Generalization. Springer (1997)

Structural Emergence with Order Independent Representations

R. Muhammad Atif Azad and Conor Ryan

Department Of Computer Science And Information Systems
University of Limerick
Ireland
{Atif.Azad,Conor.Ryan}@ul.ie

Abstract. This paper compares two grammar based Evolutionary Automatic Programming methods, Grammatical Evolution (GE) and Chorus. Both systems evolve sequences of derivation rules which can be used to produce computer programs, however, Chorus employs a position independent representation, while GE uses polymorphic codons, the meaning of which depends on the context in which they are used.
We consider issues such as the order in which rules appear in individuals, and demonstrate that an order always emerges with Chorus, which is similar to that of GE, but more flexible.
The paper also examines the final step of evolution, that is, how *perfect* individuals are produced, and how they differ from their immediate neighbours.
We demonstrate that, although Chorus appears to be more flexible structure-wise, GE tends to produce individuals with a higher neutrality, suggesting that its representation can, in some cases, make finding the perfect solution easier.

1 Introduction

Chorus [8] is a Genetic Programming (GP) [4] type evolutionary algorithm that evolves programs. It belongs to the same family of algorithms as Grammatical Evolution (GE) [9] and like [2,3,5], uses grammars to evolve the programs.

As with GE, Chorus recognizes a distinction between genotype and phenotype. The genotype is a string of 8 bit integers. It is translated into a high level language program described by a context free grammar given in Backus Naur Form (BNF) using a genotype-phenotype mapping process.

The nature of the mapping process is such that the functionality of the genes is almost never tied with their exact locations on the chromosome. This is the crucial difference between Chorus and GE, as they share many similar traits, as described in Sect. 4. Position independence not only allows the system to evolve the constituents of its individuals, but it also gives the flexibility to evolve their order.

This study is an investigation into any underlying trends that the system may have in forming its individuals and compares it with GE. The study examines the patterns that emerge in the chromosomes over the course of evolution, and considers their consequences. In particular, we wish to investigate if an order can emerge using the position independent representation of Chorus. Emergence of an order in a position independent system can also be interesting if it can be correlated with building block formation. The individuals are also tested for robustness to see the amount of neutrality available in the search space surrounding them. Finally, we also look into the leap the system has to make to reach the ideal fitness. For this purpose we analyse the parents of the ideal solutions and compare them with their perfect offsprings.

The paper first gives a brief introduction to Backus Naur Form. Next, a brief introduction to GE is followed by a description of Chorus. Section 5 then describes the genetic operators used by the systems. The following section describes the experimental setup to carry out this study and comments on the results obtained from the experiments.

2 Backus Naur Form

Backus Naur Form (BNF) is a notation for describing grammars. A grammar is represented by a tuple $\{N, T, P, S\}$, where T is a set of terminals, i.e. items that can appear in legal sentences of the grammar, and N is a set of non-terminals, which are interim items used in the generation of terminals. P is the set of production rules that map the non-terminals to the terminals, and S is a start symbol, from which all legal sentences may be generated.

Below is a sample grammar, which produces individuals similar to those used by Koza [4] in his symbolic regression and integration problems.

```
S = <expr>

<expr>    ::=  <expr> <op> <expr>      (0)
          |  ( <expr> <op> <expr>)(1)
          |  <pre-op> ( <expr> )       (2)
          |  <var>                     (3)
<op>      ::= + (4) | - (5) | % (6)
          | * (7)
<pre-op>  ::= Sin (8) | Cos (9)
          | Exp (A)| Log (B)
<var>     ::= 1.0 (C) | X (D)
```

3 Grammatical Evolution

Grammatical Evolution (GE)[9] evolves computer programs in an arbitrary language. Unlike GP, as described in [4], which typically uses abstract syntax trees, GE recognizes a distinction between genotype and phenotype. Genotypes are 8 bit integer (typically referred to as a codon) strings of varying length. Translation to the phenotype is dictated by a context free grammar.

One of the distinguishing features of GE is *intrinsic polymorphism*, that allows it to possess a genotype without non coding regions(introns). Whenever a rule has to be chosen for a non-terminal, a codon is **mod**ed against the number of rules pertaining to that non-terminal. This ensures that every codon produces a rule that is immediately consumed. A complete description of GE can be found in [6].

4 The Chorus System

In a manner similar to GE, Chorus uses 8 bit integer strings to represent the genotypes. However, unlike the intrinsically polymorphic codons of GE, each codon in Chorus corresponds to a particular rule in the grammar. In this way it is similar to GAGS[5], but it does not have the same issue with introns.

Fixed valued codons are different from GE, where the meaning of every codon is dependent upon its predecessors. The 8 bit codon is **mod**ed with the total number of rules in the grammar so as to point to a rule from the grammar. This behaviour differs from GE, where the interpretation of a codon depends upon the state of the mapping process. As described earlier, GE **mod**es a codon with the number of rules relevant to the non-terminal being processed. Depending upon the state of the mapping process, the same codon can be read in for different non-terminals. This suggests that the behaviour of a codon is determined by its situation in the chromosome.

As the total number of rules in the grammar remains constant irrespective of the state of the mapping process, a codon in Chorus behaves in a fixed manner regardless of its location in the genome.

When the mapping begins, the genome is read from left to right, to pick the rules from the grammar so as to arrive at a legal sentence comprising of all terminals. However, fixed valued codons may not necessarily be in the order the mapping process may demand. We keep track of all the rules that we come across in a *concentration table*. The concentration table has an entry for every rule in the grammar and is initialised to all zeros when the mapping begins. Every time a codon is read, concentration of the corresponding rules is incremented in the table.

Consider a sample individual, which has been **mod**ed to give the codons: 8 D D C 2 3. If we consider the grammar given in Sect. 2, the start symbol is <expr> which maps to any of the rules 0 through 3 . While we traverse through the genome to find an *applicable* rule, we can find rules that may be used later. We keep track of all the rules that we come across in the concentration table.

The table is divided into different sections, each pertaining to a particular non-terminal. As the grammar in the Sect. 2 has four non-terminals, the table has four different sections. The first section contains entries for rules 0 through 3, the second consists of rules 4 through 7 and so on. When we have to find a rule corresponding to a non-terminal, we consult the *relevant* section from the table. The rule with the highest concentration at that time is chosen. In case of a tie (e.g at the beginning of the mapping) we read the genome left to right incrementing the concentration of every rule we read in. The reading stops when one of the rules from the relevant section becomes a clear winner. When required, subsequent scanning of the genome will continue from that point.

For the current example, we stop when we encounter rule 2 so that next reading position points to rule 3. The rule with the maximum count in the relevant section (rule 2 in this case) is considered dominant and is chosen, so that the start symbol is expanded into <pre-op>(<expr>). We always stick to the left most non-terminal, a <pre-op> in this case. Rule 8 is a clear winner for this case, as it is the only applicable rule that has been read in. We do not need to read from the genome in this case. Thus, we see that a codon may not be consumed the moment it is read in. This delay in the expression of a codon, combined with the requirement to be in majority to make a claim brings the position independence in the system. The important thing is the presence or otherwise of a codon and its location is less so. For more information, consult [8][1].

5 Genetic Operators

The binary string representation of individuals effectively provides a separation of search and solution spaces. This permits us to use all the standard genetic operators at the string level. Crossover is implemented as a simple, one point affair, the only restriction being that it takes place at the codon boundaries.

Bit mutation is employed at a rate of 0.01, whereas crossover occurs with a probability of 0.9. Steady state replacement is used with roulette wheel selection.

As with GE, if an individual fails to map after a complete run through the genome, Chorus uses a wrapping operator to reuse the genetic material. However, the exact implementation of this operator has been kept different from the traditional GE implementation [9]. Repeated reuse of the same genetic material effectively makes the wrapped individual behave like multiple copies of the same genetic material stacked on top of each other in layers. When such an individual is subjected to crossover, the stack is broken into two pieces. When linearized, the result of crossover is different from one or both of its parents at regular intervals. In order to minimize such happenings, Chorus limits wrapping to the initial generation. If an individual undergoes wrapping, it is then *unrolled* by appending all the genetic material at the end of the genome the number of times it is reused. The unrolled individual then replaces the original individual in the population. This altered use of wrapping in combination with position flexibility, promises to maintain the exploitative effects of crossover. Unlike the traditional implementation of GE, those individuals that fail to map in second and subsequent generations are not wrapped, and are simply considered unfeasible.

All the experiments described in this paper employ the altered wrapping both for GE and Chorus, except where mentioned otherwise.

6 Experiments

This paper makes a comparison between the structures of the individuals that are evolved by the two grammar based evolutionary algorithms. To facilitate the cause, we have employed a few different approaches. In the first approach, termed *Segment Rule Profiles*, the individuals are divided into ten segments to see if there are certain kinds of rules dominating certain regions in the genomes. This can help us understand the relationship between the genomes and the derivation trees they encode.

As is the nature of the mapping in both the algorithms, all the genetic material available in the genomes may not be required for the mapping. Instead, a certain percentage of the chromosome may be read in. This study only monitors the *effective* lengths of the genomes that contribute towards the mapping. Considering only effective length is also motivated by the fact that, even though fixed valued codons found in the *tails* or unused portions of Chorus individuals can be sampled, intrinsically polymorphic codons in GE can not be interpreted so as the mapping terminates before reading them in.

We have used two problems to make such a comparison. The first problem is Koza's[4] symbolic regression of the quartic polynomial $(x^4 + x^3 + x^2 + x)$. In order to evolve such a function, rules 0,1,3,4,7 and D from the grammar in Sect. 2 play a significant role. However, if we can use a problem whose solution requires less number of rules, we can have a clearer view of any underlying trends. For this purpose we have used an abbreviated form of the same problem, where the objective function is $x^4 + x^3$.

We then move on to compare the robustness of the individuals produced by each of the algorithms. In this test we compare the perfect individuals produced for the two problems against their nearest neighbourhoods. Bit string representation means that any individual having a hamming distance of 1 exists in the nearest neighbourhood of the individual under observation. Percentage fall off in the fitness is recorded by generating all the nearest neighbours. This provides us with a test of neutrality around the perfect solutions found by the systems.

We also compare the similarity in the rule history of the perfect individuals with their immediate neighbours. It gives us a measure of similarity in structure in the space immediately surrounding those individuals.

We then use the notion of *productive parents* [7], that is, parents that produce perfectly fit offspring, and considers the relationship between these parents and the offspring in terms of the leap in fitness required to get to the ideal fitness.

All the experiments involve 100 independent runs with a population size of 500 spanning 250 generations. Runs terminated upon finding an ideal fitness. All fitness comparisons involve normalized fitness as described in [4]. The experimental setup is primarily the same as used by Koza[4] for the symbolic regression problem.

6.1 Segment Rule Profiles

In GE, every sampled codon decodes to an applicable rule, so the effective length of an individual can be seen as a history of the rules selected.

Position independence in Chorus, however, may not permit such minimal individuals. Effective length in a Chorus individual may possess introns as well as high counts of the competing rules trying to maximise their concentrations to stake a claim. Thus, we have also recorded rule selection histories of the Chorus Individuals to see what kind of derivation trees are being produced.

The figures reported in this section show a snapshot of the 50th generation. It was observed that the rule percentiles reasonably stabilised upon reaching the 50th generation.

Figure 1 shows the rule percentiles in the effective lengths of the Chorus individuals for both the problems. As mentioned earlier, the figures show that the system has figured out the necessary constituents of the ideal solutions, depicted by their higher percentages in different segments. However, some of these significant rules show clearly dominating percentiles in certain segments,

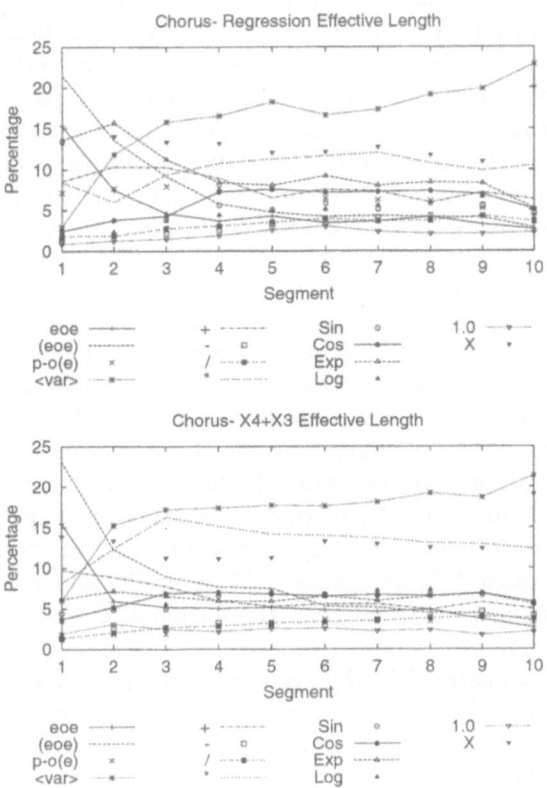

Fig. 1. *Rule percentages across* the 10 segments for the effective length of Chorus Individuals

and a low profile in the others. Let's consider the regression problem first. If we consider segment 1, the rules <expr><op><expr> (rule 0 in the grammar in Sect. 2) and (<expr><op><expr>) (rule 1 in the grammar) constitute about 38% of the profile. However, as we move to the right, the percentages of these rules decrease consistently up until segment 5, maintaining a relatively stable proportion thereafter. <var>, that links the aforementioned rules to the terminals, has only 2.935% share in segment 1, but as the percentiles of those rules decrease, <var> shows a clear tendency to increase attaining its maximum value in the last segment.

The terminal symbol X scores about 13% in segment 1, and does not show as much variation as the rules discussed earlier, except in the last segment.

The two operators + and ∗ start off at the same level. + then takes over briefly for the earlier segments, before ∗ takes the lead in later segments, emphasizing its increased role towards the end of the derivation tree.

Figure 1 shows that the artificial problem shows a trend very similar to the regression problem. The reduced effect of the + operator gives us a much clearer picture of the same phenomenon.

It appears from the findings that the system arranges its genomes in a way that the earlier part primarily consists of the rules that define the structure of the derivation tree and lead to increase in its size. This dominance is further enhanced by the low percentage of the rule <var> in the earlier segments, thus ensuring that even a high percentage of the terminal symbol X, will not hinder the structure formation. This is a consequence of the property of the system, which not only allows it to evolve the values of its genes but also the order in which the chromosome may possess them. This can be helpful in forming building blocks by allowing the rules of conflicting roles towards the derivation tree formation to exist in close vicinity.

Figure 2 shows a comparison of the rule selection histories of the two algorithms on the regression problem. The figure for Chorus correlates well with the earlier figure showing a high percentile (about 80%) of the structure imposing rules in earlier segments. <var> has a much lower appearance with terminal symbol X being almost non-existent with a share of 0.257%.

GE also shows a clear dominance of the structure forming rules in the first segment. However, there are two noticeable exceptions. The combined percentage of rule 0 (<expr><op><expr>) and rule 1 ((<expr><op><expr>) is much lower compared to Chorus(about 48% as against 80%). Also, the percentage of <var> is much higher, about 18% as against 7.277% in Chorus. The figure is unable to show clearly that the percentage of the terminal X is also higher for GE in segment 1 at 7.312%.

The differences observed in the regression problem become even clearer in the artificial problem (Fig. 3). There is an increase in the difference in combined percentages of rule 0 and rule 1 between GE and Chorus. Also, <var> has an increased percentile with GE.

The results of the rule selection history seem to suggest that, while, the individuals produced by Chorus tend to create thinner and deeper derivation trees, GE individuals encode broader trees (see Fig. 4). This fact can be attributed

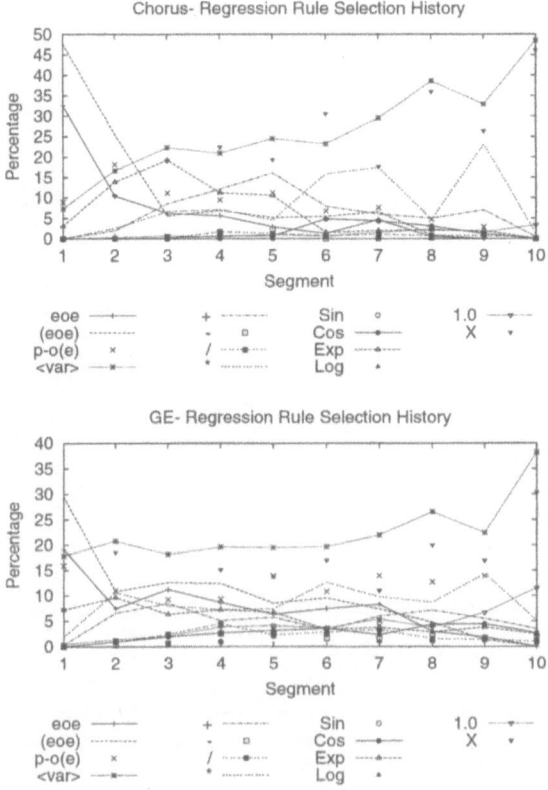

Fig. 2. A comparison of the choices made by GE and Chorus across the 10 segments for the Regression problem

to the depth first mapping employed by both the algorithms. As, earlier on, rules 0 and 1 dominate in Chorus, depth first mapping uses them to increase the depth of the tree. The phenomenon is also visible in GE, but given that it possesses <var> with a higher percentage in comparison with Chorus, it may possess broader trees as depicted by Fig. 4. This may have implications towards crossover. We get an impression that Chorus tends to preserve the structure in the earlier part of the genome. Also, the earlier portions possess the terminals that may not have been expressed so far. Thus, while the incoming fragment may contain terminals of its own, it may also be expected to trace a path to the terminals already contained by the receiving individual. Crossover, thus, can potentially behave in a more context preserving fashion.

GE, however, can have more irregularly shaped trees exchanged, possibly leading to a more exploratory behaviour. However, a study focused on the derivation trees can be more conclusive in uncovering such trends.

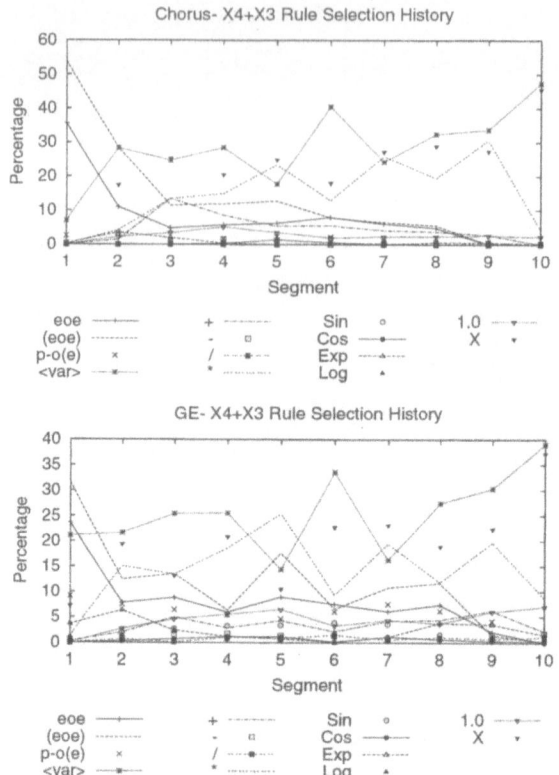

Fig. 3. A comparison of the choices made by GE and Chorus across the 10 segments for the artificial problem

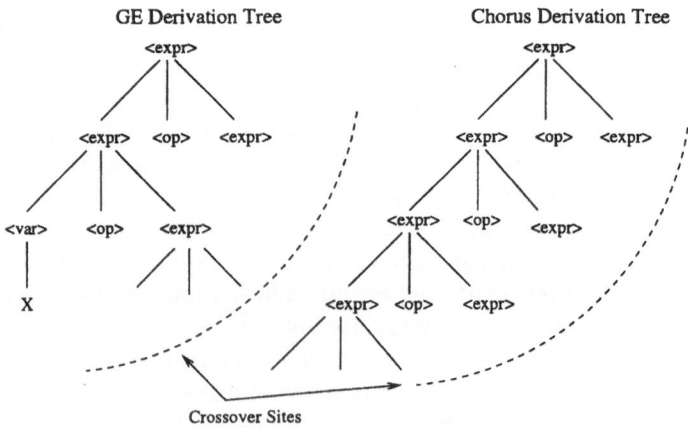

Fig. 4. Partial derivation trees for Chorus and GE

6.2 Fitness Fall Off and Rule History Similarity in the Neighbourhood

In this section we analyse the robustness of the individuals produced by the two algorithms. We consider 32 perfect individuals produced against each problem by GE and Chorus, and generate all of their immediate neighbours. As mentioned earlier, the neighbours are generated only for the effective lengths. This is so because mutations in the tails, will give us exactly the same rule selection histories thus leading to the same phenotypes. This will give a false sense of neutrality in terms of percentage fall off in fitness as well as the similarity in the rule histories. However, the neighbours may require a longer genome length in order to map completely. Thus, we have done two kinds of test. In one case, we allow the neighbours to use the tails if they require them. In the other case we do not allow them to go beyond the effective length of the corresponding perfect individual.

Table 1 explores the neighbourhoods for neutrality in fitness values. The leftmost column shows the fall off in the fitness. The entries in the table show the percentage of immediate neighbourhood lying in a range of fitness fall off. If we consider the regression problem, we can see that Chorus shows 35.017% of the immediate neighbours having perfect fitness even without being allowed to use the tails. With the use of tails, the figure slightly rises to 37.71%. GE shows higher fitness neutrality at about 41.059%. Allowing the use of tails increases this figure very slightly.

The bottom of the table also shows better figures for GE, with Chorus showing high percentage of individuals getting a zero fitness.

The artificial problem also shows that GE seems more robust to handle bit mutation. The allowance to use the tails helps Chorus to decrease the difference in performance. GE, however, remains largely unaffected by it.

Table 2 shows the similarity between the rule histories of the perfect individuals and their neighbours. It shows that most of the neutrality shown in fitness

Table 1. Percentage fall off in Fitness of Immediate Neighbours

%age	Chorus-Regression		GE-Regression		Chorus-$X^4 + X^3$		GE-$X^4 + X^3$	
	No Tails	Tails	No Tails	Tails	No Tails	Tails	No Tails	Tails
0	35.017	37.71	41.059	41.332	33.789	39.129	41.334	41.334
10	0	0	0	0	0	0	0	0
20	0	0	0	0	0	0	0	0
30	0	0	0	0	0	0	0.018	0.018
40	0	0	0	0	0	0	0.263	0.263
50	0	0	0	0	0	0	0.105	0.105
60	0	0	0	0	0	0	0	0
70	0.0060	0.0060	0.174	0.186	0.018	0.072	0	0
80	6.423	7.188	5.543	5.63	10.391	13.523	9.506	9.664
90	4.031	5.187	4.849	5.146	8.789	10.961	9.979	10.662
100	54.524	49.909	48.375	47.706	47.013	36.314	38.796	37.955

Table 2. Percentage similarity in the Rule Selection History of Immediate Neighbours

%age	Chorus-Regression		GE-Regression		Chorus-$X^4 + X^3$		GE-$X^4 + X^3$	
	No Tails	Tails	No Tails	Tails	No Tails	Tails	No Tails	Tails
0	3.81	3.815	0.409	0.409	4.861	4.843	0.35	0.35
10	2.625	2.613	0.31	0.31	3.829	3.811	0.683	0.683
20	4.303	4.32	0.608	0.608	3.738	3.729	0.735	0.735
30	4.734	4.722	0.57	0.57	3.376	3.222	1.296	1.296
40	4.649	4.546	0.471	0.471	4.001	3.747	0.63	0.63
50	7.347	7.381	0.818	0.818	6.725	6.988	1.576	1.576
60	6.803	6.429	1.128	1.128	7.902	7.476	2.714	2.714
70	5.72	5.811	3.137	3.137	4.435	4.797	7.143	7.143
80	6.281	5.726	11.471	11.471	10.889	7.594	9.191	9.191
90	19.881	18.974	45.3	45.3	18.773	18.049	41.877	41.877
100	33.849	35.663	35.776	35.776	31.472	35.744	33.806	33.806

is basically due to the similar derivation sequences. For instance, in regression problem 33.849% of the neighbours show complete neutrality for Chorus, as against a fitness neutrality of 35.017% (see Table 1), when tails are not allowed. The bulk of fitness neutrality for GE is also derived from neutrality in derivation sequences. However, the difference in 100% fitness neutrality (41.059%) (see Table 1) and 100% rule history similarity(35.776%) is higher compared to Chorus. It suggests that GE has more individuals in the neighbourhood which may be somewhat different in appearance, yet they are able to reach the ideal fitness. A similar trend is observable in the artificial problem.

The results suggest that GE is more tolerant to bit mutation as compared to Chorus. It is helped by intrinsic polymorphism that provides more neutrality in GE codons. A codon is moded only with the number of rules pertaining to the non-terminal which has to be mapped. Every non-terminal in the grammar given in Sect. 2 has 4 rules except <var>, which has 2. This means an 8 bit codon can represent the same rule in 64 different ways. The figure rises to 128 for <var>.

Chorus, on the other hand, uses fixed valued codons by moding them with the total number of rules in the grammar(14 in this case). This means every codon can represent a rule in about 18 ways, a figure considerably less than that of GE.

6.3 Productive Parents

In this section we compare the fitness of the parents of the ideal solutions to their offsprings. This gives us an idea about the relationship between the parents and their offsprings in terms of fitness.

Table 3 shows that on average the parents are quite distant from their offsprings in the fitness landscape. As is the case in neighbourhood tests, GE does better by evolving parents that are closer to the ideal solutions in terms of fit-

Table 3. Percentage distance between the fitness of productive parents and the ideal fitness. A low score indicates a higher fitness

	Chorus		GE	
	Regression	$X^4 + X^3$	Regression	$X^4 + X^3$
Maximum	91.74	92.95	83.55	82.86
Average	77.11	79.07	66.53	73.8
Minimum	49.31	41.46	21.87	31.7

ness. This suggests that GE has a relatively smoother approach towards the ideal solutions.

Artificial problem reported that 33% of the Chorus parents and 59% of GE parents encoded for the building block $x^3 + x^2$ which is 75% away from the ideal fitness. While it shows the ruggedness of the search space, it is interesting to see how the systems managed to find their way through low fitness area instead of getting trapped into highly fit but possibly misleading areas of search space.

7 Conclusions

Chorus and Grammatical Evolution are from the same family of algorithms, and clearly have much in common as a result. This paper investigates and compares the two algorithms in terms of the individuals they produce. There appears to be a convergence in behaviour, yet some differences have been pointed out. The segment profile tests show that, despite Chorus possessing position independence, an order seems to settle down as the evolution progresses. In particular, the genes that deal with structural information, tend to be at the start of individuals.

This emergent order shows that the system has the potential for building block formation, and produces individuals more conducive for crossover operations. A study focused on derivation trees can shed further light in exploring the effects of crossover.

Intrinsic polymorphism enables GE to have more tolerance towards bit mutation compared to fixed valued codons, while Chorus individuals require significantly larger codons to achieve the same neutrality.

A result of this is that GE also seems to enjoy a smoother passage towards the ideal fitness in terms of fitness distance between the productive parents and the offsprings. Further investigation into the roles of the genetic operators can help us understand it better.

References

1. Azad, R.M.A., Ryan C., Burke, M.E., and Ansari, A.R., A re-examination of The Cart Centering Problem using The Chorus System. In the proceedings of *Genetic and Evolutionary Computation Conference (GECCO), 2002* (pp. 707–715), Morgan Koffman, NYC, NY, 2002.

2. Freeman, J.J., "A Linear Representation for GP using Context Free Grammars" in *Genetic Programming 1998: Proc. 3rd Annu. Conf.*, J.R. Koza, W. Banzhaf, K. Chellapilla, K. Deb, M. Dorigo, D.B. Fogel, M.H. Garzon, D.E. Goldberg, H. Iba, R.L. Riolo, Eds. Madison, Wisconsin: MIT Press, 1998, pp. 72–77.
3. Keller, R. and Banzhaf, W., "GP using mutation, reproduction and genotype-phenotype mapping from linear binary genomes into linear LALR phenotypes" in *Genetic Programming 1996: Proc. 1st Annu. Conf.*, J.R. Koza, D.E. Goldberg, D.B. Fogel, and R.L. Riolo, Eds. Stanford, CA: MIT Press 1996, pp. 116–122.
4. Koza, J.R., Genetic Programming: On the Programming of Computers by Means of Natural Selection, MIT Press, Cambridge, MA, 1992.
5. Paterson, N., and Livesey, M., "Evolving caching algorithms in C by GP" in *Genetic Programming 1997: Proc. 2nd Annu. Conf.*, MIT Press, 1997, pp. 262–267. MIT Press.
6. O'Neill M. and Ryan C. Grammatical Evolution. *IEEE Transactions on Evolutionary Computation.* 2001.
7. O'Sullivan, J., and Ryan, C., An Investigation into the Use of Different Search Strategies with Grammatical Evolution. In the proceedings of *European Conference on Genetic Programming (EuroGP2002)* (pp. 268–277), Springer, 2002.
8. Ryan, C., Azad, A., Sheahan, A., and O'Neill, M., No Coercion and No Prohibition, A Position Independent Encoding Scheme for Evolutionary Algorithms – The Chorus System. In the Proceedings of *European Conference on Genetic Programming (EuroGP 2002)* (pp. 131–141), Springer, 2002.
9. Ryan, C., Collins, J.J., and O'Neill, M., Grammatical Evolution: Evolving Programs for an Arbitrary Language, in *EuroGP'98: Proc. of the First European Workshop on Genetic Programming* (Lecture Notes in Computer Science 1391, pp. 83–95), Springer, Paris, France, 1998.

Identifying Structural Mechanisms in Standard Genetic Programming

Jason M. Daida and Adam M. Hilss

Center for the Study of Complex Systems and the Space Physics Research Laboratory
The University of Michigan, 2455 Hayward Avenue, Ann Arbor, Michigan 48109-2143

Abstract. This paper presents a hypothesis about an undiscovered class of mechanisms that exist in standard GP. Rather than being intentionally designed, these mechanisms would be an unintended consequence of using trees as information structures. A model is described that predicts outcomes in GP that would arise solely from such mechanisms. Comparisons with empirical results from GP lend support to the existence of these mechanisms.

1 Introduction

Aside from crossover and selection, are there other mechanisms that might determine a broad variety of phenomena in standard GP?

At first glance, the answer might be "probably not." GP uses Darwinian natural selection and genetic crossover as metaphors for computation. Its design lineage stems from genetic algorithms (GA). Its processes are more similar with other algorithms in genetic and evolutionary computation than they are different. The question suggests the unusual possibility that there are additional mechanisms at work beyond that which has been intentionally engineered. In other words, given a standard GP system that has the mechanisms of crossover and selection and not much else, the question insinuates that other fundamental mechanisms might exist.

Even upon closer scrutiny, the answer might still be "probably not," although there has been previous work that suggests that there are other considerations that need to be made. The line of investigation by Soule and Foster (e.g., [1]) implied that something else is at work, while Langdon et al. has hypothesized that the causes for what Soule and Foster have found was a consequence of a random walk on the distribution of tree shapes (see [2, 3]).

However, according to this paper, the answer would be "yes, additional mechanisms do exist." We offer a hypothesis that identifies several of these mechanisms, all of which exist at a layer of abstraction below that of the biological metaphors that have given rise to GP. We further argue that not only do these mechanisms exist, but that these mechanisms can profoundly determine the extent to which GP can search.

This paper is organized as follows. Section 2 qualitatively describes our hypothesis. Section 3 describes a model that is a consequence of this hypothesis. Section 4 details the consequences of applying the model to GP search space and offers predictions of GP behavior in this search space. Section 5 compares the predictions with empirical data. Section 6 concludes.

2 A Hypothesis

2.1 Trees as Information Structure

Our hypothesis involves the existence of mechanisms that are, on one hand, pervasive and are fundamental to standard GP and many of its derivatives. On the other hand, our hypothesis requires the discovery of mechanisms that are unknown. This discovery would need to take place in an engineered system in which every single aspect of that system has been open to inspection, and has been so for over a decade. How is it possible that mechanisms exist if we never put them there in the first place?

The answer, surprisingly, stems from the choice of using trees as an information structure. In evolutionary computation, the issue of trees would typically fall under the category of *problem representation* (e.g., [4]). One of the basic considerations in evolutionary computation is the matter of choosing a representation that is suited for solving a particular problem [5]. The manner in which one represents an appropriate solution—say, a vector of real values—in turn drives the type of operators, the type of objective function, even the type of evolutionary algorithm to be used [6]. Representations, and their corresponding evolutionary algorithms, are graded according to their worth in solving a particular problem. In the post-NFL era, one takes an egalitarian view of representation and evolutionary algorithms: there is no single representation or evolutionary algorithm that is universally superior to the rest when it comes to optimization [7].

However, the matter of trees as information structure is distinct from the matter of trees as an issue of problem representation. When one does an analysis in problem representation, one implicitly assumes that information from a problem's domain is a factor throughout that analysis. When one does an analysis in information structure, that assumption is not a given. Instead, it is common to treat trees as mathematical entities apart from the information such trees would carry. Consequently, it is not unusual to treat information structure as a level of abstraction below that of problem representation (as in [8]).

Whereas there are several problem representations that have been of interest in evolutionary computation (i.e., binary strings, real-valued vectors, finite-state representations, and parse trees), there are primarily two information structures that are fundamental to computer science: linear lists and trees. Each type of information structure has its own mathematical heritage. Each type of information structure also has a set of techniques for representing and manipulating such a structure within a computer ([8], p. 232).

Trees as information structures represent an extensive body of literature that has been, for the most part, separate from work in GP.[1] Trees were likely first formalized as a mathematical entity in [13]. The term *tree* and the beginnings of an extensive treatment of trees as a mathematical entity likely started with [14]. Trees overtly represented information in computer memory in some of the earliest machine-language

---

[1] Langdon has used results from [9-10] in his analyses of GP performance. His incorporation of the mathematical literature seems to be exception and not the rule in GP theory. In evolutionary computation at large, incorporation of the literature has been mostly with respect to finding alternative representations for use in various evolutionary algorithms (e.g., [11-12]).

computers for the purpose of algebraic formula manipulation ([8], 459). Because tree structures were well suited for formula manipulation, these structures were used in some of the 1950's work in automatic programming [15].

2.2 Visualizing Tree Structures

To appreciate the role of tree structure in forming solutions in GP, it helps to visualize the kinds of structures that occur. Visualizing structure, particularly for typical trees encountered in GP, can be done by mapping m-ary trees (binary trees for this paper) to a circular grid. For binary trees, this grid can be derived by starting with a full binary tree, which we designate as C.

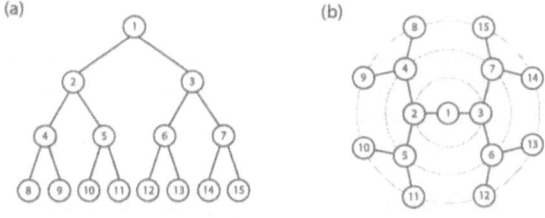

The nodes of this labeled, full binary tree C can be mapped to a circular grid in polar coordinates (r, θ) by first assigning depth levels to a set of concentric rings. The exact mapping of depth levels to ring radii is arbitrary, so long as depth level 0 remains at the center and increasing depths are assigned to rings of increasing radius. For convenience, we let depth level n be mapped to a ring of radius n. Figure 1 gives an example of a full binary tree of depth 3 that has been labeled in this manner.

Fig. 1. Mapping a full binary tree to a circular grid. (a) Full binary tree of depth 3. (b) Corresponding circular grid

An arbitrary binary tree A can be mapped to labels of a full binary tree by showing that a set of labels corresponding to A is a subset of labels corresponding to C. (Note: a formal description of our visualization method is found in [16].)

Figure 2 shows an example of a full binary tree of depth 10 (2047 nodes) and an example of binary tree of depth 26 (461 nodes) from GP. From the standpoint of typical measures of tree structure (i.e., *number of nodes* and *depth level*), the GP tree is unremarkable. The GP binary tree was generated under circumstances that were also unremarkable and is a solution from a symbolic regression problem that has been documented elsewhere in the literature [17–20].

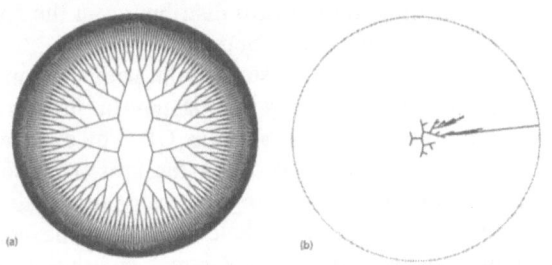

Fig. 2. Two examples of binary trees. (a) Full binary tree, depth 10. (b) Structure of a GP-generated solution (using arity-2 functions), depth 26

Figure 3 shows the individual in Fig. 2b in the context of 11 other individuals that have been taken from the same population of 500. Visualization of this population (not shown) depicts much similarity between tree shapes in a population.

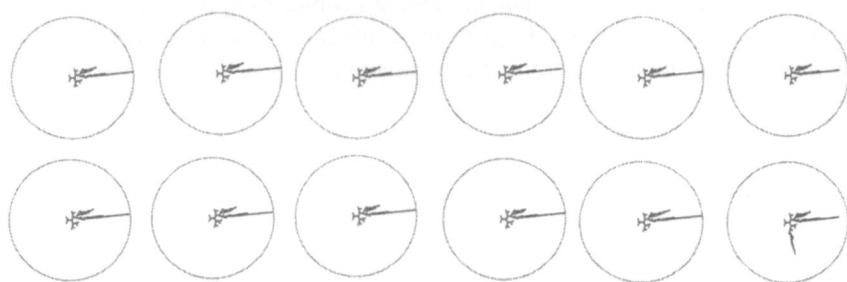

Fig. 3. Sample of a population. These are 12 representatives out of a population of 500. The individual that is shown in Fig. 2b is the second from the top left corner. The remaining 488 individuals look similar to these

In the following section, we pose a hypothesis that may account for the patterns hinted at in these figures.

2.2 Trees and Structural Mechanisms: A Hypothesis

We hypothesize that there exists a class of structural mechanisms that "piggyback" on the operators that generate variation in GP (e.g., crossover and mutation). These structural mechanisms incrementally modify the shape of the tree structures during iteration. In so doing, these mechanisms incrementally modify the probabilities of choosing a subtree that are distributed across a particular (m-ary) tree structure. In standard GP, probabilities of choosing a subtree are distributed across nodes as either uniform or piecewise uniform. (In many GP systems, there is a user-determined bias — sometimes called *internal node bias* – that weights the probability of selecting an internal node against that of terminal node. The conditional probability of choosing for either type of node is uniform within that type.) As a result, for (disjoint) subtree partitions at a given depth, the probabilities of choosing a subtree at that depth are biased toward those subtrees with more terminal nodes. This would generally hold true even under internal node bias, since subtrees with more terminal nodes often have more internal nodes. Consequently, this shift in probability, together with mechanisms that add structure, can feedback positively for the growth of deeper subtrees. This type of random growth would subsequently result in a non-uniform distribution in the space of allowable tree structures. In other words, the trees are inherently sparse.

We can illustrate various aspects of this hypothesis by starting out with an example of "piggybacking" in crossover and mutation. Figure 4 shows how a structurally equivalent operation can result from two different operators in GP. For crossover: Fig. 4a depicts Parent 1 binary tree with an arrow indicating the crossover point; Fig. 4b, Parent 2 binary tree with an arrow indicating the crossover point; Fig. 4c, Child binary tree; Fig. 4d, the oriented equivalent with bold indicating the added structure relative to Parent 1. For mutation: Fig. 4e depicts Parent 1 binary tree with an arrow indicating the point at which mutation starts; Fig. 4f, Child plane tree; Fig. 4g, the oriented equivalent with bold indicating the added structure relative to Parent 1. Both 4d and 4g show structural equivalents of adding a three-node binary

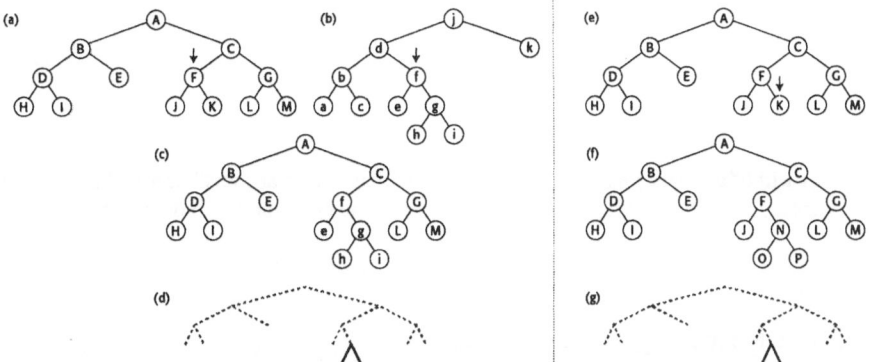

Fig. 4. Crossover and mutation. Crossover and mutation can result in a structurally equivalent operation that adds a common (oriented) subtree

subtree to Parent 1. The degree and content of each node is arbitrary, as is the location of crossover and mutation points, so long as their orientation remains fixed.

We would consider structurally equivalent operators—like the one that has been discussed in Fig. 4 – as belonging to the class of mechanisms that we have hypothesized. Namely, here is an example of a mechanism that "piggybacks" on the operators like crossover and mutation.

Such operators can explain the preferential growth of some parts of a tree over others. Figure 5 indicates two subtrees that are labeled 1 and 2 at their respective roots. Assume that the probability of picking any node is uniform. Further assume, for illustrative purposes only, that whatever the variational operator, the net effect is to increase a subtree by a three-node binary tree. In Fig. 5a, the probability of picking a node from either subtree is equal (p = 1/3). In Fig. 5b, the probability of picking a node in subtree 2 is 3 times greater than in picking a node from subtree 1 (p = 3/5 v. p = 1/5). In Fig. 5c, the probability of picking a node in subtree 2 is 5 times greater (p = 5/7 v. p = 1/7). In Fig. 5d, the probability of picking a node in subtree 2 is 7 times greater (p = 7/9 v. p = 1/9). The longer a subtree is not selected in this process of incremental growth, the less likely it is for this subtree to grow.

We can extend this thought experiment by continuing this iteration. To simplify matters, we can presume to replace a randomly selected terminal node with a three-node binary subtree. Figure 6 shows a result of this.

This iterative process results in a tree that compares well to the visualization of actual trees in Figs. 2 and 3. However, this thought experiment still falls short of having a verifiable model. Towards this end, then, the next section describes our model.

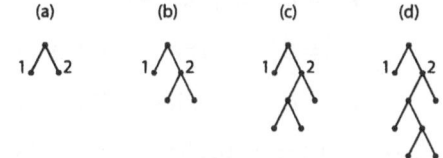

Fig. 5. Preferential growth between subtrees

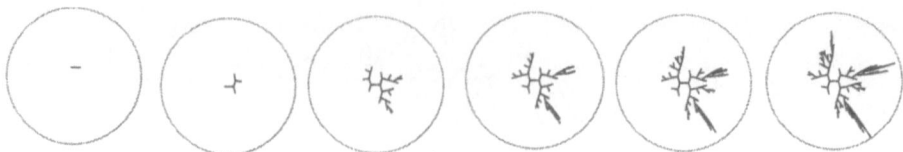

Fig. 6. Growth of a binary plane tree. This figure depicts six "snapshots" during the growth of a binary tree. The outer gray ring is for reference only and corresponds to depth 15

3 The Lattice-Aggregate Model

The lattice-aggregate model formalizes the hypothesis set forth in the previous section. It uses set-theoretic operations on ordered sets of integers to describe the iterative, stochastic growth of m-ary trees as encountered in processes like GP. As in Sect. 2, we simplify our discussion by considering binary trees only.

3.1 Basic Description

Let the root node be defined at depth $d = 0$. Assuming that tree T has depth $d > 0$, the set A of nodes of T can be decomposed into two mutually exclusive, non-empty sets J and K such that

- Set J corresponds to the internal nodes of T
- Set K corresponds to the leaves of T

We define a set B to correspond to a subtree of T. Note that set B is a function of k, whereupon $k \in A$. The smallest possible subtree, a leaf, can be described as

$$B^1 = \{k\}, k \in \mathfrak{I}^+, k \in A. \tag{1}$$

The next smallest subtree, a parent and two nodes, can be described as

$$B^3 = \{k, 2k, 2k+1\}. \tag{2}$$

For the purposes of modeling GP behaviors[2] for depths 0 – 26, we arbitrarily define B^5, B^7, B^9, and B^{11} to correspond to 5-, 7-, 9-, and 11-node subtrees, respectively.[3]

$$B^5 = \{k, 2k, 2k+1, 4k+2, 4k+3\}. \tag{3}$$

$$B^7 = B^5 \cup \{8k+4, 8k+5\}. \tag{4}$$

$$B^9 = B^7 \cup \{16k+10, 16k+11\}. \tag{5}$$

$$B^{11} = B^9 \cup \{32k+20, 32k+21\}. \tag{6}$$

(Note that the particular selection of elements for B^5, B^7, B^9, and B^{11} seems somewhat arbitrary. These subtrees represent a minimal set of subtrees that can account for

---

[2] Depth 26 was chosen largely on the basis of comparison with known empirical data sets.
[3] For depths larger than 26, we would need to consider larger and deeper subtrees, since the opportunity to pick a deeper subtree exists for a larger and deeper tree.

much of the behavior that has been observed in GP at depths 1–26. A fuller account for how these subtrees were chosen has been deferred to a future paper.)

Now let $k \in K$. We can then represent the growth of a tree by B^5 as

$$A' = A \cup B^5(k) = A \cup \{2k, 2k+1, 4k+2, 4k+3\}. \tag{7}$$

Likewise, we can do the same for B^7, B^9, and B^{11}.

Consequently, we can represent a stochastic model of tree growth for depths 0 – 26 as a recursive operation upon integer sets, namely:

$$A' = A \cup B^i(k), \tag{8}$$

where i is a discrete random variable with sample space $S_B = \{5, 7, 9, 11\}$ and k is a discrete, uniformly distributed random variable with sample space $S_B = K$. It can be demonstrated that an appropriate probability distribution function corresponding to i entails the following relationship[4]:

$$P(i=5) = 2 P(i=7) = 4 P(i=9) = 8 P(i=11). \tag{9}$$

Example. Given $A = \{1, 2, 3, 6, 7, 12, 13, 14, 15\}$. Set A decomposes into $J = \{1, 3, 6, 7\}$ and $K = \{2, 12, 13, 14, 15\}$. Assuming that the second element of K and that B^5 have been chosen, $A' = \{1, 2, 3, 6, 7, 12, 13, 14, 15, 24, 25, 50, 51\}$.

3.2 Variations

There are several additional variations that need to be considered in the modeling of tree growth in GP. The first set of variations assists in identifying the upper and lower density bounds of tree growth, while the second set of variations address methods of population initialization.

The density of a set A can be defined as follows:

$$\text{density} \equiv \frac{N(A)}{2^{\log_2(\max(A)+1)} - 1}, \tag{10}$$

where N(A) is the number of elements in A and max(A) identifies the maximum value in A. This definition corresponds to a ratio that is the number of nodes of a tree that is normalized by the number of nodes in a full tree of identical depth.

To identify the upper density bound, equation (8) can be restated as

$$A' = A \cup B^3(k). \tag{11}$$

Equation 11 corresponds to the case in which tree growth is entirely determined by three-node subtrees. Note that if k was deterministic (instead of stochastic) such that all $k \in K$ is selected for replacement by B^3, the resulting tree would approach being full.

To identify a lower density bound, equation (8) can be restated as

---

[4] This assumes that the comparison is with standard GP, in which the probability of choosing an internal node for crossover is uniform. These probabilities were determined through the following coarse approximation: given a full binary tree, what is the probability of selecting a subtree of depth j relative to the probability of selecting a subtree of depth $j-1$?

$$A' = A \cup B^n(k), \tag{12}$$

where B^n is the least dense set of those sets B that are used in modeling growth. It is assumed that the density for sets B are determined at $k = 1$. For the proposed model for depths 0 – 26, the set that is least dense is B^{11}.

It is possible to modify equation (8) to account for varying methods of population initialization. While such modifications have been done to model Koza's ramped half-and-half for depths 2–6, the exposition of these modifications have been left to a future paper.

4 Model Predictions

4.1 Determination of Search Space Boundaries

The lattice-aggregate model was subsequently used to derive boundaries in the size-shape search space of trees from depths 0 – 26. This derivation consisted of four steps, namely:
- Used Monte Carlo methods to sample the proposed lattice-aggregate model corresponding to equation (11).
- Extracted depth and number of nodes from each sample.
- Computed the cumulative distribution of the numbers of nodes per tree at a given depth d. Repeat for $d = \{0, 1, 2, \ldots 26\}$.
- Determined isopleths in size-shape space that correspond to contours of constant distribution.

This process is shown in Fig. 7 for 50,000 samples. Isopleths were generated for 99%, 75%, median, 25%, and 1% distributions. Given the relatively steep fall-offs in distribution, the 99% and the 1% isopleths do approximate boundaries that specify where trees do or do not occur in this search space.

A similar procedure was applied to determine isopleths for equations (11) and (12). Again, given relatively steep fall-offs in distribution, the 99% isopleth for equation (11) approximated the uppermost bound of search, while the 1% isopleth for equation (12) approximated the lowermost bound of search.

4.2 Model Results

Figure 8 summarizes the isopleths for equations (8), (11), and (12). The isopleths suggest the existence of at least four distinct regions for depths 0 – 26. The regions are as follows:
- *Region I*. This is the region where most solutions in standard GP would occur (for binary trees). Full mixing of various size / shape subtrees in the derivation of solutions occurs here. The width of Region I is driven largely by population initialization.
- *Region II*. These are the regions (II_a and II_b) where increasingly fewer individuals would appear the further these individuals are from Region I. Only partial mixing of size / shape subtrees occurs here, with mixing becoming non-existent towards the boundaries furthest away from Region I. Region II_a is delineated by the boundaries that are approximately located by the 99% isopleth for equation (11)

and the 99% isopleth for equation (8). The transitions between Regions II and III are pronounced.
- *Region III*. These are the regions (III$_a$ and III$_b$) where even fewer individuals would typically appear. Region III$_a$ is delineated by the boundaries that are approximately located by the 99% isopleth for equation (11) and the limit for full trees. Region III$_b$ is delineated by the boundaries that are approximately located by the 1% isopleth for equation (12) and the limit for minimal trees.
- *Region IV*. These are the regions (IV$_a$ and IV$_b$) that are precluded from binary trees.

5 Empirical Comparison

We compared the map of predicted regions against results obtained from common test problems. In particular, we considered *quintic, sextic, 6-input multiplexer*, and *11-input multiplexer*. Furthermore, we placed the outcomes from these problems in the context of both our theory and in the prevailing theory [3, 25]. The setup and implementation of *quintic* and *sextic* are from [21]; the *multiplexer* problems that use arity-2 functions, [22].

The GP parameters used were Koza's [21]: population 4,000; tournament selection $M = 7$; maximum generations 50. The maximum depth was set to 512. We used a modified version of lilgp [23] as was used in [17]. Most of the modifications were for bug fixes and for the replacement of the random number generator with the Mersenne Twister [24]. We configured lilgp to run as a single thread. Internal node bias was set at default (0.9).

Figure 9 shows the results for these test problems. Each dot represents a best-of-trial individual out of a

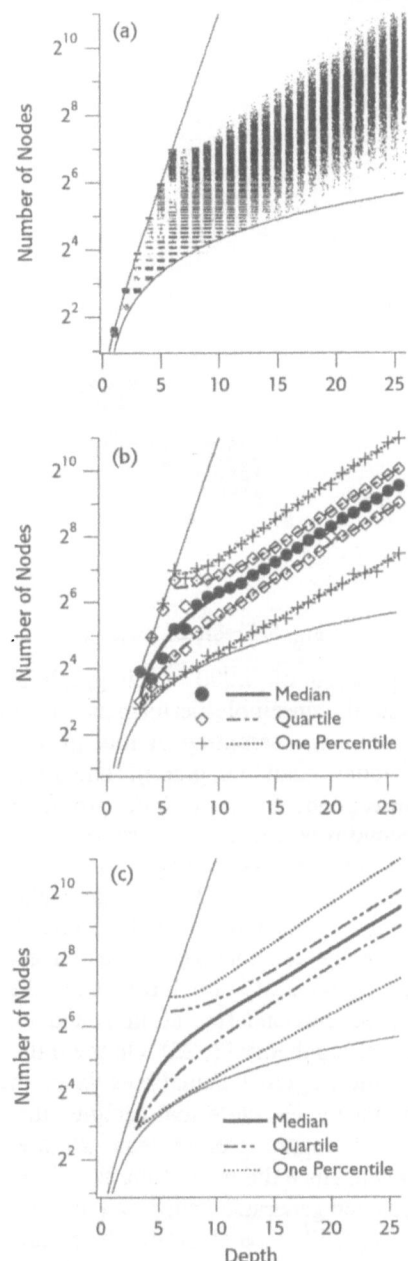

Fig. 7. Derivation of isopleths (Region I). Top plot shows the Monte Carlo results. Middle shows the numerical values of constant distribution and fitted isopleths. Bottom shows just the fitted isopleths

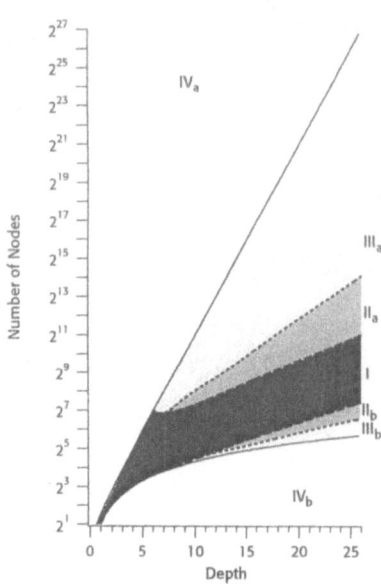

Fig. 8. Predicted regions

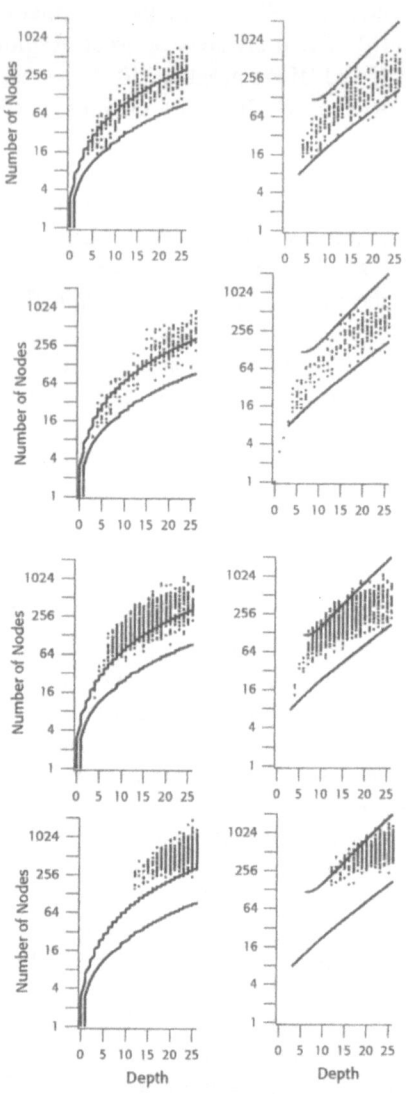

Fig. 9. Empirical comparison

population of 4,000. Each graph represents the ensemble performance of 1,000 trials (i.e., a sampling of four million individuals total per graph). The first row corresponds to the *quintic* problem; the second row, *sextic*; the third, *6-input multiplexer*; and the fourth, *11-input multiplexer*. The data in the left- and right-side graphs of each row are the same. What differ are the boundaries that have been superimposed. On the left side are the 5% and 95% boundaries as indicated by the prevailing theory [3, 25]. On the right side are the Region I boundaries as predicted by our model. Note that samples that occur at depths greater than 26 are not shown, since the map of predicted regions has been generated using an approximation that is not valid for depths greater than 26 (i.e., B^5, B^7,..., B^{11}). At depths greater than 26, a better approximation would need to be used and is left for future work.

Most of the data shown in Fig. 9 do fall within the Region I boundaries. In particular, for *quintic*, 97% of the depicted samples fall within the Region I boundaries in comparison to 61% for the prevailing theory. For *sextic*, about 95% fall within the Region I boundaries, as in comparison to 43% for the prevailing theory. For *6-input multiplexer*, Region I accounts for 93%, in comparison to the prevailing theory (15%).

For *11-input multiplexer*, Region I accounts for 92%, in comparison to the prevailing theory (0.62%).

Another comparison with empirical results is given in this proceedings [27]. This work introduces a test called the *Lid* problem, which is anticipated by our hypothesis and model. Indeed, if structure alone accounts for significant GP behavior, it should be possible to devise a structure-only problem in GP that demonstrates this. Other empirical comparisons can be found in [28].

6 Conclusions

This paper described a hypothesis of an undiscovered class of mechanisms that exist in standard GP. Rather than being intentionally designed, these structural mechanisms have been hypothesized to be an unintended consequence of using trees as information structures. To test this hypothesis, a model has been described that predicts outcomes in GP that would arise solely from such mechanisms.

There are several other conclusions that can be made.
- *The model is a logical progression in the consideration of tree structure in GP*. Although the idea of undiscovered mechanisms may seem farfetched, they are consistent with the idea of random, diffusion processes in GP. The difference, however, between our hypothesis and the prevailing theory [3, 25] is the inclusion of iterative growth in this otherwise random process. So instead of a random walk on the distribution of tree shapes, the random walk itself determines tree shape distribution because GP trees grow iteratively. This seemingly minor distinction is analogous to the distinction in physics between purely diffusive processes and diffusion-limited aggregation [26].
- *The hypothesis and model have suggested a method of visualizing tree structures in GP*. The method was briefly mentioned in this paper and in [28]. A detailed account is given in this proceedings [16].
- *The empirical results support our hypothesis and model for the existence of structural mechanisms in GP*. The model predicts a region where most outcomes of GP would be found (i.e., Region I). This would occur because Region I is where all of the available subtree shapes could be used to form a solution. Because the derivation for this region is based only on structure and not programmatic content, this region should be applicable to a broad variety of GP problems that use just arity-2 functions. The empirical results given in this paper do support the existence of Region I, with that region accounting for better than 90% of all solution outcomes of the experiment.
- *The hypothesized mechanisms provide a simple explanation of the evolution of size and shape in standard GP*. Arguably, this explanation is simpler than that provided under the prevailing theory. The prevailing theory implies that selection pressure on content accounts for the discrepancies. While the results suggest that selection pressure on content is still a factor within Region I, such pressure remains within the context of a consistent structural account for size and shape.
- *The existence of the other predicted regions is not supported by the empirical results*. Although the empirical results do provide support for our hypothesis and model, the results fall short of what is needed to demonstrate the existence of those regions. The empirical results subsequently fall short of validating the exis-

tence of structural mechanisms. However, as it has turned out, obtaining such evidence is nontrivial. Another paper [27] more directly addresses this issue than has been possible here.

Acknowledgments. We thank the following individuals and organizations for their help: CSCS U-M / Santa Fe Institute Fall Workshops, L.M. Sander, S. Stanhope, J. Polito 2, P. Litvak, S. Yalcin. D. Maclean, W. Worzel, M. Samples, and UMACERS teams Magic, Meta-Edge, Royal, Borges, and Binomial-3. We appreciate the critiques extended to us from the anonymous reviewers, E. Korkmaz, K. Seo, and N. McPhee. The first author also acknowledges I. Kristo and S. Daida.

References

[1] T. Soule, et al., "Code Growth in Genetic Programming," in *GP 1996*, J.R. Koza, et al., Eds. Cambridge: The MIT Press, 1996, pp. 215–223.
[2] W.B. Langdon, et al.; "The Evolution of Size and Shape," in *Advances in Genetic Programming 3*, L. Spector, et al. Eds. Cambridge: The MIT Press, 1999, pp. 163–190.
[3] W.B. Langdon and R. Poli, *Foundations of Genetic Programming*. Berlin: Springer-Verlag, 2002.
[4] P.J. Angeline, "Parse Trees," in *Handbook of Evolutionary Computation*, T. Bäck, et al., Eds. Bristol: Institute of Physics Publishing, 1997, pp. C1.6:1–C1.6:3.
[5] K. Deb, "Introduction [to Issues in Representations]," in *Handbook of Evolutionary Computation*, T. Bäck, et al., Eds. Bristol: Institute of Physics Publishing, 1997, pp. C1.1:1–C1.1:4.
[6] D.B. Fogel and P.J. Angeline, "Guidelines for a Suitable Encoding," in *Handbook of Evolutionary Computation*, T. Bäck, et al., Eds. Bristol: Institute of Physics Publishing, 1997, pp. C1.7:1–C1.7:2.
[7] D.H. Wolpert and W.G. Macready, "No Free Lunch Theorems for Optimization," *IEEE Transactions on Evolutionary Computation*, vol. 1, pp. 67–82, 1997.
[8] D.E. Knuth, *The Art of Computer Programming: Volume 1: Fundamental Algorithms*, vol. 1, Third ed. Reading: Addison–Wesley, 1997.
[9] L. Alonso and R. Schott, *Random Generation of Trees*. Boston: Kluwer Academic Publishers, 1995.
[10] P. Flajolet and A. Odlyzko, "The Average Height of Binary Trees and Other Simple Trees," *Journal of Computer and System Sciences*, vol. 25, pp. 171–213, 1982.
[11] J. Knowles and D. Corne, "A New Evolutionary Approach to the Degree Constrained Minimum Spanning Tree Problem," *IEEE Transactions on Evolutionary Computation*, vol. 4, pp. 125–134, 2000.
[12] G.R. Raidl, "A Hybrid GP Approach for Numerically Robust Symbolic Regression," in *GP 1998*, J. R. Koza, et al., Eds. San Francisco: Morgan Kaufmann Publishers, 1998, pp. 323–328.
[13] G. Kirchoff, *Annalen der Physik und Chemie*, vol. 72, pp. 497–508, 1847.
[14] A. Cayley, *Collected Mathematical Papers of A. Cayley*, vol. 3, 1857.
[15] H. G. Kahrimanian, presented at Symposium on Automatic Programming, Washington, D.C., 1954.
[16] J.M. Daida and A. Hilss,"Visualizing Tree Structures in Genetic Programming," *GECCO 2003*.
[17] J.M. Daida, et al., "Analysis of Single-Node (Building) Blocks in Genetic Programming," in *Advances in Genetic Programming 3*, L. Spector, et al., Eds. Cambridge: The MIT Press, 1999, pp. 217–241.

[18] J.M. Daida, et al., "What Makes a Problem GP-Hard? Analysis of a Tunably Difficult Problem in GP," in *GECCO 1999*, W. Banzhaf, et al., Eds. San Francisco: Morgan Kaufmann Publishers, 1999, pp. 982–989.
[19] O.A. Chaudhri, et al., "Characterizing a Tunably Difficult Problem in Genetic Programming," in *GECCO 2000*, L. D. Whitley, et al. Eds. San Francisco: Morgan Kaufmann Publishers, 2000, pp. 395–402.
[20] JM. Daida, et al., "What Makes a Problem GP-Hard? Analysis of a Tunably Difficult Problem in Genetic Programming," *Journal of Genetic Programming and Evolvable Hardware*, vol. 2, pp. 165–191, 2001.
[21] J. R. Koza, *Genetic Programming II: Automatic Discovery of Reusable Programs*. Cambridge: The MIT Press, 1994.
[22] W.B. Langdon, "Quadratic Bloat in Genetic Programming," in *GECCO 2000*, L. D. Whitley, et al., Eds. San Francisco: Morgan Kaufmann Publishers, 2000, pp. 451–458.
[23] D. Zongker and W. Punch, "lilgp," v. 1.02 ed. Lansing: Michigan State University Genetic Algorithms Research and Applications Group, 1995.
[24] M. Matsumoto and T. Nishimura, "Mersenne Twister: A 623-Dimensionally Equidistributed Uniform Pseudorandom Number Generator," *ACM Transactions on Modeling and Computer Simulation*, vol. 8, pp. 3–30, 1998.
[25] W.B. Langdon, "Size Fair and Homologous Tree Crossovers for Tree Genetic Programming," *Journal of Genetic Programming and Evolvable Machines*, vol. 1, pp. 95–119, 2000.
[26] T. A. Witten and L. M. Sander, "Diffusion-Limited Aggregation: A Kinetic Critical Phenomenon," *Physics Review Letters*, vol. 47, pp. 1400–1403, 1981.
[27] J.M. Daida, et al., "What Makes a Problem GP-Hard? Validating a Hypothesis of Structural Causes," *GECCO 2003*.
[28] J.M. Daida, "Limits to Expression in Genetic Programming: Lattice-Aggregate Modeling," *CEC 2002*, Piscataway: IEEE Press, pp. 273–278.

Visualizing Tree Structures in Genetic Programming

Jason M. Daida, Adam M. Hilss, David J. Ward, and Stephen L. Long

Center for the Study of Complex Systems and the Space Physics Research Laboratory
The University of Michigan, 2455 Hayward Avenue, Ann Arbor, Michigan 48109-2143

Abstract. This paper presents methods to visualize the structure of trees that occur in genetic programming. These methods allow for the inspection of structure of entire trees of arbitrary size. The methods also scale to allow for the inspection of structure for an entire population. Examples are given from a typical problem. The examples indicate further studies that might be enabled by visualizing structure at these scales.

1 Introduction

"Structure always affects function." ([1], p. 268)

In saying this, the author was commenting on the nature of complex networks. To many who study the anatomy of such networks, the topology of interconnections can have a pronounced effect on function. True, the author was thinking along the lines of complex networks like an electrical power grid or the Internet. However, he could have just as well been referring to the kinds of solution outcomes that happen in GP.

In particular, one might say that networks are prevalent in GP. Each individual can be framed as a topology of interconnections between functions and terminals. Each population can be framed as a topology of interconnections not only between individuals at time t_i, but also of the times preceding t_i.

One could also suspect that such networks in GP are complex. After all, the metaphor from which GP draws from is generally considered to be complex. One might even speculate that the mathematical properties of complex networks (e.g., scale-free networks) might also apply to GP. To say, however, that GP *produces* complex networks is a matter of conjecture. No one in the field has proven this. Nevertheless, there has been evidence that suggests that the topology of interconnections has a significant effect on the outcomes for GP.

For example, some of the early theoretical work in GP has alluded to the consequences of structure in determining solution outcomes. Rosca and Ballard [2, 3] hypothesized that solution outcomes are determined, in part, by rooted-tree schema. They argued that during a run, GP identifies these structures first and subsequently builds upon these structures to form solutions.

Other researchers have independently supported Rosca and Ballard's theory, which highlighted the importance of root structure. This support has come largely in the form of empirical studies that featured selected benchmark problems. These would include [4, 5]. Of these two, McPhee and Frietag's work made the strongest empirical case that root structures play a key role in determining solution outcomes for GP in general.

Structure, of course, has played a significant role in other studies in GP as well. For example, structure was a key determinant of fitness in one of the earliest adaptations of the Royal Road problem [6] (in genetic algorithms) to GP [7]. Work by Goldberg and O'Reilly [8, 9] resulted in the development of the ORDER and MAJORITY fitness problems in which the structures of solution outcomes were driven by the problem at hand. Structure has played a key role in the analysis by Soule, Foster et al. [10-13], who has looked at the evolution of shape in GP outcomes. Soule and Foster's work on the evolution of tree shapes has since been carried forward and extended by Langdon [14, 15].

Structure plays a key role, too, in the prevailing theory on GP. Poli's work on a GP schema theory (and later Langdon and Poli) [15-18] presumes schemata that are structurally based (i.e., has a certain size and depth). Bloating has also been associated with structure; it has been Langdon and Poli's contention that bloating occurs partly as a result of a random walk on a landscape of tree-shape distributions (e.g., [19, 20]).

However, in spite of this longstanding interest in the role of structure, the direct visualization of such structures has *not* progressed far. Nearly all visualizations of structure have been along the lines of "for illustrative purposes only," as was first done by Koza [21]. There have not been a viable means of inspecting structures for typical trees that have been discussed in GP, let alone a viable means of inspecting structures for an entire population. Consequently, our investigations have focused on the visualization of tree structures and other topologies of interconnections between nodes. The purpose of this paper, then, is to describe our methods for visualizing whole tree structures, which would be about two orders of magnitude larger in tree size than what has typically been published. The purpose of this paper is to also describe our methods for visualizing structures for entire populations, which would be another two orders of magnitude beyond that of an individual tree.

This paper is organized as follows. Section 2 provides the mathematical context and summarizes the current conventions in visualizing trees. Section 3 describes our method for visualizing the structure for an entire tree and provides examples of these visualizations. Section 4 extends the methods described for individual trees and scales them to encompass populations. Section 5 concludes this work.

2 Background

2.1 Mathematical Preliminaries

The conventions and terms that are used in this paper are those of Knuth [22] and Stanley [23, 24].

A *tree T* is formally defined as a finite set of nodes such that:
a) There is one node that is designated as the *root* of the tree, and
b) The remaining nodes (excluding the root) are partitioned into $m \geq 0$ disjoint, non-empty sets $T_1,...,T_m$, each of which is a tree. The trees $T_1,...,T_m$ are called *subtrees* of the root.

The number of subtrees associated with a node is called the *degree* of that node. A node of degree zero is called a *terminal node* or *leaf*. Each root is considered as the *parent* of the roots of its subtrees; the roots of these subtrees are said to be the *children* of their parent and *siblings* to each other.

A *plane tree* or *ordered tree* is defined similarly to a *tree*, except that (b) becomes

b´) The remaining nodes (excluding the root) are placed into an ordered partition $(T_1,...,T_m)$ of $m \geq 0$ disjoint, non-empty sets $T_1,...,T_m$, each of which is a plane tree.

If only the relative orientation of nodes is considered and not their order, a tree is said to be *oriented*.

Let $m \geq 2$ in the definition of a tree. An *m*-ary tree is defined similarly to a *tree*, except that (a) and (b) become

a´´) Either *T* is empty, or else there is one node that is designated as the *root* of *T*, and

b´´) The remaining nodes (excluding the root) are put into an ordered partition of exactly *m* disjoint (possibly empty) sets $T_1,...,T_m$, each of which is an *m*-ary tree.

An *m*-ary tree is said to be *complete* if every node that is not a leaf has *m* children. In deference to GP (and not Knuth), we use term *full* to refer to a complete *m*-ary tree that has m^d terminal nodes, where *d* is the depth of that tree and the root node is considered at depth 0. A *binary* tree is a 2-ary tree.

Note that an *m*-ary tree is *not* a special case of the formal definition of a tree. Consequently, a binary tree is not assumed to be an instance of a *tree T*. By definition, however, a complete binary tree is such an instance.

The treatment of trees in this paper differs from the current treatment of trees in GP. GP uses conventions that tend to focus on node content. (For example, internal nodes in GP are assumed to be functions of a specified arity. In turn, arity determines the number of children for that node.) In contrast, the treatment of trees in this paper uses the conventions that tend to focus on structure.

2.2 Current Conventions

In GP, the convention for drawing trees starts with the root node at the top and grows downward with the terminal nodes at the bottom. This is a prevalent convention, starting with [21] through current works like [25]. *Right*, *left*, *top*, and *bottom* are subsequently denoted. *Depth*, as opposed to *height*, is also indicated. This convention follows the standard practice in computer science at large ([22], p. 311). Stanley explicitly states this convention ([23], p. 295): that for *m*-ary trees (which computer scientists typically use), lines (edges) joining a node (vertex) to the roots of its subtrees are drawn at equal angles that are symmetric with respect to a vertical axis. This convention works if both *m* and the number of nodes remain small (i.e., small *m*-ary trees). For example, Knuth depicts two small *m*-ary trees that still consume a page of illustration (for one, *m* is 2 and number of nodes is 63; for the other, *m* is 11 and the number of nodes is 61).[1]

For many problems, drawing a large *m*-ary tree is not needed, since a small *m*-ary tree would suffice as a representative sample. For GP, however, we claim that it is instructive to visualize *m*-ary trees of a few thousand nodes. Our reasoning is that such visualizations would show broader patterns of solution structure that are not readily seen with smaller samples or with structural metrics (like *number of nodes* or

---

[1] Even so, Knuth adjusted convention slightly by laying each trees on its side.

depth level). Unfortunately for these larger trees, there is little by way of a drawing convention. Consequently, the next section introduces a method for drawing large, *m*-ary trees. To simplify matters, we restrict our attention to plane binary trees.

3 Visualizing Structure for a Tree

3.1 Visualization Method

Our method for visualizing trees involves mapping *m*-ary (binary trees for this paper) to a circular grid. For binary trees, this grid can be derived by starting with a full binary tree, which we designate as C. The nodes of C are labeled by a positive integer k ($k \in \mathfrak{I}^+$) in the following way:

1. The root node is designated as $k = 1$ and at depth level 0.
2. The leftmost node for each depth level n is designated as $k = 2^n$.
3. Nodes at depth level 1 are labeled from left to right as $k = 2$ and $k = 3$, respectively.
4. Nodes at each subsequent depth level n are labeled from left to right using the following sequence: $\{2^n, 2^n+1, 2^n+2, \ldots 2^{n+1}-1\}$.

The nodes of this labeled full binary tree can be mapped to a circular grid in polar coordinates (r, θ) by first assigning depth levels to a set of concentric rings. The exact mapping of depth levels to ring radii is arbitrary, so long as depth level 0 remains at the center and increasing depths are assigned to rings of increasing radius. For convenience, we let depth level n be mapped to a ring of radius n. Figure 1a gives an example of a full binary tree of depth 3 that has been labeled in this manner.

Ring radius is subsequently determined by using the following mapping $\rho(k): k \rightarrow r$, such that $k \in \mathfrak{I}^+$, $r \in \mathfrak{I}^+$, and

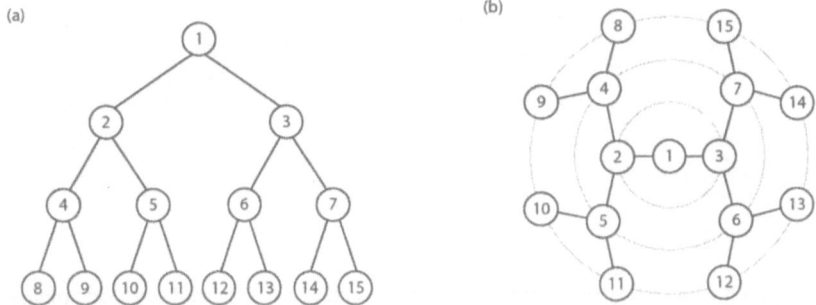

Fig. 1. Mapping a Full binary tree to a circular grid. (a) Full binary tree of depth 3. (b) Corresponding circular grid

$$\rho(k) = \begin{cases} 0, & k=1. \\ \lfloor \log_2 k \rfloor & k>1. \end{cases} \quad (1)$$

Nodes on a ring are positioned by using the following mapping $\phi(k): k \to \theta$, such that $k \in \mathfrak{I}^+, r \in \mathfrak{R}$, and

$$\phi(k) = \begin{cases} 0, & k=1. \\ \pi\left(\dfrac{1}{2} + \dfrac{1}{2^{\rho(k)}} + \dfrac{k \bmod 2^{\rho(k)}}{2^{\rho(k)-1}}\right) & k>1. \end{cases} \quad (2)$$

For the moment, we designate set L_C of plane binary trees to be defined by the collection of labeled nodes (vertices) of C in polar coordinate space. Figure 1b gives an example of a full binary tree of depth level 3 that has been mapped to set L_C.

We can show that an arbitrary plane binary tree A (or for that matter, an arbitrary binary tree) can be mapped to L_C by showing that a set of labels corresponding to A is a subset of L_C. To do so, we traverse A in preorder[2] and label each node visited in the following manner:

1. The root node of tree A is designated as $k = 1$ and at depth level 0.

2. The root node of the left subtree is labeled $2l$, where l is the label of that node's parent.

3. The root node of the right subtree is labeled $2l+1$, where l is the label of that node's parent.

We designate this set of labels for A as L_A. Figure 2 depicts an example of an arbitrary plane binary tree A that is mapped to L_C. If A is a full plane binary tree, it can be readily shown that this preorder labeling does result in an identical labeling that corresponds to C.

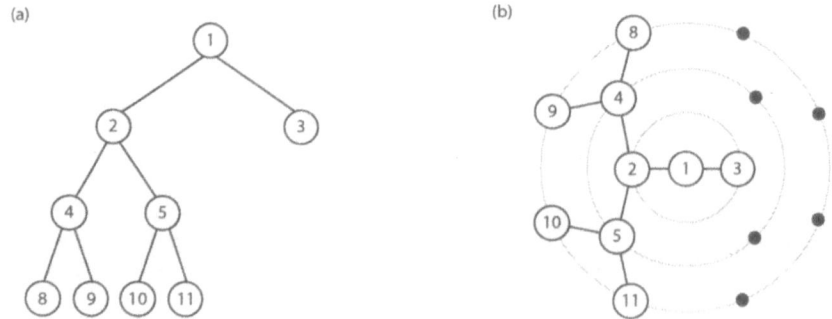

Fig. 2. Mapping an arbitrary plane binary tree to a circular grid. (a) Arbitrary plane binary tree of depth 3. (b) Corresponding circular grid

---

[2] Preorder traversal is defined recursively and is described in [22], p. 319. There are three steps that are taken: visit the root; traverse the left subtree; traverse the right subtree.

If one forgoes the information content in each node and treats each node as a vertex, one can use the described visualization method to examine the gross structure of (plane binary) trees. The compact representation allows for the depiction of structures consisting of one to several thousands of vertices.

3.2 Visualization Examples

Figure 3 shows an example of a full binary tree of depth level 10 (2047 nodes) and an example of plane binary tree of depth level 26 (461 nodes) that result from GP. From the standpoint of typical metrics that quantify aspects of tree structure (i.e., *number of nodes* and *depth level*), the GP binary tree is unremarkable. The GP tree was generated under circumstances that were also unremarkable and is a solution from a symbolic regression problem that has been documented elsewhere in the literature [26-29].[3]

In comparison to a full binary tree, the GP plane tree that is shown in Fig. 3 is sparse and is characterized by a few radial spikes in which much of the nodes are found. There are significant areas of the lattice that are not populated. It is asymmet-

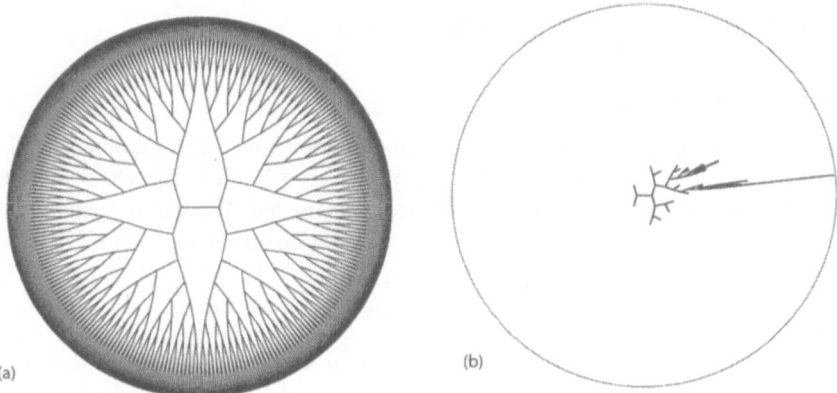

Fig. 3. Two examples of plane binary trees. (a) Full binary tree, depth 10. (b) Structure of a GP-generated solution (using arity-2 functions), depth 26. The gray circle around (b) represents depth 26 and is shown for reference

ric. As it turned out, though, this gross pattern is representative of the population from which it was taken.

---

[3] The problem is an instance taken from symbolic regression and involves solving for the function $f(x) = 1 + 3x + 3x^2 + x^3$. Fitness cases are 50 equidistant points generated from $f(x)$ over the interval [-1, 0). The function set was {+, −, ×, ÷}, which correspond to arithmetic operators of addition, subtraction, multiplication, and protected division. A terminal set was {X, **R**}, where X is the symbolic variable and **R** is the set of ephemeral random constants that are distributed uniformly over the interval [-1, 1]. Most of the GP parameters were similar to those mentioned in Chapter 7 in [21]. Unless otherwise noted: population size = 500; crossover rate = 0.9; replication rate = 0.1; population initialization with ramped half-and-half; initialization depth of 2–6 levels; and tournament selection ($n_{tournment}$ = 7). Other parameter values were maximum generations = 200 and maximum tree depth = 26.

Figure 4 shows that individual tree structure in the context of 47 other individuals. All of these trees have been taken from the same population of 500. Figure 5 shows another 48 individuals out of a population of 500, except that these individuals were generated with a slightly different set of GP parameters (i.e., fitness-proportionate, rather than tournament selection). In both figures, although many of the plane binary trees structures look similar (even identical) they are, in fact, distinct individuals. Close inspection shows minor structural differences between even the most similar-looking individuals.

Figures 4 and 5 pose a number of questions. What drives the shape of these structures? Are these figures representative of GP as a whole? Are there limits to the kinds of structures that can be generated in GP? How do structures evolve over time? Why is there a lack of structural diversity? Does the type of selection influence the kinds of structures that are generated? How does content correlate with structure? Which metrics best characterize these structural features? How do structures vary between trials? Clearly, there is a high degree of structural correlation in this small sample. Unfortunately, the questions that can be raised by a suggestion of structural pattern are more than what can be adequately answered by this paper.

4 Visualizing Structure in a Population

4.1 Visualization Method

Although only 9.6% of the their respective populations have been depicted in Figs. 4 and 5, the high degree of structural correlation depicted in those figures suggests that the rest of their populations are also structurally correlated. We can test this possibility by starting with a cumulative distribution on L_C for an entire population, i.e.,

$$L_P = \sum_{\forall A \in P} L_A, \quad (3)$$

where A is a tree in a population P, $L_P \supset L_C$ and $\mathbf{L}_i$ is a vector corresponding to L_i such that

$$\mathbf{L}_i \equiv \sum_{\forall a \in L_i} \mathbf{i}_a. \quad (4)$$

Note that $\mathbf{i}_a$ specifies a unit component vector, and that a is a label in L_i.

In other words, the (un-normalized) cumulative distribution of a population P can be treated as a sum of the vectors corresponding to the trees in that population. The vector that corresponds to each tree A is defined as a sum of unit vectors, where each unit vector corresponds to an occupied grid-point in L_C. Consequently, the tree that spans the population is described by L_P.

For example, suppose a population P consists of four binary trees that have the labels $\{1, 2, 3, 6, 7, 14, 15\}$, $\{1, 2, 3\}$, $\{1, 2, 3, 4, 5\}$, and $\{1\}$. The corresponding un-normalized cumulative distribution for this population would be $\mathbf{L}_P = 4\mathbf{i}_1 + 3\mathbf{i}_2 + 3\mathbf{i}_3 + \mathbf{i}_4 + \mathbf{i}_5 + \mathbf{i}_6 + \mathbf{i}_7 + \mathbf{i}_{14} + \mathbf{i}_{15}$, with $L_P = \{1, 2, 3, 4, 5, 6, 7, 14, 15\}$.

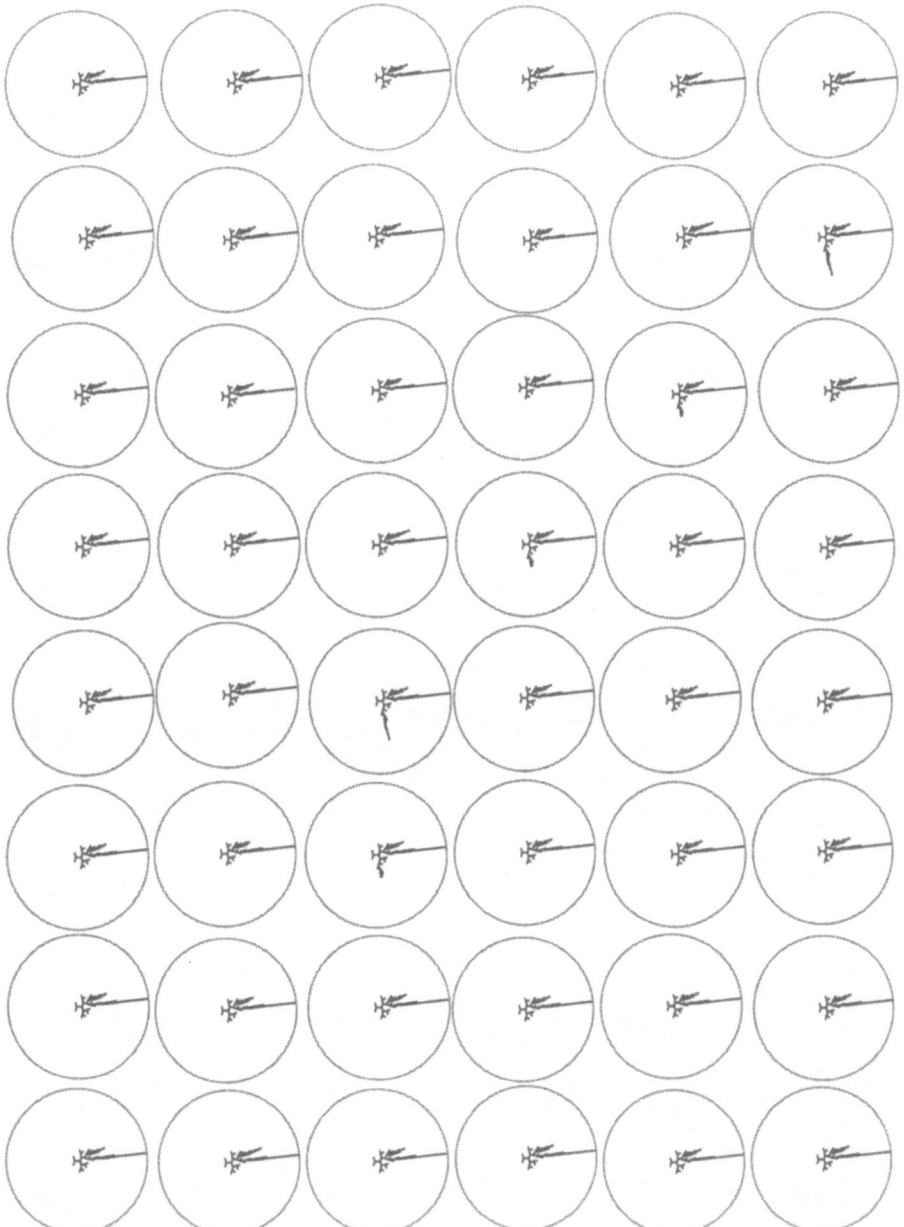

Fig. 4. Sample of a population. These are 48 representatives out of a population of 500. The individual that is shown in Figure 3b is the second from the top-left corner. These results are for a symbolic regression problem (binomial-3) using tournament selection ($n_{tournment} = 7$). As reference, a gray circle around each individual represents depth level 26

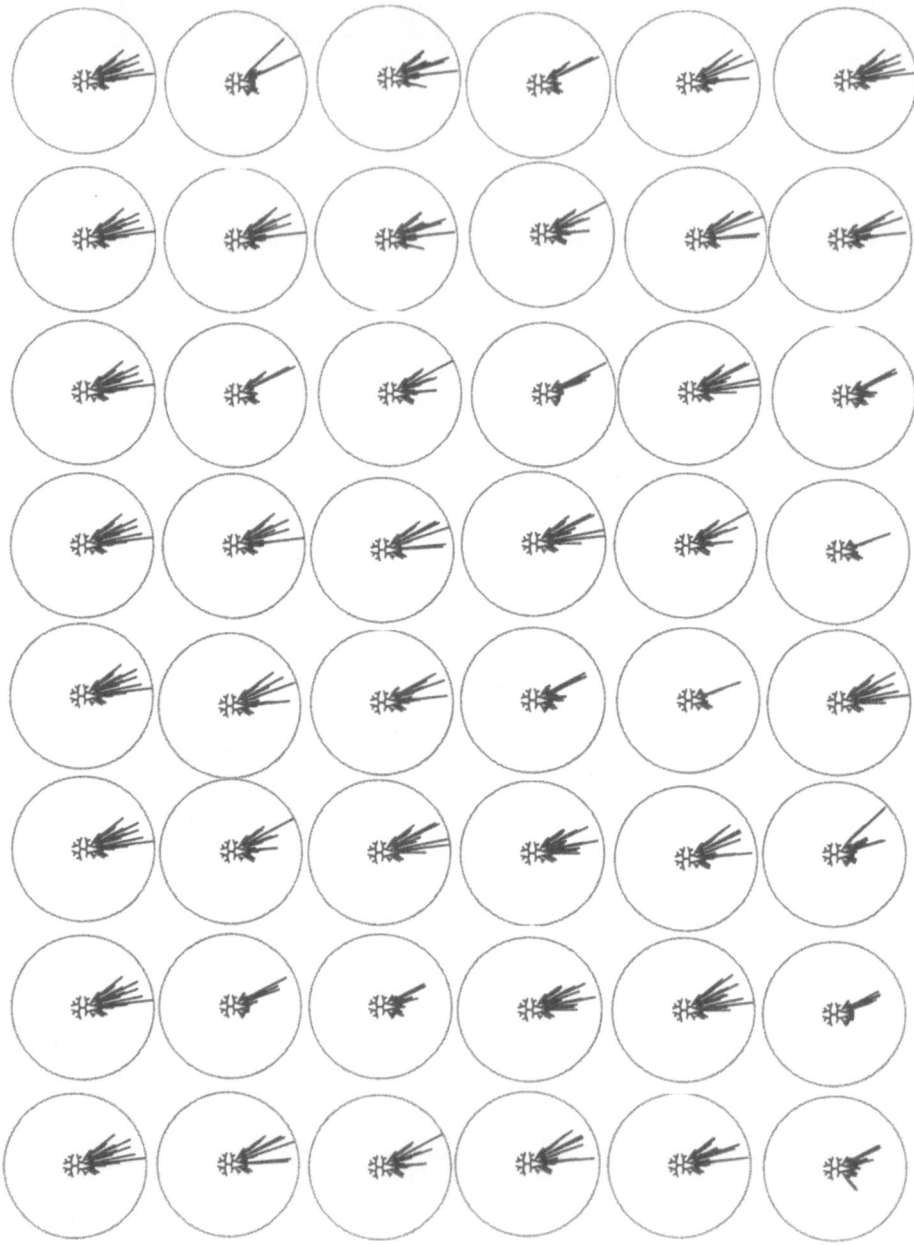

Fig. 5. Another population sample. These are 48 representatives out of a population of 500. These results are for the same symbolic regression problem and GP parameters that were used to generate the results of Fig. 5. The exception was that fitness-proportionate instead of tournament selection was used to generate these results. As reference, a gray circle around each individual represents depth level 26

Visualization of a population can subsequently be done in three steps:

1. Compute L_P.
2. Normalize each component of L_P by the number of individuals in population P.
3. Construct the tree that corresponds to L_P, except that the grayscale value of each line from parent to child is proportionate to the magnitude of the component vector corresponding to that child. (Assume that higher magnitudes map to darker grays).

The resulting visualization is tantamount to overlaying all of the trees in a population. Structures that are used most often show up darker; the least used structures, lighter.

4.2 Visualization Examples

The left-side graph of Fig. 6 corresponds to the population of which Fig. 4 is a subset. The right-side graph of Fig. 6 corresponds to the population of which Fig. 5 is a subset. Each graph summarizes the structure of 500 trees, which represents an entire population.

While individual trees provide an insight to the degree of structural similarity within a population, they are but a hint. This limitation exists even though our method of visualization of single trees does enables one to see more structure than has been possible by orders of magnitude. It is still not enough for understanding structural trends at a population level.

For those reasons, we would argue that the visualizations of Fig. 6 do help in seeing some of these larger trends. In particular, we can observe the following:

- The darkest regions of each graph occur in the center and radiate outward. (Darker regions represent locations of higher structural correlation.) This indicates that structures near the root are highly correlated within a population. This observation is consistent with theory and observations about the rapid convergence of root structure within GP (e.g., [2, 3, 5]).
- The lightest regions tend to occur near the edges of each graph. This means that

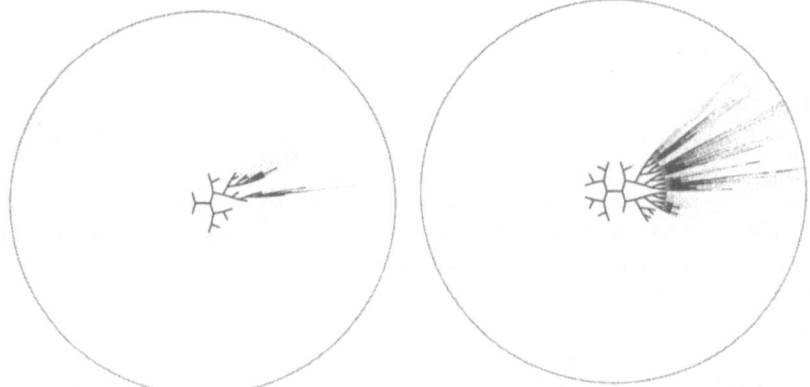

Fig. 6. Visualization of populations. The left graph corresponds to the population for Fig. 4; the right graph, the population for Fig. 5

the structures proximal to the terminals are the least likely to be found from individual to individual within a population. That they tend to be the lightest parts of each graph in Fig. 6 is also consistent with theory and observations (e.g., [4]).
- The graph on the left seems to exhibit less structural diversity than the graph on the right. Both graphs in Fig. 6 correspond to the same problem and to mostly the same GP parameters. The exception is that the graph on the left corresponds to a trial using tournament selection, while the one on the right corresponds to a trial using fitness-proportionate selection.

Of these three observations, the last one requires more effort to interpret. It is somewhat arguable to judge by visual inspection alone whether a population is more or less structurally diverse than another. Fortunately, the visualization method for populations requires the computation of a cumulative distribution. We can use the information from Equations 3 and 4 to plot cumulative distribution by rank (i.e., from grid points that are the most used to grid points that are the least used). The result of this is shown in Fig. 7.

The top graph of Fig. 7 corresponds to the cumulative distribution under tournament-selection, while the bottom graph corresponds to the cumulative distribution under fitness-proportionate selection. The plots of cumulative distributions show distinctly different curves. The plot for tournament selection shows that grid points are either occupied by most of the population or by nearly none at all—which is what one would expect for a structurally similar population. In comparison, the fitness proportionate case is heavy-tailed and indicates that many

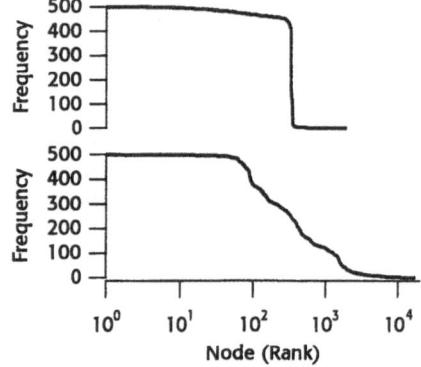

Fig. 7. Distribution curves

more structures are being preserved. This finding is consistent with theoretical work in tournament selection, which argues that diversity is lost under tournament selection [30].

5 Conclusions

"Structure always affects function."
Regardless of how strongly stated was this sentiment, work done in the research community does suggest that structure plays a significant role in understanding how GP works. This paper has reviewed a number of investigations that involve the structure of solution outcomes under GP. In spite of these efforts in understanding the role of structure in GP, there has been a deficit in visualization methods that can depict these structures at the scales that would be of interest to the GP community. Consequently, this paper has described two methods to visualize the structure of trees that occur in GP. The first method allows for the inspection of structure of entire trees and can be scaled for trees consisting of several thousand nodes. The second method allows for the visual inspection of structure at the level of an entire population. Both methods are amenable for further quantitative analysis.

Examples were given from a typical problem in GP. We used these examples as a case study that demonstrates how these visualization methods can be used. We further indicated how such visualizations could either raise new questions or complement existing investigations.

References

[1] S.H. Strogatz, "Exploring Complex Networks," *Nature*, vol. 410, pp. 268–276, 2001.
[2] J.P. Rosca, "Analysis of Complexity Drift in Genetic Programming," in *GP 1997 Proceedings*, J. R. Koza, et al., Eds. San Francisco: Morgan Kaufmann Publishers, 1997, pp. 286–94.
[3] J.P. Rosca and D.H. Ballard, "Rooted-Tree Schemata in Genetic Programming," in *Advances in Genetic Programming 3*, L. Spector, et al. Eds. Cambridge: The MIT Press, 1999, pp. 243–271.
[4] C. Gathercole and P. Ross, "An Adverse Interaction Between Crossover and Restricted Tree Depth in Genetic Programming," in *GP 1996 Proceedings*, J. R. Koza, et al., Eds. Cambridge: The MIT Press, 1996, pp. 291–96.
[5] N.F. McPhee and N.J. Hopper, "Analysis of Genetic Diversity through Population History," in *GECCO '99 Proceedings*, vol. 2, W. Banzhaf, et al., Eds. San Francisco: Morgan Kaufmann Publishers, 1999, pp. 1112–1120.
[6] M. Mitchell, S. Forrest, and J.H. Holland, "The Royal Road for Genetic Algorithms: Fitness Landscapes and GA Performance," in *Proceedings of the First European Conference on Artificial Life.*, F.J. Varela and P. Bourgine, Eds. Cambridge: The MIT Press, 1992, pp. 245–254.
[7] W. Punch, et al., "The Royal Tree Problem, A Benchmark for Single and Multiple Population GP," in *Advances in GP*, vol. 2, P. J. Angeline and J.K.E. Kinnear, Eds. Cambridge: The MIT Press, 1996, pp. 299–316.
[8] D.E. Goldberg and U.-M. O'Reilly, "Where Does the Good Stuff Go, and Why?," in *Proceedings of the First European Conference on Genetic Programming*, W. Banzhaf, et al., Eds. Berlin: Springer-Verlag, 1998.
[9] U.-M. O'Reilly and D.E. Goldberg, "How Fitness Structure Affects Subsolution Acquisition in GP," in *GP 1998 Proceedings*, J. R. Koza, et al. Eds. San Francisco: Morgan Kaufmann Publishers, 1998, pp. 269–77.
[10] T. Soule, J.A. Foster, and J. Dickinson, "Code Growth in Genetic Programming," in *GP 1996 Proceedings*, J.R. Koza, et al., Eds. Cambridge: The MIT Press, 1996, pp. 215–223.
[11] T. Soule and J.A. Foster, "Code Size and Depth Flows in Genetic Programming," in *GP 1997 Proceedings*, J.R. Koza, et al., Eds. San Francisco: Morgan Kaufmann Publishers, 1997, pp. 313–320.
[12] T. Soule and J.A. Foster, "Removal Bias: A New Cause of Code Growth in Tree Based Evolutionary Programming," in *ICEC 1998 Proceedings*, vol. 1. Piscataway: IEEE Press, 1998, pp. 781–786.
[13] W.B. Langdon, et al., "The Evolution of Size and Shape," in *Advances in Genetic Programming 3*, L. Spector, et al., Eds. Cambridge: The MIT Press, 1999, pp. 163–190.
[14] W.B. Langdon, "Size Fair and Homologous Tree Crossovers for Tree Genetic Programming," *Genetic Programming and Evolvable Machines*, vol. 1, pp. 95–119, 2000.
[15] W.B. Langdon and R. Poli, *Foundations of Genetic Programming*. Berlin: Springer-Verlag, 2002.
[16] R. Poli and W.B. Langdon, "A New Schema Theory for GP with One-Point Crossover and Point Mutation," in *GP 1997 Proceedings*, J.R. Koza, et al., Eds. San Francisco: Morgan Kaufmann Publishers, 1997, pp. 279–285.
[17] R. Poli and W. B. Langdon, "Schema Theory for Genetic Programming with One-Point Crossover and Point Mutation," *Evolutionary Computation*, vol. 6, pp. 231–252, 1998.

[18] R. Poli, "Exact Schema Theorem and Effective Fitness for GP with One-Point Crossover," in *GECCO 2000 Proceedings*, L. D. Whitley, et al., Eds. San Francisco: Morgan Kaufmann Publishers, 2000, pp. 469–476.
[19] W.B. Langdon and R. Poli, "Fitness Causes Bloat," in *Soft Computing in Engineering Design and Manufacturing*, P.K. Chawdhry, R. Roy, and R.K. Pant, Eds. London: Springer-Verlag, 1997, pp. 23–27.
[20] W.B. Langdon, "Quadratic Bloat in Genetic Programming," in *GECCO 2000 Proceedings*, L.D. Whitley, et al., Eds. San Francisco: Morgan Kaufmann Publishers, 2000, pp. 451–458.
[21] J.R. Koza, *Genetic Programming: On the Programming of Computers by Means of Natural Selection*. Cambridge: The MIT Press, 1992.
[22] D.E. Knuth, *The Art of Computer Programming: Volume 1: Fundamental Algorithms*, vol. 1, Third ed. Reading: Addison–Wesley, 1997.
[23] R.P. Stanley, *Enumerative Combinatorics I*, vol. 1. Cambridge: Cambridge University Press, 1997.
[24] R.P. Stanley, *Enumerative Combinatorics II*, vol. 2. Cambridge: Cambridge University Press, 1999.
[25] C. Jacob, *Illustrating Evolutionary Computation with Mathematica*. San Francisco: Morgan Kaufmann, 2001.
[26] J.M. Daida, et al., "Analysis of Single-Node (Building) Blocks in GP," in *Advances in Genetic Programming 3*, L. Spector, W.B. Langdon, U.-M. O'Reilly, and P. J. Angeline, Eds. Cambridge: The MIT Press, 1999, pp. 217–241.
[27] J.M. Daida, et al., "What Makes a Problem GP-Hard? Analysis of a Tunably Difficult Problem in Genetic Programming," in *GECCO '99 Proceedings*, vol. 2, W. Banzhaf, et al., Eds. San Francisco: Morgan Kaufmann Publishers, 1999, pp. 982–989.
[28] O.A. Chaudhri, et al. "Characterizing a Tunably Difficult Problem in Genetic Programming," in *GECCO 2000 Proceedings*, L.D. Whitley, et al., Eds. San Francisco: Morgan Kaufmann Publishers, 2000, pp. 395–402.
[29] J.M. Daida, et al., "What Makes a Problem GP-Hard? Analysis of a Tunably Difficult Problem in Genetic Programming," *Journal of Genetic Programming and Evolvable Hardware*, vol. 2, pp. 165–191, 2001.
[30] T. Bickle and L. Thiele, "A Mathematical Analysis of Tournament Selection," in *ICGA95 Proceedings*, L.J. Eshelman, Ed. San Francisco: Morgan Kaufmann Publishers, 1995, pp. 9–16.

What Makes a Problem GP-Hard? Validating a Hypothesis of Structural Causes

Jason M. Daida, Hsiaolei Li, Ricky Tang, and Adam M. Hilss

Center for the Study of Complex Systems and Space Physics Research Laboratory
The University of Michigan, 2455 Hayward Avenue, Ann Arbor, Michigan 48109-2143

Abstract. This paper provides an empirical test of a hypothesis, which describes the effects of structural mechanisms in genetic programming. In doing so, the paper offers a test problem anticipated by this hypothesis. The problem is tunably difficult, but has this property because tuning is accomplished through changes in structure. Content is *not* involved in tuning. The results support a prediction of the hypothesis – that GP search space is significantly constrained as an outcome of structural mechanisms.

1 Introduction

What makes a problem GP-hard? A common hypothesis of why a problem is hard lies with a metaphor of a rugged landscape. Deception, ruggedness, epistasis – all these terms have been evoked at one time or another to describe attributes of a fitness landscape that might be encountered in genetic programming (GP). These concepts have been developed in the genetic algorithms and evolutionary computation communities over the past several decades. There is evidence to suggest that these concepts have legitimacy for GP, as well.

However, there may be more to problem difficulty in GP than what may have been thought. In [1, 2], we presented a hypothesis that argues for the existence of structural mechanisms. The existence of these mechanisms is predicted to have significant consequences. One significant consequence is that GP is greatly limited to the kinds of structures it can create.

In particular, some structures would be easy for GP to produce. Other structures would be difficult. With our theory, it has been possible to create a map of where either difficult or easy structures might exist. The map indicates that only a small portion of the total possible structures would be considered easy. The rest are difficult.

Such a map raises two questions. What happens if GP needs to produce a solution that would best be served by a "difficult" structure? Can there be a purely structural basis to problem difficulty?

We have been investigating ways to validate the predictions that have been set forth by our hypothesis. This paper represents one of those efforts and describes a link between structure and problem difficulty. To start, we have narrowed the scope of the first question. The question becomes this: What happens if GP needs to produce a solution that can *only* be served by a particular set of structures?

If, indeed, there is a purely structural basis for problem difficulty in GP, it should be possible to craft a problem that tests for this. Admittedly, such a problem would be unusual. Content in functions and terminals would be rendered irrelevant. Fitness would be based on measurements of structure (i.e., like *number of nodes* and *depth of tree*). While GP was not designed for a problem like this, the hypothesis suggests its existence. In essence, GP would be asked to evolve *nothing*, as far as substance is concerned. However, the hypothesis predicts that some *nothing*s would be harder to achieve than others. It further predicts that most of *nothing* is difficult for GP. The problem that we present in this paper is called the *Lid*, a tunably difficult problem that is based on structure.

In this paper, we use the *Lid* as a probe that empirically tests for locations in the map that identifies where "hard" or "easy" structures exist. This paper presents the first of these results.

Section 2 summarizes both current theories and our hypothesis of how tree structure might determine problem difficulty. Section 3 describes our proposed tunably difficult problem. Section 4 describes our experimental procedure while Section 5 shows the results. Section 6 discusses the results in the context of the theory. Section 7 concludes.

2 Theory

2.1 Previous Work

That structure has any influence on solution outcomes was alluded to in some of the earliest references to "bloat" (i.e., [3, 4]). For example, Altenberg (in [3]) suggested "[P]rogram bloating with junk code may simply be driven by the recombination operator, a neutral diffusion process with upward drift." He further stated, "[P]rograms might be expected to bloat because of the asymmetry between addition and deletion of code at the lower boundary of program size." Although he did not directly implicate tree structures, Altenberg was probably the first to articulate the idea that there might be a neutral, random diffusive process that occurs regardless of whatever selection pressure there might be on content.

Altenberg's conjecture was independently picked up two years later by Soule and Foster. In a series of papers [5-8], they described bloating as a pervasive GP phenomenon that happens in a number of problems. While they did not go so far as to say the phenomenon is caused by a neutral process, they did propose a "fix" that was largely structural (i.e., invoke a penalty on solution size). Soule and Foster's work helped to initiate compelling theories that explain phenomena that occur across a broad variety of problems encountered in GP (e.g., [9, 10]).

Soule and Foster eventually teamed with Langdon and Poli in [11]. Together they furnished the linkage between the generation of random trees and bloating. In later work [10, 12], Langdon further elaborated on this linkage and hypothesized that the distribution of solution shapes follows that of the theoretically derived distribution of tree shapes. He posited that GP solutions are a result of a random walk on the "surface" of the distribution of tree shapes in the search space defined by tree size and depth.

Whether the distribution of tree shapes has sufficient explanatory power to account for problem difficulty is discussed later in this paper.

There are two other noteworthy investigations.

Goldberg and O'Reilly [13, 14] have isolated and examined how program dependencies may affect GP tree structures. Their work helped to establish a link between the form of a solution and the function that this solution is to embody.

Punch et al. published a paper describing a tunably difficult problem that they deemed the *Royal Tree* [15]. The *Royal Tree* problem was one of the earliest instances of borrowing the *Royal Road* function, which had been used in the GA community for some time. It is also significant in that it was the first instance of a tunably difficult problem that was structurally based. Its program dependencies were arguably straightforward and easily quantifiable. The solutions to the problem were easy to visualize. What was surprising was that it did not take much to turn the *Royal Tree* problem from a "hard" problem into an "impossible" one, in spite of the significant computational resources that were used. For the most part, however, the community has apparently dismissed the problem as anomalous – "real" coding solutions do not appear like *Royal Tree* programs; it's "obvious" that the *Royal Tree* problem is hard because the solutions need to be full trees; and so on.

2.2 Theoretical Context of This Paper

With the exception of the *Royal Tree*, nearly all of the previous work presumes a role of content in contributing toward the dynamics observed in GP. Content is presumed to be a necessary condition. In contrast, the theoretical context of this paper does not presume this. Rather, our hypothesis examines dynamics that can be brought about by considering just structural causes. If there is a presumption, it would be this: that much of the dynamics driven by structure would occur at a lower level than dynamics driven by content.

Although we describe our hypothesis in other papers [1, 2], it is worthwhile to summarize their findings here.

- The iterative growth of trees in standard GP is analogous to a physical process of diffusion-limited aggregation (also known as ballistic accretion).
- Objects that are created by diffusion-limited aggregation have a characteristic density (i.e., because they are a kind of fractal).
- If the iterative growth of trees in standard GP is analogous to diffusion-limited aggregation, it is possible that trees also have a characteristic "density" or range of "densities" (which can be measured by tree sizes and depths).
- We can create a model of iterative tree growth to map out which structures are likely.
- The model predicts for four regions in the space of number of nodes and tree depth: I (easy), II (transitional), III (hard), IV (out-of-bounds). A map of these regions is shown in Fig. 1. This particular map presumes GP with an arity-2 function set.

One can think of these regions as a kind of "bottle." Region I would denote the interior of this "bottle." Most GP trees would be found in here. Region II would denote the walls of this "bottle." Fewer trees would be found here. Region III would denote the outside of this "bottle." Even fewer trees would be found here. Region IV represents the out-of-bounds, since these regions are impossible for an arity-2 tree to exist.

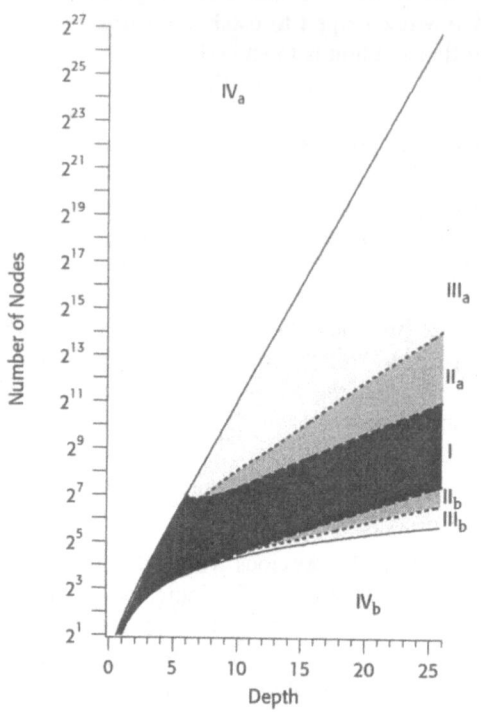

Fig. 1. Predicted regions of difficulty (from [2])

Note that the model itself focuses only on the role of tree structure on determining problem difficulty. We do *not* imply that structure alone accounts for all the phenomena observed in determining problem difficulty. Rather, the model was designed with the intent of isolating the effect of structure away from content. We view Regions I–IV, then, as representing limiting conditions within GP search space. Evidence presented in [1,2] supports the notion that content-driven mechanisms (which also determine problem difficulty) occur within these limiting conditions.

Note, too, that this type of iterative random growth does result in a substantially different "surface" than would be suggested by the distribution of binary tree shapes, as suggested in [10, 16].

Finally, we note that although this surface is different, our reasoning has drawn heavily from those works mentioned in Sect. 2.1. In a sense, we show what happens when one extends Langdon's model by including consequences that stem from iterative random growth.

2.3 Motivation for a New Test Problem

A partial verification of our hypothesis would include a comparison with GP results on problems with an arity-2 function set. (A limited comparison was given in [1, 2].) However, known problems – like *multiplexer* or *sextic* – do not account just for structural effects. Known problems inherently involve content or syntax. Even the structure-based *Royal Tree* problem (see [15]) involves some knowledge of node syntax, if only to properly compute a fitness score.

There is no known problem in GP that isolates information structure away from information content. Furthermore, there is no known problem that tests for variations in problem difficulty due only to variations in information structure.

Given that the hypothesis has predicted for significant variations in problem difficulty that would arise just from information structure, it has been incumbent upon us to show that such a problem exists. The remainder of this paper describes such a problem.

3 The *Lid* Problem

The objective of the *Lid* problem is to derive a tree of a given size and depth. If our hypothesis is correct, the *Lid* problem should be tunably difficult simply by specifying particular sizes and depths. This would occur because trees of certain sizes and depths would be more likely than others.

Given this fairly broad objective, there are a number of variants of the *Lid* problem that can be described. However, for the purposes of this paper, we describe one variant that should naturally "fit" standard versions of GP. The described variant is a depth-first version, which means that GP would need to solve for the depth of a target tree first before solving for the number of nodes of that target tree. This ordering suits standard GP, since meeting a depth criterion has been a relatively easy task.

We leave for future work descriptions and results from other variants. As far as we have been able to determine, however, all variants have displayed tunable difficulty.

3.1 Problem Specification

As in the *ORDER* and *MAJORITY* problems [13, 14], the function set for the *Lid* problem is the single primitive function JOIN {J}, which is of arity-2. As in the *Royal Tree* problem [15], the terminal set for the *Lid* problem is the single primitive {X}.

The *Lid* problem requires the specification of a target tree depth (i.e., d_{target}) and a target number of tree terminals (i.e., t_{target}). We assume that *Lid* trees are rooted, and that the convention for measuring depth presumes that the node at the root lies at null (0) depth.

How an arbitrary *Lid* tree measures against its target specifications is given by the following metrics for depth and terminals[1] – Eqs. (1) and (2), respectively.

$$metric_{depth} = W_{depth}\left(1 - \left(\frac{|d_{target} - d_{actual}|}{d_{target}}\right)\right) \quad (1)$$

$$metric_{terminals} = \begin{cases} W_{terminals}\left(1 - \left(\frac{|t_{target} - t_{actual}|}{t_{target}}\right)\right) & \text{if } metric_{depth} = W_{depth} \\ 0, & \text{otherwise} \end{cases} \quad (2)$$

where d_{actual} and t_{actual} correspond to the depth and number of terminals for the measured *Lid* tree, respectively. Weights W_{depth} and $W_{terminals}$, are chosen arbitrarily such that

$$W \equiv W_{depth} + W_{terminals} = 100. \quad (3)$$

---

[1] There are several variants of these metrics. For example, one could have GP solve for terminals first, instead of depth. Considerations of these variants are deferred to future work.

Raw fitness is defined as

$$fitness_{raw} = metric_{depth} + metric_{terminals}. \qquad (4)$$

3.2 Tunability Considerations

There are two parameters that are suggested by our hypothesis that can be used for tuning problem difficulty: d_{target} and t_{target}.

Varying t_{target} while fixing d_{target} results in solutions that vary in size, but not in depth. Note that it can be shown that t_{target} is directly related to the size of a target Lid tree, i.e.,

$$Size_{target} = 2t_{target} - 1. \qquad (5)$$

In this case, the initial combinatorial search space in J and X for each target tree remains invariant across t_{target}, although the size of each specified tree grows exponentially.

Varying d_{target} while fixing t_{target} results in solutions that should vary in depth but not in size. In this case, the combinatorial search space in J and X for each target tree remains invariant across d_{target}. We note, however, that the numbers of trees that satisfy a particular 2-tuple of (d_{target}, t_{target}) will vary as d_{target} varies. For example, there are 64 instances that satisfy $(d_{target}, t_{target}) = (7, 8)$; however, just one instance satisfies $(d_{target}, t_{target}) = (3, 8)$.

4 Experimental Procedure

This paper considers an empirical test of information structure driving problem difficulty. In particular, if the shape of a tree is a major factor in determining problem difficulty, we should be able to detect this by changing the target shape of a tree. We would subsequently note how difficult it is for GP to attain that shape. For this paper, we used a simple measure of problem difficulty. Out of N trials, we noted how many of these trials were ones in which GP successfully derived the target shape.

We subsequently considered the case of taking two one-dimensional slices along the x- and y-axes of the map of theoretically derived regions shown in Fig. 1. In essence, we used the Lid problem as a kind of "probe" that "sounds" for the predicted regions in GP search space. We basically set the tunable Lid parameters and measured the difficulty at that point. We sampled for changes in difficulty along two slices across these regions. A horizontal slice was determined by varying d_{target} while fixing t_{target}. A vertical slice was determined by varying t_{target} while fixing d_{target}.

We used a modified version of lilgp [17] as was used in [18]. Most of the modifications were for bug fixes and for the replacement of the random number generator with the Mersenne Twister [19]. We configured lilgp to run as a single thread.

Most of the GP parameters were identical to those mentioned in Chapter 7 [20]. Unless otherwise noted: population size = 500; crossover rate = 0.9; replication rate = 0.1; population initialization with ramped half-and-half; initialization depth of 2–6 levels; and fitness-proportionate selection. Other parameter values were maximum generations = 200 and the maximum depth = 512 (assuming a root node to be at depth

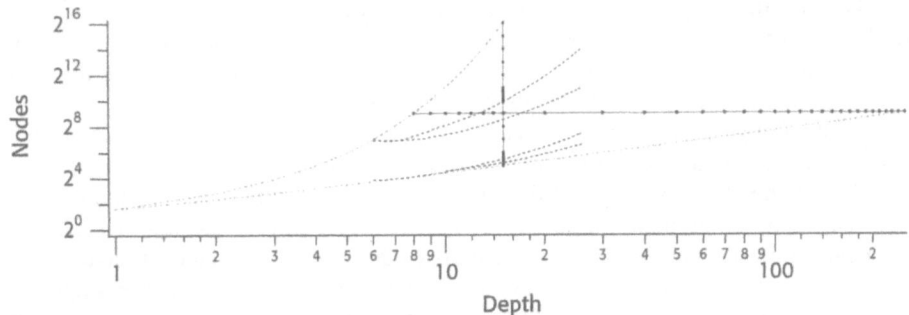

Fig. 2. Sampling of search space using *Lid*

= 0). Note that as in stock lilgp, crossover for this experiment was biased towards the internal nodes of an individual (i.e., 90% internal, 10% terminals).

The parameters of the *Lid* problem were as follows: $W_{depth} = 30$ and $W_{terminals} = 70$. Although sensitivity studies of varying W_{depth} and $W_{terminals}$ have been conducted, the results of these studies have been set aside for a future paper. Those studies have suggested that while varying the *Lid* parameters can influence results, those results are still consistent with the findings of this paper.

Figure 2 shows both slices superimposed over the map of theoretically derived regions. Lines indicate the locations of the slices. Filled circles indicate the locations of the sample taken with *Lid*. For the horizontal cut, the target number of terminals t_{target} was fixed at 256 (which corresponds to a tree that is 511 nodes large), while target depth d_{target} was varied. For the vertical cut, the target depth d_{target} was fixed at 15, while the target number of terminals t_{target} was varied. Sampling along either cut was higher where the rate of change in measured difficulty was higher.

Note that for clarity in Figure 2, the *x*-axis has been changed from a linear scale to a logarithmic one. Note also that the map of regions stops at depth 26. For depths greater than 26, model results have been left pending for a future paper.

For each 2-tuple of (d_{target}, t_{target}), a total of 1,000 trials were run. Consequently, there were 32 data sets collected of 1,000 trials apiece for the horizontal cut. There were 60 data sets of the same number of trials for the vertical cut. A total of 92,000 trials were completed. Trials were run on Linux and Sun UNIX workstations.

5 Results

Horizontal Cut
Figure 3 shows the results from the horizontal cut. This figure shows two graphs that share target depth d_{target} as a common abscissa. Each curve represents the ensemble performance of a total of 16 billion individuals that were distributed among 1,000 trials in 32 data sets. The sampling of these individuals was such that only the best individual per trial was depicted.

The domain of these graphs is the span of allowable depths for a binary tree (which are also *Lid* individuals) for a specified number of terminals. For the target terminal specification $t_{target} = 256$, the minimum allowable depth is 8 (a full tree); the maximum allowable depth, 255 (a sparse tree).

The top graph represents the success rate of identifying a perfect individual in a trial per 1,000 trials as a function of target depth d_{target}, which serves as the horizontal cut's tuning parameter. This plot is analogous to a curve of tunable difficulty as was shown in [21, 22]. The harder a problem is, the lower the success rate.

The bottom graph depicts a scatter plot of the size (i.e., number of nodes) of an individual as a function of target depth d_{target}. As with the previous plots, there are 32,000 data points shown, with 1,000 points per target depth d_{target}. For the horizontal cut discussed in this paper, a size of 511 nodes represents a successful trial. A line delineates this size for this graph. When all 1,000 points for a data set have the same number of nodes, those points are delineated with an 'x.' The 'x' markers appear for d_{target} = 12, 13, 14, 15, 20, 30, 40, 50, which means that *all* of the trials at those depths solved for the correct solution. This scatter plot subsequently provides a measure of how close individuals were to their corresponding target. (As it turned out, no trial in either horizontal or vertical cuts failed to reach its corresponding target depth. This means, then, that target depth and actual depth can be treated as the same.) The harder a problem is, the further away the sizes are from the target size of 511 nodes.

Vertical Cut

Figure 4 shows the results from the vertical cut. This figure shows two graphs that share target size (nodes) as a common abscissa. Note that target size (Eq. 5) was used instead of target terminals t_{target} so that the results can be compared to Fig. 1 (which has an ordinate in nodes, as opposed to terminals). Each curve represents the ensemble performance of a total of 30 billion individuals that were distributed among 1,000 trials in 60 data sets. The sampling of these individuals was such that only the best individual per trial is depicted.

The domain of these graphs is the span of allowable sizes for a binary tree (which are also *Lid* individuals) for a specified depth. For the target depth specification d_{target} = 15, the minimum allowable size is 31 (a sparse tree); the maximum allowable size, 65,535 (a full tree).

As in the horizontal cut, the top graph represents the success rate of identifying a perfect individual in a trial per 1,000 trials as a function of target size, which serves as the vertical cut's tuning parameter.

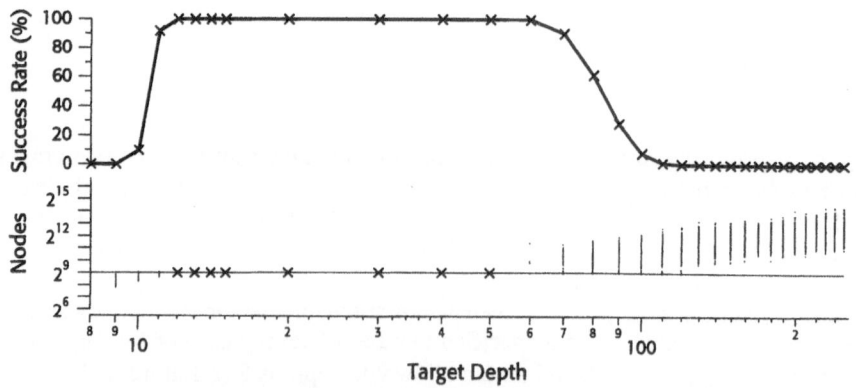

Fig. 3. Horizontal cut results

The bottom graph depicts a scatter plot of the size (i.e., number of nodes) of an individual as a function of target size. There are 60,000 data points shown, with 1,000 points per target size. For the vertical cut discussed in this paper, a successful trial is represented when the target size equals the measured size. A 45° line delineates this condition for this graph. When all 1,000 points for a data set have resulted in success, those points are delineated with a '+.' The '+' markers appear for target size = 43, 45, 47, 49, 51, 53, 55, 57, 59, 61, 63, 127, 255, 511, 915, which means that *all* of the trials at those sizes solved for the correct solution. This scatter plot subsequently provides a measure of how close individuals were to their corresponding target. The harder a problem is, the further away the measured sizes are from the 45° line.

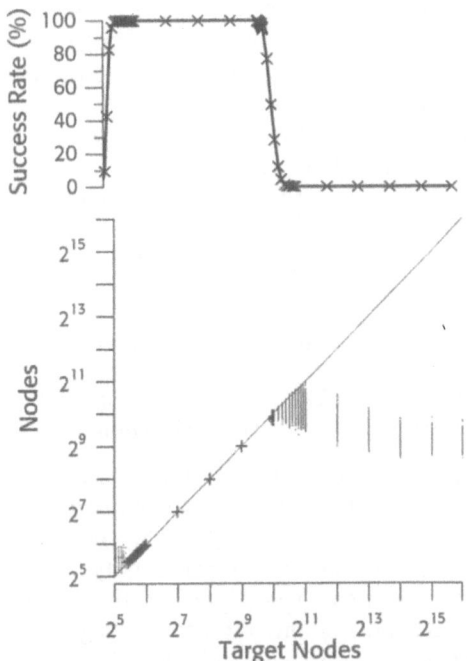

Fig. 4. Vertical cut results

6 Discussion

6.1 Regions

Observation: The empirical results support our predicted outcome of four regions of problem difficulty.

The curves for success rate for Figs. 3 and 4 indicate the existence of three regions of problem difficulty (with the fourth – "out-of-bounds" – as being impossible to achieve with binary trees). Presented in order of increasing difficulty, these regions are:

Region I. ("Easy") This region would be characterized by a plateau at or near 100% in the success-rate curve. For the horizontal cut, it measured to a plateau that spans from about depth 12 to 60. For the vertical cut, it measured to a plateau that spans from about 41 to 969 nodes. (Note: as in [1,2], this region was measured by using the upper one-percentile isopleth relative to the absolute maximum of the region, which in this case was 100%.)

Regions II. ("Transitional") These regions (i.e., IIa and IIb) would be characterized by a transition from Region I to Regions III (i.e., IIIa and IIIb). For the horizontal cut, the upper-bound transition was measured from about depth 60 to 110, while the lower-bound transition was measured from about depth 10 to 11. For the vertical cut, the upper-bound transition was measured from about 969 to 1,639 nodes, while the lower-bound transition was from about 31 to 41 nodes.

Regions III. ("Hard") These regions (i.e., IIIa and IIIb) would be practically categorized as those regions that might not be accessible to GP, even though these regions still represent legitimate tree structures. For the horizontal cut, the upper-bound region measured from about depth 110 to the end of the allowable domain (i.e., 255), while the lower-bound region measured from depths 8 to 10. For the vertical cut, the upper-bound transition was measured from about 1,639 nodes to the end of the allowable range (i.e., 65,535 nodes) while the lower-bound transition was not measurable.

Region IV. ("Out-of-Bounds") Region IV consists of those depths that are not attainable by means of a binary tree.

6.2 Contrast with Current Theory

Observation: The frequency distribution of shapes corresponding to binary trees does not have sufficient explanatory power to account for the results.

The current theory (*cf.* [10, 16]) posits that solution size and shape are the results of a random walk on the node-depth surface of the frequency distribution of tree shapes for a given size and depth. The current theory also posits that tree shapes tend to be biased towards growth, particularly in the direction of the most likely shape for a tree of a given size and depth. Finally, the current theory predicts that this random walk would most commonly occur along the "ridgeline" of this surface, with most individuals being found within the upper- and lower-5% boundaries for a given number of terminals.

If this were true, a problem like the *Lid* should show a positive correlation for problem difficulty with this surface. To determine this, we independently derived and recomputed the surface as described in [10] so that those results can be compared to our results (particularly the success-rate curves in Figs. 3 and 4).

The top graph in Fig. 5 shows a view of this surface as depicted in [10]. It shows the "out-of-bounds" boundary for binary trees, the "ridgeline," and the upper- and lower-five percentile boundaries. The top graph represents the common visualization [10, 16] of a binary tree distribution.

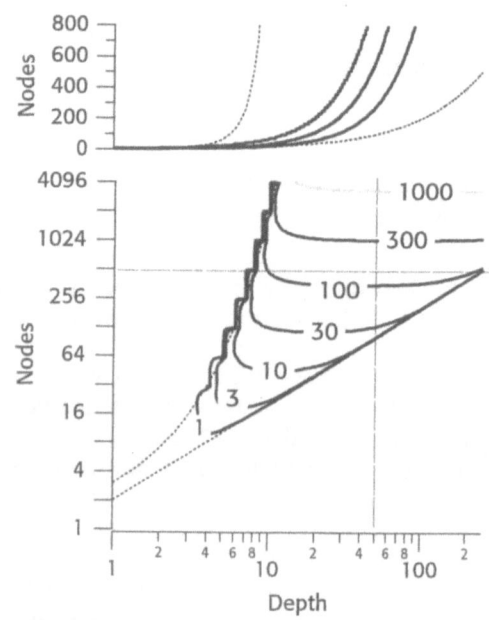

Fig. 5. Two views of the distribution of binary trees

What Makes a Problem GP-Hard? Validating a Hypothesis of Structural Causes

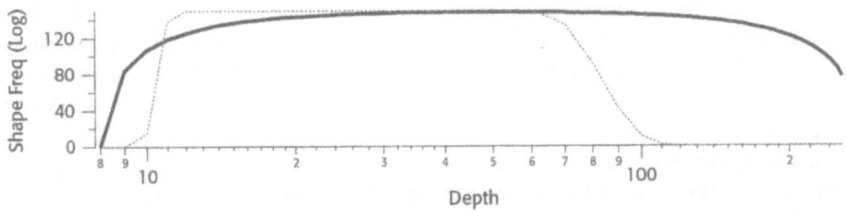

Fig. 6. Distribution of binary trees for 256 terminals

One might guess that the upper- and lower-5% boundaries are analogous to contour lines. They are not. The bottom graph in Fig. 5 shows the same surface, except visualized with the contour lines of constant frequency distribution. The contours are given in $\log_{10}$ quantities. If anything, the surface is dominated by the peak. Just in terms of what is shown in the bottom figure, the peak (located near the upper right corner) is greater than $10^{1,000}$. This value is 700 orders of magnitude greater than most of the surface that has been shown.

Also for Fig. 5, lines have been superimposed on the bottom graph to indicate the location of the horizontal and vertical cuts of this paper.

Figure 6 shows the frequency distribution of trees for the horizontal cut. The corresponding success-rate curves from the *Lid* problem are superimposed in light gray. While Fig. 6 loosely has the shape as shown in the horizontal-cut results (Fig. 3), the locations of the transition zones differ considerably from the empirical data.

Figure 7 shows the frequency distribution of trees for the vertical cut. The corresponding success-rate curves from the *Lid* problem are superimposed in light gray. If the current hypothesis is sufficient in explaining the *Lid* results, the frequency distribution curves should track these results. Figure 7 shows that the frequency distribution curve has a significantly different shape than the vertical-cut results (Fig. 4). For the range shown, the frequency of binary tree shapes is exponentially increasing. The rate of that increase is dramatic: we computed the distributions for only a sixteenth of the total span of the cut, but the frequency distribution of trees had surpassed $10^{1,200}$. In other words, the frequency distribution of trees should predict (in this range) that the *Lid* problem would become easier for the vertical cut when the number of target nodes is greater. Our results have shown otherwise.

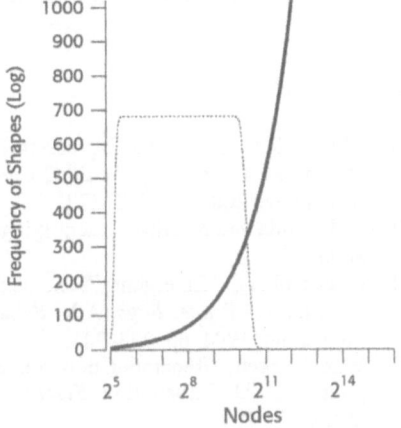

Fig. 7. Distribution of binary trees for depth 15

7 Conclusions

As this paper has shown, tree structure can have a substantial impact in determining problem difficulty. It has provided empirical support for our hypothesis that the iterative random growth of information structures alone can be a significant and limiting factor that determines problem difficulty. As we have described in [1,2], iterative random growth is a result of structural mechanisms that "piggyback" on the operators that generate variation in GP. This paper lends further support for the existence and significance of these structural mechanisms.

In working towards validating our hypothesis, we developed a tunably difficult problem called the *Lid*. It was designed, in part, to serve as a kind of "probe" that "sounds" for the predicted regions in GP search space. Its tunable difficulty stems from a predicted outcome of our theory: that some structures are easier for GP to attain than others. As the results indicate, most structures are difficult for GP.

We have shown that the current hypothesis (*cf.* [16]) does not have sufficient explanatory power to account for the results.

Finally, we note that this paper does *not* say that program dependency, fitness landscapes, epistasis, etc. are invalid. Rather, that context-free mechanisms may also need to be considered when determining what makes a problem GP-hard.

References

[1] J.M. Daida, "Limits to Expression in Genetic Programming: Lattice-Aggregate Modeling," in *Proceedings of the 2002 Congress on Evolutionary Computation*, vol. 1. Piscataway: IEEE, 2002, pp. 273–278.

[2] J.M. Daida and A. Hilss, "Identifying Structural Mechanisms in Standard GP," in *GECCO 2003*.

[3] L. Altenberg, "Emergent Phenomena in Genetic Programming," in *Evolutionary Programming – Proceedings*, A.V. Sebald and L.J. Fogal, Eds. Singapore: World Scientific Publishing, 1994, pp. 233–241.

[4] W.A. Tackett, "Recombination, Selection and the Genetic Construction of Computer Programs," Ph.D. Dissertation, *Electrical Engineering*. Los Angeles: University of Southern California, 1994.

[5] T. Soule, et al., "Code Growth in Genetic Programming," in *GP 1996 Proceedings*, J. R. Koza, et al., Eds. Cambridge: The MIT Press, 1996, pp. 215–223.

[6] T. Soule, et al., "Using Genetic Programming to Approximate Maximum Clique," in *GP 1996 Proceedings*, J.R. Koza, et al. Eds. Cambridge: The MIT Press, 1996, pp. 400–405.

[7] T. Soule and J.A. Foster, "Code Size and Depth Flows in Genetic Programming," in *GP 1997 Proceedings*, J.R. Koza, et al. Eds. San Francisco: Morgan Kaufmann Publishers, 1997, pp. 313–320.

[8] T. Soule and J.A. Foster, "Removal Bias: A New Cause of Code Growth in Tree Based Evolutionary Programming," in *1998 IEEE ICEC Proceeding*, vol. 1. Piscataway: IEEE Press, 1998, pp. 781–786.

[9] W.B. Langdon, "Scaling of Program Fitness Spaces," *Evolutionary Computation*, vol. 7, pp. 399–428, 1999.

[10] W.B. Langdon, "Size Fair and Homologous Tree Crossovers for Tree Genetic Programming," *GP and Evolvable Machines*, vol. 1, pp. 95–119, 2000.

[11] W.B. Langdon, et al., "The Evolution of Size and Shape," in *Advances in Genetic Programming 3*, L. Spector, et al., Eds. Cambridge: The MIT Press, 1999, pp. 163–190.

[12] W.B. Langdon, "Size Fair and Homologous Tree Genetic Programming Crossover," in *GECCO 1999 Proceeding*, vol. 2, W. Banzhaf, et al., Eds. San Francisco: Morgan Kaufmann Publishers, 1999, pp. 1092–1097.

[13] U.-M. O'Reilly and D.E. Goldberg, "How Fitness Structure Affects Subsolution Acquisition in Genetic Programming," in *GP 1998 Proceedings*, J. R. Koza, et al., Eds. San Francisco: Morgan Kaufmann Publishers, 1998, pp. 269–277.

[14] D. E. Goldberg and U.-M. O'Reilly, "Where Does the Good Stuff Go, and Why?," in *Proceedings of the First European Conference on GP*, W. Banzhaf, et al., Eds. Berlin: Springer-Verlag, 1998.
[15] W. Punch, D. Zongker, and E. Goodman, "The Royal Tree Problem, A Benchmark for Single and Multiple Population Genetic Programming," in *Advances in Genetic Programming*, vol. 2, P. J. Angeline and J. K.E. Kinnear, Eds. Cambridge: The MIT Press, 1996, pp. 299-316.
[16] W. B. Langdon and R. Poli, *Foundations of Genetic Programming*. Berlin: Springer-Verlag, 2002.
[17] D. Zongker and W. Punch, "lilgp," v. 1.02 ed. Lansing: Michigan State University Genetic Algorithms Research and Applications Group, 1995.
[18] J. M. Daida, et al., "Analysis of Single-Node (Building) Blocks in Genetic Programming," in *Advances in Genetic Programming 3*, L. Spector, et al., Eds. Cambridge: The MIT Press, 1999, pp. 217-241.
[19] M. Matsumoto and T. Nishimura, "Mersenne Twister: A 623-Dimensionally Equidistributed Uniform Pseudorandom Number Generator," *ACM Transactions on Modeling and Computer Simulation*, vol. 8, pp. 3-30, 1998.
[20] J. R. Koza, *Genetic Programming: On the Programming of Computers by Means of Natural Selection*. Cambridge: The MIT Press, 1992.
[21] O. A. Chaudhri, et al., "Characterizing a Tunably Difficult Problem in Genetic Programming," in *GECCO 2000 Proceedings*, L. D. Whitley, et al., Eds. San Francisco: Morgan Kaufmann Publishers, 2000, pp. 395-402.
[22] J. M. Daida, et al., "What Makes a Problem GP-Hard? Analysis of a Tunably Difficult Problem in Genetic Programming," *Journal of GP and Evolvable Hardware*, vol. 2, pp. 165-191, 2001.

Generative Representations for Evolving Families of Designs

Gregory S. Hornby

Mail Stop 269-3, NASA Ames Research Center
Moffett Field, CA, 94035-1000
hornby@email.arc.nasa.gov

Abstract. Since typical evolutionary design systems encode only a single artifact with each individual, each time the objective changes a new set of individuals must be evolved. When this objective varies in a way that can be parameterized, a more general method is to use a representation in which a single individual encodes an entire class of artifacts. In addition to saving time by preventing the need for multiple evolutionary runs, the evolution of parameter-controlled designs can create families of artifacts with the same style and a reuse of parts between members of the family. In this paper an evolutionary design system is described which uses a generative representation to encode families of designs. Because a generative representation is an algorithmic encoding of a design, its input parameters are a way to control aspects of the design it generates. By evaluating individuals multiple times with different input parameters the evolutionary design system creates individuals in which the input parameter controls specific aspects of a design. This system is demonstrated on two design substrates: neural-networks which solve the 3/5/7-parity problem and three-dimensional tables of varying heights.

1 Introduction

Evolutionary algorithms have been used in a variety of different design domains, with each individual in the evolutionary design system typically encoding a single design. With this type of representation, each time the objective changes (such as the desired lift of an aircraft wing or the receptive properties of an antenna) a new evolutionary run must be performed. While one option is to use previous results to seed a new run – as Gruau did in evolving parity networks [1] – these additional evolutionary runs can be avoided by evolving individuals which take an input parameter that controls some feature of the resulting design.

One method for an individual to encode a family of designs is for each member of the family to be encoded separately in the genotype. Yet with such a representation the size of the genotype grows with the number and size of each family member and it does not easily generalize to produce a design not already encoded. The alternative is to encode a family of designs with an algorithm which reuses parts of the genotype for different family members.

In addition to efficiencies of space, the reuse of genotypic elements for multiple members in the design family has two other advantages. For consumer products it is often desirable to have different versions of a product that vary in some way, yet have the same style. Whereas the individuals produced by different evolutionary runs usually have a different structure to them, a single individual that generates a family of designs with a reuse of assemblies of components will produce designs with a similar style. This reuse of parts leads to a second advantage of evolving design families, which is improved manufacturability. Since the members of these design families have more components in common than designs produced by multiple runs, there are fewer different parts to test and having assemblies of parts in common across the entire family should result in lower manufacturing costs.

An algorithm for encoding families of designs can be described as a *program* for mapping a *seed* to a design. Using these definitions existing work in evolutionary design can be classified as evolving either a single seed, a program, or both together. For example, the evolution of a vector of parameters with Dawkins' Biomorphs [2] and Ventrella's stick creatures [3] is the evolution of a seed for a pre-defined creature-building program. More common is the evolution of programs for fixed seeds, such as the evolution of cellular automata rules for a fixed starting state [4,5] and the evolution of Lindenmayer systems (L-systems) with a fixed axiom [6,7]. Finally, both seeds and programs have been evolved together, such as Frazer's evolution of both starting condition and cellular-automata-style rules [8] and the evolution of the axiom and rules of L-systems by Jacob [9] and Hornby [10].

Previously we defined *generative representations* as the class of representations in which elements of the genotype are reused in the translation to the phenotype and demonstrated a generative representation in which the genotype contained both a program for creating designs and the input parameters for the starting rule [10]. Here we describe an extension of this work from evolving a single design, with reuse of modules within the design, to evolving design families, with a reuse of modules across different members of the family. To produce individuals which represent a family of designs we now encode in the genotype only the program for constructing a design, and then evaluate an individual multiple times by compiling the program with different starting parameters. For each of these evaluations a specific phenotypic property of the resulting design is then compared with the desired result and the individual's fitness is adjusted based on how closely they match. By testing an individual with different starting parameters in this way, individuals are evolved to be responsive to the parameter values. We demonstrate the generality of this approach by evolving families of designs on two different design substrates: neural-networks which correctly calculate 3/5/7-parity, and three-dimensional tables of varying height.

The rest of this paper is organized as follows. First the generative representation for encoding families of designs is described, followed by a description of the overall evolutionary design system. The next two sections describe the evolution of a family of networks for calculating the 3/5/7 odd-parity function and the evolution of three-dimensional tables of varying heights. Finally we close with a *summary of this paper.*

2 L-Systems as a Generative Representation

The generative representation for our design families is based on parametric Lindenmayer systems (PL-systems) [11]. PL-systems are a grammar consisting of a set of production rules for string rewriting. Production rules are composed of a predecessor, which is the symbol to be replaced, followed by a number of condition-successor pairs. The condition is a boolean expression on the parameters to the production-rule, and the successor consists of a sequence of characters that replace the predecessor. For example in the production:

$$A(n_1, n_1) : n_2 > 5 \rightarrow B(n_2+1)cD(n_2+0.5, n_1-2)$$

the predecessor is $A(n_1, n_2)$, the condition is $n_2 > 5$ and the successor is $B(n_2+1)$ c $D(n_2+0.5, n_1$-2). Predecessor symbols are re-written by testing each of their conditions sequentially and then replacing the predecessor symbol with the successor of the first condition that succeeds.

To generate strings with the grammar a starting string is required. For example the following grammatical rules,

$$A(n_1, n_2) : (n_1 > 0) \rightarrow a(n_1) \; B(n_2, n_1) \; A(n_1 - 1, n_2)$$
$$A(n_1, n_2) : (n_1 \leq 0) \rightarrow a(0)$$
$$B(n_1, n_2) : (n_1 > 1) \rightarrow b(n_2) \; B(n_1 - 1, n_2)$$
$$B(n_1, n_2) : (n_1 \leq 1) \rightarrow b(n_2)$$

when compiled starting with the string, $A(3, 2)$, produce the sequence,

$$A(3,2)$$
$$a(3)B(2,3)A(2,2)$$
$$a(3)b(3)B(1,3)a(2)B(2,2)A(1,2)$$
$$a(3)b(3)b(3)a(2)b(2)B(1,2)a(1)B(2,1)A(0,2)$$
$$a(3)b(3)b(3)a(2)b(2)b(2)a(1)b(1)B(1,1)a(0)$$
$$a(3)b(3)b(3)a(2)b(2)b(2)a(1)b(1)b(1)a(0)$$

The combination of $A(3,2)$ and the grammatical rules is a generative representation for producing the final string in the sequence. In this case the seed consists of $A(3,2)$ and the program is the grammar. Alternatively, by using the starting string $A(n_1, n_2)$, the grammar is a program for creating a family of designs: the first parameter, n_1, controls the number of blocks of b's that are created and the second parameter, n_2, controls how many b's are in each block.

By assigning a meaning to each symbol, the strings produced by a PL-system can be used to construct artifacts. Consider the following PL-system:

$$P0(n_1) : \; n_1 > 1.0 \rightarrow [\; P1(n_1 * 1.5) \;] \; up(1) \; forward(3)$$
$$down(1) \; P0(n_1 - 1)$$

$$P1(n_1) : \; n_1 > 1.0 \rightarrow \{\; [\; forward(n_1) \;] \; left(1) \;\}(4)$$

If this PL-system is started with the string P0(4), it produces the following sequence of strings,

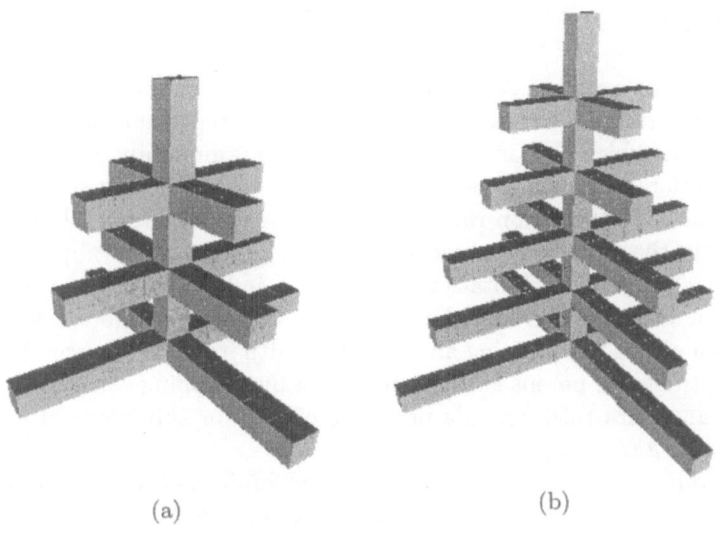

Fig. 1. Two example structures

1. *P0(4)*
2. *[P1(6)] up(1) forward(3) down(1) P0(3)*
3. *[{ [forward(6)] left(1) }(4)] up(1) forward(3) down(1) [P1(4.5)] up(1) forward(3) down(1) P0(2)*
4. *[{ [forward(6)] left(1) }(4)] up(1) forward(3) down(1) [{ [forward(4.5)] left(1) }(4)] up(1) forward(3) down(1) [P1(3)] up(1) forward(3) down(1) P0(1)*
5. *[{ [forward(6)] left(1) }(4)] up(1) forward(3) down(1) [{ [forward(4.5)] left(1) }(4)] up(1) forward(3) down(1) [{ [forward(3)] left(1) }(4)] up(1) forward(3) down(1)*
6. *[[forward(6)] left(1) [forward(6)] left(1) [forward(6)] left(1) [forward(6)] left(1)] up(1) forward(3) down(1) [[forward(4.5)] left(1) [forward(4.5)] left(1) [forward(4.5)] left(1) [forward(4.5)] left(1)] up(1) forward(3) down(1) [[forward(3)] left(1) [forward(3)] left(1) [forward(3)] left(1) [forward(3)] left(1)] up(1) forward(3) down(1) forward(3)*

By interpreting the final string as a sequence of commands to a LOGO-style turtle, this PL-system creates the tree in Fig. 1a.

To encode families of designs with this generative representation, the starting string (seed) is set to the predecessor of the first production rule with variables for starting parameters instead of numerical values. In this example the starting string is $P0(n_1)$ and different values of n_1 will produce different trees: the tree in Fig. 1b is created from this system by starting it with n_1 equal to six.

3 Method

The system for evolving design families uses a canonical evolutionary algorithm (EA) with variation operators customized for the representation. A generational EA is used in which each individual encodes a design family using the generative representation described in Sect. 2. Parents are selected with stochastic remainder selection [12] based on rank, using exponential scaling [13]. To create new individuals, the mutation and recombination operators of Hornby [10] are applied with equal probability. Mutation modifies an individual by changing one symbol with another, perturbing the parameter value of a symbol, adding/deleting some symbols, or recombining an individual with itself. With recombination, one parent is the main parent and it is modified by swapping some genetic material – either an entire rule, a single production body or substrings of a production body – with a second parent.

To produce individuals which encode for families of designs, individuals are evolved such that the value(s) of the input parameter(s) controls a certain feature of a design in a specific way. Each individual is tested with a range of different input values and each design's score is modified by how well the feature in the design matches the desired result. The following two sections will describe the application of this system for two design substrates.

4 Evolution of Parameter-Controlled n-Parity Networks

The first substrate for which families of designs are evolved is neural networks which calculate the odd-parity function. The odd-n-parity function returns true if the number of true inputs is odd and returns false otherwise. This function is difficult because the correct output changes for every change of an input value. In addition, the even/odd-n-parity functions have become a standard benchmark function in genetic programming (GP) and past experiments have shown that GP does not solve the five-parity (or higher) problem without automatically defined functions [14].

The method for using generative representations to encode neural networks is the same as our earlier work [15], which we now summarize. First the generative representation (Sect. 2) is compiled into an assembly procedure and each neural network is constructed from an initial graph by executing this assembly procedure. The initial graph consists of a single neuron which has a single edge from itself to itself and the assembly procedure is a sequence of commands from the following command set, for which the current link connects from neuron A to neuron B:

- add-input(n), creates an input neuron with a link from it to neuron B with weight n.
- add-output(n), creates an output neuron with a link from B to it with weight n.
- decrease-weight(n), subtracts n from the weight of the current link. If the current link is a virtual link, it creates it with weight $-n$.

- duplicate(n), creates a new link from neuron A to neuron B with weight n.
- increase-weight(n), adds n to the weight of the current link. If the current link is a virtual link, it creates it with weight n.
- loop(n), creates a new link from neuron B to itself with weight n.
- merge(n), merges neuron A into neuron B by copying all inputs of A as inputs to B and replacing all occurrences of neuron A as an input with neuron B. The current link then becomes the nth input into neuron B.
- next(n), changes the from-neuron in the current link to its nth sibling.
- output(n), creates an output-neuron, with a linear transfer function, from the current from-neuron with weight n. The current-link continues to be from neuron A to neuron B.
- parent(n), changes the from-neuron in the current link to the nth input-neuron of the current from-neuron. Often there will not be an actual link between the new from-neuron and to-neuron, in which case a virtual link of weight 0 is used.
- [, pops the top state off the stack and makes it the current state.
-], pushes the current state to the stack.
- reverse, deletes the current link and replaces it with a link from B to A with the same weight as the original.
- set-function(n), changes the transfer function of the to-neuron in the current link, B, with: 0, for sigmoid; 1, linear; and 2, for oscillator.
- split(n), creates a new neuron, C, with a sigmoid transfer function, and moves the current link from C to B and creates a new link connecting from neuron A to neuron C with weight n.

The design problem is to evolve an individual that specifies a family of networks with three/five/seven inputs which calculates the three/five/seven-odd-parity problem. To specify which network to construct, the first input parameter is set to 3.0, 5.0 and 7.0 to solve the three, five and seven parity problem. Input values are 1.0 for true and -1.0 for false. Networks are updated four times and then the value of the output neuron is examined to determine the parity value calculated for that set of input values. If the value of the output neuron is > 0.9 then the output of the network is taken as true and if the value of the output neuron is < 0.9 then the output of the network is taken as false. If the absolute value of the output neuron is < 0.9, the network is iteratively updated until its output value is either > 0.9 or < -0.9, for a maximum of six updates. The network receives a score of 2.0 for returning the correct parity value and a score of -1 for an incorrect answer. If the absolute value of the output neuron is less than 0.9 after six network updates, the network receives a score of 1.0 if the value of the output neuron is positive and the parity was true or if the value of the output neuron is negative and the parity is false. No penalty is given for having an incorrect value in this case. The fitness value of a network is the sum of its scores on all possible inputs and an individual's fitness score is the sum of its scores for the three networks.

Using the fitness function described in the previous paragraph, the graph in Fig. 2 contains a plot of the fitness of the best individual in the popula-

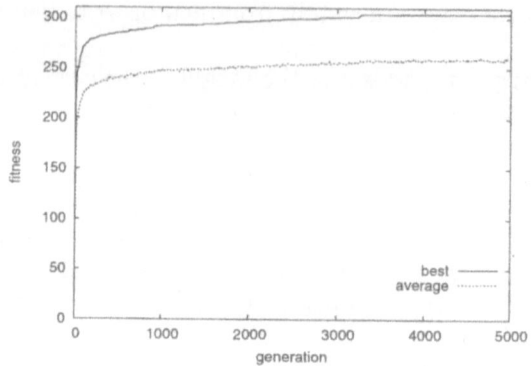

Fig. 2. Best fitness and average population fitness, averaged over fifty trials, for solving 3/5/7-parity. The maximum possible score is 336

tion averaged over fifty trials. For these trials the generative representation was set to a maximum of fifteen productions, each with two parameters and three sets of condition-successor pairs. The maximum number of commands in each condition-successor pair is fifteen and the maximum length of an assembly procedure generated by the representation is ten thousand commands. The evolutionary algorithm used a population of five hundred individuals and was run for five thousand generations. Out of these fifty runs, the generative representation found a solution that produced correct 3/5/7-parity networks twelve times. For those runs that were successful it took 1800 generations, on average, to find a solution that produced correct networks. The smallest networks produced by a single individual that correctly calculates the 3/5/7-parity problems are shown in Fig. 3 (the genotype of this individual is listed in Appendix B of [10]).

5 Evolution of Parameter-Controlled Table Designs

The second design problem is that of evolving families of tables in which the input parameter controls the height of the table. With this substrate, the command set consists of commands for controlling a LOGO-style turtle in a three-dimensional grid [16]. As the turtle moves it fills in voxels, creating a three-dimensional object. The commands for this substrate are:

- back(n), move in the turtle's negative X direction n units.
- clockwise(n), rotate heading $n \times 90°$ about the turtle's X axis.
- counter-clockwise(n), rotate heading $n \times -90°$ about the turtle's X axis.
- down(n), rotate heading $n \times -90°$ about the turtle's Z axis.
- forward(n), move in the turtle's positive X direction n units.
- left(n), rotate heading $n \times 90°$ about the turtle's Y axis.
- [, pops the top state off the stack and makes it the current state.
-], pushes the current state to the stack.

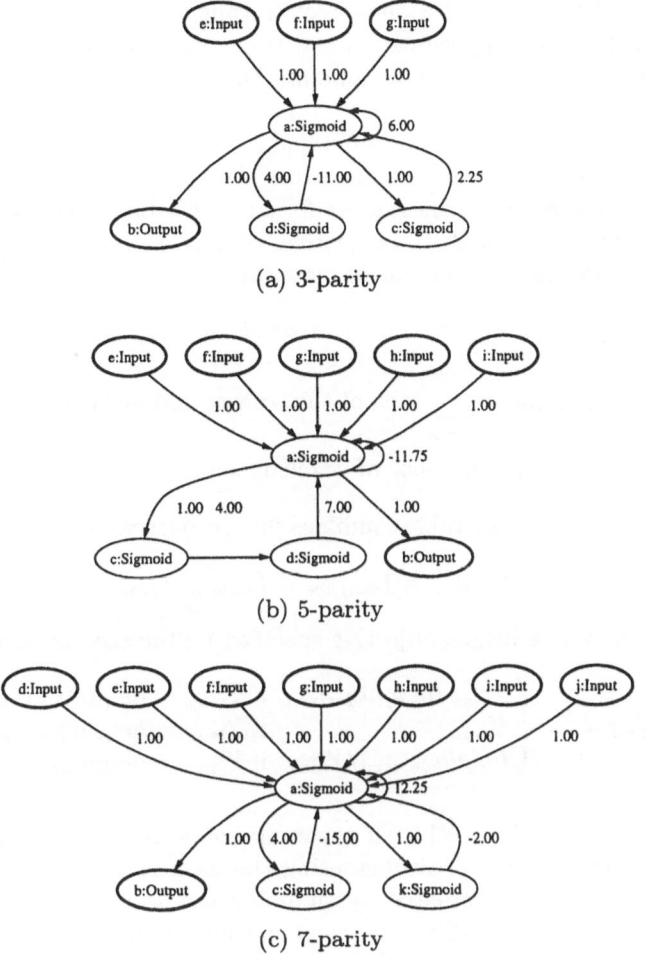

Fig. 3. Networks constructed to solve 3, 5, 7-parity from the same evolved network encoding

- right(n), rotate heading $n \times -90°$ about the turtle's Y axis.
- up(n), rotate heading $n \times 90°$ about the turtle's Z axis.

Section 2 contains an example of a design family encoded using this command set and the assembly procedure it compiles into.

Rather than have the input parameter exactly specify the height of the table, we evolve tables whose height is four times higher than the value of the input parameter. Using a volume of 40×40×40 voxels, the maximum height of a table is forty units, so the valid range of input parameters is from one to ten. To allow us to later determine if an evolved generative representation will interpolate between tested input values and extrapolate beyond tested input values, we evaluate an individual using four input values that cover the range: 2.0, 4.0, 6.0, and 8.0. As with the previous set of experiments, individuals are encoded with the generative representation of Sect. 2 using a maximum of fifteen production rules, each with two parameters and three condition-successor pairs. Since production rules take two input parameters, the second input value is an evolved value and is fixed for all trials.

To score the sensitivity of an individual to its input parameters, an individual's fitness is a combination of scores from the four designs created using four different input values. The fitness score for a single table design is based on that of our earlier work [16], with the objectives of maximizing stability and surface area while minimizing the amount of material used:

$$f_{surface} = \text{the number of voxels at } Y_{max}$$

$$f_{stability} = \sum_{y=0}^{Y_{max}} \text{area of the convex hull at height } y$$

$$f_{material} = \text{number of voxels not on the surface}$$

The overall score of a single table combines these objectives into a single function:

$$score(table) = f_{surface} \times f_{stability}/f_{material} \qquad (1)$$

In addition, there is a height objective specified by the seed parameter:

$$f_{height} = \begin{cases} Y_{max}/height_{desired} & \text{if } Y_{max} < height_{desired} \\ Y_{max} & \text{if } Y_{max} = height_{desired} \\ height_{desired}/Y_{max} & \text{if } Y_{max} > height_{desired} \end{cases}$$

This height objective is a value in the range of zero to one that penalizes a design for under-shooting or overshooting the desired height. A single fitness value for an individual is created by summing the scores for each of the four tables created by the four different seeds and multiplying them by the sum of the height penalties for all four tables:

$$fitness = \left(\sum_{i=1}^{4} f_{height}(table_i) \right) \times \left(\sum_{i=1}^{4} score(table_i) \right) \qquad (2)$$

The reason for summing all the height penalties and applying them to the scores for all tables is to put pressure on the EA to evolve individuals which are sensitive to the seed parameter. With an early version of this test function in which the height penalty for a given table was applied only to that table,

$$fitness = \left(\sum_{i=1}^{4} f_{height}(table_i) \times score(table_i) \right) \qquad (3)$$

evolved individuals tended to produce tables with high fitness for some of the seed parameters and had low fitnesses for others.

Using the fitness function of Eq. (2), Fig. 4 contains a graph plotting the fitness of the best individual in the population and average fitness of the entire

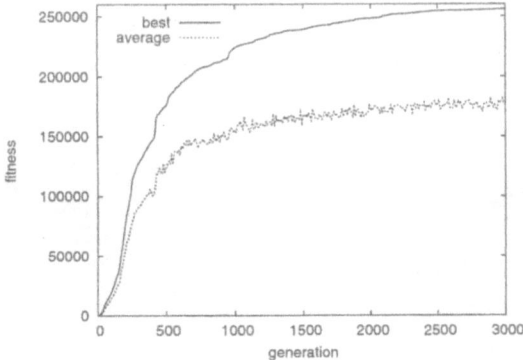

Fig. 4. Best fitness and average population fitness, averaged over fifty trials, for the table design problem

population averaged over fifty trials. Evolved tables were responsive to the seed values and in all fifty trials the individuals in the final generation created tables within one voxel of the desired height for all four seed values. Figure 5 contains six tables in the design family for one of the evolved parameter-controlled tables. The tables in 5a–d are the tables that are generated with tested input parameters 2.0, 4.0, 6.0 and 8.0 – the second parameter is evolvable and in this case is 2.0. The table in 5e is an example with an input of 7.0, demonstrating that this individual can interpolate and generate a table with a seed value that is inside the tested range, and the table in 5f is an example with an input of 10.0, which demonstrates that this individual can produce designs that extrapolate beyond the range tried during evolution.

6 Summary

Typical evolutionary design systems must evolve a new set of individuals each time the design objective changes. Here we presented an evolutionary design system in which individuals use a generative representation to encode a family of designs. By encoding designs as a program and not directly, the generative representation uses an input parameter to control a design feature. Individuals are evolved to be sensitive to this input parameter by evaluating each one multiple times with different input values and combining the scores for the resulting designs into a single fitness function.

Using this system, families of designs were evolved on two different problem domains. On the first design problem, individuals were evolved that encoded three networks that calculated 3, 5 and 7 parity. On the second domain, individuals were evolved such that an input parameter controlled the height of a table. In addition, it was demonstrated that one evolved individual produces tables of the correct height for an input value in between those tested during evolution and for an input value greater than those tested during evolution: examples of

Fig. 5. Parameter-controlled tables: (a)-(d) are the four trial parameters, (e) is an interpolation example, and (f) is an extrapolation example

interpolation and extrapolation. In general, evolving families of designs with a generative representation produced individuals with a reuse of modules among members of the design family. This reuse produced designs in a similar style and should lead to improved manufacturability.

References

1. Gruau, F.: Neural Network Synthesis Using Cellular Encoding and the Genetic Algorithm. PhD thesis, Ecole Normale Supérieure de Lyon (1994)
2. Dawkins, R.: The Blind Watchmaker. Harlow Longman (1986)
3. Ventrella, J.: Explorations in the emergence of morphology and locomotion behavior in animated characters. In Brooks, R., Maes, P., eds.: Proceedings of the Fourth Workshop on Artificial Life, Boston, MA, MIT Press (1994)
4. de Garis, H.: Artificial embryology : The genetic programming of an artificial embryo. In Soucek, B., the IRIS Group, eds.: Dynamic, Genetic and Chaotic Programming, Wiley (1992)
5. Bonabeau, E., Gu rin, S., Snyers, D., Kuntz, P., Theraulaz, G.: Three-dimensional architectures grown by simple 'stigmergic' agents. BioSystems **56** (2000) 13–32
6. Ochoa, G.: On genetic algorithms and lindenmayer systems. In Eiben, A., Baeck, T., Schoenauer, M., Schwefel, H.P., eds.: Parallel Problem Solving from Nature V, Springer-Verlag (1998) 335–344
7. Coates, P., Broughton, T., Jackson, H.: Exploring three-dimensional design worlds using lindenmayer systems and genetic programming. In Bentley, P.J., ed.: Evolutionary Design by Computers. (1999)
8. Frazer, J.: An Evolutionary Architecture. Architectural Association Publications (1995)
9. Jacob, C.: Genetic L-system Programming. In Davidor, Y., Schwefel, P., eds.: Parallel Problem Solving from Nature III, Lecture Notes in Computer Science. Volume 866. (1994) 334–343
10. Hornby, G.S.: Generative Representations for Evolutionary Design Automation. PhD thesis, Michtom School of Computer Science, Brandeis University, Waltham, MA (2003)
11. Prusinkiewicz, P., Lindenmayer, A.: The Algorithmic Beauty of Plants. Springer-Verlag (1990)
12. Bäck, T.: Evolutionary Algorithms in Theory and Practice. Oxford University Press, New York (1996)
13. Michalewicz, Z.: Genetic Algorithms + Data Structures = Evolution Programs. Springer-Verlag, New York (1992)
14. Koza, J.R.: Genetic Programming: on the programming of computers by means of natural selection. MIT Press, Cambridge, Mass. (1992)
15. Hornby, G.S., Pollack, J.B.: Creating high-level components with a generative representation for body-brain evolution. Artificial Life **8** (2002) 223–246
16. Hornby, G.S., Pollack, J.B.: The advantages of generative grammatical encodings for physical design. In: Congress on Evolutionary Computation. (2001) 600–607

Evolutionary Computation Method for Promoter Site Prediction in DNA

Daniel Howard and Karl Benson

Software Evolution Centre
Knowledge and Information Systems
QinetiQ, St Andrews Road, Malvern
WORCS WR14 3PS, United Kingdom
{dhoward,kabenson}@qinetiq.com
http://www.qinetiq.com

Abstract. This paper develops an evolutionary method that learns inductively to recognize the makeup and the position of very short consensus sequences, which are a typical feature of promoters in eukaryotic genomes. This class of method can be used to discover candidate promoter sequences in primary sequence data. If further developed, it has the potential to discover genes which are regulated together.

1 Introduction

Sequencing the complete genome for a variety of organisms is taking place at an unprecedented pace. Meanwhile, computer programs are being developed to scan genomes and to identify likely genes, the *trans*-acting sites. Yet, genomic sequencing data also presents an opportunity for pattern recognition of *cis*-acting sites in the genome whose purpose is to attract proteins, the enzymes, which bind DNA to initiate and regulate gene transcription.

By analogy, the *trans*-acting part of the genome represents the memory of a computer and the *cis*-acting part of the genome represents a computer program that acts on this memory. Understanding *cis*-acting sites is essential to providing models of what is the makeup of this computer program and how it operates. Physical and functional differences between two living forms such as a human and a chimpanzee have much more to do with differences in the execution of this computer program than with differences in the building blocks (both human and chimp use essentially the same proteins). Another example is *Drosophila*, where the proteins used for antenna development are essentially similar to those used for the development of legs. The difference between the appearance and function of these parts is controlled by the sequence of gene expression (mutation of the antennapedia gene transforms the antenna into a leg). It is also interesting to note that this computer program executes in a sophisticated and parallel fashion, as exemplified by polymerase pausing (transcription of a gene halts part way through, and remains in this halted state until a protein, a message, arrives at the scene as a trigger to resume transcription).

An important *cis*-acting region is the *promoter* region located typically upstream of the gene's transcription start site (TSS). This *cis*-acting region attracts proteins essential to the binding of RNA polymerase to DNA strands. RNA polymerase is the enzyme that catalyzes production of RNA from the DNA template.

Promoters are of interest to drug discovery research. It has been argued that Huntington's chorea, a monogenetic disease, occurs when a mutant gene product blocks a crucial promoter region elsewhere in the body, in turn reducing expression of a necessary protein, which results in damage to the brain over time [Cattaneo et al., 2002]. Over-expression of an enzyme by a variation in the promoter plays a crucial role in pharmacokinetics (or how a patient's body handles a particular drug). Genetic variants of the enzymes responsible for absorption, inactivation, and excretion may affect the amount of drug that reaches the biological target significantly, with over-expression causing the drug to be absorbed too rapidly and the body to eliminate it before it can take effect [RGEP, 2002].

Computer programs that identify the *cis*-acting control regions have potential for enabling a researcher to use a computer to scan large sequence databases for specific classes of genes which possess a desired program of regulation.

1.1 Eukaryotic Promoter Prediction

In prokaryotes, transcription is regulated by one RNA polymerase while in eukaryotes there exists three forms. RNA polymerase II (pol-II), is of particular interest because it transcribes heterogeneous RNA (the precursors of mRNAs, and also small nuclear RNAs), and also because the regulation of class II genes is the most complex of the three.

Promoters used by pol-II are recognized principally by separate accessory factors rather than principally by the polymerase enzyme as in prokaryotes. A basal transcription machinery is responsible for the recruitment of pol-II.

An "accessory factor" or "transcription factor" is defined to be any protein needed for the initiation of transcription regardless of the mechanics. Important basal transcription factors are: TFIID, TFIIA, TFIIB, TFIIF, TFIIE, TFIIH, TFIIJ and TFIIS. These factors are all general and responsible only for low basal levels of transcription. There also exist many gene-specific transcriptional factors (activators) which recognize elements in the promoter or other regions of the gene, and binding to them are able to recruit the general factors.

As a rule, promoters have elements or nucleotide sequence motifs that are important to their utility, these are present in all promoters and conserved with some variation. In a prokaryote or bacteria, consensus sequences are well known and established owing to the simple form of the initiation of transcription machinery (the overwhelming role of the σ transcription factor). Eukaryotes also have sequence motifs important to the basal apparatus (see below, Table 6). However, a consensus is not clear, and promoters can function without many of these elements (without important ones such as TATA).

In eukaryotes there must be tissue specific gene transcription because many different cell types need to be created. Regulation must be selective in different cells and tissues, and is unique for more than 200 different cell types, (see chapter 6, [Tsonis, 2003]). This gives rise to a complex interaction of many factors which causes the *cis*-acting regions in eukaryotes to contain a far more complicated distribution of motifs than do promoters in bacteria for instance. Characterizing promoter regions in a prokaryote is far simpler than in eukaryotes.

It is important to note that the gene-specific factors do recognize specific motifs, and that these can affect transcription rates by many orders of magnitude (50 times). Mutations that result in a promoter that is less identical to the basal transcription apparatus "consensus" should lead to lower levels of transcription but this is offset dramatically by the presence of the gene-specific motifs in the promoter or elsewhere.

Commonly accepted definitions are the 'core' and 'proximal' promoter regions of length 50 bp and 500 bp respectively [Bucher, 1990]. A length of 250 bp falls between both of these and is commonly used. With eukaryotes the default state of a gene is inactive and pol-II can only initiate transcription when the general transcription factors recruit it and the machinery binds to *cis*-acting sites, which are usually, but not always, located upstream of the gene. However, as already discussed, many transcription factors will bind outside of the promoter region to greatly assist with basal transcription, and in some cases there may not be enough "pattern" in the promoter region alone to recognize it (and a much longer sequence would be needed). Therefore, expect a lower sensitivity than otherwise as it becomes difficult to eliminate false negatives.

By its definition, the promoter region is located in the general vicinity of TSS, but there is also a distinct region known as the enhancer. Known enhancer regions are several kilobases (kb) upstream of TSS and about 100 base pairs (bp) in length. They contain several closely arranged sequences that stimulate initiation, some of which are quite similar to those of the promoter. Proteins bind to enhancer regions and interact with proteins bound to promoter regions to form a large protein complex. DNA must be coiled or rearranged to allow this interaction which complicates the *in vivo* modelling of transcription. Enhancers assist initiation from great distances, functioning in either orientation, and from either side of the gene [Lewin, 2000]. For this reason, the value of promoter prediction software is to propose putative *cis*-acting sequences, potential promoter regions in eukaryotic genomes, to prioritize the costly experimental effort. This argues for not aiming for highest specificity (tolerating certain false positives).

1.2 Evolutionary Computation and Promoter Prediction

A solution for promoter prediction is to use Evolutionary Computation (EC) to discover the predictor by learning the classification rule from a number of known promoter and non-promoter regions. The aim is to predict whether or not the sequence is a promoter (other research predicts TSS, see [Pedersen et al., 1995]).

A typical feature of promoters is a collection of very short *cis*-acting sequences or motifs. This is an important promoter characteristic that can be used for recognizing promoters. It seems sensible to use this knowledge to design the architecture of the solution which is evolved. That is, analyze the problem in advance using human insight to determine that a certain decomposition is likely to be helpful in solving the problem; establishing an advantageous architecture for the yet to be evolved computer program [Koza, 1999]. One desirable decomposition is to:

1. automatically discover meaningful short sequences;
2. match them to locations;
3. construct meaningful relationships between the location of the sequences;

Additionally, it is desirable to maintain algorithmic simplicity and human-readability for interpretation of the resulting algorithm.

2 Algorithm Description

The combination of Genetic Programming within finite state automata (FSA) or GP-Automata (FSA-GP) was introduced by [Ashlock, 1997]. This representation and architecture was enhanced to include the explicit gathering of state information via the use of state logical functions and successfully applied to machine vision problems [Benson, 2000b]. For the promoter prediction problem the authors enhanced the algorithm further as described in this section.

2.1 Genetic Programming Modules

The objective of the GP tree structure within the GP-Automata is to find motifs within the promoter and non-promoter regions. The function set consists of a single two valued function called GLUE, which concatenates the terminals to produce a motif, and behaves as a logical AND returning true if both input conditions are met.

The terminal set is composed of nucleotides A, C, G, and T. For example the tree (GLUE G (GLUE A T)) is the motif GAT. This tree will return true if in the genomic sequence there is a G in the first position, an A in the second, and a T in the third; else it returns false. To increase flexibility the terminal set is enriched to include terminals which return true for more than one given nucleotide, using the IUPAC ambiguity codes for nucleic acid sequences listed in Table 1. For example (GLUE Y R) will match the patterns CA CG TA TG.

2.2 Finite State Automaton Component

The FSA acts as the main program and each state q_i possesses one GP tree or 'function' F_i which gets called as and when the FSA tour invokes it. Each GP tree returns a boolean that indicates which of two transitions τ_i^T or τ_i^F it should follow to a new state, and consequently deciding what GP-tree function to call next. The FSA possesses a start state and an end state and can be halted after a pre-determined number of moves when it cycles.

A *logical function*, e.g. NAND, is associated with each transition τ_i and an integer value that moves a positional pointer over the DNA sequence, e.g. +10, is

Table 1. Terminals used by GP trees at states of the FSA. Wildcards (left) identify a match to two or more nucleotides (right) using the standard IUPAC convention

M	A C
R	A G
W	A T
S	C G
Y	C T
K	G T
V	A C G
H	A C T
D	A G T
B	C G T
N	A C G T

associated with each state q_i of the FSA. Both undergo genetic modification. As a tour is underway and state visits and transitions between states take place, the main program builds a logical relationship between the boolean output of state functions. The logical function stored at the forthcoming transition combines: (a) the current cumulative boolean result with (b) the boolean output of the function that is executed at the current state (see Table 4).

During transition, the positional pointer to the 300 bp long DNA sequence is moved from left to right (downstream) as many bp as indicated by the positional integer, which is held at the destination state. The convention followed being that sequences are ordered such that TSS is at vector index 250. By definition TSS is at sequence position +1 ([Lewin, 2000], pg 234):

vector index: 0 1 2 .... 249 250 252 .... 298 299
TSS sequence position: -250 -249 -248 .... -1 +1 +2 .... +49 +50

The algorithm can be clarified with the aid of Fig. 1. By default, the machine initiates execution at vector index 0 (corresponding to TSS sequence position -250). Execution commences at the start state q_0. The positional integer at this state is 0 meaning that the positional pointer need not be moved. Pattern SNBHGW is tested and assuming it does not match, the state returns "false" and the machine transits to state q_1 (F in the arrow joining q_0 to q_1). The positional integer at state q_1 is -79 and this means that the pointer is moved 79 bp to the left (it must now subtract 78 places from vector index 299, periodicity is assumed). This places the pointer at vector index 221 (corresponding to TSS sequence position -29). Next it applies pattern WW of state q_1. If a match does not occur the machine transits to state q_2 where the pointer is moved 111 bp to the left, which places the pointer at vector index 110 (corresponding to TSS sequence position -140). Pattern DV corresponding to this state is applied there. Assume that there is no match. This causes a recursion at state q_2 and the pointer is once again moved 111 bp to the left which places the pointer at vector index 299 (corresponding to TSS sequence position +50). Pattern DV corresponding to this state is applied: letter D at TSS sequence position +50 and letter V at

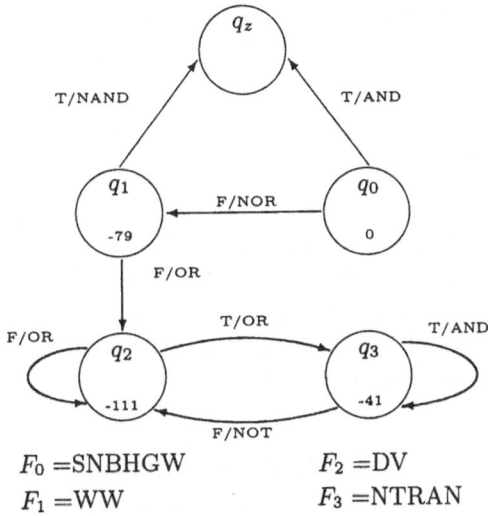

Fig. 1. One of the resulting algorithms from the 5-fold cross validation

TSS position -250 because of the periodicity[1]. Now a match occurs causing the machine to follow the transition to state q_3 where the pointer is moved 41 bp to the left placing it near TSS at TSS sequence position +9 where pattern NTRAN is applied.

State q_z is the so-called end state. Each FSA-GP individual carries with it an integer denoting *maximum number of transitions*. While recursion is not a very serious problem because the pointer gets moved along, this maximum number of states that can be visited limit avoids long tours through the graph.

As the machine executes it 'writes' a logical statement with the logical functions which are associated with each transition. This follows a particular convention (Table 4) which illustrates possible paths through the FSA-GP. The FSA-GP is capable of establishing quite complex decision rules for classification of patterns and has several interesting properties:

1. Since the motifs are constructed via GP tree structures, they can be of various lengths. Since IUPAC wildcards are used, they can match more than one pattern in promoters and non-promoters.
2. The algorithm also incorporates a decision-type process. In the example F_2 did not match the current location being inspected so it was looked for upstream. Only after this motif had been found was the next motif (F_3) searched for. A given motif may be looked for in many places accounting for the possibility that it may occur in more than one place.

---

[1] This does not correspond to the physical reality because the motif is meant to represent a *short sequence* capable of attracting a protein and it is only biologically meaningful for it to occur all together and at one place.

3. The classification of promoter or non-promoter is not based on only one criterion. Instead, the decision is based on a logical combination of the applied motif responses.

2.3 Evolutionary Scheme

Modification of the GP-Automata is achieved using the various mutation operators given in Table 2. Since these are many, selecting the frequency at which they should be applied is difficult. Self-adaptive mutation rates are used to overcome this difficulty [Benson, 2000a]. The particular results presented in this paper restricted the size of the GP-Automata arbitrarily to a maximum of 5 states (in this case to encourage a simple solution).

Table 2. Mutation operators

1. Add a State.	9. Exchange state GPs.
2. Delete a state.	10. Replace a GP with randomly created.
3. Mutate start of reading site.	11. Headless chicken crossover on a GP.
4. Mutate a transition.	12. Grow a subtree.
5. Cycle two states.	13. Shrink a subtree.
6. Mutate a state logical function.	14. Mutate a GP terminal.
7. Headless chicken crossover on a state.	15. Mutate a GP function.
8. Mutate the maximum number of transitions.	16. Mutate the integer on the transition.

Selection was achieved using a $(\mu+\lambda)$-EP strategy and the fitness function $f =$ sensitivity+specificity, i.e. the negative of the sum of equation 1 and equation 2 (see below). In this paper $\mu = 125$ and $\lambda = 125$. At each generation a tournament of size 4 took place and the top μ individuals produce one offspring each via mutation giving a total of λ children. These $\mu+\lambda$ individuals were then carried over into the next generation. The *maximum number of transitions* for each machine was set to be at least 3 and never to exceed 30.

Obviously, many of these choices were arbitrary and the algorithm could be evolved in many other different ways.

3 Application

A data set of promoters and non-promoters held at fruitfly.org[2] was used to evaluate the effectiveness of the algorithm at identifying promoters. Each sequence was 300 bp long extracted 250 bp upstream and 50 bp downstream of TSS.

The data set already comes arranged to facilitate 5-fold cross validation, thus allowing algorithms to be trained and tested on five sets of data. Each file consists of 565 promoters and 890 non-promoters.

---
[2] http://www.fruitfly.org/seq_tools/datasets/Human/promoter/

Table 3. Parameters used in the experimentation

	Parameters
Number of runs	50
Number of Generations	1000
Population size	250
Max states	5 + halt state
Min states	3 + halt state
Max motif length	16
Min motif length	1
Tree terminals	A, C, G, T, M, R, W, S, Y, K, V, H, D, B, N
Tree functions	GLUE

The non-promoters were taken from the coding sequence (CDS) data component of this data set. Each training data set file contains 452 promoters and 712 non-promoters, and each test data set file contains 113 promoters and 178 non-promoters.

The best individual found on each of the 5 training sets was then applied to its test set, as this is the standard way to evaluate the likely performance of an algorithm.

The evolutionary process, however, produces many such algorithms at different generations but in this instance, no attempt was made to generalize (to prevent over-training). No attempt was made to carry out a search for the simplest model that fits the data. Such procedure would judge when to stop the evolution of the different parallel independent runs to avoid over-training, a procedure which may have selected more generalist individuals than those used to generate the results reported here.

Table 3 lists parameters used in the experiments and Table 5 lists the results of the 5-fold cross validation on the five test sets.

Statistical calculations on the results in Table 5 provide a measure of algorithm performance. In the context of this work, sensitivity is the likelihood that a promoter is detected given that it is present, and specificity is the likelihood that a non-promoter is detected given that it is present. Equations 1 and 2 are mathematical definitions of sensitivity and specificity. Here true positives (TP) are correctly classified promoters; false negatives (FN) are promoters classified as non-promoters; true negatives (TN) are correctly classified non-promoters; and false positives (FP) are non-promoters classified as promoters.

$$\text{sensitivity} = \frac{TP}{TP + FN} = 0.79 \tag{1}$$

$$\text{specificity} = \frac{TN}{TN + FP} = 0.92 \tag{2}$$

Table 4. Paths through the machine and resulting logical decision formulae, i.e. fusion of pattern matches. For simplicity of exposition in this table, the logical formulae presented assumes that 4 is the maximum number of state visits allowed for any FSA tour. The logical function fusing two states is stored in the next transition. Note that the logical function NOT is used differently from the other logical functions. NOT negates the output at state q_i and the running formula thus far is discarded

path 1	$F_0(-250) =$ T $\rightarrow$ HALT
path 2	$F_0(-250) =$ F $\rightarrow$ $F_1(-29) =$ T $\rightarrow$ HALT
path 3i, 3ii,...	$F_0(-250) =$ F $\rightarrow$ $F_1(-29) =$ F $\rightarrow$ $F_2(-140) =$ F $\rightarrow$ $F_2(+50) =$ F or T $\rightarrow$ ...
path 4i, 4ii,...	$F_0(-250) =$ F $\rightarrow$ $F_1(-29) =$ F $\rightarrow$ $F_2(-140) =$ T $\rightarrow$ $F_3(-181) =$ F or T $\rightarrow$ ...
path 1	T = PROMOTER
path 2	F NAND T = PROMOTER
path 3i	((F OR F) OR F) OR F = NON PROMOTER
path 3ii	((F OR F) OR F) OR T = PROMOTER
path 4i	((F OR F) OR T) NOT F = PROMOTER
path 4ii	((F OR F) OR T) AND T = PROMOTER

Table 5. Results of five fold cross validation tests (for the best individual from each set of 50 runs). TP, TN, FP, FN defined in text

	TP	TN	FP	FN
Test 1	92	159	19	21
Test 2	87	165	13	26
Test 3	86	168	10	27
Test 4	91	167	11	22
Test 5	90	160	18	23
$\sum$	446	819	71	119

As already stated the fitness measure that obtained these results combined sensitivity and specificity objectives equally.

4 Discussion

Inductive learning by evolution was used to discover knowledge about promoter classificational identification. This can be compared with the straightforward application of knowledge about promoters. Several texts and papers [Lewin, 2000] [Bucher, 1990] establish that the following patterns are useful for recognizing eukaryotic promoters:

1. TATA pattern located approximately 30 bp upstream of TSS;
2. CAAT or TAAC located approximately 75 bp upstream of TSS;
3. GGGCGG or GGCGGG located approximately 90 bp upstream of TSS;
4. YYANWYY located at TSS (known as Inr pattern);

Table 6 presents TP, FP, specificity and sensitivity values obtained by applying the patterns listed above individually and together, to the data from one of the

Table 6. Application of known basal transcription patterns to the data set

	patterns	location (buffer)	TP	FP	sensitivity	specificity
Train 1	all of below		260	69	0.58	0.90
	TATA	-30 (10:10)	203	14	0.45	0.98
	CAAT or TAAC	-75 (10:10)	58	32	0.13	0.95
	GGGCGG or GGCGGG	-90 (10:10)	20	6	0.04	0.99
	Inr	0 (3:3)	45	20	0.10	0.97
Test 1	all of below		66	20	0.59	0.89
	TATA	-30 (10:10)	50	5	0.45	0.97
	CAAT or TAAC	-75 (10:10)	16	11	0.14	0.94
	GGGCGG or GGCGGG	-90 (10:10)	9	1	0.08	0.99
	Inr	0 (3:3)	6	6	0.05	0.97

5 folds. It appears that nearly half of the promoters are either TATA-less or have variations of the patterns listed and thus cannot be predicted easily. The low statistical significance of these consensus sequences and the better success rate of the TATA-box among them is a typical finding (the position-weight matrices obtained by [Audic and Claviere, 1997] on vertebrate sequences from the Eukaryotic Promoter Database (EPD)). Finally, it was not possible to discover the so called TATA downstream promoter pattern in this data set which sometimes appears in DNA.

Figure 1 is a successful algorithm arising in one of the 5-fold cross validation experiments. The four discovered sequences are very short, quite general, and in themselves cannot be discriminatory. Note from Table 4, however, that pattern WW, which stands for AA, AT, TA or AA, gets applied at -29 and that this seems plausible as this is the same TSS sequence location where the TATA box is expected. Yet, it would seem that the evolved FSA relied on a lot of movement to achieve the promoter recognition, most of them evolving to use 30 for the maximum number of transitions.

It is much harder to discriminate promoter regions from non-coding sequences than from CDS. A future implementation will evolve the FSA-GP using non-coding sequences (rather than CDS) for examples of non-promoter regions.

5 Conclusions

Ashlock's GP-Automata [Ashlock, 1997] with logical functions [Benson, 2000b] was enhanced with positional pointers to evolve an FSA-GP promoter predictor. Early results show that it has potential to discover candidate promoter sequences in primary sequence data. EC was used to: (a) discover motifs of various lengths in automata states; (b) discover integer values stored at states to position the testing of the motifs on the DNA string; and (c) combine the motif matches using logical functions to arrive at a promoter identification decision. The resulting FSA-GP could be said to operate comparably to an explicit sequence similarity-based method which looks for the local accumulation in the 250 bp window of matches with previously recognized transcriptional elements [Pestridge, 1995]. It is argued that this scheme is consistent with domain knowledge. For example, it is known that there exist a number of short *cis*-acting sequences in promoters (popular ones appear in Table 6) and that these tend to occur at fixed locations. Some are expected to occur more than once but many occur only once and at a particular location.

The scheme had the required simplicity to enable analysis of the resulting programs to potentially facilitate discovery of putative and meaningful *cis*-acting sites. In spite of the small size imposed (a maximum of 5 transitions) it had the required complexity to arrive at context sensitive decisions for classification of sequences.

The more successful FSA-GPs detected promoters by relying on a great number of visits, and on building an involved truth table of numerous matches to very small (2 or 3 bp long) patterns. More analysis of results on test cases is needed to determine whether the emerging promoter prediction strategy is correct biologically. If it is not, future research should establish how to design the FSA-GP to reflect molecular biology more faithfully than with the current approach.

Simpler EC schemes than the one presented in this paper can be applied to predict core and proximal promoters (weighted matrices with evolution of the weights, a weight is associated with each nucleotide type at every location in the sequence). The scheme presented in this paper could become useful to identify *cis*-acting motifs for gene regulation *in vivo* as part of a computational system that processes much longer DNA sequences (including enhancers and promoters upstream, downstream and inside of the gene). A property of the FSA-GP is code reuse and the creation of structure (function subprograms at the GP states of the FSA-GP). Both can become important for discovery of subtle mechanisms. The algorithm presented applies motifs at precise locations in a sequential testing tour and, if required, a more powerful variant could be less precise when applying its motifs or work in a holistic (parallel) fashion.

Acknowledgements. The authors would like to thank the anonymous reviewers for their careful reading and suggestions, and David Corne for bringing the promoter prediction problem to their attention.

References

[Ashlock, 1997] Ashlock, D. (1997). GP-Automata for dividing the dollar. In Koza et al. (Eds) *Genetic Programming: Proceedings of the Second Annual Conference*, 18–26, Stanford University.

[Audic and Claviere, 1997] Audic, S. and Claviere, J.-M. (1997). Detection of eukaryotic promoters using Markov transition matrices. Computers Chem. Vol 21, No 4, pp 223–227, 1997.

[Benson, 2000a] Benson, K.A. (2000). Evolving Finite State Machines with Embedded Genetic Programming for Automatic Target Detection within SAR Imagery. In *Proceedings of the Congress on Evolutionary Computation*, 1543–1549, La Jolla, San Diego, USA.

[Benson, 2000b] Benson, K.A. (2000). Performing Automatic Target Detection with evolvable finite state automata. Journal of Image and Vision Computing, Volume 20, Issue 9–10, Elsevier.

[Bucher, 1990] Bucher, P. (1990). Weight matrix description of four eukaryotic RNA Polymerase II promoter elements derived from 502 unrelated promoter sequences. Journal of Molecular Biology, 212, pp. 563–578.

[Cattaneo et al., 2002] Cattaneo E., Rigamonti D.,Zuccato C. The Enigma of Huntington's Disease. *Scientific American*, December 2002.

[Fessele et al., 2002] Fessele S., Maier H. Zischek C., Nelson P.J. and Werner T. Regulatory context is a crucial part of gene function. *Trends in Genetics*, vol 18, issue 2, pp 60–63, 2002.

[Handley, 1995] Handley S. Predicting whether or not a nucleic acid sequence is an *E. coli* promoter region using Genetic Programming. Proc. of First International Symposium on Intelligence in Neural and Biological Systems (INBS'95), pp. 122–127, IEEE Comp. Soc. Press, 1995.

[Hannenhalli and Levy, 2001] Hannenhalli S. and Levy S. (2001). Promoter prediction in the human genome. In *Proceedings of the 9th International Conference on Intelligent Systems for Molecular Biology*. Copenhagen, Denmark, July 21–25, 2001, Bioinformatics, Vol 17, Supplement 1, pp. S90–S96. ISSN: 1367–4803.

[Koza, 1999] John R. Koza (1999). *Genetic Programming III: Darwinian Invention and Problem Solving*, MIT Press.

[Lewin, 2000] Benjamin Lewin (2000). Genes VII, Oxford University Press.

[Orphanides et al., 1996] Orphanides, G., Lagrange, T., Reinberg, D.. The general transcription factors of RNA polymerase II. *Genes. Dev.*, vol. 10, 2657–2683, 1996.

[Tsonis, 2003] Tsonis P.S. Anatomy of Gene Regulation: A Three-Dimensional Structural Analysis. Cambridge University Press, 2003.

[Pedersen et al., 1995] Pedersen, A.G. and Engelbrecht, J. Investigations of *Escherichia coli* promoter sequences with ANN. Proceedings: Third International Conference on Intelligent Systems for Molecular Biology, pp. 292–299, 1995.

[Pedersen et al., 1999] Pedersen, A.G., Baldi, P., Chauvin, Y., Brunak, S. The biology of eukaryotic promoter prediction – a review. *Computers and Chemistry*, vol. 23, 191–207, 1999.

[Pestridge, 1995] Pestridge D.S. (1995) Predicting pol-II promoter sequences using transcription factor binding sites. *Journal of Molecular Biology*, vol. 249, 923–932, 1995.

[RGEP, 2002] Roche Genetics Education Program CD-ROM. *Scientific American*, December 2002.

Convergence of Program Fitness Landscapes

W.B. Langdon

Computer Science, University College, London, Gower Street, London, UK
W.Langdon@cs.ucl.ac.uk
http://www.cs.ucl.ac.uk/staff/W.Langdon

Abstract. Point mutation has no effect on almost all linear programs. In two genetic programming (GP) computers (cyclic and bit flip) we calculate the fitness evaluations needed using steepest ascent and first ascent hill climbers and evolutionary search. We describe how the average fitness landscape scales with program length and give general bounds.

1 Introduction

Fitness landscapes are an attractive metaphor. Easy problems are supposed to have "smooth" landscapes, while hard problems are supposed to be caused by "rugged" landscapes. Much analysis of landscapes is empirical, but these have not led to general results in GP. [Kinnear, Jr., 1994] found little relationship between GP difficulty and ruggedness on a number of GP benchmarks, while [Nikolaev and Slavov, 1998,Clergue et al., 2002] give counter examples. However [Daida et al., 2001] agrees the landscape metaphor may be deceptive for GP. We conclude the empirical picture is not clear. As an alternative to amassing yet more data, we have analysed the general properties of GP, giving general results.

In [Langdon and Poli, 2002] we showed for both linear and tree genetic programming (GP) that in general the space of programs which GP searches converges as the programs get bigger. That is, beyond a threshold further increase in size makes little difference. Since then, for simplicity, we have concentrated on linear GP [Banzhaf et al., 1998]. In linear GP, a program consists of a linear sequence (i.e. no loops) of instructions which manipulate memory registers. In [Langdon, 2002a] we defined five computer models (any, average, cyclic, bit flip and Boolean) and provided quantitative bounds on how long programs have to be so that the distribution of their outputs is near its limit. [Langdon, 2002b] deals quantitatively not only with program's outputs but also with the relationship between a program's outputs given different inputs, i.e. the function it implements. In [Langdon, 2003] we show that reversible programs tend to rather different limits. So far we have gathered formal results about the distribution of fitness of programs. This earlier analysis tells us about the search space but not how search operators connect it into a fitness landscape. Using the same computer models we use simple mutation in conjunction with two simple hill climbing strategies and population based search to yield results on the convergence of fitness landscapes and expected solution times.

In Sects. 3 and 4 we consider in detail the fitness landscape of two simple computers and calculate how long various search techniques will take to find programs of specified fitness. Section 5 shows we can put bounds on the fitness landscape for any computer, while Sect. 6 considers average behaviour across all computers. Some experimental measurements for the fifth model (Boolean linear GP) are given in Fig. 3.

2 Size of Neighbourhood – Point Mutation

Point mutation uniformly at random selects one instruction in a program and changes it. (So the mutant is always genetically different from its parent). Assume there are I different instructions, so a point mutation at a chosen point will convert the program to one of $I-1$ other programs. If the program contains l instructions, there are $l(I-1)$ other programs that can be created from it by a single point mutation. I.e., under point mutation each point in the fitness landscape has $l(I-1)$ distinct neighbours. Note the point mutation neighbourhood size increases directly in proportion to program size.

However since point mutation does not change program size, only a tiny fraction of all the programs can be reached by point mutation. Even as a fraction of programs of the same size, the neighbourhood is only $l(I-1)/I^l \approx l\, I^{1-l}$ of the total.

3 Cyclic Computer – Point Mutation Neighbourhood

The cyclic computer has only three instructions ($I = 3$) add one to memory, subtract one from memory, do nothing. Therefore the point neighbourhood size under point mutation is $l(I-1) = 2l$. The cyclic computer is obviously an unrealistic model of real computation, nonetheless by studying it we learn about real computers.

Suppose our program contains n_+ increments, n_0 no ops and n_- decrements. The complete point mutation neighbourhood contains n_- programs whose answers are two more (allowing wrap around), $n_0 + n_-$ whose answers are one more, $n_+ + n_0$ whose answers are one less and n_+ whose answers are two less. Note there are no neutral changes. The proportions of these four changes depends only upon the relative numbers of the three types of instructions in the original program and not directly on the length of the program. If we look at changes in program output, the point mutation landscape converges immediately to 1/6 −2, 1/3 −1, 1/3 +1 and 1/6 +2. Whereas if we consider the actual outputs exponentially large programs are needed for convergence [Langdon, 2002a].

If we start from a very unusual part of the search space ($n_+ \neq n_0 \neq n_-$) then this will distort the point mutation fitness landscape by changing the fractions of the four changes. If high fitness is associated with output values very different from those input, high fitness programs will have n_+ very different from n_-. This means *point mutation of high fitness programs will on average produce offspring with lower fitness.*

3.1 Hill Climbing with Point Mutation – Cyclic Computer

The simple instruction set means the output of any program is $(x+p) \bmod 2^m$. (Where x is its input, m is the number of bits in the output register and $p = n_+ - n_-$. So p is a constant for the program.) Note given the output for an input, the program's output for any other input can be inferred. Indeed we need only define one fitness case (e.g. input is zero). Define $F = |\text{output} - \text{target}|$, so the goal is to minimise F. We can calculate how long it will take two hill climbing strategies to find a solution.

In both cases, we start from a single randomly chosen program. For simplicity assume $n_+ = n_-$. (This is equivalent to assuming random fluctuations are small compared to the target. I.e. $\sqrt{l/3} \ll \text{target}$. Also assume target $< 2^{m-1}$, so the fastest route is to increase p.) $p = 0$ initialy and the task is to find a sequence of point mutations which will increase p to target. If we use steepest ascent, we need only replace target/2 decrement instructions by increment instructions. This will take $2l \times \text{target}/2 = l \times \text{target}$, fitness evaluations.

With first ascent, the fraction of mutations incrementing n_+ is $n_0/2l + n_-/2l$, while the fraction leading to a fall in n_- is n_-/l. Define $x = n_+/l$ and $y = n_-/l$ and treat x and y as continuous variables of the expected case. So

$$\frac{dx}{dt} = \frac{1}{l}\left(\frac{n_0}{2l} + \frac{n_-}{2l}\right) \Rightarrow x = 1 - \tfrac{2}{3}e^{-t/2l} \text{ and } \frac{dy}{dt} = -\tfrac{1}{l}\frac{n_-}{l} \Rightarrow y = \tfrac{1}{3}e^{-t/l}$$

The number of steps expected to be needed to first find a solution is given by the value of t for which target $= l(x - y)$. So target $= l(1 - \tfrac{2}{3}e^{-t/2l} - \tfrac{1}{3}e^{-t/l})$. Rearranging gives

$$t = -2l \log(2\sqrt{1 - 3\,\text{target}/4l} - 1) \quad \text{If target} \ll l \quad t \approx 3/2 \text{ target} \qquad (1)$$

That is, fitness initialy rises linearly but the rate of increase slows as the maximum fitness, given by the length of the program l, is approached.

If target $= l$ the problem becomes very similar to the OneMax problem. The approximation of treating x and y as continuous variables needs to be treated with care. Set target $= l - \epsilon$ in (1) and assuming $l \gg \epsilon$ and then let $\epsilon = 1$ gives t being in the region of $2l \log(2l/3)$. (Cf. $O(l \log l)$ for One-Max [Muhlenbein and Schlierkamp-Voosen, 1993].) In contrast steepest ascent requires l^2 fitness evaluations.

3.2 Population Approaches – Cyclic Computer

In the following analysis for simplicity we allow an offspring produced by point mutation into the population only if it is fitter than its parent. In which case, it replaces its parent.

Parallel Steepest Ascent. With steepest ascent each step explores the complete neighbourhood ($2l$ fitness evaluations) but, unless the problem has been solved, steepest ascent is guaranteed to find a better child, which will be accepted by the population. Therefore the solution found will be the direct descendent of the fittest program (i.e. smallest F) in the initial population (fitness F_0). Since fitness is given by difference between a programs output and the required output and steepest ascent reduces the difference by 2 each generation, a solution will be found in generation $F_0/2$.

The number of n_+ in the initial (random) programs is given by a binomial distribution $C_{n_+}^{l-n_+}(\frac{1}{3})^{n_+}(\frac{2}{3})^{(l-n_+)}$. Even for modest l, this can be approximated by a Gaussian distribution with mean $l/3$ and variance $\frac{1}{3}\frac{2}{3}l = \frac{2}{9}l$. The initial distribution of n_0 and n_- are the same as that of n_+. Provided none of them is near zero, we can treat the distribution of any two as being independent, so the distribution of $p = n_+ - n_-$ can also be approximated by a Gaussian (whose mean is the difference in the means (0) and variance is sum of the variances $\frac{4}{9}l$, i.e. standard deviation $\sqrt{4/9l} = \frac{2}{3}\sqrt{l}$).

The likely fitness of the fittest program in the initial population is given by the population size, M_l, $F_{0l} = \text{target} - \frac{2}{3}\sqrt{l}\,\Phi^{-1}(1-1/M_l)$. ($\Phi^{-1}$ is the inverse of the integral of the Gaussian distribution, it gives the number of standard deviations for a particular probability.) Note that both larger initial programs and a larger population can be expected to give a fitter initial best program. The expected number of generations to find a solution is $\text{target}/2 - \max_l 1/3\sqrt{l}\,\Phi^{-1}(1-1/M_l)$. If the initial population is large, so that there are a large number of programs of each length, we can be confident that the fittest program in the initial population is also one of the longest. Further that $\Phi^{-1}(1-1/M_l) \approx 3$. Therefore the expected number of generations to find the first solution will be about $\text{target}/2 - \sqrt{l_{\max}}$.

The number of fitness evaluations depends upon the spread of initial fitness values and the selection technique used. We assume the selection pressure is strong enough to ensure at least one copy of the fittest program is copied to the mating pool. With tournament selection and tournaments of size T there will be on average T copies of the best. This leads to rapid convergence of the population (in $\approx \log_T M$ generations[1]). If $\log_T M \ll \text{target}/2 - \sqrt{l_{\max}}$ then the number of fitness evaluations expected to be required to find a solution will be about $Ml_{\max}(\text{target} - 2\sqrt{l_{\max}})$. Rather fewer fitness evaluations will be needed if the population still retains shorter programs.

At the other extreme is to have no selection pressure and instead to give each member of the current population exactly one child. With steepest ascent, each offspring will be exactly 2 fitter than its parent and so all children will be inserted into the next generation. Therefore the number of fitness evaluations expected to be required to find a solution will be about $M\bar{l}(\text{target} - 2\sqrt{l_{\max}})$. Where $\bar{l}$ is the mean length of programs, in the initial and hence every generation.

Notice how the temporal granularity of having fixed non-overlapping generations gives rise to a simple population dynamics. Suppose instead of forcing each steepest ascent in the population to synchronise by waiting for every new child, we allow each child to be compared with its parent immediately. This gives a speed advantage to shorter programs. Using tournament selection we now get a race. A few of the longer programs initialy have an advantage and we can

---

[1] [Goldberg and Deb, 1991, p74 and p80] includes an additional $\log_T \log M$ term.

expect the average length of programs to start to increase. What happens next depends in a complicated way on the distribution of program lengths and the selection pressure. If the selection pressure is very high and the programs are of similar lengths then we expect shorter programs to be removed from the population before they can catch up with the fittest (longest) program and the average program size will continue to increase. However if the shorter programs are very much shorter and the selection pressure is not so great, one of them can increase its fitness much faster than the longest and so then will be selected for, causing the average program length to decrease.

Since the second selection scheme (where every program gets exactly one child, and hence is replaced by it) is simpler, we can analyse it in more detail. Assume all programs are at least long enough to be able to solve the problem. After tM fitness evaluations the fitness of the best program of length l will be about target $- 2\sqrt{l} - t/l$. (Remember its initial fitness $\approx$ target $- 2\sqrt{l}$ and it improves by $+2$ every $2lM$ fitness evaluations.) Initialy the best program in the population will also be one of the longest but the shortest will catch it up. We can calculate the expected number of fitness evaluations t needed by setting the expected best fitness of the longest and shortest programs to be equal. Define $r = l_{\max}/l_{\min}$ then

$$2\sqrt{l_{\min}} + t/l_{\min} = 2\sqrt{l_{\max}} + t/l_{\max} \Rightarrow t = 2\frac{1-r^{-0.5}}{r-1} l_{\max}^{3/2}$$

If target is small then it will be found first by one of the longest programs, otherwise by one of the shortest. Substituting t we get the critical target value target$_{\text{crit}} = 2\left(\frac{r-r^{-0.5}}{r-1}\right)\sqrt{l_{\max}}$. If target is less than the critical value the number of fitness evaluations expected to solve the problem is about $Ml_{\max}(\text{target} - 2\sqrt{l_{\max}})$ and $Ml_{\min}(\text{target} - 2\sqrt{l_{\min}})$ otherwise.

Parallel First Ascent. With first ascent mutation in a population it is natural to consider a generational approach in which every offspring is produced by exactly one point mutation. The M new individuals then become candidates to be members of the new population. Note this finer level of granularity removes the speed advantage of shorter programs seen with steepest ascent (previous section). In fact longer programs now have a modest advantage since as fitness climbs the chance of making a successful point mutation falls more slowly than it does for the shorter programs, cf. derivation of Equation (1).

With tournament selection we would expect rapid convergence of program sizes towards that of the fittest individual in the initial population. (With a large population we expect this to be the longest length in the initial population.) I.e. the average program size will grow towards $l_{\max}$ in $\approx \log_T M$ generations.

If the selection pressure is high ($T \gg 2l/n_- \approx 6$) then we can be reasonably sure each generation at least one of the T children of the best individual in the population will have been formed by mutating a decrement instruction into an increment instruction, increasing its output by 2 compared to its parent. Thus the same number of generations, target$/2 - \sqrt{l_{\max}}$, will be needed to solve the problem, as are needed by steepest ascent. The number of fitness evaluations will be $M(\text{target}/2 - \sqrt{l_{\max}})$. If the fittest is not quite so dominant, after each improvement is found their may be a gap generation where the next improvement is not found or a smaller improvement in fitness is found. The existing best will spread through the population, giving it about T^2 copies on average in the next generation, making it much more likely a +2 fitness improvement will be found. Hence $\text{target}/2 - \sqrt{l_{\max}}$ is a reasonable estimate even for more modest tournament sizes.

In the second selection scheme (in which every program gets exactly one child, which is the same length as it) there will be no change in average program size. The second selection scheme means each slot in the population is independent and so hill climbs in parallel but in isolation. Taking into account the expected best fitness in the initial generation ($\approx 2\sqrt{l_{\max}}$) in Equation (1), gives the average number of fitness evaluations required to solve the problem as $\approx -2Ml_{\max}\log(2\sqrt{1 - 3\,(\text{target} - 2\sqrt{l_{\max}})/4l_{\max}} - 1)$. With M searches in parallel we can expect one lucky one to find a solution before the others. If this were included, the number of fitness evaluations would be reduced by $O(M\sqrt{t})$. Where $\text{target} \ll l_{\max}$ the number of fitness evaluations to reach target is about $3/2\; M(\text{target} - 2\sqrt{l_{\max}})$, cf. Approximation (1).

Notice that even though we have used a single genetic operator on the same fitness function, i.e. a single fitness landscape, we have seen many different behaviours. Small differences in the sequence of fitness evaluation, selection and replacement can lead to macroscopic changes. "One operator, One landscape" [Jones, 1995] is not sufficient to explain evolution.

4 Bit Flip Computer

Our second example is the bit flip computer. It contains N bits of memory and $N+1$ instructions, one no op and N instructions which read their memory cell and invert it. All N bits can be used during a program's execution but only the m bits of the output register (which overlap the input register) are used for output when the program stops. Like the cyclic computer, it can only implement 2^m functions and in the long program limit each are equally likely. However they need only contain $\frac{1}{4}(N+1)(\log(m)+4)$ random instructions to be close to the uniform limit, rather than an exponentially large number for the cyclic computer [Langdon, 2002a].

If we follow any program by $\frac{1}{4}(N+1)(\log(m)+4)$ randomly chosen instructions its new fitness will effectively be uncorrelated with its original fitness. Adjacent pairs of instructions flipping the same bit can be stripped out of the random addition with no effect. Each remaining random i,j pair has the same effect as replacing an i instruction in the original program with a j instruction. (If the program did not contain any i instructions, two can be created by mutating a pair of two other instructions, which need not be adjacent, into i instructions.) I.e. no more than $\frac{1}{4}(N+1)(\log(m)+4)$ independent point mutations are needed to scramble any bit flip computer program's fitness.

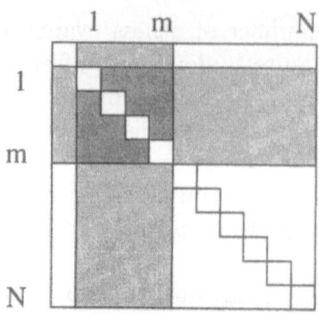 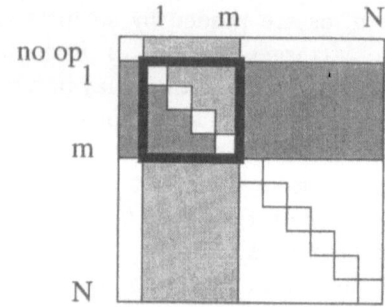

Fig. 1. Effect of point mutation on bit flip computer programs. Original instruction (rows) v. new instruction (columns). White – not a mutation or no effect, light grey – 1 bit flipped, dark – 2 output bits flipped.

Fig. 2. Acceptance by first ascent hill climber. White – not a mutation or not accepted, light grey – accepted if no. new instructions is even, dark – accepted if no. old instructions was even. Inside bold square mutation will change fitness by two bits, elsewhere only one.

Unlike the cyclic computer, many point mutations have no effect. Of the $N(N+1)$ mutations from one instruction to another, $(N-m)(N+1-m)$ affect only the $N-m$ bits of memory not used by the output register and so have no effect on the value output by the program. $2m(N+1-m)$ invert one bit of the output register, while the remaining $m(m-1)$ invert two output bits (see Fig. 1).

For concreteness we define the fitness function to be given uniquely by the function implemented by each program. For this computer, this is equal to the value returned by the program when given input zero. Which in turn is given by counting the number of instructions flipping bits $1\ldots m$. Call each of these C_i, and then the formula for fitness is fitness $= \sum_{i=1}^{m}(C_i \bmod 2)2^{i-1}$. In the long program limit, each fitness is equally likely, so the mean fitness is $2^{m-1} - 0.5$ with variance $\sqrt{(2^{2m}-1)/12}$ (SD $\approx 0.5774\ 2^{m-1}$).

The fitness neighbourhood under point mutation of a program depends upon the relative number of each of the instructions from which is made, i.e. not just on its fitness. For example, a program of $2N$ no ops will have fitness 0 and can be mutated to fitness $0, 1, 2, 4, \ldots 2^m$. (Looking at the top row of Fig. 1 we see the probability of no fitness change is $(N-m)/N$ and that of each change is $1/N$.) While another program with N pairs of each bit flip operations will also have fitness 0. It too can be mutated to fitness 0 (cf. both lower white areas in Fig. 1, probability $((N-m)/N)^2$), and to $1, 2, 4, \ldots 2^m$. (Each 1 bit change has probability $(1 + 2(N-m))/N^2$, cf. light grey in Fig. 1). But also to fitnesses $3, 5, 6, 9, 10, \ldots (2^{m-1}+2^m)$. (Each 2 bit change has probability $2/N^2$, dark grey in Fig. 1). These two examples show, that the current fitness of a program is not sufficient to define either its neighbours or the probability of moving between particular points on the fitness landscape.

4.1 Bit Flip Computer – Steepest Ascent

We start with a randomly chosen program of length l ($l \geq m$). About half the C_i will be odd. Guided by the fitness function, steepest ascent hill climbing will take about $m/2$ complete steps to find a program with maximum fitness $2^m - 1$. Each step takes lN fitness evaluations. I.e. we expect to reach the optimum in $\frac{1}{2}lmN$ fitness evaluations. If we chose an initial program of exactly the minimum length ($l = m$) then the effort is minimised $\frac{1}{2}m^2 N$.

4.2 Bit Flip Computer – First Ascent

Again we start with a randomly chosen program of length l. For simplicity we will assume exactly $m/2$ of C_i ($1 \leq i \leq m$) are odd. Otherwise the instructions are uniformly chosen. We make random point mutations one at a time. After each the offspring's fitness is calculated and the mutation is accepted if its fitness is greater than before. See Fig. 2.

If $l \gg m$ we can assume that even as the search proceeds and changes are made to the program, the proportion of each instruction remains nearly uniform. Therefore each of the $N(N+1)$ possible mutations (cf. Fig. 1) remain equally likely. However the chance of a mutation being accepted falls as more instructions become correctly paired. Unless both the instruction being replaced and the one replacing it affect the output register (i.e. both lie inside the inner square of Fig. 2) the chance of acceptance will remain proportional to the number bits unset in the fitness value. Inside the square, mutations affect two bits of the fitness. Accepted mutations may set them both or set the larger one but clear the smaller. For simplicity we ignore the second possibility, this means we will slightly over estimate the average number of mutations needed to reach a solution.

The probability of an accepted mutation which increases the number of $C_{1 \leq i \leq m}$ which are odd by one is $\frac{\text{even}}{m} \frac{2m(N+1-m)}{N(N+1)}$. Where even is the number of $C_{1 \leq i \leq m}$ which are even. While the chance of an accepted mutation increasing the number of C_i which are odd by two is about $\left(\frac{\text{even}}{m}\right)^2 \frac{m(m-1)}{N(N+1)}$. Combining gives the average reduction in the number of mismatched I/O flips per mutation as $\geq 2 \frac{\frac{\text{even}}{m} m(N+1-m) + \left(\frac{\text{even}}{m}\right)^2 m(m-1)}{N(N+1)}$. So the expected number of fitness evaluations required to find a solution is

$$\leq \frac{N(N+1)}{2} \sum_{\text{even}=m/2}^{1} \frac{1}{\frac{\text{even}}{m}m(N+1-m) + \left(\frac{\text{even}}{m}\right)^2 m(m-1)}$$

$$= \frac{N(N+1)}{2} \sum_{\text{even}=m/2}^{1} \frac{1}{\frac{\text{even}}{m}m(N+1-m)} - \frac{1}{\frac{\text{even}}{m}m(N+1-m) + \frac{m^2(N+1-m)^2}{m(m-1)}}$$

$$\approx \frac{N(N+1)}{2(N+1-m)} \left(H\left(\frac{m}{2}\right) - H\left(\frac{m}{2} + \frac{m(N+1-m)}{(m-1)}\right) + H\left(1 + \frac{m(N+1-m)}{(m-1)}\right) \right)$$

H is the Harmonic Number, $H(x) = \sum_{i=1}^{x} 1/i$. If $x \gg 1$, $H(x) \approx \log x + \gamma$, where γ is Euler's constant (≈ 0.57721566). So assuming $N \gg m \gg 1$, the expected number of fitness evaluations required to find a solution is about $\frac{1}{2}N\log(0.8905362\,m)$. Note this means on average first ascent requires fewer fitness evaluations than than steepest ascent.

5 Any Irreversible Computer

By an "irreversible" computer we mean that there is a program which when run with two or more inputs yields the same answer. I.e. it is impossible to run the program backwards from its output to uniquely determine what input it was initialy given. Practical computers are irreversible[2].

Consider two or more identical copies of a computer. They run copies (or mutated copies) of the same program and their clocks are synchronised. They may have different initial conditions but we shall show after running a long randomly chosen program they will all become synchronised. Once two such computers are synchronised they will remain synchronised (unless one of them strikes a mutation).

We strengthen our definition of "irreversible" to require that for every pair of states there exists a computer program which when run on two computers, each starting in one of the states, will eventually cause both to be in the same state at the same time. Define a as being the length of the longest (over all pairs of states) such minimal program.

Suppose we chose a program at random and evaluate its fitness, i.e. run it on each of T fitness cases. Then we mutate it and evaluate the fitness of its offspring. In both cases the computer goes through a random sequence of states until the program reaches its last instruction and halts. For an arbitrary mutation, after a instructions past the mutation site, it is possible that both programs will arrive at the same state. (By definition, at least one of the I^a, sequences of a instructions, will do this.) In which case, they will remain synchronised and so output the same value. If for each fitness case, they always synchronise then they will have the same overall fitness.

If the mutation is followed by a instructions the chance the two programs will finish in the same state is at least I^{-a}. Suppose that just before the mutation site, over the T pairs of runs, the computers are in T' distinct states ($T' \leq T$). If the mutation is followed by $T'a$ instructions, the chance the two programs will finish together in the same state when run in T pairs is at least $T'!I^{-T'a}$. If the mutation is followed by i random instructions the chance of being in different states for at least one input is no more than $(1-T'!I^{-T'a})^{\lfloor i/T'a \rfloor}$. For a program of length l the average chance of a mutation causing a fitness change is no more than

$$\leq \tfrac{1}{l}\sum_{i=T'a}^{l-1}(1-T'!I^{-T'a})^{\lfloor i/T'a \rfloor} < \tfrac{T'aI^{T'a}}{l}\left(1-(1-T'!I^{-T'a})^{l/T'a}\right)$$

The average chance of a point mutation changing the fitness of a long average program falls at least in proportion to its length. For any irreversible computer,

---

[2] In contrast quantum computers are reversible.

and any fitness function, almost all (i.e. at least 90%) point mutations on programs longer than $10\ TaI^{Ta}$ have no effect.

5.1 Average Fitness on Any Computer

We shall show on any irreversible computer almost all programs are useless. Again we run each of the T fitness cases on T copies of the same program running on identical computers. If the program is chosen at random and is at least a instructions long then there is a chance of at least $(T-1)I^{-a}$ that two of the computers will yield the same output. If we make it aT instructions long the chance all T computers are synchronised is at least $(T-1)!I^{-Ta}$. So the chance of not returning a constant is less than $1 - (T-1)!I^{-Ta}$. For programs of length $l \geq Ta$ the chance of not returning a constant is $\leq (1 - (T-1)!I^{-Ta})^{\lfloor l/Ta \rfloor}$. Setting this to 10%, taking logs and rearranging gives a lower bound, meaning almost all programs longer than $2.3\ TaI^{Ta}/(T-1)!$ when run all T fitness cases the program will yield the same output. If the fitness testing is exhaustive, (all 2^n tests are run) then on any irreversible computer almost all programs longer than $2.3\ 2^n aI^{2^n a}/(2^n - 1)! \approx 0.85\ a(2.718\ I^a/2^n)^{2^n}$ have zero fitness.

6 Average Computer

In the previous sections we have been treating the computer as a machine whose state is given by its memory and movement between those states is controlled by instructions in its instruction set. In two cases we have considered a specific class of computers and this prescribed their instruction set, while in the other we have set very loose restrictions on the computer to derive general limits for all computers. We define an average computer as one that is representative of all the computers with linear programs (no loops), a fixed memory N bits and fixed number of instructions I. The trick is to realise that a random change is by far the most likely of all the possible ways an instruction could change the computer's memory. I.e. the average computer contains I instructions each of which makes fixed but random changes to the value held in its memory.

On the average computer point mutation is very disruptive. Changing a single instruction means that just after the mutation the state of a computer running the parent program and that of one running the mutation are totally uncorrelated.

6.1 Fitness of the Average Computer Program

Here we take a very high level definition of fitness. We say if a program is run on T test cases and yields T' different answers then its fitness is T'. As before we set up T copies of the computer and program, all running in lock step. Assume $2^N \gg 2^m \gg T \gg 1$. If $l \leq 2^N/(T-1)$ then on average all of the T computers are in different states and so the expected fitness $\overline{f}$ is T. If l is between $2^N/(T-1)$ and $2^N/(T-1) + 2^N/(T-2)$ then on average two of them are in the same state so $\overline{f}$ is $T-1$. In general, if $l \leq 2^N \sum_{i=T'-1}^{i=T-1} 1/i$ then $\overline{f} \geq T'$. Approximating this sum (the Harmonic number) with the logarithm and rearranging gives the expected fitness $\overline{f} \approx 1 + (T-1)e^{-l/2^N}$, indicating on average fitness falls exponentially with program length. (This is consistent with the expected length at which programs become independent of their inputs $(\log(T-1) + \gamma)2^N$ [Langdon, 2002b, 4.2].)

6.2 Average Computer – Point Mutation

Suppose we set up T pairs of identical computers running in lock step. Initialy all run the same program. Now suppose we make the same point mutation to one program of each pair. As the computer pairs run, initialy they will be in the same state but overall we expect the total number of different states to start at T and then fall exponentially (as described in Sect. 6.1). Suppose at the point of mutation, the total number of different states of the $2T$ computers is T'. Just after the point where the programs run into the mutation, it is very likely that each pair will separate so doubling the total number of different states to $2T'$. However if the programs run on after the mutation site, the number of different states falls exponentially. Averaging across all l mutation sites, we can approximate the expected number of different states following a point mutation (at instruction x) when a program of length l terminates as

$$\frac{1}{l}\int_0^l 1 + \left(1+2(T-1)e^{-x/2^N}\right)e^{-(l-x)/2^N} dx = 1+2(T-1)e^{-l/2^N} + \frac{2^N}{l}\left(1-e^{-l/2^N}\right)$$

$$\approx 1 + \frac{2^N}{l} \qquad \text{if } l \gg 2^N$$

In small programs ($l \ll 2^N$) this is approximately $2Te^{-l/2^N}$. I.e. on average a single point mutation to a short program is very disruptive but for very long programs its impact on the program's fitness falls as $O(l^{-1})$.

6.3 Non-reversible Computer – Crossover

Two point crossover between two average programs essentially means inserting a randomly selected code fragment into one program and the corresponding fragment into the other parent. The situation is slightly more complex than with point mutation. For example the length of the code changed aught to be considered. However on the average machine, inserting a random code fragment will have much the same effect as a single random instruction change.

With uniform crossover even with very long programs there will be coding changes near each end of the program. The effect becomes very similar to replacing the whole of the parent program with another randomly selected one. Therefore we would expect no correlation between the outputs of the parent and child program. In the case of the average computer, the expected number of different outputs they produce with T input test cases will be the same. (Since

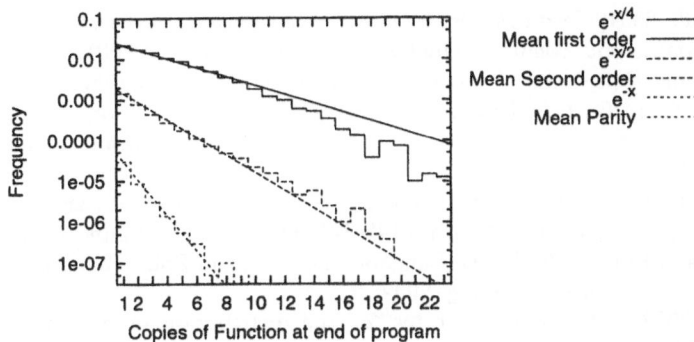

Fig. 3. Proportion of programs (1000 instructions, 128 bits) containing multiple copies of each function. The functions fall into four classes, constants (not shown), copies of inputs, second order and parity.

it is given by their length, which, depending upon the crossover operator, will be the same.) I.e. with this fitness function, we expect their fitness to be about the same.

7 Conclusions

We have proved general results. Fitness landscapes do converge. Most programs are useless, and mutating them is unlikely to improve them. How do we reconcile this with GP? How do we progress? Here the results have been for linear non reversible programs, which loose information. However other representations, which do not (e.g. tree GP, reversible computing and linear GP with inputs write protected) also converge. Nor should we hope to concentrate of programs smaller than the convergence threshold, since this can still be a very large number of programs.

How does Nature do it? When we look at evolved organisms we see tremendous reuse of partial solutions, hierarchies and modularity. Most programs are amorphous. The fitness landscape as a whole is dominated by these useless programs. If the fitness landscape metaphor is to help, like any map, it needs to concentrate on routes to where we want to be. May be future analysis will shed light on the structure of good programs and the route map created by crossover.

Acknowledgements. I would like to thank Ben Dias.

References

[Banzhaf et al., 1998] Wolfgang Banzhaf, Peter Nordin, Robert E. Keller, and Frank D. Francone. *Genetic Programming – An Introduction.* Morgan Kaufmann.

[Clergue et al., 2002] Fitness distance correlation and problem difficulty for genetic programming. In *GECCO 2002*, pp 724–732, New York, 9-13 July 2002.

[Daida et al., 2001] Jason M. Daida, et al.. What makes a problem GP-hard? *Genetic Programming and Evolvable Machines*, 2(2):165–191, June 2001.

[Goldberg and Deb, 1991] A comparative analysis of selection schemes used in genetic algorithms. In G. J. E. Rawlins, editor, *FOGA*, pp 69–93. Morgan Kaufmann.

[Jones, 1995] Terry Jones. One operator, one landscape. Technical Report SFI TR 95-02-025, Santa Fe Institute, January 1995.

[Kinnear, Jr., 1994] Kenneth E. Kinnear, Jr. Fitness landscapes and difficulty in genetic programming. In *WCCI*, pp 142–147, Orlando, 27-29 June 1994. IEEE Press.

[Langdon and Poli, 2002] W. B. Langdon and Riccardo Poli. *Foundations of Genetic Programming*. Springer-Verlag, 2002.

[Langdon, 2002a] Convergence rates for the distribution of program outputs. In *GECCO 2002*, pp 812–819, New York, 9-13 July 2002. Morgan Kaufmann.

[Langdon, 2002b] W. B. Langdon. How many good programs are there? How long are they? In Jonathan Rowe, et al. editors, *FOGA VII*. Morgan Kaufmann.

[Langdon, 2003] W. B. Langdon. The distribution of reversible functions is Normal. In Rick Riolo, editor, *GP Theory and Practise*. 2003. Forthcoming.

[Muhlenbein and Schlierkamp-Voosen, 1993] Predictive models for the breeder genetic algorithm. *Evolutionary Computation*, 1(1):25–49, 1993.

[Nikolaev and Slavov, 1998] Concepts of inductive genetic programming. In W. Banzhaf, et al. editors, *EuroGP*, *LNCS* 1391, pp 49–60. Springer-Verlag.

Multi-agent Learning of Heterogeneous Robots by Evolutionary Subsumption

Hongwei Liu[1,2] and Hitoshi Iba[1]

[1] Graduate School of Frontier Science, The University of Tokyo
Hongo 7-3-1, Bunkyo-ku, Tokyo 113-8656, Japan
[2] School of Computer and Information, Hefei University of Technology
Hefei 230009 China
{Lhw,Iba}@miv.t.u-tokyo.ac.jp

Abstract. Many multi-robot systems are heterogeneous cooperative systems, systems consisting of different species of robots cooperating with each other to achieve a common goal. This paper presents the emergence of cooperative behaviors of heterogeneous robots by means of GP. Since directly using GP to generate a controller for complex behaviors is inefficient and intractable, especially in the domain of multi-robot systems, we propose an approach called Evolutionary Subsumption, which applies GP to subsumption architecture. We test our approach in an "eye"-"hand" cooperation problem. By comparing our approach with direct GP and artificial neural network (ANN) approaches, our experimental results show that ours is more efficient in emergence of complex behaviors.

1 Introduction

Genetic Programming (GP) has proven successful in designing robots capable of performing a variety of non-trivial tasks [7,11]. However, the fields' focus is almost exclusively on single-robot systems. Many tasks can be solved more efficiently when a multi-robot system is used; while some tasks cannot be solved at all with single-robot systems. Therefore, recently more and more researchers have applied evolutionary computation techniques to the design of various types of multi-robot/agent systems [3,4,5,6,8,9].

In a multi-robot system several robots simultaneously work to achieve a common goal via interaction; their behaviors can only emerge as a result of evolution and interaction. How to learn such behaviors is a central issue of Distributed Artificial Intelligence, which has recently attracted much attention. It is very important and interesting to study the emergence of robots' behaviors in multi-robot systems by means of artificial evolution.

Most of the aforementioned researches are on homogeneous systems. Although D. Floreano et al. [3] presented a heterogeneous system, the relationship between the two robots is competitive. In this paper we address the issue in the context of a heterogeneous multi-robot system, in which two real robots, i.e., Khepera, are evolved using GP to solve a cooperative task.

Since directly using GP to generate a program of complex behaviors is difficult, a number of extensions to basic GP have been proposed to solve these control problems of the robot. For instance, J. Koza employed GP to generate a subsumption architecture control program [7]. W.F. Punch et al. proposed an approach to solve robot navigation problems, it incorporated subsumption principles into the Echo Augmented Genetic Programming approach [12]. H. Iba et al. studied the emergence of the cooperative behavior in multiple robots/agents by means of GP and proposed three types of strategies, i.e., homogeneous breeding, heterogeneous breeding, and co-evolutionary breeding, for the purpose of evolving the cooperative behavior [4]. They used a heterogeneous breeding approach of GP, evolving a multi-agent learning system, to solve robot navigation and Tile World problems [5]. They also applied the proposed GP system to a homogeneous cooperative multi-robot system and tested their approach in an "escape problem" [8]. These researches showed that GP is efficient in multi-robot/agent learning.

We report an improvement of GP, called Evolutionary Subsumption–which combines the GP with Brooks' subsumption architecture [1] and compare our approach with direct GP and ANN approaches. Our experiments show that this method is effective in solving such complex problems of robot control.

The rest of this paper is organized as follows: in Sect. 2 we will analyse the target system and its complexity, our approaches will be presented in Sect. 3, and in Sect. 4 the experimental result with comparison of evolutionary subsumption and direct GP will be reported. Finally, discussion and some empirical conclusions are presented.

2 Task Domain and Complexity

The approaches are evaluated in an "eye"-"hand" cooperation task. In this task two heterogeneous robots learn complex robotic behaviors by cooperation. One of them, which is mounted with a digital camera, acts as the "eye" and the other, which is mounted with a gripper, acts as the "hand" (Fig. 1). Their task is: the "eye" tries to find a cylindrical object[1], and then navigates the "hand" to pick it up and then navigates it to carry the cylinder to the goal. The two robots are heterogeneous–they have different sensors and actuators, and have different roles in the system. Their behaviors are complex: including tracking, path planning, and communication, etc.

We classify the similar problems into three difficulty levels, according to the relationship of the observer–"eye" and actor–"hand":

---

[1] There are two cylinders in our system, one is the object that the "hand" needs to grip in the first stage and the other is the goal, which the "hand" needs to put the first object near in the second stage. In the following text, in order to ease the depiction, we use the word 'cylinder' to indicate the object or the goal according to the stage, except where we distinguish them explicitly.

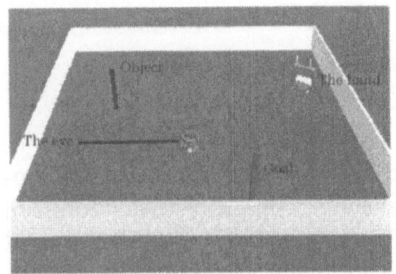 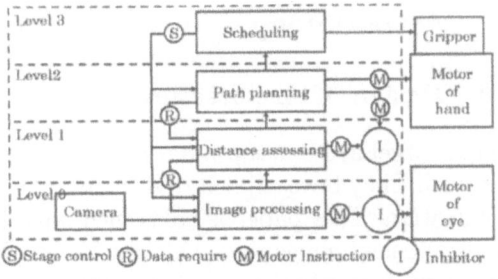

Fig. 1. "Eye"-"hand" cooperation problem

Fig. 2. Evolutionary subsumption approach's layered architecture

Difficulty 1 Fixed "Eye": the "eye" is fixed and usually acts as a bird's eye view; that is, it can see the whole environment from its fixed position. The navigation method is the most simple; but if there are obstacles in the environment it involves route selection problems.

Difficulty 2 Semi-fixed "Eye": although the "eye" can move, the relative position of "eye"-"hand" is fixed or restricted.

Difficulty 3 Unfixed "Eye": the "eye" and the "hand" can move freely, and the relative position of "eye"-"hand" is variable. Usually the "eye" can only see part of the environment.

Our target system belongs to difficulty 3. There are rather simple strategies in difficulty 1 and 2. For instance, in the "escape problem" the navigation of robots belongs to difficulty 2, the strategy is to keep the image of the button in the centre of its view field and to enlarge the image though movement, by getting closer to the object, and finally, touching it [8]. In difficulty 3 the situation is much more complicated. Since the relative position between "eye" and "hand" is variable, the "eye" must track two objects simultaneously. The search space of difficulty 3 has one more dimension than difficulty 2. Specifically, in difficulty 2, the only object that the "eye" needs to observe is fixed; but, in difficulty 3 one of the two objects, the hand, is movable. Therefore, the search space of a system which belongs to difficulty 3 is large and the emergence of robots' rational strategies is very difficult. The "eye" must select suitable viewpoints, observe the environment, and send correct instructions to "hand". Along with the moving of the "hand", the "eye" must be able to adjust its position and send new instructions according to the new situation.

3 Methodology

3.1 Design of Architecture

In our target system, the two robots need to coordinate their behaviors to achieve the goal. They have explicit division of roles and need to be synchronized by communication. This paper is concerned with how such cooperative behaviors can be established efficiently; that is, what kind of architecture should be employed and how to synthesize such an architecture. We employ the evolutionary subsumption approach and compare it with other approaches, such as the direct GP approach.

According to analysis in Sect. 2, this problem belongs to difficulty 3 and its search space is very large, it is intractable to search for a direct solution using Genetic Programming. The divide-and-conquer approach is an intuitive and efficient method when we encounter complex problems. Being a divide-and-conquer approach, the subsumption architecture decomposes the problem into a set of levels [1] and each level implements a task-achieving behavior. We employed the subsumption architecture, dividing the whole behavior into several simple behaviors. Then each level is automatically generated by Genetic Programming respectively; the lower level is formed by Genetic Programming at first, and then uses lower levels' output as nodes of the next level of Genetic Programming.

3.2 Evolutionary Subsumption

The control system is divided into 4 levels: level0 image processing, level1 distance assessing, level2 path planning, and level3, scheduling. See Fig. 2. The rest of this section will introduce each level of the architecture.

Level0 Image Processing. This level gets an input image, detects whether the "hand" and cylinders appear in the view or not, and calculates the width of their image. In order to fix our attention on the task of coordination and not immerse ourselves in the field of machine vision, we use particular colors to identify the "hand" and cylinder. See Fig. 3, input at this level is one scan line of the image and outputs are W_{hand}, D_{hand}, W_{obj}, D_{obj} (i.e., the offset from center of image and the width), and two Boolean variables B_{hand} and B_{obj}, they indicate whether the "hand" and the cylinder are within the image.

Level1 Distance Assessing. Level1 takes level0's output as its input and assesses the distance of "eye"–"hand" and "eye"–cylinder. Therefore the task of level1 is a symbolic regression problem:

$$f(W_{hand}, D_{hand}, W_{obj}, D_{obj}, B_{hand}, B_{obj}) = \{Dis_{hand}, V_{hand}, Dis_{obj}, V_{obj}\} \quad (1)$$

Where Dis_{hand} and Dis_{obj} are the assessed distances, and V_{hand} and V_{obj} indicate whether the assessed distance is valid or not. These values will be used by the higher levels. If the objects, i.e., the "hand" and cylinder, do not appear in

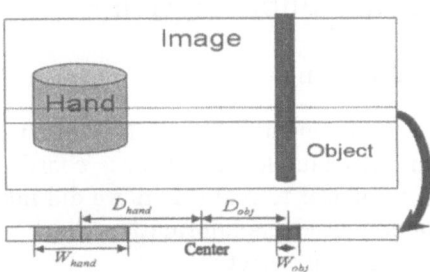
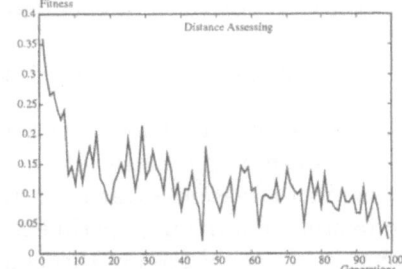

Fig. 3. Image processing approach of level0

Fig. 4. Fitness of distance assessing of level1

viewfield of "eye" or are too far from the "eye" then V_{hand} and V_{obj} will be set to "False", otherwise they will be set to "True".

This level is trained separately. For each generation, before training we generate 10 maps, which randomly specify the position and orientation of the "eye", the "hand", and the cylinder. These 10 maps will be kept constant within one generation; in the next generation they will be reformed, i.e., will be different from the prior generations. The fitness is defined as the average error between the assessed value and the real value in the 10 maps.

The function set consists of F={IFLTE, PROGN2, Data_req, IFhand, IFobj}, the terminal set consists of T={W_{hand}, D_{hand}, W_{obj}, D_{obj}, $Scan$, $Const$}. The IFLTE and PROGN2 have the same functionality as in LISP and Data_req calls the level0 to refresh its output. IFhand and IFobj are based on the B_{hand} and B_{obj} defined by level0, they take two arguments and evaluate their first argument if B_{hand}/B_{obj} is true otherwise they evaluate their second argument. The $Scan$ in the terminal set makes the "eye" rotate to scan the environment, $Const$ means constant number. The result of evolution is shown in Fig. 4. After 100 generations' evolution the error is less than 0.05, this means that the average error of assessed distance in 10 maps is less than 5cm.

Level2 Path Planning. The task of this level is to generate rational motor instructions. In our approach we used central-control architecture. It generates motor instructions for both "eye" and "hand". The rational instructions for the "eye" are to get a better viewpoint and the rational instructions for the "hand" are to drive it closer to the cylinder. As we will observe in our experimental results in Sect. 4 the two robots will learn to coordinate with each other gradually and the rational strategy will emerge along with the evolutionary procedure.

Level3 Scheduling. This level determines when the "hand" should pick up the cylinder and when it should put the cylinder down. Since the procedure of this level is fixed, we can write the program for this level manually.

4 Experiments with Evolutionary Subsumption

4.1 Environment and Experimental Setting

We used Webots of Cyberbotics for the experiments. Although in simulation the robots can obtain extra information, for example the absolute coordinates etc., in order to ease the transition from simulator to real robots we did not use such information. In our experiments we only used the information which a real khepera robot could acquire through its sensor or turret.

The size of the environment is 100 × 100 cm with high 10 cm white walls, so that the "eye" can recognize "hand" and cylinder easily. There are no obstacles in the environment. In order to keep things simple we used special colors to identify the cylinder and the goal. The "eye" robot was equipped with a k6300 digital camera turret and the "hand" robot was equipped with a Gripper turret. There is a wireless channel through which the "eye" and the "hand" can exchange messages (Fig. 1). The initial positions of "eye", "hand", cylinder, and goal are placed randomly. The limit of steps is 2 times the linear distance between "hand" and cylinder. The actions of a robot are simplified to 4 actions: MF move forward, MB move backward, MR turn right about 30 degree and move forward, and ML turn left about 30 degree and move forward.

We used a layered training method to train each level of the subsumption architecture sequentially. As described in Sect. 3.2 we first evolve the lower levels then the higher levels. When evolving, the higher levels "subsume" the lower levels. Thus the performance of level2 is the performance of the whole system.

At the beginning of each generation we generate maps by randomly placing the "eye", "hand", and cylinders. These maps are kept constant only within one generation. They will be regenerated before evolution to the next generation. The number of maps, i.e., the fitness cases, increase with the generations from 1 to 10. That is, along with evolution the difficulty of the task is increased, finally each individual must be evaluated on 10 maps. This method is able to prevent the robots from accomplishing their task by fluke. Fitness is defined as the maximum distance in all fitness cases and the distance is between the "hand" and the cylinder after the "hand" runs out of its steps.

We used function set F={IFLTE, PROGN2, Data_req, IFVhand, IFVobj} and terminal set T={D_{hand}, D_{obj}, Scan, MFe, MBe, MLe, MRe, MFh, MBh, MLh, MRh}, where the function set is very similar to level1; in the terminal set D_{hand} and D_{obj} are the output of level1, the MFe, MBe, MLe, MRe are the motor instructions of "eye", the others are the motor instructions of "hand". The other parameters are shown in Table 1.

4.2 Result

Figure 5 shows result of level2, which plots the averaged fitness values over generations for 10 runs. Note that the horizontal line at 0.1 indicates whether the "hand" can accomplish task or not. Since the fitness function is defined as the final distance between the "hand" and the cylinder, if the fitness is below that

Table 1. Parameters of Genetic Programming

Population size	2000
Crossover rate	0.85
Mutation rate	0.1
Elite rate	0.1
Maximum depth	15

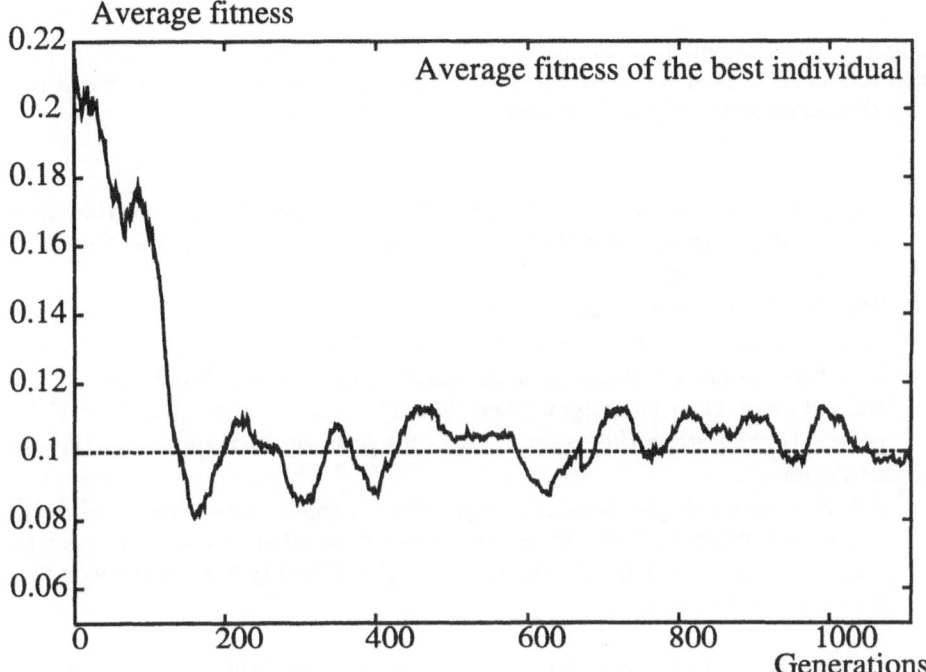

Fig. 5. Average fitness (10 runs) of the best individuals of the system

line it means finally the "hand" approached within 10 cm of the cylinder and the "hand" can detect the position of cylinder or goal, with its infrared sensors, and pick up the cylinder with its gripper, or put the cylinder down on the goal. Along with the increasing of generations the fitness cases increased from 1 to 10. This means that finally the fitness is the maximum value of 10 fitness cases. If we take this into account we can find that in this system the cooperative behaviors have been established by GP. See Fig. 8 for the emergence of cooperative behavior.

Although the concrete behaviors are various, they always display a clear *strategy*. The *robots* usually demonstrate a limited set of strategies, by analyzing the emerging behaviors we roughly classify the strategies into 3 types:

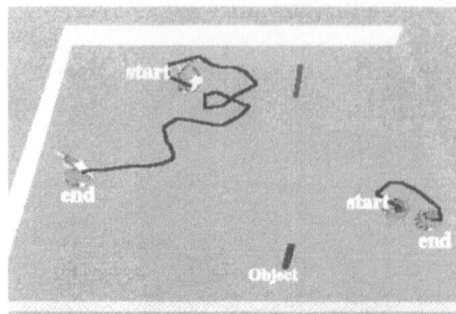
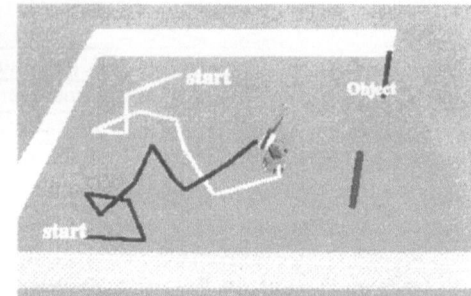

Fig. 6. At the beginning of evolution the two robots show poor coordination. Usually they move separately and aimlessly

Fig. 7. They learned cooperation along with the evolution and the "observation"–"action" rhythm emerged

1. The "eye" tries to find the "hand" and then maintains its position to it; meanwhile, searching for the cylinder and finally directing the "hand" to close to the cylinder.
2. The "eye" finds the cylinder at first, moves close to it, stays near it, and then navigates the "hand" close to the cylinder.
3. The "eye" finds a suitable position neither near to the "hand" nor near to the cylinder, then it navigates the "hand" to move. After the "hand" has moved several steps the "eye" adjusts its position in response to the new situation.

All of these strategies have the same effect, namely reduce the level of difficulty. By using such strategies the relative position of the "eye" and the "hand" becomes roughly fixed; therefore, the difficulty level is reduced from 3 to 2 (refer to Sect. 2).

Figures 6, 7, and 8 show the course of emergence of the rational strategy. At the beginning of evolution the two robots show poor coordination. Usually the "eye" and the "hand" move separately; the "hand" moves aimlessly before the "eye" surveys the environment and soon it runs out of its steps unnecessarily. Even in generation 0, there are some individuals better than others, they approach the cylinder more closely (Fig. 6).

Along with the evolution the two robots gain more skill in cooperation, they show clear rhythm of "observation"–"action"–"observation"... The "hand" never moves before the "eye" because it must save its limited steps (Fig. 7).

Finally, the two robots are more skillful. They have averaged more than 60% probability to accomplish the task. Figure 8 shows their trajectory. As shown in Fig. 8, the two robots show favorable coordination. At first the "eye" observes the environment and directs the "hand" to move and then the "eye" observes again adjusting its position and directing the "hand" to move again... We can also observe that the trajectory of "hand" is getting more and more smooth along with their interaction. These phenomena indicate that the rational strategy has emerged.

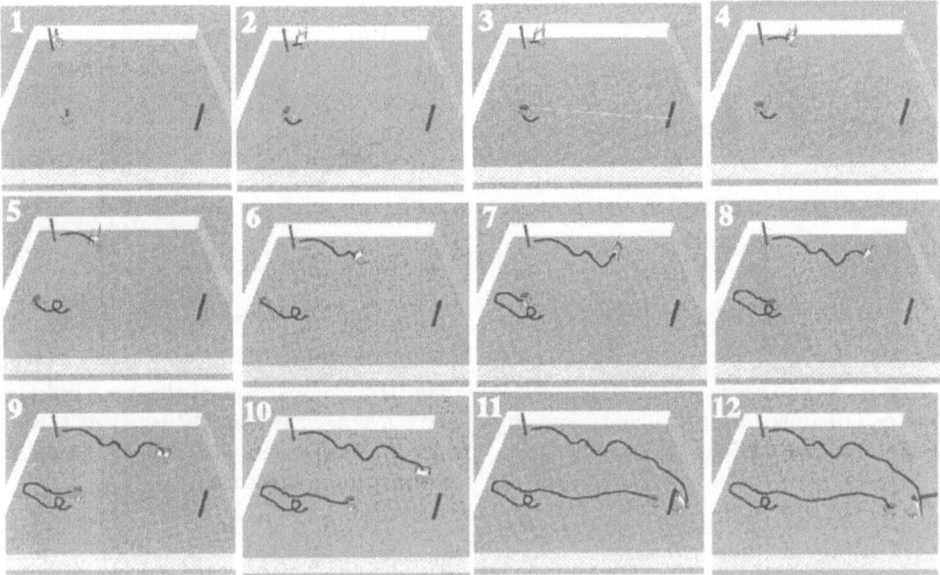

Fig. 8. Finally the skillful cooperative behaviors emerged

In our target task, due to the complexity, the two robots could not ensure accomplishment of the task in all cases. Although, along with the evolution the success rate increased. We test the success rate by the following method: for each generation we select out the best individual to test its success rate as the success rate of the generation. In the test stage we generate n maps in advance, giving the position of the "eye", "hand", cylinder, and goal as different from the training environment; but, keeping them constant to provide a fair condition to all the individuals, which are taken to test the success rate. If the "hand" can approach the cylinder within 10cm and then approach the goal also within 10 cm, then we mark the individual as able to accomplish the task on this map. The success rate is defined as the ratio of the number of accomplished maps to total maps n. In Fig. 9 the upper curve shows the trend of the average success rate along with the evolutionary process.

4.3 Comparison with Direct GP and ANN

For comparison, we also employed direct GP approaches to solve the problem. In the direct GP approach, in order to keep things simple, we did not use the input image directly; instead, we kept the level0 fixed and just used GP to generate programs for level1 and level2. The definition of fitness and the other parameters are the same as in the evolutionary subsumption approach. In direct GP approach, although the "eye" and the "hand" can find the rational strategies and produce cooperative behaviors, they take almost 3 times the number of generations of the evolutionary subsumption approach and often failed to converge

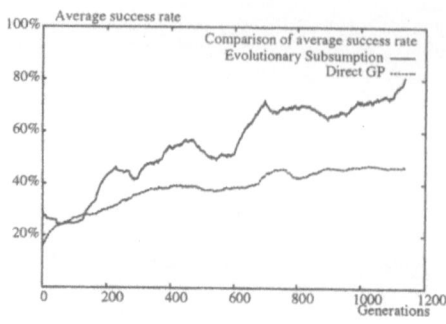 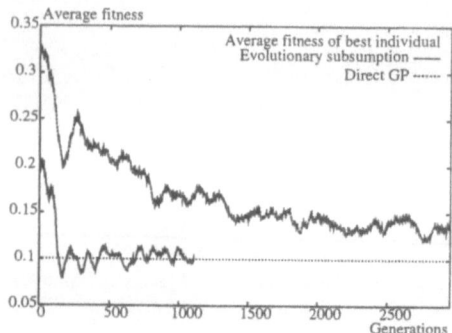

Fig. 9. Comparison of averaged success rates (10 runs) of evolutionary subsumption and direct GP

Fig. 10. Comparison of the best individual's averaged fitness (10 runs) of the direct GP approach and the evolutionary subsumption approach

due to premature convergence. Furthermore, the final fitness of the evolutionary subsumption approach is much better than that of the direct GP. Figure 10 shows the comparison of the best individual's fitness over the generations and in Fig. 9, we can find the success rate of the evolutionary subsumption approach superior to the direct GP approach. This seems due to the reasonable subsumption architecture, we have designed the suitable framework for the whole system and the GP need only search the optimal solution for each layer in a relatively small search space. On the other hand, the direct GP approach has to search in a large search space and often times it can not, or it must take a number of generations' evolution to decompose the problem into rational components. Therefore, the final results are inferior to the subsumption architecture.

In the ANN approach we used a 3 layer feed-forward neural network and used a generic BP algorithm to train the neural network. Since the BP algorithm is a supervised approach, when using the BP algorithm we have to provide a set of correct input-output pairs as training samples. However for "eye" there are too many possible strategies. It is impossible to foresee the action of "eye"; therefore, we can not provide standard input and output and use an error value to train the "eye". In other words, the reasonable strategies can not emerge automatically, they must be specified by the designer. So as an alternative we restricted the movement of the "eye": fixed the position of the "eye" and ensured that the "hand" and the cylinder both appear in its view-field. We used "batch training mode", i.e., weights are changed only after the "hand" ran out of its steps and stopped. We used the distance of this moment as the error to train the weight. Still the cooperative behaviors could not emerge within reasonable training times. Occasionally they succeeded in one fitness case, but it could not be generalized to other cases. For this approach the success rate approximates to 0%.

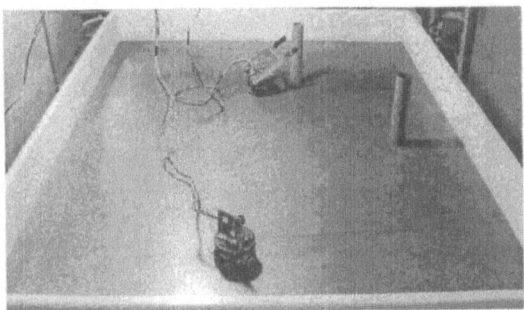

Fig. 11. The preliminary environment of the real robots

4.4 From Simulation to Real Robots

Usually the simulators are too ideal and abstract compared to the real world and gaps exist when moving from the simulator to real robots. In our ongoing experiment (Fig. 11) we employ two Khepera robots, they are connected to a desktop workstation though aerial cable. The aerial cable provides the robots with data communication from the desktop.

Since our approach employed subsumption architecture, in which the layers can be redesigned and added incrementally, when moving from the simulator to real robots we just need to redesign level0 and part of level1. Moreover, we did not use any extra information that only can be obtained via the simulator. This alleviates the complexity of moving to real robots.

5 Discussion

When designing an intelligent robot it is impossible to foresee all the potential situations a robot will encounter and specify all behaviors in advance. Especially in multi-robot systems, the situations that each robot encounters are much more complex and unpredictable. Therefore, the basic issue is how to design a control program for such systems. In this paper we studied such an issue and evaluated our evolutionary approach in an "eye"-"hand" cooperative problem. The experimental result shows that by applying GP to subsumption, our approach, efficient emergence of a rational strategy for multi-robot systems is possible.

Subsumption architecture decomposes the control system into a set of layers, each layer implements an asynchronous task achieving behavior [1]. Though it can produce a robust and flexible robot control system, the design of each layer is still a burdensome task, especially the high level layers, for instance path planning and reasoning etc. Furthermore, the robot's behavior is specified manually by adding corresponding layers, namely the reasonable behaviors can not emerge automatically just like in neural network architecture (see Sect. 4.3). In a multi-robot system it is impossible to specify the strategy of each robot in advance, the rational strategy of a multi-robot system can only emerge as a result of interaction of robots in the system.

Artificial evolutionary approaches can establish rational strategy automatically, especially in multi-robot systems. Dario Floreano et al. applied a co-evolutionary neural network in a predator-prey problem and proposed that the chase and evasion strategy would emerge automatically by the interaction of the two robots [3]. H. Iba et al. discussed the emergence of cooperative behavior for a multi-robot/agent system [4,8]. Their research showed that sophisticated strategies can be produced efficiently by the GP approach.

However, it is also intractable to search for a direct solution using evolutionary approaches for complex multistage behaviors, such as our target system. In this paper we proposed an *Evolutionary Subsumption* method, which combines the above two approaches. Compared with direct GP and classical subsumption architecture, our experimental result shows that it has been endowed with the advantages of the two approaches. By generating task achieving behaviors automatically, it alleviates the burden of design of subsumption. Furthermore, by employing an evolutionary approach, rational behaviors can emerge along with the evolution. On the other hand, when the search space is too large, subsumption architecture can restrict the search space and make evolutionary approaches converge within a practicable time.

In summary, comparing subsumption architecture, direct GP and ANN approaches:

1. It is difficult for subsumption architecture to search for a rational strategy, the behaviors should be specified in advance. Also subsumption architecture claimed that complex behaviors are simply the reflection of a complex environment, but the behaviors are specified by the designer, manually with corresponding layers.
2. Direct GP converges several times slower than the evolutionary subsumption approach.
3. BP algorithm ANN approach, due to its supervised nature, has the same problem as the subsumption architecture; that is, the cooperative behaviors can not emerge automatically.

The other essential issue for a multi-robot system is what kind of architectures should be employed, i.e., central control or distributed control. Chern H.Y. et al. employed GA to evolve a team of neural networks in a cooperative multi-robot system and studied the tradeoff between central and distributed controllers [2]. They found that co-evolutionary architecture can accelerate the convergence. Since our target system is different from the systems in literature [2, 3,8], that is the "hand" is not completely autonomous, the convergent speed of co-evolutionary architecture is slower than central-control architecture. We will report the comparison of distributed control and central control architecture in another paper.

6 Conclusion

We examined the emergence of cooperative behavior and proposed that the two robots can find several quite reasonable strategies. According to the results in

Sect. 4 these strategies are different, but have the same effect; that is by interacting with each other, the "eye" and the "hand" determine their relationship, thereby reducing the difficulty level efficiently from 3 to 2, and finally accomplishing the task. Our results lead us to deduce the following empirical conclusions:

1. Evolutionary Subsumption is efficient in emergence of a heterogeneous multi-robot system. It shows superiority to both classical subsumption architecture and GP approach. As a divide-and-conquer method this approach can be applied to a number of other robot control problems.
2. The direct GP approach can also be used to deal with some complex systems; but compared with the evolutionary subsumption approach the efficiency is much lower and easily falls into local optimum.
3. Since a standard feed-forward neural network with BP training algorithm needs samples to train the network, it is awkward in emergence of novel behavior.

However, the path of the "hand" to approach the cylinder is not optimal and the success rate is much lower for some high-reliability applications. For future research we want to extend this approach to co-evolutionary architecture and hope to improve the robustness and efficiency of our target system.

Acknowledgement. This work was partially supported by the Grants-in-Aid for Scientific Research on Priority Areas (C), "Genome Information Sciences" (No. 12208004) from the Ministry of Education, Culture, Sports, Science and Technology in Japan.

References

1. Brooks R., "A Robust Layered Control System for a Mobile Robot", IEEE Journal of Robotics and Automation 2(1), 14–23 (1986).
2. Chern Han Yong and Risto Miikkulainen, Cooperative Coevolution of Multi-Agent Systems, Technical Report AI01-287
3. D. Floreano, S. Nolfi, and F. Mondada, Competitive Co-Evolutionary Robotics: From Theory to Practice. In R. Pfeifer, editor, From Animals to Animats V: Proceedings of the Fifth International Conference on Simulation of Adaptive Behavior, MIT Press-Bradford Books, Cambridge, MA, 1998.
4. Hitoshi Iba, Emergent cooperation for Multiple Agents using Genetic Programming, in Parallel Problem Solving form Nature IV (PPSN96), 1996, 32–41
5. Hitoshi Iba, Evolving Multiple Agents by Genetic Programming, Genetic Programming 3, Spector, L.Langdon, W.B., O'Reilly, U.-A., and Angeline, P.A.,(eds), MIT Press, pp. 447–466, 1999.
6. M. Terao and Hitoshi Iba Controlling Effective Introns for Multi-Agent Learning by Genetic Programming, in Proc. of the Genetic and Evolutionary Computation Conference (GECCO2000), pp. 419–426, 2000
7. J.R. Koza, Evolution of subsumption using genetic programming, in F.J. Varela and P. Bourgine (Eds.) Toward a Practice of Autonomous Systems, pp. 110–119 Cambridge, MA: MIT Press. 1992.

8. K. Yanai and H. Iba Multi-agent Robot Learning by Means of Genetic Programming: Solving an Escape problem, in Evolvable Systems: From Biology to Hardware, 4th International Conference (ICES2001), pp. 192–203. 2001
9. Matt Quinn, A comparison of approaches to the evolution of homogeneous multi-robot teams, In Proceedings of the Congress on Evolutionary Computation (GECCO2001) pp. 128–135, Seoul, S. Korea. IEEE Press, 2001
10. Nolfi S. and Floreano D., Evolutionary Robotics: The Biology, Intelligence, and Technology of Self-Organizing Machines. Cambridge, MA: MIT Press/Bradford Books, 2000.
11. P. Nordin and W. Banzhaf., An on-line method to evolve behavior and to control a miniature robot in real time with genetic programming, Adaptive Behavior, 5(2):107–140, 1997
12. W.F. Punch and W.M. Rand,GP+Echo+Subsumption – Improved Problem Solving, in Proc. of the Genetic and Evolutionary Computation Conference (GECCO2000), pp. 411–418, 2000

Population Implosion in Genetic Programming

Sean Luke, Gabriel Catalin Balan, and Liviu Panait

Department of Computer Science, George Mason University
4400 University Drive MSN 4A5, Fairfax, VA 22030, USA
{sean,gbalan,lpanait}@cs.gmu.edu
http://www.cs.gmu.edu/~eclab

Abstract. With the exception of a small body of adaptive-parameter literature, evolutionary computation has traditionally favored keeping the population size constant through the course of the run. Unfortunately, genetic programming has an aging problem: for various reasons, late in the run the technique become less effective at optimization. Given a fixed number of evaluations, allocating many of them late in the run may thus not be a good strategy. In this paper we experiment with gradually decreasing the population size throughout a genetic programming run, in order to reallocate more evaluations to early generations. Our results show that over four problem domains and three different numbers of evaluations, decreasing the population size is always as good as, and frequently better than, various fixed-sized population strategies.

1 Introduction

In generational EC, the choice of how to allocate evaluations may be cast as a tradeoff between exploration and exploitation. A maximally large population, run for a single generation, is an extreme in exploration, approximating random search. On the other hand, depending on the locality of the modification operator, a minimal (2 person) population exploits local features to the point of hill-climbing.

Choosing a good population size vs. runlength tradeoff is a long-explored topic in evolutionary computation. This choice has traditionally hinged on two factors: theoretical justifications for a given population size; and practical necessity for capping population size, primarily because of memory capacity. Recent advances in computer technology have obviated much of the second factor: genetic programming has occasionally seen population sizes around a million individuals divided among multiple machines in Beowulf clusters [1,2]. Contrast this to early 1 + 1 evolution strategies techniques which used a population size of 2!

The most common runlength for a genetic programming problem is a very short 50 generations, and the standard population sizes have ranged from 500 to 2000. The reason for these layouts is mostly tradition: they were the layouts popularized by Koza in [3]. Koza established these layouts through trial and error, and his decisions seem to be justified for GP problems. As was discussed in [4], it appears that many of the canonical test problem domains for GP have a surprisingly short maximum useful runlength. In the Symbolic Regression domain, for example, beyond 32 generations it becomes advantageous to split the run into two or more shorter runs.

This paper examines the effects of decreasing the population size towards zero during the course of a genetic programming run. Early in the run the system would evolve with a large population of individuals, but late in the run the system would be effectively reduced to hill-climbing. This idea was inspired by simulated annealing, which also explores early and exploits (hill-climbs) late in the run, albeit in a very different fashion. We imagined that this "population implosion" would be effective because it would reallocate evaluations away from late generations where the population would have mostly stagnated anyway.

Decreasing the population size has another unforseen advantage in genetic programming: it counters the effects of bloating on memory consumption. The total memory consumed by the system is a function largely of the size of the average individual times the number of individuals in the population. As the first factor goes up, a drop in the other factor may maintain current memory consumption levels. However, this issue is not considered in the presented experiments.

We define a *layout* as a choice of population size and runlength in order to allocate N evaluations for a run. In generational evolutionary computation, layouts are almost always rectangular, with the minor exception of the $\mu + \lambda$ evolution strategy, which may have a small initial generation. In this paper we have chosen to decrease the population using a *diagonal* layout. The diagonal layout starts with a large population, then linearly decreases the population size at each generation until it reaches zero.

The paper discusses previous approaches to modifying the population size during the search process, followed by three experimental investigations. The first experiment compares a diagonal layout and three traditional layouts over several GP problem domains and evaluation lengths. The second experiment broadens the comparisons with a wider range of standard layouts. The third experiment applies the new layout to two GA domains. The paper concludes with a brief discussion and directions for future work.

2 Previous Work

Setting the population size has been a challenging issue in the EC domain for a very long time. Initial work concentrated on deciding the optimal number of individuals in fixed-sized populations. Interesting research has recently suggested methods for adapting the size of the population during the search process, depending on various online parameters like fitness improvement or size of individuals.

In [5], De Jong investigated the influence of population size, mutation, and crossover rates on the efficiency of the search process. He also suggested parameter values that showed good performance in his experiments. Later, Grefenstette used a meta-GA to find good parameters for the search process and recommended a smaller population size than the one suggested by De Jong [6]. Grefenstette's results were later supported by theoretical investigations [7]. [8] provides a theoretical and empirical investigation on the relation between population size and crossover probability, while [9] presents a theoretical analysis meant to determine how to set the population size in order to promote the selection of correct building blocks. More recently, [10] describes another theoretical investigation of the impact of the population size on the performance of the GA algorithms in the OneMax problem domain, and [11] finds statistical significant differences when using different population sizes and classifier lengths in an SCS/LCS system.

Some studies have also raised the possibility of modifying the population size during the search process. [12] introduces the GAVaPS algorithm, which assigns fitness-dependent lifetimes to individuals in a steady-state EC system. Individuals are removed from the population only when their lifetime is exceeded. Another interesting approach is presented in [13], where an adaptive mechanism adjusts the population size as a way to control selective pressure.

Population size has been studied in the GP domain as well. Koza ([3]) advocated using large population sizes, but small populations have also been espoused [14]. An interesting approach is reported in [15], where the number of *tree nodes*, not individuals, in the population is kept constant in order to prevent bloating. This approach is reported to give similar results to the standard method, however it reduces the use of computational resources.

3 First Experiment

In our first experiment, we compared four layouts against three different numbers of evaluations and four different genetic programming problem domains. The problem domains were Symbolic Regression, Artificial Ant, 5-bit Parity, and 11-bit Boolean Multiplexer. We chose to include three choices of number of evaluations: 26600, 52224, and 102400 evaluations (hereafter referred to as, inaccurately, 25K, 50K, and 100K).[1] For each number of evaluations, we picked rectangular layouts with population sizes of 1024, 1448, and 2048. We compared these against a diagonal layout starting at a population size of 2048, and decreasing linearly towards zero.

1448 was included because we were concerned that the diagonal might outperform other choices simply because it struck a middle-ground between large population sizes and long runlengths. Therefore we included our own rectangular middle-ground. Note however that the 1448 rectangular layout cannot have *exactly* the same number of evaluations as the others: we chose to err on the side of very slightly fewer evaluations. For the 50K evaluations runs, for example, the 1448 layout had 520 fewer evaluations than the others did.

Figure 1 illustrates the four layouts used.

---

[1] These slightly odd evaluation sizes are due to our decision to run for 51 generations.

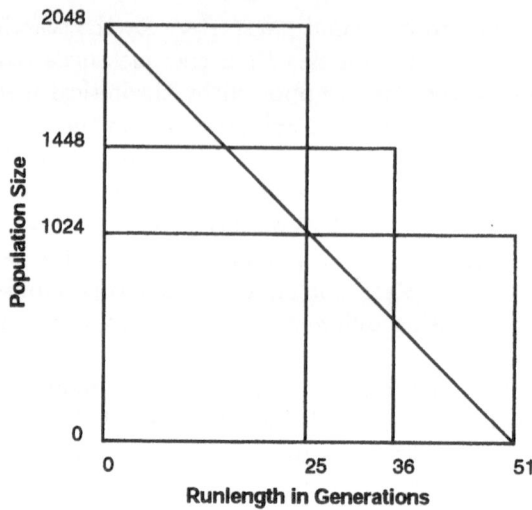

Fig. 1. Four layouts for allocating ∼50K (52224) evaluations: 2048x26; 1448x36; 1024x51; and Diagonal 2048x51

3.1 Other Parameters

Our evolutionary computation system was ECJ [16]. Unless stated otherwise, parameters have values as described in [3]. Symbolic Regression used no ephemeral constants and a function of $x^4 + x^3 + x^2 + x$. Artificial Ant used the Santa Fe trail. We used 7-tournament selection. All other initialization, modification, and representation parameters are the same as those used in [3].

We compared mean best fitness of run among experiments, and performed 200 runs for each experiment setup. Our difference of means test was an ANOVA at 95% confidence, plus a Tukey post-hoc comparison.

3.2 Results

We show the mean fitness results in Tables 1 through 4. Layouts are ordered left-to-right in worsening fitness. Horizontal bars above and connecting different layouts indicate no statistically significant difference between them.

The four problems vary significantly in problem difficulty; some prefer much larger populations, others prefer longer runlengths. The diagonal layout often outperformed all the other layouts, though not always by a statistically significant margin. Nonetheless, what was interesting was that the diagonal layout consistently appeared in the best class of layouts for every single problem domain and every single number of evaluations attempted; and it was the only layout to do so. The diagonal layout did particularly well with smaller numbers of evaluations; for two of the domains (Multiplexer and Ant) it had no peer.

There is an interesting trend among the rectangular layouts. In all four problems, for smaller numbers of evaluations, smaller-population layouts are pre-

Table 1. Statistical significance groupings for mean best fitness of run for the Symbolic Regression domain (First Experiment)

Problem and Evals	Layouts			
Regression 25K Mean Best Fitness	Diagonal 0.09131543	1024 0.103177731	1448 0.145182262	2048 0.169305242
Regression 50K Mean Best Fitness	Diagonal 0.05270353	2048 0.057168302	1448 0.075613984	1024 0.079054647
Regression 100K Mean Best Fitness	Diagonal 0.037023631	2048 0.038560673	1448 0.039290056	1024 0.060902268

Table 2. Statistical significance groupings for mean best fitness of run for the Artificial Ant domain (First Experiment)

Problem and Evals	Layouts			
Ant 25K Mean Best Fitness	Diagonal 22.325	1024 24.63	1448 24.79	2048 25.99
Ant 50K Mean Best Fitness	Diagonal 17.71	2048 18.505	1024 20.015	1448 21.87
Ant 100K Mean Best Fitness	2048 15.865	Diagonal 16.18	1024 18.16	1448 18.775

Table 3. Statistical significance groupings for mean best fitness of run for the 5-bit Parity domain (First Experiment)

Problem and Evals	Layouts			
5-bit Parity 25K Mean Best Fitness	1024 6.68	Diagonal 6.885	1448 7.785	2048 8.85
5-bit Parity 50K Mean Best Fitness	1024 4.77	Diagonal 4.835	1448 4.835	2048 5.825
5-bit Parity 100K Mean Best Fitness	Diagonal 3.35	2048 3.44	1448 3.46	1024 3.55

ferred; and for larger numbers of evaluations, larger-population layouts are preferred. This suggests that there is an optimal useful runlength for each of the problems. As the number of evaluations increases, the smaller-population layouts are extended beyond this runlength and are essentially wasting time. Similarly, for small numbers of evaluations, the large population layouts do not run for enough generations to produce competitive results.

Table 4. Statistical significance groupings for mean best fitness of run for the 11-bit Multiplexer domain (First Experiment)

Problem and Evals	Layouts			
11-bit Multiplexer 25K Mean Best Fitness	Diagonal 197.16	1024 230.475	1448 274.695	2048 352.72
11-bit Multiplexer 50K Mean Best Fitness	Diagonal 92.38	1448 119.33	2048 122.43	1024 128.32
11-bit Multiplexer 100K Mean Best Fitness	2048 46.59	Diagonal 53.32	1448 70.43	1024 90.41

Why did the diagonal perform well? We include Fig. 2, showing results in the Artificial Ant domain, to illustrate the general trend during the course of evolution. We think that the diagonal layout is effectively taking advantage of the maximal-runlength phenomenon by providing *both* the runlength of the longer layouts *and* the emphasis on population size of the larger layouts. We note that in both cases, up until 10,000 evaluations, the diagonal layout closely resembles the 2048-population layout in performance, and then in the 20,000-30,000 range the diagonal layout "speeds up" and surpasses the smaller 1024-population layout, which has mostly converged.

4 Second Experiment

Our second experiment added additional layouts and repeated the 50K evaluation runs for each problem domain, to see if the diagonal was outperforming other layouts simply because it started at 2048K or was able to eke out long numbers of generations in its tail. The additional layouts included ones with both larger and smaller population sizes (and correspondingly shorter and longer runlengths).

Our additional layouts brought the total rectangular layouts to 4096, 2048, 1448, 1024, 501, 251, and 125 population sizes. The diagonal layout again started at 2048 and decreased to 0.

We again compared mean best fitness of run among experiments, and performed 100 runs for each experiment setup. Our difference of means test was an ANOVA at 95%, plus a Tukey post-hoc comparison.

4.1 Results

In the second experiment we changed the random seeds from the first experiment; hence the means for the original layouts are slightly different. We show the mean fitness results in Table 5. Layouts are again ordered left-to-right in worsening fitness, and horizontal bars above and connecting different layouts indicate no statistically significant difference between them.

Keep in mind that due to outlier biases from the worst layout results, the conservative Tukey test now cannot discern statistically significant differences

Table 5. Statistical significance groupings for mean best fitness of run for the Symbolic Regression, Artificial Ant, 5-bit Parity, and 11-bit Multiplexer domains (Second Experiment)

Problem and Evals	Layouts							
Regression 50K Mean Best Fitness	Diagonal 0.051	2048 0.063	1024 0.078	1448 0.084	4096 0.102	251 0.198	502 0.232	125 0.373
Ant 50K Mean Best Fitness	Diagonal 17.7	2048 20.23	1448 21.25	4096 21.31	1024 22.49	251 26.63	502 28.45	125 29.68
Parity 50K Mean Best Fitness	Diagonal 4.67	1448 4.93	1024 4.95	251 5.32	2048 5.78	125 6.01	502 6.16	4096 8.48
Multiplexer 50K Mean Best Fitness	Diagonal 54.68	2048 54.72	1448 69.64	1024 81.38	251 247.26	4096 296.98	502 301.84	125 363.03

among the original four layouts. This is as expected: the goal of the second experiment was only to see if the original four layouts were good choices, and it appears that they were. The new rectangular layouts were usually poor performers, and the diagonal layout usually outperformed all of them by a statistically significant margin.

5 Third Experiment

We also performed a similar experiment using two well-known GA problem domains, the Rastrigin and Rosenbrock problems. Each GA individual's representation was a vector of 100 floating-point genes each ranging from -5.12 to 5.12. We used one-point GA crossover and a gene-independent mutation probability of 0.01, where mutation consisted of gene randomization. We applied tournament selection of size 2, plus one-individual elitism.

Both functions are minimization functions. The Rosenbrock problem [5] computes fitness over a genome of size n using the function

$$\text{Rosenbrock}(x_1, ..., x_n) = \sum_{i=1}^{n} 100(x_i^2 - x_{i+1})^2 + (1 - x_i^2)$$

Similarly, the difficult Rastrigin problem [17] computes fitness using the function

$$\text{Rastrigin}(x_1....x_n) = \sum_{i=1}^{n} x_i^2 + a(1 - \cos(2\pi x_i))$$

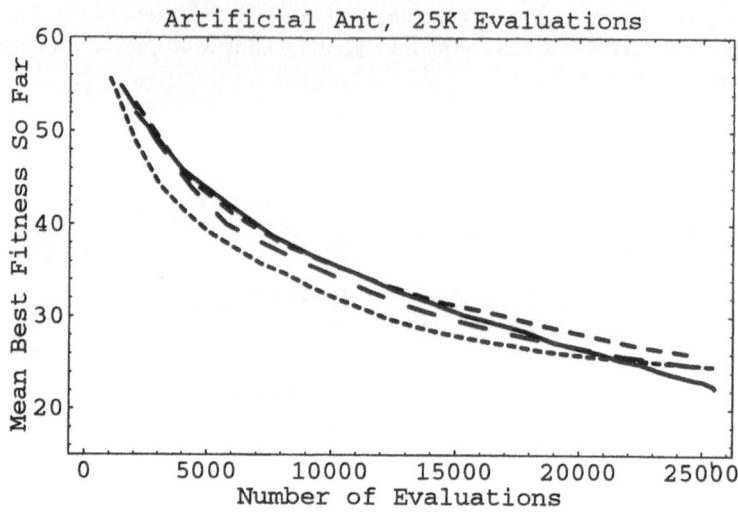

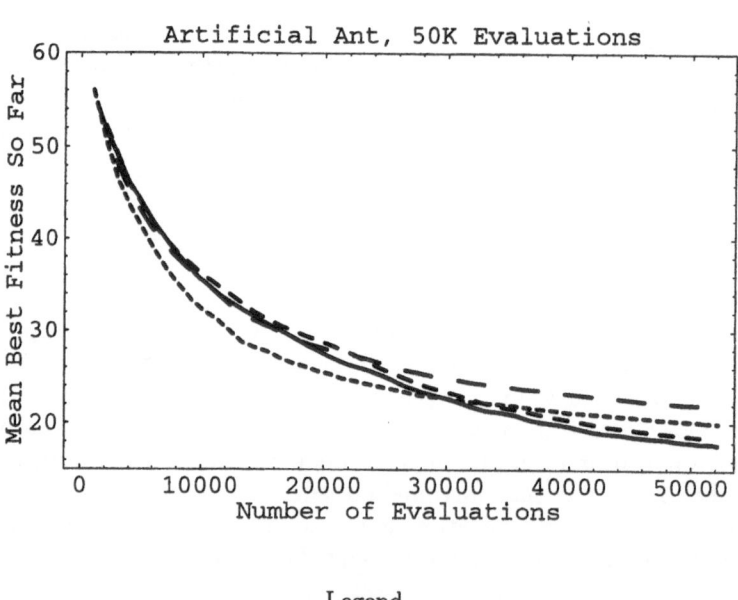

Fig. 2. Best-So-Far Curves of Four Layouts, Artificial Ant Problem, at 25K and 50K Evaluations

Table 6. Statistical significance groupings for mean best fitness of run for the Rosenbrock and Rastrigin domains (Third Experiment)

Problem and Evals		Layouts		
Rosenbrock	32	Diagonal	45	64
Mean Best Fitness	2256.585	2415.601	3281.273	5131.429
Rastrigin	32	Diagonal	45	64
Mean Best Fitness	136.0726	142.2995	155.8251	184.689

We adjusted the size of layouts to make them more appropriate to the GA realm: 32x1024, 45x724, 64x512, and a diagonal layout with initial population size of 64, running for 1024 generations. We performed 100 independent trials for each problem. As shown in Table 6, for both problem domains the ordering was the same, namely: 32x1024 outperformed diagonal, which in turn outperformed 45x724, which in turn outperformed 64x512. However, on the Rastrigin domain the difference between the 32x1024 and the diagonal layout was not statistically significant.

This is a mixed result. The result is similar to that of the 5-bit Parity GP domain, namely that longer runs are consistently preferred over shorter ones. But for the first time in this paper, diagonal has come in second in one of the experiments (Rastrigin). We note that the number of evaluations (32K, a typically size for GAs) is nonetheless similar to the smaller GP experiments. Further runs may suggest that GAs too have "diminishing returns" late in the run: but this is not yet borne out from the evidence here.

6 Conclusions and Future Work

In this paper we examined the possibility of imploding the population — gradually decreasing it towards zero as the run progressed — in the context of genetic programming. A linear decrease in population size proved effective regardless of the number of evaluations used. A diagonal layout was consistently in the top tier in every GP experiment, and usually gave the best results in the experiments. In initial GA experiments however, the results were mixed.

From this we can draw two conclusions: the primary conclusion is of course that non-rectangular layouts may yield better results than rectangular ones in environments like GP. But the second more troubling conclusion is that our experiments add to existing evidence that GP may have an aging problem. Whether due to premature convergence, bloat, or other factors, GP does not appear to use large-population resources effectively late in the run. While methods like diagonal layouts may serve to work around the issue, there is need for a closer examination as to *why* GP has diminishing returns, and how the representation or breeding methods may be changed to alleviate the problem.

Acknowledgements. This research was partially supported through a gift from SRA International and through Department of Army grant DAAB07-01-9-L504.

References

1. Streeter, M.J., Keane, M.A., Koza, J.R.: Iterative refinement of computational circuits using Genetic Programming. In Langdon, W.B., Cantú-Paz, E., Mathias, K., Roy, R., Davis, D., Poli, R., Balakrishnan, K., Honavar, V., Rudolph, G., Wegener, J., Bull, L., Potter, M.A., Schultz, A.C., Miller, J.F., Burke, E., Jonoska, N., eds.: GECCO 2002: Proceedings of the Genetic and Evolutionary Computation Conference, New York, Morgan Kaufmann Publishers (2002) 877–884
2. Koza, J.R., Keane, M.A., Yu, J., Mydlowec, W.: Automatic synthesis of electrical circuits containing a free variable using Genetic Programming. In Whitley, D., Goldberg, D., Cantu-Paz, E., Spector, L., Parmee, I., Beyer, H.G., eds.: Proceedings of the Genetic and Evolutionary Computation Conference (GECCO-2000), Las Vegas, Nevada, USA, Morgan Kaufmann (2000) 477–484
3. Koza, J.: Genetic Programming: on the Programming of Computers by Means of Natural Selection. MIT Press (1992)
4. Luke, S.: When short runs beat long runs. In: Proceedings of the Genetic and Evolutionary Computation Conference (GECCO) 2001, Morgan Kaufmann Publishers (2001) 74–80
5. De Jong, K.: An Analysis of the Behavior of a Class of Genetic Adaptive Systems. PhD thesis, University of Michigan, Ann Arbor, MI (1975)
6. Grefenstette, J.J.: Optimization of control parameters for Genetic Algorithms. IEEE Transactions on Systems Man and Cybernetics **16** (1986) 122–128
7. Schaffer, J.D., Caruana, R.A., Eshelman, L.J., Das, R.: A study of control parameters affecting online performance of Genetic Algorithms for function optimization. In: Proceedings of the Third International Conference on Genetic Algorithms, Morgan Kaufmann Publishers Inc. (1989) 51–60
8. De Jong, K., Spears, W.M.: An analysis of the interacting roles of population size and crossover in Genetic Algorithms. In Schwefel, H.P., Männer, R., eds.: Parallel Problem Solving from Nature - Proceedings of 1st Workshop, PPSN 1. Volume 496., Dortmund, Germany, Springer-Verlag, Berlin, Germany (1991) 38–47
9. Goldberg, D.E., Deb, K., Clark, J.H.: Genetic Algorithms, noise, and the sizing of populations. Complex Systems **6** (1992) 333–362
10. Giguère, P., Goldberg, D.E.: Population sizing for optimum sampling with Genetic Algorithms: A case study of the onemax problem. In Koza, J.R., Banzhaf, W., Chellapilla, K., Deb, K., Dorigo, M., Fogel, D.B., Garzon, M.H., Goldberg, D.E., Iba, H., Riolo, R., eds.: Genetic Programming 1998: Proceedings of the Third Annual Conference, University of Wisconsin, Madison, Wisconsin, USA, Morgan Kaufmann (1998) 496–503
11. Federman, F., Dorchak, S.F.: A study of classifier length and population size. In Koza, J.R., Banzhaf, W., Chellapilla, K., Deb, K., Dorigo, M., Fogel, D.B., Garzon, M.H., Goldberg, D.E., Iba, H., Riolo, R., eds.: Genetic Programming 1998: Proceedings of the Third Annual Conference, University of Wisconsin, Madison, Wisconsin, USA, Morgan Kaufmann (1998) 629–634
12. Arabas, J., Michalewicz, Z., Mulawka, J.J.: GAVaPS - A Genetic Algorithm with varying population size. In: Proceedings of IEEE Conference on Evolutionary Computation. Volume 1. (1994) 73–78
13. Balazs, M.E., Richter, D.L.: A Genetic Algorithm with dynamic population: Experimental results. In Brave, S., Wu, A.S., eds.: Late Breaking Papers at the 1999 Genetic and Evolutionary Computation Conference, Orlando, Florida, USA (1999) 25–30

14. Fuchs, M.: Large populations are not always the best choice in Genetic Programming. In Banzhaf, W., Daida, J., Eiben, A.E., Garzon, M.H., Honavar, V., Jakiela, M., Smith, R.E., eds.: Proceedings of the Genetic and Evolutionary Computation Conference. Volume 2., Orlando, Florida, USA, Morgan Kaufmann (1999) 1033–1038
15. Wagner, N., Michalewicz, Z.: Genetic Programming with efficient population control for financial time series prediction. In Goodman, E.D., ed.: 2001 Genetic and Evolutionary Computation Conference Late Breaking Papers, San Francisco, California, USA (2001) 458–462
16. Luke, S. ECJ 9: A Java EC research system. http://www.cs.umd.edu/projects/plus/ec/ecj/ (2002)
17. Cervone, G., Michalski, R., Kaufman, K., Panait, L.: Combining Machine Learning with Evolutionary Computation: Recent results on LEM. In: Proceedings of the Fifth International Workshop on Multistrategy Learning. (2000) 41–58

Methods for Evolving Robust Programs

Liviu Panait and Sean Luke

Department of Computer Science, George Mason University
4400 University Drive MSN 4A5, Fairfax, VA 22030, USA
{lpanait,sean}@cs.gmu.edu
http://www.cs.gmu.edu/~eclab

Abstract. Many evolutionary computation search spaces require fitness assessment through the sampling of and generalization over a large set of possible cases as input. Such spaces seem particularly apropos to Genetic Programming, which notionally searches for computer algorithms and functions. Most existing research in this area uses ad-hoc approaches to the sampling task, guided more by intuition than understanding. In this initial investigation, we compare six approaches to sampling large training case sets in the context of genetic programming representations. These approaches include fixed and random samples, and adaptive methods such as coevolution or fitness sharing. Our results suggest that certain domain features may lead to the preference of one approach to generalization over others. In particular, coevolution methods are strongly domain-dependent. We conclude the paper with suggestions for further investigations to shed more light onto how one might adjust fitness assessment to make various methods more effective.

1 Introduction

The bulk of evolutionary computation has been applied to non-stochastic problems with a finite set of inputs. Because the problem input space is fixed, the quality of a candidate solution can often be determined precisely, and often rapidly. More formally, given a set A of candidate problem solutions, much of evolutionary computation is typically trying to maximize some function f over a single fixed context c : Arg Max$_{a \in A} f(a,c)$.

There are important and notable exceptions to this general trend in EC. Interestingly, genetic programming has not been one of them. We say that this is interesting because genetic programming's notional goal is the development of *computer programs* or *algorithms* which are human competitive. But perhaps because of the difficulty of the search space, much of the GP community has focused instead on simple problems with little computational complexity, and thus needing only a finite, small input context. Even problems like Symbolic Regression, which technically have an infinite input space, are reduced to a fixed set of samples. Nonetheless, as computer power increases we expect to see the community more and more trying to tackle bigger computational challenges, demanding fixed and variable amounts of internal state, iteration, and recursion.

Often, such more "challenging" algorithmic problems range over an *infeasibly large* set of inputs. By "infeasibly" we mean that the set is so large that there is no way that the algorithm may be exhaustively tested on every possible input. In fact many, if not most, common algorithms operate over an *infinite* sized set of inputs, and inputs in these sets often differ in size or difficulty. Compare, for example, a function which sorts a single vector of predefined numbers to a computer algorithm which can sort any vector of any size and content. More formally we might describe these latter ones as optimization problems: Arg Max$_{a \in A} f(a, \langle c_0, c_1, ..., c_\infty \rangle)$. In many cases this optimization may be described as a summation: Arg Max$_{a \in A} \sum_{i=0}^{\infty} f(a, c_i)$ This class of problems is challenging because there is no way to prove through empirical exhaustion that such an algorithm is correct or optimal, simply because the input space is too large.

Certain "non-algorithms" can fall into this class as well. For example, real-valued feed-forward systems such as neural networks or symbolic regression trees may operate over an infinite set of numbers. Additionally, some stateless functions are intended to be repeatedly pulsed to iteratively manipulate an external world state: a soccer robot might have a simple set of boolean classification functions to be tested against an infinite number of possible opponent contexts.

As EC has been applied to problems in this class, various techniques for dealing with generalization over the input space have been proposed and tested. One such approach, *coevolution* has also attracted some theoretical attention. In this paper we will discuss and compare several common approaches to doing evolutionary computation in the face of large sets of inputs, and will cast them in the context of genetic programming.

We admit up front two ways in which our methodology will seem odd given the justifications described earlier. First, none of the problem domains we used actually has iteration, recursion, or internal state. We chose them instead because the goal in the study is to compare methods for large input spaces, rather than introduce new domains. As such we felt it more useful to use common and readily implementable problem domains relevant to the genetic programming literature. Second, certain problem domains tested (such as 4-parity) do not have large input spaces: however these problems do have a property central to the experiment, namely that a small sample of the input space does not provide many clues about the input space as a whole.

2 Robustness

In this paper we adopt the term *robust* and presume, as did [1], that it is synonymous with *generalizable*. In the machine learning community both terms imply the ability of a hypothesis induced from a set of examples (in our case, the learned genetic program) to adequately model the entire universe of exemplars (inputs). In [2], the notion of robustness is also attached to the ability to continue to perform well despite mutations in the evolved program code (distinguishing such "genotypic robustness" from generalizability or "phenotypic robustness"). We will not consider this issue in our study.

How does one go about searching for robustness? If possible, one may begin by attempting to reduce the input space to a set of "prototypical inputs" which can be proven to be a sufficient set to learn on. For example, when evolving a sorting network of size n it is not necessary to sort vectors of all possible integers; instead it is sufficient to sort all vectors consisting of only zeros and ones [3]. In many cases, however, this reduction is not possible, not obvious, or insufficient to reduce the input space to manageable sizes. Beyond this, a learning system has no choice but to sample the space.

The space may be sampled in a variety of ways. The obvious approach is to establish a fixed initial random sample which is repeatedly presented until the system has learned it. A naive variation is to fix the sample to a few input cases and hope for the best: this is the approach common in Symbolic Regression, for example. One interesting statistical issue not considered in this paper is the size of the sample: an extreme approach is to use one input case and evaluate as many candidate solutions as possible. At the other extreme, excessively large samples may waste evaluation resources that could be better spent on testing other potential solutions. We realize that such results may be sensitive to the sample size and plan to investigate this aspect in the future. For this work, we decided to fix the sample size to a relatively small fraction of the input space as we think this will be representative of typical problems.

Another approach is to randomly and uniformly resample the space at each presentation. This has the benefit of hindering convergence to a predefined set of exemplars, but a constantly changing input sample can also prevent the candidate solutions from having any search gradient. Sampling may also be done adaptively. There are two popular adaptive methods in the literature. *Sharing* methods discount the value of a given input based on how easy it is for the population as a whole to solve it. *Coevolutionary* methods evolve inputs along with the candidate solutions: the fitness of an input is based on how many candidate solutions it stumps.

It has been often the case that robustness was a desirable property for the end-result of the search process. This property comes however with overwhelming drawbacks. First, there is little understanding on how to proceed about searching for individuals that exhibit this feature. Second, because a large number of input cases may be necessary, the search process may require an increased computational time. Third, it is not trivial to test the presence of the property in the results of the evolutionary process (especially for prohibitively large numbers of possible inputs). In this paper, we plan to shed some light on the first and last concerns, while also investigating whether small samples of input cases can provide enough information for robust programs.

[4] presents a good survey of existing work, as well as an excellent study on evolving robust programs for the Artificial Ant problem by modifying the ant's food trails and creating new ones. [3] coevolves sorting networks with training sets and obtains a remarkable success in a difficult domain. [5] reports obtaining more robust programs when using coevolution, but, as we will see later in the experimental section, this conclusion is strongly domain-dependent. [6] suggests that adding the inverse size to the fitness increases the generality of evolved

programs, while [7] states that less robust results are obtained when a preference for smaller programs biases the GP system; [8] further investigates the relation between evolved program size, generality, and modularity. [9] presents a method for producing highly parsimonious DAG-based boolean programs which can generalize using only a small set of the subfunctions and variables in a problem. Adding noise to the inputs [10] or using multiple fitness sets, augmentation, and refinement [1] are other approaches to improving robustness of evolved programs.

3 Evolving Robust Programs

We selected six methods as candidates for our study. All six methods are either widely used or were reported to have yielded robust results. In our description of the methods below, we define Err(p, c) to be the error of a program p on the training case c.

Fixed-Random-Initial (FRI) randomly selects a set $c_1, ..., c_{N_c}$ of training cases beforehand and uses them during the entire run (reported in [11], for example); program p's fitness is assessed using the formula $\frac{1}{N_c} \sum_{i=1}^{N_c} \text{Err}(p, c_i)$. The individuals are therefore evaluated on just a fixed small subset of all possible input cases. This has advantages and disadvantages: all individuals are evaluated on the same inputs, so they can be directly compared in terms of performance; however, overfitting to a poorly chosen set of inputs could yield poorly performing individuals. We decided to select a small number of samples and to use the average for combining the results.

Sample Randomization involves modifying the training cases during the EC run. Again, program p's fitness is assessed using the formula $\frac{1}{N_c} \sum_{i=1}^{N_c} \text{Err}(p, c_i)$, however the set $c_1, ..., c_{N_c}$ is randomized for each individual (*Random-Per-Individual* or *RPI*) or once per generation (*Random-Per-Generation* or *RPG*). The main advantage of the two methods is that new training cases are introduced all the time in the search process. However, because of the randomization, it may be difficult to compare and rank individuals: when performance is assessed on different training cases, a better individual may do worse just because its training case was more difficult. Random-Per-Generation was introduced in [12]; Random-Per-Individual was used, for example, when noise was added to improve robustness [10].

Coevolution (CVL) helped obtain very robust results for the particular domain of evolving sorting networks [3]; instead of having fixed or randomized training cases, Hillis experimented with coevolving them in a different population. Our implementation differs somewhat from the one presented in [3]: rather than having grid-worlds and locality notions, we decided for a much simpler implementation. More specifically, there are two populations: one with programs and one with sets of training cases. The fitness for program p is calculated as

$\frac{1}{N_c}\sum_{i=1}^{N_c}\text{Err}(p,C_i)$, where $C_1,...,C_{N_c}$ represents the fittest set of training cases from the previous generation. Similarly, the fitness of the set of training cases $c_1,...,c_{N_c}$ is calculated as $\frac{1}{N_c}\sum_{i=1}^{N_c}\text{Err}(P,c_i)$, where P is the best program from the previous generation. After assessing the fitness of programs and training cases, selection and breeding operators are used to create new populations of each kind. The coevolution of the two populations is simultaneous, and a random program and a random set of training cases are selected for evaluating the first generation.

Coevolution With Opponent Sharing (Coshare or CSH) is another coevolutionary method used when multiple populations are present. It has been first described in [13] under the name "competitive fitness sharing". The main idea is to treat every individual in the population as a resource and reward individuals that defeat opponents defeated by just few individuals. In this paper we consider two populations (one with programs and one with training cases) and we assume a complete mixing (each individual is tested on each training case). The fitness of a program p is calculated using the formula $\frac{1}{N_c}\sum_{i=1}^{N_c}\text{Err}(p,c_i)\frac{\text{Err}(p,c_i)}{\sum_{j=1}^{N_p}\text{Err}(p_j,c_i)}$, where N_c and N_p are the sizes of the populations of training cases, respectively of programs, and c_i and p_j represent individuals in the training case and program populations.

Fitness Sharing (FSH) implies that the weight of the performance of individuals on different training cases depends on the performance of the population on the specific training cases. The fitness of program p is calculated similarly as in the *Coshare* method by using the formula $\frac{1}{N_c}\sum_{i=1}^{N_c}\text{Err}(p,c_i)\frac{\text{Err}(p,c_i)}{\sum_{j=1}^{N_p}\text{Err}(p_j,c_i)}$, however $c_1,...,c_{N_c}$ represents a set of training cases that is randomized every generation (as in the *Random-Per-Generation* method), rather than coevolved. A related approach is reported in [14].

4 Testing Robustness

One challenge to sampling is that the assessed quality of an individual is sensitive to its evaluation context (the set of input cases used). Thus to study the robustness of the programs, we used a variation of the *Train-Test-Validate* methodology [15,16]. In this methodology, individuals' fitnesses are assessed using a *training set*. The best-of-generation individuals are gathered, and the best-of-run individual is chosen from among them by testing each individual against a different *test set*. The final quality of the best-of-run individual is assessed using yet another different *validation set*. The size of these sets is relative to the frequency in which they are applied. Thus the validation set is larger than the test set, which in turn is larger than the training set. The use of the three separate sets means that the training, testing, and final comparison phases of the experiment are statistically independent. This is standard procedure in the Machine Learning and Data Mining communities, and as [17] argues, should be used more often in EC.

In the Artificial Ant, Multiplexer, and Even Parity problems, we also included results against an exhaustive sampling rather than a randomized validation set; our interest was in whether or not a large validation set was sufficient to justify not doing exhaustive sampling (particularly with infinite-sized input sets!). The validation sets, as it turns out, produced very similar results to the exhaustive samples.

We conducted the experiments on several common domains, all described below. More details on the domains and their standard settings can be found in [11]. All experiments were performed with the ECJ system [18]. Unless stated otherwise, the populations had 128 individuals and used 90% crossover probability, 10% reproduction probability, tournament selection with size 7, and one individual elitism. Other parameters were taken from [11]. Each experiment consisted of 50 independent runs. As we were testing generalization accuracy and not learning speed, we chose to give runs ample time (500 generations) to converge.

The problem domains used in this investigation are:

Symbolic Regression. The task is to learn the function $f(x) = x^4 + x^3 + x^2 + x$ from a set of pairs $\langle x, f(x) \rangle$ selected with x varying from -1 to 1. No ephemeral constants were used. The full input space is infinite in this problem. The training, testing and validation sets contained 20, 500, and 2000 random input cases respectively. The error on an input case was defined as the absolute distance between the real value of the generator function and the value estimated by the evolved program.

For the *Coevolution* method, the second population contained 128 training sets, each with 20 training cases; the training sets bred using Gaussian mutation ($\mu = 0, \sigma = 0.1$) with 100% probability, one-point crossover with 90% probability, and tournament selection of size 7. For the *Coshare* method, the second population had 20 individuals, each consisting of a single training case (a number between -1 and 1). The tournament selection had size 2, and Gaussian mutation ($\mu = 0, \sigma = 0.1$) was the only breeding operator.

11 Bit Multiplexer. The task is to search for boolean multiplexers that receive as input eight data and three address bits, and output the data bit corresponding to the specific address. The full input space has 2048 cases. The training, testing and validation sets contained 16, 32, and 64 random input cases respectively. The error on an input case was defined as 0 if the correct output was presented and 1 if not.

For the *Coevolution* method, the second population contained 128 training sets, each with 16 training cases. The training sets used bit-flip mutation (on average 2 bits were modified per training set) with 100% probability, one-point crossover with 90% probability, and tournament selection of size 7. For the *Coshare* method, the second population had 16 individuals, each consisting of a single training case. *Coshare* used tournament selection of size 2, followed by bit-flip mutation (on average, 3 bits per training case were modified).

Artificial Ant. The standard GP Artificial Ant problem consists of a two-dimensional discretized toroidal environment which contains a broken trail of

Table 1. Statistical significance groupings for the best-of-run performances for all problem domains. Horizontal bars at the same level indicate techniques with statistically insignificant differences in means. Methods are ordered according to the performance of the best-of-run individuals from best (leftmost) to worst (rightmost) for each domain. The Validation Ranking is obtained by evaluating the best-of-run individuals on randomly generated validation sets. The Exhaustive Ranking is obtained by evaluating the best-of-run individuals on all possible input cases

	Validation Ranking	Exhaustive Ranking
Regression	CVL RPG FSH FRI RPI CSH	
Multiplexer	RPI FSH FRI RPG CSH CVL	RPI FSH FRI CSH RPG CVL
Ant	CVL CSH RPG FSH RPI FRI	CVL CSH RPG FSH RPI FRI
4 Bit Parity	CSH FSH RPG RPI FRI CVL	CSH FSH RPG RPI FRI CVL

food pellets (we used the "Santa Fe" trail map). The GP program controls an ant starting at an initial position and orientation, which tries to collect as much food as possible in a fixed number of steps. Our experiment deviates from the standard by allowing the ant to start from any position and have any initial orientation. The full input space has 4096 possible input cases (triplets $\langle x, y, o \rangle$ where x and y range between 0 and 31, and o may take any of the four orientations). The training, testing and validation sets contained 10, 100 and 500 input cases each. The error on an input case was defined as the amount of food not consumed.

For the *Coevolution* method, the second population contained 128 training sets, each with 10 training cases; this population used a 3.4% mutation (about one gene per training set was randomized), 90% one-point crossover, and tournament selection of size 7. For the *Coshare* method, the second population had 10 individuals, each consisting of an training case (3 genes). *Coshare* again used tournament selection with size 2, this time followed by mutation with 34% probability of randomizing a gene, plus one-point crossover with 90% probability.

4-Bit Even Parity. The task consists of learning the parity of a bit string without counting the bits, using only the four boolean functions AND, OR, NAND, and NOR. 4-bit Even Parity has a small input space (16 cases). However the task is interesting in that it is difficult to generalize from a reduced sample (in our case, 8 cases in the training and 16 in the testing sets). The small input size also gave us a chance to see what would happen if the training, testing, and validation sets did not uniformly sample the space: in particular, the validation set was done on 24 random input cases. This is larger than the input space and by design cannot uniformly sample it.

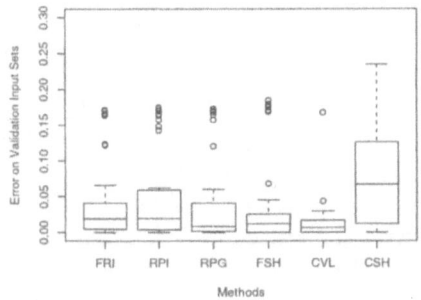

Fig. 1. Symbolic Regression: Boxplot of errors of methods on validation sets.

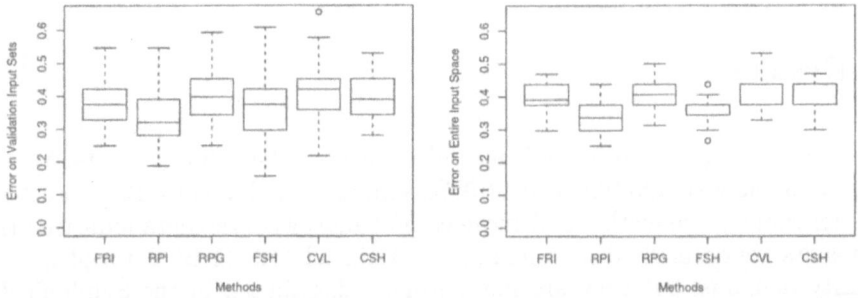

Fig. 2. 11-Bit Multiplexer: Boxplots of errors of methods on validation sets (left) and on all possible input cases (right).

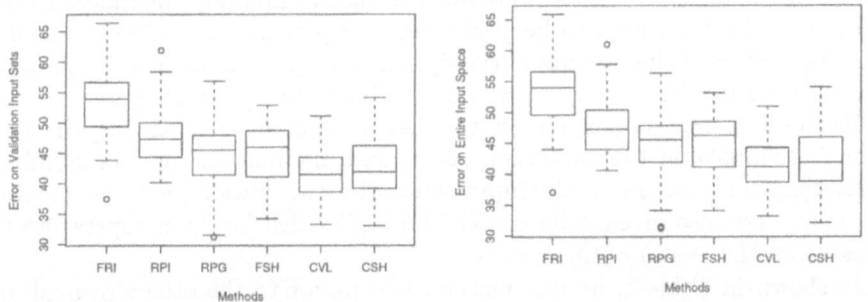

Fig. 3. Artificial Ant: Boxplots of errors of methods on validation sets (left) and on all possible input cases (right).

The *Coevolution* method contained 128 tests sets in the the second population, each with 8 training cases, using tournament selection of size 7, and bit-flip mutation that changed on average three bits per training set and one-point crossover with 90% probability. In the *Coshare* method, the second population had 8 individuals, each consisting of an training case. *Coshare* used tournament selection of size 2, followed by mutation flipping 2 bits per training case on average.

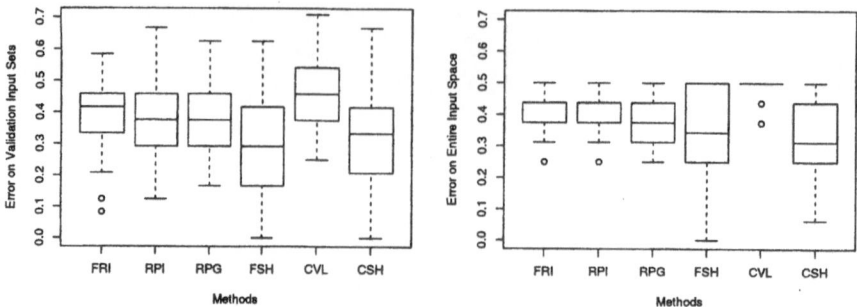

Fig. 4. 4-Bit Even Parity: Boxplots of errors of methods on validation sets (left) and on all possible input cases (right).

5 Results

Our results compare the best-of-run individuals on the validation sets. We initially used one-way ANOVA tests at 95% confidence, followed by Tukey post-hoc tests for ranking the methods. A more careful analysis of the data indicated that the results are normally distributed in the Artificial Ant, 11-Bit Multiplexer and 4-Parity domains, but they are not normally distributed in the Symbolic Regression domain. Additionally, the results have different variances, which may pose problems for the Tukey tests which assume equal variances. We compared the methods by pairwise comparisons using Welch's two-sample tests combined with Boole's inequality[1]. For 95% confidence when comparing 6 methods, 15 two-sample pairwise tests need to be performed at confidence $1 - \frac{1-0.95}{15} = 99.67\%$ each. We confirmed the ranking in the Symbolic Regression domain by performing comparisons using a nonparametric test due to Steel and Dwass.

Figures 1–4 present boxplots of the distributions of error among the best-of-run individuals on the validation set of each method. Except for Symbolic Regression, which has an infinite input space, the distributions on the exhaustive case spaces are also given. The center line inside each box plot represents the median (not the mean) of the sample.

As shown in Table 1, no one method is superior to the others over all domains in the study. In the Symbolic Regression domain, *Coevolution*, *Random-Per-Generation*, and *Fitness-Sharing* performed best. *Coevolution* and *Coshare* method did best in the Artificial Ant problem[2]. In the boolean function domains, the results are different. *Random-Per-Individual* and *Fitness-Sharing* perform

---

[1] Welch's two-sample test is similar to the standard t-test, but the possibly unequal variances of the two distributions are separately approximated. Boole's inequality is used for repeated testing of the same data sets, and it mainly increases the confidence requirements for each individual test proportional to the number of tests performed.

[2] In the Ant domain, *Random-Per-Individual* has significantly worse performance than *Fitness-Sharing*, but there is not enough confidence to state it is also worse than *Random-Per-Generation* which has a large variance in results.

best in the 11-Bit Multiplexer domain, and *Coshare* and *Fitness-Sharing* perform best in the 4-Bit Even Parity.

We analyzed the average performance of the best-of-generation and best-of-run individuals on the training, test and validation sets (graphs not shown). In general, *Coevolution* and *Coshare* both have higher errors on the training sets than on the validation sets; this is indicative of coevolution's relative success in discovering and promoting difficult training sets. We also observed overfitting to the testing sets with *Random-Per-Generation*, *Random-Per-Individual* and *Fixed-Random-Initial*. By comparing the validation and test results with a t-test at 95%, we found that in the Artificial Ant domain, these three methods all performed significantly worse on the validation sets than on the test sets.

We found that the Train-Test-Validate methodology as used in this paper gives a good approximation of the robustness of the evolved programs over the entire input space. As shown in Table 1, the ordering is relatively similar between validation and exhaustive rankings.

We finish with some speculation as to why various methods performed the way they did.

The four domains can be divided into two categories: in the boolean problem domains, it is easy to discover a difficult input case; whereas in the Artificial Ant and Regression problems, it is relatively more challenging to discover one. This is because Regression does not usually have large jumps with slightly modified X values, and small changes in orientation or location do not usually affect Artificial Ant solutions radically. We believe this may explain the behavior of *Coevolution*, which did poorly in the boolean problems but well in the other two. Specifically, we suspect that *Coevolution*'s population of training cases adapted too rapidly to difficult problems for the GP population to solve, thus contributing to a loss of gradient.

To quickly test this hypothesis, we changed the population of training cases in the 4-bit Parity domain to contain just two individuals (effectively hillclimbing), with size 2 tournament selection, and a mutation rate smaller by one order of magnitude. The goal was to prevent the training cases from improving too rapidly. The resulting solutions had statistically significantly better performance on both the validation and exhaustive sets of cases.

The Symbolic Regression domain differs from other problems in that each input case can cause an arbitrarily large error. This particularly affects the *Coshare* and *Fitness-Sharing* methods, in which a moderately large, but unique error in a single test case can ruin the fitness of an individual even if it significantly outperforms mediocre peers in the population on all the other cases.

The *Fixed-Random-Initial* method never fell into the top tier, suggesting that it should not be used as the first choice method when robust results are desired.

6 Conclusions and Future Work

Methods for sampling large input spaces may have particular utility to the evolution of computer algorithms and functions. In this initial investigation of the issue, we compared several approaches to sampling this space in the context of four genetic programming domains. However, we observed that *which* approaches did best was dependent on problem domain features. For example, *Coevolution* performed well in the Symbolic Regression and Artificial Ant domains, but it had the worst results in the 11-Bit Multiplexer and 4-Bit Even Parity. *Fitness-Sharing* fell in the first tier in the Symbolic Regression and 4-Bit Even Parity, ant fairer reasonably well (second tier) in the 11-Bit Multiplexer and the Artificial Ant problems.

Part of this may be due to unforeseen consequences of how fitness is assessed (such as Symbolic Regression). This brings up an important question to be addressed in future work: when the performance of an individual is assessed over several input cases, how can its overall performance be approximated? Possible alternatives to averaging may use the mode, median, maximum or minimum values, or include standard deviation or interquartile information ([10] in particular suggests using the minimum of several noisy evaluations). Additional research is also required to identify when overfitting occurs and how it can be avoided.

Acknowledgements. This research was partially supported through a gift from SRA International and through Department of Army grant DAAB07-01-9-L504. The authors would like to thank Dr. Daniel Menasce, R. Paul Wiegand, Elena Popovici, Gabriel Balan and Marcel Barbulescu for their help in conducting the research investigation. We would also like to thank Dr. Clifton Sutton for his assistance in analyzing the data.

References

1. Bersano-Begey, T.F., Daida, J.M.: A discussion on generality and robustness and a framework for fitness set construction in Genetic Programming to promote robustness. In Koza, J.R., ed.: Late Breaking Papers at the 1997 Genetic Programming Conference, Stanford University, CA, USA, Stanford Bookstore (1997) 11–18
2. Pagie, L., Hogeweg, P.: Evolutionary consequences of coevolving targets. Evolutionary Computation **5** (1997) 401–418
3. Hillis, D.: Co-evolving parasites improve simulated evolution as an optimization procedure. Artificial Life II, SFI Studies in the Sciences of Complexity **10** (1991) 313–324
4. Kushchu, I.: Genetic Programming and evolutionary generalization. IEEE Transactions on Evolutionary Computation **6** (2002) 431–442
5. Juille, H., Pollack, J.: Coevolutionary arms race improves generalization. In Koza, J.R., ed.: Late Breaking Papers at the Genetic Programming 1998 Conference, University of Wisconsin, Madison, Wisconsin, USA, Stanford University Bookstore (1998)
6. Kinnear, Jr., K.E.: Generality and difficulty in Genetic Programming: evolving a sort. In Forrest, S., ed.: Proceedings of the 5th International Conference on Genetic Algorithms, ICGA-93, University of Illinois at Urbana-Champaign, Morgan Kaufmann (1993) 287–294

7. Cavaretta, M.J., Chellapilla, K.: Data mining using Genetic Programming: The implications of parsimony on generalization error. In Angeline, P.J., Michalewicz, Z., Schoenauer, M., Yao, X., Zalzala, A., eds.: Proceedings of the Congress on Evolutionary Computation. Volume 2., Mayflower Hotel, Washington D.C., USA, IEEE Press (1999) 1330–1337
8. Rosca, J.: Generality versus size in Genetic Programming. In Koza, J.R., Goldberg, D.E., Fogel, D.B., Riolo, R.L., eds.: Genetic Programming 1996: Proceedings of the First Annual Conference, Stanford University, CA, USA, MIT Press (1996) 381–387
9. Droste, S.: Efficient Genetic Programming for finding good generalizing boolean functions. In Koza, J.R., Deb, K., Dorigo, M., Fogel, D.B., Garzon, M., Iba, H., Riolo, R.L., eds.: Genetic Programming 1997: Proceedings of the Second Annual Conference, Stanford University, CA, USA, Morgan Kaufmann (1997) 82–87
10. Reynolds, C.W.: Evolution of corridor following behavior in a noisy world. In: Simulation of Adaptive Behaviour (SAB-94). (1994)
11. Koza, J.: Genetic Programming: on the programming of computers by means of natural selection. MIT Press (1992)
12. Moore, F.W., Garcia, O.N.: New methodology for reducing brittleness in Genetic Programming. In Pohl, E., ed.: Proceedings of the National Aerospace and Electronics 1997 Conference (NAECON-97), IEEE Press (1997)
13. Rosin, C., Belew, R.: New methods for competitive coevolution. Evolutionary Computation 5 (1997) 1–29
14. Forrest, S., Smith, R.E., Javornik, B., Perelson, A.S.: Using Genetic Algorithms to explore pattern recognition in the immune system. Evolutionary Computation 1 (1993) 191–211
15. Rowland, J.: On model selection in supervised learning: Do we really know when to stop? In: Evolutionary and Neural Computation in Bioinformatics: A PPSN VII Workshop. (2002)
16. Brameier, M., Banzhaf, W.: A comparison of Linear Genetic Programming and Neural Networks in medical data mining. IEEE Transactions on Evolutionary Computation 5 (2001) 17–26
17. Eiben, A.E., Jelasity, M.: A critical note on experimental research methodology in EC. In: Proceedings of the 2002 Congress on Evolutionary Computation (CEC 2002). (2002) 582–587
18. Luke, S. ECJ 9: An Evolutionary Computation research system in Java. Available at http://www.cs.umd.edu/projects/plus/ec/ecj/ (2002)

On the Avoidance of Fruitless Wraps in Grammatical Evolution

Conor Ryan[1], Maarten Keijzer[2], and Miguel Nicolau[1]

[1] Department Of Computer Science And Information Systems
University of Limerick, Ireland
{Conor.Ryan,Miguel.Nicolau}@ul.ie
[2] CS Dept., Free University, Amsterdam
mkeijzer@cs.vu.nl

Abstract. Grammatical Evolution (GE) is an evolutionary system that employs variable length linear chromosomes to represent computer programs. GE uses the individuals to produce derivation trees that adhere to a Backus Naur Form grammar, which are then mapped onto a program. One unusual characteristic of the system is the manner in which chromosomes can be "wrapped", that is, if an individual has used up all of its genes before a program is completely mapped, the chromosome is reread. While this doesn't guarantee that an individual will map, prior work suggested that wrapping is beneficial for the system, both in terms of increased success rates and a reduced number of invalid individuals. However, there has been no research into the number of times an individual should be wrapped before the system gives up, and an arbitrary upper limit is usually chosen.
This paper discusses the different types of grammars that could be used with this system, and indicates the circumstances under which individuals will fail. It then presents a heuristic to minimize the number of wraps that have to be made before the system can determine that an individual will fail. It is shown that this can drastically reduce the amount of wrapping on a pathologically difficult problem, as well as on two *classes* of grammar often used by the system.

1 Introduction

Grammatical Evolution(GE) [10][6] is an Evolutionary Automatic Programming system that uses a variable length Genetic Algorithm to evolve programs in any language. The key to the system is the manner in which a Backus Naur Form (BNF) grammar is employed to specify the target language, and is used to map the linear genomes into syntactically correct programs.

One unusual characteristic of GE is the manner in which an individual's genome is reused, using a technique known as *wrapping*. That is, if, when the end of the genome is reached, an individual hasn't fully mapped to a program, another pass is made through the genome. The *intrinsically polymorphic* [3], that is, the manner in which the meaning of a codon is dependant on the context in which it is used, nature of the codons on the genome means that it is possible that, on the second and subsequent passes, a different interpretation will be produced. This can lead to a complete mapping for an individual that would otherwise have failed.

Although previous work [5] illustrated that wrapping often helps evolution and, at worst, does not harm it, no research has been conducted into the how many times an individual should wrap. An upper limit is chosen, but there is no way to tell in advance how this number should be set. If it is too low, then potentially useful individuals will wrongly be dubbed as invalid, while if it is too high, the system will waste time wrapping individuals that will *never* map.

This paper takes a formal look at the process of wrapping, and devises a heuristic to determine as early as possible when the wrapping of an individual should stop. We show that this heuristic reduces the effort required by GE without effecting the performance, because only those individuals that would fail to wrap are removed from the population.

The paper is laid out as follows. The next section introduces Grammatical Evolution and Backus Naur Form, while Sect. 3 discusses some of the properties of the simplest type of context free grammar(CFG). Section 4 expands this to include CFGs with two non-terminals, and Sect. 5 expands the work further again to the general case. Some experiments are conducted in Sect. 6, before the paper concludes with a brief summary in Sect. 7.

2 Grammatical Evolution

Grammatical Evolution is a genotype-phenotype mapping system that can map variable length binary strings into sentences of an arbitrary language. To use the system, one specifies the target language using a Backus Naur Form (BNF) grammar. BNF is a convenient representation for describing grammars. A grammar is represented by a tuple $\{N, T, P, S\}$, where T is a set of terminal symbols, i.e., items that can appear in legal sentences of the grammar, and N is a set of non-terminal symbols, which are interim items used in the generation of terminals. P is a set of production rules that map the non-terminal symbols to a sequence of terminal (or non-terminal) symbols, and S is a start symbol, from which all legal sentences must be generated.

Below is a sample grammar, similar to that used by Koza [2] in his symbolic regression and integration problems. Although Koza did not use grammars, the terminals in this grammar are similar to his function and terminal sets.

```
S = <expr>
<expr>     ::=  <expr> <op> <expr>
            |  ( <expr> <op> <expr> )
            |  <pre-op> ( <expr> )
            |  <var>
<op>       ::= + | - | / | *
<pre-op>   ::= Sin | Cos | Exp | Log
<var>      ::= 1.0
```

2.1 The Mapping Process

Rather than encoding actual programs, GE encodes *choices*. At many stages during the derivation of a legal sentence, one is required to make a choice. For example, in the previous grammar, there are four possible choices when mapping the <expr> non-terminal. To decode the choices, one first looks upon the genome as being divided into 8-bit codons, each of which codes for a single choice. Consider the individual below, written in decimal for brevity.

220 203 51 123 2 45

Initially, the *goal-stack*, that is, the stack of non-terminals, contains <expr>, as this is the start symbol. The first codon is read and **mod**ed by the number of choices giving, in this case, 0, so the first choice is made, causing the top of the stack to be replaced by the newly produced non-terminals. The individual continues to be read, codon by codon, until either the individual has mapped, or the all the codons have been exhausted.

In the latter case, the individual is *wrapped*. That is, the genome is read again. Clearly, if the first non-terminal to be mapped is the same as the *start* symbol, the individual will keep growing, and will never terminate. However, if the non-terminal is different, then this time around, the first codon will be used differently. This change will then ripple down through the rest of the codons, as the meaning of each codon is dependent on the context in which it is used. This property is referred to as *intrinsic polymorphism*. Even with wrapping, however, there is no guarantee that an individual will map to completion, so an upper limit is placed on the number of wraps permitted. Once this number has been reached, an individual is terminated.

Although the choice of this limit is clearly very important, given that if it is too low, individuals may be terminated before they get a chance to map, while an

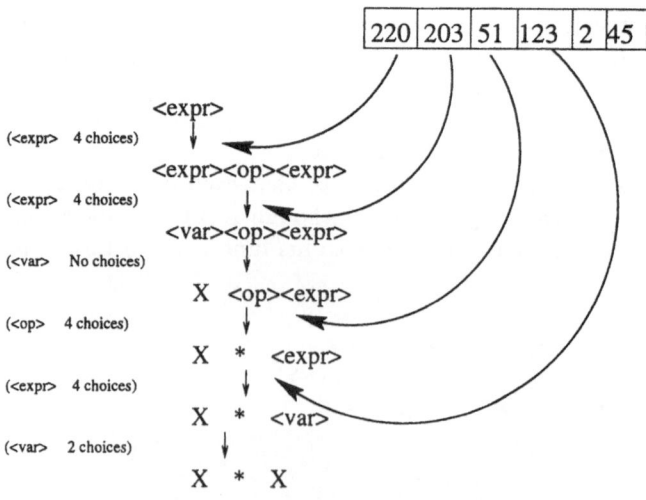

Fig. 1. Example mapping outline

excessively high limit leads to excessive wraps and memory usage (as individuals doomed to failure from excessive wraps tend to keep growing), no investigation has been carried out into determining a suitable number. An upper limit of ten was suggested in [4], although in a related system [9] a different approach is used, in which wrapping is only permitted in the first generation. These individuals are then "unrolled" so that they map in a single pass, and no wrapping is permitted in subsequent generations. In all cases, individuals that fail to map are given zero fitness, and are not permitted to engage in any reproductive activity.

Investigations into wrapping [5] suggested that, in the worst case, it doesn't adversely affect the performance of the system, while in the best case, it improves the performance.

These results suggest that, in general, most of the wrapping occurs early on in a run, and that most individuals (regardless of whether they wrap or not) map to completion after the first few generations. That said, there is clearly a lot of wasted evaluation expended on fruitless wraps. If there was some way to identify individuals that will *never* map without performing a large number of wraps, then the system could expect to enjoy a speed up, without any cost in performance, as none of these individuals contribute their genetic material through crossover.

3 Single Non-terminal Grammars

The simplest type of grammars are single non-terminal grammars. These grammars adhere to the "closure" principle of GP, that is, all functions (non-terminals for GE) can take all other functions and terminals as arguments. This can be illustrated with the following grammar, the productions of which are labelled for later reference :

```
E ::= (+ E E) | (* E E) | (- E E) | (% E E) | x | 1
        0         1         2         3        4   5
```

This grammar produces the same individuals as a GP set up with a function set of { +, *, -, %} (each of which has an arity of 2) and a terminal set of {x, 1}. In this case there is just *one* non-terminal, which ensures that any production can be applied at any time.

Furthermore, as there is only one non-terminal, codons effectively have *absolute* values, as they will never be re-interpreted. Thus, an individual that doesn't successfully map on the first pass will *never* map.

Single non-terminal grammars are, however, still useful in this context, as their simplicity helps us provide some definitions. We define *producers* to be those rules which increase the stack size, by adding one more more non-terminals to the stack, e.g. rules 0-3 above. Similarly, *consumers* are those rules which reduce the stack size, e.g. rules 4 and 5 above. Further, those rules which, after application, leave the size of the stack unchanged are referred to as *neutrals*.

We assign each production rule a **PCN** (Producer/Consumer/Neutral) number. That is, the effect it has on the stack size. From the grammar above, the

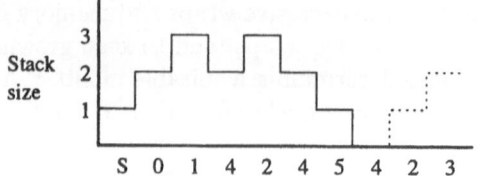

Fig. 2. Example shape graph

first four rules each have a PCN# of 1, while the remaining two have a PCN# of -1. Consider the (decoded) individual **0 1 4 2 4 5 4 2 3**. If we align the codons as follows, we can determine how the stack will grow and contract:

Codon	0	1	4	2	4	5	4	2	3
PCN#	1	1	-1	1	-1	-1	-1	1	1

The stack size always starts at 1, to reflect the presence of the start symbol. As each codon is successively applied, its PCN# is added to the size. If the size reaches zero, the individual has completely mapped.

A useful visualisation method is that of shape graphs [1]. These can be used to indicate the manner in which a stack grows and contracts over time. Figure 2 gives an example of the individual above.

The mapping stops when the size goes to zero, so the unexpressed codons are represented by the dashed line. Because this is a single non-terminal grammar, each codon has an absolute value, so the meaning is not influenced by its predecessors. Thus, we can examine the PCN numbers to determine whether or not an individual will map. That is, the PCN number for string s is

$$pcn(s) = \sum_{i}^{N} pcn(s_i)$$

If this number is less than zero, the mapping process will necessarily terminate. Furthermore, a string will terminate if and only if there is a k such that

$$\sum_{i}^{k} pcn(s_i) = -1$$

and, if a string doesn't terminate the first time around, it never will, as wrapping is essentially a double sum over PCN.

In particular, we can assert that, if the total PCN of consumers is greater than that of producers in an individual, then that individual will map.

Early work [4] indicated that the number of invalid individuals in GE dropped off from an average of 20% in the initial generation to almost zero by the fourth generation. We postulate that this is due to the evolution of a *stop sequence* of codons which, *almost always*, regardless of what productions were performed at the start of the mapping process, will complete an individual. Notice, however,

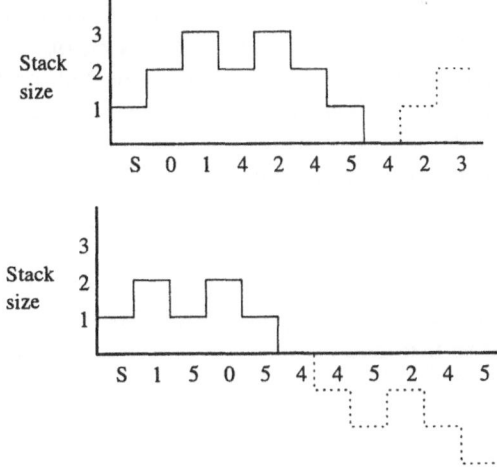

Fig. 3. Example shape graph of two GE individuals, with the second individual containing a particularly strong stop sequence

that one cannot expect a stop sequence to guarantee to complete all individuals, as individuals that grow without bounds initially may not map.

This is illustrated by the shape graphs in Fig. 3, in which the second parent has a tail of unexpressed codons, most of which are consumers. We describe any sequence of codons that, on average, consumes non-terminals as a stop sequence. Clearly, not all stop sequences are equal, with some being stronger than others as they consume more, and so have a steeper slope when plotted on a shape graph. Any individual crossed over with the second parent will be more likely to terminate due to the relatively strong stop sequence in the unexpressed tail of that individual.

4 Dual Non-terminal Grammars

The simplest case in which an individual may usefully be mapped is that in which there are two non-terminals. Consider the simple grammar:

```
A -> aAA | B
B -> a | b
```

where A is the start symbol. Typically, GE will wrap an individual if the stack is non-empty after the first pass. However, if the non-terminal A is on top of the stack, there is no point, as the system has just established that, *for this individual*, applying it to non-terminal A will (a) cause the stack to grow; and (b) leave another A on top of the stack. Clearly, this individual will keep growing.

If, on the other hand, we change context to the other non-terminal, we must *go through* the individual again. There are three possible results, (a) the individual completes; (b) an A is left on top; and (c) a B is left on top.

If an *A* is on top, then there is clearly no point in continuing, as the stack will continue growing. On the other hand, if there is a *B*, our next course of action depends on what happened to the stack. If it increased, or remained the same, then we should stop, while if it decreased, then we should continue wrapping.

Consider the individual 0 0 1, which is expressed using such low numbers to avoid having to perform the **mod** rule. After the first pass, the individual will have mapped to:

aaBAA

In this case, there is a different non-terminal at the top of the stack, so we wrap the individual to produce :

aaaaBAA

Although the individual has increased in size due to the introduction of the terminals, the stack has remained the same. Thus, we can assume that this individual will never map.

5 General Case

Most grammars have more than two non-terminals. Unfortunately, in these cases, one cannot be guaranteed that the behaviour of an individual will remain constant even in the same context. Consider the grammar :

```
A -> A | BC
B -> b | b
C -> BBB | BBB
```

where upper case letters denote non-terminals, and lower case letters terminals. If we try to map the individual 01, the following results:

Pass #	Individual
1	BC
2	bBBB
3	bbbB
4	bbbb

The result of pass #2 suggests that, with *B* on top of the stack, this particular individual will cause the stack to continue to increase in size. However, this is not the case, because, on the following pass, the stack decreases, and mapping is complete on the next pass.

This means that a simple watch on the size of the stack isn't enough to guarantee that an individual is going to fail. In general, an individual that will fail to map will enter into a cyclical behaviour. That is, the individual will keep being applied to the same non-terminal, or *set* of non-terminals. That is, although the individual may keep changing contexts, there is still a cycle. The difficulty, of course, is spotting these cycles.

Consider the individual above. The non-terminals on top of the stack at the start of each pass (including the first) are A,B,B and B. One could claim that the Bs form a repetitive pattern, but there are enough changes occurring in the stack to prevent it from becoming a pathological pattern. Clearly, only examining the top of the stack doesn't present enough information. We suggest the following heuristic : **if the entire stack from the last pass is at the top of the stack from this pass, then stop.**

This will indicate that the system is stuck in a cycle because, when the same stack from the last time around is on top, the exact same derivation sequence will be repeated. Notice that, while this heuristic isn't guaranteed to stop all wraps, it is guaranteed not to terminate an individual too early.

6 Experiments

To test how well the heuristic performed, an artificial problem was created that actively promotes individuals that wrap. This is achieved by making the fitness function the number of wraps an individual requires to map to completion. An individual that fails to wrap is given a score of zero.

The grammar used in this problem is designed to make it difficult to identify individuals that will fail. In particular, the mutually recursive relationship between the two non-terminals, coupled with the fact that both have productions that produce no non-terminals means that the stack can grow and contract several times throughout a derivation.

The actual grammar used is :

```
A -> BB | a
B -> BB | AA | c
```

Notice that, while this grammar is not designed to solve any particular problem, simply to promote wrapping, this turned out to be a nontrivial task, and individuals that reached the maximum fitness were extremely complicated.

A population size of 500 was run for 100 generations, with steady state tournament selection, and a tournament size of three. All results are averaged over 50 independent runs. The maximum fitness was 99, and the maximum number of wraps allowed was set to 100. Figure 4 shows the average fitness over time, and compares the number of illegal individuals identified by the heuristic to the number that exceeded the maximum number of wraps. In no case did an individual deemed a failure by the heuristic go on to map successfully.

Clearly, the heuristic fails to catch all the individuals that fail to map. However, it does identify an average of 35% of them after just two wraps. This represents a saving of 765,576 wraps for this problem. Given that this grammar *was* specifically designed to encourage wrapping, this is quite a considerable saving.

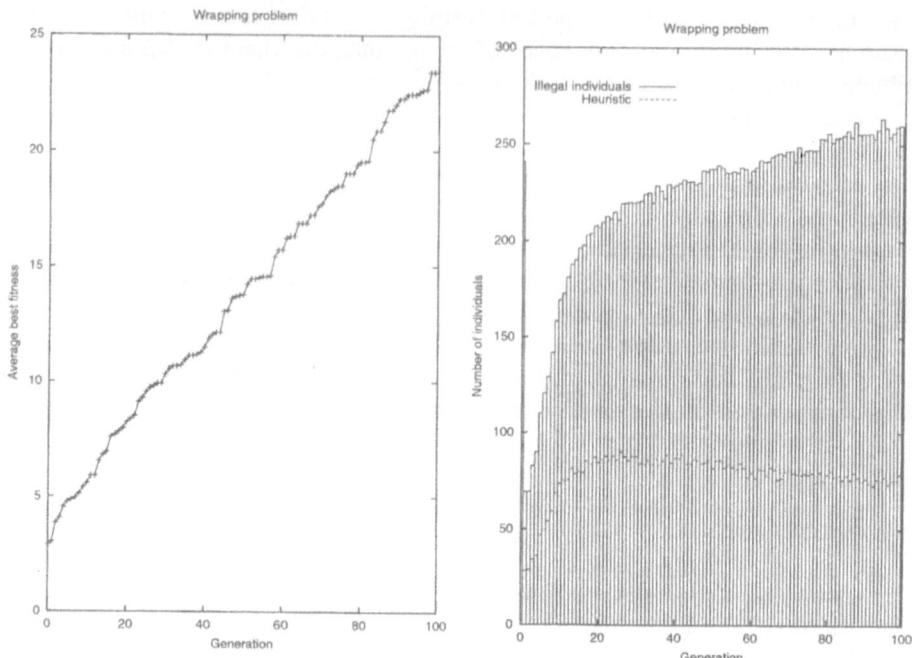

Fig. 4. Average fitness for the artificial wrapping problem (left) and a comparison of the number of individuals flagged by the heuristic, and those that timed out(right)

6.1 Application Grammars

The actual grammars employed by GE do not, in general, have such inherently pathological properties that facilitate wrapping. This section looks at two commonly used grammars, one suited for Symbolic Regression type problems, as shown in Sect. 2, and one for the Santa Fe Ant Trail, as described below.

```
<code> :: =    <line> |<code><line>
<line> :: =    <if-statement>|<op>
<if-statement> :: = if(food_ahead()) {<line>} else {<line>}
<op> :: =      left();  | right();  | move();
```

We are now more interested in the behaviour of the heuristic on a *class* of grammars, rather than on particular problems. To this end, we again use the number of wraps as a fitness function, to create a worst case scenario which may give us a lower bound on how well one can expect the heuristic to perform.

These problems use the same parameters normally associated with GE. That is, a population of 500 evolving for 50 generations, with steady state selection and a mutation probability of 0.01.

Figure 5 shows the performance of each of the grammars using the number of successful wraps as a measure of success. The Santa Fe results immediately leap out, showing that no individual in any of the runs attained a fitness greater

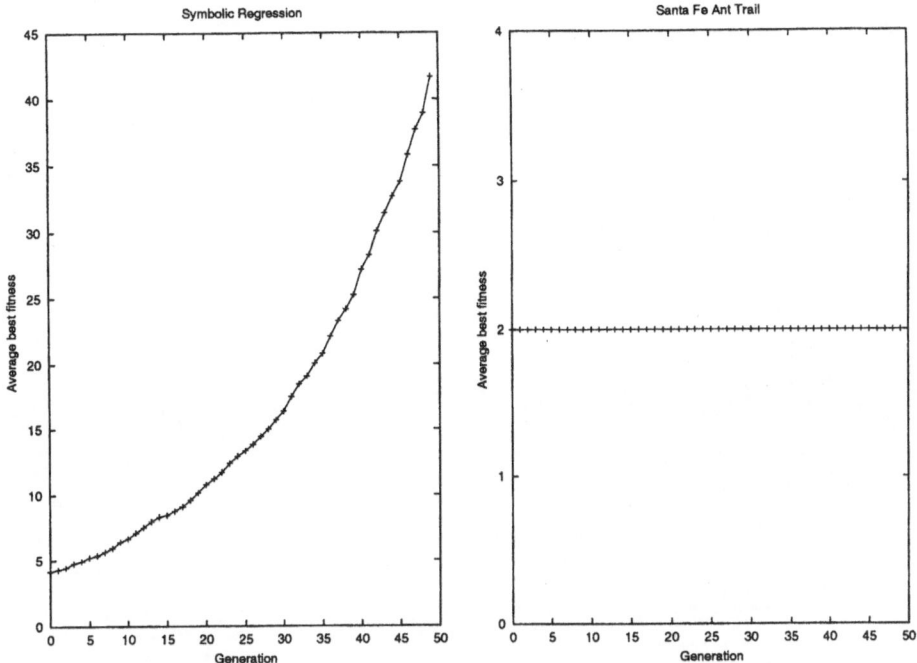

Fig. 5. Performance using number of successful wraps as a measure for Symbolic Regression (left) and Santa Fe Ant (right)

than two. It is an inherent property of the grammar that it will either complete after two wraps or not at all, but the heuristic is still capable of identifying individuals, and does so after an average of 1.3 wraps.

On the other hand, the Symbolic Regression grammar climbs continuously right up until the final generation, clearly indicating that legal individuals in this case are capable of considerably more wraps.

The numbers of individuals identified by the heuristic as being illegal in each case are shown in Fig. 6. In the case of the Symbolic Regression grammar, the heuristic performs very well at the start, but, after about 15 generations, starts to fail to identify failures. Across the 50 generations, the heuristic, on average, identifies 33% of invalid individuals, with an average of 2.06 wraps.

In the case of the Santa Fe grammar, the heuristic performs much better, correctly identifying around 84% of failures, with an average of just 1.3 wraps.

The fact that these problems specifically encourage wrapping should be kept in mind, so, particularly with the symbolic regression grammar, in which the fitness keeps rising, it shouldn't be surprising that the number of invalid individuals keeps increasing. This was done specifically to create a difficult situation for the heuristic, but in usual GE runs, the number of invalid individuals drops to near zero by the fourth or fifth generation.

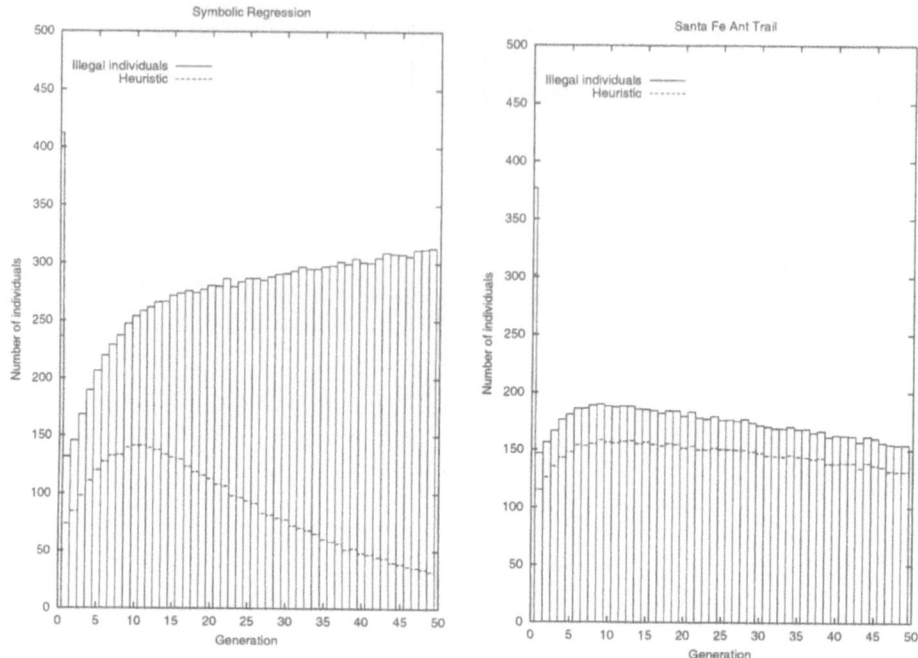

Fig. 6. The number of individuals identified by the heuristic for the Symbolic Regression grammar (left) and the Santa Fe grammar (right)

7 Conclusions

This paper has investigated the phenomenon of wrapping in Grammatical Evolution. We have demonstrated circumstances under which individuals will fail to map using a number of different types of context free grammars, namely, single non-terminal grammars, dual non-terminal grammars and multiple non-terminal grammars. Using shape graphs, we describe a stop sequence, which is any sequence of genes that is likely to terminate another string after crossover.

We showed a simple algorithm to determine whether or not an individual using a dual non-terminal grammar will fail to map, and illustrated the difficulty in extending this algorithm to multiple non-terminal grammars. However, a very cheap heuristic has been shown that, on a problem specifically designed to produce individuals that will fail, successfully identifies 35% of individuals that won't map.

The paper also considered two classes of grammars, a symbolic regression type, and a Santa Fe Ant Trail type. For both of these, the worst case was examined, that is, where individuals are actually rewarded for wrapping. In these cases, the heuristic successfully identified 33% and 84% of failures for the symbolic regression grammar and the Santa Fe Ant Trail grammar respectively.

References

1. Keijzer M. Scientific Discovery using Genetic Programming PhD Thesis, Danish Hydraulic Institute, 2001.
2. Koza, J.R., Genetic Programming: On the Programming of Computers by Means of Natural Evolution, MIT Press, Cambridge, MA, 1992.
3. Keijzer M., Ryan C., O'Neill M., Cattolico M., and Babovic V. Ripple crossover in genetic programming. In *Proceedings of EuroGP 2001*, 2001.
4. M. O'Neill. Automatic Programming in an Arbitrary Language: Evolving Programs with Grammatical Evolution. PhD thesis, University Of Limerick, 2001.
5. O'Neill M. and Ryan C. Genetic code degeneracy: Implications for grammatical evolution and beyond. In *ECAL'99: Proc. of the Fifth European Conference on Artificial Life*, Lausanne, Switzerland, 1999.
6. O'Neill M. and Ryan C. Grammatical Evolution. *IEEE Transactions on Evolutionary Computation*. 2001.
7. O'Sullivan, J., and Ryan, C., An Investigation into the Use of Different Search Strategies with Grammatical Evolution. In the proceedings of *European Conference on Genetic Programming (EuroGP2002)* (pp. 268–277), Springer, 2002.
8. Ryan, C., and Azad, R.M.A., Sensible Initialisation in Chorus. Accepted for *European Conference on Genetic Programming (EuroGP 2003)*.
9. Ryan, C., Azad, A., Sheahan, A., and O'Neill, M., No Coercion and No Prohibition, A Position Independent Encoding Scheme for Evolutionary Algorithms - The Chorus System. In the Proceedings of *European Conference on Genetic Programming (EuroGP 2002)* (pp. 131–141), Springer, 2002.
10. Ryan, C., Collins, J.J., and O'Neill, M., Grammatical Evolution: Evolving Programs for an Arbitrary Language, in *EuroGP'98: Proc. of the First European Workshop on Genetic Programming* (Lecture Notes in Computer Science 1391, pp- 83–95), Springer, Paris, France, 1998.

Dense and Switched Modular Primitives for Bond Graph Model Design

Kisung Seo[1], Zhun Fan[1], Jianjun Hu[1], Erik D. Goodman[1], and Ronald C. Rosenberg[2]

[1]Genetic Algorithms Research and Applications Group (GARAGe)
Michigan State University
[2]Department of Mechanical Engineering, Michigan State University
East Lansing, MI 48824, USA
{ksseo,fanzhun,hujianju,goodman,rosenber}@egr.msu.edu

Abstract. This paper suggests dense and switched modular primitives for a bond-graph-based GP design framework that automatically synthesizes designs for multi-domain, lumped parameter dynamic systems. A set of primitives is sought that will avoid redundant junctions and elements, based on pre-assembling useful functional blocks of bond graph elements and (optionally) using a switched choice mechanism for inclusion of some elements. Motivation for using these primitives is to improve performance through greater search efficiency and thereby to reduce computational effort. As a proof of concept for this approach, an eigenvalue assignment problem, which is to find bond graph models exhibiting minimal distance errors from target sets of eigenvalues, was tested and showed improved performance for various sets of eigenvalues.

1 Introduction

Design of interdisciplinary (multi-domain) dynamic engineering systems, such as mechatronic systems, differs from design of single-domain systems, such as electronic circuits, mechanisms, and fluid power systems, in part because of the need to integrate the several distinct domain characteristics in predicting system behavior (Youcef-Toumi [1]). However, most current research for evolutionary design has been optimized for a single domain (see, for example, Koza et. al., [2,3]).

In order to overcome this limitation and enable open-ended search, the Bond Graph / Genetic Programming (BG/GP) design methodology has been developed, based on the combination of these two powerful tools (Seo *et al.* [4,5] and tested for a few applications – an analog filter (Fan *et al.* [6]), printer drive mechanism (Fan et. al., [7]), and air pump design (Goodman *et al.* [8]). BG/GP worked efficiently for these applications. The search capability of this system has been improved dramatically by introduction of a new form of parallel evolutionary computation, called Hierarchical Fair Competition GP (HFC-GP, Hu, *et al.*, [9]), which can strongly reduce premature convergence and enable scalability with smaller populations.

However, two issues still arise: one is the need for much stronger synthesis capability arising from the complex nature of multi-domain engineering design, and the other is the desire to minimize computational demands. While we have made inroads in improving of GP search by introducing HFC-GP, we want to exploit the notion of modularity of GP function primitives to make additional gains. Much useful modularity can be discovered during an evolutionary process, as is done, for example, by the ADF (Koza [10]). However, in many cases, we believe that explicit introduction of higher-level modules as function primitives, based on domain knowledge, will yield faster progress than requiring their recognition during the evolutionary process. Some research has been devoted to choice or refinement of the function set in GP. Soule and Heckendorn [11] examined how the function set influences performance in GP and showed some relationship between performance and GP functions sets, but their work was limited to generating simple sine functions varying only arithmetic and trigonometric operators (e.g., +, -, *, /, tan, ...). We will try to exploit higher-level function sets, rather than simply choosing different sets at the same level.

In this paper, a generic type of primitive is introduced, and specialized here to capture specific domain knowledge about bond graphs – the *dense switched modular primitive*.

First, we introduce the *dense* module concept to generate compact bond graph models with fewer operations. It replaces several operations in the basic (original) set with one operation, yielding a smaller tree attainable with less computational effort.

Second, the *switched* module concept creates a small function set of elements with changeable forms, which can assist in evolving complex functionality, while eliminating many redundant bond graph structures evolved if it is not used. Elements eliminated include "dangling" junctions that connect to nothing and many one-port components (such as resistors, capacitors, inductors, etc.). Their elimination makes the resulting bond graph simpler and the speed of evolution faster.

A careful design of a dense and switched modular primitive should considerably increase the efficiency of search and also, for the bond graph case, the efficiency of fitness assessment, as is illustrated in this paper.

As a test class of design problems, we have chosen one in which the objective is to realize a design having a specified set of eigenvalues. The eigenvalue assignment problem is well defined and has been studied effectively using linear components with constant parameters. Section 2 discusses the inter-domain nature, efficient evaluation, and graphical generation of bond graphs, including the design methodology used in approaching such problems. Section 3 explains the basic set and redundancy problem and Sect. 4 describes the dense switched modular primitive set. Section 5 presents results for 6-, 10- and 16-eigenvalue design problems, and Sect. 6 concludes the paper.

2 Evolutionary Bond Graph Synthesis for Engineering Design

2.1 The BG/GP Design Methodology

There is a strong need for a unified design tool, able to be applied across energy domains – electrical, mechanical, hydraulic, etc. Most design tools or methodologies require user interaction, so users must make many decisions during the design process. This makes the design procedure more complex and often introduces the need for trial-and-error iterations. Automation of this process – so the user sets up the specifications and "pushes a button," then receives candidate design(s) – is also important.

A design methodology that combines bond graphs and genetic programming can serve as an automated and unified approach (Fig. 1). The proposed BG/GP (Bond Graph with Genetic Programming) design methodology requires only an embryo model and fitness (performance) definition in its initial stage; the remaining procedures are automatically executed by genetic programming search. However, due to the complexity of the engineering design problem, the need for efficiency in the design search is very high. It is this problem that is addressed here.

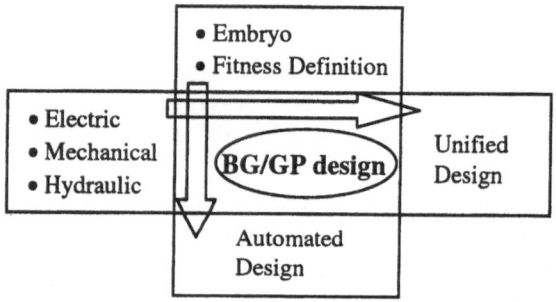

Fig. 1. Key features of the BG/GP design methodology

2.2 Bond Graphs

Topologically, bond graphs consist of *elements* and *bonds*. Relatively simple systems include passive one-port elements C, I, and R, active one-port elements S_e and S_f, and two-port elements TF and GY (transformers and gyrators). These elements can be attached to *0- (or 1-) junctions*, which are multi-port elements, using bonds. The middle of Fig. 2 consists of S_e, 1-junction, C, I, and R elements, and that same bond graph represents, for example, either a mechanical mass, spring and damper system(left), or an RLC electrical circuit. S_e corresponds with force in mechanical systems, or voltage in electrical (right). The 1-junction implies a common velocity for 1) the force source, 2) the end of the spring, 3) the end of the damper, and 4) the mass in the mechanical system, or implies that the current in the RLC loop is common. The R, I, and C represent the damper, inertia (of a mass), and spring in the mechanical system, or the resistor, inductor, and capacitor in the electrical circuit.

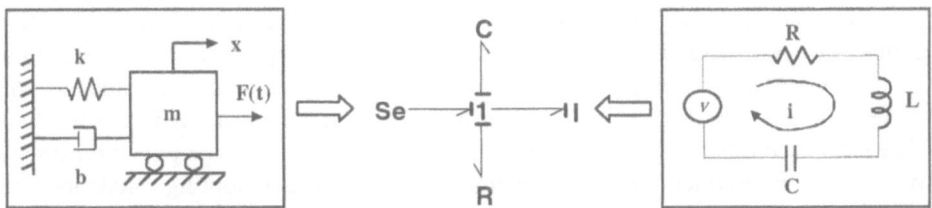

Fig. 2. The same bond graph model for two different domains

3 Basic Set and Redundancy

The initial BG/GP system used GP functions and terminals for bond graph construction as follows. There are four types of functions: *add* functions that can be applied only to a junction and which add a C, I, or R element; *insert* functions that can be applied to a bond and which insert a 0-junction or 1-junction into the bond; *replace* functions that can be applied to a node and which can change the type of element and corresponding parameter values for C, I, or R elements; and *arithmetic* functions that perform arithmetic operations and can be used to determine the numerical values associated with components (Table 1). Details of function definitions are illustrated in Seo *et al.* [5].

Table 1. Functions and terminals in Basic set

Name	Description
add_C	Add a C element to a junction
add_I	Add an I element to a junction
add_R	Add an R element to a junction
insert_J0	Insert a 0-junction in a bond
insert_J1	Insert a 1-junction in a bond
replace_C	Replace current element with C element
replace_I	Replace current element with I element
replace_R	Replace current element with R element
+	Sum two ERCs
-	Subtract two ERCs
endn	End terminal for add element operation
endb	End terminal for insert junction operation
endr	End terminal for replace element operation
erc	Ephemeral random constant (ERC)

Many redundant or unnecessary junctions and elements were observed in experiments with this basic set. Such unnecessary elements can be generated by the free combinatorial connection of elements, and, while they can be removed without any change in the physical meaning of the bond graph, their processing reduces the efficiency of processing and of search. At the same time, such a "universal" set guarantees that all possible topologies can be generated. However, many junctions "dangle" without further extension and many arrangements of one-port components (C, I, R) that can be condensed are generated. Figure 3 illustrates redundancies that are marked with dotted circles in the example. First, the dangling 0- and 1-junctions in the left-hand figure can be eliminated, and then three C, I, and R elements can be joined together at one 1-junction. Furthermore, two R elements attached to neighboring 0-junctions can be merged to a single equivalent R. Avoiding these redundant junctions and elements improves search efficiency significantly.

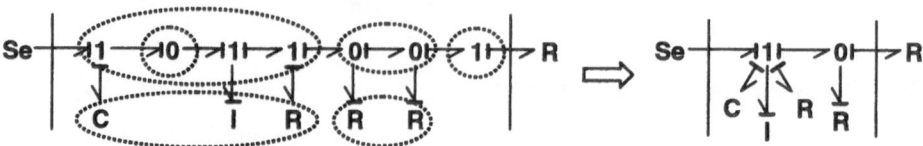

Fig. 3. Example of redundant 0- and 1-junctions and R elements (left) in generated bond graph model, and equivalent model after simplification (right). The dotted lines represent the boundary of the embryo

4 Construction of Dense Switched Modular Primitives

The redundancy problem is closely related with the performance and computational effort in the evolutionary process. The search process will be hastened by eliminating the redundancy, and it is hypothesized that this will happen without loss of performance of the systems evolved. It is obvious that computational resources can be saved by removal of the redundancy. To reduce the redundancy noted above and to utilize the concept of modularity, a new type of GP function primitives has been devised – the *dense switched* modular primitives ("DSMP"). Roughly speaking, a *dense* representation (eliminating redundant components at junctions, guaranteeing causally well-posed bond graphs, and avoiding adjacent junctions of the same type) will be combined with a *switched* structure (allowing components that do not impact causal assignment at a junction to be present or absent depending on a binary switch).

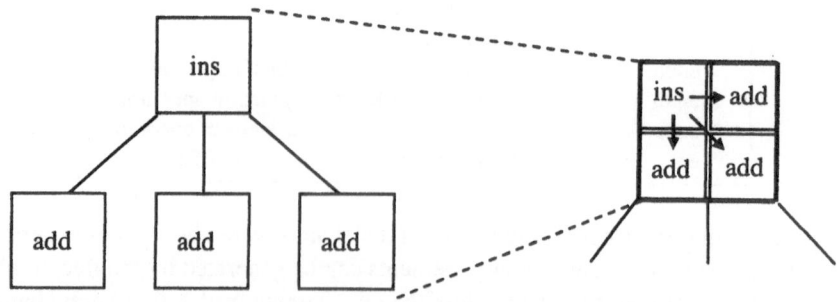

Fig. 4. The dense modular primitive

The major features of the modular primitives are as follows. First, a single dense function replaces all add, insert, and replace functions of the basic set. This concept is explained in Fig. 4, in which mixed *ins* and *add* operations can be merged into one operation. Therefore, a GP tree that represents a certain bond graph topology can be much smaller than attainable with the basic set. This dense function not only incorporates multiple operations, but also reflects design knowledge of the bond graph domain, such as causality (discussed later).

Second, any combination of C, I, and R components can be instantiated according to the values of a set of on/off *switch* settings that are evolved by mutation. This modularity also helps to relieve the redundancy of C, I, and R components, giving them fewer places to proliferate that appear to be different, but are functionally equivalent. This new set introduces further modularity through a controllable switching function for selection of C, I, R combinations (Fig. 5). The function set of the dense switched modular primitives is shown in Table 2. It consists of two functions that replace all *ins*, *add*, and *replace* functions in the basic set (Table 1).

Table 2. New functions in the switched modular primitive set

Name	Description
insert_JPair_SWElements	Insert a 0-1 (or 1-0) junction pair in a bond and attach switched C, I, R elements to each junction
add_J_ SWElements	Add a counter-junction to a junction and attach switched C, I, R elements

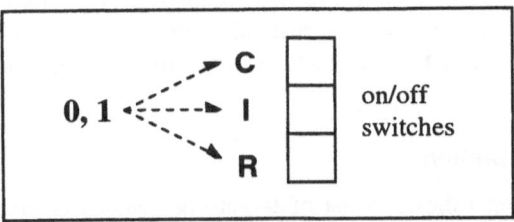

Fig. 5. Switched modular primitive

Third, the proper typing of 0-junctions and 1-junctions is determined by an implicit genotype-phenotype mapping, considering the neighbor junction to which the primitive is attached. This allows insertion of only "proper pairs" of junctions on bonds, preventing generation of consecutive junctions of the same type that are replaceable by a single one.

Fourth, we insure that we generate only feasible individuals, satisfying the causally well-posed property, so automatic state equation formulation is simplified considerably. One of the key advantages of BG/GP design is the efficiency of the evaluation. The evaluation stage is composed of two steps: 1) causality analysis, and, when merited, 2) dynamic simulation. The first, causal analysis, allows rapid determination of feasibility of candidate designs, thereby sharply reducing the time needed for analysis of designs that are infeasible. In most cases, all bonds in the graph will have been assigned a causal stroke (determining which variables are assigned values at that point, rather than bringing to it pre-assigned values) using only integral causality of C or I and extension of causal implication. Some models can have all causality assigned without violation – the causally satisfied case. Other models are assigned causality, but with violations – the causally violated case. If one has to continue to use an arbitrary causality of an R, it means that some algebraic relationships must be solved if the equations are to be put into standard form. This case can be classified as causally undetermined. Detail causality analysis is described in Karnopp et al. [12].

The dense switched modular primitives with implicit genotype-phenotype mapping and the guaranteed feasibility of the resulting causally well-posed bond graphs can speed up the evolution process significantly.

5 Experiments and Analysis

To evaluate and compare the proposed approach with the previous one, the eigenvalue assignment problem, for which the design objective is to find bond graph models with minimal distance errors from a target set of eigenvalues, is used. The problem of eigenvalue assignment has received a great deal of attention in control system design. Design of systems to avoid instability and to provide specified response characteristics as determined by their eigenvalues is often an important and practical problem.

5.1 Problem Definition

In the example that follows, a set of target eigenvalues is given and a bond graph model with those eigenvalues must be generated, in a classic "inverse" problem. The following sets (consisting of various 6-, 10- and 16-eigenvalue target sets, respectively) were used for the genetic programming runs:

- Eigenvalue sets used in experiments:
    1) $\{-1\pm 2j, -2\pm j, -3\pm 0.5j\}$
    2) $\{-10\pm j, -1\pm 10j, -3\pm 3j\}$
    3) $\{-20\pm j, -1\pm 20j, -7\pm 7j\}$
    4) $\{-1, -2, -3, -4, -5, -6\}$
    5) $\{-20\pm j, -1\pm 20j, -7\pm 7j, -12\pm 4j, -4\pm 12j\}$
    6) $\{-1, -2, -3, -4, -5, -6, -7, -8, -9, -10\}$
    7) $\{-20\pm 1j, -1\pm 20j, -7\pm 7j, -12\pm 4j, -4\pm 12j, -15\pm 2j, -9\pm 5j, -5\pm 9j\}$

The fitness function is defined as follows: pair each target eigenvalue one:one with the closest one in the solution; calculate the sum of distance errors between each target eigenvalue and the solution's corresponding eigenvalue, divide by the order, and perform hyperbolic scaling as follows. Relative distance error (normed by the distance of the target from the origin) is used.

$$Fitness(Eigenvalue) = 0.5 + \frac{1}{(2 + \sum Error/Order)}$$

We used a strongly-typed version (Luke, [13]) of lilgp (Zongker and Punch [14]) with HFC (Hierarchical Fair Competition, Hu, et al., [9]) GP to generate bond graph models. These examples were run on a single Pentium IV 2.8GHz PC with 512MB RAM. The GP parameters were as shown below.

Number of generations : 500
Population sizes : 100 in each of ten subpopulations for multiple population runs
Initial population: half_and_half
Initial depth : 3-6
Max depth : 12 (with 800 max_nodes)
Selection : Tournament (size=7)
Crossover : 0.9
Mutation : 0.1

The tabular results of 6- and 10-eigenvalue runs are provided in Tables 3-4, with statistics including mean relative distance error (averaged across each target eigenvalue) and mean tree size, for each set of 10 experiments.

Table 3 illustrates the comparison between the basic set and the DSMP (dense switched modular primitive) set on typical complex conjugate and real six-eigenvalue target sets. In the first set, {-1±2j, -2±j, -3±0.5j}, the average error of the basic set (0.151) is larger than that of the DSMP set (0.043). The second and third sets, for two different target eigenvalue sets that have larger norms from the origin, show average distance errors of the basic set that are also larger. The numbers in parentheses regarding distance error of the DSMP set represent their ratio to the basic set distance errors.

Table 3. Results for 6 eigenvalues

6-Eigenvalue Placement Problem (10 runs)				
	Basic set		DSMP set	
Eigenvalue set	Dist error	Tree Size	Dist error	Tree Size
{-1±2j, -2±j, -3±0.5j}	0.151	513.6	0.043(28%)	237.0
{-10±1j, -1±10j, -3±3j}	0.068	451.8	0.026(38%)	296.8
{-20±1j, -1±20j, -7±7j}	0.056	399.4	0.021(37%)	285.6
{-1, -2, -3, -4, -5, -6}	0.144	445.7	0.009(6%)	307.1

In a fourth example, an all-real set of target eigenvalues {-1, -2, -3, -4, -5, -6} is tested and shows that the ratio of errors between the approaches is more than ten (0.144 for the basic set vs. 0.009 for the DSMP set, only 6% of the basic set error). Also, mean tree sizes of all basic set runs are much larger than those of DSMP set.

Results for a 10-eigenvalue assignment problem are shown in Table 4. The results for a complex conjugate 10-eigenvalue set {-20±1j, -1±20j, -7±7j, -12±4j, -4±12j} show that the average error of the basic set (0.210) is three times larger than that of the DSMP set (0.064). The results for a real 10-eigenvalue set also show the average error of the basic set (0.267) is more than ten times larger than that of the DSMP set (0.023). As with 6 eigenvalues, the mean tree sizes of the basic set are larger than those of the DSMP set.

Table 4. Results for 10 eigenvalues

10-Eigenvalue Placement Problem (10 runs)				
	Basic set		DSMP set	
Eigenvalue set	Dist error	Tree size	Dist error	Tree size
{-20±1j, -1±20j, -7±7j, -12±4j, -4±12j}	0.210	564.9	0.064 (30%)	385.6
{-1, -2, -3, -4, -5, -6, -7, -8, -9, -10}	0.267	564.5	0.023 (9%)	425.8

Results for a 16-eigenvalue assignment problem – a much more difficult problem – are shown in Table 5. The results for a complex conjugate 16-eigenvalue set {-20±1j, -1±20j, -7±7j, -12±4j, -4±12j, -15±2j, -9±5j, -5±9j} show that the average error of the basic set (0.279) is twice as large as that of the DSMP set (0.132). Mean size of the GP tree, BG size, and computation time are also given in Table 5. BG size represents the mean number of junctions and C, I, R elements in each individual. All mean tree sizes, BG sizes, and computation times of the DSMP set are less, respectively, than their basic set counterparts

Table 5. Results for 16 eigenvalues

16-Eigenvalue Placement Problem (10 runs) {-20±1j, -1±20j, -7±7j, -12±4j, -4±12j, -15±2j, -9±5j, -5±9j}							
Basic set				DSMP set			
Dist error	Mean Tree Size	BG Size	Compu. Time (min)	Dist error	Mean Tree Size	BG Size	Compu. Time (min)
0.279	663.1	62.2	72.4	0.132 (47%)	592.6	37	56.1

Although the experiments run to date are not sufficient to allow making strong statistical assertions, it appears that the search capability of the DSMP set is superior to that of the basic set for bond graph design. The superiority of the DSMP set seems very clear. Although the difference may be not seem large, it is very significant considering that the results of the basic set runs are already taking advantage of HFC (Hierarchical Fair Competition, Hu, et al., [9]).

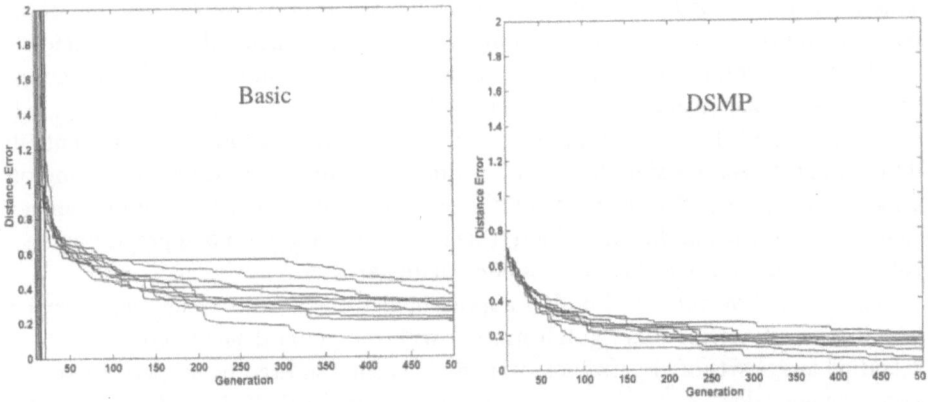

Fig. 6. Distance error for 16 eigenvalues

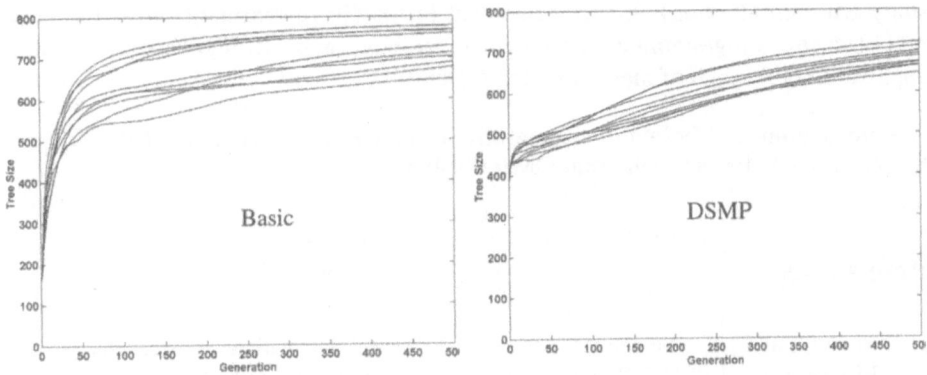

Fig. 7. Mean tree size for 16 eigenvalues

The distance errors (vs. generation) in 10 runs of the 16-eigenvalue problem are shown in Fig. 6. The distance errors of the DSMP set in Fig. 6 have already decreased rapidly within 50 generations, because only causally feasible (well-posed) individuals appear in the population. Figure 7 gives the mean tree sizes for each approach on the

16-eigenvalue problem. The DSMP set clearly obtains better performance using smaller trees. This bodes well for the scalability of the approach.

7 Conclusion

This paper has introduced the dense switched modular primitive for bond graph/GP-based automated design of multi-domain, lumped parameter dynamic systems. A careful combination is made of a *dense* representation (eliminating redundant components at junctions, guaranteeing causally well-posed bond graphs, and avoiding adjacent junctions of the same type) and a *switched* structure (allowing components that do not impact causal assignment at a junction to be present or absent depending on a binary switch). The use of these primitives considerably increases the efficiency of fitness assessment and the search performance in generation of bond graph models, to solve engineering problems with less computational effort.

As a proof of concept for this approach, the eigenvalue assignment problem, which is to synthesize bond graph models with minimum distance errors from pre-specified target sets of eigenvalues, was used. Results showed better performance for various eigenvalue sets when the new primitives were used. This tends to support the conjecture that a carefully tailored, problem-specific representation and operators that generate only feasible solutions with smaller amounts of redundancy and fewer genotypes that map to the same effective phenotype will improve the efficiency of GP search. This, in turn, offers promise that much more complex multi-domain systems with more detailed performance specifications can be designed efficiently. Further study will aim at extension and refinement of the GP representations for the bond-graph/genetic programming design methodology, and at demonstration of its applicability to design of more complex systems.

Acknowledgment. The authors gratefully acknowledge the support of the National Science Foundation through grant DMII 0084934.

References

1. Youcef-Toumi, K.: Modeling, Design, and Control Integration: A necessary Step in Mechatronics. IEEE/ASME Trans. Mechatronics, vol. 1, no.1, (1996) 29–38
2. Koza, J.R., Bennett, F. H., Andre, D., Keane, M.A., Dunlap, F.: Automated Synthesis of Analog Electrical Circuits by Means of Genetic Programming. IEEE Trans. Evolutionary Computation, vol. 1. no. 2. (1997) 109–128.
3. Koza, J.R., Bennett F.H., Andre D., Keane M.A., Genetic Programming III, Darwinian Invention and Problem Solving. Morgan Kaufmann Publishers, (1999)
4. Seo K., Goodman, E., Rosenberg, R.C.: First Steps toward Automated Design of Mechatronic Systems Using Bond Graphs and Genetic Programming. Proc. Genetic and Evolutionary Computation Conference, GECCO-2001, San Francisco (2001) 189

5. Seo K., Hu, J., Fan, Z., Goodman E.D., Rosenberg R.C.: Automated Design Approaches for Multi-Domain Dynamic Systems Using Bond Graphs and Genetic Programming. Int. Jour. of Computers, Systems and Signals, vol. 3. no. 1. (2002) 55–70
6. Fan Z., Hu, J., Seo, K., Goodman E.D., Rosenberg R.C., Zhang, B.: Bond Graph Representation and GP for Automated Analog Filter Design. Genetic and Evolutionary Computation Conference Late-Breaking Papers, San Francisco (2001) 81–86
7. Fan Z., Seo, K., Rosenberg R.C., Hu J., Goodman E.D.: Exploring Multiple Design Topologies using Genetic Programming and Bond Graphs. Proceedings of the Genetic and Evolutionary Computation Conference, GECCO-2002, New York (2002) 1073–1080.
8. Goodman E.D., Seo, K., Rosenberg R.C., Fan Z., Hu, J.: Automated Design of Mechatronic Systems: Novel Search Methods and Modular Primitives to Enable Real-World Applications. Proc. 2003 NSF Design, Service and Manufacturing Grantees and Research Conference, January, Birmingham, Alabama, (2003)
9. Hu J., Goodman E.D., Seo, K., Pei, M.: Adaptive Hierarchical Fair Competition (AHFC) Model for Parallel Evolutionary Algorithms. Proc. Genetic and Evolutionary Computation Conference, GECCO-2002, New York (2002) 772–779.
10. Koza, J.R., Genetic Programming II: Automatic Discovery of Reusable Programs. The MIT Press, (1994)
11. Soule, T., Heckendorn, R.B.: Function Sets in Genetic Programming.: Proc. Genetic and Evolutionary Computation Conference, GECCO-2001, San Francisco (2001) 190
12. Karnopp, D.C., Rosenberg R.C., Margolis, D.L., System Dynamics, A Unified Approach, 3rd ed., John Wiley & Sons (2000)
13. Luke S., Strongly-Typed, Multithreaded C Genetic Programming Kernel. http://www.cs.umd.edu/users/seanl/gp/patched-gp/. (1997)
14. Zongker, D., Punch, W., lil-gp 1.1 User's Manual. Michigan State University, (1996)

Dynamic Maximum Tree Depth
A Simple Technique for Avoiding Bloat in Tree-Based GP

Sara Silva[1] and Jonas Almeida[1,2]

[1] Biomathematics Group, Instituto de Tecnologia Química e Biológica
Universidade Nova de Lisboa, PO Box 127, 2780-156 Oeiras, Portugal
sara@itqb.unl.pt
http://www.itqb.unl.pt:1111

[2] Dept Biometry & Epidemiology, Medical Univ South Carolina, 135 Cannon Street
Suite 303, PO Box 250835, Charleston SC 29425, USA
almeidaj@musc.edu
http://bioinformatics.musc.edu

Abstract. We present a technique, designated as dynamic maximum tree depth, for avoiding excessive growth of tree-based GP individuals during the evolutionary process. This technique introduces a dynamic tree depth limit, very similar to the Koza-style strict limit except in two aspects: it is initially set with a low value; it is increased when needed to accommodate an individual that is deeper than the limit but is better than any other individual found during the run. The results show that the dynamic maximum tree depth technique efficiently avoids the growth of trees beyond the necessary size to solve the problem, maintaining the ability to find individuals with good fitness values. When compared to lexicographic parsimony pressure, dynamic maximum tree depth proves to be significantly superior. When both techniques are coupled, the results are even better.

1 Introduction

Genetic Programming (GP) solves complex problems by evolving populations of computer programs, using Darwinian evolution and Mendelian genetics as inspiration. Its search space is potentially unlimited and programs may grow in size during the evolutionary process. Code growth is a healthy result of genetic operators in search of better solutions. Unfortunately, it also permits the appearance of pieces of redundant code, called introns, which increase the size of programs without improving their fitness. Besides consuming precious time in an already computationally intensive process, introns may start growing rapidly, a situation known as bloat. Several plausible explanations for this phenomenon have been proposed (reviews in [9,12]), and a combination of factors may be involved, but whatever the reasons may be, the fact remains that bloat is a serious problem that may stagnate the evolution, preventing the algorithm from finding better solutions.

Several techniques have been used in the attempt to control bloat (reviews in [7,11]), most of them based on parsimony pressure. This paper presents a technique especially suited for tree representations in GP, based on tree depth limitations. The idea is to introduce a dynamic tree depth limit, very similar to the Koza-style strict limit [3] except in two aspects: it is initially set with a low value; it is increased when needed to accommodate an individual that is deeper than the limit but is better than any other individual found during the run. Unlike some of the most recent successful approaches, this technique does not require especially designed operators [6,14] and its performance is not reduced in the presence of populations where all individuals have different fitness values [7]. It is also simple to implement and can easily be coupled with other techniques to join forces in the battle against bloat.

2 Introns

In GP, the term *intron* usually refers to a piece of redundant code that can be removed from an individual without affecting its fitness [1]. The term *exon* refers to all non-intron code. However, on designing a procedure for detecting introns in GP trees, a more precise definition arises.

Our procedure for detecting introns works recursively in a tree from its root to its terminals, trying to maximize the number of introns detected. It can be described by the following pseudo-code, where *nintrons* designates the number of introns detected:

If tree is a terminal node:

nintrons = 0

Otherwise:

Evaluate tree and all its branches
If none of the branches returns the same evaluation as tree:

nintrons = sum of the number of introns in each branch

Otherwise:

Count nodes and introns in branches with same evaluation as tree
Pick branch with lowest number of non intron nodes: *ibranch*
nintrons = all nodes in tree minus non intron nodes in *ibranch*

This does not detect introns in code like not(not(X)) or -(-(X)), although removing pieces of this code, leaving only the underlined parts, would not affect the fitness of the individual. We use the term *redundant complexity* to designate these and other cases where a tree could be replaced by a smaller tree (but not by one of its branches) without affecting the fitness of the individual. Redundant complexity is also a problem in GP [10].

We use the term *bloat* loosely to designate a rapid growth of the mean population size in terms of the number of nodes that make up the trees, whether it happens in the beginning or in the end of the run, and whether the growth is caused by introns or exons.

3 Dynamic Maximum Tree Depth

Dynamic maximum tree depth is a technique that introduces a dynamic limit to the maximum depth of the individuals allowed in the population. It is similar to the Koza-style depth limiting technique, but it does not replace it – both dynamic and strict limits are used in conjunction. There is also an initial depth limit imposed on the trees that form the initial population [3], which must not be mistaken for the initial value of the dynamic limit. The dynamic limit should be initially set with a low value at least as high as the initial tree depth. It can increase during the run, but it always lies somewhere between the initial tree depth and the strict Koza limit. Whenever a new individual is created by the genetic operators, these rules apply:

— if the new individual is deeper than the strict Koza depth limit, reject it and consider one of its parents for the new generation instead;
— if the new individual is no deeper than the dynamic depth limit, consider it acceptable to participate in the new generation;
— if the new individual is deeper than the dynamic limit (but no deeper than the strict Koza limit) measure its fitness and:
  – if the individual has better fitness than the current best individual of the run, increase the dynamic limit to match the depth of the individual and consider it acceptable to the new generation;
  – if the individual is not better than the current best of run, leave the dynamic level unchanged and reject the individual, considering one of its parents for the new generation instead.

Once increased, the dynamic level will not be lowered again. If and when the dynamic limit reaches the same value as the strict Koza limit, both limits become one and the same. By setting the dynamic limit to the same value as the strict Koza limit, one is in fact switching it off, and making the algorithm behave as if there were no dynamic limit.

The dynamic maximum depth technique is easy to implement and can be used with any set of parameters and/or in conjunction with other techniques for controlling bloat. Furthermore, it may be used for another purpose besides controlling bloat. In real world applications, one may not be interested or able to invest a large amount of time in achieving the best possible solution, particularly in approximation problems. Instead, one may only consider a solution to be acceptable if it is sufficiently simple to be comprehended, even if its accuracy is known to be worse than the accuracy of other more complex solutions. Choosing less stringent stop conditions to allow the algorithm to stop earlier is not enough to ensure that the resulting solution will be acceptable, as it cannot predict its complexity. By starting with a low dynamic limit for tree depth and repeatedly raising it as more complex solutions prove to be better than simpler ones, the dynamic maximum tree depth technique can in fact provide a series of solutions of increasing complexity and accuracy, from which the user may choose the most adequate one. Once again, it is important to choose a low value for the initial dynamic depth limit, to force the algorithm to look for simpler solutions before adopting more complex ones.

4 Experiments

To test the efficacy of the dynamic maximum tree depth technique we have decided to perform the same battery of tests using six different approaches: (1) Koza-style tree depth limiting alone [3]; (2) lexicographic parsimony pressure (coupled with the Koza-style limit, which yielded better results than lexicographic parsimony pressure alone, with no bucketing) [7]; (3) dynamic maximum tree depth (also coupled with the Koza-style limit, as described in the previous section) with initial dynamic limit 9; (4) dynamic maximum tree depth, with initial dynamic limit 9, coupled with lexicographic parsimony pressure; (5) dynamic maximum tree depth with initial dynamic limit 6 (the same as the initial tree depth); (6) dynamic maximum tree depth, with initial dynamic limit 6, coupled with lexicographic parsimony pressure.

Two different problems were chosen for the experiments: Symbolic Regression of the quartic polynomial ($x^4 + x^3 + x^2 + x$, with 21 equidistant points in the interval -1 to $+1$), and Even-3 Parity. These problems are very different in the number of possible fitness values of the evolved solutions – from a potentially infinite number in Symbolic Regression, to very few possible values in Even-3 Parity. This facilitates the comparison with lexicographic parsimony pressure because it covers the two domains where this technique, due to its characteristics, behaved most differently [7].

A total of 30 runs were performed with each technique for each problem. All the runs used populations of 500 individuals evolved during 50 generations, even when an optimal solution was found earlier. The parameters adopted for the experiments were essentially the same as in [3] and [7], to facilitate the comparison between the techniques, but a few differences must be noted. Although some effort was put into promoting the diversity of the initial population, the tree initialization procedure used (Ramped Half-and-Half [3]) did not guarantee that all individuals were different. For 500 individuals and initial trees of maximum depth 6, the mean diversity of the initial population was roughly 75% for Symbolic Regression and 80% for Even-3 Parity, where diversity is the percentage of individuals in the population that account for the total number of different individuals (based on variety in [5]). Standard tree crossover was used, but with total random choice of the crossover points, independently of being internal or terminal nodes.

As stated in the previous section, the dynamic maximum tree limit can be switched on and off without affecting the rest of the algorithm, and switching it off is setting it to the same value as the Koza-style limit, which was always set to depth 17. Likewise, lexicographic parsimony pressure can be switched on by using a modified tournament selection that prefers smaller trees when their fitness values are the same, or switched off by using the standard tournament selection. Table 1 summarizes the combinations of tournament and initial dynamic limit that lead to each technique.

All the experiments were performed with GPLAB [8], a GP Toolbox for MATLAB [13]. Statistical significance of the null hypothesis of no difference was determined with ANOVAs at $p = 0.01$.

Table 1. Tournament and initial dynamic limit settings for each technique used

Technique	Tournament	Initial dynamic limit
(1) Koza-style tree depth limiting (K)	standard	17
(2) Lexicographic parsimony pressure (L)	modified	17
(3) Dynamic maximum tree depth 9 (S9)	standard	9
(4) L+S9	modified	9
(5) Dynamic maximum tree depth 6 (S6)	standard	6
(6) L+S6	modified	6

5 Results

The results of the experiments are presented as boxplots and evolution curves concerning the mean size of the trees (Fig. 1), where size is the number of nodes, the mean percentage of introns in the trees (Fig. 2), the population diversity as defined in the last section (Fig. 3), and best (lowest) fitness achieved (Fig. 4). The boxplots are based on the mean values (except Fig. 4, based on the best value) of all generations of each run, and each technique is represented by a box and pair of whiskers. Each box has lines at the lower quartile, median, and upper quartile values, and the whiskers mark the furthest value within 1.5 of the quartile ranges. Outliers are represented by + and × marks the mean. The evolution curves represent each generation separately (averaged over all runs), one line per technique. Throughout the text, we use the acronyms introduced in Table 1 to designate the different techniques.

5.1 Mean Size of Trees

Figure 1 shows the results concerning the mean size of trees. In the Symbolic Regression problem, techniques S9 and S6 performed equally well and were able to significantly lower mean tree sizes of run when compared to K and L. None of the compound techniques (L+S9, L+S6) proved to be significantly different from its counterpart (S9 and S6, respectively). The evolution curves show a clear difference in growth rate between the four techniques that use dynamic maximum tree depth (S9, L+S9, S6, L+S6) and the two that do not (K, L).

In the Even-3 Parity problem, techniques S9 and S6 were significantly different from each other. Both outperformed K, but only S6 performed better than L (no significant difference between S9 and L). However, both compound techniques (L+S9, L+S6) were able to outperform L, as well as their respective counterparts S6 and S9. The evolution curves show a tendency for stabilization of growth in all three techniques that use lexicographic parsimony pressure (L, L+S9, L+S6).

5.2 Mean Percentage of Introns

Figure 2 shows the results concerning the mean percentage of introns. Starting with Symbolic Regression, the striking fact about this problem is that it prac-

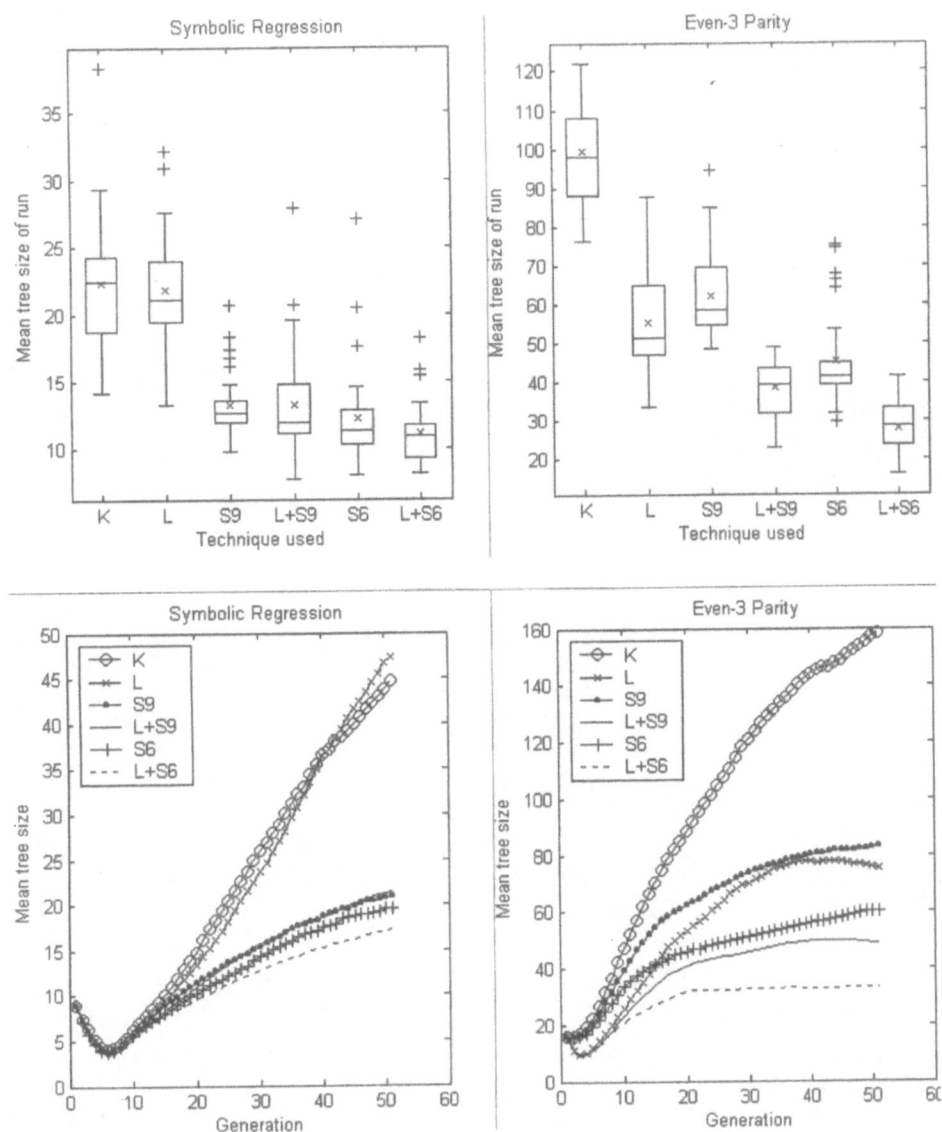

Fig. 1. Boxplots and evolution curves of mean tree size. See Table 1 for the names of the techniques

tically does not produce introns, redundant complexity being the only apparent cause of bloat. The differences between techniques in mean intron percentage of run were not significant. The evolution curves show that the percentage of introns tends to slowly increase after an initial reduction.

In the Even-3 Parity problem the percentage of introns is generally high. Techniques S9 and S6 outperformed K, and S6 also outperformed L, with no

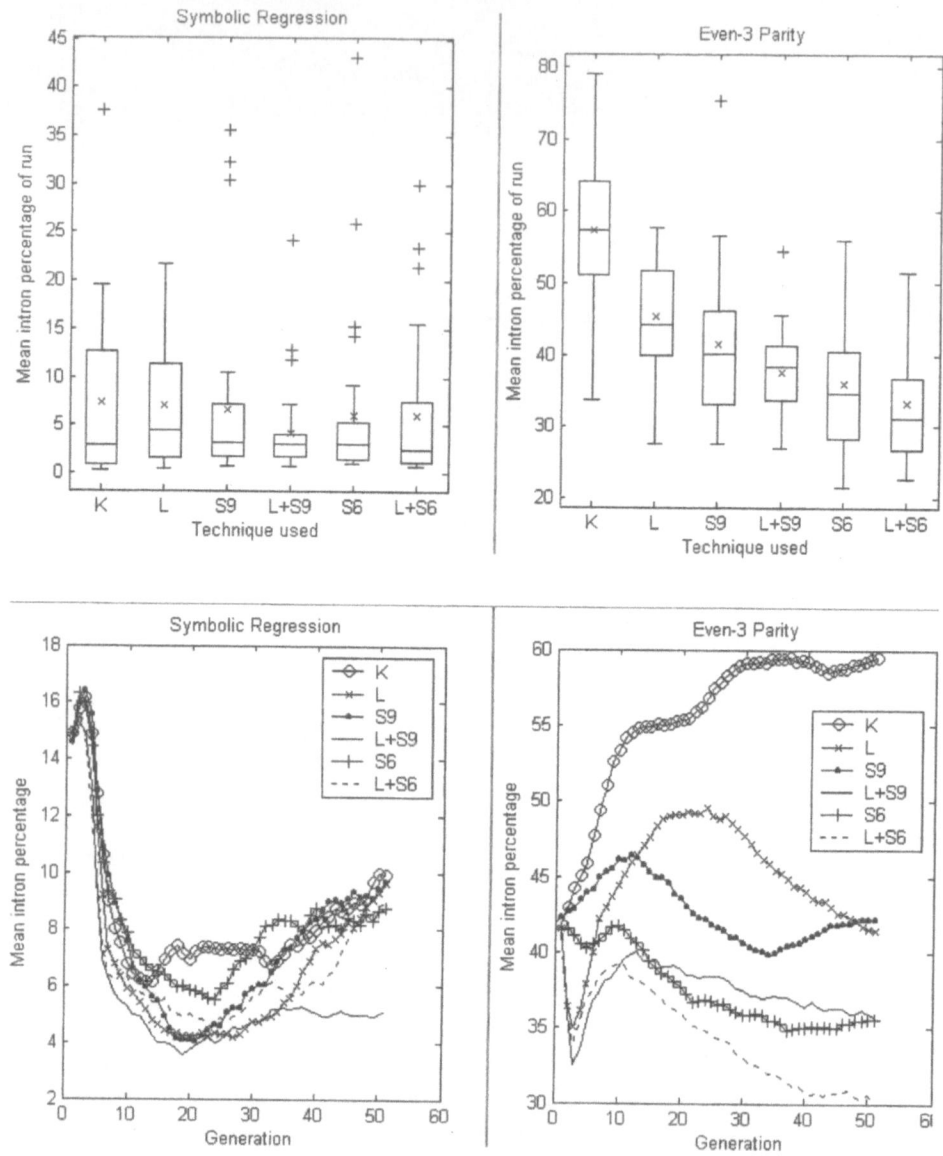

Fig. 2. Boxplots and evolution curves of mean intron percentage. See Table 1 for the names of the techniques

significant difference between S9 and L (similarly to what had been observed in mean tree size). The compound techniques (L+S9, L+S6) showed no significant differences from their counterparts (respectively S9 and S6), but both were able to significantly outperform L. The evolution curves show that, by the end of the run, the mean intron percentage has not increased from its initial value, in all techniques except K.

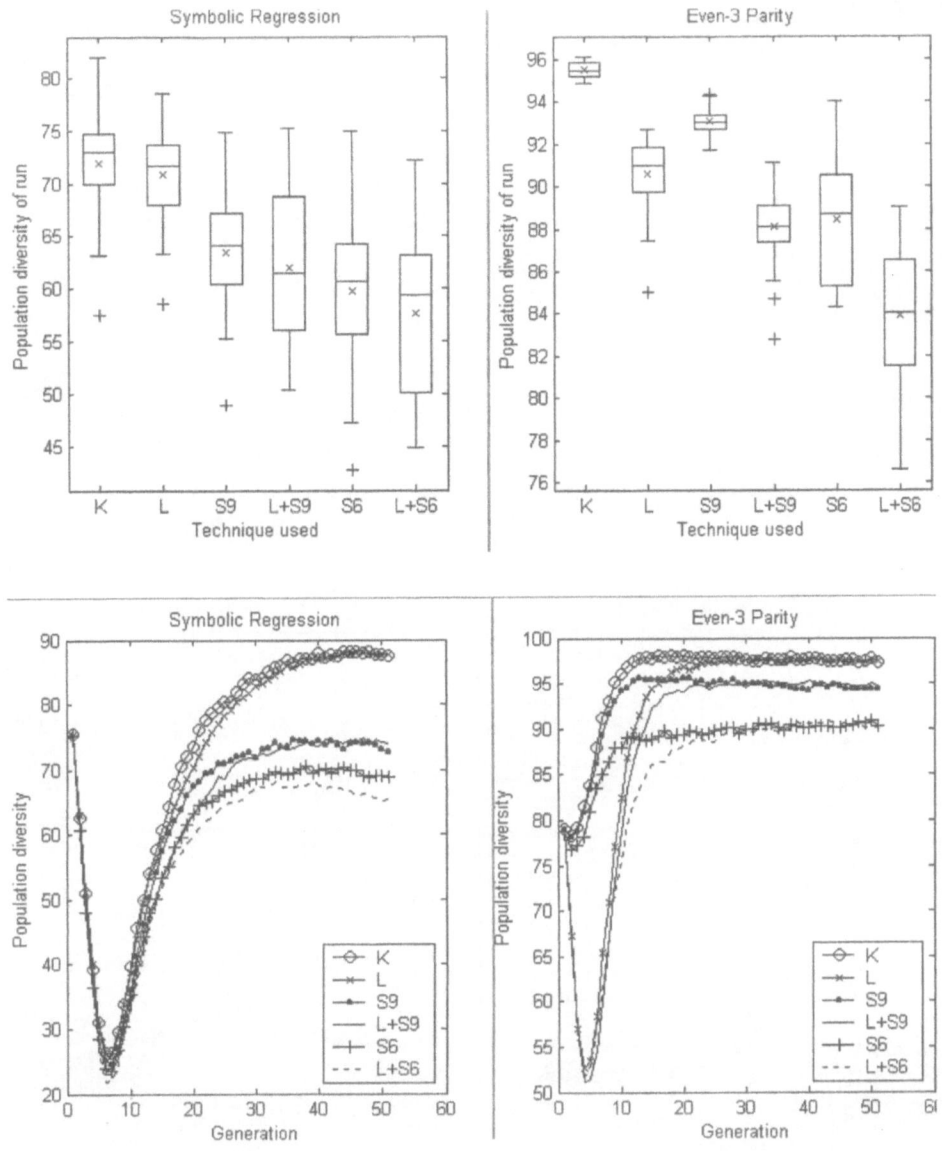

Fig. 3. Boxplots and evolution curves of population diversity. See Table 1 for the names of the techniques

5.3 Population Diversity

Figure 3 shows the results concerning the population diversity. In Symbolic Regression, techniques S9 and S6 caused a significant decrease in population diversity when compared to K and L. None of the compound techniques (L+S9, L+S6) was significantly different from its counterpart (S9 and S6, respectively). The evolution curves show a conspicuous decrease in population diversity in the

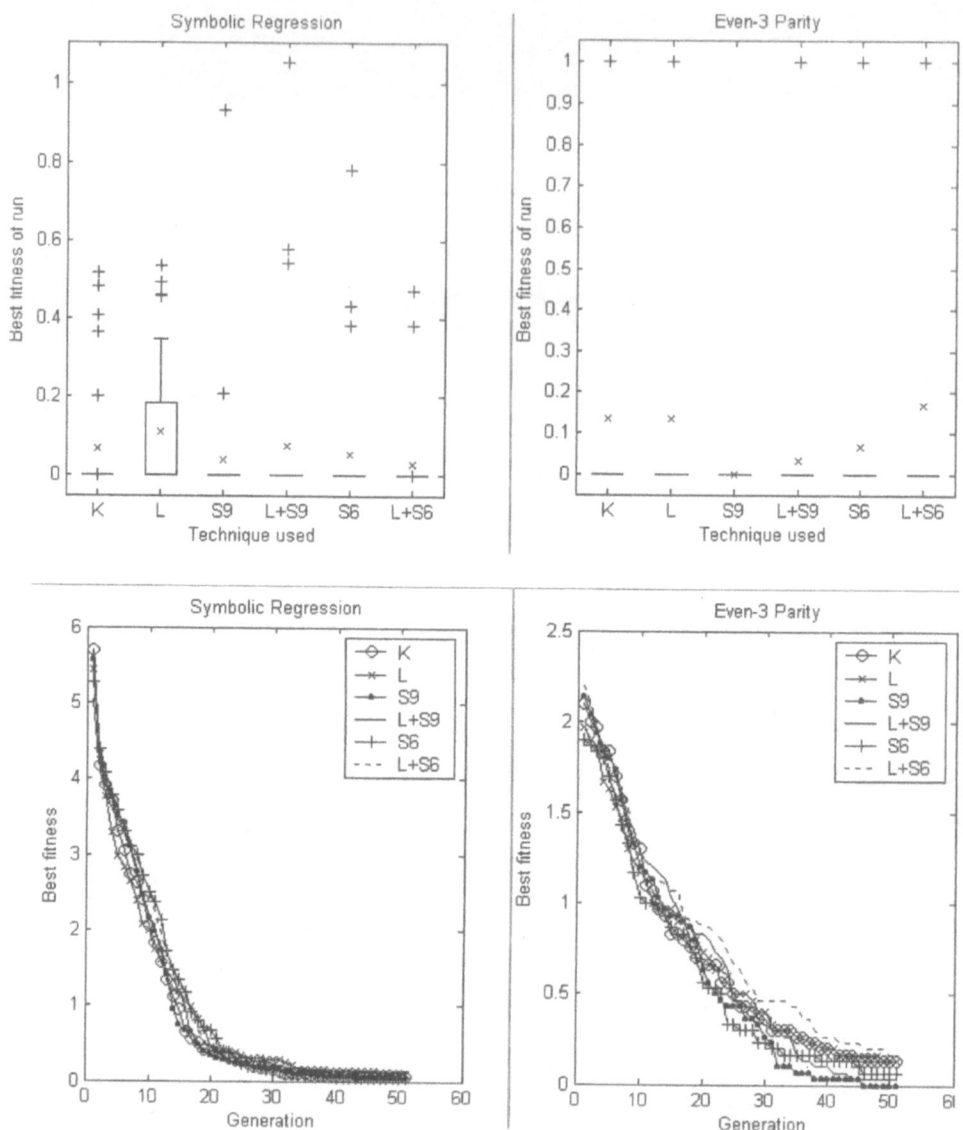

Fig. 4. Boxplots and evolution curves of best fitness. See Table 1 for the names of the techniques

beginning of the run, promptly recovered afterwards. The two techniques that do not use dynamic maximum tree depth (K, L) are able to sustain much higher population diversity in later generations.

In the Even-3 Parity problem, once again S9 and S6 caused a significant decrease in population diversity when compared to K, but showed significantly different performances when compared to each other. While S9 outperformed L, maintaining significantly higher population diversity, S6 was outperformed by L. Both compound techniques (L+S9, L+S6) significantly lowered population diversity when compared to L and to their counterparts S9 and S6. The evolution curves for this problem also show an abrupt loss of diversity in the beginning of the run, but only in the techniques that use lexicographic parsimony pressure (L, L+S9, L+S6). In this problem, techniques using dynamic maximum tree depth (S9, L+S9, S6, L+S6) also result in lower population diversity in later generations, with both S6 and L+S6 yielding the worst results.

An initial decrease in population diversity is expected in problems where the number of possible fitness values is unlimited, as in Symbolic Regression. Since most of the random initial trees (particularly the larger ones) have very poor fitness, selection quickly eliminates them from the population in favor of a more homogeneous group of better (and usually smaller) individuals. It can be observed that the abrupt loss of diversity shown in Fig. 3 occurs at the same time in the run as the decrease in mean tree size shown in Fig. 1 (evolution curve on the left). In the Even-3 Parity problem, the number of possible fitness values is very limited, so the initial trees never have very poor fitness. However, several different trees have the same fitness value, and lexicographic parsimony pressure quickly eliminates the larger trees in favor of the smaller ones, thus producing the same effect (which also explains the initial decrease of the intron percentage that can be observed in Fig. 2, evolution curve on the right). This behavior has not been reported before.

A higher number of clonings resulting from unsuccessful crossovers, or simply the reduction of the search space, caused by the introduction of the dynamic limit, may account for the persistent lower diversity observed in the later stages of the run, with all techniques using dynamic maximum tree depth.

5.4 Best Fitness

Figure 4 shows the results concerning the best fitness. Although there has been some concern regarding whether depth restrictions may affect GP performance negatively when using crossover [2,4], in both Symbolic Regression and Even-3 Parity none of the techniques was significantly different from the others. The evolution curves indicate that convergence to good solutions was not a problem.

6 Conclusions and Future Work

The dynamic maximum tree depth technique was able to effectively control code growth in both Symbolic Regression and Even-3 Parity problems. In spite of a noticeable decrease in population diversity, the ability to find individuals with good fitness values was not compromised in these two problems. Dynamic maximum tree depth was clearly superior to lexicographic parsimony pressure in the Symbolic Regression problem. In the Even-3 Parity problem, where the differences between both techniques were not always significant, the combination of both outperformed the use of either one separately.

There seems to be no disadvantage in using the lower, and most limitative, value for the initial depth limit. The size of the trees was kept low without losing convergence into good solutions. However, the persistent lower population diversity observed when using the dynamic level, particularly with the more limitative setting, may become an impediment to good convergence in smaller populations, which may require a higher value for the initial depth limit in order to maintain the convergence ability. Further work must be carried out in order to confirm or reject this hypothesis. The performance of the dynamic maximum tree depth technique must also be checked in harder problems.

Acknowledgements. This work was partially supported by grants QLK2-CT-2000-01020 (EURIS) from the European Commission and POCTI/1999/BSE/34794 (SAPIENS) from Fundação para a Ciência e a Tecnologia, Portugal. We would like to thank Ernesto Costa (Universidade de Coimbra), Pedro J.N. Silva (Universidade de Lisboa), and all the anonymous reviewers for carefully reading the manuscript and providing many valuable suggestions.

References

1. Banzhaf, W., Nordin, P., Keller, R.E., Francone, F.D.: Genetic programming – an introduction, San Francisco, CA. Morgan Kaufmann (1998)
2. Gathercole, C., Ross, P.: An adverse interaction between crossover and restricted tree depth in genetic programming. In Koza, J.R., Goldberg, D.E., Fogel, D.B., Riolo, R.L., editors, Proceedings of GP'96, Cambridge, MA. MIT Press (1996) 291–296
3. Koza, J.R.: Genetic programming – on the programming of computers by means of natural selection, Cambridge, MA. MIT Press (1992)
4. Langdon, W.B., Poli, R.: An analysis of the MAX problem in genetic programming. In Koza, J.R., Deb, K., Dorigo, M., Fogel, D.B., Garzon, M., Iba, H., Riolo, R.L., editors, Proceedings of GP'97, San Francisco, CA. Morgan Kaufman (1997) 222–230
5. Langdon, W.B.: Genetic Programming + Data Structures = Automatic Programming!, Boston, MA. Kluwer (1998)
6. Langdon, W.B.: Size fair and homologous tree crossovers for tree genetic programming. Genetic Programming and Evolvable Machines, 1 (2000) 95–119
7. Luke, S., Panait, L.: Lexicographic parsimony pressure. In Langdon, W.B., Cantú-Paz, E., Mathias, K., Roy, R., Davis, D., Poli, R. Balakrishnan, K., Honavar, V., Rudolph, G., Wegener, J., Bull, L., Potter, M.A., Schultz, A.C., Miller, J.F., Burke, E., Jonoska, N., editors, Proceedings of GECCO-2002, San Francisco, CA. Morgan Kaufmann (2002) 829–836
8. Silva, S.: GPLAB – a genetic programming toolbox for MATLAB. (2003) http://www.itqb.unl.pt:1111/gplab/
9. Soule, T.: Code growth in genetic programming. PhD thesis, University of Idaho (1998)
10. Soule, T.: Exons and code growth in genetic programming. In Foster, J.A., Lutton, E., Miller, J., Ryan, C., Tettamanzi, A.G.B., editors, Proceedings of EuroGP-2002, Berlin. Springer (2002) 142–151

11. Soule, T., Foster, J.A.: Effects of code growth and parsimony pressure on populations in genetic programming. Evolutionary Computation, 6(4) (1999) 293–309
12. Soule, T., Heckendorn, R.B.: An analysis of the causes of code growth in genetic programming. Genetic Programming and Evolvable Machines, 3 (2002) 283–309
13. The MathWorks. (2003) http://www.mathworks.com
14. Van Belle, T., Ackley, D.H.: Uniform subtree mutation. In Foster, J.A., Lutton, E., Miller, J., Ryan, C., Tettamanzi, A.G.B., editors, Proceedings of EuroGP-2002, Berlin. Springer (2002) 152–161

Difficulty of Unimodal and Multimodal Landscapes in Genetic Programming

Leonardo Vanneschi[1], Marco Tomassini[1], Manuel Clergue[2], and Philippe Collard[2]

[1] Computer Science Institute, University of Lausanne, Lausanne, Switzerland
[2] I3S Laboratory, University of Nice, Sophia Antipolis, France

Abstract. This paper presents an original study of fitness distance correlation as a measure of problem difficulty in genetic programming. A new definition of distance, called structural distance, is used and suitable mutation operators for the program space are defined. The difficulty is studied for a number of problems, including, for the first time in GP, multimodal ones, both for the new hand-tailored mutation operators and standard crossover. Results are in agreement with empirical observations, thus confirming that fitness distance correlation can be considered a reasonable index of difficulty for genetic programming, at least for the set of problems studied here.

1 Introduction

The fitness distance correlation (*fdc*) coefficient has been used as a tool for measuring problem difficulty in genetic algorithms (GAs) and genetic programming (GP) with controversial results: some counterexamples have been found for GAs [15], but *fdc* has been proven an useful measure on a large number of GA (see for example [3] or [9]) and GP functions (see [2,16]). In particular, Clergue and coworkers ([2]) have shown *fdc* to be a reasonable way of quantifying problem difficulty for GP for a set of functions.

In this paper, we use a measure of structural distance for trees (see [7]) to calculate *fdc*. Then, we employ *fdc* to measure problem difficulty for two kinds of GP processes: GP using the standard Koza's crossover as the only genetic operator (that we call from now on *standard GP*) and GP using two new mutation operators based on the transformations on which structural distance is defined (*structural mutation genetic programming* from now on). The test problems that we have chosen are unimodal and multimodal trap functions, royal trees and two versions of the MAX problem. The present study is the first attempt to quantify problem difficulty of functions with multiple global optima with the same fitness in evolutionary algorithms by the *fdc*.

This paper is structured as follows: the next section gives a short description of the structural tree distance used in the paper, followed by the definition of the basic mutation operators that go hand-in-hand with it. Section 4 presents the main results and their discussion for a number of GP problems. Finally, Sect. 5 gives our conclusions and hints to future work.

2 Distance Measure for Genetic Programs

In GAs individuals are represented as strings of digits and typical distance measures are Hamming distance or alternation. Defining a distance between genotypes in GP is much more difficult, given the tree structure of the individuals. In [2] an *ad hoc* distance between trees was used. In the spirit of the *fdc* definition [9], it would be better to use GP operators that have a direct counterpart in the tree distance. For that reason, we adopt the structural distance for trees proposed in [7] and we define corresponding operators (see next section). According to this measure, given the sets $\mathcal{F}$ and $\mathcal{T}$ of functions and terminal symbols, a coding function c must be defined such that $c : \{\mathcal{T} \cup \mathcal{F}\} \to \mathbb{N}$. One can think of many ways for the specification of c, for example the "complexity" of the primitives or their arity. The distance between two trees T_1 and T_2 is calculated in three steps: (1) T_1 and T_2 are overlapped at the root node and the process is applied recursively starting from the leftmost subtrees (see [7] for a description of the overlapping algorithm). (2) For each pair of nodes at matching positions, the difference of their codes (eventually raised to an exponent) is computed. (3) The differences computed in the previous step are combined in a weighted sum. This gives for the distance between two trees T_1 and T_2 with roots R_1 and R_2 the following expression:

$$dist(T_1, T_2) = d(R_1, R_2) + k \sum_{i=1}^{m} dist(child_i(R_1), child_i(R_2)) \quad (1)$$

where: $d(R_1, R_2) = (|c(R_1) - c(R_2)|)^z$, $child_i(Y)$ is the ith of the m possible children of a generical node Y, if $i \leq m$, or the empty tree otherwise, and c evaluated on the root of an empty tree is 0. Constant k is used to give different weights to nodes belonging to different levels and z is a constant usually chosen in such a way that $z \in \mathbb{N}$. In most of this paper, except for the MAX function, individuals will be coded using the same syntax as in [2] and [14], i.e. considering a set of functions A, B, C, etc. with increasing arity (i.e. $arity(A) = 1$, $arity(B) = 2$, and so on) and a single terminal X (i.e. $arity(X) = 0$) as follows: $\mathcal{F} = \{A, B, C, D, \ldots\}, \mathcal{T} = \{X\}$ and the c function will be defined as follows: $\forall x \in \{\mathcal{F} \cup \mathcal{T}\}\ c(x) = arity(x) + 1$. In our experiments we will always set $k = \frac{1}{2}$ and $z = 2$. By keeping $0 < k < 1$, the differences near the root have higher weight. This is convenient for GP as it has been noted that programs converge quickly to a fixed root portion [11].

3 Structural Mutation Operators

In [13] O'Reilly used the edit distance, which is related to, but not identical with the tree distance employed here. She also defined suitable edit operators and related them to GP mutation. She used the edit distance for a different purpose: how to measure and control the step size of crossover in order to balance explotation and exploration in GP. She also suggested that edit distance could be useful in the study of GP fitness landscapes, but did not develop the issue. Here we take up exactly this last point.

The following operators have been inspired by the distance definition presented in the last section and by the work in [13]. Given the sets $\mathcal{F}$ and $\mathcal{T}$ and the coding function c defined in Sect. 2, we define c_{max} (respectively, c_{min}) as the maximum (respectively, the minimum) value taken by c on the domain $\{\mathcal{F} \cup \mathcal{T}\}$. Moreover, given a symbol n such that $n \in \{\mathcal{F} \cup \mathcal{T}\}$ and $c(n) < c_{max}$ and a symbol m such that $m \in \{\mathcal{F} \cup \mathcal{T}\}$ and $c(m) > c_{min}$, we define: $succ(n)$ as a primitive such that $c(succ(n)) = c(n) + 1$ and $pred(m)$ as a primitive such that $c(pred(m)) = c(m) - 1$. Then we can define the following editing operators on a generic tree T:

- **inflate mutation**. A primitive labelled with a symbol n such that $c(n) < c_{max}$ is selected in T and replaced by $succ(n)$. A new random terminal node is added to this new node in a random position (i.e. the new terminal becomes the ith son of $succ(n)$, where i is comprised between 0 and $arity(n)$).
- **deflate mutation**. A primitive labelled with a symbol m such that $c(m) > c_{min}$, and such that at least one of his sons is a leaf, is selected in T and replaced by $pred(m)$. A random leaf, between the sons of this node, is deleted from T.

The terms *inflate* and *deflate* have been used to avoid confusion with the similar and well known *grow* and *shrink* mutations that have already been proposed in GP. The following property holds (the proof appears in [17]):

Property 1. Distance/Operator Consistency.
Let's consider the sets $\mathcal{F}$ and $\mathcal{T}$ and the coding function c defined in Sect. 2. Let T_1 and T_2 be two trees composed by symbols belonging to $\{\mathcal{F} \cup \mathcal{T}\}$ and let's consider the k and z constants of definition (1) to be equal to 1. If $dist(T_1, T_2) = D$, then T_2 can be obtained from T_1 by a sequence of $\frac{D}{2}$ editing operations, where an editing operation can be an inflate mutation or a deflate mutation.

From this property, we conclude that the operators of inflate mutation and deflate mutation are completely coherent with the notion of distance defined in Sect. 2, i.e. an application of these operators allow us to move on the search space from a tree to its neighbors according to that distance. We call the new GP process based on these operators structural mutation genetic programming (SMGP).

4 Experimental Results

In all experiments shown in the following, fdc has been calculated via a sampling of 40000 randomly chosen individuals. All the experiments have been done with generational GP, a total population size of 100 individuals, ramped half and a half initialization, tournament selection of size 10. When standard GP has been used, the crossover rate has been set to 95% and the mutation rate to 0%, while when SMGP has been employed, no crossover has been used and the rate of the two new mutation operators has been set to 95%. The GP process has been stopped either when a perfect solution has been found (global optimum) or when 500 generations have been executed. All experiments have been performed 100 times.

4.1 Fitness Distance Correlation

An approach proposed for GAs [9] states that an indication of problem hardness is given by the relationship between fitness and distance of the genotypes from known optima. Given a sample $F = \{f_1, f_2, ..., f_n\}$ of n individual fitnesses and a corresponding sample $D = \{d_1, d_2, ..., d_n\}$ of the n distances to the nearest global optimum, fdc is defined as: $fdc = \frac{C_{FD}}{\sigma_F \sigma_D}$, where: $C_{FD} = \frac{1}{n}\sum_{i=1}^{n}(f_i - \overline{f})(d_i - \overline{d})$ is the covariance of F and D and σ_F, σ_D, $\overline{f}$ and $\overline{d}$ are the standard deviations and means of F and D. Given the tree structure of genotypes, the normalization problem is not a trivial one in GP. Dividing all the distances in the sampling by the maximum of all the possible distances between two trees in the search space is not a practically applicable method, since it would give too large a number for the typically used maximum tree depth and operator arity values. This problem has been tackled in [4]. In [2], Clergue and coworkers used a "sufficiently large" integer constant to obtain the normalisation of distances. Here, similarly to [4], we obtain the normalized distance between two trees by dividing their distance by the maximum value between the distances of the same two trees from the empty one. As suggested in [9], GA problems can be empirically classified in three classes, depending on the value of the fdc coefficient: **misleading** ($fdc \geq 0.15$), in which fitness increases with distance, **unknown** ($-0.15 < fdc < 0.15$) in which there is virtually no correlation between fitness and distance and **straightforward** ($fdc \leq -0.15$) in which fitness increases as the global optimum approaches. The second class corresponds to problems for which the difficulty can't be estimated, because fdc doesn't bring any information. In this case, examination of the fitness-distance scatterplot may give information on problem difficulty (see [9]).

4.2 Unimodal Trap Functions

Trap functions [5] allow one to define the fitness of the individuals as a function of their distance from the optimum, and the difficulty of trap functions can be changed by simply modifying some parameters. A function $f : distance \rightarrow fitness$ is a unimodal trap function if it is defined in the following way:

$$f(d) = \begin{cases} 1 - \dfrac{d}{B} & \text{if } d \leq B \\ \dfrac{R \cdot (d - B)}{1 - B} & \text{otherwise} \end{cases}$$

where d is the distance of the current individual from the *unique* global optimum, normalized so as to belong to the range $[0, 1]$, and B and R are constants

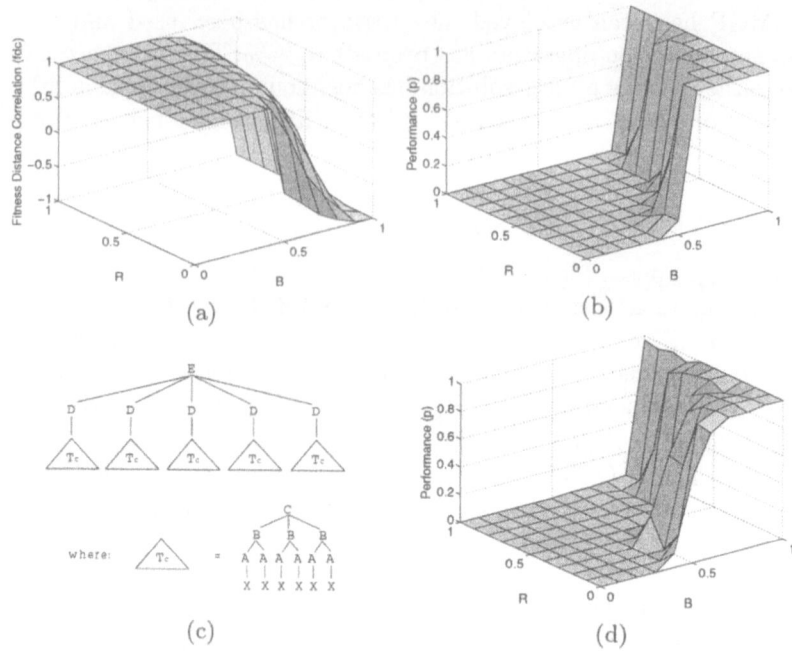

Fig. 1. (a): *fdc* values for some trap functions obtained by changing the values of the constants B and R. (b): Performance (p) values of SMGP for traps. (d): Performance (p) values of standard GP for traps. (c): Stucture of the tree used as optimum in the experiments reported in (a), (b) and (d).

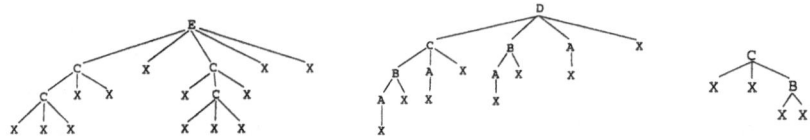

Fig. 2. The trees used as optima in experiments analogous to the ones of Fig. 1.

belonging to $[0,1]$. B allows to set the width of the attractive basin for each one of the optima and R sets their relative importance. By construction, the difficulty of trap functions decreases as the value of B increases, while it increases as the value of R decreases. For a more detailed explanation of trap functions see for instance [2]. Figure 1 show values of the performance p (defined as the proportion of the runs for which the global optimum has been found in less than 500 generations over 100 runs) and of *fdc* for various trap functions obtained by changing the values of the constants B and R. These experiments have been performed considering the tree represented in Fig. 1c as the global optimum. The same experiments have been performed using as global optimum the trees shown in Fig. 2 and the results (not shown here for lack of space) are qualitatively analogous to the ones of Fig. 1. In all cases *fdc* is confirmed to be a reasonable measure for quantifying problem difficulty both for SMGP and standard GP for unimodal trap functions.

4.3 "W" Multimodal Trap Functions

Unimodal trap functions described in Sect. 4.2 are characterized by the presence of a unique global optimum. The functions used here and first proposed in [6] (informally called "W" trap functions, given their typical shape shown in Fig. 3) are characterized by the presence of several global optima. They depend on 5 variables called B_1, B_2, B_3, R_1 and R_2 and they can be expressed by the following formula:

$$f(d) = \begin{cases} 1 - \dfrac{d}{B_1} & \text{if } d \leq B_1 \\[2mm] \dfrac{R_1 \cdot (d - B_1)}{B_2 - B_1} & \text{if } B_1 \leq d \leq B_2 \\[2mm] \dfrac{R_1 \cdot (B_3 - d)}{B_3 - B_2} & \text{if } B_2 \leq d \leq B_3 \\[2mm] \dfrac{R_2 \cdot (d - B_3)}{1 - B_3} & \text{otherwise} \end{cases}$$

where B_1, B_2, B_3, R_1 and R_2 are constants belonging to the interval $[0, 1]$ and the property $B_1 \leq B_2 \leq B_3$ must hold. A systematic study of problem difficulty for multimodal landscapes via an algebraic indicator such as *fdc* has, to our knowledge, never been performed before in evolutionary computation. Here is how our study has been performed: we choose a particular tree belonging to the search space and we call it *origin*. Then we artificially assign a maximum fitness $(= 1)$ to this tree, so as to make it a global optimum. Then, if we set the

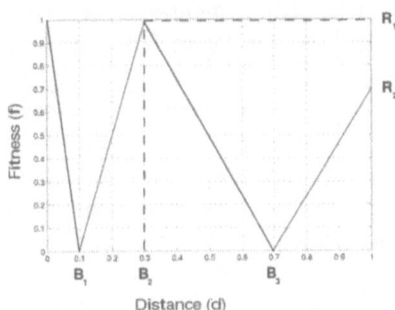

Fig. 3. Graphical representation of a "W" trap function with $B_1 = 0.1$, $B_2 = 0.3$, $B_3 = 0.7$, $R_1 = 1$, $R_2 = 0.7$. Note that distances and fitness are normalized into the interval $[0, 1]$.

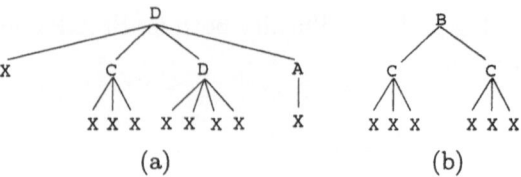

Fig. 4. Two trees having a normalized distance equal to 0.5 between them

R_1 constant to 1, all the trees having a normalized distance equal to B_2 from the origin, given the definition of "W" trap functions, have a fitness equal to 1 and thus are global optima too. Two series of experiments have been performed, using both SMGP and standard GP. In the first one, the tree represented in Fig. 4a is considered as the origin (let's call this tree T_1).

Since the tree in Fig. 4b (that we call T_2) has a normalized distance of 0.5 from T_1, if we set R_1 to 1 and B_2 to 0.5 then T_2 is a global optimum too. To calculate the *fdc*, the minimum of the distances from T_1 and T_2 to each tree in the sampling is considered (as suggested for GAs by Jones in [9], even though never practically experimented). The distribution of these "minimum distances" has been studied on a sample of 40000 individuals and it appears to have a regular shape, as it is the case for unimodal trap functions (results not shown here for lack of space).

Since these series of experiments represent a first attempt to use *fdc* to measure the difficulty of fitness landscapes with more than one global optimum, and since we want to be able to study the dynamics of the GP process in detail, we have decided, as a first step, to use fitness landscapes containing only two global optima. Thus, a random normalized fitness different from 1 has been arbitrarily assigned to each tree having a distance equal to 0.5 from T_1 and a genotype different from T_2. This choice, of course, alters the "W" trap functions definition and influences the fitness landscape, even though only marginally. Anyway, we consider this choice perfectly "fair", given that we are not interested in testing *fdc* on standard benchmarks, but rather on artificially defined fitness landscapes. Table 1 shows a subset of the experimental results that we have obtained with this first series of tests, by setting B_2 to 0.5 and R_1 to 1 and by varying the values of B_1, B_3 and R_2.

Given the enormous number of experiments performed (about 10000 GP runs) and the limited space available, we couldn't show all the results. Thus, we have decided to proceed as follows: for each trap function (identified by particular values of B_1, B_3 and R_2), we have calculated the *fdc*. Then, we have discarded all the traps for which the *fdc* is comprised between -0.15 and 0.15, because no experimental result would give us any useful information for these functions (see Sect. 4.1). For a subset of the other trap functions, 100 runs have been performed with both SMGP and standard GP. Results shown in Table 1 are encouraging and suggest that *fdc* could be a reasonable measure to predict problem difficulty for some typical "W" trap functions. The results not shown in Table 1 are qualitatively similar and lead us to the same conclusions.

Difficulty of Unimodal and Multimodal Landscapes in Genetic Programming

Table 1. Results of *fdc* using SMGP and standard GP for a set of "W" trap functions where the origin is the tree shown in Fig. 4a and the second global optimum is the one shown in Fig. 4b; p stands for performance

	fdc	*fdc* prediction	p (SMGP)	p (stGP)
$B_1 = 0.4, B_3 = 0.9, R_2 = 0.5$	-0.62	straightf.	0.75	0.81
$B_1 = 0.5, B_3 = 0.8, R_2 = 0.4$	-0.88	straightf.	0.98	0.94
$B_1 = 0.3, B_3 = 0.9, R_2 = 0.7$	-0.61	straightf.	0.80	0.77
$B_1 = 0.2, B_3 = 0.9, R_2 = 0.1$	-0.69	straightf.	0.72	0.91
$B_1 = 0.1, B_3 = 0.9, R_2 = 0.3$	-0.72	straightf.	0.85	0.98
$B_1 = 0.5, B_3 = 0.6, R_2 = 0.9$	0.34	misleading	0.33	0.20
$B_1 = 0.4, B_3 = 0.6, R_2 = 0.9$	0.36	misleading	0.14	0.30
$B_1 = 0.3, B_3 = 0.6, R_2 = 0.9$	0.33	misleading	0.31	0.13

Table 2. Results of *fdc* using SMGP and standard GP for a set of "W" trap functions where the origin is the tree shown in Fig. 5a and the second global optimum is the one shown in Fig. 5b; p stands for performance

	fdc	*fdc* prediction	p (SMGP)	p (stGP)
$B_1 = 0.1, B_3 = 0.1, R_2 = 0.1$	-0.76	straightf.	0.91	0.88
$B_1 = 0.1, B_3 = 0.9, R_2 = 0.1$	-0.93	straightf.	0.97	0.99
$B_1 = 0.05, B_3 = 0.7, R_2 = 0.1$	-0.90	straightf.	0.98	0.92
$B_1 = 0.1, B_3 = 0.8, R_2 = 0.5$	-0.81	straightf.	0.90	0.83
$B_1 = 0.1, B_3 = 0.5, R_2 = 0.1$	-0.71	straightf.	0.91	0.90
$B_1 = 0.05, B_3 = 0.2, R_2 = 0.9$	0.89	misleading	0.02	0.13
$B_1 = 0.05, B_3 = 0.4, R_2 = 0.9$	0.74	misleading	0.35	0.17
$B_1 = 0.05, B_3 = 0.1, R_2 = 0.6$	0.78	misleading	0.25	0.23

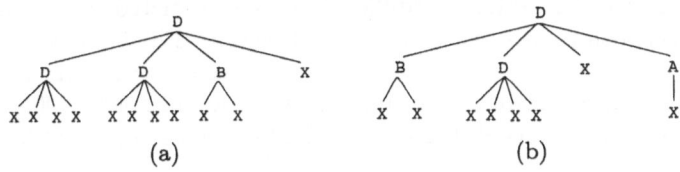

Fig. 5. Two trees having a normalized distance equal to 0.1 between them

Analogous experiments have also been performed by considering the tree in Fig. 5a as the origin and the tree in Fig. 5b as the second global optimum. Since the normalized distance between these two trees is equal to 0.1, this time B_2 is set to 0.1 and R_1 to 1. Once again, a random normalized fitness different from 1 is assigned to all the trees having a distance equal to 0.1 from the one in Fig. 5a and a genotype different from the one in Fig. 5b. A subset of the results of this second series of experiments are shown in Table 2. These results are encouraging too and seem to confirm the validity of the *fdc* both for SMGP and standard GP.

In any event, the amount of results shown in this section doesn't allow us to draw definitive conclusions: many more cases should be studied. For example, in the case of fitness landscapes containing two global optima, a larger set of differently shaped trees should be considered as origin and as second global optimum. Moreover, fitness landscapes with more than two global optima should be investigated. However, a methodology for the study of difficulty of multimodal landscapes has been proposed, and some non-trivial choices have been performed, as the one of calculating *fdc* by considering the minimum of the distances from the global optima for every individual of the sampling. These choices needed an experimental confirmation. Results shown here are encouraging and should pave the way for a deeper study of problem difficulty for multimodal lanscapes.

4.4 Royal Trees

The next functions we study are the royal trees proposed by Punch et al. [14]. The language used is the same as in Sect. 2, and the fitness of a tree (or any subtree) is defined as the score of its root. Each function calculates its score by summing the weighted scores of its direct children. If the child is a perfect tree of the appropriate level (for instance, a complete level-C tree beneath a D node), then the score of that subtree, times a *FullBonus* weight, is added to the score of the root. If the child has a correct root but is not a perfect tree, then the weight is *PartialBonus*. If the child's root is incorrect, then the weight is *Penalty*. After scoring the root, if the function is itself the root of a perfect tree, the final sum is multiplied by *CompleteBonus* (see [14] for a more detailed explanation). Values used here are as in [14] i.e. *FullBonus* = 2, *PartialBonus* = 1, *Penalty* = $\frac{1}{3}$, *CompleteBonus* = 2. Different experiments, considering different nodes as the node with maximum arity allowed, have been performed. Results are shown in Table 3.

Predictions made by *fdc* for level-A, level-B, level-C and level-D functions are correct. Level-E function is "difficult" to be predicted by the *fdc* (i.e. no correlation between fitness and distance is observed). Finally, level-F and level-G functions are predicted to be "misleading" (in accord with Punch in [14]) and they really are, since the global optimum is never found before 500 generations. Royal trees problem spans all the classes of difficulty as described by the *fdc*.

4.5 MAX Problem

The task of the MAX problem for GP, defined in [8] and [10], is "to find the program which returns the largest value for a given terminal and function set

Table 3. Results of *fdc* for the Royal Trees; p stands for performance

Root	fdc	fdc prediction	p (SMGP)	p (stGP)
B	-0.31	straightf.	1	1
C	-0.25	straightf.	1	1
D	-0.20	straightf.	0.76	0.70
E	0.059	unknown	0	0.12
F	0.44	misleading	0	0
G	0.73	misleading	0	0

Table 4. Results of *fdc* for the MAX problem using SMGP and standard GP. The first column shows the sets of functions and terminals used in the experiments; p stands for performance

MAX problem	fdc	fdc prediction	p (SMGP)	p (stGP)
{+} {1}	-0.87	straightf.	1	1
{+} {1,2}	-0.86	straightf.	1	1

with a depth limit d, where the root node counts as depth 0". We set d equal to 8 and we use the set of functions $\mathcal{F} = \{+\}$ and the set of terminals $\mathcal{T}_1 = \{1\}$ or $\mathcal{T}_2 = \{1, 2\}$. When using $\mathcal{T}_1$, we specify the coding function c as: $c(1) = 1$, $c(+) = 2$, when using $\mathcal{T}_2$, we pose: $c(1) = 1$, $c(2) = 2, c(+) = 3$. The study of standard GP for these MAX functions comports no particular problem, while for the case of SMGP, the definitions of the operators of inflate and deflate mutation given in Sect. 3 must be slightly modified, since we are considering a different coding language. The inflate and deflate mutations are now defined in such a way that, when using $\mathcal{T}_1$, a terminal symbol 1 can be transformed in the subtree $T_1 = +(1,1)$ by one step of inflate mutation and the *vice-versa* can be done by the deflate mutation. When using $\mathcal{T}_2$, the inflate mutation can transform a 1 node into a 2 node and a 2 node into the subtrees $T_2 = +(2,1)$ or $T_3 = +(1,2)$ (with a uniform probability). On the other hand, the deflate mutation can tranform T_1 or T_2 into a leaf labelled by 2, and a 2 node into a 1 node. Table 4 shows the *fdc* and p values for these test cases. Both problems are correctly classified as straightforward by *fdc*, both for SMGP and standard GP.

5 Conclusions and Future Work

Two new kinds of tree mutations corresponding to the operations of the structural distance are defined in this paper. Fitness distance correlation (calculated using this structural distance between trees) has been shown to be a reasonable way of quantifying the difficulty of unimodal trap functions, of a restricted set of multimodal trap functions, of royal trees, and of two MAX functions for GP using these mutations as genetic operators, as well as for GP based on standard crossover. The results show that, for the functions studied, using crossover or our mutation operators, does not seem to have a marked effect on difficulty as measured by *fdc*. Thus the remarks of [1] and [12], and other work where it is claimed that the standard GP crossover does not seem to markedly improve performance with respect to other variation operators seem to be confirmed. Although *fdc* has confirmed its validity in the present study, in view of some counterexamples for GAs mentioned in the text, it remains to be seen whether the use of *fdc* extends to other classes of functions, such as typical GP benchmarks. In the future we plan to improve the study of problem difficulty of multimodal landscapes and MAX functions (e.g. when $\mathcal{F} = \{+, *\}$ and $\mathcal{T} = \{0.5\}$), and look for a measure for GP difficulty that can be calculated without prior knowledge of the global optima, thus eliminating the strongest limitation of *fdc*. Moreover, since *fdc* is surely not an infallible measure, we plan to build a counterexample for *fdc* in GP. Another open problem consists in taking into account in the distance definition the phenomenon of introns (whereby two different genotypes can lead to the same phenotypic behavior). Finally, we intend to look for a better measure than performance to identify the success rate of functions, possibly independent from the maximum number of generations chosen.

Acknowledgments. This work has been partially sponsored by the Fonds National Suisse pour la recherche scientifique.

References

1. K. Chellapilla. Evolutionary programming with tree mutations: Evolving computer programs without crossover. In J.R. Koza, K. Deb, M. Dorigo, D.B. Fogel, M. Garzon, H. Iba, and R.L. Riolo, editors, *Genetic Programming 1997: Proceedings of the Second Annual Conference*, pages 431–438, Stanford University, CA, USA, 1997. Morgan Kaufmann.
2. M. Clergue, P. Collard, M. Tomassini, and L. Vanneschi. Fitness distance correlation and problem difficulty for genetic programming. In *Proceedings of the genetic and evolutionary computation conference GECCO'02*, pages 724–732. Morgan Kaufmann, San Francisco, CA, 2002.
3. P. Collard, A. Gaspar, M. Clergue, and C. Escazut. Fitness distance correlation as statistical measure of genetic algorithms difficulty, revisited. In *European Conference on Artificial Intelligence (ECAI'98)*, pages 650–654, Brighton, 1998. John Witley & Sons, Ltd.
4. E.D. de Jong, R.A. Watson, and J.B. Pollak. Reducing bloat and promoting diversity using multi-objective methods. In L. Spector et al., editor, *Proceedings of the Genetic and Evolutionary Computation Conference GECCO '01*, San Francisco, CA, July 2001. Morgan Kaufmann.
5. K. Deb and D.E. Goldberg. Analyzing deception in trap functions. In D. Whitley, editor, *Foundations of Genetic Algorithms, 2*, pages 93–108. Morgan Kaufmann, 1993.
6. K. Deb, J. Horn, and D. Goldberg. Multimodal deceptive functions. *Complex Systems*, 7:131–153, 1993.
7. A. Ekárt and S.Z. Németh. Maintaining the diversity of genetic programs. In J.A. Foster, E. Lutton, J. Miller, C. Ryan, and A.G.B. Tettamanzi, editors, *Genetic Programming, Proceedings of the 5th European Conference, EuroGP 2002*, volume 2278 of *LNCS*, pages 162–171, Kinsale, Ireland, 3–5 April 2002. Springer-Verlag.
8. C. Gathercole and P. Ross. An adverse interaction between crossover and restricted tree depth in genetic programming. In J.R. Koza, D.E. Goldberg, D.B. Fogel, and R.L. Riolo, editors, *Genetic Programming 1996: Proceedings of the First Annual Conference*, pages 291–296, Stanford University, CA, USA, 28–31 July 1996. MIT Press.
9. T. Jones. *Evolutionary Algorithms, Fitness Landscapes and Search*. PhD thesis, University of New Mexico, Albuquerque, 1995.

10. W.B. Langdon and R. Poli. An analysis of the max problem in genetic programming. In J.R. Koza, K. Deb, M. Dorigo, D.B. Fogel, M. Garzon, H. Iba, and R.L. Riolo, editors, *Genetic Programming 1997: Proceedings of the Second Annual Conference on Genetic Programming*, pages 222–230, San Francisco, CA, 1997. Morgan Kaufmann.
11. N.F. McPhee and N.J. Hopper. Analysis of genetic diversity through population history. In W. Banzhaf, J. Daida, A.E. Eiben, M.H. Garzon, V. Honavar, M. Jakiela, and R. E. Smith, editors, *Proceedings of the Genetic and Evolutionary Computation Conference*, volume 2, pages 1112–1120, Orlando, Florida, USA, 13–17 July 1999. Morgan Kaufmann.
12. U.-M. O'Reilly. *An Analysis of Genetic Programming*. PhD thesis, Carleton University, Ottawa, Ontario, Canada, 1995.
13. U.-M. O'Reilly. Using a distance metric on genetic programs to understand genetic operators. In *Proceedings of 1997 IEEE International Conference on Systems, Man, and Cybernetics*, pages 4092–4097. IEEE Press, Piscataway, NJ, 1997.
14. B. Punch, D. Zongker, and E. Goodman. The royal tree problem, a benchmark for single and multiple population genetic programming. In P. Angeline and K. Kinnear, editors, *Advances in Genetic Programming 2*, pages 299–316, Cambridge, MA, 1996. The MIT Press.
15. R.J. Quick, V.J. Rayward-Smith, and G.D. Smith. Fitness distance correlation and ridge functions. In *Fifth Conference on Parallel Problems Solving from Nature (PPSN'98)*, pages 77–86. Springer-Verlag, Heidelberg, 1998.
16. V. Slavov and N.I. Nikolaev. Fitness landscapes and inductive genetic programming. In *Proceedings of International Conference on Artificial Neural Networks and Genetic Algorithms (ICANNGA97)*, University of East Anglia, Norwich, UK, 1997. Springer-Verlag KG, Vienna.
17. L. Vanneschi, M. Tomassini, P. Collard, and M. Clergue. Fitness distance correlation in structural mutation genetic programming. In C. Ryan et al., editor, *Genetic Programming, 6th European Conference, EuroGP2003*, Lecture Notes in Computer Science. Springer-Verlag, Heidelberg, 2003.

Ramped Half-n-Half Initialisation Bias in GP

Edmund Burke, Steven Gustafson*, and Graham Kendall

School of Computer Science & IT
University of Nottingham
{ekb,smg,gxk}@cs.nott.ac.uk

Tree initialisation techniques for genetic programming (GP) are examined in [4,3], highlighting a bias in the standard implementation of the initialisation method *Ramped Half-n-Half* (RHH) [1]. GP trees typically evolve to random shapes, even when populations were initially full or minimal trees [2]. In canonical GP, unbalanced and sparse trees increase the probability that bigger subtrees are selected for recombination, ensuring code growth occurs faster and that subtree crossover will have more difficulty in producing trees within specified depth limits. The ability to evolve tree shapes which allow more legal crossover operations, by providing more possible crossover points (by being bushier), and control code growth is critical. The GP community often uses RHH [4]. The standard implementation of the RHH method selects either the **grow** or **full** method with 0.5 probability to produce a tree. If the tree is already in the initial population it is discarded and another is created by **grow** or **full**. As duplicates are typically not allowed, this standard implementation of RHH favours **full** over **grow** and possibly biases the evolutionary process.

The **full** and **grow** methods are similar algorithms for recursively producing GP trees. The **full** algorithm makes trees with branches extending to the maximum initial depth. The **grow** algorithm does not require this and allows branches of varying length (up to the maximum initial depth). As many more unique trees exist which are full (as full trees contain more nodes), there is a tendency, especially with particular function and terminal sets, to produce more duplicate trees with the **grow** method.

To estimate the bias of the RHH method with a particular function and terminal set, we use the results from Luke [3] (Lemma 1). The expected number of nodes E_d at depth d is defined as: $E_d = \{1 \text{ if } d = 0, \text{else } E_{d-1}pb \text{ if } d > 0\}$ where pb is the expected number of children of a new node (p is the probability of picking a nonterminal, and b is the expected number of children of a nonterminal). Luke [3] uses this to calculate the expected size of trees in the infinite case. Here we bound the calculation to depth $d = 4$. For our analysis, E_d' correctly predicted the **full** method would contribute more trees to the initial population whenever the expected size of the two methods was not similar (i.e. **grow** made smaller trees, causing more duplicates and rejected trees).

Canonical GP trees grow in size to maximum depth limits, making the initial trees seeds for the evolutionary process. As the **full** algorithm is more likely to evolve larger and more bushier trees with more nodes than the **grow** method, we conduct an experimental study to observe these differences in the evolutionary

---

* corresponding author

process on standard problems. We use RHH and the **full** and **grow** methods exclusively. Initial depths of 4 and 6, with maximum tree depths of 10 and 15, respectively, a population size of 500, a total of 51 generations, and standard subtree crossover is used for recombination on the artificial ant, even-5-parity, and symbolic regression problem with the quartic polynomial. 50 random runs were performed for each of the 6 experiments for each problem.

On the ant problem, GP produced the best fitness with smaller initial trees (smaller initial depth or those produced by the **grow** method). Initial trees in the ant problem are particularly important as they encode the initial path that the ant takes. The parity problem experiments produced better fitness with bigger and more fuller trees; the populations created only by **grow** had the worst fitness, followed by the smaller depth limit populations. The RHH method causes 70% of the initial ant and parity population to be created by the **full** method, a negative bias for the ant problem and a positive bias for the parity problem. The regression problem always produced trees which were sparse due to the number of unary functions in the regression function set (log, exp, sin, cos), which is typical in function sets for more complex regression problems. While the RHH method was not overly biased (**full** produced 55% of initial trees in the RHH experiments and **full** and **grow** had similar fitness), the sparseness and unbalancedness of trees caused a significant loss of genotypic diversity (the number of unique trees).

The bias of the RHH method can be positive or negative, but it can also effect diversity. A loss of genotypic diversity can result from two factors here: the creation of trees by crossover already in the population, or the failure of crossover to find acceptable crossover points (leading to the duplication of one of the parents). The ant and parity populations produced with **grow** were sparser, more unbalanced and lost diversity. All the regression populations had similar behaviour due to the function set.

A disadvantage exists for GP trees which are unable to grow effectively. GP trees which grow sparse and unbalanced will cause more code growth, less genotypic diversity, and search a smaller space of possible programs. These populations will be less effective in the evolutionary process. Current research is investigating various ways to adapt and detect program shapes for improved performance of fitness and recombination operators.

References

1. J.R. Koza. *Genetic Programming: On the Programming of Computers by Means of Natural Selection.* MIT Press, Cambridge, MA, USA, 1992.
2. W. Langdon, T. Soule, R. Poli, and J.A. Foster. The evolution of size and shape. In Lee Spector et al., editors, *Advances in Genetic Programming 3*, chapter 8, pages 163–190. MIT Press, Cambridge, MA, USA, June 1999.
3. S. Luke. Two fast tree-creation algorithms for genetic programming. *IEEE Transactions on Evolutionary Computation*, 4(3):274–283, September 2000.
4. S. Luke and L. Panait. A survey and comparison of tree generation algorithms. In L. Spector et al., editors, *Proceedings of the Genetic and Evolutionary Computation Conference*, pages 81–88, San Francisco, USA, 7-11 July 2001. Morgan Kaufmann.

Improving Evolvability of Genetic Parallel Programming Using Dynamic Sample Weighting

Sin Man Cheang, Kin Hong Lee, and Kwong Sak Leung

Department of Computer Science and Engineering
The Chinese University of Hong Kong
{smcheang,khlee,ksleung}@cse.cuhk.edu.hk

Abstract. This paper investigates the sample weighting effect on Genetic Parallel Programming (GPP) that evolves parallel programs to solve the training samples captured directly from a real-world system. The distribution of these samples can be extremely biased. Standard GPP assigns equal weights to all samples. It slows down evolution because crowded regions of samples dominate the fitness evaluation and cause premature convergence. This paper compares the performance of four sample weighting (SW) methods, namely, Equal SW (ESW), Class-equal SW (CSW), Static SW (SSW) and Dynamic SW (DSW) on five training sets. Experimental results show that DSW is superior in performance on tested problems.

1 Introduction

Genetic Programming (GP) has been used to learn computer programs for data classification. Linear GP (LGP) uses a linear representation which is directly corresponding to the program statements based on a register machine. In this paper, experiments are performed on a novel LGP paradigm, Genetic Parallel Programming (GPP) [1], which evolves parallel programs based on a multiple ALU register machine, Multi-ALU Processor (MAP). The MAP employs a tightly coupled, shared-register, MIMD architecture. It can perform multiple operations in parallel in each processor clock cycle. GPP has been used to evolve parallel programs for symbolic regression, recursive function regression and artificial ant problems. Besides of evolving parallel programs to make use of the parallelism of the specific parallel architecture, the GPP accelerating phenomenon [3] reveals the efficiency of evolving algorithms in parallel form.

Sample Weighting (SW) is a technique to change the samples' weights in order to increase the efficiency of fitness-based evolutionary algorithms. The basic principle of SW is to balance the biased distribution of sample points in a solution space. Researchers have proposed different SW methods which determine samples' weights in runtime. For example, we have applied Dynamic SW (DSW) to Genetic Algorithms (GA) to evolve Finite State Machines [2]. In this paper, we show that DSW can also be applied to GP to increase evolutionary efficiency of Boolean function regression and data classification problems. We compare DSW to other three SW methods,

Equal SW (ESW), Class-equal SW (CSW) and Static SW (SSW). In ESW, all samples' weights are set to equal without considering the distribution of samples. In CSW, the calculation of samples' weights is based on the number of classes. In DSW, samples' weights are recalculated periodically based on the contributions of individual samples to the population. SSW is a special case of DSW that does not modify samples' weights after the initialization.

2 Experiments, Results, and Conclusions

We have applied four SW methods on five training sets. Three training sets are Boolean functions, 3-even-parity (*3ep*), 5-even-parity (*5ep*) and 5-symmetery (*5sm*). The remainings are UCI data classification databases, Wisconsin Breast Cancer (*bcw*) and Bupa Liver Disorder (*bld*). Three Boolean function sets (F_1={and,or,xor,nop}, F_2={and,or,nand,nor,nop} and F_3={and,or,not,nop}) are used for Boolean functions. One arithmetic function set (F_4={+,−,×,%,move,nop}) is used for UCI databases. Table 1 below shows the relative computation effort of ESW versus DSW. Obviously, DSW boosts the evolutionary efficiency of all tested problems (all values are greater than 1.0). In the *5sm*_F_3 experiment, DSW reduces the computational effort by 744.5 times. This shows that DSW is superior in evolutionary speedup on this problem. Even if the improvements are not significant on the *bcw* and *bld* problems, DSW is a helpful SW method to speed up evolution because it is easily to be implemented on the original population-based evolutionary algorithms.

Research in the near future will be focused on extending the usage of DSW to different applications areas, e.g. multi-class classification and numeric function regression problems. The main concern is to apply DSW to non-count based algorithm, e.g. floating-point function regression problems in which the fitness of an individual program is based on the total error of the expected outputs and the program outputs.

Table 1. Relative computation effort (E_{ESW}/E_{DSW})

	3ep_F_3	5ep_F_1	5ep_F_2	5sm_F_1	5sm_F_2	5sm_F_3	bcw	bld
E_{ESW}/E_{DSW}	11.8	3.7	1.8	51.4	226.7	744.5	1.2	1.1

References

1. Leung, K.S., Lee, K.H., Cheang, S.M.: Evolving Parallel Machine Programs for a Multi-ALU Processor. Proc. of IEEE Congress on Evolutionary Computation (2002) 1703-1708
2. Leung, K.S., Lee, K.H., Cheang, S.M.: Balancing Samples' Contributions on GA Learning. Proc. of the 4th Int. Conf. on Evolvable Systems: From Biology to Hardware (ICES), Lecture Notes in Computer Science, Springs-Verlag (2001) 256–266
3. Leung, K.S., Lee, K.H., Cheang, S.M.: Parallel Programs are More Evolvable than Sequential Programs. Proc. of the 6th Euro. Conf. on Genetic Programming (EuroGP), Lecture Notes in Computer Science, Springs-Verlag (2003)

Enhancing the Performance of GP Using an Ancestry-Based Mate Selection Scheme

Rodney Fry and Andy Tyrrell

Department of Electronics
University of York, UK
{rjlf101,amt}@ohm.york.ac.uk

Abstract. The performance of genetic programming relies mostly on population-contained variation. If the population diversity is low then there will be a greater chance of the algorithm being unable to find the global optimum. We present a new method of approximating the genetic similarity between two individuals using ancestry information. We define a new diversity-preserving selection scheme, based on standard tournament selection, which encourages genetically dissimilar individuals to undergo genetic operation. The new method is illustrated by assessing its performance in a well-known problem domain: algebraic symbolic regression.

1 Ancestry-Based Tournament Selection

Genetic programming [1] in its most conventional form, suffers from a well-known problem: the search becomes localised around a point in the search space which is not the global optimum. This may be a result of not taking steps to preserve the genetic diversity of the population.

A selection scheme in GP may be made to promote diversity if one could select individuals not only based on their fitness, but also based on how dissimilar the individuals' genotypes are. Previous work in measuring how genotypically similar individuals are has concentrated their phenotypic relationship [2] and structural difference [3]. A good measure of how similar two individuals are can be attained by looking at the individuals' ancestries. For example, siblings are far more likely to have similar genotypes than two randomly chosen individuals. If each individual in a GP population had the capability of knowing who its parents and grandparents were, and chose a mate based partly on dissimilarity between their respective ancestries, then search of deceptive local optima would be reduced. Previous work in the area includes the incest prevention scheme [4] and crossover prohibition based on ancestry [5].

In our scheme, a binary tree of ancestry is associated with each individual. The information stored relates to which previous individuals were used in the creation of the current solution and the proportion of each parent individual that was donated to create the current solution: the *Parental Contribution Fraction* or PCF. The first of the two individuals that will take part in the crossover operation is selected using standard tournament selection. A second parent individual must then be selected based on how genetically dissimilar it is when compared to the first. This can be achieved by modifying the fitness values of the candidate individuals to reflect their *apparent* fitness from the first parent's point of view, before they enter the tournament.

2 Results

The modified tournament selection mechanism was tested and results were compared with the standard tournament selection scheme and the ancestry-based crossover prohibition scheme. Tests were carried out on the symbolic regression problem outlined in Section 10.2 of Koza [1]. As Fig. 1 shows, our 'PCF Tree' implementation outperforms the standard tournament selection scheme and the crossover prohibition scheme.

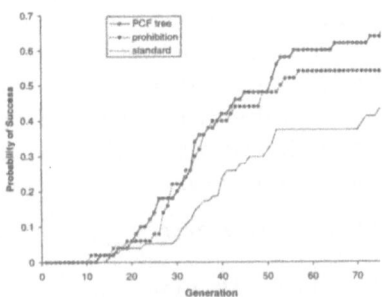

Fig. 1. Comparison of the performance of the PCF ancestry tree, the crossover prohibition scheme and standard tournament selection

References

1. John R. Koza. *Genetic Programming: On the Programming of Computers by Means of Natural Selection*. MIT Press, 1992.
2. S.W. Mahfoud. Crowding and preselection revisited. In Reinhard Männer and Bernard Manderick, editors, *Parallel problem solving from nature 2*, pages 27–36, Amsterdam, 1992. North-Holland.
3. R. Keller and W. Banzhaf. Explicit maintenance of genetic diversity on genospaces, 1994.
4. L.J. Eshelman and J.D. Schaffer. Preventing premature convergence in genetic algorithms by preventing incest. In *Proceedings of the Fourth International Conference on Genetic Algorithms*, pages 115–122, San Diego, CA, 1991. Morgan Kaufmann.
5. Rob Craighurst and Worthy Martin. Enhancing ga performance through crossover prohibitions based on ancestry. In Larry J. Eshelman, editor, *Proceedings of the Sixth International Conference on Genetic Algorithms*, pages 130–135. Morgan Kaufmann, 1995.

A General Approach to Automatic Programming Using Occam's Razor, Compression, and Self-Inspection

Extended Abstract

Peter Galos, Peter Nordin, Joel Olsén, and Kristofer Sundén Ringnér

Complex Systems Group
Department of Physical Resource Theory
Chalmers University of Technology
S-412 96 Göteborg, Sweden

This paper describes a novel general method for automatic programming which can be seen as a generalization of techniques such as genetic programming and ADATE. The approach builds on the assumption that data compression can be used as a metaphor for cognition and intelligence. The proof-of-concept system is evaluated on sequence prediction problems. As a starting point, the process of inferring a general law from a data set is viewed as an attempt to compress the observed data. From an artificial intelligence point of view, compression is a useful way of measuring how deeply the observed data is understood. If the sequence contains redundancy it exists a shorter description i.e. the sequence can be compressed. Consider the sequence:

$$ABCDEABCDEABCDEABCDEABC$$

This sequence could be described with a program.

```
for i = 1..5 loop            for i = 1..infinity loop
    print 'ABCDE'                print 'ABCDE'
end loop                     end loop
print 'ABC'
```

The principle of Occam's Razor states that from two possible solutions (cf. the programs above) to a problem the simpler one (the program to the right) should be chosen. According to algorithmic information theory, the information content of the simplest program that without any prior information describes a data series is called the Kolmogorov complexity. It has been shown that prediction of the continuation of a sequence is more likely by a simpler program [VL1]. Solomonoff's induction method [So1] is used to produce the target data as well as a continuation the target data. Since it is not known whether a program will halt or not before execution, the Kolmogorov complexity is in general not computable. However, it can be approximated, in this case by programs consisting of recursively applied compression methods. The learning problem is therefore turned into a search problem in the space of all possible permutations of the set of compression methods. The fitness criterion is based on algorithmic information theory [Sh1].

It is possible to search the space of hypotheses randomly in order to find a solution. However, a method that takes into account previous search results before suggesting a new hypothesis can give rice to an accelerated search process. In order to achieve self-inspection, a prediction algorithm is used by examining the algorithm's own progress. This pattern search in the history of previous programs' performance, enables the algorithm to predict the most likely way to search for future hypotheses. In a way, this is a reinforcement problem and thus just another sequence prediction problem. An advantage of this self-inspection is that no additional theory or learning algorithm is needed. To achieve an accurate measure of the information content of a program a registry machine is used. Table 1 shows some examples of predictions[1] using the approach proposed in this paper, were the "least complex programs found" column presents combinations of different compression methods.

Table 1. Results of predicting data series using the algorithm

Data sequence	Least complex program found	Prediction
1. 5, 10, 15, 20, 25	Delta→SRP	30, 35, 40, 45,...
2. 2, 4, 8, 16, 32	LCB→Delta→SRP	64, 128, 256, 512,...
3. 0, 9, 3, 4, 13, 7	Delta→SRP	8, 17, 11, 12, 21,...
4. 1, 2, 2, 3, 3, 3, 4, 4, 4, 4	RL→Delta→SRP	5, 5, 5, 5, 5, 6,...
5. 0, 3, 12, 21, 48, 75, 102	Delta→LCB→RL→Delta→SRP	183, 264, 345,...
6. 2, 2, 3, 2, 2, 4, 2, 2, 5, 2, 2, 6	N-gram→PWC→Delta→SRP	2, 2, 7, 2, 2, 8, 2, 2,...
7. 0, 1, 3, 6, 10, 15, 21, 28	Delta→Delta→SRP	36, 45, 55, 66, 78,...
8. 1, 1, 2, 1, 2, 3, 1, 2, 3	LZSS→PWC→Delta→SRP	4, 1, 2, 3, 4, 5, 1, 2, 3,...
9. 0, 1, 2, 4, 5, 7, 10, 11, 13, 16	Delta→LZSS→PWC→Delta→SRP	20, 21, 23, 26, 30,...

In conclusion the algorithm is simple and requires little domain specific input which makes it able to predict a wide range of input sequences. Furthermore, no prior knowledge is necessary. In addition, the paper pioneers the implementation and verification of the theory [VL1] concerning the connection between compression and learning. Much work on theory and evaluation is still needed in order to transform the method into a practical tool with the same wide applicability as for instance genetic programming. However, the firm theoretical foundation holds promise for a very efficient method and advantageous scaling properties.

References

[Sh1] Shannon, C.E.: A Mathematical Theory of Communication The Bell System Technical Journal Vol. 27, (1948) pp. 13
[So1] Solomonoff, R., J.: A preliminary report on a general theory of inductive inference. (1960) pp. 11–14
[VL1] Vitányi, P., Li, M.: Simplicity, Information, Kolmogorov Complexity, and Prediction. (1998) pp. 3

---

[1] Computed within 4 minutes on an Intel PII/133MHz

Building Decision Tree Software Quality Classification Models Using Genetic Programming

Yi Liu and Taghi M. Khoshgoftaar

Florida Atlantic University
Boca Raton, Florida, USA
{yliu,taghi}@cse.fau.edu

Abstract. Predicting the quality of software modules prior to testing or system operations allows a focused software quality improvement endeavor. Decision trees are very attractive for classification problems, because of their comprehensibility and white box modeling features. However, optimizing the classification accuracy and the tree size is a difficult problem, and to our knowledge very few studies have addressed the issue. This paper presents an automated and simplified genetic programming (GP) based decision tree modeling technique for calibrating software quality classification models. The proposed technique is based on multi-objective optimization using strongly typed GP. Two fitness functions are used to optimize the classification accuracy and tree size of the classification models calibrated for a real-world high-assurance software system. The performances of the classification models are compared with those obtained by standard GP. It is shown that the GP-based decision tree technique yielded better classification models.

A timely quality prediction of the software system can be useful in improving the overall reliability and quality of the software product. Software quality classification (SQC) models which classify program modules as either fault-prone (*fp*) or not fault-prone (*nfp*) can be used to target the software quality improvement resources toward modules that are of low quality. With the aid of such models, software project managers can deliver a high-quality software product on time and within the allocated budget. Usually software quality estimation models are based on software product and process measurements, which have shown to be effective indicators of software quality. Therefore, a given SQC technique aims to model the underlying relationship between the available software metrics data, i.e., independent variables, and the software quality factor, i.e., dependent variable.

Among the commonly used SQC models, the decision tree (DT) based modeling approach is very attractive due to the model's white box feature. A DT-based SQC model is usually a binary tree with two types of nodes, i.e., query nodes and leaf nodes. A query node is a logical equation which returns either true or false, whereas a leaf node is a terminal which assigns a class label. Each query node in the tree can be seen as a classifier which partitions the data set into subsets.

All the leaf nodes are labelled as one of the class types, such as *fp* or *nfp*. An observation or case in the given data set will "trickle" down, according to its attribute values, from the root node of the decision tree to one of its leaf nodes.

This study focuses on calibrating GP-based decision tree models for predicting the quality of software modules as either *fp* or *nfp*. In the context of what constitutes a good DT-based model, two factors are usually considered: classification accuracy and model simplicity. Classification accuracy is often measured in terms of the misclassification error rates, whereas model simplicity of decision trees is often expressed in terms of the number of nodes. Therefore, an optimal DT is one that has low misclassification error rates and has a (relatively) few number of nodes. GP is a logical solution to the problems that require a multi-objective optimization, primarily because it is based on the process of natural evolution which involves the simultaneous optimization of several factors.

Very few studies have investigated GP-based decision tree models, and among those none has investigated the GP-based decision trees for the SQC problem. Previous works related to GP-based classification models have focused on the standard GP process, which requires that the function and terminal sets have the *closure property*. This property implies that all the functions in the function set must accept any of the values and data types defined in the terminal set as arguments. However, since each decision tree has at least two different types of nodes, i.e., query nodes and leaf nodes, the closure property requirement of standard GP does not guarantee the generation of a valid individual(s).

Strongly Typed Genetic Programming (STGP) has been used to alleviate the closure property requirement (Montana [1]), by allowing each function to define the different kinds of data types it can accept. Moreover, each function, variable, and terminal are specified by certain types in STGP. When the STGP process generates an individual or performs a genetic operation during the simulated evolution process, it considers additional criteria that are not part of standard GP. For example, in our study of calibrating SQC models, a function in the function set can only be in query nodes, while the terminal variables such as *fp* and *nfp* can only be in the leaf nodes. Therefore, in our study we use STGP to build decision trees.

In this study we investigated a simplified GP-based multi-objective optimization method for automatically building optimal SQC decision trees that have a high classification accuracy rate and a relatively small tree size. The GP-based decision trees were build to optimize two fitness functions: the average weighted cost of misclassification, and the tree size which is expressed in terms of the number of tree nodes. Moreover, the relative classification performances of the GP-based DT model and the that of standard GP were evaluated in the context of a real-world industrial software project. It was observed that the proposed approach yielded useful decision trees.

References

1. Montana, D.J.: Strongly typed genetic programming. Evolutionary Computation (1995) 199–230

Evolving Petri Nets with a Genetic Algorithm

Holger Mauch

University of Hawaii at Manoa, Dept. of Information and Computer Science,
1680 East-West Road, Honolulu, HI 96822
hmauch@hawaii.edu

Abstract. In evolutionary computation many different representations ("genomes") have been suggested as the underlying data structures, upon which the genetic operators act. Among the most prominent examples are the evolution of binary strings, real-valued vectors, permutations, finite automata, and parse trees. In this paper the use of place-transition nets, a low-level Petri net (PN) class [1,2], as the structures that undergo evolution is examined. We call this approach "Petri Net Evolution" (PNE). Structurally, Petri nets can be considered as specialized bipartite graphs. In their extended version (adding inhibitor arcs) PNs are as powerful as Turing machines. PNE is therefore a form of Genetic Programming (GP). Preliminary results obtained by evolving variable-size place-transition nets show the success of this approach when applied to the problem areas of boolean function learning and classification.[1]

1 Overview

PNs are a nice formalisms that allow the complex modeling of numerous real world problems. Their strength is the "built in" concurrency mechanism, which allows a formal description of systems containing parallel processes. For this study small sample problems have been used to test the feasibility of PNE. The PNs evolved belong to a class of low level PNs called *place-transition nets*.

This work has been inspired by the generalization of the recombination operator from trees to bipartite graphs. Since recombination seems to be an important genetic operator in PNE, this approach follows the GA stream within EC.

The common feature of traditional GP and PNE is the variable length genotype. This is in contrast to the standard GA which employs fixed-length strings. In contrast to traditional tree-based GP, there is no need to specify a function set explicitly for PNE. Functions are implicitly built into the semantics of the chosen PN model. Because the place-transition PN model used is integer-based our focus is on discrete problems rather than continuous ones.

PNE evolves a population of PNs in an evolution-directed search of the space of possible PNs for ones, which, when executed, will exhibit high fitness. To the best of our knowledge the idea of evolving a population of individuals which are variable-sized PNs does not appear in previous literature.

---

[1] This research was supported in part by DARPA grant NBCH1020004.

2 Elements of Petri Net Evolution

In experiments PNs have been evolved for boolean function learning and classification. We partition the set of places $P = P_{in} \cup P_{center} \cup P_{out}$. For a classification problem such as *PlayTennis* [3, p.59] the input places P_{in} contain the values of the attributes of a fitness case in its initial marking. After firing a suitable number of transitions the PN is evaluated on its output places P_{out} according to the induced submarking M_{out}. Every PN individual is executed on several fitness cases to determine its fitness.

One problem with the evaluation of a PN is its potentially nondeterministic behavior due to conflicts (i.e. the situation when two transitions are enabled, but firing one of the two disables the other one.) To solve this difficulty the fitness measure works with the mean value of a sample of simulation runs. This introduces a random component (the firing order) into the fitness evaluation process.

The PNE implementation uses the incidence matrix representation as described in [2] to represent the directed, weighted, bipartite graph data structure of a place-transition net. The incidence matrix and the initial marking vector M_0 are implemented by a low-level data structure that can grow dynamically.

After random generation of generation 0 PNs evolutionary computation enters the cycle of evaluation, selection and recombination/mutation. Several mutation operators to manipulate the various components (i.e. arcs, places, transitions, initial marking) of a PN have been designed. Mutation also allows a PN to grow and shrink in size. At the heart of PNE is the specifically designed recombination operator which creates 2 offspring PNs from 2 mating parent PNs. The parents do not need to match in size, so every individual in the population can mate with any other individual. A conventional tournament selection with tournament size 2 is used.

In addition to the usual parameters (population size, maximum number of generations, parameters for recombination, mutation and selection) PNE uses a parameter for the maximum duration of a single PN execution, since termination is not guaranteed when a PN is executed.

First results in the area of classification and boolean function learning show that numerous runs produced highly fit PN individuals. However, the examples examined are very simple in nature. Future work will involve larger sized problem instances. That will allow the evaluation of the robustness of PNE. What has been shown is the feasibility of successfully evolving PNs when suitable procedures for the evaluation, mutation and recombination are employed.

References

1. Reisig, W.: Petri Nets: an Introduction. Springer-Verlag, Berlin, Germany (1985)
2. Murata, T.: Petri nets: Properties, analysis and applications. Proceedings of the IEEE **77** (1989) 541–580
3. Mitchell, T.M.: Machine Learning. WCB/McGraw-Hill, Boston, MA (1997)

Diversity in Multipopulation Genetic Programming

Marco Tomassini[1], Leonardo Vanneschi[1], Francisco Fernández[2], and Germán Galeano[2]

[1] Computer Science Institute, University of Lausanne, Lausanne, Switzerland
[2] Computer Science Institute, University of Extremadura, Spain

In the past few years, we have done a systematic experimental investigation of the behavior of multipopulation GP [2] and we have empirically observed that distributing the individuals among several loosely connected islands allows not only to save computation time, due to the fact that the system runs on multiple machines, but also to find better solution quality. These results have often been attributed to better diversity maintenance due to the periodic migration of groups of "good" individuals among the subpopulations. We also believe that this might be the case and we study the evolution of diversity in multi-island GP. All the diversity measures that we use in this paper are based on the concept of *entropy* of a population P, defined as $H(P) = -\sum_{j=1}^{N} F_j \log(F_j)$. If we are considering phenotypic diversity, we define F_j as the fraction n_j/N of individuals in P having a certain fitness j, where N is the total number of fitness values in P. In this case, the entropy measure will be indicated as $H_p(P)$ or simply H_p. To define genotypic diversity, we use two different techniques. The first one consists in partitioning individuals in such a way that only identical individuals belong to the same group. In this case, we have considered F_j as the fraction of trees in the population P having a certain genotype j, where N is the total number of genotypes in P and the entropy measure will be indicated as $H_G(P)$ or simply H_G. The second technique consists in defining a distance measure, able to quantify the genotypic diversity between two trees. In this case, F_j is the fraction of individuals having a given distance j from a fixed tree (called *origin*), where N is the total number of distance values from the origin appearing in P and the entropy measure will be indicated as $H_g(P)$ or simply H_g. The tree distance used is Ekárt's and Németh's definition [1].

Figure 1 depicts the behavior of H_G, H_g and H_p during evolution for the symbolic regression problem, with the classic ploynomial equation $f(x) = x^4 + x^3 + x^2 + x$, an input set composed of 1000 fitness cases and a set of functions equal to F={*,//,+,-}, where // is like / but returns 0 instead of *error* when the divisor is equal to 0. Fitness is the sum of the square errors at each test point. Curves are averages over 100 independent runs for generational GP, crossover rate: 95%, mutation rate: 0.1%, tournament selection of size: 10, ramped half and half initialization, maximum depth of individuals for the creation phase: 6, maximum depth of individuals for crossover: 17, elitism. Genotypic diversity for the panmictic case tends to remain constant over time, and to have higher values than the distributed case. On the contrary, the average phenotypic entropy for the multipopulation case tends to remain higher than in the panmictic case. The oscillating behavior of the multipopulation curves when groups of individuals are sent and received is not surprising: it is due to the sudden change in diversity when new individuals enter a subpopulation. Finally, we remark that the behavior of the two measures used to calculate genotypic diversity (H_G and H_g) is qualitatively equivalent. Analogous results

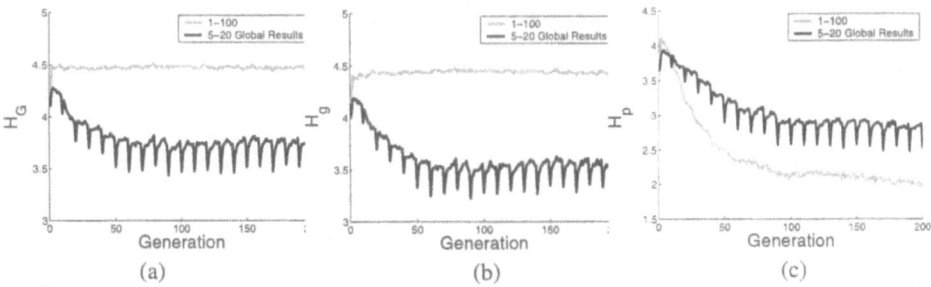

Fig. 1. Symbolic Regression Problem. 100 total individuals. (a): Genotypic diversity calculated using the H_G measure (see the text). Gray curve: panmictic population. Black curve: aggregated subpopulations. (b): Genotypic diversity calculated with the H_g measure (c): Phenotypic diversity (H_p)

Table 1. Numbers represent the average amount of time spent by a solution or a part thereof in the corresponding population. Results refer to the Ant problem with a population size of 1000

Pop 1	Pop 2	Pop 3	Pop 4	Pop 5
20	18.62	21.5	15.62	24.5

(not shown here for lack of space) have been obtained for the Even Parity 4 problem and for the Artificial Ant on the Santa Fe Trail problem.

It is also interesting to study how solutions originate and propagate in the distributed system. Table 1 is a synthesis of a few runs of the Artificial Ant problem. Although data are not statistically significant (too few runs have been done), they do indicate that all the islands participate in the formation of a solution.

By defining genotypic and phenotipic diversity indices and by monitoring their variation over a large number of runs on three standard test problems, we have empirically studied how diversity evolves in the distributed case. While genotypic diversity is not much affected by splitting a single population into multiple ones, phenotypic diversity, which is linked to fitness, remains higher in the multipopulation case for all problems studied here.

We have also studied how solutions arise in the distributed case, and we have seen that all the subpopulations contribute in the building of the right genetic material, which again seems to confirm that smaller communicating populations are more effective than a big panmictic one. In conclusion, using multiple loosely coupled populations is a natural and easy way for maintaining diversity and, to some extent, avoiding premature convergence in GP.

References

1. A. Ekárt and S.Z. Németh. Maintaining the diversity of genetic programs. In J.A. Foster et al., editor, *Genetic Programming, Proceedings of the 5th European Conference, EuroGP 2002*, volume 2278 of *LNCS*, pages 162–171. Springer-Verlag, 2002.
2. F. Fernández, M. Tomassini, and L. Vanneschi. An empirical study of multipopulation genetic programming. *Genetic Programming and Evolvable Machines*, March 2003. Volume 4. Pages 21–51. W. Banzhaf Editor-in-Chief.

An Encoding Scheme for Generating λ-Expressions in Genetic Programming

Kazuto Tominaga, Tomoya Suzuki, and Kazuhiro Oka

Tokyo University of Technology, Hachioji, Tokyo 192-0982 Japan
tomi@acm.org, {tomoya,oka}@ns.it.teu.ac.jp

Abstract. To apply genetic programming (GP) to evolve λ-expressions, we devised an encoding scheme that encodes λ-expressions into trees. This encoding has closure property, i.e., any combination of terminal and non-terminal symbols forms a valid λ-expression. We applied this encoding to a simple symbolic regression problem over Church numerals and the objective function $f(x) = 2x$ was successfully obtained. This encoding scheme will provide a good foothold for exploring fundamental properties of GP by making use of lambda-calculus.

In this paper, we give an encoding scheme that encodes λ-expressions into trees, which is appropriate for evolving λ-expressions by genetic programming (GP). A λ-expression is a term in lambda-calculus [2], composed of *variables*, *abstraction* and *application*. Here are some examples of λ-expressions.

(λx.(λy.(x(xy))))   (let *2* denote this λ-expression)
(λx.(λy.(λz.((xy)(xy)z))))   (let *D* denote this λ-expression)

A λ-expression can be regarded as a program. Giving the input *2* to the program *D* is represented by applying *D* to *2* as the following.

(*D* *2*) ≡ ((λx.(λy.(λz.((xy)(xy)z))))(λx.(λy.(x(xy)))))

By several steps of β-reduction, the above λ-expression is rewritten into

(λx.(λy.(x(x(x(xy)))))) .   (let *4* denote this λ-expression)

β-reduction steps can be regarded as the execution of the program, and *4* can be regarded as the output of the execution. If λ-expressions can be represented as trees, GP can evolve λ-expressions.

We devised an encoding scheme to handle λ-expressions in GP. This encoding scheme encodes a λ-expression into a tree. The non-terminal symbol L represents abstraction, and the non-terminal symbol P represents application. A terminal symbol is a positive integer and it represents a variable. A variable labeled by an integer n is bound by n-th abstraction in the path from the variable to the root. Example trees are (L 1), (L (L 1)) and (L (L (P 2 1))), which represent λ-expressions (λx.x), (λx.(λy.y)) and (λx.(λy.(xy))), respectively.

This encoding scheme has the following good properties. One is that any tree encoded by this scheme is syntactically and semantically valid as a λ-expression. Crossover and mutation always produce valid individuals. Secondly the meaning of a subtree does not change by crossover since relation of a bound variable and the abstraction that binds the variable is preserved under crossover.

We applied this encoding to a simple symbolic regression problem. The objective function is $f(x) = 2x$ over Church numerals [2]. Church numerals represent natural numbers; examples are *2* and *4* shown above. The function f can be expressed as a λ-expression in many ways, and D is such a λ-expression. We implemented a β-reduction routine in C and incorporated it to lil-gp [3] to make the whole GP system. Main GP parameters are: population size 1000; maximum generation 1000; 10 % reproduction with elites 1 %, 89.9 % crossover and 0.1 % mutation. In the experiment, the objective function D was obtained in 13 out of 25 runs. The transition of the best individuals in a successful run is shown in Table 1 in their canonical forms. The fitness was measured based on structural difference between the correct answer and the output of the individual.

The experimental results indicate that the encoding scheme is appropriate for evolving λ-expressions by GP, and thus it will provide a good foothold for exploring fundamental property of GP by using lambda-calculus.

Table 1. Example transition of best individuals

Generation	Fitness	Best individual
0	153	$(\lambda x.x)$
233	150	$(\lambda x.(\lambda y.(\lambda z.(z((xy)(yy))))))$
346	147	$(\lambda x.(\lambda y.(\lambda z.(z((xy)(y(z(yy))))))))$
365	146	$(\lambda x.(\lambda y.(\lambda z.(z((xy)(y(y(yy))))))))$
376	145	$(\lambda x.(\lambda y.(\lambda z.(z((xy)(y(y(zy))))))))$
401	143	$(\lambda x.(\lambda y.(\lambda z.(z((xy)(y(y(z(yz)))))))))$
403	82	$(\lambda x.(\lambda y.(\lambda z.(z((xy)(y(y((xz)(\lambda w.w)))))))))$
414	81	$(\lambda x.(\lambda y.(\lambda z.(z((xy)(y((xz)(\lambda w.w))))))))$
420	80	$(\lambda x.(\lambda y.(\lambda z.((xy)(y(y((xz)(\lambda w.w))))))))$
422	9	$(\lambda x.(\lambda y.(\lambda z.(z((xy)(y(y((xy)(\lambda w.w)))))))))$
425	6	$(\lambda x.(\lambda y.(\lambda z.((xy)(y(y((xy)(\lambda w.w))))))))$
433	2	$(\lambda x.(\lambda y.(\lambda z.((xy)((xy)(\lambda w.w))))))$
439	0	$(\lambda x.(\lambda y.(\lambda z.((xy)((xy)z)))))$

References

1. Koza, J.: Genetic Programming. MIT Press (1992)
2. Revesz, G.: Lambda-Calculus, Combinators, and Functional Programming. Cambridge University Press (1988)
3. lil-gp Genetic Programming System.
   http://garage.cps.msu.edu/software/lil-gp/lilgp-index.html

AVICE: Evolving Avatar's Movernent

Hiromi Wakaki and Hitoshi Iba

Graduate School of Frontier Science
The University of Tokyo
7-3-1 Hongo, Bunkyou-ku, Tokyo, 113-8656, Japan
{wakaki,iba}@miv.t.u-tokyo.ac.jp
http://www.miv.t.u-tokyo.ac.jp/~wakaki/HomePage.html

Abstract. In recent years, the widespread penetration of the Internet has led to the use of diverse expressions over the Web. Among them, many appear to have strong video elements. However, few expressions are based on human beings, with whom we are most familiar. This is deemed to be attributable to the fact that it is not easy to generate human movements. This is attributable to the fact that the creation of movements by hand requires intuition and technology.

With this in mind, this study proposes a system that supports the creation of 3D avatars' movements based on interactive evolutionary computation, as a system that facilitates creativity and enables ordinary Users with no special skills to generate dynamic animations easily, and as a method of movement description.

1 Introduction

The creation of 3D-CG animations requires a tool that can easily and accurately change the details of the movement rather than joining the parts of the movement. inverse kinematics (IK) is good at balancing the movements of joints relative to other joints, but a great deal of knowledge, experience and intuition is required t o describe a movement.

In consideration of the above, we realized a System called AVICE that generates movements through Interactive Evolutionary Computation (1EC) [3,2,1]. AVICE Supports the user creativity in designing motions, which is empirically shown.

2 Outline of AVICE System

AVICE system uses H-Anim to express avatar. The Humanoid Animation Working Group (http://h-anim.org/) specifies how to express standard human bodies in VRML97, which is called Humanoid Animation (hereinafter referred to as "H-Anim").

AVICE system integrates Creator, Mixer and Remaker modules (See Fig. 1). Each relationship is that Mixer module take in the Output of Creator module and other Softwares. Creator module makes VRML files suited user's taste from initialization used IEC [4]. Mixer module makes VRML files suited user's taste from inputted files used IEC.

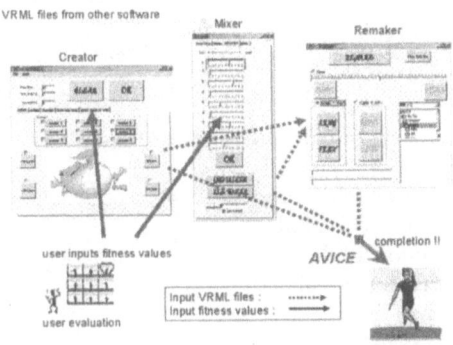 

Fig. 1. outline of AVICE system   **Fig. 2.** Example of created movements

3 Demonstration of Created Movements

Movements shown in the Fig. 2 were acquired as a result of the experiment of AVICE System. In this movement demo, avatars are dancing to two types of music. The movements were created by joining movements deemed to suitable to the music. Even if no music data has been imported, it shows that they are in rhythm. The difference between the two also shows that their dance is not only in rhythm to the music but also suits the atmosphere thereof.

4 Conclusion

Few expressions are based on human beings, with whom we are most familiar. This is deemed to be attributable to the fact that it is not easy to generate human movements. If movements of avatars can be created based on VRML, which can be displayed by a browser, it should increase the ways in which individuals express themselves, as they do through paintings and music.

References

1. P.J. Bentley. Evolutionary Design by Computers. Morgan Kaufmann Publishers Inc., 1999.
2. H. Takagi. Interactive evolutionary computation – cooperation of computational intelligence and human kansei. In *Proceeding of 5th International Conference on Sofi Computing und Information/Intelligent Systems*, pp. 41–50, Oct. 1998.
3. T. Unemi. Sbart2.4:breeding 2d cg images and movies, and creating a type of Collage. In *Proceedings of The Third International Conference on Knowledge-based Intelligent Information Engineering Systems*, pp. 288-291, Adelaide, Australia, Aug. 1999.
4. H. Wakaki and H. Iba. Motion design of a 3d-cg avatar using interactive evolutionary computation. In *2002 IEEE International Conference on Systems, Man und Cybernetics (SMC02)*. IEEE Press, 2002.

Evolving Multiple Discretizations with Adaptive Intervals for a Pittsburgh Rule-Based Learning Classifier System

Jaume Bacardit and Josep Maria Garrell

Intelligent Systems Research Group
Enginyeria i Arquitectura La Salle
Universitat Ramon Llull
Psg. Bonanova 8, 08022-Barcelona
Catalonia, Spain, Europe
{jbacardit,josepmg}@salleURL.edu

Abstract. One of the ways to solve classification problems with real-value attributes using a Learning Classifier System is the use of a discretization algorithm, which enables traditional discrete knowledge representations to solve these problems. A good discretization should balance losing the minimum of information and having a reasonable number of cut points. Choosing a single discretization that achieves this balance across several domains is not easy. This paper proposes a knowledge representation that uses several discretization (both uniform and non-uniform ones) at the same time, choosing the correct method for each problem and attribute through the iterations. Also, the intervals proposed by each discretization can split and merge among them along the evolutionary process, reducing the search space where possible and expanding it where necessary. The knowledge representation is tested across several domains which represent a broad range of possibilities.

1 Introduction

The application of Genetic Algorithms (GA) [1,2] to classification problems is usually known as Genetic Based Machine Learning (GBML) or Learning Classifier Systems (LCS), and traditionally it has been addressed from two different points of view: the Pittsburgh approach (also known as Pittsburgh LCS), and the Michigan approach (or Michigan LCS). Some representative systems of each approach are *GABIL* [3] and XCS [4]. The classical knowledge representation used in these systems is a set of rules where the antecedent is defined by a prefixed finite number of intervals to handle real-valued attributes. The performance of these systems is tied to the right election of the intervals through the use of a discretization algorithm.

There exist several good heuristic discretization methods which have good performance on several domains. However, they lack robustness on some other domains because they loose too much information from the original data. The alternative of a high number of simple uniform-width intervals usually expands the size of the search space without a clear performance gain.

In a previous paper [5] we have proposed a representation called *adaptive discrete intervals (ADI) rule representation* where several uniform-width discretizations are used at the same time. Thus, allowing the GA choose the correct discretization for each rule and attribute. Also, the discretization intervals were split and merged through the evolutionary process. This representation has been used in our Pittsburgh approach *LSC*.

In this paper we generalize the ADI representation approach (proposing *ADI2*) by also using heuristic non-uniform discretization methods. In our case the well-known Fayyad & Irani discretization algorithm [6]. In our previous work se reported that using only the Fayyad & Irani discretization can lead to a non-robust system in some domains, but using it together with some uniform-width discretizations improves the performance in some domains and, most important, presents a robust behavior across most domains. Also, the probability that controls the split and merge operators is redefined in order to simplify the tuning needed to use the representation.

This rule representation is compared across different domains against the results of the original representation and also the *XCSR* representation [7] by Wilson which uses rules with real-valued intervals in the well known XCS Michigan LCS [4]. We want to state clearly that we have integrated the *XCSR* representation into our Pittsburgh system, instead of using it inside *XCS*.

The paper is structured as follows. Section 2 presents some related work. Then, we describe the framework of our classifier system in Sect. 3. The revision of the ADI representation is explained in Sect. 4. Next, Sect. 5 describes the test suite used in the comparison. The results obtained are summarized in Sect. 6. Finally, Sect. 7 discusses the conclusions and some further work.

2 Related Work

There are several approaches to handle real-valued attributes in the LCS field. These approaches can be classified in a simple manner into two groups: Systems that discretize or systems that work directly with real values. Reducing the real-valued attributes to discrete values let the systems in the first group use traditional symbolic knowledge representations. There are several types of algorithms which can perform this reduction. Some heuristic examples are the Fayyad & Irani method [6] which works with information entropy or the χ^2 statistic measures [8].

Some specific GBML applications of discretization algorithms were presented by Riquelme and Aguilar [9] and are similar to the representation proposed here. Their system evolves conjunctive rules where each conjunct is associated to an attribute and is defined as a range of adjacent discretization intervals. They use their own discretization algorithm, called *USD* [10].

Lately, *several alternatives to the discrete rules have been presented.* There are rules composed by real-valued intervals (XCSR [7] by Wilson or COGITO

[11] by Aguilar and Riquelme). Also, Llorà and Garrell [12] proposed a knowledge independent method for learning other knowledge representations like instance sets or decision trees. Most of these alternatives have better accuracy than the discrete rules, but usually they also have higher computational cost [11].

3 Framework

In this section we describe the main features of our classifier system which is a Pittsburgh LCS based on GABIL [3]. Directly from GABIL we have borrowed the semantically correct crossover operator and the fitness computation (squared accuracy).

Matching Strategy: The matching process follows a "if ... then ... else if ... then..." structure, usually called *Decision List* [13].

Mutation Operators: The system manipulates variable-length individuals, making more difficult the tuning of the classic gene-based mutation probability. In order to simplify this tuning, we define p_{mut} as the probability of mutating an individual. When an individual is selected for mutation (based on p_{mut}), a random gene is chosen inside its chromosome for mutation.

Control of the Individuals Length: Dealing with variable-length individuals arises some serious considerations. One of the most important ones is the control of the size of the evolving individuals [14]. This control is achieved using two different operators:

- *Rule deletion:* This operator deletes the rules of the individuals that do not match any training example. This rule deletion is done after the fitness computation and has two constraints: (a) the process is only activated after a predefined number of iterations, to prevent a massive diversity loss and (b) the number of rules of an individual never goes below a lower threshold.
- *Selection bias using the individual size:* Selection is guided as usual by the fitness (the accuracy of the individual). However, it also gives certain degree of relevance to the size of the individuals, having a policy similar to multi-objective systems. We use tournament selection because its local behavior lets us implement this policy. The criterion of the tournament is given by our own operator called "hierarchical selection" [15]. This operator considers two individuals "similar" if their fitness difference is below a certain threshold (d_{comp}). Then, it selects the individual with fewer number of rules. Our previous tests showed that sizing d_{comp} to 0.01 was quite good for real problems and 0.001 was quite good for synthetic problems. Although a fine tuning of this parameter probably would improve the performance for each test domain, for the sake of simplicity we will use these values.

4 The Adaptive Discretization Intervals (ADI) Rule Representation

4.1 The Original ADI Representation

The general structure of each rule is taken from *GABIL*. That is, each rule consists of a condition part and a classification part: $condition \rightarrow classification$. Each condition is a Conjunctive Normal Form (CNF) predicate defined as:

$$((A_1 = V_1^1 \vee \ldots \vee A_1 = V_m^1) \bigwedge \ldots \bigwedge (A_n = V_2^n \vee \ldots A_n = V_m^b))$$

Where A_i is the ith attribute of the problem and V_i^j is the jth value that can take the ith attribute.

In the *GABIL* representation this kind of predicate can be encoded into a binary string in the following way: if we have a problem with two attributes, where each attribute can take three different values {1,2,3}, a rule of the form "If the first attribute has value 1 or 2 and the second one has value 3 then we assign class 1" will be represented by the string 110|001|1. For real-valued attributes, each bit (except the class one) would be associated to a discretization interval.

The intervals of the rules in the ADI representation are not static, but they evolve splitting and merging among them (having a minimum size called *micro-intervals*). Thus, the binary coding of the *GABIL* representation is extended as represented in Fig. 1, also showing the split and merge operations.

The ADI representation is defined in depth as follows:

1. Each individual, initial rule and attribute term is assigned a number of "low level" uniform-width and static discretization intervals (called *micro-intervals*).
2. The intervals of the rule are built joining together adjacent *micro-intervals*.
3. Attributes with different number of *micro-intervals* can coexist in the population. The evolution will choose the correct number of *micro-intervals* for each attribute.

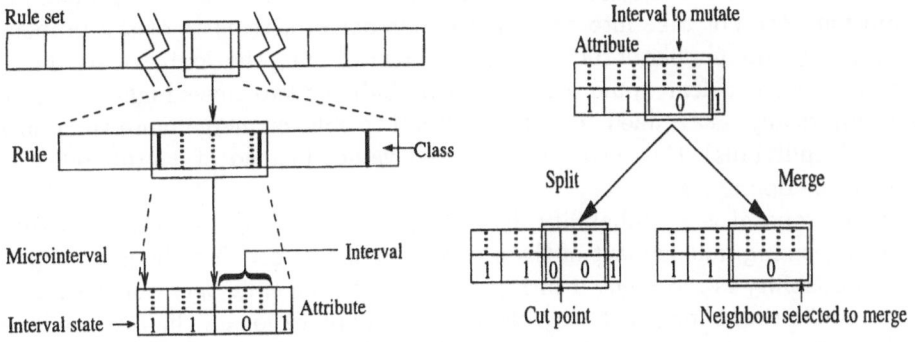

Fig. 1. Adaptive intervals representation and the split and merge operators

4. For computational cost reasons, we will have an upper limit in the number of intervals allowed for an attribute, which in most cases will be less than the number of *micro-intervals* assigned to each attribute.
5. When we split an interval, we select a random point in its *micro-intervals* to break it.
6. When we merge two intervals, the state (1 or 0) of the resulting interval is taken from the one which has more *micro-intervals*. If both have the same number of *micro-intervals*, the value is chosen randomly.
7. The number of *micro-intervals* assigned to each attribute term is chosen from a predefined set.
8. The number and size of the initial intervals is selected randomly.
9. The cut points of the crossover operator can only take place in attribute terms boundaries, not between intervals. This restriction takes place in order to maintain the semantical correctness of the rules.
10. The *hierarchical selection* operator uses the length of the individuals (defined as the sum of all the intervals of the individual) instead of the number of rules as the secondary criteria. This change promotes simple individuals with more reduced interval fragmentation.

In order to make the interval splitting and merging part of the evolutionary process, we have to include it in the GAs genetic operators. We have chosen to add to the GA cycle two special stages applied to the offspring population after the mutation stage. For each stage (split and merge) we have a probability (p_{split} or p_{merge}) of applying the operation to an individual. If an individual is selected for split or merge, a random point inside its chromosome is chosen to apply the operation.

4.2 Revisions to the Representation: ADI2

The first modification to the ADI representation affects the definition of the split and merge probabilities. These probabilities were defined at an individual-wise level, and needed an specific adjust for each domain that we want to solve. The need of this fine-adjusting is motivated by the fact that this probability definition does not take into account the number of attributes of the problem nor its optimum number of rules. Thus, a problem having the double of attributes than another problem would also need a probability two times higher. Also, it was empirically determined that it was useful to split or merge more than once for each individual, thus using an expected values instead of a probability to control the operators.

Our proposal is a probability defined for each attribute term of each rule. Thus, becoming independent of these two factors. The code for the merge operator probability is represented in Fig. 2. Code for the split operator is similar. This redefinition allows us to use the same probability for all the test domains with similar results to the original probability definition. The probability value has been empirically defined as 0.05.

```
ForEach Individual i of Population
    ForEach Rule j of Population individual i
        ForEach Attribute k of Rule j of Population individual i
            If random [0..1] number < p_merge
                Select one random interval of attribute term k
                    of rule j of individual i
                Apply a merge operation to this interval
            EndIf
        EndForEach
    EndForEach
EndForEach
```

Fig. 2. Code of the application of the merge operator

Our second modification is related to the kind of discretization that we use. The experimentation reported of the original representation included a comparison with a discrete representation where the discretization intervals where given by the Fayyad & Irani method [6]. This discrete representation performed better than the adaptive intervals rule representation in some domains, but it was significantly outperformed in some other domains, showing a lack of robustness.

Our aim is to modify the representation in order to include non-uniform discretization into it, in a way that improves the performance of the system in the problems where it is possible, but maintains a robust behavior in the kind of problems where the systems using only heuristic representations fail. A graphical representation of the two types of rules is in Fig. 3. The changes introduced to the definition of the representation are minimum. A revised definition follows:

1. **A set of static discretization intervals (called *micro-intervals*) is assigned to each attribute term of each rule of each individual.**
2. The intervals of the rule are built joining together *micro-intervals*.
3. **Attributes with different number and sizes of *micro-intervals* can coexist in the population. The evolution will choose the correct discretization for each attribute.**
4. For computational cost reasons, we will have an upper limit in the number of intervals allowed for an attribute, which in most cases will be less than the number of *micro-intervals* assigned to each attribute.
5. When we split an interval, we select a random point in its *micro-intervals* to break it.

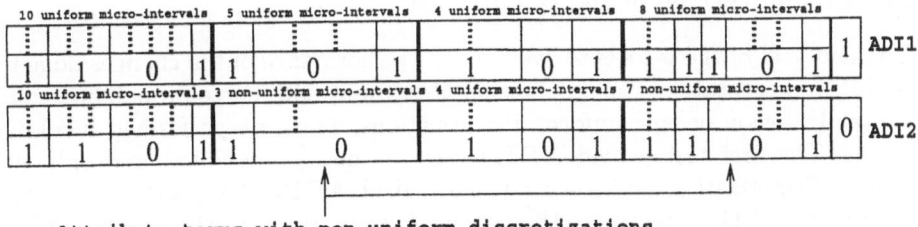

Fig. 3. Example of the differences between ADI1 and ADI2 rules

6. When we merge two intervals, the state (1 or 0) of the resulting interval is taken from the one which has more *micro-intervals*. If both have the same number of *micro-intervals*, the value is chosen randomly.
7. **The discretization assigned in the initialization stage to each attribute term is chosen from a predefined set.**
8. The number and size of the initial intervals is selected randomly.
9. The cut points of the crossover operator can only take place in attribute terms boundaries, not between intervals. This restriction takes place in order to maintain the semantical correctness of the rules.
10. The *hierarchical selection* operator uses the length of the individuals (defined as the sum of all the intervals of the individual) instead of the number of rules as the secondary criteria. This change promotes simple individuals with more reduced interval fragmentation.

5 Test Suite

This section summarizes the tests done in order to evaluate the accuracy and efficiency of the method presented in this paper. We also compare it with some alternative methods. The tests were conducted using several machine learning problems which we also describe.

5.1 Test Problems

The selected test problems are the same that were used in the original version of the representation being studied in this paper. The first problem is a synthetic problem (*tao* [12]) that has non-orthogonal class boundaries. We also use several problems provided by the University of California at Irvine (UCI) repository [16]. The problems selected are: Pima Indians Diabetes (*pim*), Iris (*irs*), Glass (*gls*) and Winsconsin Breast Cancer (*breast*). Finally we will use three real problems from private repositories. The first two deal with the diagnosis of breast cancer based on biopsies (*bps* [17]) and on mammograms (*mmg* [18]), whereas the last one is related to the prediction of student qualifications (*lrn* [19]). The characteristics of the problems are listed in Table 1. The partition of the examples into the train and test sets was done using *stratified ten-fold cross-validation* [20].

5.2 Configurations of the GA to Test

The main goal of the tests is to evaluate the performance of the changes done to the ADI representation: the redefinition of the split and merge probabilities and also the inclusion of non-uniform discretizations. Thus, we performed two kind of tests: The first test (called **ADI2**) uses the new probabilities and only the uniform discretizations. The second one (called **ADI2+Fayyad**) adds to the configuration **ADI2** the use of the Fayyad & Irani discretization method [6].

This revision of the ADI representation is compared to three more configurations:

Table 1. Characteristics of the test problems

Name	ID.	Instances	Real attr	Discrete attr	Type	Classes
Tao	tao	1888	2	-	synthetic	2
Pima-Indians-Diabetes	pim	768	8	-	real	2
Iris	irs	150	4	-	real	3
Glass	gls	214	9	-	real	6
Winsconsin Breast Cancer	bre	699	-	9	real	2
Biopsies	bps	1027	24	-	real	2
Mammograms	mmg	216	21	-	real	2
Learning	lrn	648	4	2	real	5

- Non-adaptive discrete rules (using the *GABIL* representation) using the Fayyad & Irani discretization method to determine the intervals used. This test is labeled **Fayyad**.
- The original ADI representation with the old split and merge probability definition and only using uniform discretizations (called *ADI1*).
- Rules with real-valued intervals using the *XCSR* representation [7] (called **XCSR**).

We have decided not to add any other non-uniform discretization to *ADI2+ Fayyad* because the comparison of these tests with the *Fayyad* test would not be fair if more non-uniform discretization methods were used.

The GA parameters are shown in Tables 2 and 3. The first one shows the general parameters and the second one the domain-specific ones. The reader can appreciate that the sizing of both p_{split} and p_{merge} is the same for all the problems except the *tao* problem. Giving the same value to p_{merge} and p_{split} produce solutions with too few rules and intervals, as well as less accurate than the results obtained with the configuration shown in Table 3.

The reason of this fact comes from the definition of this synthetic problem with only two attributes and rules with intervals as small as one 48th of an attribute domain in the optimum rule set. The probability of generating these very specific rules is very low because of the *hierarchical selection* operator used to apply generalization pressure. The operator is still used because it is beneficial for the rest of problems and we can fix this bad interaction by increasing p_{split}. The characteristics of this problem explained here are also the reason of the bad performance of the test using only the Fayyad & Irani intervals because the discretization generates too few intervals and, as a consequence, loses too much information from the original data.

6 Results

In this section we present the results obtained. The aim of the tests was to compare the methods studied in this paper in three aspects: accuracy and size of the solutions as well as the computational cost. For each method and test

Table 2. Common parameters of the GA

Parameter	Value
General parameters	
Crossover probability	0.6
Selection algorithm	Tournament
Tournament size	3
Population size	300
Probability of mutating an individual	0.6
Rule Deletion operator	
Iteration of activation	30
Minimum number of rules before disabling the operator	number of classes + 3
Hierarchical Selection	
Iteration of activation	10
ADI rule representation	
Number of intervals of the uniform discretizations	4,5,6,7,8,10,15,20,25
XCSR rule representation	
Maximum radius size in initialization	0.7 of each attribute domain
Maximum center offset in mutation	0.7 of each attribute domain
Maximum radius offset in mutation	0.7 of each attribute domain

Table 3. Problem-specific parameters of the GA

Code	Parameter
#iter	Number of GA iterations
d_{comp}	Distance parameter in the "size-based comparison" operator
p_{split} orig	Expected value interval split in *ADI1*
p_{merge} orig	Expected value of interval merging in *ADI1*
p_{split}	Probability of splitting an interval in *ADI2*
p_{merge}	Probability of merging an interval in *ADI2*

Problem	Parameter					
	#iter	d_{comp}	p_{merge} orig	p_{split} orig	p_{merge}	p_{split}
tao	900	0.001	1.3	2.6	0.05	0.25
pim	250	0.01	0.8	0.8	0.05	0.05
irs	275	0.01	0.5	0.5	0.05	0.05
gls	900	0.01	1.5	1.5	0.05	0.05
bre	300	0.01	3.2	3.2	0.05	0.05
bps	275	0.01	1.7	1.7	0.05	0.05
mmg	225	0.01	1.0	1.0	0.05	0.05
lrn	700	0.01	1.2	1.2	0.05	0.05

problem we show the average and standard deviation values of: (1) the cross-validation accuracy, (2) the size of the best individual in number of rules and number of intervals and (3) the execution time in seconds. Obviously, the *XCSR* results lack the intervals per attribute column. The tests were executed in an AMD Athlon 1700+ using Linux operating system and C++ language. Runs for each method, problem and fold has been repeated 15 times using different random seeds and the results averaged. The full detail of the results is shown in Table 4 and a summary of them taking the form of a methods ranking is in Table 5. The ranking for each problem and method is based on the accuracy. The global rankings are computed averaging the problem rankings. The results were also analyzed using two-sided Student t-tests [20] with a significance level of 1% in order to determine if there were significant outperformances between the methods tested. The results of the t-tests are shown in Table 6. None of the *ADI* was outperformed.

The results were also analyzed using the two-sided t-test [20] with a significance level of 1% in order to determine if there were significant outperformances between the methods tested.

The first interesting fact from the results is the comparison between *ADI1* and *ADI2*, that is, the comparison between the split and merge probabilities definitions. *ADI2* the method that uses the same probabilities for all the domains is always better (with only one exception) than *ADI1*, the method that needed domain-specific probabilities. Thus, the gain is double, performance and simplicity of use.

The performance of *ADI2+Fayyad* is also quite good, being the second method in the overall ranking but very close to *ADI2*, the first one. The objective of increasing the *ADI* accuracy in the problems where the *Fayyad* discrete rules alone performed well and, at the same time, maintain the accuracy in the domains where *Fayyad* presented very poor performance has been achieved.

Also, the results of the *XCSR* representation show that the representations based on a discretization process, if well used, present good performance compared to representations that evolve real-valued intervals. This observation matches the ones reported by Riquelme and Aguilar [11].

The computational cost continues being the main drawback of the representation. The comparison of the ADI representation run time with the *Fayyad* discrete rules is clearly favorable to *Fayyad* test. The comparison with *XCSR*, however, is not so clear. In some domains one representation is faster and in other domains it is the reverse situation.

7 Conclusions and Further Work

This paper focused on a revision and generalization of our previous work on representations for real-valued attributes: the Adaptive Discretization Intervals (ADI) rule representation. This representation evolves rules that can use multiple discretizations, letting the evolution choose the correct discretization for each rule and attribute. Also, the intervals defined in each discretization can split or merge among them through the evolution process, reducing the search space where it is possible and expanding it where it is needed. This revision has consisted of two parts: redefining the split and merge operators probabilities and using non-uniform discretizations in the representation in addition of the existing uniform ones. The redefinition of the operators has simplified enormously

Table 4. Mean and deviation of the accuracy, number of rules and intervals per attribute for each method tested. Bold entries show the method with best results for each test problem

Problem	Configuration	Accuracy	Number of Rules	Intervals per attribute	Run time
tao	Fayyad	87.8±1.1	3.1±0.3	3.4±0.1	37.3±2.1
	ADI1	94.3±1.0	19.5±4.9	6.0±0.6	145.6±20.9
	ADI2	**94.7±0.9**	17.5±4.4	8.7±0.6	162.1±20.9
	ADI2+Fayyad	94.0±0.9	15.6±4.4	7.6±1.0	150.5±23.4
	XCSR	91.1±1.4	12.9±3.1	—	110.6±14.5
pim	Fayyad	73.6±3.1	6.6±2.6	2.3±0.2	**13.2±1.5**
	ADI1	74.4±3.1	5.8±2.2	**1.9±0.4**	29.9±4.5
	ADI2	**74.5±3.8**	3.7±1.2	2.1±0.3	30.2±3.2
	ADI2+Fayyad	**74.5±3.3**	3.6±0.9	2.0±0.3	32.1±3.0
	XCSR	74.2±3.3	**3.4±1.0**	—	18.7±2.9
irs	Fayyad	94.2±3.0	**3.2±0.6**	2.8±0.1	**4.1±0.1**
	ADI1	96.2±2.2	3.6±0.9	**1.3±0.2**	6.2±0.6
	ADI2	96.0±2.6	3.9±0.7	1.5±0.3	6.0±2.0
	ADI2+Fayyad	**96.3±2.4**	3.7±0.7	1.5±0.2	6.4±2.1
	XCSR	95.2±2.3	3.5±0.6	—	**3.5±0.1**
gls	Fayyad	65.7±6.1	8.1±1.4	2.4±0.1	**16.8±1.3**
	ADI1	65.2±4.1	**6.7±2.0**	**1.8±0.2**	46.1±6.0
	ADI2	65.6±3.7	7.6±1.5	2.6±0.2	41.8±3.8
	ADI2+Fayyad	**66.2±3.6**	7.7±1.6	2.5±0.2	42.7±3.5
	XCSR	65.2±5.2	8.4±1.2	—	37.5±4.9
bre	Fayyad	95.2±1.8	4.1±0.8	3.6±0.1	**8.3±0.9**
	ADI1	95.3±2.3	2.6±0.9	**1.7±0.2**	25.0±1.4
	ADI2	**95.9±2.3**	**2.4±0.6**	1.8±0.3	19.6±1.4
	ADI2+Fayyad	95.7±2.4	**2.4±0.6**	1.8±0.3	22.8±1.6
	XCSR	95.8±2.7	3.2±0.9	—	21.4±2.9
bps	Fayyad	80.0±3.1	7.1±3.8	2.4±0.1	**31.1±5.0**
	ADI1	80.1±3.3	5.1±2.0	**2.0±0.3**	99.6±17.0
	ADI2	**80.6±2.9**	**3.6±1.0**	2.1±0.2	113.8±9.9
	ADI2+Fayyad	80.3±2.9	**3.6±1.0**	**2.0±0.2**	119.2±7.7
	XCSR	79.7±3.1	4.3±1.1	—	151.6±13.4
mmg	Fayyad	65.3±11.1	**2.3±0.5**	2.0±0.1	**3.8±0.3**
	ADI1	65.0±6.1	4.4±1.9	**1.9±0.2**	12.3±2.5
	ADI2	**65.6±5.7**	4.9±1.0	2.3±0.1	13.7±1.2
	ADI2+Fayyad	65.2±4.3	4.7±1.1	2.2±0.1	14.4±1.2
	XCSR	63.1±4.1	5.6±1.6	—	14.9±2.4
lrn	Fayyad	67.5±5.1	14.3±5.0	4.4±0.1	**26.5±3.4**
	ADI1	66.7±4.1	11.6±4.1	**3.4±0.2**	53.9±7.2
	ADI2	67.4±4.3	7.2±1.7	3.5±0.2	46.6±5.1
	ADI2+Fayyad	**67.6±4.6**	6.9±1.4	**3.4±0.2**	45.9±5.0
	XCSR	67.1±4.4	9.1±2.5	—	52.6±7.5

the tuning needed to use the representation because we have converted an operator which needed domain-specific tuning into an operator which uses the same probability for all the domains. The addition of non-uniform discretizations to the representation has improved its performance in some domains while main-

Table 5. Performance ranking of the tested methods. Lower number means better ranking

Problem	Fayyad	ADI1	ADI2	ADI2+Fayyad	XCSR
tao	5	2	1	3	4
pim	5	3	1	1	4
irs	5	2	3	1	4
gls	2	4	3	1	4
bre	5	4	1	3	2
bps	4	3	1	2	5
mmg	2	4	1	3	5
lrn	2	4	3	1	4
Average	3.75	3.25	1.75	1.875	4
Final rank	4	3	1	2	5

Table 6. Summary of the statistical two-sided t-test performed at the 1% significance level. Each cell indicates how many times the method in the row outperforms the method in the column

Method	Fayyad	ADI1	ADI2	ADI2+Fayyad	XCSR	Total
Fayyad	-	0	0	0	0	0
ADI1	1	-	0	0	2	3
ADI2	1	0	-	0	2	3
ADI2+Fayyad	1	0	0	-	2	3
XCSR	1	0	0	0	-	1
Total	4	0	0	0	6	

taining the robust behavior that presented the original version, according to the statistical t-tests done.

The tests done show that the new version is always better than the original one and also better than the other representations included in the comparison when used in a Pittsburgh LCS. However, the computational cost continues being a major drawback, compared to the discrete rules.

As a further work, other non-uniform discretization methods beside the Fayyad & Irani one should be tested, to increase the range of problems where this representation performs well. This search for more discretization methods, however, should always have in mind maintaining the robustness behavior that the representation has showed so far. On the other hand, the ADI representation should be compared with other kind of representations dealing with real-valued attributes, for example non-orthogonal ones.

Acknowledgments. The authors acknowledge the support provided under grant numbers 2001FI 00514, TIC2002-04160-C02-02 and TIC 2002-04036-C05-03 and 2002SGR 00155. Also, we would like to thank Enginyeria i Arquitectura La Salle for their support to our research group. The Wisconsin breast cancer databases was obtained from the University of Wisconsin Hospitals, Madison from Dr. William H. Wolberg.

References

1. Holland, J.H.: Adaptation in Natural and Artificial Systems. University of Michigan Press (1975)
2. Goldberg, D.E.: Genetic Algorithms in Search, Optimization and Machine Learning. Addison-Wesley Publishing Company, Inc. (1989)
3. DeJong, K.A., Spears, W.M.: Learning concept classification rules using genetic algorithms. Proceedings of the International Joint Conference on Artificial Intelligence (1991) 651–656
4. Wilson, S.W.: Classifier fitness based on accuracy. Evolutionary Computation **3** (1995) 149–175
5. Bacardit, J., Garrell, J.M.: Evolution of multi-adaptive discretization intervals for A rule-based genetic learning system. In: Proceedings of the VIII Iberoamerican Conference on Artificial Intelligence (IBERAMIA'2002), LNAI vol. 2527, Springer (2002) 350–360
6. Fayyad, U.M., Irani, K.B.: Multi-interval discretization of continuous-valued attributes for classification learning. In: IJCAI. (1993) 1022–1029
7. Wilson, S.W.: Get real! XCS with continuous-valued inputs. In Booker, L., Forrest, S., Mitchell, M., Riolo, R.L., eds.: Festschrift in Honor of John H. Holland, Center for the Study of Complex Systems (1999) 111–121
8. Liu, H., Setiono, R.: Chi2: Feature selection and discretization of numeric attributes. In: Proceedings of 7th IEEE International Conference on Tools with Artificial Intelligence, IEEE Computer Society (1995) 388–391
9. Aguilar-Ruiz, J.S., Riquelme, J.C., Valle, C.D.: Improving the evolutionary coding for machine learning tasks. In: Proceedings of the European Conference on Artificial Intelligence, ECAI'02, Lyon, France, IOS Press (2002) pp. 173–177
10. Giráldez, R., Aguilar-Ruiz, J.S., Riquelme, J.C.: Discretización supervisada no paramétrica orientada a la obtención de reglas de decisión. In: Proceedings of the CAEPIA2001. (2001) 53–62
11. Riquelme, J.C., Aguilar, J.S.: Codificación indexada de atributos continuos para algoritmos evolutivos en aprendizaje supervisado. In: Proceedings of the "Primer Congreso Español de Algoritmos Evolutivos y Bioinspirados (AEB'02)". (2002) 161–167
12. Llorà, X., Garrell, J.M.: Knowledge-independent data mining with fine-grained parallel evolutionary algorithms. In: Proceedings of the Genetic and Evolutionary Computation Conference (GECCO-2001), Morgan Kaufmann (2001) 461–468
13. Rivest, R.L.: Learning decision lists. Machine Learning **2** (1987) 229–246
14. Soule, T., Foster, J.A.: Effects of code growth and parsimony pressure on populations in genetic programming. Evolutionary Computation **6** (1998) 293–309
15. Bacardit, J., Garrell, J.M.: Métodos de generalización para sistemas clasificadores de Pittsburgh. In: Proceedings of the "Primer Congreso Español de Algoritmos Evolutivos y Bioinspirados (AEB'02)". (2002) 486–493
16. Blake, C., Keogh, E., Merz, C.: Uci repository of machine learning databases (1998) Blake, C., Keogh, E., & Merz, C.J. (1998). UCI repository of machine learning databases (www.ics.uci.edu/mlearn/MLRepository.html).

17. Martínez Marroquín, E., Vos, C., et al.: Morphological analysis of mammary biopsy images. In: Proceedings of the IEEE International Conference on Image Processing (ICIP'96). (1996) 943–947
18. Martí, J., Cufí, X., Regincós, J., et al.: Shape-based feature selection for microcalcification evaluation. In: Imaging Conference on Image Processing, 3338:1215-1224. (1998)
19. Golobardes, E., Llorà, X., Garrell, J.M., Vernet, D., Bacardit, J.: Genetic classifier system as a heuristic weighting method for a case-based classifier system. Butlletí de l'Associació Catalana d'Intel.ligència Artificial **22** (2000) 132–141
20. Witten, I.H., Frank, E.: Data Mining: practical machine learning tools and techniques with java implementations. Morgan Kaufmann (2000)

Limits in Long Path Learning with XCS

Alwyn Barry

Department of Computer Science
University of Bath
Claverton Down
Bath, BA2 7AY, UK
A.M.Barry@bath.ac.uk

Abstract. The development of the XCS Learning Classifier System [26] has produced a stable implementation, able to consistently identify the accurate and optimally general population of classifiers mapping a given reward landscape [15,16,29]. XCS is particularly powerful within direct-reward environments, and notably within problems suitable for commercial application [3]. The application of XCS within delayed reward environments has also shown promise, although early investigations were within enviroments with a comparatively short delay to reward (e.g. [28, 19]). Subsequent systematic investigation [19,20,1,2] have suggested that XCS has difficulty accurately mapping and exploiting even simple environments with moderate reward delays. This paper summarises these results and presents new results that identify some limits and their implications. A modification to the error computation within XCS is introduced that allows the minimum error parameter to be applied relative to the magnitude of the payoff to each classifier. First results demonstrate that this modification enables XCS to successfully map longer delayed-reward enviroments.

1 Background

Learning Classifier Systems (*'LCS'*) are a class of machine learning techniques that utilise evolutionary computation to provide the main knowledge induction algorithm. They are characterised by the representation of knowledge in terms of a population of simplified production rules (*'classifiers'*) in which the conditions are able to cover one or more inputs. The *Michigan* LCS [13] maintain a single population of production rules with a Genetic Algorithm (*'GA'*) operating within the population ... each rule maintains its own fitness estimate. LCS are general machine learners, primarily limited by the constraints in the representation adopted for the production rules (see, for example, [29]) and by the complexity of solutions that can be maintained under the action of the GA [10].

LCS have been successfully applied to many application areas – most notably for Data Mining [22,14,3] but also in more complex control problems (e.g. [12,24])[1]. Although LCS performance has been competitive with the most effective machine learning techniques, it is notable that many of these cases (and all

---
[1] see [21] for further details of LCS applications

of those cited earlier in this paragraph) have been with LCS implementations that use direct reward allocation. In contrast, the application of LCS to complex delayed reward tasks has, until recently, been more problematic[2]. Recently a number of new LCS implementations appear to have overcome some of the instabilities of LCS when learning in delayed reward situations. ZCS [25] is a simplification of earlier LCS implementations and was designed for academic study, but [4,6] have shown that ZCS can *perform* optimally with appropriate parameter settings. [16] has argued persuasively that strength-based LCS implementations will inevitably lead to over-generalisation in tasks with biased reward functions. XCS [26] is an accuracy-based LCS derived from ZCS which overcomes many of these limitations, and the *Optimality Hypothesis:* [15] suggests that XCS will always identify a sub-population of accurate optimally general classifiers that occupy a larger proportion of the population than other classifiers. Bull argues that the fitness sharing mechanism of ZCS acts as a mechanism to prevent over-generalisation within ZCS, making ZCS a competitive strength-based LCS [5].

The effectiveness of XCS in its application to direct reward environments has been empirically demonstrated by many workers (for example, [26,15,29,3,11]). Research into the performance of XCS within delayed reward environments has been more limited. [26,27] provided a proof-of-concept demonstration of the operation of XCS within the Woods2 environment. [17] identified that within certain Woods-like environments XCS was unable to identify optimum generalisations. This was attributed to two major factors: an inequality in exploration of all states in the environment allowing over-general classifiers to appear accurate, and an input encoding which meant that certain generalisations were not explored as often as others. [18] sought to apply these lessons to more complex Woods-based environments and discovered that XCS was additionally unable to establish a solution to the long chain Woods-14 problem [9]. This was due in part to the number of possible alternatives to explore in each state that prevented XCS from attributing equal exploration time to later states within the chain. It has been shown that XCS is able to learn this environment using an alternative explore-exploit strategy [6].

Whilst other work has investigated some of the more complex problems that delayed reward environments present, such as perceptual aliasing, there had been little investigation of the comparative performance of traditional LCS implementations and XCS in delayed reward environments beyond the simple Woods environments. Much more importantly, no work had attempted to identify the limits of XCS learning with increasing environment complexity or length when applied to delayed reward environments. [2] presented the results of an investigation of the abililty of XCS to form the population of optimally general classifiers mapping the payoff of environments of incrementally increasing length. It was shown that XCS performance was very good within the GREF-1 environment (100% performance within 1000 explorations – CFS-C [23] achieved 90% performance in 10,000 explorations). It was also shown that XCS can reliably learn the optimal route in a corridor environment (an extension of Fig. 1) of length

---

[2] There are many reasons for this ... for a review of the issues see [1].

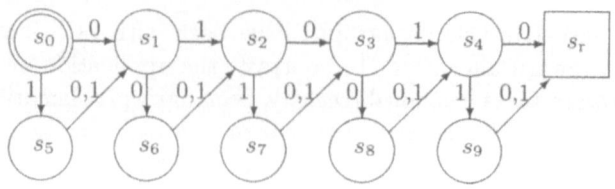

Fig. 1. A length 5 corridor Finite State World for delayed reward experiments

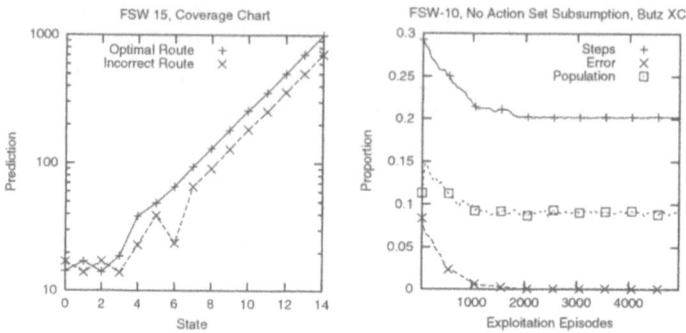

Fig. 2. (a) State × Prediction (log scale) coverage of the populations from 10 runs of Barry's XCS in a length 15 corridor environment after 15000 iterations; (b) Average of 10 runs of Butz's XCS in a length 10 corridor environment showing errors in the 'steps' plot

40 where generalisation was not used. However, once XCS was required to learn the optimal *generalisations* XCS was unable to reliably select the optimal route even within the length 10 environment. Learning performance deteriorated sufficiently thereafter to make investigation of performance in corridors longer than 15 steps superfluous (see Fig. 2a – the payoff prediction values of classifiers in the early states are highly confused). Subsequent investigations using an alternative XCS implementation [7] have confirmed these findings (see the perturbation in the 'steps' plot in Fig. 2b[3]).

Analysis of the population coverage of the environment indicated that XCS was unable to learn appropriate *generalisations* for early states in the environment. It was hypothesised that this was partially due to the reduction in payoff prediction values in classifiers covering these early states, making the prediction values sufficiently similar that XCS can identify a few over-general classifiers to cover the early states (see Fig. 2a). This conclusion caused some other workers to suggest that the problem might be resolved through the use of Wilson's modified accuracy measure or the use of an alternative subsumption algorithm (both introduced within [8]). These suggestions, though ill-founded, led to fur-

---

[3] XCS should rapidly identify the optimum number of steps to the reward, but instead often chooses at least one sub-optimal action in each corridor traversal

ther investigations into the cause of the over-generalisation within these early states.

2 XCS Learning in Delayed-Reward Environments

The *match set* ('[M]') is the set of all classifiers whose conditions *match* the current environmental input. XCS selects one of the actions proposed by [M] and all classifiers in [M] which advocate that action become members of the *action set* ('[A]') in the current iteration. When reward R is obtained by XCS in a delayed reward environment the reward is allocated to all classifiers in [A]. Clearly the reward could only have been reached as a result of earlier actions. So that the optimal path to the reward can be established XCS allocates payoff to classifiers within earlier action sets. In each exploration iteration the maximum action set prediction of [M] discounted by a discount factor γ is used as the payoff P in the update of the predictions p of the classifiers in the previous action set ($[A]_{t-1}$) using the following update scheme ($0.0 < \beta \leq 1.0$):

$$p' = p + \beta(P - p) . \quad (1)$$

Over time all action sets will converge upon an estimate of the discounted payoff that will be received if the action was to be taken. The discount allows the estimate to take account of distance to reward as well as the magnitude of the reward so that a trade-off of effort to reward is inherent in the action selection. Unfortunately the discount also means that the payoff prediction becomes much smaller with distance from the reward. With the 'standard' XCS discount ($\gamma = 0.71$) the payoff reduces from the reward of 1000 to less than 10 within 14 steps ... the payoff in state n of an N state path to reward R is:

$$p_n = R\gamma^{N-n} . \quad (2)$$

XCS is an accuracy-based LCS. Accuracy is a steep logarithmic function[4] of the error in the classifier's prediction of payoff:

$$\kappa = \begin{cases} \ln(\alpha) \frac{\varepsilon - \varepsilon_0}{\varepsilon_0} m & (\varepsilon > \varepsilon_0) \\ 1.0 & (\text{otherwise}) . \end{cases} \quad (3)$$

(α and m are constants), where the error ε is calculated as:

$$\varepsilon' = \varepsilon + \beta(\varepsilon_{\text{abs}} - \varepsilon) \quad (4)$$

$$\text{where } \varepsilon_{\text{abs}} = \frac{|P - p|}{R_{\text{max}} - R_{\text{min}}} . \quad (5)$$

($R_{\text{max}} - R_{\text{min}}$ is the reward range). Fitness is based on the *relative accuracy* of the classifiers appearing within each [A].

---
[4] [8] introduce an alternative accuracy function which is calculated using a power function.

A cut-off parameter in the accuracy function, ε_0 (typically 0.01), is used to identify when the accuracy (κ) of a classifier is sufficiently close enough to 1.0 to be considered fully accurate (see Eq. 3). The conclusions in [2] failed to take into account the effect of this cut-off calculation of accuracy on the generalisation behaviour of XCS. Once the error in payoff prediction falls below $\varepsilon_0 R$ all predictions will be considered accurate. This is normally useful, filtering out noise in the payoff algorithm. However, in the early states the payoff prediction is sufficiently small that ε_0 is a large proportion of the prediction. This will allow classifiers which advocate sub-optimal actions with a payoff prediction that is variant from the payoff by less than $\varepsilon_0 R$ to be considered accurate. When the difference in stable payoff values for the same action in two neighbouring states falls below this threshold it is possible to identify a single classifier to cover that action in both states. This classifier will be considered accurate even though there is a fluctuation in the payoff to the classifier. XCS will use this false accuracy to proliferate classifiers that generalise over successive states, so producing accurate over-general classifiers. That this is the case can be seen within Fig. 2a. In this environment the optimal route alternates between action 0 and 1. A single classifier is covering action 0 and another single classifier covers action 1 in states s_0 to s_3 so that the optimal action is not selected in states s_1 and s_3.

Although this might suggest that a solution to the problem would be to reduce ε_0, this is problematic because the noise created whilst seeking to identify appropriate generalisations will also prevent the identification of accurate classifiers. An alternative is to change the value of γ to reduce the amount of discount and keep the difference in neighbouring payoffs above $\varepsilon_0 R$. This may increase the length of paths that can be learnt, but as the level of discount is reduced the difference between neighbouring prediction values will decrease. It has been noted that where prediction values are close and there is an area of the environment where over-generals are encouraged, it is possible for large numerosity over-general classifiers to develop [2]. These classifiers use their numerosity to dominate the action-sets, reducing the payoff of the action-sets to a value similar to their prediction, thereby making themselves more accurate and giving themselves more breeding opportunities. It is therefore hypothesised that:

Hypothesis *Reducing the discount level will increase the number of steps over which XCS can learn the optimal state × action × payoff mapping, but there will be a point beyond which further reduction will cause the mapping to be disrupted by over-general classifiers.*

3 Experimental Approach

[1] introduced a test environment designed for testing the ability to learn an optimal policy as the distance to reward increases, whilst controlling other potentially confounding variables. This environment, depicted in Fig. 1 has many useful properties for these experiments – see [1]. In the experiments that follow the environment used will be labelled 'FSW-N', where N is the length of the optimal path from the start to the reward state.

Unless stated otherwise, each experiment uses the following XCS parameters: $\beta = 0.2, \theta = 25, \varepsilon = 0.01, \alpha = 0.1, \chi = 0.8, \mu = 0.04, p_r = 0.5, P_\# = 0.33, p_i = 10.0, \varepsilon_i = 0.0, f_i = 0.01, m = 0.1, s = 20$ (see [26] for a parameter glossary). Each experiment was repeated 30 times and the results presented are the average of 30 runs unless otherwise stated.

4 Investigating Length Limits

It was argued in §2 that the reason for the failure to learn the optimal solution in FSW-10 and FSW-15 was due to the small difference in action-set payoff as a result of the high discount value (γ). To demonstrate that this is the case XCS was run within the FSW-15 environment, changing γ to 0.75, 0.80, 0.85, 0.90, 0.95 in each successive batch of 30 experiments. For these experiments the maximum population size (N) was 1200 and the message size was 7 bits (for comparability with [2]). For each exploitation iteration the System Relative Error [1] was calculated, in addition to the number of steps taken in the environment and the population size. The results of each batch of 30 runs were averaged and the averages compared. Figures 3 and 4 show the results and coverage chart at discount values 0.80 and 0.95.

From 0.75 through to 0.90 the maximum System Relative Error reduces and the number of steps taken to achieve the reward moves towards the optimal route, with $\gamma = 0.95$ allowing the optimal route to be reliably selected. The identification that XCS can learn the optimal route in the FSW-15 environment given an appropriate discount factor provides an initial verification of the hypothesis. However, it is useful to question why a discount of 0.8 or 0.85 was not effective. The answer is partially revealed by an examination of Table 1. The difference between the predictions in s_0 and s_1 for discount 0.8 is 11 and the difference in payoff between the optimal and sub-optimal route in s_0 is 8.8 – below the ε_0 error boundary and therefore a candidate for generalisation. This is reflected in the results for $\gamma = 0.8$ – the 'iterations' plot shows one incorrect decision is taken in each episode (see Fig. 3). At $\gamma = 0.85$ the difference between

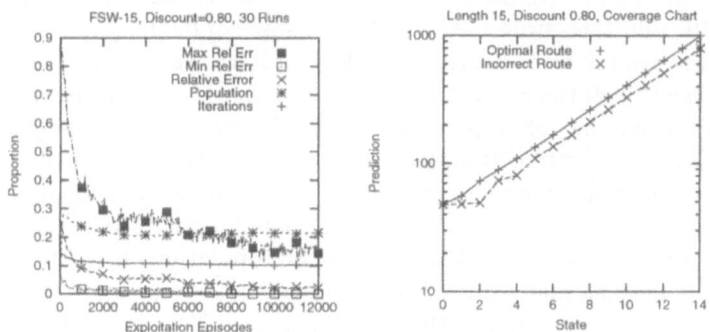

Fig. 3. Averaged results from 30 runs of XCS in FSW-15 at $\gamma = 0.80$

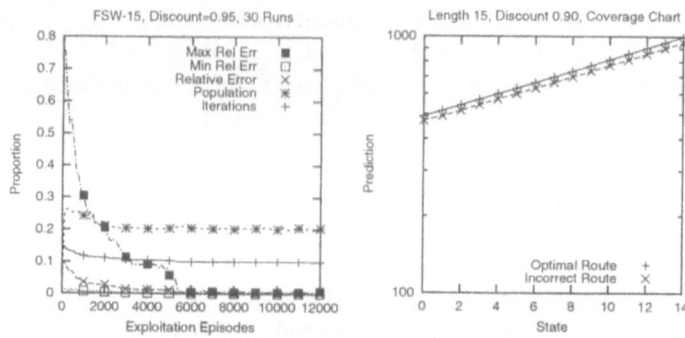

Fig. 4. Averaged results from 30 runs of XCS in FSW-15 at $\gamma = 0.95$

Table 1. Payout predictions in states within FSW-15 with different values of γ

γ	s_{0err}	s_0	s_1	s_2	s_3	s_4	s_5
0.71	5.87	8.27	11.65	16.41	23.11	32.55	45.85
0.75	13.36	17.82	23.76	31.68	42.24	56.31	75.08
0.8	35.18	43.98	54.98	68.72	85.90	107.37	134.21
0.85	87.35	102.77	120.91	142.24	167.34	196.87	231.62
0.9	205.89	228.77	254.19	282.43	313.81	348.68	387.42
0.95	463.29	487.68	513.34	540.36	568.80	598.74	630.25

the predicted payoff in s_0 and s_1 is increased. Unfortunately the coverage graph (not shown) indicates that difference of 15 between the optimal and non-optimal routes in s_0 is still sufficiently small to adversely influence the coverage of the sub-optimal route in s_0.

Now that it is clear that $\gamma = 0.95$ allows XCS to learn the optimal coverage of FSW-15, the second part of the hypothesis must be tested. This suggests that as the discount becomes small over-general classifiers will develop. It is worth noting that the proximity of the payoff values produced by $\gamma = 0.95$ were assumed by the author prior to the investigation to be sufficient to start to produce this generalisation. To investigate further, γ was systematically reduced to 0.99 in steps of 0.01, and then from 0.99 to 0.999 in steps of 0.001.

When the results were analysed, a clear pattern was evident. As γ was increased towards 0.99 the System Relative Error increased and more errors were evident in the number of steps taken to s_r. However, as it drew near to 0.99 and thereafter, the System Relative Error reduced even though the number of steps taken gradually moved towards that expected for a random selection of action in each state. An examination of the populations revealed that at 0.995 fully general classifiers dominated 7 of the 30 populations (see Fig. 5a) and at 0.999 all populations maintained high numerosity fully general classifiers. This is no surprise – the proximity of the predictions at 0.999 causes the difference in prediction to be well below ε_0 and the range of predictions is sufficiently small to allow a fully general classifier to easily gain sufficient numerosity to suppress the

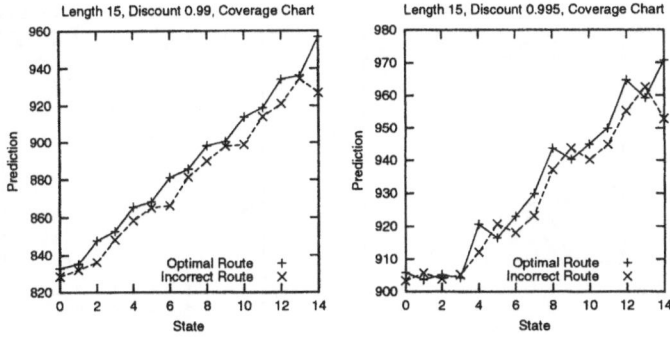

Fig. 5. Coverage charts for FSW-15 at $\gamma = 0.99$ and $\gamma = 0.995$

predictions and always be classed as accurate. What was surprising was the revelation that XCS was able to maintain such a reasonable coverage at a discount level of $\gamma = 0.99$.

Having demonstrated that XCS can perform optimally within FSW-15, it was appropriate to investigate how much the environment could be extended before XCS would once more perform sub-optimally. XCS was run with $\gamma = 0.95$ in FSW of length 20, 25, 30 and 40. If errors in coverage are guaranteed when the difference between payoff in s_n and s_{n+1} is 10.0 then a simple mathematical exercise would suggest that the practical limit is an environment of length 30. However it is now known that a difference of 15 was sufficient to prevent appropriate early state coverage, and this magnitude difference is seen between optimal and sub-optimal routes only 22 states from s_r. When the experiments were completed and the results analysed, is was found that XCS was able to find an optimal solution to FSW-20 at $\gamma = 0.95$. In FSW-25 the coverage chart remained almost optimal, and a careful analysis of each run showed that a single erroneous step was taken in 2.3% of the episodes after episode 3000. Within FSW-30 up to three additional steps were taken in each episode after episode 3000, and the coverage chart indicated generalisation over states up to s_6 (23 states from s_r).

These limitations present some difficulties. XCS has shown itself to be powerful in application to direct-reward problems, and yet apparently fundamentally limited in relatively small delayed-reward environments. However, it is important to understand that the problems arise not because of the basic mechanisms of XCS but due to the use of an *absolute* measure of error. As equation 4 indicates, error is computed relative to the reward range without taking account of the discount mechanism. A way to tackle this problem may be to identify an error measure that is independent of the position of the classifier in the action chain to the reward. A first attempt at such a solution was devised. This involved a simple modification to XCS to retain an estimate of the distance d from the *reward within each classifier*. This was used to calculate an action set estimate of distance to the nearest reward and to calculate the error as a proportion of

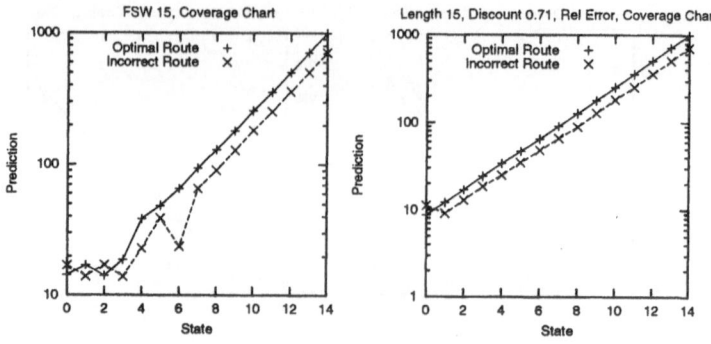

Fig. 6. Comparison of coverage charts for absolute error calculation and relative error calculation in FSW-15 at $\gamma = 0.71$

$\gamma^{N-d}R$. Although this technique resulted in a reduction in the System Relative Error, considerable error remained within the calculations, due to the magnification of errors in the estimate of d by the calculation. It is possible that reducing d to an integer value within this expression may mask these errors.

An alternative approach replaced the absolute calculation of error with:

$$\varepsilon = \begin{cases} \frac{|P-p|}{\max(P,p)} & (\max(P,p) > 0) \\ 0.0 & (\text{otherwise}) \end{cases} \qquad (6)$$

As the prediction becomes more accurate, the relative difference between P and p will reduce and so the error will become small independently of the magnitude of the payoff P. This, and an alternative error calculation: $\varepsilon = \frac{|P-p|}{p}$ (capped to 1.0 if $\varepsilon > 1.0$), have been the subject of recent investigations. Figure 6 shows that the use of the relative error update method reduces the over-generalisation in states above the state s_0. In FSW-15 at $\gamma = 0.71$ the two relative error expressions appear to produce similar results, although the second should provide less variance in the initial updates.

The results of applying a relative error calculation are encouraging, though further investigations are now required to identify any penalties that may be present in the formulation of the optimal sub-population of accurate classifiers. It would appear that the use of relative error makes the identification of accurate classifiers more problematic leading to greater divergence in the population, although this does not appear to dramatically affect the time taken for XCS to identify the optimal route. The use of a more focused GA selection technique, such as Tournament Selection, may resolve this problem.

5 Discussion

The limitations on the length of delayed reward environments are highly constraining. Delayed reward environments are commonly of much greater size within the laboratory, let alone within 'real-world' applications. In defence of XCS it should be noted that the corridor environment used within these tests is highly artificial. Richer environments may provide more distinct payoff gradients which will enable XCS to work over a larger range. It should also be noted that changes in parameters alongside modification to γ may affect the performance. For example, in experiments conducted alongside this investigation it was found that using the within-niche mutation scheme of [8] produced a much weaker performance because it encouraged an early decision on the most accurate generalisation and so aided the formation of over-general classifiers. Mutation schemes that encourage more diversity will act as a pressure against generalisation, and so enable XCS to map the environment using more specific classifiers. This is hardly a desirable solution, however.

The results are interesting in the light of [16]. This identifies that all non-trivial delayed reward environments are environments which encourage the development of over-general classifiers. Whilst XCS, as an accuracy-based LCS, has some protection against over-general classifiers, it is clear that the formation of over-generals will be encouraged as soon as there is a failure of the accuracy function to distinguish between payoff boundaries. The lack of provision for discount within the error calculation leads to an inequity of accuracy computation that encourages such a failure.

Whilst the results presented on the use of relative error are promising, more work is required in order to identify the dynamics of the calculated error in various parts of the action chain, and to identify the effect of the measure on the ability of XCS to satisfy the Optimality Hypothesis. The use of a relative measure of error should allow the length of paths that can be optimally mapped to be extended, but new limits must be established.

It is recognised that any discounted payoff scheme will cause hard limits in the length of path learning. Therefore work towards autonomous problem-space subdivision, the autonomous identification of sub-goals and their use in hierarchical planning remain important research objectives. Many other Reinforcement Learning methods face related problems in long path learning, and lessons can be drawn from these areas. However, the limits this paper has sought to address are those generated by the requirement to identify the minimum set of input generalisations that produce an accurate condition × action × payoff prediction mapping. The combination of accuracy-based learning and the requirement for the production of an optimally compact and accurate mapping is unique to XCS.

References

1. Alwyn Barry. *XCS Performance and Population Structure within Multiple-Step Environments.* PhD thesis, Queens University Belfast, 2000.
2. Alwyn M. Barry. The stability of long action chains in xcs. *Journal of Soft Computing*, 6(3-4):183-199, 2002.

3. Ester Bernadó, Xavier Llorà, and Josep M. Garrell. XCS and GALE: a Comparative Study of Two Learning Classifier Systems with Six Other Learning Algorithms on Classification Tasks. In *Proceedings of the 4th International Workshop on Learning Classifier Systems (IWLCS-2001)*, pages 337–341, 2001. Short version publishe in Genetic and Evolutionary Compution Conference (GECCO2001).
4. Larry Bull. On using ZCS in a Simulated Continuous Double-Auction Market. In Wolfgang Banzhaf, Jason Daida, Agoston E. Eiben, Max H. Garzon, Vasant Honavar, Mark Jakiela, and Robert E. Smith, editors, *Proceedings of the Genetic and Evolutionary Computation Conference (GECCO-99)*, pages 83–90. Morgan Kaufmann, 1999.
5. Larry Bull. On accuracy-based fitness. *Journal of Soft Computing*, 6(3–4):154–161, 2002.
6. Larry Bull, Jacob Hurst, and Andy Tomlinson. Mutation in Classifier System Controllers. In J.A. Meyer et al., editor, *From Animals to Animats 6: Proceedings of the Sixth International Conference on Simulation of Adaptive Behavior*, pages 460–467, 2000.
7. Martin V. Butz. An Implementation of the XCS classifier system in C. Technical Report 99021, The Illinois Genetic Algorithms Laboratory, 1999.
8. Martin V. Butz and Stewart W. Wilson. An Algorithmic Description of XCS. Technical Report 2000017, Illinois Genetic Algorithms Laboratory, 2000.
9. Dave Cliff and Susi Ross. Adding Temporary Memory to ZCS. *Adaptive Behavior*, 3(2):101–150, 1994. Also technical report: ftp://ftp.cogs.susx.ac.uk/pub/reports/csrp/csrp347.ps.Z.
10. M. Compiani, D. Montanari, R. Serra, and P. Simonini. Learning and Bucket Brigade Dynamics in Classifier Systems. *Special issue of Physica D (Vol. 42)*, 42:202–212, 1990.
11. Lawrence Davis, Chunsheng Fu, and Stewart W. Wilson. An incremental multiplexer problem and its uses in classifier system research. In Pier Luca Lanzi, Wolfgang Stolzmann, and Stewart W. Wilson, editors, *Advances in Learning Classifier Systems*, volume 2321 of *LNAI*, pages 23–31. Springer-Verlag, Berlin, 2002.
12. Marco Dorigo and Marco Colombetti. *Robot Shaping: An Experiment in Behavior Engineering*. MIT Press/Bradford Books, 1998.
13. John H. Holland, Keith J. Holyoak, Richard E. Nisbett, and P.R. Thagard. *Induction: Processes of Inference, Learning, and Discovery*. MIT Press, Cambridge, 1986.
14. John H. Holmes. A genetics-based machine learning approach to knowledge discovery in clinical data. *Journal of the American Medical Informatics Association Supplement*, 1996.
15. Tim Kovacs. Evolving Optimal Populations with XCS Classifier Systems. Master's thesis, School of Computer Science, University of Birmingham, Birmingham, U.K., 1996.
16. Tim Kovacs. *A Comparison and Strength and Accuracy-based Fitness in Learning Classifier Systems*. PhD thesis, University of Birmingham, 2002.
17. Pier Luca Lanzi. A Study of the Generalization Capabilities of XCS. In Thomas Bäck, editor, *Proceedings of the 7th International Conference on Genetic Algorithms (ICGA97)*, pages 418–425. Morgan Kaufmann, 1997.
18. Pier Luca Lanzi. Solving Problems in Partially Observable Environments with Classifier Systems (Experiments on Adding Memory to XCS). Technical Report 97.45, Politecnico di Milano. Department of Electronic Engineering and Information Sciences, 1997.

19. Pier Luca Lanzi. Generalization in Wilson's XCS. In A.E. Eiben, T.Bäck, M. Shoenauer, and H.-P. Schwefel, editors, *Proceedings of the Fifth International Conference on Parallel Problem Solving From Nature – PPSN V*, number 1498 in LNCS. Springer Verlag, 1998.
20. Pier Luca Lanzi. An Analysis of Generalization in the XCS Classifier System. *Evolutionary Computation*, 7(2):125–149, 1999.
21. Pier Luca Lanzi and Rick L. Riolo. A Roadmap to the Last Decade of Learning Classifier System Research (from 1989 to 1999). In Pier Luca Lanzi, Wolfgang Stolzmann, and Stewart W. Wilson, editors, *Learning Classifier Systems. From Foundations to Applications*, volume 1813 of *LNAI*, pages 33–62, Berlin, 2000. Springer-Verlag.
22. Alexandre Parodi and P. Bonelli. The Animat and the Physician. In J.A. Meyer and S.W. Wilson, editors, *From Animals to Animats 1. Proceedings of the First International Conference on Simulation of Adaptive Behavior (SAB90)*, pages 50–57. A Bradford Book. MIT Press, 1990.
23. Rick L. Riolo. The Emergence of Coupled Sequences of Classifiers. In J. David Schaffer, editor, *Proceedings of the 3rd International Conference on Genetic Algorithms (ICGA89)*, pages 256–264, George Mason University, June 1989. Morgan Kaufmann.
24. Robert E. Smith, B.A. Dike, R.K. Mehra, B. Ravichandran, and A. El-Fallah. Classifier Systems in Combat: Two-sided Learning of Maneuvers for Advanced Fighter Aircraft. In *Computer Methods in Applied Mechanics and Engineering*. Elsevier, 1999.
25. Stewart W. Wilson. ZCS: A zeroth level classifier system. *Evolutionary Computation*, 2(1):1–18, 1994.
26. Stewart W. Wilson. Classifier Fitness Based on Accuracy. *Evolutionary Computation*, 3(2):149–175, 1995.
27. Stewart W. Wilson. Generalization in evolutionary learning. Presented at the Fourth European Conference on Artificial Life (ECAL97), Brighton, UK, July 27-31., 1997.
28. Stewart W. Wilson. Generalization in the XCS classifier system. In John R. Koza, Wolfgang Banzhaf, Kumar Chellapilla, Kalyanmoy Deb, Marco Dorigo, David B. Fogel, Max H. Garzon, David E. Goldberg, Hitoshi Iba, and Rick Riolo, editors, *Genetic Programming 1998: Proceedings of the Third Annual Conference*, pages 665–674. Morgan Kaufmann, 1998.
29. Stewart W. Wilson. Mining Oblique Data with XCS. Technical Report 2000028, University of Illinois at Urbana-Champaign, 2000.

Bounding the Population Size in XCS to Ensure Reproductive Opportunities

Martin V. Butz and David E. Goldberg

Illinois Genetic Algorithms Laboratory (IlliGAL)
University of Illinois at Urbana-Champaign
104 S. Mathews, 61801 Urbana, IL, USA
{butz,deg}@illigal.ge.uiuc.edu

Abstract. Despite several recent successful comparisons and applications of the accuracy-based learning classifier system XCS, it is hardly understood how crucial parameters should be set in XCS nor how XCS can be expect to scale up in larger problems. Previous research identified a *covering challenge* in XCS that needs to be obeyed to ensure that the genetic learning process takes place. Furthermore, a *schema challenge* was identified that, once obeyed, ensures the existence of accurate classifiers. This paper departs from these challenges deriving a *reproductive opportunity bound*. The bound assures that more accurate classifiers get a chance for reproduction. The relation to the previous bounds as well as to the specificity pressure in XCS are discussed as well. The derived bound shows that XCS scales in a machine learning competitive way.

1 Introduction

The XCS classifier system has recently gained increasing attention. Especially in the realm of classification problems, XCS has been shown to solve many typical machine learning problems with a performance comparable to other traditional machine learning algorithms [15,1,7]

On the theory side, however, XCS is still rather poorly understood. There are few hints for parameter settings and scale-up analyses do not exist. Wilson [16,17] suggested that XCS scales polynomially in the number of concepts that need to be distinguished and in the problem length. However, this hypothesis has not been investigated to date.

This paper derives an important population size bound that needs to be obeyed to ensure learning. The derivation departs from the previously identified *covering challenge*, which requires that the genetic learning mechanism takes place, and the *schema challenge*, which requires that important (sufficiently specialized) classifiers are present in the initial population [3]. The two challenges were shown to result in specificity-dependent population size bounds.

In addition to these requirements, the population must be large enough to ensure that more accurate classifiers have reproductive opportunities. Following this intuition, we derive a *reproductive opportunity bound* that bounds the population size of XCS in $O(l^{k_d})$ where l denotes the problem length and k_d denotes the minimal schema order in the problem. Since k_d is usually small, Wilson's hypothesis holds that XCS scales polynomially in problem length l. In general, our analysis provides further insights on scale-up behavior, necessary parameter settings, and basic XCS functioning.

After a short overview of the XCS system, we review the previous identified problem bounds, identify a time dimension in the schema challenge, and then analyze the additional *reproductive opportunity bound*. Empirical confirmation of the derived bound is provided. Summary and conclusions complete the paper.

2 XCS in Brief

The XCS classifier system investigated herein is based on the work in [16,17, 10] and derived from the algorithmic description in [6]. This XCS overview provides only the details necessary for the rest of the paper. The interested reader is referred to the cited literature. For simplicity, we introduce XCS as a pure classifier. The results, however, should readily carry over to multi-step problems in which reward needs to be propagated.

We define a classification problem as a binary problem that provides problem instances $\sigma \in \{0,1\}^l$ (denoting problem length by l). After an instance is classified, the problem provides feedback in terms of scalar payoff $\rho \in \Re$ reflecting the quality of the classification. The task is to maximize this feedback effectively always choosing the correct classification.

XCS is basically a rule learning system. A population of rules, or *classifiers*, is evolved and adapted. Each rule consists of a condition part and a classification part. The condition part C specifies when the rule is active, or *matches*. It is coded by the ternary alphabet $\{0, 1, \#\}$ (i.e. $C \in \{0, 1, \#\}^l$) where a #-symbol matches both zero and one. The proportion of non-don't care symbols in the condition part determines the *specificity* of a classifier. In addition to condition and classification parts, classifiers specify a *reward prediction* p that estimates resulting payoff, *prediction error* ϵ that estimates the mean absolute deviation of p, and *fitness* F that measures the average relative scaled accuracy of p with respect to all competing classifiers. Further classifier parameters are *experience exp*, *time stamp ts*, *action set size estimate as*, and *numerosity num*.

XCS continuously evaluates and evolves its population. Rule parameters are updated by the Widrow-Hoff rule [14] and reinforcement learning techniques [13]. Effectively, the parameters approximate the average of all possible cases.

The population is evolved by the means of a covering mechanism and a genetic algorithm (GA). Initially, the population of XCS is empty. Given a problem instance, if there is no classifier that matches the instance for a particular classification, a covering classifier is generated that matches in that instance and specifies the missing classification. At each position of the condition, a #-symbol is induced with a probability of $P_\#$. A GA is applied if the average time in the action set [A] since the last GA application, recorded by the time stamp ts, is greater than the threshold θ_{GA}. If a GA is applied, two classifiers are selected in [A] for reproduction using fitness proportionate selection with respect to the fitness of the classifiers in [A]. The classifiers are reproduced and the children undergo mutation and crossover. In mutation, each attribute in C of each classifier is changed with a probability μ to either other value of the ternary alphabet (niche mutation is not considered herein). The action is mutated to any other possible action with a probability μ. For crossover, two-point crossover is applied with a probability χ. The parents stay in the population and compete with their offspring. The classifiers are inserted applying *subsumption deletion* in which classifiers are absorbed by experienced, more general, accurate classifiers.

If the number of (micro-)classifiers in the population exceeds the maximal population size N, excess classifiers are deleted. A classifier is chosen for deletion with roulette wheel selection proportional to its action set size estimate as. Further, if a classifier is sufficiently experienced ($exp > \theta_{del}$) and significantly less accurate than the average fitness in the population ($f < \delta * \sum_{cl \in [P]} f_{cl} / \sum_{cl \in [P]} num_{cl}$), the probability of being selected for deletion is further increased. Note that the GA is consequently divided into a reproduction process that takes place inside particular action sets and a deletion process that takes place in the whole population.

3 Specificity Change and Previous Problem Bounds

Previous investigations of XCS have proven the existence of intrinsic generalization in the system. Also, a covering challenge was identified that requires that the GA takes place as well as a schema challenge that requires that important classifiers are present in the initial population. This section gives a short overview of these bounds. Moreover, assuming that none of the important classifiers are present in the initial population, a time estimate for the generation of the necessary classifiers is derived.

3.1 Specificity Pressure

Since XCS reproduces classifiers in the action set but deletes classifiers from the whole population, there is an implicit generalization pressure inherent in XCS's evolutionary mechanism. The change in specificity due to reproduction can be derived from the expected specificity in an action set $s_{[A]}$ given current specificity in the population $s_{[P]}$ which is given by [2]:

$$s_{[A]} = \frac{s_{[P]}}{2 - s_{[P]}} \quad (1)$$

Additionally, mutation pushes towards an equal number of symbols in a classifier so that the expected specificity of the offspring s_o can be derived from parental specificity s_p [4]:

$$\Delta_m = s_o - s_p = 0.5\mu(2 - 3s_p) \quad (2)$$

Taken together, the evolving specificity in the whole population can be asserted considering the reproduction of two offspring in a reproductive event and the impact on the whole population [4] (ignoring any possible fitness influence):

$$s_{[P(t+1)]} = f_{ga}\frac{2(s_{[A]} + \Delta_m - s_{[P(t)]})}{N} \tag{3}$$

Parameter f_{ga} denotes the frequency the GA is applied on average which depends on the GA threshold θ_{ga} but also on the current specificity $s_{[P]}$ as well as the distribution of the encountered problem instances. While the specificity is initially determined by the don't care probability $P_\#$, the specificity changes over time as formulated by Equation 3. Thus, parameter $P_\#$ can influence XCS's behavior only early in a problem run.

From equation 3 we can derive the steady state value of the specificity $s_{[P]}$ as long as no fitness influence is present. Note that the speed of convergence is dependent on f_{ga} and N, the converged value is independent of these values.

$$s_{[A]} + \Delta_m - s_{[P]} = 0$$

$$\frac{s_{[P]}}{2 - s_{[P]}} + \frac{\mu}{2}\left(2 - 3\frac{s_{[P]}}{2 - s_{[P]}}\right) = s_{[P]} \tag{4}$$

Solving for $s_{[P]}$:

$$s_{[P]}^2 - (2.5\mu + 1)s_{[P]} + 2\mu = 0$$

$$s_{[P]} = \frac{1 + 2.5\mu - \sqrt{6.25\mu^2 - 3\mu + 1}}{2} \tag{5}$$

This derivation enables us to determine the converged specificity of a population given a mutation probability μ. Table 1 shows the resulting specificities in theory and empirically determined on a random Boolean function (a Boolean function that returns randomly either zero or one and thus either 1000 or zero reward). The empirical results are slightly higher due to higher noise in more specific classifiers, the biased accuracy function, the fitness biased deletion method, and the parameter initialization method [4]. Although mutation's implicit specialization pressure (as long as $s_{[P]} < 2/3$) can be used to control the specificity of the population, too high mutation most certainly disrupts useful problem substructures. Thus, in the future other ways might be worth exploring to control specificity in $[P]$.

Table 1. Converged Specificities in Theory and in Empirical Results

μ	0.02	0.04	0.06	0.08	0.10	0.12	0.14	0.16	0.18	0.20
$s_{[P]}$, theory	0.040	0.078	0.116	0.153	0.188	0.223	0.256	0.288	0.318	0.347
$s_{[P]}$, empirical	0.053	0.100	0.160	0.240	0.287	0.310	0.329	0.350	0.367	0.394

3.2 Covering and Schema Bound

Before specificity change and evolutionary pressure in general apply, though, it needs to be assured that the GA takes place in the first place. This is addressed by the *covering challenge* [3]. To meet the challenge, the probability of covering a given random input needs to be sufficiently high (assuming uniformly distributed #-symbols):

$$P(cover) = 1 - \left(1 - \left(\frac{2 - s_{[P]}}{2}\right)^l\right)^N > 0 \qquad (6)$$

Once the covering bound is obeyed, the GA takes place and specificity changes according to Equation 3. To evolve accurate classifiers, however, more accurate classifiers need to be present in a population. This leads to the *schema challenge* which determines the probability of the existence of a *schema representative* in a population with given specificity [3].

Order of Problem Difficulty. To understand the schema challenge, we use the traditional schema notation [9,8] of GAs to define a *schema representative*. Hereby, a schema can be characterized by its *order* k, which denotes the number of specified attributes, and its *defining length* δ, which is the distance between the outermost specified attributes. For example, schema **11***0*000* has order six and defining length nine whereas schema **1*0*0****** has order three and defining length four. As proposed in [3], a classifier is said to be a *schema representative* if its condition part has at least all k attributes of the corresponding order k schema specified and its action corresponds to the schema corresponding action. For example, a classifier with condition C=##1#0#0##000# would be a representative of the second schema but not of the first one. Consequently, the specificity of a representative is at least equal to the order k of the schema it represents. If the specificity of a representative cl exactly equals k, then it is said to be *maximally general* with respect to the schema.

A problem is now said to be of *order of problem difficulty* k_d if the smallest order schema that supplies information gain is of order k_d. Thus, if a problem is of order of problem difficulty k_d any schema of order less than k_d will have the same (low) accuracy and essentially the same accuracy as the completely general schema (order $k = 0$). Similarly, only classifiers that have at least k_d attributes specified can have higher accuracy, higher fitness, and can thus provide *fitness guidance*. Thus, a problem of order of difficulty k_d can only be solved by XCS if representatives of the order k_d schema exist (or are generated at random) in the population.

Given a particular problem, any schema can be assigned a probability of being correct. It has been shown in [3] that the more consistently a classifier is correct or incorrect, the more accurate the classifier will be. Thus, a problem provides fitness guidance from the over-general side, if classifiers that specify parts of the relevant attributes are more consistently correct/incorrect.

An extreme case of this is the *parity problem* in which all k relevant bits need to be specified to reach any fitness guidance. The specification of any subset of those k bits will result in a 50/50 chance for correct classification and consequently a base accuracy which is equal to the completely general classifier. Thus, in the parity problem there is no fitness guidance before reaching full accuracy (effectively going from zero to completely accuracy in one step) so that the order of problem difficulty in the parity problem is equal to the size of the parity.

Note that the order of problem difficulty measure also affects many other common machine learning techniques. For example, the inductive decision tree learner C4.5 [12] totally relies on the notion of information gain. If a problem has an order of difficulty $k_d > 1$, C4.5 would basically decide at random on which attribute to expand first. Consequently, C4.5 would generate an inappropriately large decision tree.

Schema Representative. With this notion we can formulate the probability that there exists an order k schema representative in a population that has specificity $s_{[P]}$ [3]:

$$P(representative\ in\ [P]) = 1 - \left(1 - \frac{1}{n}\left(\frac{s_{[P]}}{2}\right)^k\right)^N > 0 \quad (7)$$

where n denotes the number of possible outcome classes.

Similar to the probability of a representative in a population, the probability that a particular classifier in the population is a representative as well as the expected number of representatives in the population can be assessed:

$$P(representative) = \frac{1}{n}(s_{[P]})^k$$
$$E(representative) = N\frac{1}{n}(s_{[P]})^k \quad (8)$$

Requiring that at least one representative exists in a particular population with specificity $s_{[P]}$ we derive the following *representative bound* on the population size:

$$E(representative) > 1$$
$$N > \frac{n}{s_{[P]}^k} \quad (9)$$

While this bound is rather easy to obey given a reasonable high specificity, the more interesting result is the constraint on the specificity given an actual population size N:

$$s_{[P]} > \left(\frac{n}{N}\right)^{1/k} \quad (10)$$

This shows that with increasing population size, the required specificity decreases.

3.3 Extension in Time

As mentioned above, initially the population has a specificity of $s_{[P]} = 1 - P_\#$. From then on, as long as the population is not over-specific and the GA takes place, the specificity changes according to Equation 3. For example, say we set $P_\# = 1.0$ so that the initial specificity will be zero, the specificity will gradually increase towards the equilibrium specificity.

The time dimension is determined from the time it takes that a representative is created due to random mutations. Given a current specificity $s_{[P]}$, we can determine the probability that a representative of a schema with order k is generated in a GA invocation:

$$P(\text{generation of representative}) = (1 - \mu)^{s_{[P]}k} \cdot \mu^{(1-s_{[P]})k} \quad (11)$$

Out of this probability, we can determine the expected number of steps until at least one classifier may have the desired attributes specified. Since this is a geometric distribution,

$$E(t(\text{generation of representative})) =$$
$$1/P(\text{generation of representative}) =$$
$$\left(\frac{\mu}{1-\mu}\right)^{s_{[P]}k} \mu^{-k} \quad (12)$$

To derive a lower bound on the mutation probability, we can assume a current specificity of zero. The expected number of steps until the generation of a representative then equals to μ^{-k}. Thus, given we start with a completely general population, the expected time until the generation of a first representative is less than μ^{-k} (since $s_{[P]}$ increases over time). Requiring that the expected time until a classifier is generated is smaller than some threshold Θ, we can generate a lower bound on the mutation μ:

$$\mu^{-k} < \Theta$$
$$\mu > \Theta^{\frac{1}{-k}} \quad (13)$$

This representative bound actually relates to the existence of a representative bound determined above in equation 10. Setting Θ equal to N/n we get the same bound (s can be approximated by $a \cdot \mu$ where $a \approx 2$, see Table 1). Assuming that no representative exists in the population, the order of difficulty of a problem k_d influences speed of learning to take place due to possible delayed supply of representatives. In the extreme case of the parity function, the bound essentially bounds learning speed since a representative is automatically fully accurate.

4 Reproductive Opportunity Bound

Given the presence of representatives and GA application, another crucial requirement remains to be satisfied: The representative needs to get the chance to reproduce. Since GA selection for reproduction takes place in action sets, reproduction can only be assured if the representative will be part of an action set before it is deleted. To derive an approximate population size bound to ensure reproduction, we can require that the probability of deletion of a representative is smaller than the probability that the representative is part of an action set.

The probability of deletion is

$$P(deletion) = \frac{1}{N} \qquad (14)$$

assuming that there are no action set size estimate influences nor any fitness influences in the deletion process. This is a reasonable assumption given that the action set size estimate is inherited from the parents and fitness is initially decreased by 0.1.

Given a particular classifier cl that has specificity s_{cl}, we can determine the probability that cl will be part of an action set given random inputs.

$$P(\text{in } [A]) = \frac{1}{n} 0.5^{l \cdot s_{cl}} \qquad (15)$$

This probability is exact if binary input strings are encountered that are uniformly distributed over the whole problem instance space $\{0,1\}^l$.

Combining equations 14 and 15 we can now derive a constraint for successful evolution in XCS by requiring that the probability of deletion is smaller than the probability of reproduction. Since two classifiers are deleted at random from the population but classifiers are reproduced in only one action set (the current one), the probability of a reproductive opportunity needs to be larger than approximately twice the deletion probability:

$$P(\text{in}[A]) > 2P(deletion)$$
$$\frac{1}{n} 0.5^{l \cdot s_{cl}} > \frac{2}{N}$$
$$N > n 2^{l \cdot s_{cl}+1} \qquad (16)$$

This bounds the population size by $O(n2^{ls})$. Note that the specificity s depends on the order of problem difficulty k_d and the chosen population size N as expressed in Equation 10. Since k_d is usually small, s can often be approximated by $s = 1/l$ so that the bound diminishes. However, in problems in which no fitness guidance can be expected from the over-general side up to order $k_d > 1$, specificity s cannot be approximated as easily.

The expected specificity of such a representative of an order k schema can be estimated given a current specificity in the population $s_{[P]}$. Given that the classifier specifies all k positions its expected average specificity will be approximately:

$$E(s(\text{representative of schema of order } k)) = \frac{k + (l-k)s_{[P]}}{l} \qquad (17)$$

Substituting s_{cl} in equation 16 with the expected specificity of a representative of a schema of order $k = k_d$ necessary in a problem of order of difficulty k_d, the population size N can be bounded by

$$N > n2^{l \cdot \frac{k_d+(l-k_d)s_{[P]}}{l}+1}$$
$$N > n2^{k_d+(l-k_d)s_{[P]}+1} \tag{18}$$

This bound ensures that classifiers necessary in a problem of order of difficulty k_d will get reproductive opportunities. Once the bound is satisfied, existing representatives of the necessary order k_d schemata are ensured to reproduce and XCS is enabled to evolve a more accurate population.

Note that this population size bound is actually exponential in schema order k_d and in string length times specificity $l \cdot s_{[P]}$. This would mean that XCS would scale exponential in the problem length which is certainly highly undesirable. However, as seen above the necessary specificity actually decreases with increasing population sizes.

Considering this, we show in the following that out of the specificity constraint a general *reproductive opportunity bound* (ROP-bound) can be derived that shows that population size needs to grow in $O(l^{k_d})$. Equations 10 and 18 can both be denoted in O-notation:

$$s_{[P]} = O((\frac{n}{N})^{1/k})$$
$$N = O(2^{l s_{[P]}}) \tag{19}$$

Ignoring additional constants, we can derive the following population size bound which solely depends on problem length l and order of difficulty k_d.

$$N > 2^{l(\frac{n}{N})^{1/k_d}}$$
$$N(\log_2 N)^{k_d} > nl^{k_d} \tag{20}$$

This *reproductive opportunity bound* essentially shows that population size N needs to grow approximately exponential in the order of the minimal building block k_d (i.e., order of difficulty) and polynomial in the string length.

$$N = O(l^{k_d}) \tag{21}$$

Note that in the usual case k_d is rather small and can often be set to one (see Sect. 3.2). As shown in the extreme case of the parity problem, though, k_d can be larger than one.

5 Bound Verification

To verify the derived reproductive opportunity bound as well as the time dimension of the representative bound, we apply XCS to a Boolean function problem of order of difficulty k_d where k_d is larger than one. The hidden-parity problem,

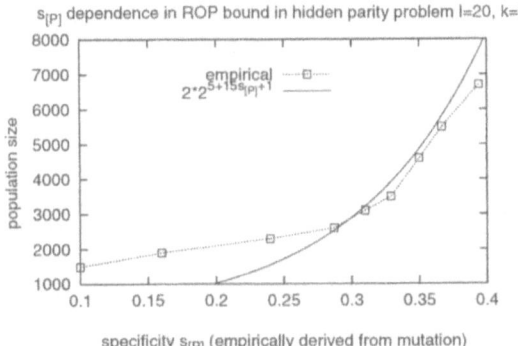

Fig. 1. Minimal population size depends on applied specificity in the population (manipulated by varying mutation rates). When mutation and thus specificity is sufficiently low, the bound becomes obsolete

originally investigated in [11], is very suitable to manipulate k_d. The basic problem is represented by a Boolean function in which k relevant bits determine the outcome. If the k bits have an even number of ones, the outcome will be one and zero otherwise. As discussed above, the order of difficulty k_d is equivalent to the number of relevant attributes k.

Figure 1 shows that the reproductive opportunity bound is well approximated by equation 16.[1] The experimental values are derived by determining the population size needed to reliably reach 100% performance. The average specificity in the population is derived from the empirical specificities in Table 1. XCS is assumed to reach 100% performance reliably if all 50 runs reached 100% performance after 200,000 steps. Although the bound corresponds to the empirical points when specificity is high, in the case of lower specificity the actual population size needed departs from the approximation. The reason for this is the time dimension of the representative bound determined above. The lower the specificity in the population, the longer it takes until a representative is actually generated (by random mutation). Thus, the representative bound extends in time rather than in population size.

To validate the $O(l^k)$ bound of equation 21, we ran further experiments in the hidden parity problem with $k = 4$, varying l and adapting mutation rate μ (and thus specificity) appropriately.[2] Results are shown in Fig. 2. The bound was derived from experimental results averaged over 50 runs determining the population size for which 98% performance is reached after 300000 steps.

---

[1] XCS with tournament selection was used [5]. If not stated differently, parameters were set as follows: $\beta = 0.2$, $\alpha = 1$, $\epsilon_0 = 1$, $\nu = 5$, $\theta_{ga} = 25$, $\tau = 0.4$, $\chi = 1.0$, uniform crossover, $\mu = 0.04$, $\theta_{del} = 20$, $\delta = 0.1$, $P_\# = 1.0$, $\theta_{sub} = 20$.
[2] Mutation was set as follows with respect to l: $l = 10, \mu = 0.10; l = 15, \mu = 0.075; l = 20, \mu = 0.070; l = 25, \mu = 0.065; l = 30, \mu = 0.060; l = 35, \mu = 0.057; l = 40, \mu = 0.055; l = 45, \mu = 0.050; l = 50, \mu = 0.050; l = 55, \mu = 0.048; l = 60, \mu = 0.048; l = 65, \mu = 0.048$.

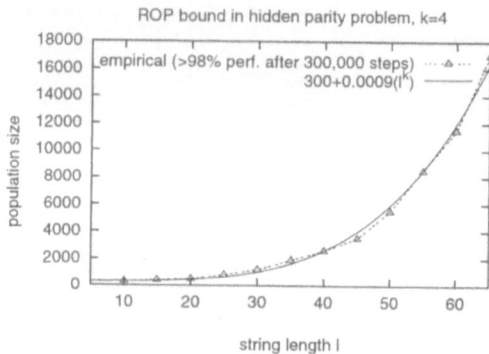

Fig. 2. The *reproductive opportunity bound* can be observed altering problem length l and using optimal mutation μ, given a problem of order of difficulty k_d. The experimental runs confirm that the evolutionary process indeed scales up in $O(l^k)$

With appropriate constants added, the experimentally derived bound is well-approximated by the theory.

6 Summary and Conclusions

In accord with Wilson's early scale-up hypothesis [16,17], this paper confirms that XCS scales-up polynomially in problem length l and exponentially in order of problem difficulty and thus in a machine learning competitive way. The empirical analysis in the hidden parity function confirmed the derived bounds. Combined with the first derived scale-up bound in XCS, the analysis provides several hints on mutation and population size settings. All bounds so far identified in XCS assure that the GA takes place (covering bound), that initially more accurate classifiers are supplied (schema bound—in population size and generation time), and that these more accurate classifiers are propagated (reproductive opportunity bound).

Many issues remain to be addressed. First, we assumed that crucial schema representatives provide fitness guidance towards accuracy. Type, strength, and reliability of this guidance require further investigation. Second, classifier parameter values are always only approximations of their assumed average values. Variance in these values needs to be accounted for. Third, although reproductive opportunity events are assured by our derived bound it needs to be assured that recombination effectively processes important substructures and mutation does not destroy these substructures. Forth, the number of to-be represented classifier niches (denoted by $[O]$ in [11]) needs to be taken into account in our model. Fifth, the distribution of classifier niches such as overlapping niches and obliqueness needs to be considered [15]. Integrating these points should allow us to derive a rigorous theory of XCS functioning and enable us to develop more competent and robust XCS learning systems.

Acknowledgment. We are grateful to Kumara Sastry and the whole IlliGAL lab for their help on the equations and for useful discussions. The work was sponsored by the Air Force Office of Scientific Research, Air Force Materiel Command, USAF, under grant F49620-00-0163. Research funding was also provided by a grant from the National Science Foundation under grant DMI-9908252. The US Government is authorized to reproduce and distribute reprints for Government purposes notwithstanding any copyright notation thereon. Additional support from the Automated Learning Group (ALG) at the National Center for Supercomputing Applications (NCSA) is acknowledged. Additional funding from the German research foundation (DFG) under grant DFG HO1301/4-3 is acknowledged. The views and conclusions contained herein are those of the authors and should not be interpreted as necessarily representing the official policies or endorsements, either expressed or implied, of the Air Force Office of Scientific Research, the National Science Foundation, or the U.S. Government.

References

1. Bernadó, E., Llorà, X., Garrell, J.M.: XCS and GALE: A comparative study of two learning classifier systems and six other learning algorithms on classification tasks. In Lanzi, P.L., Stolzmann, W., Wilson, S.W., eds.: Advances in Learning Classifier Systems: 4th International Workshop, IWLCS 2001. Springer-Verlag, Berlin Heidelberg (2002) 115–132
2. Butz, M.V., Kovacs, T., Lanzi, P.L., Wilson, S.W.: Theory of generalization and learning in XCS. IEEE Transaction on Evolutionary Computation (submitted, 2003) also available as IlliGAL technical report 2002011 at http://www-illigal.ge.uiuc.edu/techreps.php3.
3. Butz, M.V., Kovacs, T., Lanzi, P.L., Wilson, S.W.: How XCS evolves accurate classifiers. Proceedings of the Genetic and Evolutionary Computation Conference (GECCO 2001) (2001) 927–934
4. Butz, M.V., Pelikan, M.: Analyzing the evolutionary pressures in XCS. Proceedings of the Genetic and Evolutionary Computation Conference (GECCO 2001) (2001) 935–942
5. Butz, M.V., Sastry, K., Goldberg, D.G.: Tournament selection in XCS. Proceedings of the Genetic and Evolutionary Computation Conference (GECCO 2003) (2003) this volume.
6. Butz, M.V., Wilson, S.W.: An algorithmic description of XCS. In Lanzi, P.L., Stolzmann, W., Wilson, S.W., eds.: Advances in Learning Classifier Systems: Third International Workshop, IWLCS 2000. Springer-Verlag, Berlin Heidelberg (2001) 253–272
7. Dixon, P.W., Corne, D.W., Oates, M.J.: A preliminary investigation of modified XCS as a generic data mining tool. In Lanzi, P.L., Stolzmann, W., Wilson, S.W., eds.: Advances in Learning Classifier Systems: 4th International Workshop, IWLCS 2001. Springer-Verlag, Berlin Heidelberg (2002) 133–150
8. Goldberg, D.E.: Genetic Algorithms in Search, Optimization and Machine Learning. Addison-Wesley, Reading, MA (1989)
9. Holland, J.H.: Adaptation in natural and artificial systems. Universtiy of Michigan Press, Ann Arbor, MI (1975) second edition 1992.

10. Kovacs, T.: Deletion schemes for classifier systems. Proceedings of the Genetic and Evolutionary Computation Conference (GECCO-99) (1999) 329–336
11. Kovacs, T., Kerber, M.: What makes a problem hard for XCS? In Lanzi, P.L., Stolzmann, W., Wilson, S.W., eds.: Advances in Learning Classifier Systems: Third International Workshop, IWLCS 2000. Springer-Verlag, Berlin Heidelberg (2001) 80–99
12. Quinlan, J.R.: C4.5: Programs for Machine Learning. Morgan Kaufmann (1993)
13. Sutton, R.S., Barto, A.G.: Reinforcement learning: An introduction. MIT Press, Cambridge, MA (1998)
14. Widrow, B., Hoff, M.: Adaptive switching circuits. Western Electronic Show and Convention **4** (1960) 96–104
15. Wilson, S.W.: Mining oblique data with XCS. In Lanzi, P.L., Stolzmann, W., Wilson, S.W., eds.: Advances in Learning Classifier Systems: Third International Workshop, IWLCS 2000. Springer-Verlag, Berlin Heidelberg (2001) 158–174
16. Wilson, S.W.: Classifier fitness based on accuracy. Evolutionary Computation **3** (1995) 149–175
17. Wilson, S.W.: Generalization in the XCS classifier system. Genetic Programming 1998: Proceedings of the Third Annual Conference (1998) 665–674

Tournament Selection: Stable Fitness Pressure in XCS

Martin V. Butz, Kumara Sastry, and David E. Goldberg

Illinois Genetic Algorithms Laboratory (IlliGAL)
University of Illinois at Urbana-Champaign,
104 S. Mathews, 61801 Urbana, IL, USA
{butz,kumara,deg}@illigal.ge.uiuc.edu

Abstract. Although it is known from GA literature that proportionate selection is subject to many pitfalls, the LCS community somewhat adhered to proportionate selection. Also in the accuracy-based learning classifier system XCS, introduced by Wilson in 1995, proportionate selection is used. This paper identifies problem properties in which performance of proportionate selection is impaired. Consequently, tournament selection is introduced which makes XCS more parameter independent, noise independent, and more efficient in exploiting fitness guidance.

1 Introduction

Learning Classifier Systems (LCSs) [12,3] are rule learning systems in which rules are generated by the means of a genetic algorithm (GA) [11]. The GA evolves a population of rules, the so-called *classifiers*. A classifier usually consists of a condition and an action part. The condition part specifies when the classifier is applicable and the action part specifies which action to execute. In contrast to the original LCSs, the fitness in the XCS classifier system, introduced by Wilson [18], is based on the accuracy of reward predictions rather than on the reward predictions directly. Thus, XCS is meant to not only evolve a representation of an optimal behavioral strategy, or classification, but rather to evolve a representation of a complete *payoff map* of the problem. That is, XCS is designed to evolve a representation of the expected payoff in each possible situation-action combination. Recently, several studies were reported that show that XCS performs comparably to several other typical classification algorithms in different standard machine learning problems [20,2,8].

Although many indicators can be found in the GA literature that point out that proportionate selection is strongly fitness dependent [9] and moreover is strongly dependent on the degree of convergence [10], the LCS community has largely ignored this insight. In XCS, fitness is a scaled estimate of relative accuracy of a classifier. Due to the scaling, it does not seem necessary to apply a more fitness-independent selection method. Nonetheless, in this paper we identify problem types that impair the efficiency of proportionate selection. We show that tournament selection makes XCS selection more reliable, outperforming proportionate selection in all investigated problem types.

We restrict our analysis to Boolean classification, or one-step, problems in which each presented situation is independent of the history of situations and classifications. Particularly, we apply XCS to several typical Boolean function problems. In these problems, feedback is available immediately so that no reinforcement learning techniques are necessary that propagate reward [13].

The aim of this study is threefold. First, to make the XCS classifier system more robust and more problem independent. Second, to contribute to the understanding of the functioning of XCS in general. Third, to prepare the XCS classifier system to solve decomposable machine learning problems quickly, accurately, and reliably.

The next section gives a short overview of the major mechanisms in XCS. Next, the Boolean multiplexer problem is introduced and noise is added which causes XCS to fail. Next, we introduce tournament selection to XCS. Section 5 provides further experimental comparisons of the two mechanisms in several Boolean function problems with and without additional noise. Finally, we provide summary and conclusions.

2 XCS in a Nutshell

Although XCS was also successfully applied in multi-step problems [18,14,15,1], we restrict this study to classification problems to avoid the additional problem of reward propagation. However, the insights of this study should readily carry over to multi-step problems. This section consequently introduces XCS as a pure classification system providing the necessary details to comprehend the remainder of this work. For a more complete introduction to XCS the interested reader is referred to the original paper [18] and the algorithmic description [6].

We define a classification problem as a problem that consists of problem instances $s \in S$ that need to be classified by XCS with one of the possible classifications $a \in A$. The problem then provides scalar payoff $R \in \Re$ with respect to the made classification. The goal for XCS is to choose the classification that results in the highest payoff. To do that, XCS is designed to learn a complete mapping from any possible $s \times a$ combination to an accurate payoff value. To keep things simple, we investigate problems with Boolean input and classification, i.e. $S \subseteq \{0,1\}^L$ where L denotes the fixed length of the input string and $A = \{0,1\}$.

XCS evolves a population $[P]$ of rules, or *classifiers*. Each classifier in XCS consists of five main components. The condition $C \in \{0,1,\#\}^L$ specifies the subspace of the problem instances in which the classifier is applicable, or *matches*. The "don't care" symbol $\#$ matches in all input cases. The action part $A \in A$ specifies the advocated action, or classification. The payoff prediction p approaches the average payoff encountered after executing action A in situations in which condition C matches. The prediction error ε estimates the average deviation, or error, of the payoff prediction p. The fitness reflects the average relative accuracy of the classifier with respect to other overlapping classifiers.

XCS iteratively updates its knowledge base with respect to each problem instance. Given current input s, XCS forms a *match set* $[M]$ consisting of all classifiers in $[P]$ whose conditions match s. If an action is not represented in $[M]$, a covering classifier is created that matches s (#-symbols are inserted with a probability of $P_\#$ at each position). For each classification, XCS forms a *payoff prediction* $P(a)$, i.e. the fitness-weighted average of all reward prediction estimates of the classifiers in $[M]$ that advocate classification a. The payoff predictions determine the appropriate classification. After the classification is selected and sent to the problem, payoff R is provided according to which XCS updates all classifiers in the current action set $[A]$ which comprises all classifiers in $[M]$ that advocate the chosen classification a. After update and possible GA invocation, the next iteration starts.

Prediction and prediction error parameters are update in $[A]$ by $p \leftarrow p + \beta(R-p)$ and $\varepsilon \leftarrow \varepsilon + \beta(|R-p|-\varepsilon)$ where β ($\beta \in [0,1]$) denotes the *learning rate*. The fitness value of each classifier in $[A]$ is updated according to its current scaled relative accuracy κ':

$$\kappa = \begin{cases} 1 & \text{if } \varepsilon < \varepsilon_0 \\ \alpha \left(\frac{\varepsilon_0}{\varepsilon}\right)^\nu & \text{otherwise} \end{cases} \qquad \kappa' = \frac{\kappa}{\sum_{x \in [A]} \kappa_x} \qquad (1)$$

$$F \leftarrow F + \beta(\kappa' - F) \qquad (2)$$

The parameter ε_0 ($\varepsilon_0 > 0$) controls the tolerance for prediction error ε; parameters α ($\alpha \in (0,1)$) and ν ($\nu > 0$) are constants controlling the rate of decline in accuracy κ when ε_0 is exceeded. The accuracy values κ in the action set $[A]$ are then converted to set-relative accuracies κ'. Finally, classifier fitness F is updated towards the classifier's current set-relative accuracy. Figure 1 shows how κ is influenced by ε_0 and α. The determination of κ, then, also causes the scaling of the fitness function and thus strongly influences proportionate selection. All parameters except fitness F are updated using the *moyenne adaptive modifiée* technique [17]. This technique sets parameter values directly to the average of the so far encountered cases as long as the experience of a classifier is still less than $1/\beta$. Each time the parameters of a classifier are updated, the experience counter exp of the classifier is increased by one.

A GA is invoked in XCS if the average time since the last GA application upon the classifiers in $[A]$ exceeds threshold θ_{ga}. The GA selects two parental classifiers using proportionate selection (the probability of selecting classifier cl ($P_s(cl)$) is determined by its relative fitness in $[A]$, i.e. $P_s(cl) = F(cl)/\sum_{c \in [A]} F(c)$). Two offspring are generated reproducing the parents and applying (two-point) crossover and mutation. Parents stay in the population competing with their offspring. We apply free mutation in which each attribute of the offspring condition is mutated to the other two possibilities with equal probability. Parameters of the offspring are inherited from the parents, except for the experience counter exp which is set to one, the numerosity num which is set to one, and the fitness F which is multiplied by 0.1. In the insertion process, *subsumption deletion* may be applied [19] to stress generalization. Due to the possible strong effects of *action-set subsumption* we apply *GA subsumption* only.

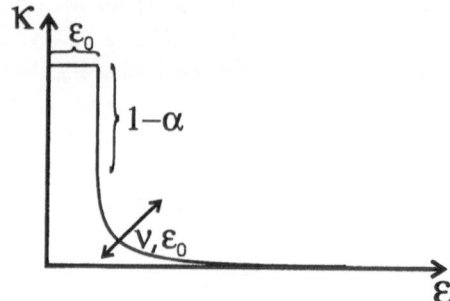

Fig. 1. The scaling of accuracy κ is crucial for successful proportionate selection. Parameters ε_0, α, and ν control tolerance, offset, and slope, respectively

The population of classifiers $[P]$ is of fixed size N. Excess classifiers are deleted from $[P]$ with probability proportional to an estimate of the size of the action sets that the classifiers occur in (stored in the additional parameter as). If the classifier is sufficiently experienced and its fitness F is significantly lower than the average fitness of classifiers in $[P]$, its deletion probability is further increased.

3 Trouble for Proportionate Selection

Proportionate selection depends on fitness relative to action set $[A]$. The fitness F of a classifier is strongly influenced by the accuracy determination of Equation 1 visualized in Fig. 1. F is also influenced by the number of classifiers in the action sets the classifier participates in. Moreover, the initial fitness of offspring is decreased by 0.1 so that it may take a while until the superiority of offspring is reflected by its fitness value. To infer these dependencies, we apply XCS to the Multiplexer problem with altered parameter settings or with additional noise in the payoff function of the problem.

The multiplexer problem is a widely studied problem in LCS research [7,18,19,4]. It has been shown that LCSs are superior compared to standard machine learning algorithms, such as $C4.5$, in the multiplexer task [7]. The multiplexer problem is a Boolean function. The function is defined for binary strings of length $k + 2^k$. The output of the multiplexer function is determined by the bit situated at the position referred to by the k position bits. For example, in the six multiplexer $f(100010) = 1$, $f(000111) = 0$, or $f(110101) = 1$. A correct classification results in a payoff of 1000 while an incorrect classification results in a payoff of 0.

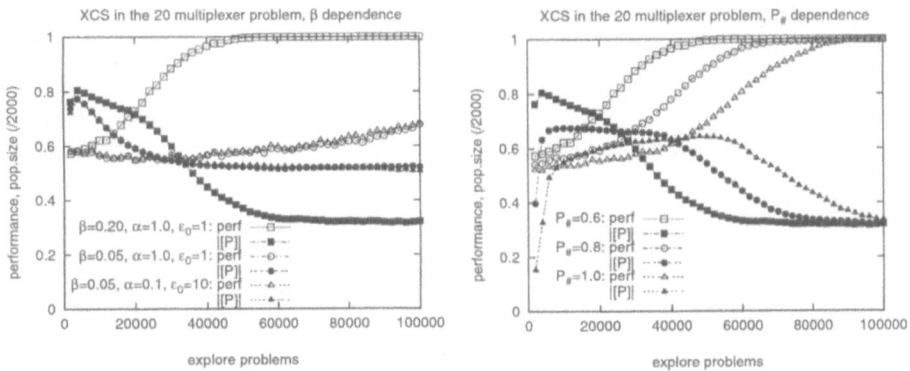

Fig. 2. Decreasing learning rate β decreases XCS learning performance (left-hand side). Accuracy function parameters have no immediate influence. Also a decrease in initial specificity strongly decreases XCS's performance (right-hand side).

Figure 2 reveals the strong dependence on parameter β.[1] Decreasing the learning rate hinders XCS from evolving an accurate payoff map. The problem is that over-general classifiers occupy a big part of the population initially. Better offspring often looses against the over-general parents since the fitness of the offspring only increases slowly (due to the low β value) and small differences in the fitness F only have small effects when using proportionate selection. Altering the slope of the accuracy curve by changing parameters α and ε_0 does not have any positive learning effect, either. Later, we show that small β values are necessary in some problems so that increasing β does not solve this problem in general. Figure 2 (right-hand side) also reveals XCS's dependence on initial specificity. Increasing $P_\#$ (effectively decreasing initial specificity) impairs learning speed of XCS facing the *schema challenge* [4].

Additional to the dependence on β we can show that XCS is often not able to solve noisy problems. We added two kinds of noise to the multiplexer problem: (1) Gaussian noise is added to the payoff provided by the environment. (2) The payoff is alternated with a certain probability, termed *alternating noise* in the remainder of this work. Figure 3 shows that when adding only a small amount of either noise, XCS's performance is strongly affected. The more noise is added, the smaller the fitness difference between accurate and inaccurate classifiers. Thus, selection pressure decreases due to proportionate selection so that the population starts to drift at random. Lanzi [15] proposed an extension to XCS that detects noise in environments and adjusts the error estimates accordingly. This approach, however, does not solve the β problem.

---

[1] All results herein are averaged over 50 experimental runs. Performance is assessed by test trials in which no learning takes place and the better classification is chosen as the classification. During learning, classifications are chosen at random. If not stated differently, parameters were set as follows: $N = 2000$, $\beta = 0.2$, $\alpha = 1$, $\varepsilon_0 = 1$, $\nu = 5$, $\theta_{GA} = 25$, $\chi = 0.8$, $\mu = 0.04$, $\theta_{del} = 20$, $\delta = 0.1$, $\theta_{sub} = 20$, and $P_\# = 0.6$.

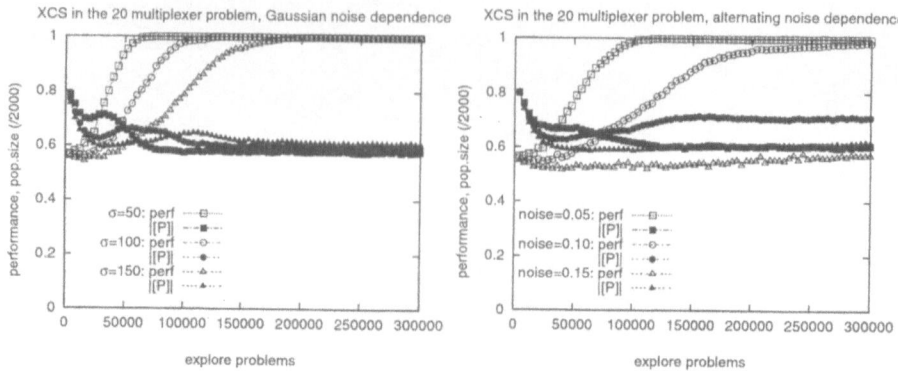

Fig. 3. Adding noise to the payoff function of the multiplexer significantly deteriorates performance of XCS. Gaussian noise (left-hand side) or alternating noise (right-hand side) is added

4 Stable Selection Pressure with Tournament Selection

By contrast to proportionate selection, tournament selection is independent of fitness scaling [9]. In tournament selection parental classifiers are not selected proportional to their fitness, but tournaments are held in which the classifier with the highest fitness wins (stochastic tournaments are not considered herein). Participants for the tournament are usually chosen at random from the population in which selection is applied. The size of the tournament controls the selection pressure. Usually, fixed tournament sizes are used.

Compared to standard GAs, the GA in XCS is a steady-state, niched GA. Only two classifiers are selected in each GA application. Moreover, selection is restricted to the current action set. Thus, some classifiers might not get any reproductive opportunity at all before being deleted from the population. Additionally, action set sizes can vary significantly. Initially, action sets are often over-populated with over-general classifiers. Thus, a relatively strong selection pressure appears to be necessary which adapts to the current action set size.

Our tournament selection process holds tournaments of sizes dependent on the current action set size $|[A]|$ by choosing a fraction τ of classifiers ($\tau \in (0,1]$) in the current action set. [2] Instead of proportionate selection, two independent tournaments are held in which the classifier with the highest fitness is selected. We also experimented with fixed tournament sizes, as shown below, which did not result in a stable selection pressure.

Figure 4 shows that XCS with tournament selection, referred to as *XCSTS* in the remainder of this work, can solve the 20-Multiplexer problem even with a low parameter value β. The curves show that XCSTS is more independent of parameter β. Even lower values of β, of course, ultimately also impair XCSTS's performance. In the case of a more general initial population ($P_\# = 1.0$) XCSTS

---

[2] If not stated differently, τ is set to 0.4 in the subsequent experimental runs.

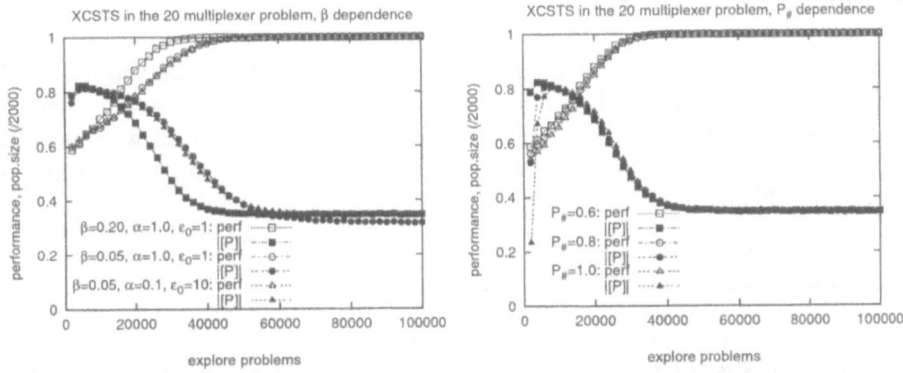

Fig. 4. XCSTS is barely influenced by a decrease of learning rate β. Accuracy parameters do not influence learning behavior. An initial over-general population ($P_\# = 1.0$) does hardly influence learning, either (right-hand side)

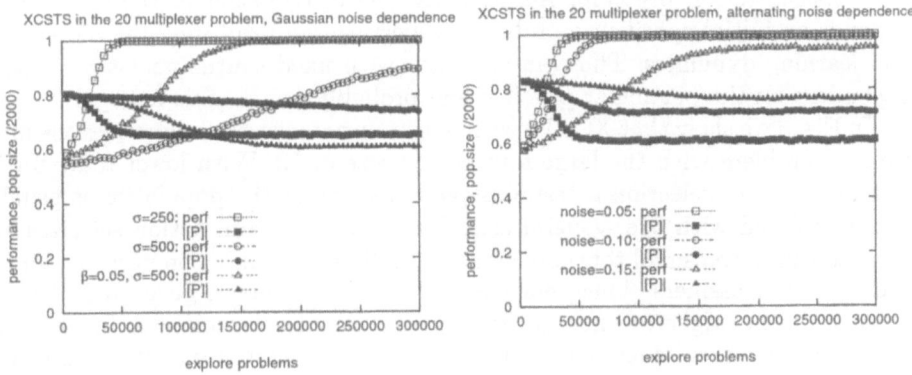

Fig. 5. XCSTS performs much better than XCS in noisy problems. For example, performance of XCSTS with Gaussian noise $\sigma = 250$ is better than XCS's performance with $\sigma = 50$ noise. Lowering β allows XCSTS to solve problems with even $\sigma = 500$

detects accurate classifiers faster and thus hardly suffers at all from the initial over-general population.

Figure 5 shows that XCSTS is much more robust in noisy problems as well. XCSTS solves Gaussian noise with a standard deviation of $\sigma = 250$ better than XCS the $\sigma = 100$ setting. In the case of alternating noise, XCSTS solves the $P_x = 0.1$ noise case faster than XCS the $P_x = 0.05$ case. As expected, the population sizes do not converge to the sizes achieved without noise since subsumption does not apply. Nonetheless, in both noise cases the population sizes decrease indicating the detection of accurate classifiers.

5 Experimental Study

This section investigates XCSTS's behavior further. We show the effect of a fixed tournament size and different proportional tournament sizes τ. We also apply XCS to the multiplexer problem with layered reward as well as to the *layered count ones problem*. In the latter problem, recombinatory events are highly beneficial.

5.1 Fixed Tournament Sizes

XCS's action sets vary in size and in distribution. Dependent on the initial specificity in the population (controlled by parameter $P_\#$), the average action set size is either large or small initially. It was shown that the average specificity in an action set is always smaller than the specificity in the whole population [5]. Replication in action sets and deletion from the whole population results in an implicit generalization pressure that can only be overcome by a sufficiently large specialization pressure. Additionally, the distribution of the specificities depends on initial specificity, problem properties, the resulting fitness pressure, and learning dynamics. Thus, an approach with fixed tournament size is quite dependent on the particular problem and probably not flexible enough.

In Fig. 6 we show that XCSTS with fixed tournament size only solves the multiplexer problem with the large tournament size of 12. With lower tournament sizes not enough selection pressure is generated. Since the population is usually over-populated with over-general classifiers early in a run, action set sizes are large so that a too small tournament size usually results in a competition among over-general classifiers. Thus, not enough fitness pressure is generated. Adding noise, an even larger tournament size is necessary for learning. A tournament size of 32, however, does not allow any useful recombinatory events anymore since the action set size itself is usually not much bigger than that. Thus, fixed tournament sizes are inappropriate for XCS's selection mechanism.

5.2 Layered Boolean Multiplexer

The layered multiplexer [18] provides useful *fitness guidance* for XCS [4]. The more position bits are specified (starting from the most significant one) the less different reward levels are available and the closer the values of the different reward levels are together so that classifiers that have more position bits specified have higher accuracy. Consequently, those classifiers get propagated. Thus, the reward scheme somewhat simplifies the problem.

The exact equation of the reward scheme is $Int(positionbits) * 200 + Int(referencedbit) * 100$. An additional reward of 300 is added if the classification was correct. For example, in the 6-multiplexer the (incorrect) classification 0 of instance (100010) would result in a payoff of 500 while the (correct) classification 1 would result in a payoff of 800.

We ran a series of experiments in the layered 20-Multiplexer increasing Gaussian noise. Figure 7 shows results for noise values $\sigma = 100$ and $\sigma = 300$. With a

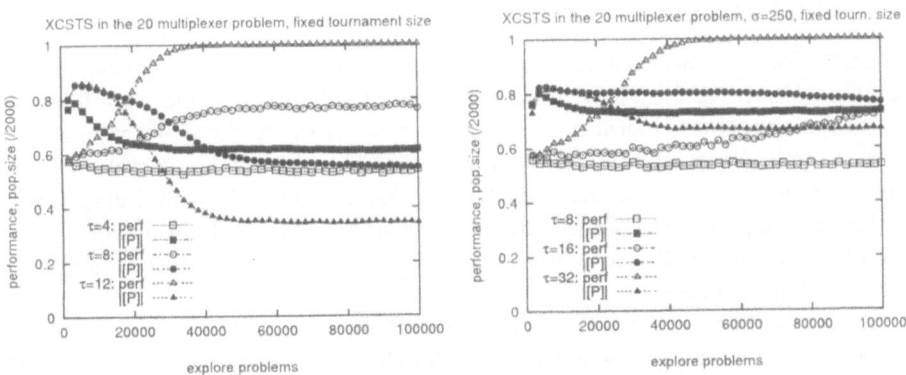

Fig. 6. Due to fluctuations in action set sizes and distributions as well as the higher proportion of more general classifiers in action sets, fixed tournament sizes are inappropriate for selection in XCSTS

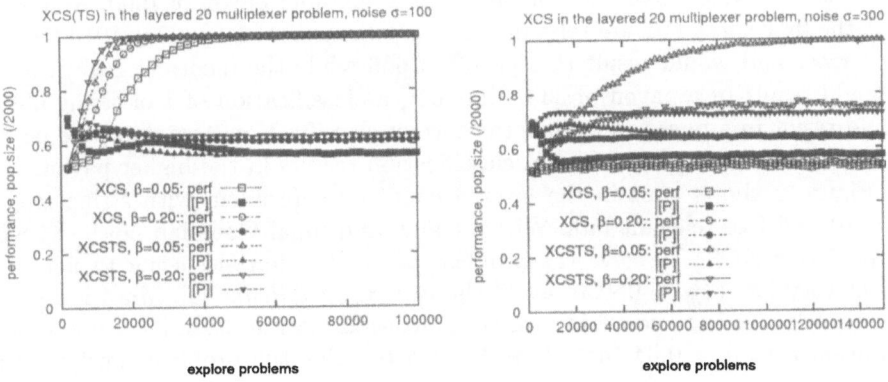

Fig. 7. Gaussian noise disrupts performance of XCS more than performance of XCSTS. While in a problem with little noise a higher value of parameter β allows faster learning, in settings with more noise high β values cause too high fluctuations and are thus disruptive. XCS fails to solve a noise level $\sigma = 300$ with either setting of parameter β while XCSTS still reaches 100% performance with parameter β set to 0.05

noise with standard deviation $\sigma = 100$, XCS as well as XCSTS have no problem to solve the task even with the lower learning rate $\beta = 0.05$ for XCS. XCSTS however already learns faster than XCS. With a noise of $\sigma = 300$ performance of XCS does hardly increase at all. XCSTS learns with either learning rate setting but reaches 100% knowledge only with a learning rate of $\beta = 0.05$. This shows that a lower learning rate is necessary in noisy problems since otherwise fluctuations in classifier parameter values can cause disruptions.

5.3 Layered Count Ones Problem

The final Boolean function in this study is constructed to reveal the benefit of recombination in XCS(TS). While in GA research whole conferences are filled with studies on crossover operators, linkage learning GAs, or the more recent probabilistic model building GAs (PMBGAs) [16], LCS research has largely neglected the investigation of the crossover operator. Herein, we take a first step towards understanding the possible usefulness of recombination in XCS by constructing a problem in which the recombination of low-fitness classifiers can increasingly result in higher fitness.

We term the problem the *layered count ones* problem. In this problem a subset of a Boolean string determines the class of the problem. If the number of ones in this subset is greater than half the size of the subset, then the class is one and otherwise zero. Additionally, we layer the payoff somewhat similarly to the layered multiplexer problem. The amount of payoff is dependent on the number of ones in the relevant bits and not the integer value of the relevant bits. For example, consider the layered count ones problem 5/3 (a binary string of length five with three relevant bits) and a maximal payoff of 1000. Assuming that the first three bits are relevant, the classification of 1 of string 10100 would be correct and would result in a payoff of 666 while the incorrect classification 0 would result in a payoff of 333. Similarly, a classification of 1 of string 00010 would result in a payoff of 0 while the correct classification 0 would give a payoff of 1000. Thus, always the correct classification results in the higher payoff.

Figure 8 shows runs in the layered count ones problem with string length $L = 70$ and five relevant bits. Without any additional Gaussian noise, XCSTS outperforms XCS. To solve the problem at all, the initial specificity needs to be set very low ($P_\# = 0.9$) to avoid the *covering challenge* [4]. Mutation needs to be set low enough to avoid over-specialization via mutation [5]. Runs with a mutation rate $\mu = 0.04$ (not shown) failed to solve the problem. Performance

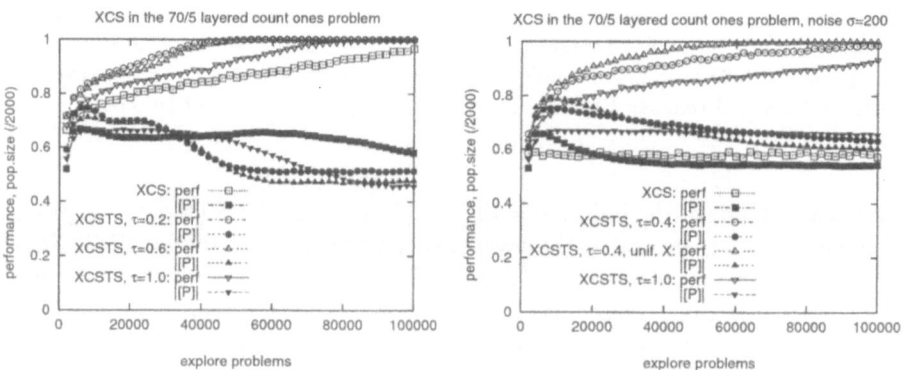

Fig. 8. Also in the layered count ones problem tournament selection has better performance. The benefit of crossover becomes clearly visible. Performance increases when uniform crossover is applied whereas it decreases when no mixing ($\tau = 1.0$) can occur

stalls at approximately 0.8 regardless of the selection type. Figure 8 (left-hand side) also shows that a higher tournament size proportion τ decreases the probability of recombining different classifiers. Thus, the evolutionary process cannot benefit from efficient recombination and learning speed decreases. Note however that proportionate selection is still worse than the selection type that always chooses the classifier with the highest fitness ($\tau = 1.0$). This indicates that recombination, albeit helpful, does not seem to be mandatory. The more important criterion for a reliable convergence is a proper selection pressure.

Adding Gaussian noise of $\sigma = 250$ to the layered count ones problem again completely deteriorates XCS's performance (Fig. 8, right-hand side). Noise hinders mutation even more from doing the trick. After 100,000 learning steps only selection with proper recombination was able to allow the evolution of perfect performance (Fig. 8). Perpetual selection of the best classifier hinders crossover from recombining classifiers that are partially specified in the relevant bits so that the evolutionary process is significantly delayed.

6 Summary and Conclusions

This paper showed that proportionate selection can prevent proper fitness pressure and thus successful learning in XCS. Applying tournament selection selection results in better (1) parameter independence, (2) noise robustness, and (3) recombinatory efficiency. No problem was found in which XCS with tournament selection with an action-set proportionate tournament size performed worse than proportionate selection. Thus, it is clear that future research in XCS should start using tournament selection instead of proportionate selection.

Despite the recent first insights in problem difficulty it remains unclear which problems are really hard for XCS. Moreover, it remains to be shown in which machine learning problems crossover is actually useful or even mandatory for successful learning in XCS and LCSs in general. In XCS the recombination of important substructures should lead to successively higher accuracy. In strength-based LCSs, on the other hand, the recombination of substructures should lead to larger payoff. Due to the apparent importance of recombinatory events in nature as well as in GAs, these issues deserve further detailed investigations in the realm of LCSs.

Acknowledgment. We are grateful to Xavier Llora, Martin Pelikan, Abhishek Sinha, and the whole IlliGAL lab. The work was sponsored by the Air Force Office of Scientific Research, Air Force Materiel Command, USAF, under grant F49620-00-0163. Research funding was also provided by a grant from the National Science Foundation under grant DMI-9908252. The US Government is authorized to reproduce and distribute reprints for Government purposes notwithstanding any copyright notation thereon. Additional support from the Automated Learning Group (ALG) at the National Center for Supercomputing Applications (NCSA) is acknowledged. Additional funding from the German research foundation (DFG) under grant DFG HO1301/4-3 is acknowledged. The views and conclusions contained herein are those of the authors and should not be interpreted as necessarily representing the official policies or endorsements, either expressed or implied, of the Air Force Office of Scientific Research, the National Science Foundation, or the U.S. Government.

References

1. Barry, A.: A hierarchical XCS for long path environments. Proceedings of the Genetic and Evolutionary Computation Conference (GECCO 2001) (2001) 913–920
2. Bernadó, E., Llorà, X., Garrell, J.M.: XCS and GALE: A comparative study of two learning classifier systems and six other learning algorithms on classification tasks. In Lanzi, P.L., Stolzmann, W., Wilson, S.W., eds.: Advances in Learning Classifier Systems: 4th International Workshop, IWLCS 2001. Springer-Verlag, Berlin Heidelberg (2002) 115–132
3. Booker, L.B., Goldberg, D.E., Holland, J.H.: Classifier systems and genetic algorithms. Artificial Intelligence **40** (1989) 235–282
4. Butz, M.V., Kovacs, T., Lanzi, P.L., Wilson, S.W.: How XCS evolves accurate classifiers. Proceedings of the Genetic and Evolutionary Computation Conference (GECCO 2001) (2001) 927–934
5. Butz, M.V., Pelikan, M.: Analyzing the evolutionary pressures in XCS. Proceedings of the Genetic and Evolutionary Computation Conference (GECCO 2001) (2001) 935–942
6. Butz, M.V., Wilson, S.W.: An algorithmic description of XCS. In Lanzi, P.L., Stolzmann, W., Wilson, S.W., eds.: Advances in Learning Classifier Systems: Third International Workshop, IWLCS 2000. Springer-Verlag, Berlin Heidelberg (2001) 253–272
7. De Jong, K.A., Spears, W.M.: Learning concept classification rules using genetic algorithms. IJCAI-91 Proceedings of the Twelfth International Conference on Artificial Intelligence (1991) 651–656
8. Dixon, P.W., Corne, D.W., Oates, M.J.: A preliminary investigation of modified XCS as a generic data mining tool. In Lanzi, P.L., Stolzmann, W., Wilson, S.W., eds.: Advances in Learning Classifier Systems: 4th International Workshop, IWLCS 2001. Springer-Verlag, Berlin Heidelberg (2002) 133–150
9. Goldberg, D.E., Deb, K.: A comparative analysis of selection schemes used in genetic algorithms. Foundations of Genetic Algorithms (1991) 69–93
10. Goldberg, D.E., Sastry, K.: A practical schema theorem for genetic algorithm design and tuning. Proceedings of the Genetic and Evolutionary Computation Conference (GECCO 2001) (2001) 328–335
11. Holland, J.H.: Adaptation in natural and artificial systems. Universtiy of Michigan Press, Ann Arbor, MI (1975) second edition 1992.
12. Holland, J.H.: Adaptation. In Rosen, R., Snell, F., eds.: Progress in Theoretical Biology. Volume 4. Academic Press, New York (1976) 263–293
13. Kaelbling, L.P., Littman, M.L., Moore, A.W.: Reinforcement learning: A survey. Journal of Artificial Intelligence Research **4** (1996) 237–258
14. Lanzi, P.L.: An analysis of generalization in the XCS classifier system. Evolutionary Computation **7** (1999) 125–149
15. Lanzi, P.L., Colombetti, M.: An extension to the XCS classifier system for stochastic environments. Proceedings of the Genetic and Evolutionary Computation Conference (GECCO-99) (1999) 353–360

16. Pelikan, M., Goldberg, D.E., Lobo, F.: A survey of optimization by building and using probabilistic models. Computational Optimization and Applications **21** (2002) 5–20
17. Venturini, G.: Adaptation in dynamic environments through a minimal probability of exploration. From Animals to Animats 3: Proceedings of the Third International Conference on Simulation of Adaptive Behavior (1994) 371–381
18. Wilson, S.W.: Classifier fitness based on accuracy. Evolutionary Computation **3** (1995) 149–175
19. Wilson, S.W.: Generalization in the XCS classifier system. Genetic Programming 1998: Proceedings of the Third Annual Conference (1998) 665–674
20. Wilson, S.W.: Get real! XCS with continuous-valued inputs. In Lanzi, P.L., Stolzmann, W., Wilson, S.W., eds.: Learning Classifier Systems: From Foundations to Applications. Springer-Verlag, Berlin Heidelberg (2000) 209–219

Improving Performance in Size-Constrained Extended Classifier Systems

Devon Dawson

Hewlett-Packard, 8000 Foothills Blvd, Roseville CA 95747-5557, USA
devon.dawson@hp.com

Abstract. Extended Classifier Systems, or XCS, have been shown to be successful at developing accurate, complete and compact mappings of a problem's payoff landscape. However, the experimental results presented in the literature frequently utilize population sizes significantly larger than the size of the search space. This resource requirement may limit the range of problem/hardware combinations to which XCS can be applied. In this paper two sets of modifications are presented that are shown to improve performance in small size-constrained 6-Multiplexer and Woods-2 problems.

1 Introduction

An extended classifier system, or XCS, is a rule-based machine learning framework developed by Stewart Wilson [3,15,16] that has been shown to posses some very appealing properties. Chiefly, XCS has been found to develop accurate, complete and compact mappings of a problem's search space [6,10,15,16]. However, to be successful XCS can require a population size large enough to allow the system to maintain, at least initially, a number of unique classifiers (i.e., macroclassifiers) constituting a significant portion of the size of the search space. This potentially prohibitive resource requirement can prevent XCS from being successful where the system is *size-constrained*, that is, where the population size is smaller than the search space.

As XCS-based systems find application in industry (and particularly in cost-constrained commercial applications), size-constrained systems are likely to be frequently encountered. In an effort to broaden the range of situations in which XCS-based systems can be utilized, this paper presents two sets of modifications that have been found to significantly improve system performance in small size-constrained problem/population configurations.

Improving upon techniques employed by Dawson and Juliano in [5], the first set of mostly minor modifications attempt to make more cautious use of the limited resources available to a size-constrained system. The second modified system builds upon the first and, among other changes, uses an effectiveness-based fitness measure (i.e., a strength/accuracy hybrid) - thus resulting in a system that develops a partial map of the payoff landscape. Experimental results are reported that demonstrate the performance improvement observed when using the proposed modifications in size-constrained 6-Multiplexer and Woods-2 problems.

2 XCS: Developing Accurate, Complete, and Compact Populations

Building upon his strength-based ZCS [14], Wilson introduced XCS in [15] as a solution to many of the problems encountered in traditional strength-based classifier systems. Namely, the problems of overgeneral classifiers, greedy classifier selection, and the interaction of the two [4,8,9,10,15]. Through the use of a simplified architecture, an accuracy-based fitness measure, a Q-Learning [12] like credit distribution algorithm and a niche Genetic Algorithm (GA) [1] – XCS is motivated towards the development of accurate, complete and compact maps of a problem's search space.

While earlier work had utilized various forms of accuracy as a factor in determining a classifier's fitness, Wilson demonstrated with XCS that the use of accuracy alone could lead the GA to develop a stable population of classifiers that accurately predict the reward anticipated as a result of their activation.

The completeness and compactness of the solution developed by XCS stems from its use of a niche GA as well as the accuracy-based fitness measure. The niche GA, introduced by Booker [1], considers classifiers to be members of specific, functionally related subsets of the population and operates upon these subsets, or niches, rather than on the population as a whole. By calculating a classifier's fitness relative to the members of the niches in which it resides and searching within these niches, the GA drives the system towards the development of the minimal number of accurate classifiers required to completely cover the search space.

3 Size-Constrained Extended Classifier Systems

Although the populations ultimately developed by XCS tend to be compact, and therefore significantly smaller than the search space (to a degree dependent upon the amount of generalization possible in the problem environment), the early stages of a run can require that the population contain a large number of classifiers, often constituting a significant percentage of the problem's search space [6,10,11,15,16]. Typically this is not a problem as the population size is set large enough to more than encompass the problem's search space. However, for practical application this may not always be possible as the size of the search space can simply exceed the physical limitations of the hardware on which the system is to execute, or the population size may be limited by the need to share system resources with other processes.

Lanzi identified in [11] that the tendency to generalize inherent in XCS may impair the system's ability to succeed in environments which afford little opportunity for generalization, and that this effect could be diminished through the use of a new *Specify* operator implemented in his XCSS. The Specify operator attempts to identify overly general classifiers and produces specialized offspring from those classifiers with some number of their don't-care bits (i.e., #'s) replaced by the corresponding bits in the current system input. Lanzi showed that as the population size was reduced, the performance of XCS and XCSS diminished in the Maze4 problem [11].

While Lanzi's work was focused on improving the performance of XCS in problems allowing little generalization, this work aims to improve performance in size-constrained problem/population combinations. For the purposes of this paper we define a size-constrained system to be one in which the population size, N, is smaller than the size of the search space, S. In order to quantify the amount of size-constraint in a system (as required by some of the modifications proposed in this work), a new size-constraint measure, Ω, is introduced and determined by:

$$\Omega = 1 + \ln((S/N)^c) . \qquad (1)$$

For all experiments reported in this work the value of the power parameter, c, was set to c=5. For configurations where $N \geq S$, the size-constraint measure is set to 1. Note that this function was selected based simply on the fact that it resulted in values that appeared reasonable for the problems and population sizes evaluated in this work. A self tuning constraint measure (e.g., one taking into account such factors as the stability of the population) would seem to be in order.

4 Modified Extended Classifier Systems

This section presents the first set of modifications that are intended to improve the performance of size-constrained systems while still allowing the development of a complete map. For the sake of comparison the system utilizing these modifications is termed the Modified Extended Classifier System, or MXCS.

4.1 Experience-Based GA Trigger Algorithm

In XCS, the GA trigger algorithm attempts to evenly distribute GA activations across all encountered match/action sets regardless of the frequency with which those input states are encountered. Evenly distributing GA activations is achieved by triggering the GA when the average amount of time (i.e., exploration episodes) since the last GA activation on the set exceeds the given threshold value (θ_{ga}).

While the overall impact of this triggering algorithm may be beneficial, its experience-independent nature becomes problematic in size-constrained systems where the brittle nature of the smaller populations can be easily disturbed by "over-clocking" the GA, that is, producing classifiers faster than they can be evaluated by the system. Though the GA trigger threshold can be increased to overcome this, it becomes difficult to determine the appropriate value due to its experience-independent nature and the increase may result in a significantly diminished learning rate.

MXCS modifies the trigger algorithm to use a simple experience-based threshold, where the GA is triggered when the given set contains any number of classifiers with experience since the GA last operated on it greater than the trigger threshold (θ_{ga}). While this experience-based algorithm appears to favor classifiers covering frequently encountered inputs, the favoritism is ameliorated by the following modifications and dissipates once a maximally-general classifier is discovered covering that niche, as the offspring will tend to be subsumed [16] into the generalist.

4.2 Focused Deletion-Selection

In XCS a classifier is selected for deletion from the population when the addition of a new classifier causes the population size to exceed the maximum size parameter (N).

The deletion-selection process begins by assigning each classifier a deletion-vote, which is the product of the classifier's numerosity (num_{cl}) and the learned action set size estimate (as_{cl}). The classifier's numerosity indicates the number of identical classifiers (i.e., *microclassifiers*) in the population represented by the given *macroclassifier*. The learned action set size estimate represents an average of the number of classifiers present in the action sets to which the classifier belongs. If the classifier's experience (exp_{cl}) is greater than the deletion threshold (θ_{del}) and its fitness is less than a constant fraction (δ) of the mean fitness of the population, then the deletion vote is scaled by the difference between the classifier's fitness and the mean population fitness. Therefore, this algorithm calculates the deletion vote as follows [3,7,15]:

$$vote_{cl} = num_{cl} * as_{cl} . \qquad (2)$$

$$\text{IF } ((exp_{cl} > \theta_{del}) \text{ AND } (f_{cl} / num_{cl} < \delta * (\Sigma f_{pop} / \Sigma num_{pop}))) \qquad (3)$$
$$vote_{cl} = vote_{cl} * (\Sigma f_{pop} / \Sigma num_{pop}) / (f_{cl} / num_{cl}) .$$

Once the deletion vote of each classifier is determined, the XCS deletion-selection algorithm typically uses *roulette-selection*, where a classifier is probabilistically selected from the population based on each classifier's deletion vote. However, in size-constrained systems roulette-selection can negatively impact system performance as its probabilistic nature often results in the deletion of effective and useful classifiers.

MXCS modifies the deletion-selection method to deterministically select the classifier with the highest deletion vote (i.e., a classifier is selected at random from the set of classifiers with the highest deletion vote). When using this *max-deletion* technique the deletion process maintains a strong downward pressure on the numerosity of classifiers, due to the deletion-vote factors of numerosity and action set size estimate.

Though not utilized in [5], the MXCS described in this work further increases the pressure on classifier numerosities as the deletion vote is modified by raising the numerosity factor by a power equivalent to the constraint measure, Ω, as follow:

$$vote_{cl} = num_{cl}^{\Omega} * as_{cl} . \qquad (4)$$

4.3 Focused GA Selection

In XCS the GA probabilistically selects parents for use in the generation of new offspring from the current set (action or match depending on placement of the GA) with the selection probability based on the value of each classifier's fitness. MXCS intensifies the focus on fit classifiers by basing the selection probability on the fitness of each classifier raised to the power of the constraint measure, Ω, as follows:

$$prob(cl) = f_{cl}^{\Omega} / \Sigma f_{[A]}^{\Omega} . \qquad (5)$$

4.4 Parameter Updates

In an effort to avoid favoring classifiers frequently chosen for activation, XCS limits the updating of classifier parameters (such as experience, error, fitness and predicted payoff), to exploration steps only. MXCS is modified to update a classifier's prediction and error estimates on both exploration and exploitation steps. Though not required for the system to succeed, this is done in order to take full advantage of all the information returned from the environment. It should be noted that a bias is introduced in that accurate classifiers chosen on exploitation steps will have more chances to refine their predictions and will tend to have more refined accuracy estimates.

5 Effectiveness-Based Extended Classifier System

This section builds upon MXCS and introduces a new effectiveness-based classifier system (EXCS) that further reduces the resource requirements of the system by focusing its resources on accurate and rewarding classifiers. This results in a far more compact *partial map* as the system discards classifiers with low payoff predictions.

Though the map developed by EXCS is an incomplete one, the use of a niche GA and the accuracy factor in the effectiveness measure allow the system to avoid the problems previously mentioned for purely strength-based systems [4,8,9,10,15].

5.1 Accuracy Function

In developing EXCS it was found that the standard accuracy function used in XCS was too sensitive to fluctuations in prediction error and too insensitive to the classifier's long term accuracy history to be used in determining the effectiveness of classifiers in highly competitive populations. This is particularly problematic in systems employing max-deletion due to the less forgiving nature of the algorithm.

The accuracy function used in EXCS uses two factors to calculate a classifier's accuracy. First a weighted average of the classifier's long-term accuracy is calculated. As shown in equations 6 and 7, this is accomplished by taking the ratio of the sum of the classifier's accuracy over time (k_sum_{cl}) and its experience (k_exp_{cl}) both discounted at each update by a constant fraction, ψ. Owing to the power series that develops with each update, the sensitivity of the function is determined by the value of $(1-\psi)^{-1}$ to which the series (i.e., $\Sigma\psi^n$) converges. Therefore, the weight attributed to an accuracy recorded n steps ago is ψ^n. A value of $\psi=0.999$ was used in all experiments reported in this paper.

Lastly, this new weighted accuracy average and the standard classifier accuracy measure are then averaged as shown in equation 8.

$$k\_sum_{cl} = (k\_sum_{cl} * \psi) + k_{cl}. \tag{6}$$

$$k\_exp_{cl} = (k\_exp_{cl} * \psi) + 1. \tag{7}$$

$$k\_ave_{cl} = ((k\_sum_{cl} / k\_exp_{cl}) + k_{cl})/2. \tag{8}$$

Note that this accuracy measure was not found to improve the performance of XCS and was therefore omitted from the MXCS implementation. It may be that accuracy-based fitness functions require more drastic near-term changes in accuracy to be effective.

5.2 Effectiveness-Based Fitness Measure

Similar to the effectiveness measure utilized by Booker's GOFER-1 [2], which was a function of a classifier's "impact" (i.e., payoff), "consistency" (i.e., prediction error) and "match score" (i.e., specificity), EXCS measures classifier effectiveness as the product of a classifier's predicted payoff (p_{cl}), average accuracy (k_ave_{cl}) and generality (gen_{cl}) – where *generality* is the percentage of the classifier's condition consisting of don't-care bits (#'s). Therefore, the effectiveness of an individual classifier is:

$$eff_{cl} = p_{cl} * k\_ave_{cl} * gen_{cl}. \qquad (9)$$

The fitness calculation in EXCS then proceeds as in XCS with two differences not found to improve the performance of MXCS. First, the GA in EXCS operates on the match set ([M]) with offspring actions determined by the crossover of the parent actions. This is done in order to cause the GA to drive the system towards a small number of effective actions per match set. Second, where XCS calculates a classifier's fitness to be the portion of the niche's accuracy sum attributed to that classifier, EXCS updates classifier fitness towards the ratio of a classifier's effectiveness and the maximum effectiveness found in the match set, as shown in equation 10.

$$f_{cl} = f_{cl} + \beta * (eff_{cl} / MAX(eff_{[M]}) - f_{cl}). \qquad (10)$$

The use of the maximum effectiveness in the match set prevents the fitness of each classifier from being impacted by the quantity of classifiers in the same match sets. This can be troublesome when comparing classifier fitness's across GA niches, as in the deletion-selection process.

It should also be noted that the action set size estimate, *as*, was changed for EXCS to measure the average size of the match sets in which the classifiers are found, rather than the action sets as in XCS and MXCS.

5.3 Conflict-Resolution

In order to accommodate the use of a partial map in EXCS, the conflict-resolution process must be modified. XCS appears to take great advantage of the fact that accurate predictions of low payoff actions can be used to direct action selection towards higher payoff actions during conflict-resolution on exploitation steps.

EXCS, on the other hand, is at a disadvantage as the incomplete map developed results in inexperienced or inaccurate classifiers with higher predictions appearing favorable to the conflict-resolution process due to the lack of accurate classifiers in the same set (e.g., the fitness weighted prediction of a classifier that is the only member of *an action set is independent* of the classifier's fitness, and therefore accuracy).

This problem is overcome in EXCS by scaling the contribution of an inexperienced classifier (i.e., $exp_{cl} < \Omega * \theta_{del}$) to the calculation of the prediction and fitness sums for an action set. As shown in equations 11 and 12 this is accomplished by scaling the prediction of an inexperienced classifier by the square of a small fraction (i.e., $.01^2$), and scaling the inexperienced classifier's contribution to the fitness sum by the same small fraction (i.e., .01).

$$\Sigma (p*f)_{[A]} = \Sigma (p*f)_{[A]} + .01^2 * p_{cl} * f_{cl}. \quad (11)$$

$$\Sigma f_{[A]} = \Sigma f_{[A]} + .01 * f_{cl}. \quad (12)$$

The use of the square of the fraction in the calculation of the payoff prediction sum helps to prevent the inexperienced classifier from overestimating its predicted payoff as well as protects against situations where the classifier is the only member of the action set. Using the fraction of the classifier's fitness in the fitness sum for the action set prevents the inexperienced classifier from distorting the payoff predicted for that action set. Note that it was found to be beneficial to performance to use these scaling factors, rather than simply disregarding inexperienced classifiers.

5.4 Population Culling

The final modification made in EXCS is the introduction of a new deletion operation, termed *culling*. The purpose of the culling operation is to rapidly remove ineffective classifiers from the population. This is accomplished by immediately removing classifiers before each invocation of the GA when the classifier's experience is greater than the deletion threshold ($exp_{cl} > \theta_{del}$), its effectiveness is below the minimum error parameter (ε_0) and the percentage of the population composed of such ineffective classifiers is above a new *cullling threshold* (θ_{cull} – in all experiments reported in this paper, $\theta_{cull}=0.1$). The process employed was to first determine the set of ineffective classifiers then randomly remove classifiers from this set until the threshold is met.

The culling threshold is used as it was found to be beneficial to performance to leave some number of "easy targets" in the population for the deletion selection process. Otherwise, should all ineffective classifiers be immediately removed, the deletion-selection process tends to be forced to remove effective classifiers, as the bulk of the population consists of experienced effective classifiers and inexperienced classifiers with low deletion votes due to their low experience levels.

6 Experimental Comparisons

This section presents experimental results that compare the performance of XCS, MXCS and EXCS in size-constrained 6-Multiplexer and Woods-2 problems.

Except for changes required for the described modifications the systems were implemented to follow the XCS definition described by Butz and Wilson in [3]. System parameters for all experiments reported in this work are as in [15] except where noted. The covering threshold parameter (θ_{mna}) was set to require one classifier per match set.

This value yields better performance in size-constrained systems. For EXCS, the deletion-vote fitness fraction (δ) was disregarded (i.e., set to infinity) as this was found to improve performance. This change had an adverse effect when used in XCS or MXCS, possibly owing to the greater amount of information encoded in the effectiveness-based fitness measure. Classifier parameters were updated in the order: experience, error, prediction, and then fitness.

Subsumption deletion [3,16] was used in all experiments, though a new subsumption experience threshold ($\theta_{sub[A]}$) was introduced for use in triggering action set subsumption. Owing to the fact that the negative impact of an erroneous action set subsumption can be far more severe than an erroneous GA subsumption, a higher subsumption threshold was required. For all experiments reported in this work the value of the action set subsumption threshold was set to be ten times the size of the problem's input space (i.e., the number of possible inputs). This gives occasionally erroneous classifiers more opportunities to be identified and removed from the system before subsuming other more accurate classifiers. Note that this level is certainly not intended to work for other problems with potentially much larger input spaces. It is merely intended to allow subsumption deletion to be used in the relatively small environments reported here.

6.1 6-Multiplexer

The multiplexer class of problems are frequently used in the literature as they present a challenging single step task with a scalable difficulty level capable of testing the systems ability to develop accurate, general and optimal rules.

Only three of the six bits in any given input string of a 6-Mulitplexer (6-Mux) are important, the two address bits and the bit at the position pointed to by the address bits (i.e., the correct output). Due to this property, the system can develop general rules with don't-care symbols in the ignored bit positions of their conditions that will accurately encode the entire set of classifiers masked by that condition. For example, while the 6-Mux problem requires 128 unique and fully defined classifiers to be complete, a smaller set of 16 *optimal* classifiers can accurately cover the entire input space given a two-level payoff landscape.

For all 6-Mux results reported in this work, the system inputs were randomly generated 6-bit strings. Rewards of 1000 and 0 were returned to the system for right and wrong answers respectively.

The results presented in this work for the 6-Mux problem use two different performance measures commonly reported in the literature: performance and optimality. *Performance* is a moving average of the percentage of the last 50 exploit steps that returned the correct output. *Optimality* measures the percentage of the optimal set of classifiers present in the population.

As stated, for the 6-Mux problem with two payoff levels the optimal set consists of 16 classifiers. However, in order to compare systems developing complete maps (i.e., XCS and MXCS) and partial maps (i.e., EXCS) the optimal set is restricted to those optimal classifiers that earn the maximum reward. Therefore, for the 6-mux problem the *optimally-effective* set of classifiers consists of 8 unique classifiers.

Furthermore, to facilitate a direct comparison of systems utilizing different GA triggering algorithms the GA trigger threshold (θ_{ga}) was selected for XCS so as to cause the GA to be invoked at the same rate as MXCS and EXCS. Therefore, θ_{ga} for XCS with N=40 was set to θ_{ga}=185 and θ_{ga} for XCS with N=20 was set to θ_{ga}=265. These values resulted in approximately the same number of GA invocations on average per run as the value of θ_{ga}=25 used in the experience-based MXCS and EXCS GA trigger algorithms. Note that the performance of XCS was not found to be greatly affected by any of the numerous other trigger values tested and results obtained with N=400 (not shown) closely approximated those reported by Wilson and others [6,10,15].

Performance. Figure 1 shows performance in the 6-Mux problem at the two size-constrained configurations of N=40 and N=20 (a size of N=400 is commonly used for the 6-Mux problem). As shown, XCS is unsuccessful at either constrained size. MXCS, on the other hand, achieves a successful population with N=40 but fails with N=20. EXCS succeeds at both N=40 and N=20.

Optimality. Figure 2 shows the optimality of the three systems. Due to the small population sizes XCS is unable to maintain much of the optimally-effective set of classifiers while MXCS succeeds with N=40 but performs poorly with N=20. EXCS performs well with both N=40 and N=20.

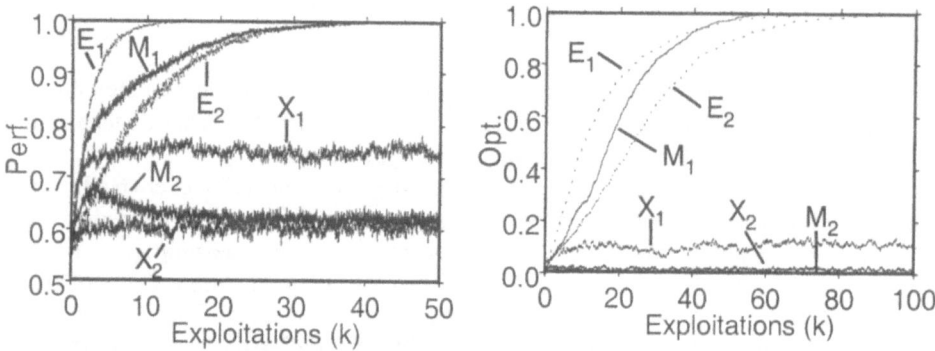

Fig. 1 (left) and Fig. 2 (right). 6-Mux performance (left) and optimality (right) of: XCS with N=40 (X_1) and N=20 (X_2), MXCS with N=40 (M_1) and N=20 (M_2), and EXCS with N=40 (E_1) and N=20 (E_2). Curves are averages of 100 runs

6.2 Woods-2

This section compares the performance of XCS, MXCS and EXCS in the Woods-2 problem. Though, as Kovacs explains in [10], Woods-2 is not a very difficult test of the abilities of a system to learn sequential decision tasks, it serves as a standard benchmark by which we can compare the system's responses to population pressure.

Woods-2, developed by Wilson [15], is a multi-step Markov problem with delayed rewards in which the system represents an "animat" [13], or organism, searching for food on a 30 by 15 grid. The Woods-2 environment is composed of five possible location types (encoded as three bit binary strings): empty space through which the animat may move, two types of "rock" that block movement, and two types of food which garner a reward when occupied by the animat. The grid contains a repeating pattern of 3 by 3 blocks which are separated by two rows of empty space and the grid "wraps around" (e.g., moving off the edge of the grid places the animat on the opposite edge of the grid), thereby creating a continuous environment.

The system input is a 24-bit binary string representing the eight space types surrounding the location currently occupied by the animat (i.e., the animat's "sensed" surroundings). The output of the system is a three-bit binary string representing in which of the eight possible directions the animat attempts to move in response to its currently sensed surroundings. At the start of each episode, the types of food and rock in each block are randomly selected but the locations remain constant. This is done to better test the generalization abilities of the systems.

For Woods-2, performance is measured as a moving average of the number of steps required for the animat to reach food on the last 50 exploitation episodes. A performance level of 1.7 steps on average is the optimal solution for Woods-2 as no population of rules can do better [10,15].

As with the 6-Mux problem, the GA trigger threshold was selected for XCS in Woods-2 in order to cause the system to invoke the GA approximately the same number of times per run as MXCS and EXCS. Therefore, values of $\theta_{ga}=500$ and $\theta_{ga}=1000$ were used for XCS at $N=80$ and $N=40$ respectively; while a value of $\theta_{ga}=25$ was used for MXCS and EXCS. As with the 6-Mux experiments, a variety of threshold values were evaluated and not found to significantly alter the performance of XCS at these population sizes.

One further change was made to the implementation of the Woods-2 problem in order to allow a comparison of systems developing complete and partial maps. The exploration phase of the Woods-2 problem involves placing the animat at a random location and ending the exploration once the random selection of actions causes the animat to move onto a food space (and thus garner a reward) or when the maximum number of steps have been taken. However, it was found that a system developing an incomplete map (i.e., EXCS) will tend to take far fewer steps on average during exploration as the majority of actions present will direct the animat towards food. This greatly reduces the number of steps taken during exploration and thus reduces the amount of "learning" taking place during each exploration episode.

In an attempt to level the playing field, the exploration episodes of the Woods-2 problems reported in this work utilize a *complete-exploration* technique. Under complete-exploration the animat is placed in a new random location after each update of the prior action set or when food is reached. This continues until the maximum number of steps have been taken, thereby ensuring that each system takes the same number of exploration steps per run. Note that this technique is not required for the system to learn, it is used to allow a direct comparison between the two types of systems.

Performance. Figure 3 shows the performance of the three systems in the Woods-2 problem at two size-constrained configurations of $N=80$ and $N=40$ (note the logarithmic vertical axis). As shown, the size constraint causes XCS to be unsuccessful at either population size, actually performing worse than would be expected from a random walk. MXCS achieves an average performance level of 1.94 steps by the end of the run with $N=80$ and 3.83 steps with $N=40$. EXCS reaches average performance levels of 1.74 steps and 1.84 steps for $N=80$ and $N=40$ respectively.

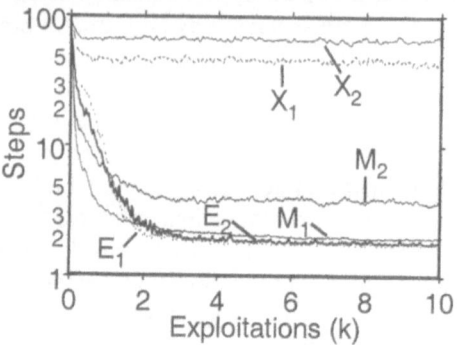

Fig. 3. Woods2 Performance of: XCS with $N=80$ (X_1) and $N=40$ (X_2), MXCS with $N=80$ (M_1) and $N=40$ (M_2), EXCS with $N=80$ (E_1) and $N=40$ (E_2). Curves are averages of 100 runs

7 Conclusions and Future Directions

This paper has attempted to broaden the range of applications to which the XCS framework can be applied by introducing modifications that may help XCS to perform in size-constrained systems. Though the solutions may take longer to reach, it makes sense to investigate alternatives that allow the system's requirements to be shifted from physical resources (i.e., population size) to system iterations as the physical resources may be limited by forces out of the implementers control.

As was demonstrated, the first modified system, MXCS, was able to outperform XCS in small size-constrained configurations by, in essence, tuning the system's algorithms to make more cautious use of the limited resources available. The second system introduced, EXCS, was shown to outperform both XCS and MXCS by taking advantage of the lower resource requirements of a partial map. Though a system developing an incomplete map such as EXCS may not be suitable for many problems, the advantages in terms of reduced resource requirements and potentially shorter time-to-solution make it appealing for those problems that allow an incomplete map.

There are a variety or areas related to this work that could be of interest for future investigation. First, a systematic evaluation of each modification is called for. Also, a more thorough investigation into the resource requirements of XCS and the relationship between population size and the ability of the system to generalize is called for. Furthermore, a comparison of the systems presented here, as well as others (such as Kovacs' SBXCS [9,10] and Lanzi's XCSS [11]), in a wider range of problems and varying size-constraint levels seems to be in order. Lastly, a comparison of these systems in terms of time-to-solution for a variety of problems and configurations may help to make an argument for the use of smaller populations, as the computational cost of an increased population size will at some point outweigh the benefit.

The author would like to thank the Hewlett-Packard Company for its support and the reviewers of this paper for their time and valuable feedback.

References

1. Booker L.B. *Intelligent behavior as an adaptation to the task environment.* Ph.D. Dissertation, The University of Michigan, 1982.
2. Booker L.B. Triggered rule discovery in classifier systems. In: J.D. Schaffer (ed.), *Proceedings of the Third International Conference on Genetic Algorithms (ICGA89)*, pages 265–274. Morgan Kaufmann, 1989.
3. Butz M.V. and Wilson S.W. An Algorithmic Description of XCS. In: Lanzi P.L., Stolzmann W. and Wilson S.W. (Eds.), *Advances in Learning Classifier Systems, LNAI 1996*, pages 253–272. Springer-Verlag, Berlin, 2001.
4. Cliff D. and Ross S., Adding Temporary Memory to ZCS. *Adaptive Behavior*, 3(2):101–150, 1995.
5. Dawson D. and Juliano B. Modifying XCS for size-constrained systems. *Neural Network World, International Journal on Neural and Mass-Parallel Computing and Information Systems, Special Issue on Soft Computing Systems*, 12(6), pages 519–531, 2002.
6. Kovacs T. *Evolving Optimal Populations with XCS Classifier Systems.* Master's Thesis, School of Computer Science, University of Birmingham, Birmingham, U.K., 1996.
7. Kovacs T. *Deletion Schemes for Classifier Systems.* In: W. Banzhaf, J. Daida, A. E. Eiben, M. H. Garzon, V. Honavar, M. Jakiela, and R. E. Smith, (Eds.), GECCO-99: Proceedings of the Genetic and Evolutionary Computation Conference, pages 329–336. Morgan Kaufmann, San Francisco (CA), 1999.
8. Kovacs T. Strength or Accuracy? Fitness calculation in learning classifier systems. In: P. L. Lanzi, W. Stolzmann, S.W. Wilson (Eds.), *Learning Classifier Systems, From Foundations to Applications*, Springer-Verlag, Berlin, 2000.
9. Kovacs T. Two Views of Classifier Systems. In: Lanzi, P.L., Stolzmann W., and Wilson, S.W., (Eds.), *Advances in Learning Classifier Systems*, pages 74–87. Springer-Verlag, Berlin, 2002.
10. Kovacs T. *A Comparison of Strength and Accuracy-Based Fitness in Learning Classifier Systems*, Ph.D. Dissertation, School of Computer Science, University of Birmingham, 2002.
11. Lanzi P.L. An Analysis of Generalization in the XCS Classifier System. *Evolutionary Computation*, 7(2):125–149. 1999.
12. Watkins C. *Learning from Delayed Rewards.* Ph.D. Dissertation, Cambridge University, 1989.
13. Wilson S.W. *Knowledge growth in an artificial animal.* Proceedings of the Tenth International Joint Conference on Artificial Intelligence. pages 217–220. Los Angeles (CA), Morgan Kaufman, 1985.
14. Wilson S.W. ZCS: A zeroeth level classifier system, *Evolutionary Computation*, 2(1), pages 1–18. 1994
15. Wilson S.W. Classifier fitness based on accuracy. *Evolutionary Computation*, 3(2), pages 149–175. 1995,
16. Wilson S.W. Generalization in the XCS classifier system. In: J.R. Koza et al. (Eds.), *Genetic Programming 1998: Proceedings of the Third Annual Conference*, pages 665–674. Morgan Kaufmann, San Francisco (CA), 1996.

Designing Efficient Exploration with MACS: Modules and Function Approximation

Pierre Gérard and Olivier Sigaud

AnimatLab (LIP6)
8, rue du Capitaine Scott
75015 PARIS

Abstract. MACS (Modular Anticipatory Classifier System) is a new Anticipatory Classifier System. With respect to its predecessors, ACS, ACS2 and YACS, the latent learning process in MACS is able to take advantage of new regularities. Instead of anticipating all attributes of the perceived situations in the same classifier, MACS only anticipates one attribute per classifier. In this paper we describe how the model of the environment represented by the classifiers can be used to perform active exploration and how this exploration policy is aggregated with the exploitation policy. The architecture is validated experimentally. Then we draw more general principles from the architectural choices giving rise to MACS. We show that building a model of the environment can be seen as a function approximation problem which can be solved with Anticipatory Classifier Systems such as MACS, but also with accuracy-based systems like XCS or XCSF, organized into a Dyna architecture.

1 Introduction

Research on Learning Classifier Systems (LCSs) has received increasing attention over the last few years. This surge of interest took two different directions.

First, a trend called "classical LCSs" hereafter comes from the simplification of Holland's initial framework [Hol85] by Wilson. The design of ZCS [Wil94] and then XCS [Wil95] resulted in a dramatic increase of LCS performance and applicability. The latter system, using the accuracy of the reward prediction as a fitness measure, has proven its effectiveness on different classes of problems such as adaptive behavior learning or data mining. XCS can be considered as the starting point of most new work along this first line of research.

Second, a new family of systems called Anticipatory Learning Classifier Systems (ALCSs) has emerged, showing the feasibility of using model-based Reinforcement Learning (RL) in the LCS framework. This second line of research is more inclined to use heuristics rather than genetic algorithms (GAs) in order to deal with the improvement of classifiers. Several systems (*e.g.* ACS [Sto98,Sto00], ACS2 [But02a] and YACS [GSS02]) have highlighted the interesting properties of this family of approaches.

The way ALCSs achieve model-based RL consists in a major shift in the classifiers representation. Instead of [condition] [action] classifiers, they use a [condition] [action] [effect] representation, where the [effect] part represent what would result from taking the action if the condition is verified. As a consequence of this representational shift, the ALCS framework is somewhat distinct from the classical LCS one.

In this paper, we want to show that one can benefit from the model-based properties of ALCSs while keeping the classical LCS framework as is, thanks to the design of a general architecture whose basic components can be either ALCSs or classical LCSs like XCS or XCSF [Wil01,Wil02], or even other kinds of function approximators.

More precisely, after presenting ALCSs in section 2, we will introduce in section 3 a new ALCS called MACS, whose generalization properties are different from those of previous ALCSs. Classifiers in MACS are only intended to predict one attribute of the next situation in the environment depending on a situation and an action, whereas classifiers in all previous systems try to predict the next situation as a whole. We will show that this new representation gives rise to more powerful generalization than previous ones, and can be realized with a modular approach. Then we will explain in section 4 how such an efficient model-based learning process can be integrated into a Dyna architecture combining exploration and exploitation criteria, giving empirical results in section 5.

Then, in section 6, we will generalize what we have learned from MACS in a wider perspective. Predicting the value of one attribute of the environment can be seen as a function approximation problem, and any system able to approximate a function can be used in the same architecture. In particular, since XCS and XCSF are such systems, we will conclude in section 7 that model-based reinforcement learning with generalization properties can be performed as well with XCS, XCSF or even other kinds of systems, provided that an adequate architecture is used.

2 Anticipatory Learning Classifier Systems

The usual formal representation of RL problems is a Markov Decision Process (MDP) which is defined by:

- a finite state space S;
- a finite set of actions A;
- a transition function $T : S \times A \to \Pi(S)$ where $\Pi(S)$ is a distribution of probabilities over the state space S;
- a reward function $R : S \times A \times S \to \mathbb{R}$ which associates an immediate reward to every possible transition.

One of the most popular RL algorithm based on this representation is Q-learning [Wat89]. This algorithm directly and incrementally updates a Q-table representing a quality function $q : S \times A \to \mathbb{R}$, without using the transition and the immediate reward functions. The quality $q(s, a)$ represents the expected payoff when the agent performs the action a in the state s, and follows the greedy policy thereafter. Then, the qualities aggregate the immediate and future expected payoffs.

The main advantage of Learning Classifier Systems with respect to other RL techniques like tabular *Q-learning* relies in their generalization capabilities. In problems such that situations are composed of several attributes, generalization makes it possible to aggregate several situations within a common description so that the model of the quality function q becomes smaller.

In [Lan00], Lanzi shows how it is possible to shift from a tabular representation of a RL problem to a classifier-based representation. While tabular Q-learning considers triples $(s, a, q) \in S \times A \times \mathbb{R}$, LCSs like XCS consider C-A-p rules [condition] [action] payoff *classifiers*). During the learning process, the LCS learns appropriate general conditions and updates the payoff prediction.

Within the classical LCS framework, the use of *don't care* symbols # in the C part of the classifiers permits generalization, since *don't care* symbols make it possible to use a single description to describe several situations. Indeed, a *don't care* symbol *matches* any particular value of the considered attribute. Therefore, changing an attribute into a *don't care* symbol makes the corresponding condition more general (it matches more situations). The main issue with generalization in classical LCSs is to organize C and A parts so that the *don't care* symbols are well placed.

XCS offers a generalization capability but, as *Q-learning*, it can only update very few measures of classifier payoff prediction at each time step, corresponding to the immediate actual previous situation and action s_{t-1} and a_{t-1}. Indeed, the model of the expected payoff can only be updated when an actual transition occurs. Sutton [Sut90] proposed the Dyna architecture to endow the system with the ability to update many qualities at the same time, in order to significantly improve the learning speed of the quality function. The Dyna architecture illustrated in figure 1 uses a model of the environment to build hypothetical transitions independently from the current experience. These simulated actions are used to update the model of the quality function more than once per time step, with a *value iteration* algorithm inspired from *Dynamic Programming*. The model of the environment is learned *latently* – i.e. independently from the reward.

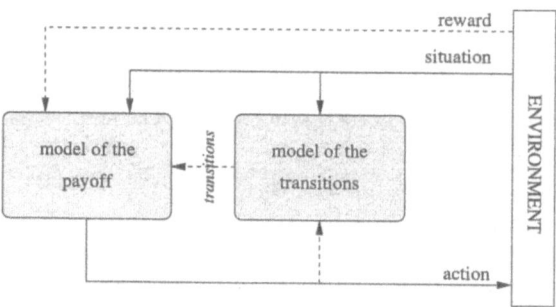

Fig. 1. A Dyna architecture to perform reinforcement learning. It combines a model of the payoff with a model of the transitions. The model of the payoff may consist in an approximation of the immediate reward function $R : S \times A \to \mathbb{R}$, and in an approximation of the quality function $q : S \times A \to \mathbb{R}$, which produces a scalar quality for each $(situation, action)$ pair. To learn the function q incrementally, the system is given $(s_{t-1}, a_{t-1}, s_t, r_t)$ tuples. The model of the transitions is an approximation of the transition function $T : S \times A \to S$. The transition function and the immediate reward functions provide hypothetical samples to the quality learning system, and subsequently speed up the reinforcement learning process. Note that the construction of both models requires a memory of the previous situation, which is not shown on the figure.

Instead of directly learning a model of the quality function q as XCS does, ALCSs such as ACS [Sto98,BGS00], ACS2 [But02a] and YACS [GS01,GSS02] learn a model of the transition function T. They take advantage of generalization capabilities to learn a model of the transition function which is more compact than the exhaustive list of all the (s_t, a_t, s_{t+1}) transitions. The transition function provides a model of the dynamics of the interactions between the agent and its environment which takes place in a Dyna architecture to speed up the plain reinforcement learning process.

In ALCSs, the classifiers are organized into [condition] [action] [effect] parts, noted C-A-E. In such classifiers, the E part represents the effects of action A in situations matched by condition C. It records the perceived changes in the environment. In ACS, ACS2 and YACS, a C part is a situation which may contain *don't care* symbols # or specific values (like 0 or 1), as in XCS. An E part is also divided into several attributes and may contain either specific values or *don't change* symbols =. Such a *don't change* symbol means that the attribute of the perceived situation it refers to remains unchanged when action A is performed. A specific value in the E part means that the value of the corresponding attribute *changes* to the value specified in that E part.

This formalism permits the representation of regularities in the interactions with the environment, like for instance *"In a grid world, when the agent perceives a wall in front of itself, whatever the other features of the current cell are, trying to move forward entails hitting the wall, and no change will be perceived in the cell's features"*.

The latent learning process is in charge of discovering C-A-E classifiers with general C parts that accurately model the dynamics of the environment. ACS and YACS generalize according to anticipated situations, and not according to the payoff, as in XCS. As a result, it does not make sense to store information about the expected payoff in the classifiers. Therefore, the list of classifiers only models environmental changes. The information concerning the payoffs must be stored separately.

3 Improved Latent Learning with MACS

3.1 Representing More Regularities with MACS

Generalization makes it possible to represent regularities in the interactions with the environment. However, while ACS and YACS are able to detect if a particular attribute is changing or not, their formalism cannot represent regularities across different attributes because it considers each situation as a whole. To make this point clear, let us consider an agent in a grid world such as those presented in figures 3, 4 and 5, where its perceptions are defined as a vector of boolean values depending on the presence or absence of walls in the eight surrounding cells. Turning right results in a two-positions left shift of the attributes. For instance, the agent may experience transitions like [11001100] [↷] [00110011].

In such a case, every attribute is changing. Thus, the formalism of ACS and YACS is unable to represent this regularity. Nevertheless, the shift in the perceived situation is actually a regularity of the dynamics of the interactions: whatever the situation is, when the agent turns clockwise, the value of the 1st attribute comes to the last value of the 3rd, the value of the 2nd becomes the 4th *etc.*

The particularity of such a regularity is that the new value of an attribute depends on the previous value of another one. Expressing generalization with *don't change* symbols forbids the representation of such regularities. In the ACS/YACS formalism, the new value of an attribute may only depend upon the previous value of the same attribute, a situation which is seldom encountered in practice. To overcome this problem, it is necessary to decorrelate the attributes in the E parts, whereas ACS and YACS classifiers anticipate all attributes at once.

To this end, our new system, MACS, describes the E parts with *don't know* symbols "?" rather than with *don't change* symbols. This way, the accurate classifier [####1###] [↷] [??1?????] means that *"just after turning right, the agent always*

Table 1. During the integration process, the LCS proposed in section 3 scans the E parts and selects classifiers whose A parts match the action and whose C part match the situation. The integration process builds all the possible anticipated situations with respect to the possible values of every attribute. Here, the system anticipates that using [11001100] as a current situation should lead either to [00110011] or to [00110111]. In deterministic environments, if all the classifiers were accurate, this process would generate only one possible anticipation.

[11001100]		←Situation
[1#######]	[↷]	[???????1?]
[#1######]	[↷]	[???????1]
[##0#####]	[↷]	[0???????]
[###0####]	[↷]	[?0??????]
[####1###]	[↷]	[??1?????]
[#####1##]	[↷]	[???1????]
[######0#]	[↷]	[????0???]
[#######0]	[↷]	[?????0??]
[#######0]	[↷]	[?????1??]
Anticipations →		[00110011]
	or	[00110111]

perceives a wall at its left when it perceived a wall behind itself, whatever the other attributes were". This classifier does not provide information about the new values of other attributes (as denoted by the ? symbol). Thus, the overall system gains the opportunity to discover regularities involving different attributes in the [condition] and the [effect] parts.

Again, this proposal for a new formalism leads to a new conception of generalization. As usual, a classifier is said to be *maximally general* if it could not contain any additional *don't care* symbol without becoming inaccurate. But it is now said to be *accurate* if, in every situation matched by its condition, taking the proposed action always sets the attributes to the values specified in the effect part, when such attributes are not *don't know* symbols.

As a result, the anticipating unit is not the single classifier anymore but the whole LCS. Given a situation and an action, a single classifier is not able to predict the whole next situation: it anticipates only one attribute. The system needs an additional mechanism which *integrates* these partial anticipations and builds a whole anticipated situation, without any *don't know* symbol in its description, as shown in Table 1.

Experimental results presented in [GMS03] demonstrated that the new formalism used by MACS actually affords more powerful generalization capacities than the formalism of YACS, without any cost in terms of learning speed.

The algorithms realizing the latent learning process in MACS will not be described in detail here, they are presented in [GMS03].

3.2 Modular Model of the Environment with MACS

In the previous section, we described how MACS represents its model of the dynamics of the environment with anticipating classifiers.

In all Dyna systems, this model consists of an exhaustive list of (s_{t-1}, a_{t-1}, s_t) triples, each specifying a whole transition, i.e. the expected value of a complete set of attributes. ACS and YACS both improve the model by adding generalization in the triples, but each classifier still specifies complete transitions.

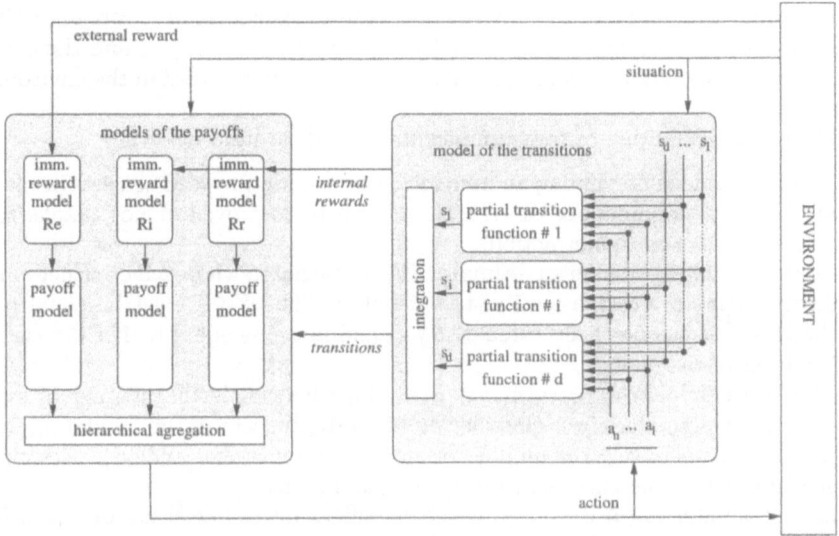

Fig. 2. MACS architecture. Each module is a simple function approximator.

Each classifier of YACS (or (s_{t-1}, a_{t-1}, s_t) triple of DynaQ+) is a subfunction of the global transition function $T : s_1 \times ... \times s_d \times a_1 \times ... \times a_e \rightarrow s_1 \times ... \times s_d{}^1$. Conversely, in MACS, each classifier is a subfunction of a partial transition function $T_i : s_1 \times ... \times s_d \times a_1 \times ... \times a_e \rightarrow s_i$. Then, it is possible to consider groups of classifiers, each group anticipating one particular attribute. Each of these groups then models a partial transition function. The global transition function can be obtained by integrating the partial functions.

The MACS architecture illustrated in figure 2 shows how the latent learning part of MACS can be considered as a modular system, each module anticipating one attribute. Each of these modules provides an approximation of one partial transition function, each predicting one single value. As we will discuss in section 7, this architecture suggests that one could replace MACS modules by some other function approximation systems like neural networks or classical LCSs like XCS or XCSF.

4 Combining Active Exploration and Exploitation

4.1 Hierarchical Aggregation of Different Criteria

The aim of active exploration is to provide the agent with a policy that maximizes the information provided by the sensori-motor loop. The agent selects actions that help improving the model according to this criterion.

---

[1] where e is the number of effectors, and d is the number of perceived attributes

As illustrated by figure 2, in order to combine active exploration with exploitation, we designed an architecture resulting from the hierarchical combination of three dynamic programming modules, each trying to maximize the reward from a different source. Indeed, we distinguish the internal reward, corresponding to a gain in information about the model of the environment, the rehearsal reward, corresponding to a measure of the time elapsed after the last visit of each transition, and the external reward, corresponding to a gain of food or whatever actual reward in the environment of the agent.

The precise definition of these immediate rewards are the following

- The general idea of defining an immediate information reward consists in measuring whether the model of the transitions can be improved or not thanks to the activation of a particular action.

More precisely, we define an estimator $El(c)$ associated with each classifier c. $El(c)$ measures the evaluation level of the classifier. Thus $El(c)$ must be equal to 0 if the classifier has not been evaluated yet, and must be equal to 1 if the classifier has been tested enough.

Thus we define $El(c) = min((b + g)/\theta_e, 1)$, where θ_e is the number of evaluations needed to classify a classifier as accurate, inaccurate or oscillating, and b and g are respectively the number of anticipation mistakes and successes already encountered by the classifier in previous evaluations.

Each estimator $El(c)$ is bound to one classifier, not to one situation. In order to compute the information gain bound to one situation $R_i(s_0)$, the process is the following.

The classifiers that match s_0 are grouped by action. For each possible action a, MACS computes the set $S_{s_0,a}$ of the possible anticipated situations as described in section 3^2. Each triple (s_0, a, s_1), where $s_1 \in S_{s_0,a}$, is one of the possible transitions that would be experienced if action a were performed in situation s_0. We define the evaluation level $El(s_0, a, s_1)$ associated with this transition as the product of the evaluation levels $El(c)$ of all the classifiers c involved in this anticipation:

$$El(s_0, a, s_1) = \prod_{c \approx (s_0,a,s_1)} El(c)$$

The classifiers c matching (s_0, a, s_1) are such that their C part matches s_0, their A part matches a, and their E part matches s_1. The less a transition has been evaluated, the greater the immediate information gain. Thus, if the transition occurs, the associated immediate information gain is:

$$R_i(s_0, a, s_1) = 1 - El(s_0, a, s_1)$$

We define the immediate information gain associated with a situation and an action as the maximum information gain over the possible associated anticipations:

$$R_i(s_0, a) = \max_{s_1 \in S_{s_0,a}} R_i(s_0, a, s_1)$$

If the model does not provide MACS with at least one anticipated situation s_1, because of incompleteness, then $R_i(s_0, a)$ is given the default value 1, which is the maximum immediate information gain.

---

[2] There may be several possible anticipated situations in the case where the classifiers are not accurate.

Finally,
$$R_i(s_0) = \max_{a \in A} R_i(s_0, a)$$

- The external reward is the usual source of reward found in any reinforcement learning framework. We define it as $Re(s_0)$ corresponding to the immediate reward obtained for visiting the situation s_0.
- The immediate rehearsal reward leads to a policy which is similar to the exploration policy described in [Sut91] or [But02b][3]. Then it grows until the situation is visited again and then drops to 0.
In order to update $R_r(s_0)$, we use the classical Widrow-Hoff equation:

$$R_r(s_0) = (1 - \beta_r)R_r(s_0) + \beta_r$$

Then, from each immediate reward, whether internal, rehearsal or external, a long term expected payoff is computed separately thanks to the *Value Iteration* algorithm (see [SB98] for a presentation). The latent learning process provides the system with the model of the transitions which is necessary for an offline computing of the qualities associated to each action, given a situation. By doing so, the values are updated independently from the actual experience of the agent, and many updates can be computed at each time step, as usual in a *Dyna* architecture.

The last component of the architecture consists of a hierarchical combination of these three criteria. Since an inaccurate model may result in an inaccurate estimation of the expected external payoff, maximizing the information gain is given the priority against the external payoff maximization. Thus, if there is a better action with respect to the information gain criterion, this action is chosen. Else, if at least two actions provide the maximal expected information gain, then the one which maximizes the external payoff is chosen. In particular, if the information about the problem is perfect, which means that no action provide any gain in information anymore, then the policy will be completely driven by the external payoff and will converge to optimal exploitation. The last criterion, rehearsal reward, is the least often used. It is only chosen if at least two actions are equally likely to be fired with respect to both previous criteria.

5 Experimental Results: Moving Sources of Reward

In order to illustrate the gain resulting from latent learning in reinforcement learning problems, we present in this paper new experiments where we tried to test MACS on problems where the sources of reward are moved after the agent succeeds in reaching them.

This problem gives the opportunity to highlight two key properties of MACS. First, the necessity to have an internal model and to be able to perform *Value Iteration* on that model as an offline mental rehearsal arises when the agent discovers that the source of reward has moved: it must forget all the payoff estimates corresponding to its previous model, and this forgetting process would be very slow without such a model. Second, the necessity of active exploration arises once the model is forgotten: the agent must re-explore the whole environment to find the new location of the source of reward. This search is much more efficient thanks to active exploration.

---

[3] The latter also uses a sort of "internal reward" biasing action selection according to the accuracy or "quality" of the predicted effects.

 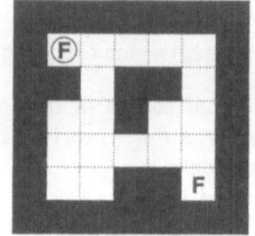 

Fig. 3. Maze216C   **Fig. 4.** Maze228C   **Fig. 5.** Maze312C

As a benchmark problem, we tested MACS on three different environments, namely Maze216C, Maze228C and Maze312C, respectively shown on figure 3, 4 and 5. These "maze-like" or "woods" problems are standard benchmarks in LCS research, they could be replaced by any finite state automaton but they provide a more intuitive view.

At the beginning of the experiments, the food is in the cell marked with a circled F. Once the agent has reached it twenty times, the food is moved to the cell marked with a plain F. This is done repeatedly, the food being moved again each time the agent has found it twenty times.

Figure 6 illustrates the number of time steps MACS needs to solve Maze216C, Maze228C and Maze312C during 250 successive trials. The results are averaged over 100 experiments. The learning rates are set to 0.1 and the memory size and number of evaluations necessary to take a specialization/generalization decision are all set to 5. The discount factor γ is set to 0.9.

The results show that, though it only performs one *Value Iteration* step per time step, MACS is able to re-adapt the policy to the new source location very fast. Each twenty trials, the longer trial corresponds to the case where the agent must find the new location of the source of reward. It is found fast thanks to active exploration and, as soon as it is found, the time necessary to reach it again converges during the next trial. The slight variations are due to the fact that the agent always starts from a random cell and the results are averaged over 100 trials only.

Note that MACS is tested here in a deterministic environment and would not perform as efficiently in a probabilistic environment, but the lessons that we will draw in the following discussion and conclusion still apply in the probabilistic case.

6 Discussion

In this paper, we have shown how a latent learning process could be combined with several dynamical programming processes into a Dyna-like architecture. As in Dyna architectures, one module learns a model of the environment and the other modules are in charge of maximizing the payoff expectation thanks to offline Dynamic Programming techniques. By using several immediate rewards, we have shown how three different policies could be obtained and how a hierarchical combination of these policies resulted in improved performance in classical problems considered difficult in the reinforcement learning framework.

A finer decomposition into modules can be obtained if one considers that the latent learning process in MACS can itself be split into one module per anticipated attribute. The global architecture of our system is shown in figure 2.

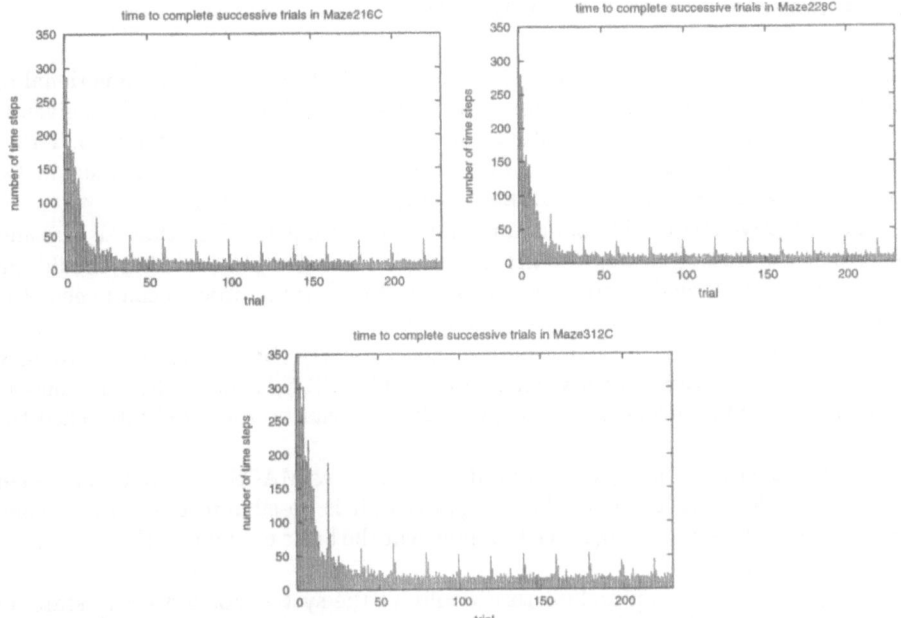

Fig. 6. MACS combining exploration and exploitation: number of time steps to reach the source of reward in successive trials in Maze216C, Maze228C and Maze312C.

As in DynaQ+, one policy relies on the external reward and a second one on the time elapsed after the last visit of each transition. But the third policy in MACS, relying on the information which can be gained by firing each classifier, is more original. It provides the agent with a systematic exploration capability which allows it to gain a perfect knowledge of its environment much faster than with a random exploration, as shown in [GMS03].

The hierarchical combination of these policies is also a distinctive feature of MACS with respect to DynaQ+. In DynaQ+, a weighted sum of the different criteria results in a compromise between the different policies. First, designing a weighted sum of different criteria while the magnitude of these criteria is not known in advance (it depends on the amount of external reward, which may vary from an environment to another) is very difficult. Thus it is likely that the weights have to be tuned for each experiment. In MACS, on the contrary, no weight parameter has to be tuned in order to deal with different levels of reward, since such weights do not exist in the hierarchical agregation.

Furthermore, since the criterion corresponding to the tendency to explore is never equal to zero, the overall behavior is always under-optimal with respect to payoff. In MACS, by contrast, giving the priority to building a correct model permits the construction of a complete model, and then the payoff maximization process naturally takes the control of the agent and gives rise to an optimal behavior.

7 Future Work and Conclusion

In the MACS architecture for latent learning presented on figure 2, the functional role of each module is to guess the value of one attribute at the next time step given the value of several attributes (conditions and actions) at the current time step. That is, each module tries to approximate a *many-to-one function*, whereas classifiers in ACS and YACS were trying to approximate a *many-to-many mapping*.

The effectiveness of this approximation results from the fact that all classifiers receive at the next time step the actual situation or attribute that they should have anticipated. I.e., though there is no supervisor, the learning process can benefit from informations on its errors as if it were supervised.

Therefore, any system relying on the same property can be used to approximate functions with the same effectiveness. In particular, XCS is such a system, since the prediction of the accuracy of each eligible classifier can be corrected after each time step.

Thus, rather than using a specialized ALCS such as MACS to solve reinforcement learning problems with a model-based approach, it is possible to use a more general system like XCS as a building block to implement the basic modules in the architecture presented in figure 2.

Using XCS in such an architecture would let the system solve discrete state and time problems verifying the Markov property as MACS does. Another possibility would be to use XCSF [Wil01,Wil02], an XCS extension devoted to the piecewise linear approximation of continuous functions, so as to address continuous state problems.

Our future work will consist in the investigation of that line of research : using XCSF modules to learn both a policy and a model of the environment. It must be mentioned that, where XCSF uses the Widrow-Hoff delta rule so as to linearly approximate a function, we envision to rather use a more powerful and efficient method coming from the field of incremental statistical estimations.

As a conclusion, we described in this paper several LCSs, casting a new light on the concept of generalization in the LCS framework. We presented MACS, a new LCS using a different formalism able to use additional regularities in its latent learning process. MACS formalism for latent learning does not consider situations as an unsecable whole, but decorrelates the attributes, making it possible to represent regularities across attributes. We have illustrated the effectiveness of our approach with MACS in the context of an active exploration problem, using a hierarchical aggregation criterion in order to tackle the exploration/exploitation tradeoff.

Beyond the presentation of MACS, we hope to have convinced the reader that solving reinforcement learning problems with a model-based approach can be seen as a conjunction of several function approximation problems, and that solving these problems can be achieved efficiently with different systems as a building block, given that they rely on a prediction process.

References

[BGS00] M. V. Butz, D. E. Goldberg, and W. Stolzmann. Introducing a genetic generalization pressure to the Anticipatory Classifier System part I: Theoretical approach. In *Proceedings of the 2000 Genetic and Evolutionary Computation Conference (GECCO 2000)*, pages 34–41, 2000.

[But02a] M. V. Butz. An Algorithmic Description of ACS2. In Lanzi et al. [LSW02], pages 211–229.
[But02b] M. V. Butz. Biasing Exploration in an Anticipatory Learning Classifier System. In Lanzi et al. [LSW02], pages 3–22.
[GMS03] P. Gérard, J.-A. Meyer, and O. Sigaud. Combining latent learning with dynamic programming. *European Journal of Operation Research*, to appear, 2003.
[GS01] P. Gérard and O. Sigaud. YACS : Combining Anticipation and Dynamic Programming in Classifier Systems. In P. L. Lanzi, W. Stolzmann, and S.W. Wilson, editors, *Advances in Learning Classifier Systems*, volume 1996 of *Lecture Notes in Artificial Intelligence*, pages 52–69. Springer-Verlag, Berlin, 2001.
[GSS02] P. Gérard, W. Stolzmann, and O. Sigaud. YACS: a new Learning Classifier System with Anticipation. *Journal of Soft Computing : Special Issue on Learning Classifier Systems*, 6(3-4):216–228, 2002.
[Hol85] J.H. Holland. Properties of the bucket brigade algorithm. In J.J. Grefenstette, editor, *Proceedings of the 1st international Conference on Genetic Algorithms and their applications (ICGA85)*, pages 1–7. L.E. Associates, july 1985.
[Lan00] P. L. Lanzi. Learning Classifier Systems from a reinforcement learning perspective. Technical Report 00-03, Dip. di Elettronica e Informazione, Politecnico di Milano, 2000.
[LSW02] P. L. Lanzi, W. Stolzmann, and S.W. Wilson, editors. *Advances in Learning Classifier Systems*, volume 2321 of *Lecture Notes in Artificial Intelligence*. Springer-Verlag, Berlin, 2002.
[SB98] R. S. Sutton and A. G. Barto. *Reinforcement Learning: An Introduction*. MIT Press, 1998.
[Sto98] W. Stolzmann. Anticipatory Classifier Systems. In J.R. Koza, W. Banzhaf, K. Chellapilla, K. Deb, M. Dorigo, D. B. Fogel, M. H. Garzon, D. E. Goldberg, H. Iba, and R. Riolo, editors, *Genetic Programming*, pages 658–664. Morgan Kaufmann Publishers, Inc., San Francisco, CA, 1998.
[Sto00] W. Stolzmann. An introduction to Anticipatory Classifier Systems. In P. L. Lanzi, W. Stolzmann, and S. W. Wilson, editors, *Learning Classifier Systems: from Foundations to Applications*, pages 175–194. Springer-Verlag, Heidelberg, 2000.
[Sut90] R. S. Sutton. Integrating architectures for learning, planning, and reacting based on approximating dynamic programming. In *Proceedings of the Seventh International Conference on Machine Learning ICML'90*, pages 216–224, San Mateo, CA, 1990. Morgan Kaufmann.
[Sut91] R. S. Sutton. Reinforcement learning architectures for animats. In J.-A. Meyer and S. W. Wilson, editors, *From animals to animats: Proceedings of the First International Conference on Simulation of Adaptative Behavior*, pages 288–296, Cambridge, MA, 1991. MIT Press.
[Wat89] C. J. Watkins. *Learning with delayed rewards*. PhD thesis, Psychology Department, University of Cambridge, England, 1989.
[Wil94] S. W. Wilson. ZCS, a zeroth level Classifier System. *Evolutionary Computation*, 2(1):1–18, 1994.
[Wil95] S. W. Wilson. Classifier fitness based on accuracy. *Evolutionary Computation*, 3(2):149–175, 1995.
[Wil01] S. W. Wilson. Function approximation with a classifier system. In L. Spector, Goodman E. D., A. Wu, W. B. Langdon, H. M. Voigt, and M. Gen, editors, *Proceedings of the Genetic and Evolutionary Computation Conference (GECCO01)*, pages 974–981. Morgan Kaufmann, 2001.
[Wil02] S. W. Wilson. Classifiers that approximate functions. *Natural Computing*, 1(2-3):211–234, 2002.

Estimating Classifier Generalization and Action's Effect: A Minimalist Approach

Pier Luca Lanzi

Artificial Intelligence and Robotics Laboratory
Dipartimento di Elettronica e Informazione
Politecnico di Milano
pierluca.lanzi@polimi.it

Abstract. We present an online technique to estimate the generality of classifiers conditions. We show that this technique can be extended to gather *some* basic information about the effect of classifier actions in the environment. The approach we present is *minimalist* in that it is aimed at obtaining as much information as possible from online experience, with as few modifications as possible to the classifier structure. Because of its plainness, the method we propose can be applied virtually to any classifier system model.

1 Introduction

Learning classifier systems are online techniques which exploit *reinforcement learning* and *evolutionary computation* to learn how to solve problems. Learning is achieved through the evolution of a set (or *population*) of condition-action-payoff rules (the *classifiers*) which represent the solution. As all the other *reinforcement learning* techniques, a learning classifier system neither assumes any knowledge about space of possible input configurations that will be encountered, nor any knowledge about the possible effect that its actions will have in the environment.

When the solutions evolved by a classifier system are analyzed, one of the main focus is the degree of generalization achieved. Classifiers that apply in many situations are studied because they provide some high level knowledge about the target problem; classifiers that apply in few situations are studied because they provide insight about some specific aspect of the target problem. In principle, since no assumptions are made about the input space, it should not be possible to know which situations classifiers have been applied to (unless we can keep track of the situations encountered as done in [3]). In practice, the analysis techniques usually applied assume that the problem space is known completely (e.g., [9]). However, this assumption becomes easily infeasible in large applications such as real-world robotics and analysis of huge amount of data. Thus it would be useful to have a technique capable of providing some information about the classifier generality which (i) does not assume that the input space is known completely, and (ii) does not store large amounts of additional information.

indicates a free cell. Classifiers conditions are 16 bits long (2 bits × 8 cells). There are eight possible actions, encoded with three bits. The system goal is to learn how to reach food position from any free position; the system is always in one free position and it can move in any surrounding free position; when the system reaches a food position (F) the problem ends and the systems is rewarded with 1000; in all the other cases the system receives zero reward.

4 Estimating Rule Generality

The method we propose is aimed at collecting as much information as possible regarding the classifier generality, making as few modifications as possible to the classifier system structure. The method is quite simple. To the classifier structure, we add (i) an array in to estimate the averages of the sensory inputs matched by the classifier; (ii) an array ε_{in} to estimate the error affecting the values in the array in. The size of in and ε_{in} is the number of sensory inputs (e.g., in the first example discussed it would be 16; in the second example it would be six). At the beginning of an experiment, the values of in and ε_{in} of classifiers in the rulebase are initialized with a predefined value (0 in our experiments, but a random value would work as well). When classifiers are matched against the current sensory inputs s_t, for every classifier cl that matches s_t, the array in and ε_{in} are updated as follows:

$$\forall cl \text{ that matches } s_t\ \forall k:$$
$$cl.in[k] \leftarrow cl.in[k] + \beta_e(s_t[k] - cl.in[k]) \tag{1}$$
$$cl.\varepsilon_{in}[k] \leftarrow cl.\varepsilon_{in}[k] + \beta_e(|s_t[k] - cl.in[k]| - cl.\varepsilon_{in}[k]) \tag{2}$$

where $s_t[k]$ represent the k-th sensory input of s_t; $cl.in[k]$ represents the k-th element of the array in of classifier cl; likewise $cl.\varepsilon_{in}[k]$ represents the k-th element of the array ε_{in} of classifier cl; β_e ($0 < \beta_e < 1$) is the learning rate. Equation 1 tries to estimate the average of all the input values that classifier cl matches. Given the average input values stored in $cl.in[k]$, Equation 2 builds an estimate of the error $cl.\varepsilon_{in}[k]$ affecting $cl.in[k]$. The reader familiar with Wilson's XCS [8] will recognize in equations 1 and 2 the update of the classifier prediction parameter p, and classifier error parameter ε used in XCS.

5 Experiments on Rule Generality

To test the proposed method for estimating classifier generality, we implemented it on an available XCS implementation [6] and applied our XCS extension to the 6-multiplexer with the following parameters (see [2] for details): $N = 400$, $\beta = 0.2$, $\alpha = 0.1$, $\epsilon_0 = 0.01$, $\nu = 5$, $\theta_{GA} = 25$, $\chi = 0.8$, $\mu = 0.04$, $\theta_{del} = 20$, $\delta = 0.1$, $\theta_{sub} = 20$, $P_{\#} = 0.6$, $\theta_{mna} = 2$, $doGASubsumption= 1$, $doActionSet$-$Subsumption= 0$ (i.e., action set subsumtption is not used). The learning rate β_e is 0.001. Table 1 reports an example of population that was evolved after 20000 learning problems; for the sake of brevity, only classifiers with non null

Table 1. Classifiers with non-zero prediction evolved by XCS for the 6-mpultiplexer. For every classifier we report: a unique identifier, the classifier condition and the classifier action separated by ":", the prediction p, the prediction error ε, and the fitness F. The array in and ε_{in} are reported as a sequence of $in[k] \pm \varepsilon_{in}[k]$ for $k \in \{0\ldots5\}$.

23331 11###1:1 $p = 1000$ $\varepsilon = 0.00$ $F = 1.00$
$\langle 1.00 \pm 0.00, 1.00 \pm 0.00, 0.53 \pm 0.50, 0.48 \pm 0.50, 0.47 \pm 0.50, 1.00 \pm 0.00 \rangle$
3485   000###:0 $p = 1000$ $\varepsilon = 0.00$ $F = 1.00$
$\langle 0.00 \pm 0.00, 0.00 \pm 0.00, 0.00 \pm 0.00, 0.48 \pm 0.50, 0.49 \pm 0.50, 0.52 \pm 0.50 \rangle$
7810   11###0:0 $p = 1000$ $\varepsilon = 0.00$ $F = 1.00$
$\langle 1.00 \pm 0.00, 1.00 \pm 0.00, 0.55 \pm 0.49, 0.47 \pm 0.50, 0.47 \pm 0.50, 0.00 \pm 0.00 \rangle$
8807   01#0##:0 $p = 1000$ $\varepsilon = 0.00$ $F = 1.00$
$\langle 0.00 \pm 0.00, 1.00 \pm 0.00, 0.54 \pm 0.50, 0.00 \pm 0.00, 0.47 \pm 0.50, 0.53 \pm 0.50 \rangle$
20897 01#1##:1 $p = 1000$ $\varepsilon = 0.00$ $F = 1.00$
$\langle 0.00 \pm 0.00, 1.00 \pm 0.00, 0.46 \pm 0.50, 1.00 \pm 0.00, 0.50 \pm 0.50, 0.46 \pm 0.50 \rangle$
373   10##1#:1 $p = 1000$ $\varepsilon = 0.00$ $F = 1.00$
$\langle 1.00 \pm 0.00, 0.00 \pm 0.00, 0.48 \pm 0.50, 0.50 \pm 0.50, 1.00 \pm 0.00, 0.50 \pm 0.50 \rangle$
7723   001###:1 $p = 1000$ $\varepsilon = 0.00$ $F = 1.00$
$\langle 0.00 \pm 0.00, 0.00 \pm 0.00, 1.00 \pm 0.00, 0.53 \pm 0.50, 0.50 \pm 0.50, 0.47 \pm 0.50 \rangle$
6108   10##0#:0 $p = 1000$ $\varepsilon = 0.00$ $F = 1.00$
$\langle 1.00 \pm 0.00, 0.00 \pm 0.00, 0.51 \pm 0.50, 0.52 \pm 0.50, 0.00 \pm 0.00, 0.52 \pm 0.5 \rangle$

---

prediction are reported. For each classifier, on the first line, we report: a unique identifier, the classifier condition and the classifier action separated by a ":", the prediction p, the prediction error ε, the fitness F; on the second line, for each input $s[k]$, we report the interval $in[k] \pm \varepsilon_{in}[k]$ which provides an estimate of the values of the k-th input that the classifier matches.

Since in Boolean multiplexer all the possible input configurations are considered, we should expect that (i) every position k of the classifier condition containing a don't care, would correspond to a $in[k]$ equal to 0.5, and to a $\varepsilon_{in}[k]$ equal to 0.5 (i.e., 0.5 ± 0.5); (ii) every position k of the classifier condition containing a 1 or a 0, would correspond to a $in[k]$ equal to 1 or 0 respectively, and to a $\varepsilon_{in}[k]$ equal to 0.0 (i.e., 0.5 ± 0.5). Table 1 confirms these expectations. All the interval corresponding to a constant input (0 or 1) have an error estimate (ε_{in}) equal to 0; all the interval corresponding to a don't care have values very near to 0.5. Of course, since in and ε_{in} are estimates they are affected by errors mainly due to the order in which the inputs appeared and to the value of β_e used. For instance, in classifier 23331, the interval corresponding to the first don't care is 0.53 ± 0.50, instead of 0.50 ± 0.50. Probably, we could obtain better approximations by tuning β_e or by using an adaptive value for β_e. On the other hand, some experiments we performed with more sophisticated techniques did not result in relevant improvements of the results; while we find that the estimates developed through our elementary technique are quite satisfactory. In fact, in Table 1 all the values of in and ε_{in} turn out to be correct when approximated to the first decimal digit.

The technique appears even more interesting if we apply it to a version of XCS involving symbolic expressions (e.g., [7]). We apply XCS with symbolic expression to the 6-multiplexer with the same parameters used in the previous experiments. In one of the final populations appears the following classifier:

((((NOT(NOT(NOT((X0 OR Y0)AND(NOT Y0)))))OR((NOT((NOT(X0 AND Y1)) AND((NOT X1) AND(Y3 AND X0)))) OR(NOT(((Y0 OR Y0)OR(Y3 OR Y1))AND((NOT Y3) AND((NOT X1)AND (Y3 AND X0)))))))AND((NOT X0)AND(NOT(NOT(Y1 AND X1))))AND((X1 AND((((Y1 AND (Y1 AND X1))AND Y1)AND(Y1 AND(NOT X0))) AND((((NOT(Y0 OR X0)) OR(((NOT X0)AND Y0)OR(Y1 AND Y0)))AND(NOT(NOT(Y1 AND X1))))AND(Y1 AND(NOT X0)))))AND X1)):1
$p = 1000\ \varepsilon = 0\ F = 1$
$\langle 0.00 \pm 0.00, 1.00 \pm 0.00, 0.55 \pm 0.50, 1.00 \pm 0.00, 0.46 \pm 0.50, 0.51 \pm 0.50 \rangle$

where AND, OR, and NOT represent the corresponding Boolean operators; X0, X1, Y0, ..., Y3 represent the six problem variables. Note that it is almost impossible to evaluate the generality of the condition (unless we simplify it), however the intervals suggest that the above classifier matches situations in which $x_0 = 0$, $x_1 = 1$, and $y_1 = 1$ while all the other variables can take any value between 0 and 1. When we simplify the condition with a tool for logic synthesis, we find that the condition corresponds to the expression $\overline{x_0}\ x_1\ y_1$, confirming what suggested by the intervals represented with in and ε_{in}.

6 Estimating Actions' Effect

The same principle used to estimate rule generality can be applied to provide some *elementary* information about the effect of classifier actions. The classifier structure is further extended, adding (i) an array ef to estimate the average values of the sensory inputs that the system will receive if the classifier action is performed; (ii) an array ε_{ef} to estimate the error affecting the values in the array ef. As in the previous case, the size of ef and ε_{ef} is the number of sensory inputs. Consider the (quite typical) performance cycle of a classifier system:

```
1.   repeat
2.       consider the current sensory input s_t ;
3.       build the set [M] of classifiers that match s_t ;
4.       select an action a_t from those in [M] ;
5.       store the classifiers in [M] with action a_t into [A] ;
6.       perform action a_t ;
7.       get the reward r_t ;
8.       get the next sensory input s_{t+1} ;
9.       distribute the reward r_t to classifiers;
10.  until stop criterion met;
```

Although the structure still resembles that of XCS, we left out some typical details of XCS (e.g., the genetic algorithm and the distribution of the reward in

[A]) to keep it as much general as possible. We can extend the cycle above by adding a step, between line 8 and line 9, in which for every classifier cl in [A], the array ef and the array ε_{ef} are updated with s_{t+1} as follows:

$$\forall cl \in [A] \; \forall k:$$
$$cl.ef[k] \leftarrow cl.ef[k] + \beta_e(s_{t+1}[k] - cl.ef[k]) \tag{3}$$
$$cl.\varepsilon_{ef}[k] \leftarrow cl.\varepsilon_{ef}[k] + \beta_e(|s_{t+1}[k] - cl.ef[k]| - cl.\varepsilon_{ef}[k]) \tag{4}$$

where $s_{t+1}[k]$ represents the k-th input of state s_{t+1}, i.e., the effect that performing action a_t in state s_t has on the k-th sensory input; as before, $cl.ef[k]$ represents the k-th element of the array ef of classifier cl; likewise $cl.\varepsilon_{ef}[k]$ represents the k-th element of the array ε_{ef} of classifier cl. Equation 3 estimates the average effect that the execution of classifier cl will have on the next k-th sensory input (i.e., on $s_{t+1}[k]$); equation 4, estimates the error affecting $cl.ef[k]$.

7 Experiments on Actions' Effect

We finally test the proposed techniques approach by applying our modified version of XCS to two grid environments: Woods1 and Maze4. The experiments are conducted as follows. First, we apply XCS to the two environments to develop some very small (i.e., very general) solutions. To achieve this we use both action-set subsumption and an additional condensation phase (see [2,5] for details). The parameters are set as in the previous experiments except: $N = 1600$, $\gamma = .7$, $doActionSetSubsumption = 1$ (i.e., AS-subsumption was used), $\theta_{nma} = 8$. For Woods1, the smallest solution consisted of 33 classifiers; for Maze4 consisted of 101 classifiers.

As a second step, we compute the exact average values that our method should be able to estimate. For every classifier cl evolved, we consider: (i) the input set I_{cl} of all the sensory inputs that cl matches; (ii) the effect set E_{cl} of all the sensory inputs that appears next to the execution of cl. Given I_{cl} and E_{cl}, for each input k we compute the average of the input values appearing in position k, $ai[k]$, the average of the input values appearing in position k after the execution of cl, $ae[k]$, and the average errors affecting $ai[k]$ and $ae[k]$, i.e., $\varepsilon_{ai}[k]$ and $\varepsilon_{ae}[k]$. More formally:

$$cl.ai[k] = (\sum_{s \in I_{cl}} s[k])/|I_{cl}| \tag{5}$$

$$cl.\varepsilon_{ai}[k] = (\sum_{s \in I_{cl}} |cl.ai[k] - s[k]|)/|I_{cl}| \tag{6}$$

$$cl.ae[k] = (\sum_{s \in E_{cl}} s[k])/|E_{cl}| \tag{7}$$

$$cl.\varepsilon_{ae}[k] = (\sum_{s \in E_{cl}} |cl.ae[k] - s[k]|)/|E_{cl}| \qquad (8)$$

The values of $cl.ai[k]$, $cl.\varepsilon_{ai}[k]$, $cl.ae[k]$, and $cl.\varepsilon_{ae}[k]$ are the average of input values for sensor k and the average effect for sensor k, i.e., what our technique should be able to estimate. Thus they serve as reference to compare the quality of the estimates developed through our approach.

Then we apply our version of XCS to Woods1. Table 2 reports three classifiers evolved for Woods1; due to space constraints, it is not possible to consider here all the 33 classifiers. For each classifier we report on the first line: a unique identifier, the condition and the action separated by a ":", the prediction p, the error ε, the fitness F. On the next lines, we report the list I of inputs that the classifier matches and the corresponding list E of inputs that appear after as effect of the classifier execution. Then, we report: (i) the array of average input values, which contains, for every k, the intervals $cl.ai[k] \pm cl.\varepsilon_{ai}[k]$ (Equation 5 and 6); (ii) the array of average effect values, which contains, for every k, the intervals $cl.ae[k] \pm cl.\varepsilon_{ae}[k]$ (Equation 7 and 8), (iii) the array representing the estimated classifier input, which contains, for every k, the intervals $cl.in[k] \pm cl.\varepsilon_{in}[k]$ (Equation 1 and 2); and (iv) the array representing the estimated classifier effect, which contains, for every k, the intervals $cl.ef[k] \pm cl.\varepsilon_{ef}[k]$ (Equation 3 and 4).

The first classifier in Table 2 (3112676) is the most general classifiers found; it matches all the possible sensory inputs of Woods1, nevertheless is very accurate and has an high fitness. Note that, since classifier 3112677 is very general, the estimate of the matching sensory inputs and the estimate of the classifier effect provide only limited information. The computed values of ai and ε_{ai} suggest that three of the sensory inputs (the second, the third, and the fifth) will be always 0; the computed values of ae and ε_{ae} suggest that no matter in which situation the classifier is applied, in the next state, nine of the inputs will be always zero. If we now consider the estimated input in and the estimated effect ef for the classifier, we find that our technique was successful in finding the same information. All the positions in ai and ae that correspond to zero values, corresponds to zero also in in and ef. While the estimated non zero values of in and ef are quite near to the corresponding computed values.

Some interesting information can be inferred from the non zero values. For instance, the values of $ae[8] = 0.06$ and $\varepsilon_{ae}[8] = 0.12$ suggests that on the subsequent state, the ninth bit will be very likely to be 0. Again, the values estimated through our approach for the same input are quite similar: the values of $ef[8] = 0.04$ and $\varepsilon_{ef}[8] = 0.8$. Note that, in Woods1 sensory inputs are not visited uniformly, accordingly the estimates evolved through our technique (e.g., in) tend to differ from the computed averages (e.g., ai), since the latter assume a uniform distribution of the inputs.

The second classifier (3112681) matches fewer situations, therefore its estimate of classifier input in and its estimate of classifier effect ef provide more information. For instance, the arrays in and ε_{in} of classifier 3112681 suggest that most of the sensory inputs matched do not change; likewise the values in ef and ε_{ef}. Again, we note that for non constant values, the estimates developed

Table 2. Three classifiers payoff evolved by our extended XCS for Woods1. For every classifier we report: a unique identifier, the classifier condition and the classifier action separated by ":", the prediction p, the prediction error ε, and the fitness F, the set I of matched inputs, the set E of inputs that arrive after the classifier execution, the computed average values of the inputs $ai \pm \varepsilon_{ai}$, and of the effects $af \varepsilon_{af}$, the *estimate* input and effect values, $in \pm \varepsilon_{in}$ and $ef \pm \varepsilon_{ef}$.

```
3112676  ###############1:001  p = 490    ε = 0.00  F = 1.00
         input set I              effect set E
         1010000000000000        1010000000000000
         0000001010000000        1010000000000010
         1010000000000010        1010000000000010
         0000001110100000        1000000000000010
         1000000000000010        0000000000001010
         0000000011100000        0000000000000010
         0000000000000010        0010100000000000
         0000000000110000        0010000000000000
         0000000000101100        0000001000000000
         0000000000101011        0000101000000000
         0000000000001010        0010101000000000
         0010000000000000        0010000000000000
         0000001000000000        1010000000000000
         0000101000000000        0000001010000000
         0010101000000000        0010101000000000
         0010100000000000        0010100000000000
```

$ai \pm \varepsilon_{ai}$ $\langle .19 \pm .30, .00 \pm .00, .31 \pm .43, .00 \pm .00, .19 \pm .30, .00 \pm .00, .31 \pm .43, .06 \pm .12,$
$.19 \pm .30, .06 \pm .12, .31 \pm .43, .06 \pm .12, .19 \pm .30, .06 \pm .12, .31 \pm .43, .06 \pm .12\rangle$
$ae \pm \varepsilon_{ae}$ $\langle .31 \pm .43, .00 \pm .00, .62 \pm .47, .00 \pm .00, .31 \pm .43, .00 \pm .00, .31 \pm .43, .00 \pm .00,$
$.06 \pm .12, .00 \pm .00, .00 \pm .00, .00 \pm .00, .06 \pm .12, .00 \pm .00, .31 \pm .43, .00 \pm .00\rangle$

$in \pm \varepsilon_{in}$ $\langle .21 \pm .33, .00 \pm .00, .37 \pm .45, .00 \pm .00, .20 \pm .32, .00 \pm .00, .24 \pm .38, .05 \pm .10,$
$.18 \pm .29, .07 \pm .12, .31 \pm .41, .05 \pm .10, .21 \pm .32, .04 \pm .09, .38 \pm .47, .09 \pm .14\rangle$
$ie \pm \varepsilon_{ie}$ $\langle .34 \pm .45, .00 \pm .00, .71 \pm .43, .00 \pm .00, .36 \pm .45, .00 \pm .00, .25 \pm .39, .00 \pm .00,$
$.04 \pm .08, .00 \pm .00, .00 \pm .00, .00 \pm .00, .09 \pm .14, .00 \pm .00, .32 \pm .42, .00 \pm .00\rangle$

```
3112681  ############1###:000   p = 700    ε = 0.00  F = 1.00
         input set I              effect set E
         0000000000101100        0000000000110000
         0000000000101011        0000000000101100
         0000000000001010        0000000000101011
```

$ai \pm \varepsilon_{ai}$ $\langle .00 \pm .00, .00 \pm .00, .00 \pm .00, .00 \pm .00, .00 \pm .00, .00 \pm .00, .00 \pm .00, .00 \pm .00$
$.00 \pm .00, .00 \pm .00, .67 \pm .44, .00 \pm .00, 1.00 \pm .00, .33 \pm .44, .67 \pm .44, .33 \pm .44\rangle$
$ae \pm \varepsilon_{ae}$ $\langle .00 \pm .00, .00 \pm .00, .00 \pm .00, .00 \pm .00, .00 \pm .00, .00 \pm .00, .00 \pm .00, .00 \pm .00$
$.00 \pm .00, .00 \pm .00, 1.00 \pm .00, .33 \pm .44, .67 \pm .44, .33 \pm .44, .33 \pm .44, .33 \pm .44\rangle$

$in \pm \varepsilon_{in}$ $\langle .00 \pm .00, .00 \pm .00, .00 \pm .00, .00 \pm .00, .00 \pm .00, .00 \pm .00, .00 \pm .00, .00 \pm .00$
$.00 \pm .00, .00 \pm .00, .63 \pm .45, .00 \pm .00, 1.00 \pm .00, .27 \pm .40, .73 \pm .40, .37 \pm .46\rangle$
$ef \pm \varepsilon_{ef}$ $\langle .00 \pm .00, .00 \pm .00, .00 \pm .00, .00 \pm .00, .00 \pm .00, .00 \pm .00, .00 \pm .00, .00 \pm .00$
$.00 \pm .00, .00 \pm .00, 1.00 \pm .00, .25 \pm .38, .75 \pm .38, .27 \pm .40, .49 \pm .49, .49 \pm .49\rangle$

```
3112677  ###############1:111  p = 1000   ε = 0.00  F = 1.00
         input set I              effect set E
         0000000000101011        0000000010101000
```

$ai \pm \varepsilon_{ai}$ $\langle .00 \pm .00, .00 \pm .00, .00 \pm .00, .00 \pm .00, .00 \pm .00, .00 \pm .00, .00 \pm .00, .00 \pm .00$
$.00 \pm .00, .00 \pm .00, 1.00 \pm .00, .00 \pm .00, 1.00 \pm .00, .00 \pm .00, 1.00 \pm .00, 1.00 \pm .00\rangle$
$ae \pm \varepsilon_{ae}$ $\langle .00 \pm .00, .00 \pm .00, .00 \pm .00, .00 \pm .00, .00 \pm .00, .00 \pm .00, .00 \pm .00, .00 \pm .00$
$1.00 \pm .00, .00 \pm .00, 1.00 \pm .00, .00 \pm .00, 1.00 \pm .00, .00 \pm .00, .00 \pm .00, .00 \pm .00\rangle$

$in \pm \varepsilon_{in}$ $\langle .00 \pm .00, .00 \pm .00, .00 \pm .00, .00 \pm .00, .00 \pm .00, .00 \pm .00, .00 \pm .00, .00 \pm .00$
$.00 \pm .00, .00 \pm .00, 1.00 \pm .00, .00 \pm .00, 1.00 \pm .00, .00 \pm .00, 1.00 \pm .00, 1.00 \pm .00\rangle$
$ef \pm \varepsilon_{ef}$ $\langle .00 \pm .00, .00 \pm .00, .00 \pm .00, .00 \pm .00, .00 \pm .00, .00 \pm .00, .00 \pm .00, .00 \pm .00$
$1.00 \pm .00, .00 \pm .00, 1.00 \pm .00, .00 \pm .00, 1.00 \pm .00, .00 \pm .00, .00 \pm .00, .00 \pm .00, \rangle$

through our technique are quite reasonable, also considering that the distribution of inputs is not considered in the computation of ai and ae.

The last classifier (3112677) is the one we discussed in the first example. Although it might appear quite general, it matches only one possible situation, accordingly its estimates are exact. To provide a rough evaluation of the estimates developed with our technique we computed the average error between the computed values for inputs and effect (ai, ε_{ai}, ae, ε_{ae}) and the corresponding estimated values (in, ε_{in}, ef, ε_{ef}). For this purpose we applied our modified version of XCS to Woods1 ten times. For each run we computed:

$$\texttt{error}(in) = \frac{\sum_{cl \in P} \sum_k |cl.ai[k] - cl.in[k]|}{|P| \times n} \quad (9)$$

$$\texttt{error}(\varepsilon_{in}) = \frac{\sum_{cl \in P} \sum_k |cl.\varepsilon_{ai}[k] - cl.\varepsilon_{in}[k]|}{|P| \times n} \quad (10)$$

$$\texttt{error}(ef) = \frac{\sum_{cl \in P} \sum_k |cl.ae[k] - cl.ef[k]|}{|P| \times n} \quad (11)$$

$$\texttt{error}(\varepsilon_{ef}) = \frac{\sum_{cl \in P} \sum_k |cl.\varepsilon_{ae}[k] - cl.\varepsilon_{ef}[k]|}{|P| \times n} \quad (12)$$

where P represents the final population evolved; n is the number of inputs (16 in Woods1). Over ten runs in Woods1, the average $\texttt{error}(in)$ is 0.0054, the average $\texttt{error}(\varepsilon_{in})$ is 0.0035, the average $\texttt{error}(ef)$ is 0.0075, the average $\texttt{error}(\varepsilon_{ef})$ is 0.0041. We applied the same technique to Maze4 with a population size $N = 2000$ for ten runs, measuring an average $\texttt{error}(in)$ of 0.0058, an average $\texttt{error}(\varepsilon_{in})$ of 0.0024, an average $\texttt{error}(ef)$ of 0.0057, an average $\texttt{error}(\varepsilon_{ef})$ of 0.0028. Since Maze4 does not allow many generalization, the prediction of the classifier effect is a little bit more accurate than that obtained for Woods1, in fact in Maze4 the values of $\texttt{error}(ef)$ and $\texttt{error}(\varepsilon_{ef})$ are a little bit smaller. Note also that, these errors measures the difference between *our estimated values* and the *best values* that can be obtained following our approach. Thus, even if the errors are small, still the accuracy of the prediction built is limited by the approach we follow which is less powerful than those methods specifically design for anticipatory behavior.

As the short example presented and the few statistics collected suggest, the method we propose provides limited information about classifiers that apply in many situations; while it provides more precise information on classifier that apply in fewer situations. In either cases, the method appears to be able to provide reliable information about classifier generality and some interesting information about the effect of classifier activation.

8 Extensions and Limitations

The proposed approach is very simple, therefore it is opened to critics and extensions.

Classifier Generality. The first limitation regards the estimates of classifier generality. Our approach, for every input k, develops an interval $in[k] \pm \varepsilon_{in}[k]$ that estimates the range of values that appears at corresponding input. The larger the range, the more general the classifier appears to be with respect to that specific input. Accordingly, we might be tempted to use these intervals to build a subsumption relation for classifier representations that usually do not allow it. Lanzi [4] noted that with symbolic conditions subsumption operators are computationally infeasible. In Sect. 7 we showed that the intervals $in[k] \pm \varepsilon_{in}[k]$ might provide some indication about the classifier generality. Thus, we might try to define a subsumption relation for symbolic conditions using our technique.

Unfortunately this is not possible. As a counter example consider the two conditions defined over a variable X∈ {0...9}: (i) (2<X) AND (X<6); (ii) (X<3) AND (X>5). Both conditions have the same average input value, $in = 4$, while the associated error ε_{in} is 0.66 for condition (i), 3 for condition (ii). Comparing the two intervals we should argue that condition (ii) is more general than condition (i) since the estimated input interval of condition (ii) subsumes the estimated input interval for condition. But the two conditions form a partition of the input domain and they cannot be compared through an inclusion operator. Thus we cannot define subsumption operators based on the estimated intervals. These in fact are developed from linear estimators, while symbolic conditions usually express non linear relations

Classifier Effect. Our approach tend to provide information on the classifier effect in terms of input values that the system will experience after the classifier activation. Other anticipatory techniques like ACS [1] provides information in term of what will change in the current sensory input after the classifier execution. It is straightforward to extend our technique to provide such kind of information. Given the current input s_t and the next input s_{t+1} we define an array $\Delta[k]$ as follows: $\Delta[k] = 1$ if $s_t[k] \neq s_{t+1}[k]$, 0 otherwise. The array Δ represents the difference that occurred to the current state s_t after the classifier execution. The value $\Delta[k]$ can now be used in Equation 3 and Equation 4 instead of $s_{t+1}[k]$ so that ef will now estimate the *changes* in the environment state rather than the next state.

In addition, note that in our approach we also take into account the final state, corresponding to the position with food. Classifier 3112677 in Table 2 has reward 1000 meaning that performing action 111 in state 0000000000101011 will take the agent to the end position, corresponding to state 0000000010101000. This information might be exploited to perform some *very elementary* planning (as done with Anticipatory Classifier Systems), although, given the basic information provided by the approach is not clear at the moment the true potential of the approach. Alternatively, we might eliminate the effect estimate from classifiers whose action takes the agent to a final state. In this case, planning should focus on reaching of classifiers with a null effect.

9 Summary

We presented a very basic method to estimate the degree of generality of classifier conditions. The method was also applied to obtain *basic* information about the effect of classifier execution. We showed some example of information that we could extract from some very elementary problems. Since the method does not provide an exact anticipation, it is currently difficult to provide some performance metrics to estimate its effectiveness. On the other hand, because of its simplicity, the method can be added virtually on any classifier system model. The code to experiment the approach discussed here is available under GNU Public License at the `xcslib` Web site [6].

References

1. Martin Butz. *Anticipatory Learning Classifier Systems*. Kluwer, 2002.
2. Martin V. Butz and Stewart W. Wilson. An algorithmic description of xcs. *Journal of Soft Computing*, 6(3-4):144-153, 2002.
3. Pierre Géard and Olivier Sigaud. Yacs: Combining dynamic programming with generalization in classifier systems. In Pier Luca Lanzi, Wolfgang Stolzmann, and Stewart W. Wilson, editors, *IWLCS*, volume 1996 of *Lecture Notes in Computer Science*, pages 52-69. Springer, 2001.
4. Pier Luca Lanzi. Extending the Representation of Classifier Conditions Part II: From Messy Coding to S-Expressions. In Wolfgang Banzhaf, Jason Daida, Agoston E. Eiben, Max H. Garzon, Vasant Honavar, Mark Jakiela, and Robert E. Smith, editors, *Proceedings of the Genetic and Evolutionary Computation Conference (GECCO-99)*, pages 345-352. Morgan Kaufmann, 1999.
5. Pier Luca Lanzi. Mining interesting knowledge from data with the xcs classifier system. In Lee Spector, Erik D. Goodman, Annie Wu, W.B. Langdon, Hans-Michael Voigt, Mitsuo Gen, Sandip Sen, Marco Dorigo, Shahram Pezeshk, Max H. Garzon, and Edmund Burke, editors, *Proceedings of the Genetic and Evolutionary Computation Conference (GECCO-2001)*, pages 958-965, San Francisco, CA 94104, USA, 7-11 July 2001. Morgan Kaufmann.
6. Pier Luca Lanzi. The xcs library. http://xcslib.sourceforge.net, 2002.
7. Pier Luca Lanzi and Alessandro Perrucci. Extending the Representation of Classifier Conditions Part II: From Messy Coding to S-Expressions. In Wolfgang Banzhaf, Jason Daida, Agoston E. Eiben, Max H. Garzon, Vasant Honavar, Mark Jakiela, and Robert E. Smith, editors, *Proceedings of the Genetic and Evolutionary Computation Conference (GECCO 99)*, pages 345-352, Orlando (FL), July 1999. Morgan Kaufmann.
8. Stewart W. Wilson. Classifier Fitness Based on Accuracy. *Evolutionary Computation*, 3(2):149-175, 1995. http://prediction-dynamics.com/.
9. Stewart W. Wilson. Compact rulesets from xcsi. In Pier Luca Lanzi, Wolfgang Stolzmann, and Stewart W. Wilson, editors, *IWLCS*, volume 2321 of *Lecture Notes in Computer Science*, pages 197-210. Springer, 2002.

Towards Building Block Propagation in XCS: A Negative Result and Its Implications

Kurian K. Tharakunnel, Martin V. Butz, and David E. Goldberg

Illinois Genetic Algorithms Laboratory (IlliGAL)
University of Illinois at Urbana-Champaign
104 S. Mathews, 61801 Urbana, IL, USA
{kurian,butz,deg}@illigal.ge.uiuc.edu

Abstract. The accuracy-based classifier system XCS is currently the most successful learning classifier system. Several recent studies showed that XCS can produce machine-learning competitive results. Nonetheless, until now the evolutionary mechanisms in XCS remained somewhat ill-understood. This study investigates the selectorecombinative capabilities of the current XCS system. We reveal the accuracy dependence of XCS's evolutionary algorithm and identify a fundamental limitation of the accuracy-based fitness approach in certain problems. Implications and future research directions conclude the paper.

1 Introduction

After Holland's introduction of *learning classifier systems* (LCSs), originally referring to a *cognitive system* [7], one of the most important steps in LCS research was the development of the accuracy-based classifier system XCS by Wilson [12]. Most of the recent work on LCSs focuses on XCS. Among the many changes from the original LCS, the accuracy-based fitness in XCS is considered as the distinctive feature and the most important reason for success.

Although crossover has always been applied in XCS, our experimental investigations showed that in most investigated classification problems to date mutation alone seems to be sufficient to evolve an appropriate solution. However, the larger a problem the more difficult it becomes to generate an accurate solution by mutation. This is the case since mutation in general results in a diversification pressure and in LCSs effectively in a specialization pressure since classifiers usually specify only a small fraction of the available features [2]. The larger the problem, however, the more general classifiers need to be to undergo sufficient evaluations and reproductive events. Thus, the larger a problem the smaller mutation needs to be. Consequently, only successful recombinations of accurate sub-structures, or *building blocks* [6], can do the trick. Additionally, as argued in [5], suitable search and ultimately innovative events are determined by effective recombination of appropriate sub-parts. As this insight led to the development of *competent GAs*—GAs that solve boundedly difficult problems quickly, accurately, and reliably—the same approach seems necessary in learning classifier system research to develop competent LCSs, that is, LCSs that solve typical (decomposable) machine learning problems efficiently.

Recently, several studies addressed how XCS works. Many of these try to explain the XCS mechanisms and help choosing appropriate parameter settings. An important aspect missing so far is an explanation for the apparent non-performance of XCS in some problems. This paper investigates a set of such hard problems. By doing that, we reveal an important interaction between the accuracy based fitness and the selectorecombinative GA [6,4] in XCS that prevents learning. Previous early hints on this interaction can be found in [1].

The role of the GA in XCS is to evolve accurate classifiers as well as to search for optimal generalizations within classification niches. The basic working of GA dictates that selection, recombination, and mutation progressively results in the discovery of better individuals. We show that the GA in XCS may not be able to achieve this in particular problems due to a lack of suitable accuracy guidance from the over-general side. We use concatenated multiplexer problems (multiplexer problems formed by concatenating the condition and action parts of more than one single multiplexer problem) to illustrate how XCS performance is affected by the combination of accuracy based fitness and selectorecombinative GAs.

After a brief review on previous analyses of XCS mechanisms, we formalize the derivation of the estimated prediction error in XCS. Since the accuracy is derived from the prediction error estimate, this estimate is crucial for XCS success. In Sect. 4 we identify a fundamental weakness in XCS's accuracy approach, illustrating successful and unsuccessful selectorecombinative events in XCS. Concluding remarks complete the paper.

2 XCS Performance

Along with the introduction of XCS, Wilson [12] illustrated the success of XCS both in single-step problems and multi-step problems with results from experiments on Multiplexer and Woods2 environments respectively. Later Wilson showed the successful performance of XCS in a standard data mining task [13]. However, the results obtained in several other test environments were not very encouraging. An interesting comparison of the performance of XCS with other LCS models in various Maze environments can be found on the web page: http://www.ai.tsi.lv/ga/lcs_performance.html

Although considerably more tractable than Holland's original LCS model, XCS is still a complex system. There have been several attempts recently to understand XCS mechanisms and their interactions. The studies by Kovacs [8] established that XCS would be able to accurately build an optimal representation of the payoff function of a problem. Kovacs also illustrated how XCS overcomes the problem of strong over-generals encountered in strength based LCSs without fitness sharing [9]. A fundamental problem difficulty dimension was characterized by Kovacs' optimal rule set size measure [O] [10]. Recently, Butz et al. [1,2] have given a more analytical treatment of XCS mechanisms, in which further dimensions of problem complexity were identified.

3 Accuracy Revisited

The most important and distinctive feature of XCS is its fitness definition. In the traditional LCS approach payoff prediction of the classifier, sometimes combined with condition specificity, is directly used as the fitness measure. In XCS, the accuracy with which a classifier predicts payoff is used as a measure of its fitness.

The actual procedure of fitness calculation involves several steps [3]. Every time a classifier takes part in an action set, first its prediction parameter is estimated using the immediate payoff received according to the standard Widrow-Hoff delta rule. Next, an estimate of the absolute error in estimate of the prediction is done using the current error (absolute difference between the immediate payoff and the current estimate of prediction). This is again done by the Widrow-Hoff delta rule. This error estimate is then used to calculate the accuracy of a classifier by using a power function on error. This ensures that classifiers with error less than a threshold value will have accuracy one and the accuracy falls steep with increase in error. Next, the accuracy values (all between zero and one) of all the classifiers in the action set are used to calculate a relative accuracy value for each classifier in the action set. These relative accuracy values are then used to update the fitness of classifiers using the Widrow-Hoff delta rule once again.

The procedure described above shows that the estimated absolute error plays a crucial role in the determination of accuracy of a classifier. To understand the full significance of this we should know what the estimates of prediction and prediction error mean to a classifier. A classifier represents a set of environmental states along with a possible action. Thus, a classifier in fact represents a set of state-action pairs for the given problem. Each state-action pair results in a particular reward (state-action value). Hence, as a classifier takes part in many action sets and with the update procedure described above, the prediction parameter of the classifier is estimated towards the average of the state-action values of the state-action pairs covered by the classifier. Similarly, the update procedure on prediction error results in an estimate of the mean absolute deviation (MAD) in the covered state-action pairs. Thus, the error estimate of a classifier in XCS is in fact an estimate of MAD of state-action values associated with that classifier. We call this value simply the MAD of the classifier.

The above definition of error can be used to calculate directly the error of a classifier and hence determine its accuracy. For example, let p_s denote the probability of correct classification with a uniform reward of r and zero reward otherwise. Then the prediction P of a classifier may be written as

$$P = rp_s + 0 * (1 - p_s) = rp_s \tag{1}$$

Similarly, the error ϵ which is the MAD of the classifier may be written as

$$\epsilon = |r - P|p_s + |0 - P|(1 - p_s) = 2r(p_s - p_s^2) \tag{2}$$

(see also [1]). We use these expressions to determine reward prediction and reward prediction error of classifiers in the following sections. Figure 1 illustrates

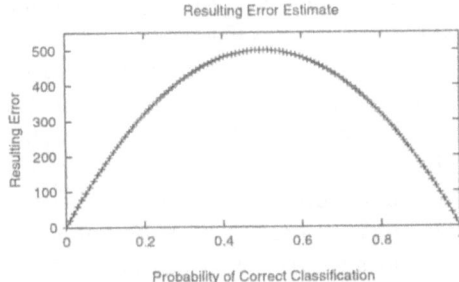

Fig. 1. Prediction-error dependence on correctness of classifier in a 1000/0 reward scheme

Equation 2. It can be seen that the more consistently a classifier classifies correctly (or incorrectly), the smaller its prediction error and thus the larger its accuracy will be.

4 Accuracy Guidance for Successful Recombination

Two fundamental processes are associated with the working of every LCS - rule discovery and rule evaluation which are executed by the GA and the credit allocation components, respectively. In XCS, these two processes work towards the objective of building a complete, accurate, and maximally general payoff map. That is, a problem representation is formed that is able to predict the resulting payoff of an action, or classification, for any possible problem instance. Important changes in XCS are the accuracy-based fitness approach and GA selection which is applied in action set niches rather than panmictically in the whole population as in traditional LCSs. Deletion, however, is done panmictically in the whole population. This results in an inherent generalization pressure as hypothesized in [12] and formulated in [2].

The question is now how the GA evolves accurate classifiers in XCS. The GA used in XCS is basically selectorecombinative. This means that the driving force of evolution towards better individuals is the dual process of selection and recombination with a continuous influence of mutation. With this point of view, accurate classifiers are evolved from selection, mutation, and recombination of good partially accurate solutions. Especially successful recombinations of different partially accurate sub-parts, or *building blocks*, should lead to higher accuracy.

4.1 Accuracy Guidance Required

Let's define an accurate, maximally general classifier as a *completely successful classifier* and the corresponding generalization a *completely successful generalization*. A perfect classifier has maximum accuracy so that any further generalization of the classifier condition will result in a reduction in accuracy. Such a classifier can be characterized by the specified positions in the condition part since all other positions are filled with don't care symbols. Let us say this classifier condition consists of three specified positions a, b and c. Based on the discussion above, this classifier can be generated through GA recombination or mutation. Hence we can reasonably assume that the specific pattern of abc can result only from recombinations or mutations of classifiers with the partial patterns a, b and c in them (since a direct generation of abc by mutation is highly unlikely). We term the classifiers that specify a subset of positions abc *partially successful classifiers* and the corresponding generalizations *partially successful generalizations*. Note that there might be more than one possible completely successful generalization and different completely successful classifiers might overlap. This issue, however, needs to be addressed separately from the concern in this paper.

The above argument leads to the hypothesis that successful generalization can be possible only if the following two conditions are satisfied. (1) Sustenance of partially successful classifiers in the population; (2) Partially successful classifiers whose conditions are more similar to completely successful classifiers have higher accuracy. The first condition says that the partially successful classifier should have some minimum accuracy (fitness) so that it will not get deleted from the population. To a great extent this is dependent on the parameter setting of XCS as well as on the problem properties. The second condition says that classifiers with conditions closer to a completely successful classifier should have higher fitness so that a proper recombination/mutation can lead to the generation of completely successful classifiers. As we investigate in the rest of this paper, this *accuracy guidance* is essential for successful evolutionary learning in XCS.

Note that we focus on when a successful recombination, i.e. a recombination that generates offspring that is more similar to a perfect classifier in its condition part than its parents, actually also results in higher accuracy. In this paper, we are not interested in the matter of a good crossover operator or even in the creation of a competent crossover operator (such as the probabilistic model building GAs field (PMBGAs) [11]), but in the satisfaction of the preconditions for such successful recombination.

4.2 Accuracy Guidance in the Six Multiplexer

To illustrate accuracy guidance further we take the Boolean multiplexer problem as an example.

Multiplexer problems are widely used test problems in LCS research. In a multiplexer problem, the agent tries to predict the output of a multiplexer function. A multiplexer function is a function defined on binary strings of length $k + 2^k$. The first k bits reference the output bit in the 2^k remaining bits. In a 3 multiplexer problem, for example, a zero in the first bit would determine the output being the second bit while a one would reference the third bit.

In the 6 multiplexer problem the input strings are of length six and the first two bits determine the output bit position. According to our definition of a completely successful classifier, it can easily be observed that one of the completely

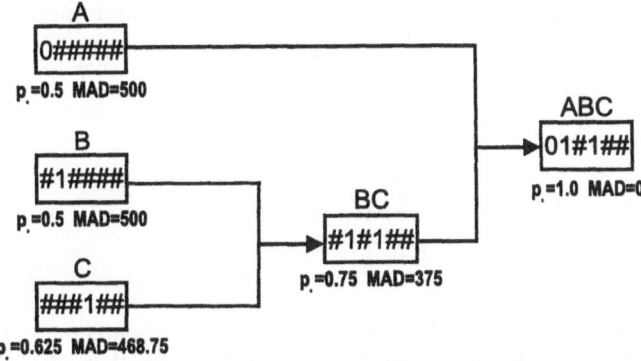

Fig. 2. Progress of successful generalization in 6 multiplexer problem

successful classifier in a 6 multiplexer problem is $01\#1\#\# \to 1$. Now consider that the condition part of this classifier consists of three consecutive patterns a, b and c involving the three specified bits respectively. Then a partially successful classifier with pattern a is $0\#\#\#\#\# \to 1$. Similarly partially successful classifiers with patterns b and c are $\#1\#\#\#\# \to 1$ and $\#\#\#1\#\# \to 1$ respectively. Here we have considered the most general version of partially successful classifiers though we can consider partially successful classifiers with more specific bits with the same effect. Now let us look at what happens to the accuracy when the partially successful classifiers combine.

Figure 2 illustrates the progressive recombination of partially successful classifiers A, B and C resulting in the completely successful classifier ABC. A 1000/0 payoff scheme is used in this example and the accuracy of classifiers is measured as MAD. It can be seen that the MAD diminishes when partially successful classifiers recombine. Eventually, the completely successful classifier with MAD 0 will be generated. Here the condition (2) of our hypothesis holds in that a successful recombination of partially successful classifiers also results in a decrease in MAD and thus in an increase in accuracy. This explains how XCS with the accuracy based fitness is able to successfully solve the 6 multiplexer problem. What happens when condition (2) is not satisfied? This is illustrated in the next section using what we call a *concatenated multiplexer problem*.

4.3 Concatenated Multiplexer Problems

We construct concatenated multiplexer problems by combining more than one multiplexer problem. In a concatenated multiplexer problem, the agent tries to predict the output of a concatenated multiplexer function. A concatenated multiplexer function is the function obtained by concatenating the input part and the output part of more than one individual multiplexer functions of the same type. The output will be the combination of outputs determined by the *individual multiplexer functions*. For example, Table 1 shows a sample of inputs and outputs of a concatenated multiplexer function made of three 3-multiplexer

Table 1. A sample input/output mapping of *3X3* multiplexer problem

Input	Output
000 000 000	000
010 100 111	101
110 101 001	010
111 110 001	100
111 111 111	111

Table 2. In the *3X3* multiplexer problem the necessary classifiers for an accurate and correct classification need to be much more specific than those for an accurate but wrong classification (classifiers that always suggest a wrong classification)

Correct Classifiers		Wrong Classifiers	
Condition	Action	Condition	Action
00# 00# 00#	000	### ### 01#	000
00# 00# 1#0	000	### ### 1#1	000
00# 1#0 00#	000	### 01# ###	000
00# 1#0 1#0	000	### 1#1 ###	000
1#0 00# 00#	000	01# ### ###	000
1#0 00# 1#0	000	1#1 ### ###	000
1#0 1#0 00#	000	### ### 00#	001
1#0 1#0 1#0	000	...	...
00# 00# 01#	001	...	...
...	...	...	...

functions. We call this concatenated problem a *3X3* multiplexer problem. In general, a *mXn* multiplexer problem is a combination of *m* problems of type *n*.

We use concatenated multiplexer problems to illustrate the hypothesis of Sect. 4.1 because we are sure of the completely successful generalizations existing in these problems. For example, in the *3X3* problem, we know the completely successful generalizations for the individual three multiplexer problems and so the completely successful generalization for the concatenated problem can be easily constructed. A part of the optimal classifier population [O] for the *3X3* problem is shown in Table 2. Note that for the wrong classifiers the wrong output of one component is sufficient for the entire problem to have a wrong output and so a greater level of generalization is possible. The size of the optimal population $|[O]|$ in an *mXn* problem where the n-multiplexer has n_p position bits can be calculated by $|[O]| = 2^m(2^{n_p^m}) + 2^m m 2_p^n = 2^{2m+n_p m} + m 2^{m+n_p}$ adding the number of correct and wrong classifiers together. In the *3X3* problem $|[O]| = 64 + 48 = 112$.

It is surprising that although an optimal population exists for the *3X3* problem, XCS fails to evolve it. We show that this is because condition (2) of our hypothesis in Sect. 4.1 is not satisfied in this problem.

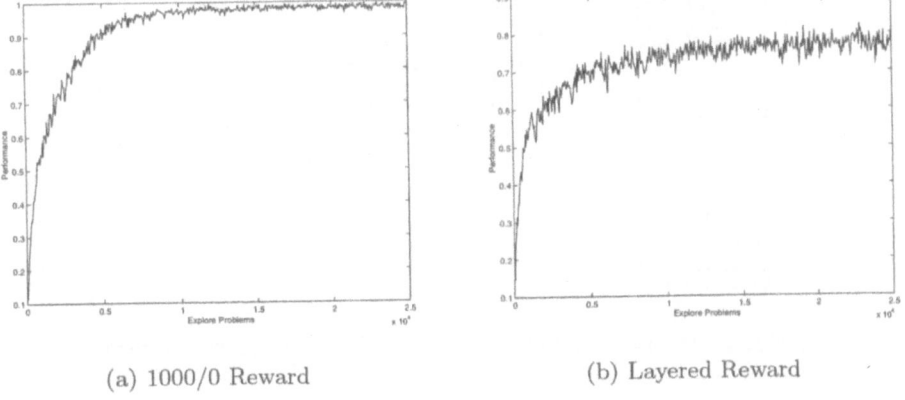

(a) 1000/0 Reward   (b) Layered Reward

Fig. 3. XCS performance in the *3X3* multiplexer problem

Before we proceed further with concatenated multiplexer problems we need to decide about the reward function to be used in these problems. We can use the usual 1000/0 reward structure which means that a correct output from all the component multiplexer problems will result in a reward of 1000 but if one of them is wrong then the output is 0. Another possibility is to use a layered reward structure, i.e. a progressive reward of 1000 with each component multiplexer giving the correct output. This will result in a reward of 1000c for the concatenated problem where c is the number of components correctly classified by the specified action. We use the 1000/0 reward function here. Later we will show that XCS performance will be worse with layered reward as the first condition (sustenance of partially successful classifiers) is violated as well in that case.

Next, we shall see why XCS is not able to perform well in the *3X3* problem. Remember that the three multiplexer is a rather trivial problem for XCS. There are only two actions and XCS will quickly find the accurate and optimal classifier population. Now, in the *3X3* problem, there are eight actions, two from each component problem. The XCS performance on the *3X3* with 1000/0 reward structure is shown in Fig. 3(a). (All the performance results shown in this paper are averages of 10 runs with parameters set as follows: $N = 2000$, $\beta = 0.2$, $\alpha = 0.1$, $\epsilon_0 = 0.01$, $\nu = 5$, $\theta_{GA} = 25$, $\chi = 0.8$, $\mu = 0.04$, $\theta_{del} = 20$, $\delta = 0.1$, $\theta_{sub} = 20$, $P_\# = 0.33$, $p_{explr} = 1$, $p_I = -10$, $\epsilon_I = 0$, and $F_I = 0.01$ [3]). It can be seen that XCS fails to reach 100% performance even after 25,000 steps.

To see why XCS is not able to perform well in this problem, let us look at the completely successful classifiers for this problem. For example, 01# 01# 01# → 111 is an accurate and optimal generalization (completely successful classifier) for the problem (the spaces between condition part of component multiplexers *are for clarity*). One possible way for this completely successful classifier to form in the population is by the recombination of the partially successful classifiers :

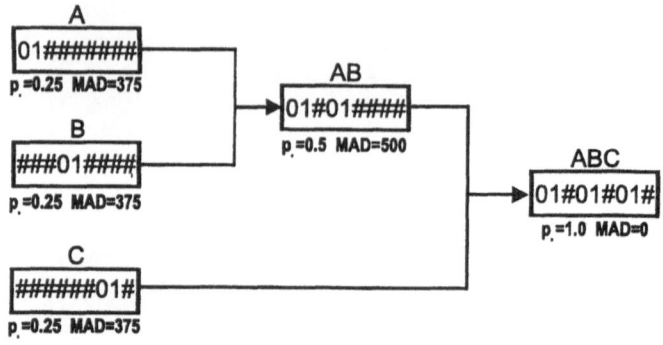

Fig. 4. Necessary learning progress in *3X3* multiplexer problem

01# ### ### → 111, ### 01# ### → 111 and ### ### 01# → 111. We have shown only the most general case of partially successful classifiers though in reality more specific forms are likely. However, this will not change the results we are going to illustrate. Also, remember that the GA occurs only in the action set specified by the action 111 and these recombinations happen progressively. In Fig. 4, a possible progressive formation of the completely successful classifier from partially successful classifiers is shown. The accuracy of classifiers are expressed as MAD. Note that the MAD of the completely successful classifier is 0.

In particular, let's look at the exact derivation of those numbers. Classifiers $cl_1 = $ 01# ### ### → 111 and $cl_2 = $ ### 01# ### → 111 classify a random input string correctly with a probability of 0.25 since there is a 50/50 chance that either of the other components is correct. Thus, the reward predictions of these classifiers will be approximately $0.25 \cdot 1000 = 250$ and consequently their reward prediction errors (MAD) will approximate 375 (see equations 1 and 2). The more specific classifier $cl_3 = $ 01# 01# ### → 111 is much closer to the desired completely successful classifier $cl_4 = $ 01# 01# 01# → 111. Thus, cl_3 should have higher fitness than cl_1 and cl_2 since a less complex recombinatory event or the mutation of the appropriate two attributes can actually lead to the generation of classifier cl_4. Classifier cl_3 has a probability of 0.5 to perform the correct classification. Thus, both its reward prediction and MAD will approximate 500. With a higher MAD, classifier cl_3 will actually have a smaller accuracy than classifiers cl_1 and cl_2. Thus, cl_3 will encounter less recombinatory events and is consequently highly likely to be deleted soon. Moreover, in order to get to classifiers cl_1 and cl_2 it is necessary that those classifiers have a lower MAD than, for example, the completely general classifier (classifier with only don't care symbols in the condition part). However, the probability of a completely general classifier to classify a random input correctly is 0.125 which results in a MAD of 218.75 so that the completely general classifier has an even higher accuracy than cl_1 or cl_2. Thus, besides the inherent generalization pressure in XCS [12,2], in this problem, fitness also pushes towards generality instead of towards specificity so that it is highly unlikely that XCS will ever evolve an accurate representation of the problem.

Indeed, looking at the evolving population in more detail revealed that none of the completely successful classifiers are present in the population. It seems somewhat surprising, though, why XCS still reaches a performance of near 100% as displayed in Fig. 3(a). A more detailed analysis revealed that XCS is actually learning the classifiers that specify inaccurate classifications accurately (shown on the right column in table 2). Thus, given a current match set, XCS eventually 'knows' that seven of the eight possible classifications will be incorrect and thus chooses the correct classification most of the time.

In Fig. 3(b), we show performance of XCS in the *3X3* multiplexer problem when the reward is $1000c$ where c is the number of components correctly classified by the specified action. It can be observed that performance is worse with this layered reward structure. This is because, with layered reward, the difference between MAD of partially successful classifiers resulting from recombination/mutation and MAD of their parent classifiers is even larger. This results in a smaller chance for the newly generated partially successful classifiers to sustain in the population and cause subsequent progress of successful generalizations. This shows that layered reward does not necessarily need to be helpful.

The concatenated multiplexer problem is essentially a problem in which the base probability of classifying an input correctly is below 0.5. Looking back at Fig. 1 we see that the resulting problem is that XCS needs to cope with the problem that more correct classifiers (those closer to 1.0 and thus initially closer to 0.5) are actually less accurate due to the MAD-based accuracy. Moreover, the problem is not symmetrical in that an accurate, maximally general classifier that predicts payoff 0 does not have the same structure as an accurate, maximally general classifier that predicts payoff 1000. If there are other problem properties that can cause the same problem or if these properties are typical in interesting problems is a subject of future research.

Note that our analysis made several simplifying assumptions. First, we assumed that reward prediction and reward prediction error of a classifier immediately take on the average values. However, the Widrow-Hoff delta rule allows only an approximation of these values with an accuracy dependent on the update rate β. Second, since classifier fitness in XCS is based on the set-relative accuracy of a classifier, overlapping classifiers (classifiers that match sometimes both and sometimes either one of them) can alter fitness determination which might cause additional problems. Third, we studied XCS in one-step, or classification, problems. In multi-step problems the additional problem of reinforcement propagation arises which can cause over-general classifiers to propagate inappropriate reinforcement values. Thus, in multi-step problems, the determination of current MAD is even harder.

5 Conclusion

This paper investigated the accuracy-based fitness approach in XCS and its implications for a successful selectorecombinative evolutionary process. Accuracy of a classifier was derived from its mean absolute deviation (MAD) which is approximated by its reward prediction error estimate. Two necessary conditions were identified for a successful evolutionary process: (1) Sustenance of partially successful classifiers; (2) classifiers closer to completely successful ones need to be more accurate. Violating either of those conditions disables the evolutionary process from evolving an accurate problem representation since there are no classifiers to propagate and/or the propagation leads towards the wrong direction. The concatenated multiplexer problem was introduced as an example problem that violates the second condition. Consequently, XCS was not able to evolve the desired complete, accurate, and maximally general payoff map of the problem.

As the concatenated multiplexer problem showed, there are problems that are inherently accuracy-misleading for the current accuracy-based fitness approach in XCS. Thus, a modification of the accuracy definition appears to be necessary to enable successful learning. Since accuracy is derived from the reward prediction error which approximates the MAD, it appears that the MAD itself is not an appropriate basis for XCS's accuracy measure. Thus, it is necessary to redefine the reward prediction error in XCS. One approach would be to change the error to an upper and lower error or a two-sided error measure. Future research needs to investigate the various possibilities for modifying the reward prediction error in XCS.

Although a sufficiently large population size with a large mutation might be able to create and maintain completely successful classifiers in our investigated concatenated multiplexer problem, mutation cannot remedy the problem in general since it does not scale up. Thus, accuracy guidance needs to be exploited for successful selectorecombinative events. Once accuracy guidance is available in a problem, the next step is to investigate and develop competent crossover operators that can further speed-up and scale-up evolutionary search in XCS.

Acknowledgments. We are grateful to Xavier Llora for the useful discussions. This work was sponsored by the Air Force Office of Scientific Research, Air Force Material Command, USAF, under grant F49620-00-0163. Research funding for this work was also provided by the National Science Foundation under grant DMI-9908252. The U.S. Government is authorized to reproduce and distribute reprints for Government purposes notwithstanding any copyright notation thereon. The views and conclusions contained herein are those of the authors and should not be interpreted as necessarily representing the official policies or endorsements, either expressed or implied, of the Air Force Office of Scientific Research, the National Science Foundation, or the U.S. Government.

References

1. Butz, M.V., Kovacs, T., Lanzi, P.L., Wilson, S.W.: How XCS evolves accurate classifiers. Proceedings of the Genetic and Evolutionary Computation Conference (GECCO-2001) (2001) 927–934

2. Butz, M.V., Pelikan, M.: Analyzing the evolutionary pressures in XCS. Proceedings of the Genetic and Evolutionary Computation Conference (GECCO-2001) (2001) 935–942
3. Butz, M.V., Wilson, S.W.: An algorithmic description of XCS. In Lanzi, P.L., Stoltzman, W., Wilson, S.W., eds.: Advances in Learning Classifier Systems: Third International Workshop, IWLCS 2000, Berlin Heidelberg, Springer-Verlag (2001) 253–272
4. Goldberg, D.E.: Genetic algorithms in search, optimization and machine learning. Addison-Wesley, Reading, MA (1989)
5. Goldberg, D.E.: The design of innovation: Lessons from and for competent genetic algorithms. Kluwer Academic Publishers, Boston, MA (2002)
6. Holland, J.H.: Adaptation in natural and artificial systems. Universtiy of Michigan Press, Ann Arbor, MI (1975) second edition 1992.
7. Holland, J.H.: Adaptation. In Rosen, R., Snell, F., eds.: Progress in Theoretical Biology. Volume 4., New York, Academic Press (1976) 263–293
8. Kovacs, T.: XCS classifier system reliably evolves accurate, complete, and minimal representations for boolean functions. In Roy, Chawdhry, Pant, eds.: Soft computing in engineering design and manufacturing. Springer-Verlag, London (1997) 59–68
9. Kovacs, T.: Towards a theory of strong overgeneral classifiers. Foundations of Genetic Algorithms 6 (2001)
10. Kovacs, T., Kerber, M.: What makes a problem hard for XCS? In Lanzi, P.L., Stolzmann, W., Wilson, S.W., eds.: Advances in Learning Classifier Systems: Third International Workshop, IWLCS 2000. Springer-Verlag, Berlin Heidelberg (2001) 80–99
11. Pelikan, M., Goldberg, D.E., Lobo, F.: A survey on optimization by building and using probabilistic models. Computational Optimization and Applications **21** (2002) 5–20
12. Wilson, S.W.: Classifier fitness based on accuracy. Evolutionary Computation **3** (1995) 149–175
13. Wilson, S.W.: Mining oblique data with XCS. In Lanzi, P.L., Stolzmann, W., Wilson, S.W., eds.: Advances in Learning Classifier Systems: Third International Workshop, IWLCS 2000. Springer-Verlag, Berlin Heidelberg (2001) 158–174

Data Classification Using Genetic Parallel Programming

Sin Man Cheang, Kin Hong Lee, and Kwong Sak Leung

Department of Computer Science and Engineering
The Chinese University of Hong Kong, Hong Kong
{smcheang,khlee,ksleung}@cse.cuhk.edu.hk

Abstract. A novel Linear Genetic Programming (LGP) paradigm called Genetic Parallel Programming (GPP) has been proposed to evolve parallel programs based on a Multi-ALU Processor. It is found that GPP can evolve parallel programs for Data Classification problems. In this paper, five binary-class UCI Machine Learning Repository databases are used to test the effectiveness of the proposed GPP-classifier. The main advantages of employing GPP for data classification are: 1) speeding up evolutionary process by parallel hardware fitness evaluation; and 2) discovering parallel algorithms automatically. Experimental results show that the GPP-classifier evolves simple classification programs with good generalization performance. The accuracies of these evolved classifiers are comparable to other existing classification algorithms.

Data Classification is a supervised learning process that learns a classifier from a training set. The learned classifier can be used to classify unseen data records. Lim *et al.* have performed a sophisticated study on 16 UCI Machine Learning Repository databases by 33 different data classification algorithms [1]. Their experimental results are used for comparison with the proposed GPP-classifier. A novel LGP paradigm – Genetic Parallel Programming (GPP) [2,3] is employed to learn data classifiers. In GPP, individual programs are represented in a sequence of parallel instructions. Each parallel instruction consists of multiple subinstructions in order to perform multiple operations in each processor clock cycle simultaneously. A parallel program is executed on a specially designed Multi-ALU Processor (MAP). The main purpose of this paper is to demonstrate that GPP can evolve data classifiers to solve real-world data classification problems. Experimental results show that GPP can evolve binary-class data classifiers with comparable generalization accuracy to the other 33 existing data classification methods presented in [1].

We adopt the 10-fold cross-validation method to estimate the classification error rate (CE) of the GPP-classifier. 10 training sets are used to learn 10 classifiers that are tested with their corresponding test sets to obtain 10 test set CE. The 10 test set CE are averaged to estimate the generalized CE. We measure the classification accuracy and the generalization performance. A good generalized classifier gives similar levels of performance on the training and test sets. Furthermore, the GPP-classifier has adopted three techniques to avoid overtraining: 1) limiting the size of genetic programs; 2) penalizing over-trained individual programs; and 3) monitoring generalization performance over the evolution. All experiments have been run on a software GPP-classifier system. It produces a parallel assembly program together with a correspondent serialized C code segment. Table 1 below shows the best, average, and standard deviation (*stddev*) of training set CE and test set CE of 10 independent runs (10-fold cross-validation on each run).

Table 1. Training set CE and test set CE of the GPP-classifier

	training set CE (%)			test set CE (%)			%ΔCE
	best	average	stddev	best	average	stddev	
bcw	2.7	2.9	0.09	3.5	3.9	0.29	25.6%
bld	27.3	28.0	0.69	29.3	31.7	1.74	11.7%
pid	22.5	22.7	0.11	23.7	24.5	0.42	7.3%
hea	14.4	14.8	0.24	16.0	18.9	1.78	21.7%
vot	3.9	4.1	0.10	4.1	4.6	0.23	10.8%
average							15.4%

In Table 1 above, the last column shows the percentage differences (%ΔCE) of the average training set CE and test set CE. The average %ΔCE of the five databases is 15.4%. It is shown that GPP can learn parallel programs to solve real-world data classification problems. Experimental results show that GPP is able to learn human understandable classifiers with comparable generalization performance to other classification algorithms. Even without tailor-making the GPP configurations for individual problem, good quality classifiers are evolved. The generalization performance of the GPP-classifier is higher than the average of the 33 benchmark algorithms in [1]. It shows that the GPP-classifier has the power to learn very simple but accurate classifiers with a suitable overtraining control strategy. Besides, the GPP-classifier automatically determines the structure of the solution program without prior knowledge of the databases. Even though the results show that classification program evolved by GPP has comparable generalization performance to other classification algorithms, further improvements can be carried out. In spite of adopting overtraining control strategies, the GPP-classifier still suffers from some overtraining, i.e. the training set CE are higher than test set CE in Table 1 above. In order to obtain a good generalization performance, we shall work out an appropriate terminating condition to detect overtraining.

References

1. Lim, T.S., Loh, W.Y., Shih, Y.S.: A Comparison of Prediction Accuracy, Complexity, and Training Time of Thirty-Three Old and New Classification Algorithms. Machine Learning Journal, Vol.40, Kluwer Academic (2000) 203–229
2. Leung, K.S., Lee, K.H., Cheang, S.M.: Evolving Parallel Machine Programs for a Multi-ALU Processor. Proc. of IEEE Congress on Evolutionary Computation (2002) 1703–1708
3. Leung, K.S., Lee, K.H., Cheang, S.M.: Genetic Parallel Programming – Evolving Linear Machine Codes on a Multiple-ALU Processor. Proc. of International Conference on Intelligence in Engineering and Technology, Univ. Malaysia Sabah (2002) 207–213

Dynamic Strategies in a Real-Time Strategy Game

William Joseph Falke II and Peter Ross

School of Computing, Napier University
10 Colinton Road, Edinburgh EH10 5DT, UK
joe@bongofury.vispa.com, P.Ross@napier.ac.uk

Abstract. Most modern real-time strategy computer games have a sophisticated but fixed 'AI' component that controls the computer's actions. Once the user has learned how such a game will react, the game quickly loses its appeal. This paper describes an example of how a learning classifier system (based on Wilson's ZCS [1]) can be used to equip the computer with dynamically-changing strategies that respond to the user's strategies, thus greatly extending the games playability for serious gamers.

1 The Game and the Classifier System

Real-time strategy (RTS) games typically involve fighting a battle against a computer opponent. This opponent typically has a very sophisticated but essentially static strategy and once the user has learned how it behaves, the game loses its playability. To get round this limitation we have used a learning classifier system (LCS) to provide the computer with the capability to dynamically change its strategy. LCSs are sometimes criticized for being slow to learn. In this example, the LCS has short conditions so as to fit in the real-time loop; success also depends on a careful choice of the types of actions.

The game we implemented, using the publicly available version of the Auran Jet game engine, involves two opposing armies each consisting of two squads of 20 soldiers, moving on hilly terrain. A squad defaults to moving in a 6-6-6-2 formation but the soldiers are individually governed by a flocking algorithm [2] so that squad shapes are very fluid in battles. The essence of the game play is to jockey for position, because it pays for a squad to attack the flank of the enemy or for two squads to attack a third if its companion squad is too far away to offer support quickly enough. Figure 1 illustrates the idea. The rectangles are identifying battle-flags. The large diamond is the cursor that the user employs to command his squads, by placing the cursor somewhere and then left- or right-clicking to summon a particular squad towards that location.

The classifier system sends commands to the computer's squads, at randomly-chosen times within a two- to four-second interval (experimentally determined to offer good playability). The condition part has 14 bits as follows, the values depend on the squad to be commanded. For enemy squad 1, two bits describe how far away it is; one bit indicates whether that squad is engaged in a fight;

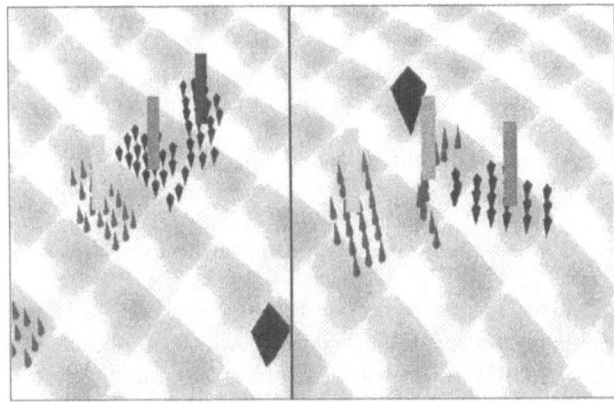

Fig. 1. Two examples of attacks

one bit indicates whether that squad is stronger than this one; one bit indicates whether this squad is positionally flanking that one. Another five bits give the same data about enemy squad 2. Bits 11 and 12 give the proximity of the friendly squad, bits 13 and 14 show whether this squad and its friend are currently engaged in battle. A battle consists of a set of soldier-to-soldier fights running in parallel, each ending with the death of one combatant. A key to the success of the system is the fairly general nature of the possible actions [3]: pursue a named enemy squad, flank a named enemy squad, flee from a named enemy squad, follow the friendly squad. Unlike in the original LCS, the reward system is invoked immediately after the environment string is built and evaluates the squad's previous action. The reward is 1000 is that action caused more kills than deaths for the squad, 0 otherwise. In experiments it took around 750 computer-vs-computer battles, at 15 seconds per battle on a 750MHz PC, to generate a strong opponent for a human. Human testers then found the computer opponent to be both hard to beat and hard to model, as it continued learning.

This demonstrates by example that an LCS, kept suitably simple and with suitably general actions, can be the basis of an effective dynamic strategy in a real-time strategy game. The frame rate of 40-50 fps is good. Moreover, acquired strategies can be saved and exchanged with other users.

References

1. Wilson, S.: ZCS: a zeroth level classifier system. Evolutionary Computation **2** (1994) 1–18
2. Reynolds, C.: Flocks, herds, and schools: A distributed behavioural model. Computer Graphics **21** (1987) 25–34
3. Smith, R., Dike, B., Ravichandran, B., El-Fallah, A., Mehra, R.: The fighter aircraft lcs: a case of different lcs goals and techniques. In Lanzi, P.L., W.Stolzmann, Wilson, S., eds.: Learning Classifier Systems, From Foundations to Applications. Number *1813* in *LNCS*. Springer-Verlag (2000) 283–300

Using Raw Accuracy to Estimate Classifier Fitness in XCS

Pier Luca Lanzi

Artificial Intelligence and Robotics Laboratory
Dipartimento di Elettronica e Informazione
Politecnico di Milano
pierluca.lanzi@polimi.it

In XCS classifier fitness is based on the *relative accuracy* of the classifier prediction [3]. A classifier is more fit if its prediction of the expected payoff is *more accurate* than the prediction given by the other classifiers that are applied in the same situations. The use of *relative accuracy* has two major implications. First, because the evaluation of fitness is based on the relevance that classifiers have in some situations, classifiers that are the only ones applying in a certain situation have a high fitness, even if they are inaccurate. As a consequence, inaccurate classifiers might be able to reproduce so to cause reduced performance; as already noted by Wilson (personal communication reported in [1]). In addition, because the computation of classifier fitness is based both (i) on the classifier accuracy and (ii) on the classifier relevance in situations in which it applies, in XCS, classifier fitness does not provide information about the problem solution, but rather an indication of the classifier relevance in the encountered situations. Accordingly, it is not generally possible to tell whether a classifier with a high fitness is accurate or not, just looking at the fitness. To have this kind of information, we need the prediction error ε which provides an indication of the *raw* classifier accuracy.

Relative Accuracy. In XCS, a classifier cl consists of a condition, an action, and four main parameters: (i) the prediction $cl.p$, which estimates the payoff that the system expects when cl is applied; (ii) the prediction error $cl.\varepsilon$, which estimates the error of the prediction $cl.p$; (iii) the fitness $cl.F$, which estimates the accuracy of the payoff prediction given by $cl.p$; and finally (iv) the numerosity $cl.num$, which indicates how many copies of classifiers with the same condition and the same action are present in the population. In XCS the update of classifier fitness consists of three steps. First, for all the classifiers cl in [A], the *raw accuracy* $cl.\kappa$ is computed: $cl.\kappa$ is 1 if $cl.\varepsilon \leq \varepsilon_0$; $\alpha(cl.\varepsilon/\varepsilon_0)^{-\nu}$ otherwise; α is usually 0.1; ν is usually 5. A classifier is considered to be *accurate* if its prediction error $cl.\varepsilon$ is smaller than the threshold ε_0; a classifier that is accurate has a *raw accuracy* $cl.\kappa$ equal to one. A classifier is considered to be *inaccurate* if its prediction error ε is larger than ε_0; the *raw accuracy* $cl.\kappa$ of an inaccurate classifier cl is computed as a potential descending slope given by $\alpha(cl.\varepsilon/\varepsilon_0)^{-\nu}$. The classifier *raw accuracy* $cl.\kappa$ is used to calculate the *relative accuracy* κ' as $(cl.\kappa \times cl.num)/\sum_{cl_i \in [A]}(cl_i.\kappa \times cl_i.num)$. Finally the *relative accuracy* κ' is used to update the classifier fitness as: $cl.F \leftarrow cl.F + \beta(cl.\kappa' - cl.F)$.

Raw Accuracy. The idea behind *raw accuracy* is quite simple: instead of updating classifier fitness $cl.F$ with the *relative accuracy* $cl.\kappa'$, we update $cl.F$ with the *raw fitness* $cl.\kappa$. The fitness update becomes: $cl.F \leftarrow cl.F + \beta(cl.\kappa - cl.F)$. Classifier fitness now conveys different information since *raw accuracy* does not provide any knowledge about the classifier relative accuracy in the encountered situations. With *relative accuracy* this information is obtained through the use of classifier numerosity, and it is *intrinsically* exploited in Wilson's XCS when fitness is used, i.e.: (i) when the system prediction is computed; (ii) when offspring classifiers are selected from [A] with probability proportional to their fitness (see [3] for details). With *raw accuracy*, this information can be obtained by combining *raw accuracy* and *numerosity* either during the computation of the system prediction, either during the selection of offspring classifiers. This gives raise to four different XCS versions, which differ in the way *raw accuracy* and *numerosity* are combined. The first model, XCSnn, implements the most basic approach possible: classifier fitness is updated using the raw accuracy κ, but classifier numerosity is never used. XCSnn lacks of any information regarding the classifier relevance in the environmental niche which is instead available with *relative accuracy*. The second model, XCSnga, extends XCSnn by introducing numerosity for selecting offspring classifiers: fitness is updated using κ, classifier numerosity is used during offspring selection, i.e., the probability of selecting a classifier cl is proportional to $cl.F \times cl.num$. The third model, XCSne, uses numerosity both for computing system prediction both for offspring selection. The fourth model, XCSnu, does not update fitness; *raw accuracy* κ is used as the measure of fitness; this is equivalent to have $cl.F \leftarrow cl.\kappa$ so as to eliminate the parameter F; classifier fitness is thus computed directly from the prediction error.

Discussion. These different versions of XCS have been compared in [2] with Wilson's XCS [3]. Lanzi [2] shows that in simple single-step problems, XCSnu learns faster than all the other XCS versions producing also the most compact solutions (see [2] for details). In more complex problems, Wilson's XCS learns faster than XCSnu and XCSne but it produces larger populations. Overall, the results reported in [2] suggest that *raw accuracy* might result in interesting performance, although Wilson's relative accuracy appears to provide the best trade-off.

References

1. Pier Luca Lanzi. An Analysis of Generalization in the XCS Classifier System. *Evolutionary Computation*, 7(2):125–149, 1999.
2. Pier Luca Lanzi. A comparison of relative accuracy and raw accuracy in XCS. Technical Report 2003.14, Dipartimento di Elettronica e Informazione. Politecnico di Milano., March 2003.
3. Stewart W. Wilson. Classifier Fitness Based on Accuracy. *Evolutionary Computation*, 3(2):149–175, 1995. http://prediction-dynamics.com/.

Towards Learning Classifier Systems for Continuous-Valued Online Environments

Christopher Stone and Larry Bull

Faculty of Computing, Engineering and Mathematical Sciences
University of the West of England
Bristol, BS16 1QY, United Kingdom
{christopher.stone,larry.bull@uwe.ac.uk}

Abstract. Previous work has studied the use of interval representations in XCS to allow its use in continuous-valued environments. Here we compare the speed of learning of continuous-valued versions of ZCS and XCS with a simple model of an online environment.

1 Introduction

Much current research is focussed on the asymptotic performance of Learning Classifier Systems (LCS). However, where the environment is ever changing, speed of learning and the ability of the system to recover from environmental change (i.e., the slope of the learning curve) may be a more relevant measure.

We are interested in continuous-valued environments and so use an interval representation to replace the $\{0, 1, \#\}$ classifier predicate with one representing an interval $[p_i, q_i)$ [1,5]. An interval is represented as an unordered tuple (p_i, q_i) and matches an environmental variable x_i if $p_i \leq x_i < q_i$.

Changes to the cover, subsumption and Genetic Algorithm (GA) operators must be made to ZCS [3] and XCS [4] to accommodate the interval representation. Action set subsumption is not used. We use single-point crossover between intervals for ZCS and two-point crossover between intervals for XCS. Both ZCS and XCS are run with an initially empty population [1,2].

2 Experiments

Experiments were performed on a six-bit real multiplexer [5]. Parameter settings used for ZCS were (using XCS terminology for consistency) $N = 800$, $\beta = 0.2$, $\tau = 0.1$, $\chi = 0.5$, $\mu = 0.04$, $f_I = 500$, $s_0 = 0.5$, $\rho = 0.25$, $\phi = 0.5$. Parameter settings for XCS were $N = 800$, $\beta = 0.2$, $\alpha = 0.1$, $\varepsilon_0 = 10$, $\nu = 5$, $\theta_{GA} = 12$, $\chi = 0.8$, $\mu = 0.04$, $\theta_{del} = 20$, $\delta = 0.1$, $\theta_{sub} = 20$, $p_I = 10$, $\varepsilon_I = 0$, $f_I = 0.01$, $\theta_{mna} = 2$, $m = 0.1$, $s_0 = 1$. All experiments used a fixed reward of 1000.

In an online environment there is no artificial distinction between exploration and exploitation trials, so we use a roulette wheel for action selection and permanently enable the GA and update mechanisms.

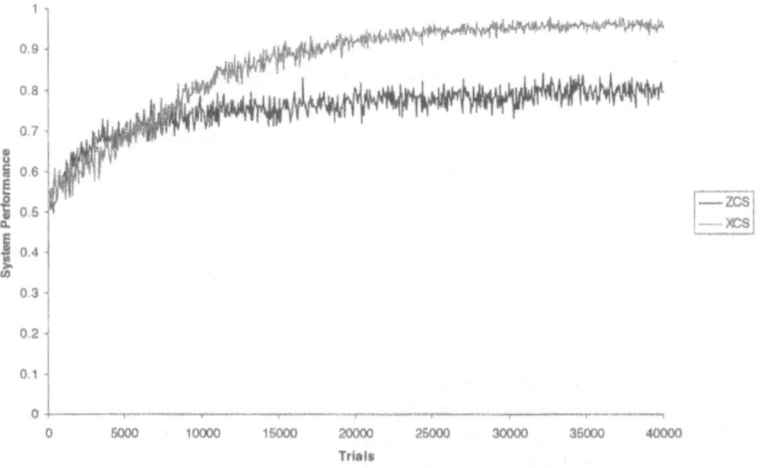

Fig. 1. Mean system performance over 10 runs of ZCS and XCS on the six-bit real multiplexer

Figure 1 shows the performance of ZCS and XCS on the real multiplexer. Similar results were obtained for a three-dimensional checkerboard problem with three divisions per dimension [2] (not shown).

We found that, although the asymptotic performance of XCS ultimately exceeds that of ZCS, ZCS performs as well as XCS during the early part of the learning curve for the problems tested. This result shows that a simple LCS architecture, such as ZCS, can compete with the more complex XCS system where speed of learning is more important than asymptotic performance.

References

1. Stone, C. & Bull, L. (2003) For real! XCS with continuous-valued inputs. To appear in *Evolutionary Computation*.
2. Stone, C. & Bull, L. (2003) Comparing Learning Classifier Systems for Continuous-Valued Online Environments. University of the West of England Learning Classifier Group Technical Report UWELCSG03-001.
   http://www.cems.uwe.ac.uk/lcsg/reports/uwelcsg03-001.ps
3. Wilson, S.W. (1994). ZCS: A zeroth order classifier system. *Evolutionary Computation*, 2(1):1-18.
4. Wilson, S.W. (1995). Classifier fitness based on accuracy. *Evolutionary Computation*, 3(2):149-175.
5. Wilson, S W. (2000). Get real! XCS with continuous-valued inputs. In P.L. Lanzi, W. Stolzmann and S.W. Wilson (eds.), *Learning Classifier Systems. From Foundations to Applications, Lecture Notes in Artificial Intelligence (LNAI-1813)*, Berlin: Springer, pages 209-219.

Artificial Immune System for Classification of Gene Expression Data

Shin Ando[1] and Hitoshi Iba[2]

[1] Dept. of Electronics, School of Engineering
University of Tokyo
ando@miv.t.u-tokyo.ac.jp
[2] Dept. of Frontier Informatics, School of Frontier Science
University of Tokyo
iba@miv.t.u-tokyo.ac.jp

Abstract. DNA microarray experiments generate thousands of gene expression measurement simultaneously. Analyzing the difference of gene expression in cell and tissue samples is useful in diagnosis of disease. This paper presents an Artificial Immune System for classifying microarray-monitored data. The system evolutionarily selects important features and optimizes their weights to derive classification rules. This system was applied to two datasets of cancerous cells and tissues. The primary result found few classification rules which correctly classified all the test samples and gave some interesting implications for feature selection.

1 Introduction

The analysis of human gene expression is an important topic in bioinformatics. The microarrays and DNA chips can measure the expression profile of thousands of genes simultaneously. Genes are expressed differently depending on its environment, such as their affiliate organs and external stimulation. Many ongoing researches try to extract information from the difference in expression profile given the stimulation or environmental change. Some of the experiments have shown promising result in diagnosis of cancer [16, 17, 4]. Our focus is on classifying gene expression data in [16] and [17]. These data are publicly available and has been applied by other classification methods.

This paper describes the implementation of Artificial Immune System (AIS) [2] for classification. The AIS simulates the human immune system, which is a complex network structure, which responds to an almost unlimited multitude of foreign pathogens. It is considered to be potent in intelligent computing applications such as detection, pattern recognition, and classification.

This implementation of AIS defines classification rules as hyperplanes, which divide the domain of sample vectors. The experiments with genomic expression data showed how the border lines are captured and how the features are selected. Compared to other classifier methods, AIS show reduced complexity of the rules, while equaling or improving on the accuracy of prediction.

2 Classification of Microarray Data

Important aspects in gene expression classification are selection of informative genes (feature selection), and optimization of strength (weight) of each gene, and generalization.

Many existing works use ranking methods to select features as a kind of dimensionality reduction on the data. Many studies [7, 15, 16], uses correlation metrics G_i (i) to rank the feature genes. Subset of genes with highest correlations is chosen as classifier genes. In (i), µ and σ are the mean and standard deviation for the expression levels of gene i, in class A or B samples. [5] compares several ranking methods and results when combined with several machine learning methods.

$$G_i = (\mu_A - \mu_B) / (\sigma_A + \sigma_B) \qquad (i)$$

For optimizing weights, machine learning methods such as weighted vote cast [16], Bayesian Network, Neural Network, RBF Network [7], Support Vector Machine [6, 15], have been applied to such data.

2.1 Classification of ALL/AML in Acute Leukemia

In [16], cancerous cells collected from 72 patients of acute leukemia are monitored over 7109 genes. Discovery and prediction of cancer class is important, as clinical outcome vary depending on its class. Each sample belongs to either ALL or AML cancer classes. Two independent data sets, training data set (38 cell samples, 27 ALL and 11 AML) to learn the cancer classes and test data set (34 samples, 20 ALL and 14 AML) to evaluate its prediction were provided in [16]. The reliable diagnoses of the samples were made by combination of clinical tests.

2.2 Colon Cancer Diagnosis

In [17], using Affymetrix oligonucleotide arrays, expression levels of 40 tumor and 22 normal colon tissues are measured for 6500 human genes. Among these genes, 2000 with the highest minimal intensity across the tissues are selected for classification purposes. Since no training data set or test dataset were classified, we randomly chosen 38 training samples and 24 test samples.

Table 1 shows comparison of performance in ALL/AML classification by various machine learning techniques. The results are cited from [6, 7, 15, 16]. Each method were trained and tested under following conditions. Weighted Vote Cast (WVC) [16] selected 50 genes with high correlation as classifier genes. Each sample is classified by the sum of each gene's correlation value. Since weights are not learned, training samples are misclassified.

Neural Network (NN) creates different classifier in every run. The success rate shown in Table 1 is the best in 10 runs, while the average and worst success rate of test data was 4.9 and 9 respectively. Bayesian Network[7] uses 4 genes with higher correlations, though how the features were chosen is not clearly stated. Support Vector Machine (SVM) is a combination of linear modeling and instance-based learning. [15] uses 50, 100, and 200 genes of high correlation. [6] uses Recursive Feature

Elimination to select informative genes. Result in Table 1 is obtained when 8, 16, 32 genes were chosen.Our implementation of AIS selects combination of genes and weights in evolutionary recombination process. The result in Table 1 is the best of 20 runs, while average number of misclassification was 0.85. More detailed result will be given in later section.

Table 1. Comparison of machine learning techniques in Leukemia data set

ALL/AML	WVC[16]	BN[7]	NN[7]	SVM[15]	SVM[6]	AIS
Training	2/38	0/38	0/38	0/38	0/38	0/38
Test	5/34	2/34	1/34	2-4/34	0/34	0/34

3 Features of Immune System

The capabilities of natural immune system, which are to recognize, destroy, and remember almost unlimited multitude of foreign pathogens, have drawn increasing interest of researchers over the past few years. Application of AIS includes fields of computer security, pattern recognition, and classification.

The natural immune system responds to and removes intruders such as bacteria, viruses, fungi, and parasites. Substances that are capable of invoking specific immune responses are referred to as antigens (Ag).

Immune system learns the features of antigens and remembers successful responses to use against invasions by similar pathogens in the future. These characteristics are achieved by a class of white blood cells called lymphocytes, whose function is to detect antigens and assist in their elimination. The receptors on the surface of a lymphocyte bind with specific epitopes on the surfaces of antigens. These proteins related to immune system are called antibodies (Ab).

Immune system can maintain a diverse repertoire of receptors to capture various antigens, because the DNA strings which codes the receptors are subject to high probability crossover and mutation, and new receptors are constantly created.

Lymphocytes are subject to two types of selection process. Negative selection, which takes place in thymus, operates by killing all antibodies that binds to any self-protein in its maturing process. The clonal selection takes place in the bone marrow. Lymphocyte which binds to a pathogen is stimulated to copy themselves. The copy process is subject to a high probability of errors, i.e., hypermutation. The combination of mutation and selection amounts to an evolutionary algorithm that produces lymphocytes that become increasingly specific to invading pathogens.

During the primary response to a new pathogen, the organism will experience an infection while the immune system learns to recognize the epitope by evolutionary process. The memory of successful receptors is maintained to allow much quicker secondary response when same or similar pathogens invade thereafter.

There are several theories of how immune memory is maintained. The AIS in this paper stores successful antibodies in permanent memory cells to store adapted results.

4 Implementation of Artificial Immune System

In the application of artificial immune system to ALL/AML classification problem, the following analogy applies. The ALL training data sets correspond to Ag, and AML training data sets to the self-proteins. Classification rules represent Ab, which captures training samples(Ag or self-proteins) when its profile satisfies the conditions of the rule. A population of Ab goes through a cycle of invasion by Ag and selective reproduction. As successful Abs are converted into memory cells, ALL/AML class is learned. Above analogy can be applied to Colon cancer data by replacing ALL with the tumor tissues and AML with normal tissue.

4.1 Rule Encoding

Classification rules are linear separators, or hyper-planes, as shown in (ii). Vector $x = (x_1, x_2, \ldots, x_i, \ldots x_n)$ represents gene expression levels of a sample, and vector $w = (w_1, w_2, \ldots, w_i, \ldots, w_n)$ represents the weight of each gene. A hyperplane $W(x)=0$ can separate the domain and the samples into two classes. W determines the class of sample x; if $W(x)$ is larger than or equal to 0, sample x is classifies as ALL. If it is smaller than 0, the sample is classified as AML.

$$W(x) \geq 0 \ \{ W(x) = w^T \cdot x \} \tag{ii}$$

Encoded rules are shown in (iii). It represents a vector w, where each loci consists of a pointer to a gene and weight value of that gene. It corresponds to a vector where unspecified gene weights are supplemented with 0. It is similar to messyGA[3] encoding of vector w.

$$(X_{123}, 0.5)\,(X_i, w_i)\,(\ldots)\,(\ldots)\,(\ldots)\ (\textit{gene index, weight}) \tag{iii}$$

4.2 Initialization

Initially, rules are created by sequential creation of locus. For each locus, a gene and a weight value is chosen randomly. With probability P_i, next locus is created. Thus, average lengths of the initial rules are $\Sigma_n n P_i n$. Empirically, the number of initial rules should be in the same order as the number of genes to ensure sufficient building blocks for classification rules.

4.3 Negative Selection

All newly created rules will first go through negative selection. Each rule is met with set of AML training samples, $x_i(i=1,2,\ldots,N_{AML})$, as self-proteins. If a rule binds with any of the samples(satisfy (iv)0, it is terminated. The new rules are created until N_{Ab} antibodies pass the negative selection. These rules constitute population of antibodies $Ab_i(i=1,2,\ldots,N_{Ab})$.

$$\prod \delta(W(x)) \neq 1 \begin{cases} \delta(t) = 0 | t > 0 \\ = 1 | t < 0 \end{cases} \tag{iv}$$

4.4 Memory Cell Conversion

The antibodies who endures the negative selection are met with invading antigens, $Ag_i(i=1,2,\ldots,N_{ALL})$, or ALL training samples. Antibodies which can capture many antigens are converted into memory cells $M_i(i=1,2,\ldots,N_{Mem})$.

Ab_i are converted to memory cell in following conditions. A set of antigens captured by Ab_i or a memory cell M_i is represented by $C(Ab_i)$, $C(M_i)$.

- M_i is removed if $C(M_i) \subset C(Ab_i)$.
- Ab_i is converted to M_{N+1} if $C(Abi) \not\subset C(M_1) \cup C(M_2) \cup \ldots \cup C(M_N)$.
- Ab_i is converted to M_{N+1}, if $C(Mi)=C(Abi)$

4.5 Clonal Selection

The memory cells and Abs which bind with Ags go through clonal selection for reproduction. This process is a cycle described as follows:
- First, an Ag is selected from the list of captured Ags. The probability of selection is proportional to $S(Ag_i)$, the concentration of Ag_i, which is initially 1.
- Randomly select two antibodies Ab_{p1} and Ab_{p2} from all the antibodies bound with the antigen.
- Ab_{p1} and Ab_{p2} are crossed over with probability P_c to produce offspring Ab_{c1} and Ab_{c2}.
The crossover operation is defined as cut and splice operation.
Crossover is followed by hypermutation, which is a series of copy mutation applied to w_{c1}, w_{c2}, and their copied offspring. There are several types of mutation. Locus deletion, deletes randomly selected locus. Locus addition, adds newly created locus to the antibody. Weight mutation changes the weight value of randomly chosen locus.
- With probability P_m, newly created antibody creates another mutated copy. Copy operation is repeated for $\Sigma n P_m^n$ times on average.
- Parents are selected from memory cells as well. Same crossover and hypermutation process is applied.
- The copied antigens go through negative selection as previously described. The reproduction processes are repeated until N_{Ab} antigens pass the negative selection.
- Finally, the score of each Ag is updated by (v). T is the score of Ag, s is the number of Ab bound to an Ag, and N is the total number of Ab. β is an empirically determined constant, 1.44 in this study. The concentration of Ag converges to 1 with appropriate β.

$$T' = \beta^{T-s/N} \qquad (v)$$

The process goes back to Negative Selection to start a new cycle.

4.7 Generalization

In a single run, many rules with same set of captured antigens, C(Mem), are stored as memory cells. After the run is terminated, one memory cell is chosen to classify the test samples. A memory cell with largest margin M (vi), is chosen.

$$M = \min_i \left| W(x_i) / W(\tilde{x}_i) \right| \qquad (vi)$$

x_i are the ALL/AML sample vectors and $\tilde{x}_i$ is the median of the samples. We try to maximize generality by choosing a hyperplane whose margin to nearest sample vector is the largest.

4.6 Summary of Experiment

AIS repeats the cycle as previously described. Flow of the system is shown in Fig. 1. Each run was terminated after $N_c(=50)$ cycles. The AIS runs on parameters shown in Table 2. The results were robust to minor tuning of these parameters.

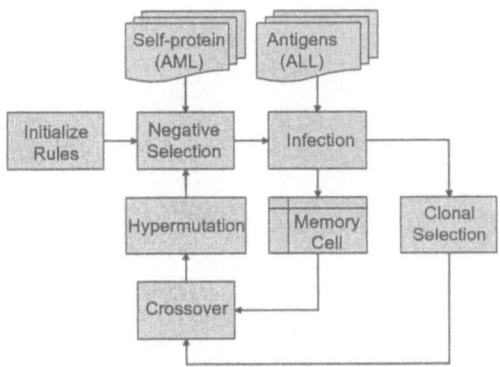

Fig. 1. Cycle of artificial immune system

Table 2. Data attributes and AIS parameters

Leukemia	$N_G = 7109$	$N_{AML} = 11$	$N_{ALL} = 27$		
Colon cancer	$N_G = 2001$	$N_{norm} = 13$	$N_{tumor} = 25$		
AIS parameters	$N_{Ab} = 7,000$	$N_c = 50$	$P_i = 0.5$	$P_c = 0.9$	$P_m = 0.6$

4.8 Results

AIS was applied to Leukemia dataset and Colon cancer dataset. Its performance was measured by average and standard deviation of 20 runs. In all runs, training data set was correctly classified. Table 3 shows average and standard deviation of the number of false positives (misclassified AML test data/ misclassified normal tissue) and false negative (misclassified ALL test data / misclassified tumor tissue) prediction on test samples for both dataset.

Table 3. The number of misclassified samples in test data set of Leukemia and Colon cancer

	#FN(Average/Standard Dev.)	#FP(Average/Standard Dev.)
Leukemia	0.3 / 0.47	0.55 / 0.51
Colon Caner	0.75 / 0.44	0.7 / 0.47

Fig. 2 and Fig. 3 show the learning process of AIS in a typical run for Leukemia dataset. Fig. 2 shows the number of Ag captured by best and worst Abs. The average number of captured Ag is also shown. It shows training samples are learned by 20th cycle. It can be read from the graph that cycles afterward are spent to derive more general rules.

Fig. 3 shows the concentration of Ags at each cycle. Most Ags converge to 1, while few are slower to converge. These samples imply the borderline of the classes. Samples near the classification border are prone to misclassification by untrained Abs, thus slower to converge.

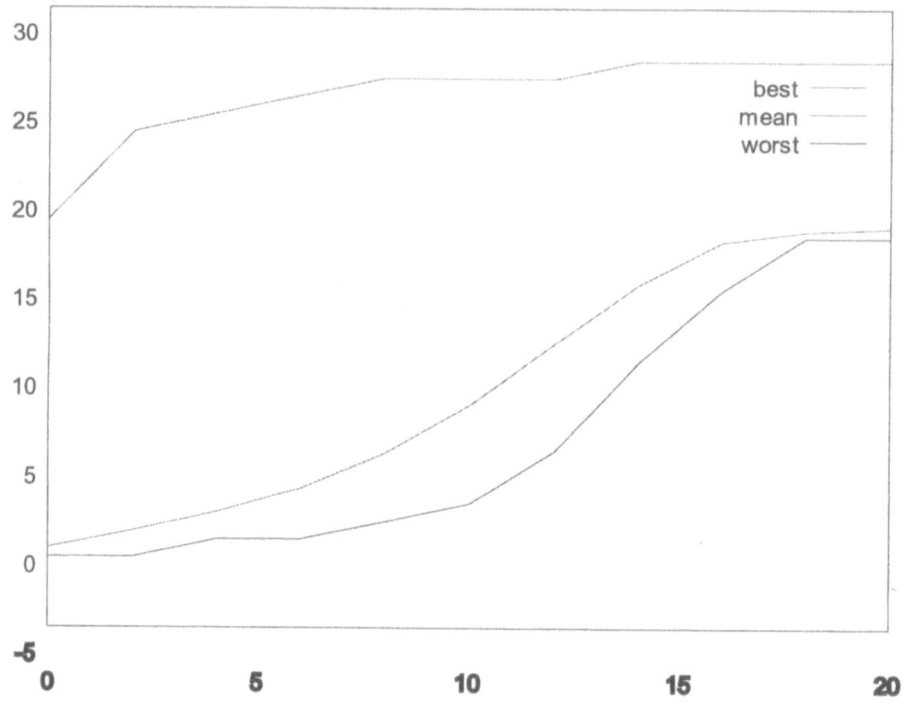

Fig. 2. Number of antigens caught by the best and worst antibodies

Empirically, the results were successful when the grasp of border is clear, i.e. all but few samples have converged. The termination criteria (number of cycle) were determined so that sufficient convergence was achieved by the end of iteration.

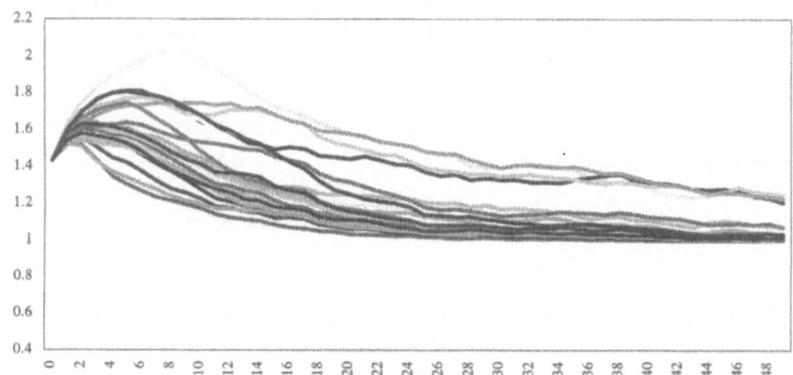

Fig. 3. Transition of antigen scores

5 Analyses of Selected Features

Fig. 4 and Fig. 5 shows some of the classification rules which correctly classified the test samples of Leukemia dataset and Colon cancer dataset respectively.

A) $1.21896X_{3675} + -1.5858X_{4474} + 1.46134X_{1540} + -1.19885X_{2105} + 1.84803X_{757} + 1.82983X_{4038}$

B) $-1.4577X_{1385} + -1.57815X_{4363} + 1.31819X_{2317} + 1.75329X_{2328}$

C) $1.00809X_{1904} + 1.7706X_{6244} + -1.41034X_{4526} + -1.14542X_{759} + 1.94696X_{2723} + -1.34382X_{4875}$

D) $1.26745X_{3110} + 1.43941X_{1190} + 1.97632 + 1.74422X_{5519} + 1.79449X_{6874} + -1.44577X_{6022}$

Fig. 4. Examples of ALL/AML classifier rules

A) $-1.678X_{1997} -1.55458X_{1516}\ 1.41783X_{1297} -1.85391X_{696}$
$1.69254X_{416} -1.39212X_{620} -1.19309X_{1672} -1.95164X_{1310}$
$1.95381X_{49}\ 1.64378X_{134}$

B) $-1.25102X_{1997} + 1.77743X_{138} + -1.42182X_{1770} + 1.3154X_{49} +$
$1.63866X_{183} + 1.17681X_{94}$

Fig. 5. Examples of colon cancer classifier rules

Each rule was different for each run. In this section, we further analyze the selected genes. Some of the genes appear repeatedly in the classification rules. Such genes have relatively high correlations (i) as shown in Table 4. These genes are fairly informative in terms of ALL/AML classification. Fig. 6 shows the expression level of those genes, and how the test data sets can be clustered with features in Table 4. The figure was created with average linkage clustering by Eisen's clustering tool and viewer [10]. It can separate ALL/AML samples with the exception of one sample.

Table 4. Correlation value of the classifier genes (GAN: Gene Accession Number)

Gene	X_{757}	X_{1238}	X_{4038}	X_{2328}	X_{1683}	X_{6022}	X_{4363}
GAN	D88270	L07633	X03934	M89957	M11722	L00634_s	X62654
Gene name	VPREB1	PSME1	CD3D	CD79B	DNTT	FNTA	CD63
Correlation	0.838	0.592	0.647	0.766	0.816	-0.837	-0.834

Many of these genes have relations to Leukemic disease which can be confirmed by biological literature. For example, CD79b is one of the surface marker molecule which could provide important additional information in leukemia cell analysis [8]. CD3D is involved in abnormal location of the genes often observed in acute leukemia [14].

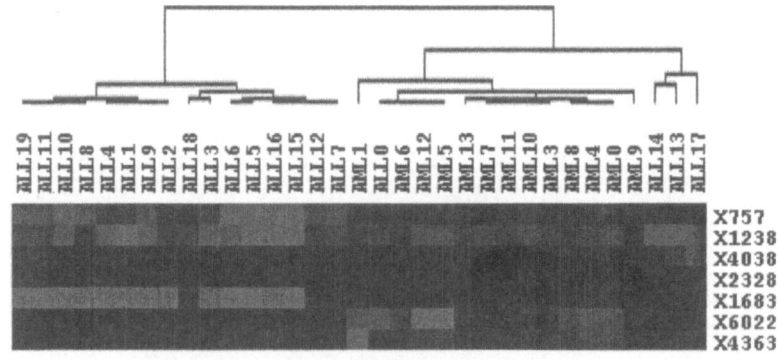

Fig. 6. Expression levels of informative genes and clustering based on those genes

The following section analyzes featured genes in rule A (Fig. 4). Each gene does not always have high correlation value as can be seen in Table 5.

Table 5. Correlation of classifier genes in rule A

Gene	X3675	X4474	X1540	X2105	X757	X4038
Weight	1.21896	-1.5858	1.46134	-1.19885	1.84803	1.82983
GAN	U73682	X69699	L38696	M62762	D88270	X03934
Correlation	0.151	0.371	0.418	-1.02	0.838	0.647

Fig. 7 shows the expression levels of feature genes in rule A. It implies that majority of the samples can be classified by few ALL(X_{2105}, X_{757}) and AML(X_{4038}) classifier genes. These classifier genes have relatively high correlation value.

Some of the samples (AML1, 6, 10, ALL11, 17, 18, 19) in Fig. 7 seem indistinguishable by those classifier genes. Functions of supplementary genes(X_{3675}, X_{4474}, X_{1540}) become evident when these samples are looked at especially.

Fig. 8 shows normalized expression levels of selected samples. In this selected group, X_{3675} and X_{4474} are highly correlated to ALL and AML respectively, while the classifier genes(X_{2105}, X_{757}, X_{4038}) became irrelevant to sample class.

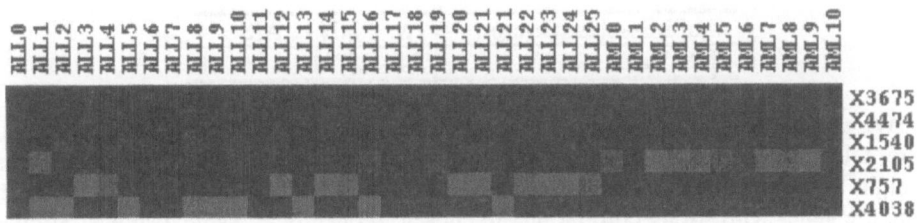

Fig. 7. Expression level of classifier genes in rule A

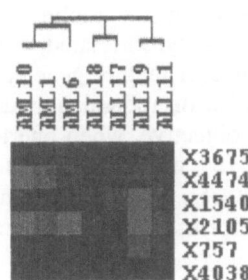

Fig. 8. Selected samples

6 Conclusion

This paper presented Artificial Immune System to classify gene expression of cancerous tissues. Despite sparseness in training data, the accuracy of prediction was satisfactory, as test data were correctly classified 8 out of 20 times in ALL/AML classification. The colon cancer classification is known to be much harder. For comparison, we implemented classification algorithms using popular machine learning methods. Table 6 shows the error rates of the algorithms when applied to the colon cancer data, using 50, 100, and 200 genes selected by correlation metric (i) as features. SVM used linear kernel and error margin of 0.001. NN was implemented as a 3-layer perceptron.

In the experiment with Colon Cancer dataset, the approximate amount of time required for each of the algorithms are: 50 min. for SVM, 3 hours for NN, and 6 hours for AIS. It is hard to compare the efficiency of the algorithms with regards to accuracy of the prediction and amount of computation required. The computational cost may depend on the conditions of the data and the algorithms, e.g. the number of iterations in SVM increase exponentially to the number of training samples, where as the AIS population must increase linear to the number of genes. Considering that preparation of the data, i.e., collecting samples from patients and performing the Microarray tests, takes months, it can be said that all algorithms require considerably small amount of time.

Table 6. Comparison of performance in colon cancer dataset

# of Missclassification	SVM	NN	AIS
Test Data	3/24	9/24	1.45/24

We presume that AIS was more effective than other methods in regards to the feature selection. AIS evolutionarily chooses informative genes whilst optimizing its weight as well. Though gene subset chosen by AIS for classification differs in each run, the genes with strong correlation are chosen frequently as Table 4 shows. Similar results can be obtained by application of Genetic Algorithm [9].

These genes with strong correlation, either selected by frequency or correlation, may not contain enough information to be a sufficient subset for classification, when many co-regulated genes are selected in the subset. Many genes are predicted to be co-expressed and those genes are expected to have similar rankings.

On the other hand, the result in Fig. 8 implies that selection of complementary genes, which are not necessary highly correlated, can be useful in classification. It might be suggested that the choice of feature gene subsets should be based not only on single ranking method, but also on redundancy and mutual information between the genes. Changing of ranking objective, when one feature is removed as a ranking criterion, has been suggested in [13], while [1937] states that performance of machine learning is naïve to choice of features. AIS selects features in the learning process and it is interesting that it can choose primary and complementary feature genes by evolutionary process.

Monitoring the convergence of the clonal selection suggested new termination criteria, as results improved when all but few genes converged. The analysis and quantitative implementation of such criteria is underway. As future work to improve classification capability, use of effective kernel functions, and expressing relations between the genes, such as combining antibodies with AND/OR functions should be addressed.

References

1. A. Ben-Dor, N. Friedman, Z. Yakini, Class discovery in gene expression data, Proc. of the 5th Annual International Conference on Computational Molecular Biology, 31–38, 2001.
2. D. Dasgupta. Artificial Immune Systems and Their Applications. Springer, 1999.
3. D. Goldberg, B. Korb and K. Deb, Messy Genetic Algorithms: Motivation, Analysis and First Results, Complex Systems, 3:493–530, 1989
4. Donna K. Slonim, Pablo Tamayo, Jill P. Mesirov, Todd R. Golub, Eric S. Lander, Class Prediction and Discovery Using Gene Expression Data, Proc. of the 4th Annual International Conference on Computational Molecular Biology(RECOMB), 263–272, 2000.
5. H. Liu, J. Li, L. Wong, A Comparative Study on Feature Selection and Classification Methods Using Gene Expression Profiles and Proteomic Patterns, in Proceeding of Genome Informatics Workshop, 2002
6. I. Guyon, J. Weston, S. Barnhill, V. Vapnik, Gene Selection for Cancer Classification using Support Vector Machines, Machine Learning Vol. 46 Issue 1–3, pp. 389–422, 2002
7. K.B. Hwang, D.Y. Cho, S.W. Wook Park, S.D. Kim, and B.Y. Zhang, Applying Machine Learning Techniques to Analysis of Gene Expression Data: Cancer Diagnosis, in Proceedings of the First Conference on Critical Assessment of Microarray Data Analysis, CAMDA2000.
8. Knapp W, Strobl H, Majdic O., Flow cytometric analysis of cell-surface and intracellular antigens in leukemia diagnosis. Cytometry 1994 Dec 15;18(4):187–98
9. L. Li, C. R. Weinberg, T. A. Darden, L. G. Pedersen, Gene selection for sample classification based on gene expression data: study of sensitivity to choice of parameters of the GA/KNN method, Bioinformatics, Vol. 17, No. 12, pp. 1131–1142, 2001
10. M.B. Eisen, P.T. Spellman, P.O. Brown, and D. Botstein. Cluster analysis and display of genome-wide expression patterns. Proceedings of the National Academy of Science, 85:14863–14868, 1998.
11. P. Baldi and A. Long, A Bayesian framework for the analysis of microarray expression data: Regularized t-test and statistical inferences of gene changes, Bioinformatics, 17:509–519, 2001.
12. P.J. Park, M. Pagano, and M. Bonetti, A nonparametric scoring algorithm for identifying informative genes from microarry data, PSB2001, 6:52–63, 2001.
13. R. Kohavi and G. H. John, Wrappers for Feature Subset Selection, Artificial Intelligence, vol. 97, 1–2, pp. 273–324, 1997
14. Rowley JD, Diaz MO, Espinosa R 3rd, Patel YD, van Melle E, Ziemin S, Taillon-Miller P, Lichter P, Evans GA, Kersey JH, et al., Mapping chromosome band 11q23 in human acute leukemia with biotinylated probes: identification of 11q23 translocation breakpoints with a yeast artificial chromosome., Proc Natl Acad Sci U S A 1990 Dec;87(23):9358–62
15. T.S. Furey, N. Cristianini, N. Duffy, D.W. Bednarski, M. Schummer, and D. Haussler. Support vector machine classification and validation of cancer tissue samples using microarray expression data. Bioinformatics, 2001
16. T.R. Golub, D.K. Slonim, P. Tamayo, Molecular classification of cancer: class discovery and class prediction by gene expression monitoring. Science, 286:531–537, 1999.
17. U. Alon, N. Barkai, D. Notterman, K. Gish, S. Ybarra, D. Mack, and A. Levine. Broad patterns of gene expression revealed by clustering analysis of tumor and normal colon cancer tissues probed by oligonucleotide arrays. Cell Biology, 96:6745–6750, 1999.

Automatic Design Synthesis and Optimization of Component-Based Systems by Evolutionary Algorithms

P.P. Angelov[1], Y. Zhang[1], J.A. Wright[1], V.I. Hanby[2], and R.A. Buswell[1]

[1] Dept of Civil and Building Engineering, Loughborough University
Loughborough, LE11 3TU, Leicestershire, United Kingdom
{P.P.Angelov,Y.Zhang,J.A.Wright}@Lboro.ac.uk

[2] Institute of Energy and Sustainable Development, De Montfort University
Leicester, LE7 9SU, United Kingdom
VHanby@dmu.ac.uk

Abstract. A novel approach for automatic design synthesis and optimization using evolutionary algorithms (EA) is introduced in the paper. The approach applies to component-based systems in general and is demonstrated with a heating, ventilating and air-conditioning (HVAC) systems problem. The whole process of the system design, including the initial stages that usually entail significant human involvement, is treated as a constraint satisfaction problem. The formulation of the optimization process realizes the complex nature of the design problem using different types of variables (real and integer) that represent both the physical and the topological properties of the system; the objective is to design a feasible and efficient system. New evolutionary operators tailored to the component-based, spatially distributed system design problem have been developed. The process of design has been fully automated. Interactive supervision of the optimization process by a human-designer is possible using a specialized GUI. An example of automatic design of HVAC system for two-zone buildings is presented.

1 Introduction

There are generally two phases in a design process:
- Conceptual design
- Detailed design

During the conceptual design phase, strategic design decisions are usually taken by the human designer based on experience and knowledge. These choices lead to one or more possible schematic representation(s) of the system. Repeated consideration of similar problems quite often results in a set of "typical" design solutions for pre-defined cases.

The conceptual design is a creative process and is highly subjective. The detailed design process identifies possible candidate solutions that meet the requirements and provide the necessary level of performance [12]; in contrast, this stage involves numerical analyses and simulation and occasionally optimization.

The development of component-based simulation techniques [3,5] made possible the investigation of changes in system operating variables and parameters. Studies considering a fixed system configuration are now relatively easy to compile [7]. Existing design options allow the designer to chose a particular configuration from a list of prescribed systems [4] or to develop a user-defined system [2,6]. The former approach presents a limited range of possible solutions and the latter is realized only by time-consuming and error-prone procedures.

An alternative approach is proposed in the paper offering a new degree of freedom in the design process. Using the technique, strategic design decisions concerning the system type and configuration are investigated as part of the design optimization process, instead of being predefined. This approach allows alternative, **novel** system configurations to be explored automatically and hence, the "best" design in terms of both the feasible configuration and optimal set of sizing parameters to be determined.

It is well known that for a given design problem, a number of plausible system configurations exist [10]. The proposed approach combines the automatic generation of alternative system configurations with the optimization. Variables and parameters describing the system configuration, the component set and the topological links, form part of the optimization search-space [13]. EAs have been selected to solve this multi-modal problem because they are particularly appropriate for tackling complex optimization problems [1].

New evolutionary operators, tailored to the specific problem of secondary HVAC system design, have been developed. The component set and the topological interconnections have been encoded into the chromosome of the EA. This has been combined with the physical variables describing the thermal and air mass transfer through the HVAC system into a "genome" (a data structure, which combines several types of chromosomes). The objective is to design a feasible and energy efficient system. The system input is a "design brief", a standardized description of the design requirements and constraints. The output from the system is a ranked list of suggested design solutions ordered by their cost. The cost is based on component size and energy consumption. The process of design is fully automated. A computational module has been produced in Java, given a performance specification and a library of system components, can synthesize functionally viable systems configurations. In addition, the specialized GUI allows interactive supervision of the design process by a human-designer. An example of automatic design of HVAC systems for two-zone buildings is given as an illustration of the technique.

2 Automatic Design Synthesis and Optimization by EA

2.1 Concept of the Approach

The concept of the proposed approach is that the design synthesis can be represented as a constraint-satisfying optimization problem. In this way, the design synthesis, including system configuration generation and parameter optimization as well as the overall system performance optimization is integrated into one search problem. This problem formulation combines the search for a feasible system configuration (comprising the list of particular components used and their interconnections) with the minimization of the energy (and possibly capital) costs. A feasible configuration is one that fulfills both the design and engineering requirements.

The computational mechanism used to solve the problem is based on EA [1,8]. This technique has proven to be powerful and robust in solving mixed variable, non-linear and discontinuous problems. The EA searches from a population of points, each representing a particular candidate solution. It needs only the value of the objective function and operates using a coding of the parameter set not the parameter values. The evaluation of a given system includes checks that the configuration satisfies the connectivity constraints and the other specific requirements of the system; in this case, that the supply air conditions satisfy the zone loads. Graph algorithms have been used to verify these requirements [14].

The approach is concerned with identifying system configurations using a library of typical components performing elementary transformations. A typical list of components of a secondary HVAC system is given in Table 1.

Table 1. List of components (example for a secondary HVAC system)

ID	Type	Description
0	Div	Divergence tee
1	Mix	Mixing tee
2	Heat	Heating coil
3	Cool	Cooling coil
4	Steam	Steam injection
5	Zone	Zone
6	Ambient	Ambient environment

The approach is generic and could be applied to other similar design synthesis and optimization problems. These include building architectural design, electronic hardware design, pipeline and sewage network design. The approach maximizes the potential for generating novel system designs, but needs to determine whether or not the system configurations conform to the connectivity constraints. It forms a highly constrained mixed integer-real multi-modal optimization problem.

Characteristics of the specific search problem are:

- the solution space is multi-modal, with the viable system configurations being distributed in a discontinuous manner throughout the search space;
- the number of performance optimization variables changes with the system configuration, and therefore, location of the solution in the search space.

2.2 Evolutionary Synthesis and Optimization of the System Configuration

An initial population is created randomly in which each member is a chromosome consisting of a coding of the components used together with their links. This population is then evolved through successive epochs using mating and mutation to improve the "fitness" of the population. Each chromosome is formed from the system connectivity [14], component capacity and air mass flow rate variables. It is possible for configurations in a given population to have different numbers of components, hence the number of capacity and flow rate variables could be different. The structure of genome, depicted in Fig.1 has been developed to handle this complexity. Essentially, a more flexible data structure is designed to combine chromosomes with variable length in an object-oriented manner. The benefit of the use of the data structure is the use of directed crossover in optimizing the duties. Unlike the loose chromosome set implementation seen in some other implementations, this structure allows relationship and interaction to be defined among its member chromosomes. It allows the consistency of the genome to be maintained although there are operators performing on chromosome level. This implementation is more generic in that:

- It does not require equal/fixed length member chromosomes;
- It can take any type of member chromosomes

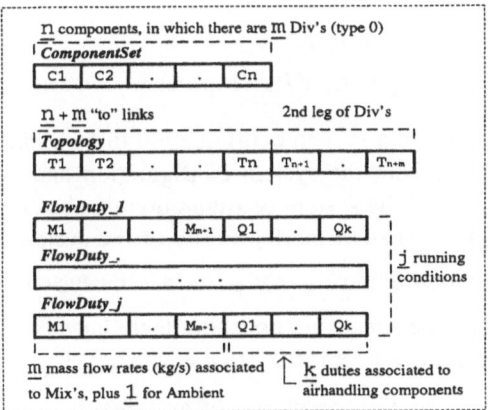

Fig. 1. Genome structure

Each problem variable is viewed as a *gene*. It can be integer (for the component ID number, its position and links (topology) or real (for the coil duty or mass flow rate). A collection of *genes* forms a *chromosome*. Examples of such chromosomes having a clear interpretation in the HVAC system design are [14]:

- *ComponentSet* (list of components present in a particular configuration);
- *Topology* (Modified adjacency list encoding of the network, where "To" links between the components are listed);
- *FlowDuty* (Real-encoded air mass flow rates and component duties for each running condition).

In short, the *genome* is a collection of chromosomes that fully describe structure and operation of a specific system. A sample of genome is presented in section 4.

2.3 Fitness and Constraints Formulation

The design synthesis and optimization problem has been formulated as a constraint satisfaction problem, which in general can be described by:

Minimize:
$$f(X,Y), \quad X = (x_1,...,x_{nl}) \in \Re^{nl}$$
$$Y = (y_1,..., y_{n2}) \in N^{n2}$$

Such that:
$$l_i \leq x_i \leq u_i, \quad \forall i \in (1,..., n1)$$
$$g_j(X) \leq 0.0, \forall j \in (1,..., q)$$
$$h_j(X) = 0, \quad \forall j \in (q+1,..., m1)$$
$$l_i \leq y_i \leq u_i, \quad \forall i \in (1,..., n2)$$
$$h_j(Y) = 0, \quad \forall j \in (q2+1,..., m2)$$

And:
$$l_i, u_i \in \Re^n$$
$$l_i \leq u_i, \forall i \in (1,..., n)$$

The EA attempts to find a system configuration that minimizes the system energy consumption subject to both the system configuration and system operation being feasible. The feasibility of the system configuration can be defined by a set of equality constraints, $h_j(Y)$, whereas the feasibility of the system operation is described by a set of inequality constraints, $g_j(X)$. In fact the system operation includes two equality constraints on the zone conditions, but these have been converted to inequality constraints by applying a tolerance, , to the equality: $|h_j(X)| - \delta \leq 0.0$. Since it is not possible to evaluate the energy use of an infeasible topology, fitness of solutions violating the topology constraints is evaluated by the degree of violations.

The fitness, $F(X,Y)$, is derived from a linear exterior penalty function and is formulated for function minimization:

$$F(X,Y) = \begin{cases} w_f f(X,Y), & \text{If feasible;} \\ w_f f(X,Y) + \sum_{j=0}^{q} w_j g_{o,j}(X) + \sum_{j=q+1}^{m} w_j (|h_{o,j}| - \delta) & \text{else if topology feasible, operation infeasible} \\ \sum_{j=q+1}^{m} w_j |h_{t,j}| & \text{else (topology infeasible).} \end{cases}$$

Where: subscript o, refers to the operation constraints and subscript t, to the topology constraints; all $g_{o,j}$ and $(|h_{o,j}| - \delta)$, have values ≥ 0.0 (are infeasible); and w_f and w_j are the objective and constraint function weights.

Several sets of fundamental and user-defined and controlled constraints are imposed on configurations generations. Generally, they consist of [14]:

- component-related rules
- connection-related rules
- process-related rules

It should be mentioned that the approach applied to the problem is **open** and allows easily to add/modify any group of constraints/rules.

3 Problem-Specific Operators

The optimization has two connected problems; search for feasible configurations and optimization of operations of the feasible configurations found. The latter can only be performed when the configuration is feasible. The effective solution of the search for feasible configurations is therefore of paramount importance. A set of problem-specific operators for mutation and recombination of topologies have been studied, to improve the chance for a candidate configuration to be feasible.

3.1 New Operators Definitions

Two operators have been defined for mutation of topology chromosomes. The "Component link mutation" (LINK) is based on the generic swap-type mutation that inverses the position of two randomly selected genes. For a topology chromosome, inversion of the integers representing the "To" links effectively exchanges two branches in the network, as illustrated in Fig.2. The number of swaps to be performed can be specified as a parameter of the operator. The advantage of the operator is its exploration power, but the probability of damaging the connectivity requirement of a topology is high. As a mutation operator, LINK increases the exploration capabilities of the algorithm by swapping the links, not only the components themselves.

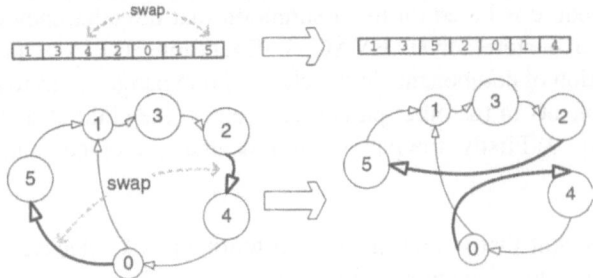

Fig. 2. Example of Component links mutation operator

The second topology operator is "component position mutation" (POS), illustrated in Fig. 3. The position of component <2> and component <0> in the network are exchanged by sequentially swapping the "To" and "From" links belonging to each component. If either component had had more than one "To" or "From" link, graph based analysis would have been applied to determine which link can be swapped to ensure preservation of the network connectivity. The advantage of this operator is that if a topology has already satisfied, the connectivity requirement, further operations do not generate topologies that violate it. The POS operator generates fewer infeasible solutions than the LINK operator, but it is weaker at exploring new topologies.

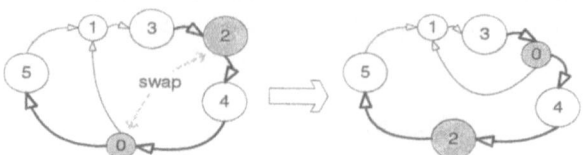

Fig. 3. Example of swapping mutation operator

The configuration of an HVAC system can be considered as a network of components in which air is flowing. Accordingly, optimization methods developed for the so-called Travelling Salesman Problem (TSP) are intuitively relevant. Goldberg [8] introduced "partially matched crossover" (PMX), which has shown its usefulness in solving the TSP. The PMX was adapted as a topology recombination operator with a minor modification to allow multiple connections to a single node. When a child chromosome is produced, the PMX for topology tries to preserve chunk of chromosome from one parent, completing the chromosome with parts of the other. In order to maintain the correct form of adjacency representation for a network, genes from the second parent often have to be repaired. This limits the inheritance of features from the second parent. In addition, any repair of the genes is executed without consideration of the preservation of connectivity. This results in a high probability that a pair of parents who themselves are already connected, will produce unconnected children. This characteristic has a significant impact on the performance of PMX in searching topologies, demonstrated by the results in the next section.

A new topology recombination operator, "adjacent component crossover" (ADJ), has been implemented in order to enhance both connectivity and propagation of features. The procedure is based on the assumption that the adjacency of specific components is useful feature for an HVAC configuration. The ADJ operator performs minimal relocation of components in the clone of one parent, so as to make the pair of adjacent components in the other parent, adjacent in the child also. The procedure is illustrated in Fig. 4. Firstly, the parent chromosomes are cloned and following steps performed:

1) A component that is present in both topologies is selected at random as the base component (component <1> in this instance).
2) The component next to the base component is identified in both topologies respectively. If this component is same in both topologies, or if it is absent from either topology, restart from step 1. The feature of parent A identified in this

example, is the adjacency of <1>→<3>, whereas the feature of parent B is the adjacency of <1>→<2>.
3) The algorithm that exchanges components in the POS operator is used to swap components in parent A in the example. This ensures the preservation of connectivity in the child A'. So for parent A, these operations are; where <3> is linked from <1>, swap <3> with <2> so that <1> and <2> become adjacent.
4) Step 3 is then repeated for parent B to achieve the adjacency feature (<1>→<3>) in parent A.

The result of these operations is that the child preserves most of the network structure of the parent, while acquiring a new feature from a second parent.

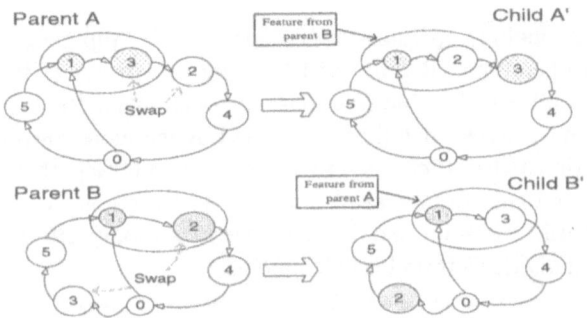

Fig. 4. Example of exchange-adjacent component crossover

3.2 Evaluation of New Operators

Two test procedures have been performed to evaluate the effectiveness of the new operators. In the first test procedure, the operators are used to generate new individuals from a pre-prepared population. The proportion of feasible topologies generated in the children of the initial population is compared to the number of new individuals. This measure gives the probability that the operator will produce feasible topologies from an existing topology.

Two initial populations are prepared, both based on a set of 15 components. The first population is filled with randomly initialized topologies and the second with feasible topologies. The feasible topologies are generated at random and tested for feasibility *a priori*. The operator under test continually selects individuals from the initial population, producing new individuals, until 10,000 new chromosomes have been produced. The same test is repeated on a given operator 20 times. The mean and standard deviation of the results are given in Table 2.

Table 2. Number of feasible solutions out of 10,000 individuals produced by mutation and crossover operators from random population and all-feasible population

Mutation	From Random Mean (stdev)	From Feasible Mean (stdev)	Crossover	From Random Mean (stdev)	From Feasible Mean (stdev)
FLIP (p = 0.1)	2.6 (1.43)	253.3 (18.1)	1P (p = 1.0)	0.9 (1.04)	118.7 (19.0)
RAND (p = 1.0)	99.1(12.0)	100.0(11.1)	PMX (p = 1.0)	95.9 (8.6)	641.0 (19.4)
LINK (swap = 1)	102.9 (20.0)	2439.7 (30.3)	ADJ (p = 1.0)	132.6 (24.1)	6221.2 (51.3)
POS (swap = 1)	141.3 (28.6)	6326.6 (35.7)			

The proposed topology operators are compared with "flip gene mutation" (FLIP), "randomization operation" (RAND) and "1-point crossover" (1P). Table 2 demonstrates that the number of feasible solutions generated by the topology-specific operators are significantly higher than those generated by the more conventional operators. Both POS mutation and ADJ crossover have more than 60% probability of producing new feasible topologies from feasible parents. Conventional operators for integer chromosomes (FLIP mutation and 1P crossover) have only 1%~3% chance of producing feasible individuals from feasible parents.

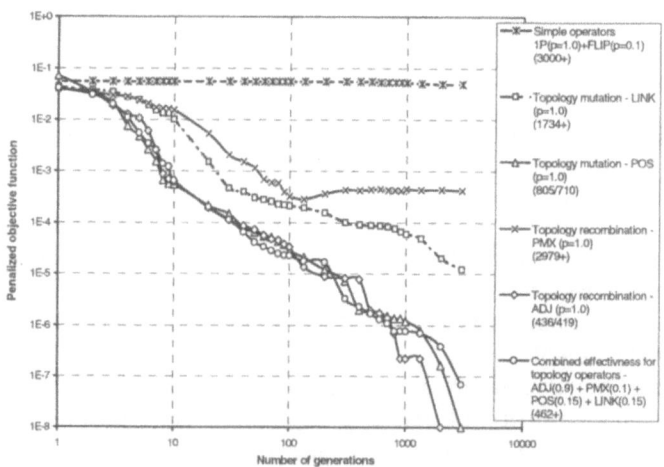

Fig. 5. Effectiveness of evolutionary operators in searching a specified topology

The second evaluation procedure tests convergence. In the procedure, a conventional configuration for a HVAC system was selected, depicted in Fig. 6. The operation of the selected configuration, including the component capacities and airflow rates, were optimized. The capacities and flow rates were then fixed and the system configuration subjected to further optimization, using the different operators. The optimal configuration in this case must be the original configuration and hence convergence on this solution can be used as a measure of operator performance. Figure 5 illustrates the rate of convergence for various operators.

The numbers given in the notation are the mean generations required to find the optimum configuration from 30 runs. The '+' sign indicates that the topology was not found in every trial run, otherwise standard deviation is given. Fig. 5 demonstrates the poor convergence using the conventional operators compared to the new problem-specific operators.

4 Automatic HVAC System Design Synthesis and Optimization

To demonstrate the design synthesis approach, consider the HVAC problem: Design the optimal HVAC system for a small office building in Oklahoma, USA. Three typical design conditions are selected to represent the operation in summer, winter, and swing season. Fig. 6 shows a conventional system with its optimal operation parameters in the summer condition. One of the auto-generated configurations is presented in Fig. 7 (encoded), and in Fig. 8 graphically.

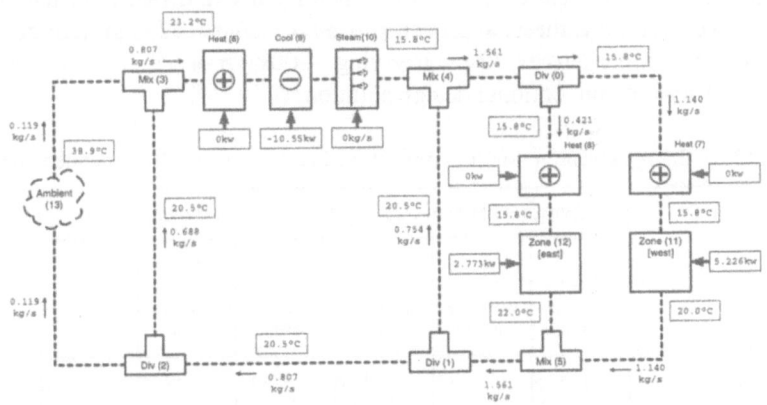

Fig. 6. A conventional HVAC system configuration (summer condition)

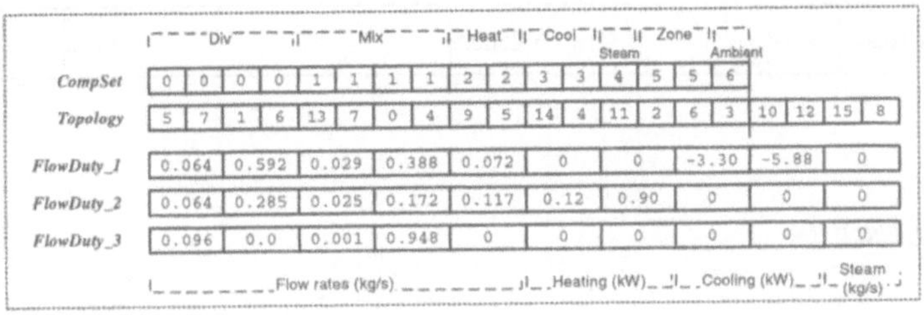

Fig. 7. The genome for the generated configuration

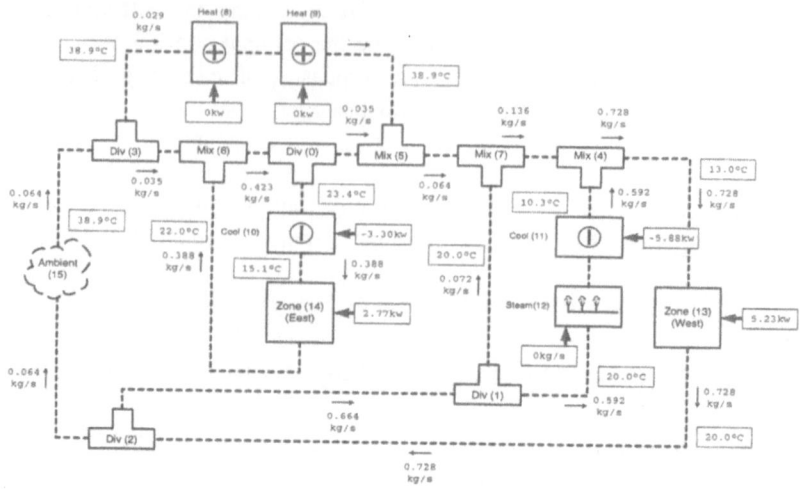

Fig. 8. Configuration generated automatically by the new approach (summer condition)

The minimum energy consumption associated with the conventional and generated configurations is compared in Table 3. The results illustrate both the ability of the automated design synthesis approach to find a viable HVAC system design and that the system is more energy efficient saving ~800kWh or 9% of the energy consumed by the optimized conventional design solution.

Table 3. Energy consumption for conventional and automatically generated configurations

Season	Conventional Configuration			Generated configuration			Total energy saving (kWh)
	Total Heat/Cooling Consumption (kWh)	Total circulation consumption (kWh)	Total energy consumption (kWh)	Total Heat/Cooling Consumption (kWh)	Total circulation consumption (kWh)	Total energy consumption (kWh)	
Summer	6857.5	396.5	7254.0	5960.5	136.5	6097.0	1157.0
Winter	858.0	71.5	929.5	611.0	65.0	676.0	253.5
Spring/Autumn	0.0	377.0	377.0	0.0	988.0	988.0	-611.0
Annual Total			8560.5			7761.0	799.5

4 Conclusions

A new approach for automatic design synthesis and optimization using EAs is introduced in the paper. The approach applies to component-based systems in general although the focus of this paper has been on secondary HVAC system problems. The whole design process is treated as a constraint satisfaction problem. The problem formulation is complex, using different variables to represent the physical and topological properties of the system. The problem objective is to design a feasible and efficient system. EAs have been shown to solve numerically complex search problems such as this. Several problem-specific operators, tailored to the component-based spatially distributed systems design, have been developed and evaluated. The results of evaluation of these operators illustrate their superiority in comparison with conventional operators for the mutation and crossover of integer genes.

The process of design can be fully automated, commencing with a design brief and resulting in a rank ordered set of optimal configurations. An example of the automated design of a secondary HVAC system for two-zone office building was presented. The synthesized system design showed energy savings of about 800 kWh (or 9%) compared to the optimal, conventional system, based on the performance over three season-typical days. This result demonstrates the potential of the approach to automatically generate viable novel HVAC system configurations that can yield improved energy performance compared to conventional design solutions.

Acknowledgement. This research has been supported by the American Society for Heating Refrigeration and Air-Conditioning Engineering (ASHRAE) through the RP-1049 research project.

References

1. Michalewicz, Z.: Genetic Algorithms + Data Structures = Evolution Programs. 3rd edn. Springer-Verlag, Berlin Heidelberg New York (1996)
2. *Blast 3.0 User Manual*, Urbana-Champaign, Illinois, Blast Support Office, Department of Machine and Industrial Engineering, University of Illinois (1992)
3. York D.A., C.C. Cappiello, *DOE-2 Engineers manual* (Version 2.1.A), Lawrence Berkeley Laboratory and Los Alamos National Lab., LBL-11353 and LA-8520-M (1981)
4. Park C., D.R. Clarke, G.E. Kelly, An Overview of HVACSIM+, a Dynamic Building /HVAC/ Control System Simulation Program, 1st *Ann. Build Energy Sim Conf*, Seattle, WA (1985)
5. Klein S.A., W.A. Beckman, J. A. Duffie, TRNSYS – a Transient Simulation Program, *ASHRAE Trans.*, **82,** Pt.2 (1976)
6. Sahlin P., *IDA – a Modelling and Simulation Environment for Building Applications*, Swedish Institute of Applied Mathematics, ITH Report No 1991:2 (1991)
7. Wright J.A., V.I. Hanby, The Formulation, Characteristics, and Solution of HVAC System Optimised Design Problems, *ASHRAE Trans.*, **93**, pt.2 (1987)
8. Goldberg G.E., *Genetic Algorithms in Search, Optimization and Machine Learning*, Addison-Wesley, Reading, M.A., USA (1989)
9. Hanby V.I., J.A. Wright, HVAC Optimisation Studies: Component Modelling Methodology, Building Services Engineering Research and Technology, **10** (1), 35–39 (1989)
10. Sowell E.F., K. Taghavi, H. Levy, D. W. Low, Generation of Building Energy Management System Models, *ASHRAE Trans.*, **90,** pt.1 (1984) 573–586.

11. Volkov A., Automated Design of Branched Duct Networks of Ventilated and air-conditioning Systems, *Proc. of the CLIMA 2000 Conference*, Napoli, Italy 15–18.09.2001.
12. P.J. Bentley (Ed.) *Evolutionary Design by Computers*, London: John Willey Ltd., 2000.
13. Angelov P.P., V.I. Hanby, J. A. Wright, An Automatic Generator of HVAC Secondary Systems Configurations, Technical Report, Loughborough University, UK (1999) 1–17.
14. Wright J.A., P.P. Angelov, Y. Zhang, R.A. Buswell, V.I. Hanby, *Building System Design Synthesis and Optimization*, Final report to 1049-RP, ASHRAE, 2002

Studying the Advantages of a Messy Evolutionary Algorithm for Natural Language Tagging*

Lourdes Araujo

Dpto. Sistemas Informáticos y Programación
Universidad Complutense de Madrid
lurdes@sip.ucm.es

Abstract. The process of labeling each word in a sentence with one of its lexical categories (noun, verb, etc) is called tagging and is a key step in parsing and many other language processing and generation applications. Automatic lexical taggers are usually based on statistical methods, such as Hidden Markov Models, which works with information extracted from large tagged available corpora. This information consists of the frequencies of the *contexts* of the words, that is, of the sequence of their neighbouring tags. Thus, these methods rely on the assumption that the tag of a word only depends on its surrounding tags. This work proposes the use of a Messy Evolutionary Algorithm to investigate the validity of this assumption. This algorithm is an extension of the fast messy genetic algorithms, a variety of Genetic Algorithms that improve the survival of high quality partial solutions or *building blocks*. Messy GAs do not require all genes to be present in the chromosomes and they may also appear more than one time. This allows us to study the kind of building blocks that arise, thus obtaining information of possible relationships between the tag of a word and other tags corresponding to any position in the sentence. The paper describes the design of a messy evolutionary algorithm for the tagging problem and a number of experiments on the performance of the system and the parameters of the algorithm.

1 Introduction

The process of labeling each word in a sentence of a text with its lexical category (noun, verb, etc) is called *tagging* and is a key step in the parsing process and many other natural language processing and generation applications: machine translation, information retrieval, speech recognition and generation, etc.

A word may have more than one possible tag (lexical ambiguity), and thus, disambiguation methods are required to proceed with the tagging. There are different approaches to automatic tagging based on statistical information, that use large amount of data to establish the probabilities of the tag assignments. Most of them are based on Hidden Markov Models (HMMs) and variants [11,4,

---

* Supported by project PR1/03-11588.

13,3] and neither require knowledge of the rules of the language nor try to deduce them. Others are rule-based approaches that apply language rules to improve the accuracy of the tagging. The Brill system [2] extracts these rules from a training corpus, obtaining competitive performance with stochastic taggers.

Evolutionary algorithms present an appealing tradeoff between efficiency and generality and for this reason they have been extensively applied to complex optimization problems. They have previously been applied to tagging [1]. In the evolutionary tagger, individuals are sequences of tags assigned to the words of a sentence. This model is a variant of the HMM in which disambiguation is introduced by assigning different probabilities to a given tag depending on which are the neighbouring tags (*context*) on both sides of the word. The computation of the fitness of the individuals is based on the data extracted of a training corpus tagged by hand. These data are organized as contexts. The tagger is able to learn from a training corpus so as to produce a table of rules (contexts) called *training table*. This table records the different contexts of each tag. The table can be computed by going through the training text and recording the different contexts and the number of occurrences of each of them for every tag in the training text.

Results indicate that the evolutionary approach for tagging texts of natural language obtains accuracies comparable to other statistical approaches, while improving the robustness of the typical algorithms used for the same purpose in other stochastic tagging approaches (such as the widely used of Viterbi). These methods typically perform at about the 96% level of correctness [3] (percentage of words correctly tagged). However, there is a limit beyond which no further improvement can be obtained, neither by enlarging the size of the context, nor the size of the training text.

This leads us to investigate the application of messy GAs to the tagging problem with the aim of studying the possible relationships between the tags of the sentence. Statistical methods for tagging rely on the assumption that the only influence to the correctness comes from the words surrounding the considered one. Therefore, in the evolutionary tagger solutions are evaluated according to the position of the genes. In a messy GA, individuals may be composed of any set of genes, therefore relaxing the previous assumption and allowing to investigate other kinds of dependencies among the words to be tagged. Messy GAs do not require all genes to be present in the chromosomes and they may also appear more than one time. This allows us to study the kind of building blocks that arise, thus obtaining information of possible relation between the tag of a word and other tags corresponding to any position in the sentence. Accordingly, this paper presents an extension of the fast messy GAs to a fast messy evolutionary algorithm (EA), which does not work with bit strings but with tag strings for the tagging problem.

1.1 Messy Genetic Algorithms

Messy Genetic Algorithms [5] are variants of Genetic Algorithms that improve the survival of high quality partial solutions or *building blocks*. This is achieved by explicitly composing increasingly longer and highly fit string from previously obtained and tested building blocks. This is a technique to tackle the problem of building block disruption — *linkage problem* [9] — mainly due to the fixed mapping from the solutions into the representations of individuals or chromosomes and to the way in which the crossover operator combine two chromosomes to obtain a new one. Classic crossover operators often break promising building block leading the algorithm to a local optimum. Furthermore, because messy GAs expedite the presence of high quality building blocks, they increase the probability of exchanging different building blocks [7].

Messy GAs use variable-length strings that are sequences of *messy genes*. A messy gene is an ordered pair composed of a position and a value. The recording of this information allows any building block to achieve tight linkage. Messy GAs do not require all genes to be present in the chromosomes. Therefore, *under* or *overspecified* chromosomes may arise. For example, in the messy string ((1 1) (3 0) (1 0)), the leftmost element specifies gene 1 and its allele value is 1. In a three-bit problem, this chromosome is underspecified, because gene 2 is absent, and overspecified, because gene 1 appears twice. Overspecification is handled by applying a first-come-first-served rule on a left to right order. As for underspecification, some problems do not require to deal with it in any special way because structures of any size can be interpreted and evaluated. Otherwise, the missed genes are filled with those of a *competitive template*, a string that is locally optimal.

A messy GA proceeds in two different phases. The first one, called *primordial* phase, aims to select tight building blocks. It is achieved by initializing the population with all possible building blocks of a specified length. The proportion of good building blocks is improved by applying selection alone for a number of generations. Furthermore, the population size is usually reduced at particular intervals. Thereafter, the next phase, the *juxtaposition* phase, proceeds applying different genetic operators, like *cut* and *splice*, with the objective of recombining the building block obtained in the first phase. In order to deal with chromosomes of variable length, the traditional crossover operator is substituted by the complementary operators cut and splice. Cut divides the chromosome with a specified probability, proportional to the length of the chromosome, while splice joins two chromosomes.

The previous description corresponds to the original messy GA, which suffers from an initialization bottleneck due to the generation of all building blocks of a particular size k: it requires a population of $O(l^k)$ individuals, l being the length of the problem strings. This problem has been dealt with by the socalled *fast* messy GAs [7,10]. Fast messy GAs use smaller populations composed of longer chromosomes. However these longer chromosomes introduce too much error places [6]. This effect is handled by the mechanism of *building block filtering*. According to this, the process begins with long chromosomes, that are reduced by applying selection and random deletion at specific intervals. Another feature of messy GAs is the use of a *thresholding selection*, which enforces any two chromosomes to share some threshold number of genes before competing for selection.

The rest of the paper proceeds as follows: Sect. 2 describes the main elements in the fast messy EA for tagging; Sect. 3 is devoted to evaluate the model and Sect. 4 draws the main conclusions of this work.

2 Messy Evolutionary Algorithm for Probabilistic Tagging

The fast messy EA for tagging works with a population of sequences of genes (chromosomes) of variable length. Each gene consists of a position indicating the word of the sentence to be tagged and a tag chosen for that word. A word may appear any number of times in the chromosome or be missing at all. Let us considered the following words and their tags: **Rice:** (NOUN), **flies:** (NOUN, VERB), **like:** (PREP, VERB), **sand:** (NOUN). Then a possible individual when tagging the sentence *Rice flies like sand* is shown in Fig. 1. The algorithm uses a *pattern* for the evaluation of the underspecified chromosomes. A pattern is a complete tagging of the sentence. Initially each tag of this pattern is randomly selected with a probability proportional to the frequency of the tag. Then the algorithm performs a number of iterations, each of which is devoted to building up increasingly longer building blocks. The pattern is updated after every iteration step, replacing its tags by those present in the best current individual. Figure 2 shows a pattern for the sentence of the previous example. When the individual of Fig. 1 is evaluated, the tag for the word *flies*, which is missing in the chromosome, is taken from the pattern. As in a messy GA, each iteration proceeds in three consecutive phases (see Fig. 3). In the *initialization* phase, a population of individuals is generated to represent the classes corresponding to each set of genes. In the original messy GA the initial population was composed of a single copy of all substrings of length k. This initialization ensured that all building blocks of the considered length were present but led to a bottleneck for a problem length l moderately large. This is avoided by a technique called *probabilistically complete initialization* [7], which creates a population of longer individuals ensuring that, with high probability, any building block appears at least once. Two parameters control the procedure: the length of the initial individuals and the population size. Let us call l' the length of the initial individuals, a value larger than k and smaller than l. The results obtained in [7,8] indicate that if $k < l/2$ and we set $l' = l - k$, a population size $O(l)$ is a good choice.

Rice	like	sand
NOUN	VERB	NOUN

Fig. 1. Example of an individual for the sentence *Rice flies like sand*

Rice	flies	like	sand
NOUN	NOUN	VERB	NOUN

Fig. 2. Example of pattern for the sentence *Rice flies like sand*

```
function Fast_messy_EA()
    pattern = random_tagging(sentence);
    for level = initial_level to final_level do{
        Probabilistic_initialization(Population, level);
        Primordial_phase(Population, pattern, level);
        Juxtaposition_phase(Population, pattern);
        pattern = update_pattern (pattern, best(Population)); }
```

Fig. 3. Scheme of the fast messy EA for tagging

These results have been obtained assuming two alleles for each gene, but they may be extended to an arbitrary number of them. Accordingly we assume these results and take, in our case, the population size as a function of the length of the sentence. Our experimental results serve as an a posteriori justification.

2.1 The Primordial Phase

After the initialization, the *primordial* phase (Fig. 4) selects the best individuals of each combination of genes of size k. Because chromosomes in a messy algorithm may contain very different sets of genes, and thus have little in common, a special selection is required to make the competition fair. This is called *thresholding selection*, and it is implemented by applying tournament selection only between two individuals that have in common a number of genes greater than a threshold value. In our case, the number of common genes is computed as the number of different words of the sentence tagged in both chromosomes.

Individual Evaluation. The fitness function is related to the total probability of the sequence of tags assigned to the sentence. The raw data to obtain this probability are extracted from the training table. The fitness of an assignment is defined as the sum of the fitness of its positions, $\sum_i (f(g_i))$. The fitness of a position g is defined as

$$f(g) = \log P(T|LC, RC),$$

where $P(T|LC, RC)$ is the probability that the tag of position g is T, given that its context is formed by the sequence of tags LC to the left and the sequence

```
function Primordial_phase(Population, pattern, level)
    level_temp = length(Population[0]) - level;
    while (level_temp > level) do{
        Selection(Population);
        Gene_deletion(Population);
        Evaluation(Population);
        level_temp--; }
```

Fig. 4. Scheme of the primordial phase

RC to the right. This probability is estimated from the training table as

$$P(T|LC, RC) \approx \frac{occ(LC, T, RC)}{\sum_{T' \in \mathcal{T}} occ(LC, T', RC)}$$

where $occ(LC, T, RC)$ is the number of occurrences of the list of tags LC, T, RC in the training table and $\mathcal{T}$ is the set of all possible tags of g_i. The contexts corresponding to the position at the beginning and the end of the sentences lack tags on the left-hand side and on the right-hand side respectively. This event is managed by introducing a special tag, NULL, to complete the context.

In our case chromosomes are evaluated by previously building a tagging of the sentence with tags coming from the individual if it contains any instance of the corresponding gene and from the pattern otherwise. For overspecified genes we take the first occurrence of the gene in a left-to-right order. For example, if we are evaluating the individual of Fig. 1 and we are considering contexts composed of one tag on the left and one tag on the right of the position evaluated, the second gene, for which there are two possible tags, NOUN (the one chosen in this individual) and VERB, will be evaluated as:

$$\frac{\#(\text{NOUN NOUN VERB})}{[\#(\text{NOUN NOUN VERB}) + \#(\text{NOUN VERB VERB})]}$$

where # represents the number of occurrences of the context. The remaining genes are evaluated in the same manner.

Another mechanism introduced in this phase is the deletion of genes to reduce the length of the individuals from l' to k. The genes to be erased are randomly selected. We expect to have enough copies of the best building blocks so that after the random deletion some of them still remain.

2.2 The Juxtaposition Phase

The last phase, called *juxtaposition* (Fig. 5), is devoted to combine the tight building blocks obtained in the previous phase. This phase also uses thresholding selection as well the *cut-and-splice* operator to combine individuals, and the mutation operator. The rate of application of the cut operator increases with the length of the individuals, while splicing is applied at a fixed rate. Thus, in

```
function Juxtaposition_phase(Population, pattern)
    generation = 0;
    while (generation < max_generation) && not convergence do{
        Selection(Population);
        Cut_splice(Population);
        Mutation(Population);
        Evaluation(Population);
        generation++; }
```

Fig. 5. Scheme of the juxtaposition phase

the beginning of the evolution process, cut-and-splice behaves almost like splice alone, increasing the length of the individuals. Later on, when the length of the individuals is long enough, cut and splice are applied together producing an effect similar to a one-point crossover operator.

Mutation is then applied to every gene of the chromosomes resulting from the cut-and-splice operation with a probability given by the mutation rate. The tag of the mutation point is replaced by another of the valid tags of the corresponding word. The new tag is randomly chosen according to its probability (the frequency it appears with in the corpus).

Chromosomes resulting from the application of genetic operators replace an equal number of individuals selected with a probability inversely proportional to their fitness.

3 Experiments

The algorithm has been implemented on C++ language on a Pentium II PC. Tables 1 and 2 compare the accuracy rate obtained with a classic EA and a fast messy EA for the tagging problem. Experiments have been carried out using a training corpus of 185000 words extracted from the *Brown* corpus [12], a test text of 500 words, population sizes once, twice or three times the length of the sentence to be tagged, a crossover rate of 30%, and a mutation rate of 1%. Table 1 presents the highest accuracy achieved in ten runs for all population sizes; Table 2 presents the average and standard deviation of these ten runs for a population size twice the sentence length, as well as the results of a t-test to asses the statistical significance of the differences in accuracy of both methods.

Table 1. Largest accuracy of ten runs obtained with the classic EA and with the fast messy EA with different numbers of generation and population sizes for the tagging problem. CEA stands for classic evolutionary algorithm and FMEA for fast messy evolutionary algorithm. In the FMEA the number of generations refers to the juxtaposition phase. The range of sizes of building blocks in the primordial phase has been $|s|/3.5 - |s|/2.5$. The threshold value to allow competition during selection has been $|s| - 2$, and the threshold value to begin to apply *cut* has been $|s|/1.5$. Two last columns show the number of fitness evaluations for $PS = |s|*2$

	Accuracy						evaluations n.	
	$PS = \|s\|$		$PS = \|s\|*2$		$PS = \|s\|*3$		$PS = \|s\|*2$	
	CEA	FMEA	CEA	FMEA	CEA	FMEA	CEA	FMEA
5 generations	95.44	96.40	95.44	96.40	95.20	95.62	98964	11521
10 generations	95.44	96.48	96.16	96.88	95.68	96.64	102944	21494
20 generations	95.68	96.88	96.16	97.12	96.16	96.68	109964	38136
30 generations	95.44	96.88	95.92	96.88	95.92	96.64	117337	51225
40 generations	95.68	96.40	95.92	96.88	95.92	96.64	123757	60728

Table 2. Study of the statistical significance of the results obtained for a population size of $2*|s|$. CEA stands for classic evolutionary algorithm and FMEA for fast messy evolutionary algorithm. Last column reports the probability that the two samples (the accuracies of the ten runs with the CEA and those of the FMEA) arise from the same probability distribution (a small number means that the difference in the means is statistically significant)

	CEA		FMEA		significance
	Mean	St. dev.	Mean	St. Dev.	t-test signif.
5 generations	95.5	0.441	96.1	0.415	0.0057
10 generations	95.5	0.411	95.8	0.343	0.17
20 generations	95.5	0.435	96.3	0.509	0.0015
30 generations	95.4	0.518	96.1	0.610	0.020
40 generations	95.5	0.499	96.5	0.495	0.0002

We can observe that the messy algorithm slightly but systematically improves the results. Although the accuracy results are limited a priori by the statistical information used in the fitness function, the systematic improvement obtained suggests the existence of some correlation between the word to be tagged and distant words, which comes to light by the nature of the messy EA. This hypothesis is supported by the shape of the building blocks obtained from the primordial phase. Figure 6 shows some instances of building blocks obtained from the primordial phase for a sentence of 29 words and a level $k = 7$. A '0' indicates that the gene corresponding to the word of that position in the sentence is absent and any other number that it is present at least once. A '1' indicates that the gene is present but the tag is wrong, '2' that the tag is right and '3' that the tag is right while it was wrong in the tagging performed with a classic EA. We can observe that the worse building blocks present very sparse genes. We can also observe that although some of the best building blocks, such as 9, present their genes very grouped, there are also some others, such as 2, with some sparse genes. Furthermore, we can also observe that some of the tags that have been

	Building blocks	Fitness
1	02200121000000102200000302	20.2156
2	00200020220200020020000302	22.1340
3	00200220002020120000020020	21.6027
4	00002000200022200202020300	21.6340
5	00232100000200102020200000	20.7156
6	02212000000022100200020022	21.1027
7	02032001200200020000200002	21.1352
8	00202120002200002000200020	21.2468
9	00200022200022000000020302	22.1340

Fig. 6. Samples of building blocks obtained from the primordial phase for a sentence of 29 words and a level $k = 7$. Average fitness is 20.92

correctly assigned by the messy EA but wrongly by the classic EA (marked '3') appear among sparse genes. The evaluation of contexts containing missing genes amounts to taking the tag of the pattern and thus, in general, at least in the primordial phase, the resulting context will be composed of tags without any relation. For the same reason, we expect that consecutive genes increase the fitness because with high probability they correspond to frequent contexts (they have appeared after the selection of the primordial phase). Therefore, the selection of chromosomes with separate genes may indicate a correlation between tags by some hidden indirect mechanisms. To illustrate this assume the following pattern for a sentence with n words:

$$P : P_1, P_2, \cdots, P_n$$

where P_i is the tag assigned to the word w_i. Let us consider now the following individuals ($\_$ denotes a missing gene), and let T represent the tag they have assigned to the word:

$$I1 : T_1, T_2, T_3 \cdots, \_$$
$$I2 : T'_1, \_, T'_3, \_, T'_5, \cdots, \_$$

Assuming that a context is composed of one tag on the left and one of the right, the evaluation of $I1$ is done according to the frequencies of the contexts (T_1, T_2), (T_1, T_2, T_3), (T_2, T_3, T_4), etc, that must have high frequencies because they have been selected. However, the evaluation of $I2$ is done with the contexts (T'_1, P_2), (T'_1, P_2, T'_3), (P_2, T'_3, P_4), etc, composed of tags not necessarily related. Therefore, in general, they will have low frequencies and will hardly appear. Accordingly, the actual occurrence of an individual as the following one

$$I : T''_1, \_, \_, T''_4, T''_5, \cdots, \_$$

may indicate a hidden relation between T''_1 and T''_4, T''_5 (e.g. they might correspond to words separated by a subordinate phrase, or by words that have only one possible tag). In this way the messy EA although less efficient than classic EAs (the number of fitness evaluations is more than twice, see two last columns of Table 1), helps to uncover hidden relations between the words of the sentence.

3.1 Study of the Algorithm Parameters

Two fundamental parameters for the probabilistic initialization phase in a messy EA are the length of the chromosomes and the size of the initial population. Some experiments have been carried out in order to determine the most appropriate values of these parameters in our problem. Because the text is tagged sentence by sentence it seems reasonable to adjust these parameters as some kind of simple function of the length of the sentence. Figure 7 shows the results obtained when varying the length of the chromosomes for different population sizes. All parameters are established as a function of the length of the sentence. We can observe that values of the order of the problem length are enough. A parameter

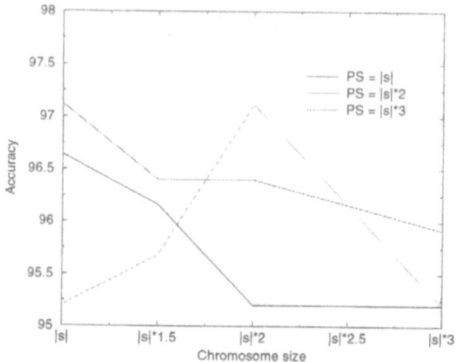

Fig. 7. Accuracy obtained with different sizes of the chromosomes. Each chart corresponds to a different population size: P1 stand for a population size of equal to the length of the sentence, P2 to one of twice the length of the sentence and P2 of three times this length

of the algorithm that needs to be studied is the threshold value for the measurement of similarities between two chromosomes, in order to decide if they are similar enough for the comparison to make sense. Table 3(a) shows the results for some values defined as a function of the length of the sentence to be tagged. The number of similarities is measured as the number of different words of the sentence that corresponds to any of the genes of the chromosome. The best result is obtained when the required similarity is equal to the length of the sentence minus 2. This indicates that only very similar chromosomes must be compared.

Another parameter to be fixed in a messy EA is the threshold value for the chromosome size to begin to apply the *cut* operator. Table 3(b) shows the results.

Table 3. Table (a) presents the accuracy obtained with different values of the threshold established to considerer two chromosomes similar enough to be compared. Table (b) shows the accuracy obtained with different values of the threshold size of the chromosomes to apply the *cut* operator. Table (c) presents the accuracy obtained with different range of sizes of building blocks in the primordial phase. |s| stands for length of the sentence. When varying one of the three parameters the other two are assigned the value marked in boldface

Thres. similarity	Acc.
|s|	96.16
|s| - 1	96.64
|s| - 2	**97.12**
|s| - 3	95.44

(a)

Thres. length	Acc.
|s| / 1.3	95.68
|s| / 1.4	96.88
|s| / 1.5	**97.12**
|s| / 1.6	96.16
|s| / 1.7	95.68

(b)

Range	Acc.
|s| / 3 - |s| / 2	96.40
|s| / 4 - |s| / 3	96.88
|s| / 3.5 - |s| / 2.5	**97.12**
|s| / 3 - |s| / 2.5	96.88

(c)

The best accuracy is obtained when cut is only applied to individuals longer than two thirds of the length of the sentence.

Finally, the range of sizes of building blocks explored in the primordial phase has also been studied. Table 3(c) shows the results obtained. In this case, the greater accuracy is obtained when the size of the building blocks explored ranges between the length of the sentence divided by 2.5 and the length divided by 3.5.

4 Conclusions

This work has investigated the kind of dependencies between words in the tagging problem, that is the assignment of lexical categories to the words of a text. This have been done by applying a fast messy evolutionary algorithm to solve the problem. This algorithm is an extension of the fast messy genetic algorithm, a variety of Genetic Algorithm that improves the survival of high quality partial solutions or *building blocks*. Thus, it is a technique to tackle the problem of building block disruption or *linkage problem* partially due to the fixed mapping from the solutions into the representations of individuals. In a messy GA individuals are variable-length strings of *messy genes*. A messy gene is an ordered pair composed of a position and a value. The recording of this information allows any building block to achieve tight linkage. These algorithms allow obtaining in a phase, called *primordial phase*, previous to the evolutionary process, a set of tight building blocks, whose features can provide information about the internal dependencies of the problem.

Results obtained for the tagging problem systematically outperform a little those obtained with a classic evolutionary algorithm, thus indicating the existence of other relation between tags apart from those between the neighbouring words. This idea is also supported by the kind of building blocks obtained from the primordial phase. Some of them present their genes grouped, but in others the genes are mainly sparse in the proximity of the group or forming others groups.

These results indicate the presence of more complex relationships between words. Accordingly, in the future those possible relationships will be investigated by studying the dependencies between the tagging and the parsing problem.

A number of parameters affecting the performance of the results have also been investigated. These parameters have been assigned values as a function of the length of the sentence. The study of the size of the chromosomes and of the initial population size in the probabilistic initialization phase indicates that values close to the length of the sentence or twice this length are enough to obtain the best results. The experiments on the threshold value to consider that two chromosomes are comparable indicates that very similar chromosomes are required (different at most by the presence of two genes).

References

1. L. Araujo. A parallel evolutionary algorithm for stochastic natural language parsing. In *Proc. of the Int. Conf. Parallel Problem Solving from Nature (PPSNVII)*, 2002.
2. E. Brill. Transformation-based error-driven learning and natural language processing: A case study in part of speech tagging. *Computational Linguistics*, 21(4), 1995.
3. E. Charniak. *Statistical Language Learning*. MIT press, 1993.
4. D. Cutting, J. Kupiec, J. Pedersen, and P. Sibun. A practical part-of-speech tagger. In *Proc. of the Third Conf. on Applied Natural Language Processing*. Association for Computational Linguistics, 1992.
5. D.E. Goldberg, Korb B., and Deb K. Messy genetic algorithms: motivation, analysis, and first results. *Complex Systems*, 3:493–530, 1989.
6. D.E. Goldberg, Korb B., and Deb K. Messy genetic algorithms revisited: Studies in mixed size and scale. *Complex Systems*, 4:415–444, 1990.
7. D.E. Goldberg, Kargupta H. Deb K., and Harik G. Rapid, accurate optimization of difficult problems using fast messy genetic algorithms. In *Proc. of the Fifth International Conference on Genetic Algorithms*, pages 56–64. Morgan Kaufmann Publishers, 1993.
8. D.E. Goldberg, Deb K., and J. H. Clark. Don't worry, be messy. In *Proc. of the Fourth International Conference in Genetic Algorithms and their Applications*, pages 24–30, 1991.
9. Georges R. Harik and David E. Goldberg. Learning linkage. In Richard K. Belew and Michael D. Vose, editors, *Foundations of Genetic Algorithms 4*, pages 247–262. Morgan Kaufmann, San Francisco, CA, 1997.
10. H. Kargupta. *Search, polynomial complexity, and the fast messy genetic algorithm*. Ph.D. thesis, Graduate College of the University of Illinois at Urbana-Champaign, 1996.
11. B. Merialdo. Tagging english text with a probabilistic model. *Computational Linguistics*, 20(2):155–172, 1994.
12. Francis W. Nelson and Henry Kucera. Manual of information to accompany a standard corpus of present-day edited american english, for use with digital computers. Technical report, Department of Linguistics, Brown University., 1979.
13. H. Schutze and Y. Singer. Part od speech tagging using a variable memory markov model. In *Proc. of the 1994 of the Association for Computational Linguistics*. Association for Computational Linguistics, 1994.

Optimal Elevator Group Control by Evolution Strategies

Thomas Beielstein[1], Claus-Peter Ewald[2], and Sandor Markon[3]

[1] Universtität Dortmund, D-44221 Dortmund, Germany
Thomas.Beielstein@udo.edu
http://ls11-www.cs.uni-dortmund.de/people/tom/index.html
[2] NuTech Solutions GmbH, Martin-Schmeisser Weg 15, D-44227 Dortmund, Germany
Ewald@nutechsolutions.de
http://www.nutechsolutions.de
[3] FUJITEC Co.Ltd. World Headquarters, 28-10, Shoh 1-chome, Osaka, Japan
markon@rd.fujitec.co.jp
http://www.fujitec.com

Abstract. Efficient elevator group control is important for the operation of large buildings. Recent developments in this field include the use of fuzzy logic and neural networks. This paper summarizes the development of an evolution strategy (ES) that is capable of optimizing the neuro-controller of an elevator group controller. It extends the results that were based on a simplified elevator group controller simulator. A threshold selection technique is presented as a method to cope with noisy fitness function values during the optimization run. Experimental design techniques are used to analyze first experimental results.

1 Introduction

Elevators play an important role in today's urban life. The elevator supervisory group control (ESGC) problem is related to many other stochastic traffic control problems, especially with respect to the complex behavior and to many difficulties in analysis, design, simulation, and control [Bar86,MN02]. The ESGC problem has been studied for a long time: first approaches were mainly based on analytical approaches derived from queuing theory, in the last decades artificial intelligence techniques such as fuzzy logic (FL), neural networks (NN), and evolutionary algorithms (EA) were introduced, whereas today hybrid techniques, that combine the best methods from the different worlds, enable improvements. Therefore, the present era of optimization could be classified as the era of computational intelligence (CI) methods [SWW02,BPM03]. CI techniques might be useful as quick development techniques to create a new generation of self-adaptive ESGC systems that can handle high maximum traffic situations.

In the following we will consider an ESGC system that is based on a neural network to control the elevators. Some of the NN connection weights can be modified, so that different weight settings and their influence on the ESGC performance can be tested. Let x denote one weight configuration. We can define the optimal weight configuration as $x^* = \arg\min f(x)$, where the performance measure $f()$ to be minimized is defined later. The determination of an optimal weight setting x^* is difficult, since it is not trivial

1. to find an efficient strategy that modifies the weights without generating too many infeasible solutions, and
2. to judge the performance or fitness $f(x)$ of one ESGC configuration. The performance of one specific weight setting x is based on simulations of specific traffic situations, which lead automatically to stochastically disturbed (noisy) fitness function values $\tilde{f}(x)$.

The rest of this article deals mainly with the second problem, especially with problems related to the comparison of two noisy fitness function values. Before we discuss a technique that might be able to improve the comparison of stochastically disturbed values in Sect. 3, we introduce one concrete variant of the ESGC problem in the next section. The applicability of the comparison technique to the ESGC problem is demonstrated in Sect. 4, whereas Sect. 5 gives a summary and an outlook.

2 The Elevator Supervisory Group Control Problem

In this section, we will consider the elevator supervisory group control problem [Bar86,SC99,MN02]. An elevator group controller assigns elevators to service calls. An optimal control strategy is a precondition to minimize service times and to maximize the elevator group capacity. Depending on current traffic loads, several heuristics for reasonable control strategies do exist, but which in general lead to suboptimal controllers. These heuristics have been implemented by Fujitec, one of the world's leading elevator manufacturers, using a fuzzy control approach. In order to improve the generalization of the resulting controllers, Fujitec developed a neural network based controller, which is trained by use of a set of the aforementioned fuzzy controllers, each representing control strategies for different traffic situations. This leads to robust and reasonable, but not necessarily optimal, control strategies [Mar95].

Here we will be concerned with finding optimal control strategies for destination call based ESGC systems. In contrast to traditional elevators, where customers only press a button to request up or down service and choose the exact destination from inside the elevator car, a destination call system lets the customer choose the desired destination at a terminal before entering the elevator car. This provides more exact information to the group controller, and allows higher efficiency by grouping of passengers into elevators according to their destinations; but also limits the freedom of decision. Once a customer is assigned to a car and the car number appears on the terminal, the customer moves away from the terminal, which makes it very inconvenient to reassign his call to another car later.

The concrete control strategy of the neural network is determined by the network structure and neural weights. While the network structure as well as many

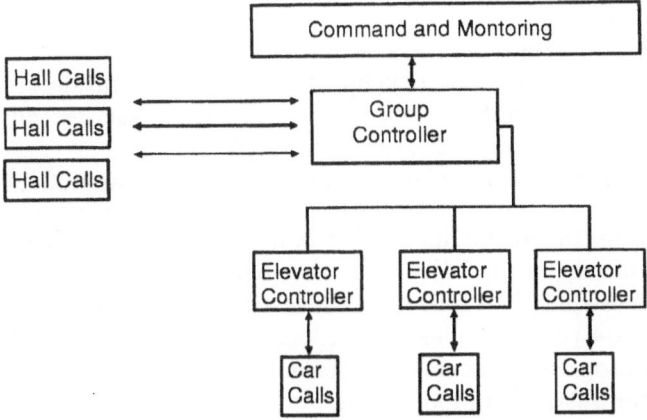

Fig. 1. The architecture of an elevator supervisory group control system [Sii97]

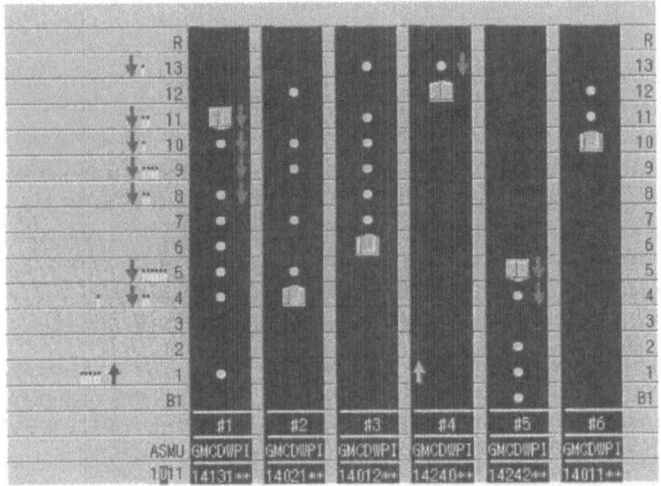

Fig. 2. Visualization of the output from the elevator group simulator

of the weights are fixed, some of the weights on the output layer, which have a major impact on the controller's performance, are variable and therefore subject to optimization. Thus, an algorithm is searched for to optimize the variable weights of the neural controller. The controller's performance can be computed by the help of a discrete-event based elevator group simulator. Figure 2 illustrates how the output from the simulator is visualized.

Unfortunately, the resulting response surface shows some characteristics which makes the identification of globally optimal weights difficult if not impossible. The topology of the fitness function can be characterized as follows:

- highly nonlinear,
- highly multi-modal,
- varying fractal dimensions depending on the position inside the search space,
- randomly disturbed due to the nondeterminism of service calls,
- dynamically changing with respect to traffic loads,
- local measures such as gradients can not be derived analytically.

Furthermore, the maximum number of fitness function evaluations is limited to the order of magnitude 10^4, due to the computational effort for single simulator calls. Consequently, efficient robust methods from the domain of black-box optimization are required where evolutionary computation is one of the most promising approaches. Therefore, we have chosen an evolution strategy to perform the optimization [BFM00,Bey01].

The objective function for this study is the average waiting time of all passengers served during a simulated elevator movement of two hours. Further experiments will be performed in future to compare this objective function to more complex ones, which e.g. take into account the maximum waiting time as well. There are three main traffic patterns occurring during a day: up-peak (morning rush hour), two-way (less intense, balanced traffic during the day), and down-peak traffic (rush hour at closing time). These patterns make up the simulation time to one third each, which forces the resulting controller to cope with different traffic situations. As described above, the objective function values are massively disturbed by noise, since the simulator calculates different waiting times on the same objective variables (network weights), which stems from changing passenger distributions generated by the simulator. For the comparison of different ES parameter settings the final best individuals produced by the ES were assigned handling capacities at 30, 35, and 40 seconds. A handling capacity of n passengers per hour at 30s means that the elevator system is able to serve a maximum of n passengers per hour without exceeding an average waiting time of 30s. These values were created by running the simulator with altering random seeds and increasing passenger loads using the network weights of the best individuals found in each optimization run. Finally, to enable the deployment of our standard evaluation process, we needed a single figure to minimize. Therefore, the handling capacities were averaged and then subtracted from 3000 pass./h. The latter value was empirically chosen as an upper bound for the given scenario. The resulting fitness function is shown in (9) and is called 'inverse handling capacity' in the following.

3 Evolution Strategies and Threshold Selection

3.1 Experimental Noise and Evolution Strategies

In this section we discuss the problem of comparing two solutions, when the available information (the measured data) is disturbed by noise. A bad solution might appear better than a good one due to the noise. Since minor differences are unimportant in this situation, it might be useful to select only much better values. A threshold value τ can be used to formulate the linguistic expression 'much better' mathematically:

$$x \text{ is much better (here: greater) than } y \Leftrightarrow \tilde{f}(x) > \tilde{f}(y) + \tau. \qquad (1)$$

Since comparisons play an important role in the selection process of evolutionary algorithms, we will use the term threshold selection (TS) in the rest of this article. TS can be generalized to many other stochastic search techniques such as particle swarm optimizers or genetic algorithms. We will consider evolution strategies in this article only. The goal of the ES is to find a vector x^*, for which holds:

$$f(x^*) \leq f(x) \quad \forall x \in \mathcal{D}, \qquad (2)$$

where the vector x represents a set of (object) parameters, and $\mathcal{D}$ is some n-dimensional search space. An ES individual is usually defined as the set of object parameters x, strategy parameters s, and its fitness value $f(x)$ [BS02].

In the following we will consider fitness function values obtained from computer simulation experiments: in this case, the value of the fitness function depends on the seed that is used to initialize the random stream of the simulator. The exact fitness function value $f(x)$ is replaced by the noisy fitness function value $\tilde{f}(x) = f(x) + \epsilon$. In the theoretical analysis we assume normal-distributed noise, that is $\epsilon \sim \mathcal{N}(\mu, \sigma^2)$. It is crucial for any stochastic search algorithm to decide with a certain amount of confidence whether one individual has a better fitness function value than its competitor. This problem will be referred to as the selection problem in the following [Rud98,AB01]. Reevaluation of the fitness function can increase this confidence, but this simple averaging technique is not applicable to many real-world optimization problems, i.e. if the evaluations are too costly.

3.2 Statistical Hypothesis Testing

Threshold selection belongs to a class of statistical methods that can reduce the number of reevaluations. It is directly connected to classical statistical hypothesis testing [BM02].

Let x and y denote two object parameter vectors, with y being proposed as a 'better' (greater) one, to replace the existing x. Using statistical testing, the selection problem can be stated as testing the null hypothesis $H_0 : f(x) \leq f(y)$ against the one-sided alternative hypothesis $H_1 : f(x) > f(y)$. Whether the null hypothesis is accepted or not depends on the specification of a critical region for the test and on the definition of an appropriate test statistic. Two kinds of errors may occur in testing hypothesis: if the null hypothesis is rejected when it is true, an alpha error has been made. If the null hypothesis is not rejected when it is false, a beta error has been made. Consider a maximization problem: the *threshold rejection probability* P_τ^- is defined as the conditional probability, that a worse candidate y has a better noisy fitness value than the fitness value of candidate x by at least a threshold τ:

$$P_\tau^- := P\{\overline{f}(y) \leq \overline{f}(x) + \tau \mid f(y) \leq f(x)\}, \qquad (3)$$

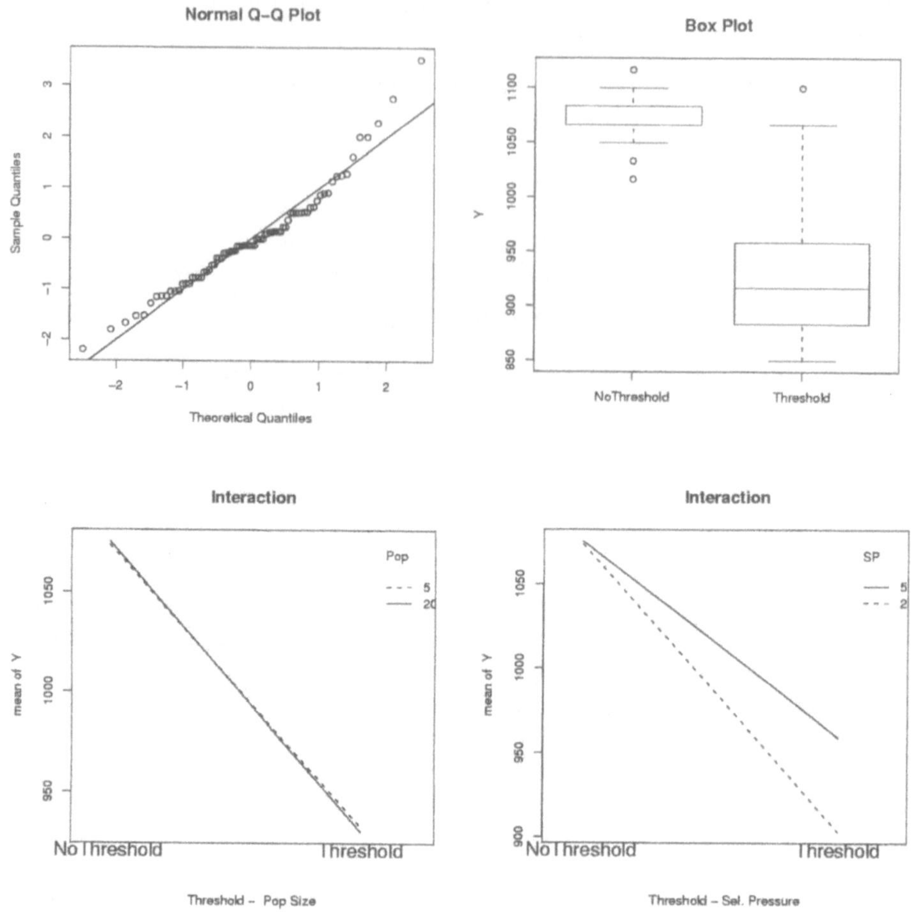

Fig. 3. Hypothesis testing to compare models: we are testing whether threshold selection has any significance in the model. The Normal QQ-plot, the box-plot, and the interaction-plots lead to the conclusion that threshold selection has a significant effect. Y denotes the fitness function values that are based on the inverse handling capacities: smaller values are better, see (9). Threshold, population size μ and selective pressure (offspring-parent ratio ν) are modified according to the values in Table 3. The function $\alpha(t)$ as defined in (6) was used to determine the threshold value

where $\overline{f}(x) := \sum_{i=1}^{n} \tilde{f}(x_i)/n$ denotes the sample average of the noisy values [BM02]. P_τ^- and the alpha error are complementary probabilities:

$$P_\tau^- = 1 - \alpha. \qquad (4)$$

(4) reveals, how TS and hypothesis testing are connected: maximization of the threshold rejection probability, an important task in ES based optimization, is equivalent to the minimization of the alpha error. Therefore, we provide tech-

niques from statistical hypothesis testing to improve the behavior of ES in noisy environments.

Our implementation of the TS method is based on the common practice in hypothesis testing to specify a value of the probability of an alpha error, the so called significance level of the test. In the first phase of the search process the explorative character is enforced, whereas in the second phase the main focus lies on the exploitive character. Thus, a technique that is similar to simulated annealing is used to modify the significance level of the test during the search process. In the following paragraph, we will describe the TS implementation in detail.

3.3 Implementation Details

The TS implementation presented here is related to doing a hypothesis test to see whether the measured difference in the expectations of the fitness function values is significantly different from zero. The test result (either a 'reject' or 'fail-to-reject' recommendation) determines the decision whether to reject or accept a new individual during the selection process of an ES. In the following, we give some recommendations for the concrete implementation. We have chosen a parametric approach, although non-parametric approaches might be applicable in this context too.

Let $Y_{i1}, Y_{i2}, \ldots Y_{in}$ ($i = 1, 2$) be a sample of n independent and identically distributed (i.i.d.) measured fitness function values. The n differences $Z_i = Y_{1j} - Y_{2j}$ are also i.i.d. random variables (r.v.) with sample mean $\overline{Z}$ and sample variance $S^2(n)$. Consider the following test on means μ_1 and μ_2 of normal distributions: if $H_0 : \mu_1 - \mu_2 \leq 0$ is tested against $H_1 : \mu_1 - \mu_2 > 0$, we have the test statistic

$$t_0 = \frac{\overline{z} - (\mu_1 - \mu_2)}{s\sqrt{2/n}}, \tag{5}$$

with sample standard deviation S (small letters denote realizations of the corresponding r.v.). The null hypothesis $H_0 : \mu_1 - \mu_2 \leq 0$ would be rejected if $t_0 > t_{2n-2, 1-\alpha}$, where $t_0 > t_{2n-2, 1-\alpha}$ is the upper α percentage point of the t distribution with $2n - 2$ degrees of freedom.

Summarizing the methods discussed so far, we recommend the following recipe:

1. Select an alpha value. The α error is reduced during the search process, e.g. the function

$$\alpha(t) = \sqrt{(1 - t/t_{\max})}/2, \tag{6}$$

   with $t_{\max}$ the maximum number of iterations and $t \in \{0, t_{\max}\}$, the actual iteration number, can been used.
2. Evaluate the parent and the offspring candidate n times. To reduce the computational effort, this sampling can be performed every k generations only.

3. Determine the sample variance.
4. Determine the threshold value $\tau = t_{2n-2, 1-\alpha} \cdot s \cdot \sqrt{2/n}$
5. A new individual is accepted, if its (perturbed) fitness value plus τ is smaller than the (perturbed) fitness value of the parent.

A first analysis of threshold selection can be found in [MAB+01,BM02]. In the following section we will present some experimental results that are based on the introduced ideas.

4 Threshold Selection in the Context of ESGC Optimization

4.1 Statistical Experimental Design

The following experiments have been set up to answer the question: how can TS improve the optimization process if only stochastically perturbed fitness function values (e.g. from simulation runs) can be obtained? Therefore, we simulated alternative ES parameter configurations and examined their results. We wanted to find out if TS has any effect on the performance of an ES, and if there are any interactions between TS and other exogenous parameters such as population size or selective pressure. A description of the experimental design (DoE) methods we used is omitted here. [Kle87,LK00] give excellent introductions into design of experiments, the applicability of DoE to evolutionary algorithms is shown in [Bei03].

The following vector notation provides a very compact description of evolution strategies [BS02,BM02]. Consider the following parameter vector of an ES parameter design:

$$\boldsymbol{p}_{\text{ES}} = (\mu, \nu, S, n_\sigma, \tau_0, \tau_i, \rho, R_1, R_2, r_0), \tag{7}$$

where $\nu := \lambda/\mu$ defines the offspring-parent ratio, $S \in \{C, P\}$ defines the selection scheme resulting in a comma or plus strategy. The representation of the selection parameter S can be generalized by introducing the parameter κ that defines the maximum age (in generations) of an individual. If κ is set to 1, we obtain the comma-strategy, if κ equals $+\infty$, we model the plus-strategy. The mixing number ρ defines the size of the parent family that is chosen from the parent pool of size μ to create λ offsprings. We consider global intermediate GI, global discrete GD, local intermediate LI, and local discrete LD recombination. R_1 and R_2 define the recombination operator for object resp. strategy variables, and r_0 is the random seed. This representation will be used throughout the rest of this article and is summarized in Table 1. Typical settings are:

$$\boldsymbol{p}_{\text{ES}} = \left(5, 7, 1, 1, 1/\sqrt{2\sqrt{D}}, 1/\sqrt{2D}, 5, GD, GI, 0\right). \tag{8}$$

Our experiment with the elevator group simulator involved three factors. A 2^3 full factorial design has been selected to compare the influence of different population sizes, offspring-parent ratios and selection schemes. This experimental

Table 1. DoE parameter for ES

Symbol	Parameter	Recommended Values
μ	number of parent individuals	$10\ldots 100$
$\nu = \lambda/\mu$	offspring-parent ratio	$1\ldots 10$
n_σ	number of standard deviations	$1\ldots D$
τ_0	global mutation parameter	$1/\sqrt{2\sqrt{D}}$
τ_i	individual mutation parameter	$1/\sqrt{2D}$
κ	age	$1\ldots\infty$
ρ	mixing number	μ
R_1	recombination operator for object variables	{discrete}
R_2	recombination operator for strategy variables	{intermediate}

Table 2. DoE parameter for the fitness function

Symbol	Parameter	Values
f	fitness function, optimization problem	ESGC problem, minimization, see (9)
D	dimension of f	36
N_{exp}	number of experiments for each scenario	10
N_{tot}	total number of fitness function evaluations	$5 \cdot 10^3$
σ	noise level	unknown

design leads to eight different configurations of the factors that are shown in Table 1. The encoding of the problem parameters is shown in Table 2. Table 3 displays the parameter settings that were used during the simulation runs: the population size was set to 5 and 20, whereas the parent-offspring ratio was set to 2 and 5. This gives four different (parent, offspring) combinations: (5,20), (5,25), (20,40), and (20,100). Combined with two different selection schemes we obtain 8 different run configurations. Each run configuration was repeated 10 times.

4.2 Analysis

Although a batch job processing system, that enables a parallel execution of simulation runs, was used to run the experiments, more than a full week of round-the-clock computing was required to perform the experiments. Finally 80 configurations have been tested (10 repeats of 8 different factor settings). The simulated inverse handling capacities can be summarized as follows:

```
Min.   : 850.0   1st Qu.:916.7    Median :1033.3
Mean   :1003.1   3rd Qu.:1083.3   Max.   :1116.7
```

A closer look at the influence of TS on the inverse handling capacities reveals that ES runs with TS have a lower mean (931.2) than simulations that used a

Table 3. ES parameter designs.

ES	ESGC-Design Model ($D = 36$)	
Variable	Low (-1)	High ($+1$)
μ	5	20
ν	2	5
κ	$+\infty$	Threshold Selection
The following values remain unchanged:		
n_σ		1
τ_0		$1/\sqrt{2\sqrt{D}}$
τ_1		$1/\sqrt{2D}$
ρ		μ
R_1		GD
R_2		GI

plus selection strategy (1075.0). As introduced in Sect. 2, the fitness function reads (minimization task):

$$g(x) = 3000.0 - \overline{f}_{p_{ES}}(x), \tag{9}$$

where the setting from Table 3 was used, $\overline{f}_{p_{ES}}$ is the averaged handling capacity (pass./h), and x is a 36 dimensional vector that specifies the NN weights. The minimum fitness function value (850.0) has been found by TS. Performing a t-test (the null hypothesis 'the true difference in means is not greater than 0' that is tested against the 'alternative hypothesis: the true difference in means is greater than 0') leads to the p-value smaller than $2.2e - 16$.

An interaction plot plots the mean of the fitness function value for two-way combinations of factors, e.g. population size and selective strength. Thus, it can illustrate possible interactions between factors. Considering the interaction plots in Fig. 3, we can conclude that the application of threshold selection improves the behavior of the evolution strategy in any case. This improvement is independent from other parameter settings of the underlying evolution strategy. Thus, there is no hint that the results were caused by interactions between other factors.

5 Summary and Outlook

The elevator supervisory group control problem was introduced in the first part of this paper. Evolution strategies were characterized as efficient optimization techniques: they can be applied to optimize the performance of an NN-based elevator supervisory group controller. The implementation of a threshold selection operator for evolution strategies and its application to a complex real-world optimization problem has been shown. Experimental design methods have been used to set up the experiments and to perform the data analysis. The obtained results gave first hints that threshold selection might be able to improve the performance of evolution strategies if the fitness function values are stochastically disturbed. The TS operator was able to improve the average passenger handling capacities of an elevator supervisory group control problem.

Future research on the thresholding mechanism will investigate the following topics:

- Developing an improved sampling mechanism to reduce the number of additional fitness function evaluations.
- Introducing a self-adaptation mechanism for τ during the search process.
- Combining TS with other statistical methods, e.g. Staggge's efficiently averaging method [Sta98].

Acknowledgments. T. Beielstein's research was supported by the DFG as a part of the collaborative research center 'Computational Intelligence' (531). R, a language for data analysis and graphics was used to compute the statistical analysis [IG96].

References

[AB01] Dirk V. Arnold and Hans-Georg Beyer. Investigation of the (μ, λ)-ES in the presence of noise. In J.-H. Kim, B.-T. Zhang, G. Fogel, and I. Kuscu, editors, *Proc. 2001 Congress on Evolutionary Computation (CEC'01)*, Seoul, pages 332–339, Piscataway NJ, 2001. IEEE Press.

[Bar86] G. Barney. *Elevator Traffic Analysis, Design and Control*. Cambridg U.P., 1986.

[Bei03] T. Beielstein. Tuning evolutionary algorithms. Technical Report 148/03, Universität Dortmund, 2003.

[Bey01] Hans-Georg Beyer. *The Theory of Evolution Strategies*. Natural Computing Series. Springer, Heidelberg, 2001.

[BFM00] Th. Bäck, D.B. Fogel, and Z. Michalewicz, editors. *Evolutionary Computation 1 – Basic Algorithms and Operators*. Institute of Physics Publ., Bristol, 2000.

[BM02] T. Beielstein and S. Markon. Threshold selection, hypothesis tests, and DOE methods. In David B. Fogel, Mohamed A. El-Sharkawi, Xin Yao, Garry Greenwood, Hitoshi Iba, Paul Marrow, and Mark Shackleton, editors, *Proceedings of the 2002 Congress on Evolutionary Computation CEC2002*, pages 777–782. IEEE Press, 2002.

[BPM03] T. Beielstein, M. Preuss, and S. Markon. A parallel approach to elevator optimization based on soft computing. Technical Report 147/03, Universität Dortmund, 2003.

[BS02] Hans-Georg Beyer and Hans-Paul Schwefel. Evolution strategies – A comprehensive introduction. *Natural Computing*, 1:3–52, 2002.

[IG96] R. Ihaka and R. Gentleman. R: A language for data analysis and graphics. *Journal of Computational and Graphical Statistics*, 5(3):299–314, 1996.

[Kle87] J. Kleijnen. *Statistical Tools for Simulation Practitioners*. Marcel Dekker, New York, 1987.

[LK00] Averill M. Law and W. David Kelton. *Simulation Modelling and Analysis*. McGraw-Hill Series in Industrial Egineering and Management Science. McGraw-Hill, New York, 3 edition, 2000.

[MAB+01] Sandor Markon, Dirk V. Arnold, Thomas Bäck, Thomas Beielstein, and Hans-Georg Beyer. Thresholding – a selection operator for noisy ES. In J.-H. Kim, B.-T. Zhang, G. Fogel, and I. Kuscu, editors, *Proc. 2001 Congress on Evolutionary Computation (CEC'01)*, pages 465–472, Seoul, Korea, May 27–30, 2001. IEEE Press, Piscataway NJ.

[Mar95] Sandor Markon. *Studies on Applications of Neural Networks in the Elevator System*. PhD thesis, Kyoto University, 1995.

[MN02] S. Markon and Y. Nishikawa. On the analysis and optimization of dynamic cellular automata with application to elevator control. In *The 10th Japanese-German Seminar, Nonlinear Problems in Dynamical Systems, Theory and Applications*. Noto Royal Hotel, Hakui, Ishikawa, Japan, September 2002.

[Rud98] Günter Rudolph. On risky methods for local selection under noise. In A. E. Eiben, Th. Bäck, M. Schoenauer, and H.-P. Schwefel, editors, *Parallel Problem Solving from Nature – PPSN V, Fifth Int'l Conf., Amsterdam, The Netherlands, September 27–30, 1998, Proc.*, volume 1498 of *Lecture Notes in Computer Science*, pages 169–177. Springer, Berlin, 1998.

[SC99] A.T. So and W.L. Chan. *Intelligent Building Systems*. Kluwer A.P., 1999.

[Sii97] M.L. Siikonen. *Planning and Control Models for Elevators in High-Rise Buildings*. PhD thesis, Helsinki Unverstity of Technology, Systems Analysis Laboratory, October 1997.

[Sta98] Peter Stagge. Averaging efficiently in the presence of noise. In A.Eiben, editor, *Parallel Problem Solving from Nature, PPSN V*, pages 188–197, Berlin, 1998. Springer-Verlag.

[SWW02] H.-P. Schwefel, I. Wegener, and K. Weinert, editors. *Advances in Computational Intelligence – Theory and Practice*. Natural Computing Series. Springer, Berlin, 2002.

A Methodology for Combining Symbolic Regression and Design of Experiments to Improve Empirical Model Building

Flor Castillo, Kenric Marshall, James Green, and Arthur Kordon

The Dow Chemical Company
2301 N. Brazosport Blvd, B-1217
Freeport, TX 77541, USA
979-238-7554
{Facastillo,KAMarshall,JLGreen,Akordon}@dow.com

Abstract. A novel methodology for empirical model building using GP-generated symbolic regression in combination with statistical design of experiments as well as undesigned data is proposed. The main advantage of this methodology is the maximum data utilization when extrapolation is necessary. The methodology offers alternative non-linear models that can either linearize the response in the presence of Lack or Fit or challenge and confirm the results from the linear regression in a cost effective and time efficient fashion. The economic benefit is the reduced number of additional experiments in the presence of Lack of Fit.

1 Introduction

The key issues in empirical model development are high quality model interpolation and its extrapolation capability outside the known data range. Of special importance to industrial applications is the second property since the changing operating conditions are more a rule than an exception. Using linear regression models based on well-balanced data generated by Design of Experiments (DOE) is the dominant approach to effective empirical modeling and several techniques have been developed for this purpose [1]. However, in many cases, due to the non-linear nature of the system and unfeasible experimental conditions, it is not possible to develop a linear model. Among the several approaches that can be used either to linearize the problem, or to generate a non-linear empirical model is Genetic Programming (GP)-generated symbolic regression. This novel approach uses simulated evolution to generate non-linear equations with high fitness [2]. The potential of symbolic regression for linearizing the response in statistical DOE when significant Lack of Fit is detected and additional experimentation is unfeasible was explored in [3]. The derived transformations based on the GP equations resolved the problem of Lack of Fit and demonstrated good interpolation capability in an industrial case study in The Dow Chemical Company. However, the extrapolation capability of the derived non-linear models is unknown and not built-in as in the case of linear models.

Extrapolation of empirical models is not always effective but often necessary in chemical processes because plants often operate outside the range of the original experimental data used to develop the empirical model. Of special importance is the use of this data since time and cost restrictions in planned experimentation are frequently encountered.

In this paper, a novel methodology integrating GP with designed and undesigned data is presented. It is based on a combination of linear regression models based on a DOE and non-linear GP-generated symbolic regression models considering interpolation and extrapolation capabilities. This approach has the potential to improve the effectiveness of empirical model building by maximizing the use of plant data and saving time and resources in situations where experimental runs are expensive or technically unfeasible. A case study with a chemical process was used to investigate the utility of this approach and to test the potential of model over-fitting. The promising results obtained give the initial confidence for empirical model development based on GP- generated symbolic regression in conjunction with designed data and open the door for numerous industrial applications.

2 A Methodology for Empirical Model Building Combining Linear and Non-linear Models

With the growing research in evolutionary algorithms and the speed, power and availability of modern computers, genetic programming offers a suitable possibility for real-world applications in industry. This approach based on GP offers four unique elements to empirical model building. First, previous modeling assumptions, such as independent inputs and an established error structure with constant variance, are not required [4]. Second, it generates a multiplicity of non-linear equations which have the potential of restoring Lack of Fit in linear regression models by suggesting variable transforms that can be used to linearize the response [3]. Third, it generates non-linear equations with high fitness representing additional empirical models, which can be considered in conjunction with linearized regression models or to challenge and confirm results even when a linear model is not significant [3]. Fourth, unlike neural networks that require significant amount of data for training and testing, it can generate empirical models with small data sets.

One of the most significant challenges of Genetic programming for empirical model building and linearizing regression models is that it produces non-parsimonious solutions with chunks of inactive terms (introns) that do not contribute to the overall fitness [4]. This drawback can be partially overcome by computational algorithms that quickly select the most highly fit and less-complex models or by using parsimony pressure during the evolution process.

Using DOE in conjunction with genetic programming results in a powerful approach that improves empirical model building and that may have significant economic impact by reducing the number of experiments.

The following are the main components of the proposed methodology.

Step 1. The Experimental Design
A well-planned and executed experimental design (DOE) is essential to the development of empirical models to understand causality in the relationship between the input

variables and the response variable. This component includes the selection of input variables and their ranges, as well as the response variable. It also includes the analysis and the development of a linear regression model for the response(s).

Step 2. Linearization via Genetic Programming

This step is necessary if the linear regression model developed previously presents Lack of Fit and additional experimentation to augment the experimental design such as a Central Composite Design (CCD) is not possible due to extreme experimental conditions. An alternative is to construct a Face-Centered Design by modification of the CCD [5]. However, this alternative often results in high correlation between the square terms of the resulting regression models. GP-generated symbolic regression can be employed to search for mathematical expressions that fit the given data set generated by the experimental design. This approach results in several analytical equations that offer a rich set of possible input transformations which can remove lack of fit without additional experimentation. In addition, it offers non-linear empirical models that are considered with the linearized regression model. The selection between these models is often a trade-off between model complexity and fitness. Very often the less complex model is easier to interpret and is preferred by plant personnel and process engineers.

Step 3. Interpolation

This step is a form of model validation in which the linearized regression model and the non-linear model(s) generated by GP are tested with additional data within the experimental region. Here again, model selection considers complexity, fitness, and the preference of the final user.

Step 4. Extrapolation

This is necessary when the range of one or more of the variables is extended beyond the range of the original design and it is desirable to make timely predictions on this range.

Step 5. Linear and Non-linear Models for Undesigned Data and Model Comparison

When the models developed in the previous sections do not perform well for extrapolation, it is necessary to develop a new empirical model for the new operating region. The ideal approach is to consider an additional experimental design in the new range but often it can not be done in a timely fashion. Furthermore, it is often desirable to use the data set already available. In these cases, while a multiple regression approach can be applied to build a linear regression model with all available data, the risk of collinearity, near linear dependency among regression variables, must be evaluated since the data no longer conforms to an experimental design. High collinearity produces ambiguous regression results making it impossible to estimate the unique effects of individual variables in the regression model. In this case regression coefficients have large sampling errors and are very sensitive to slight changes in the data and to the addition or deletion of variables in the regression equation. This affects model stability, inference and forecasting that is made based on the regression model. Of special interest here is a model developed with a GP-generated symbolic regression algorithm because it offers additional model alternatives.

One essential consideration in this case is that the models generated with this new data *can not* be used to infer *causality*. This restriction comes from the fact that the data no longer conforms to an experimental design. However, the models generated (linear regression and non-linear GP-generated models) *can* be used to *predict* the response in the new range. Both types of models are then compared in terms of complexity, and fitness, allowing the final users to make the decision.

The proposed methodology will be illustrated with an industrial application in a chemical reactor.

3 Empirical Modeling Methodology for a Chemical Reactor

3.1 The Experimental Design and Transformed Linear Model (Steps 1 and 2 in Methodology)

The original data set corresponds to a series of designed experiments that were conducted to clarify the impact on formation of a chemical compound as key variables are manipulated. The experiments consisted of a complete 2^4 factorial design in the factors x_1, x_2, x_3, x_4 with three center points. Nineteen experiments were performed. The response variable, S_k, was the yield or selectivity of one of the products. The factors were coded to a value of −1 at the low level, +1 at the high level, and 0 at the center point. The complete design in the coded variables is shown in Table 1, based on the original design in [3].

The selectivity of the chemical compound (S_k), was first fit to the following first-order linear regression equation considering only terms that are significant at the 95% confidence level:

$$S_k = \beta_o + \sum_{i=1}^{k} \beta_i x_i + \sum\sum_{i<j} \beta_{ij} x_i x_j \tag{1}$$

Table 1. 2^4 factorial design with three center points

Runs	x_1	x_2	x_3	x_4	S_k	P_k
1	1	-1	1	1	1.598	0.000
2	0	0	0	0	1.419	0.000
3	0	0	0	0	1.433	0.016
4	-1	1	1	1	1.281	0.016
5	-1	1	-1	1	1.147	0.009
6	1	1	-1	1	1.607	0.012
7	-1	1	1	-1	1.195	0.019
8	1	1	1	-1	2.027	0.015
9	-1	-1	-1	1	1.111	0.009
10	-1	1	-1	-1	1.159	0.007
11	-1	-1	-1	-1	1.186	0.006
12	1	-1	-1	1	1.453	0.013
13	1	1	-1	-1	1.772	0.006
14	-1	-1	1	-1	1.047	0.018
15	-1	-1	1	1	1.175	0.009
16	1	1	1	1	1.923	0.023
17	1	-1	-1	-1	1.595	0.007
18	1	-1	1	-1	1.811	0.015
19	0	0	0	0	1.412	0.017

Lack of Fit was induced ($p = 0.046$) by omission of experiment number 1 of the experimental design. This was done to simulate a common situation in industry in which LOF is significant and additional experimental runs are impractical due to the cost of experimentation, or because it is technically unfeasible due to restrictions in experimental conditions.

The GP algorithm (GP-generated symbolic regression) was applied to the same data set. The algorithm was implemented as a toolbox in MATLAB. Several models of the selectivity of the chemical compound as a function of the four experimental variables (x_1, x_2, x_3, x_4) were obtained by combining basic functions, inputs, and numerical constants. The initial functions for GP included: addition, subtraction, multiplication, division, square, change sign, square root, natural logarithm, exponential, and power. Function generation takes 20 runs with 500 population size, 100 number of generations, 4 reproductions per generation, 0.6 probability for function as next node, 0.01 parsimony pressure, and correlation coefficient as optimization criteria.

The functional form of the equations produced a rich set of possible transforms. The suggested transforms were tested for the ability to linearize the response without altering the necessary conditions of the error structure needed for least-square estimation (uncorrelated and normally distributed errors with mean zero, and constant variance). The process consisted of selecting equations from the simulated evolution with correlation coefficients larger than 0.9. These equations were analyzed in terms of the R^2 between model prediction and empirical response. Equations with R^2 higher than 0.9 were chosen and the original variables were transformed according to the functional form of these equations. The linear regression model presented in equation (1) was fitted to the data using the transformed variables and the fitness of this transformed linear model was analyzed considering Lack of Fit and R^2. The error structure of the models not showing lack of fit was then analyzed. This process ensured that the transformations given by GP not only linearized the response but also produced the adequate error structure needed for least square estimations.

Using the process previously described, the best fit between model prediction and empirical response from the set of potential non-linear equations was found for the following analytical function with R^2 of 0.98[3]:

$$S_k = \frac{3.13868\times10^{-17} e^{\sqrt{2x_1}} \ln[(x_3)^2] x_2}{x_4} + 1.00545 \qquad (2)$$

The corresponding input/output sensitivity analysis reveals that x_1 is the most important input.

The following transformations were then applied to the data as indicated by the functional form of the GP function shown in equation (2).

Table 2. Variable transformations suggested by GP model.

Original Variable	Transformed Variable
x_1	$Z_1 = \exp(\sqrt{2x_1})$
x_2	$Z_2 = x_2$
x_3	$Z_3 = \ln[(x_3)^2]$
x_4	$Z_4 = x_4^{-1}$

The transformed variables were used to fit a first-order linear regression model shown in equation (1). The resulting model is referred to as the Transformed Linear Model (TLM). The TLM had an R^2 of 0.99, no evidence of Lack of Fit (p=0.3072) and retained the appropiate randomized error structure indicating that the GP-generated transformations were succesful in linearizing the response. The model parameters in the transformed variables are given in [3].

3.2 Interpolation Capability of Transformed Linear Model (TLM) and Symbolic Regression model (GP) (Step 3 of Methodology)

The interpolation capabilities of the TLM and the GP model shown in equation (2) were tested with nine additional experimental points within the range of experimentation. A plot of the GP and TLM models for the interpolation data is presented in Fig. 1.

The models, GP and TLM, perform similarly in terms of interpolation with sum square errors (SSE) being slightly smaller for the TLM (0.1082) than for the GP model (0.1346). However, the models are comparable in terms of prediction with data within the region of the design. The selection of one of these models would be driven by the requirements of a particular application. For example, in the case of process control, the more parsimonious model would generally be preferred.

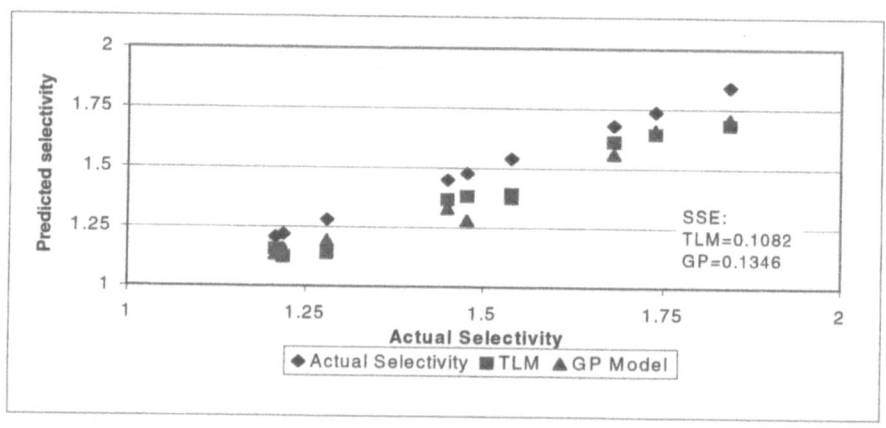

Fig. 1. Predicted selectivity for GP and TLM using interpolation data set

3.3 Extrapolation Capability of Transformed Linear Model (TLM) and Symbolic Regression Model (GP) (Step 4 of Methodology)

It was necessary to evaluate the prediction of the TLM and the GP model with available data outside the region of experimentation (beyond −1 and 1 for the input variables) that were occasionally encountered in the chemical system. The most appropriate approach would be to generate an experimental design in this new region of operability. Unfortunately this could not be completed in a timely and cost effective fashion due to restrictions in operating conditions. The data used for extrapolation in coded form is shown in Table 3. Figure 2 shows the comparison between actual and predicted selectivity for this data set.

Table 3. Data used for extrapolation for GP and TLM models

Run	X1	X2	X3	X4	Selectivity S.	GP	TLM	SSE GP	SSE TLM
1	1.0	-	-	-	1.614	1.63	1.61	0.000	0.000
2	1.0	-	-	-	1.331	1.32	1.21	0.000	0.013
3	1.0	-	-	-	1.368	1.36	1.40	0.000	0.001
4	2.0	-	-	-	1.791	2.27	2.02	0.231	0.056
5	2.0	-	-	-	1.359	1.65	1.24	0.086	0.013
6	2.0	-	-	-	1.422	1.72	1.47	0.092	0.003
7	3.0	-	-	-	1.969	3.52	2.63	2.412	0.446
8	3.0	-	-	-	1.398	2.29	1.28	0.798	0.013
9	3.0	-	-	-	1.455	2.43	1.57	0.960	0.014
10	3.0	0.67	-	-	1.480	2.62	2.10	1.318	0.384
11	3.0	0.76	-	0.47	1.343	2.30	1.48	0.923	0.020
Data is coded based on conditions of the original design								6.822	0.963

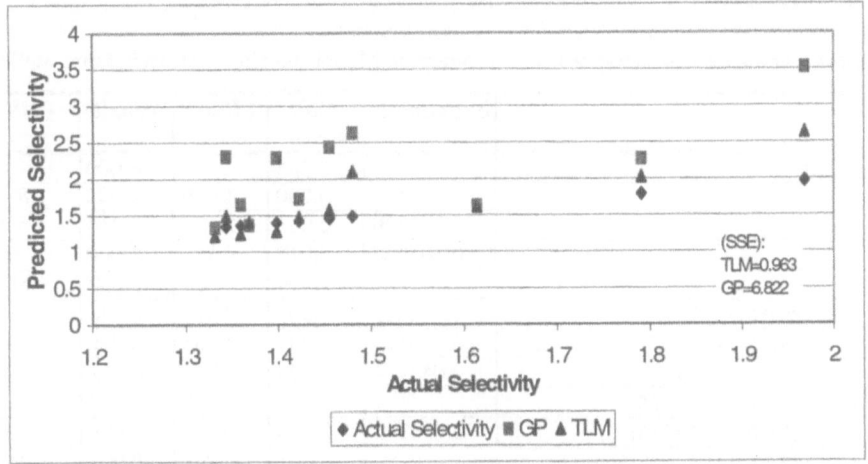

Fig. 2. Predicted selectivity for GP and TLM models using extrapolation data

The deviations between model predictions and actual selectivity (SSE) are larger for the GP model (6.822) than for the TLM (0.963) suggesting that the TLM had better extrapolation capabilities. Table 3 shows that the sum squares errors (SSE) gets larger for both models as the most sensitive input x_1 increases beyond 1, which is the region of the original design. This finding confirms that extrapolation of a GP model in terms of the most sensitive inputs is a dangerous practice. This is not a disadvantage of the GP or TLM per se, it is a fact that has challenged and will always challenge empirical models, whether linear or non-linear in nature. This is because, unlike mechanistic models that are *deduced* from the laws of physics and which apply in in general for the phenomena being modeled, empirical models are *developed* from data in a limited region of experimentation.

3.4 Linear Regression and Symbolic Regression for Undesigned Data (Step 5 of Methodology)

Given that extrapolation of the previously developed models was not effective, a different approach was explored by combining all of the data sets previously presented (DOE, interpolation and extrapolation data sets) to build a linear regression model and a GP model. However since the combined data sets do not conform to an experimental DOE, the resulting models are only to be used for prediction within the new region of operability and not to infer causality.

The Linear Regression Model. The linear regression model was built treating the three sets as one combined data set and treating them as undesigned data, or data not collected using a design of experiments. Selectivity was fit to the first-order linear regression shown in equation (1). The resulting model is referred to as Linear Regression Model (LRM). The analysis of variance revealed a significant regression equation (F ratio <0.0001) with R^2 of 0.94, and no evidence of Lack of Fit ($p = 0.1520$). The parameter estimates are presented in Table 4.

Table 4. Parameter Estimates for Linear Regression Model showing significant terms at 95%

Term	β Estimate	Std Error	t Ratio	Prob>|t|	VIF
Intercept	-	0.365	-	<.000	
x1	0.008585	0.000	17.49	<.000	1.871
x2	0.086230	0.033	2.61	0.014	1.506
x3	0.105926	0.010	9.72	<.000	1.995
x4	-	0.119	-4.21	0.000	1.688
(x1-615.789)*(x2-3.73)	0.002362	0.001	2.17	0.038	1.748
(x1-615.789)*(x3-4.37447)	0.003111	0.000	10.63	<.000	1.537
(x2-3.73)*(x3-4.37447)	0.045254	0.023	1.94	0.062	1.791
(x1-615.789)*(x4-1.18474)	-	0.003	-2.47	0.020	1.695
(x3-4.37447)*(x4-1.18474)	0.036342	0.074	0.49	0.628	1.605
(x1-615.789)*(x3-4.37447)*(x4-	-	0.001	-2.80	0.009	1.766

effect included to preserve model precedence given that third order interaction is significant

Because the LRM was built for undesigned data, multicollinearity was analyzed using Variance Inflation Factors (VIF) [6]. As previously discussed, the presence of severe multicollinearty can seriously affect the precision of the estimated regression coefficients, making them very sensitive to the data in the particular sample collected and producing models with poor prediction.

The VIF for each model parameter (VIF_j) were compared to the overall Model (VIF_{model}). The VIF for the LRM are listed in Table 4. No evidence of severe multicollinearity was detected ($VIF_j < 16.7526$). A subsequent residual analysis did not show indication of violations of the error structure required for least square estimation indicating that an acceptable LRM was achieved with the combined data set.

The GP Model. The GP model was developed for the combined data set by using a toolbox in MATLAB. This model is referred to as the GP2 model to differentiate it from the GP model that was developed using only the DOE data. The initial functions for GP2 included addition, subtraction, multiplication, division, square, change sign, square root, natural logarithm, exponential, and power. Several runs were performed with various function generation (20 to 40); population size (500 to 900); number of generations (100 to 500); parsimony pressure (0.01 to 0.1); and 40 reproductions per generation with 0.6 probability for function as next node, and two different optimization criteria (correlation coefficient and sum of squares).

Several hundred equations were obtained. The selection of the best equations was based on the value of R^2. The following equation resulted in the highest value of R^2 (0.84).

$$S_k = -4.5333 - 0.0081 * \left(-e^{\left[2x_4 e^{-2x_3 *(-x_2+x_3)} \right]} * \left| 1.9997 1 + \sqrt{|x_1|} \right|^2 - \sqrt{|x_1|} + e^{\left[2Re \frac{x_3+x_4}{\log_e x_3} \right]} * |x_4|^2 - x_2(5.0521 + x_3 - x_4) \right) \quad (3)$$

Unlike the model of equation (1), the GP2 model shown in equation (3) is a non-linear model, with a complex functional form that shows relationships between the different variables (x_1, x_2, x_3, x_4).

Comparing the Linear and Non-linear Empirical Models. Figure 3 shows the graph of the GP2 model and the Linear Regression Model (LRM) for the combined data set. Comparing the two empirical models, the LRM performs better that the GP2 model in terms of R^2 and the sum square errors.

In terms of influential observations, the linear regression model allows determination of influential observations by calculating the Cook's D influence, D_i [7]. Observations with large values of D_i have considerable influence on the least square estimates β_i in equation (1) making these estimates very sensitive to these observations. None of the observations in the combined data set was considered influential (all calculated $D_i < 1$). This information is helpful identifying potentially interesting observations that may be replicated to confirm results.

The GP algorithm does not allow the statistical determination of influential observations but it allows the determination of the most sensitive inputs. This process is somehow similar to the testing of significant parameters in linear regression but it is not based on *statistical hypothesis testing*. The GP algorithm finds the most sensitive inputs by determining how the fitness function of the GP-generated equations improves by adding the specific input. The input/output sensitivity analysis had shown x_1 as the most important input. This is in agreement with the LRM, which shows x_1 as significant (Table 4).

The simpler form of the LRM would generally be preferred by plant engineers because of the larger R^2 and because it is more parsimonious than the GP2 model shown in equation (3). Nevertheless, the functional form of the non-linear GP2 model reveals relationships among the variables that account for a large percentage of the variation of the data.

Therefore, designed and undesigned data are opportunities for consideration of an alternative GP generated model. Even when the regression model is not significant, a GP model can still be built to confirm or challenge the result. This is illustrated in the following section.

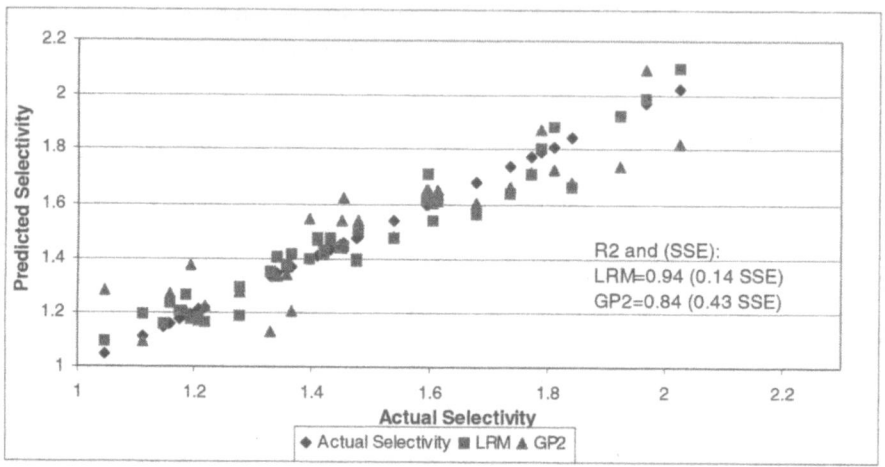

Fig. 3. Predicted selectivity for GP2 and LRM models for combined data set data

3.5 The GP Model – LRM Not Significant

A known weakness of some non-linear algorithms is the risk of model over-fitting (modeling unexplainable variation instead of real relationships that exist between inputs and outputs). This risk can significantly hinder the potential of GP models.

In order to test model over-fitting, data for the selectivity of a second chemical compound (P_k) was selected from the same 2^4 DOE experiments shown in Table 1 (P_k selectivity data is shown in the fourth column in Table 1). The standard linear regression model shown in equation (1) was used to fit to the data considering selectivity P_k as the response variable; and x_1, x_2, x_3, x_4 as the independent variables.

The corresponding analysis of variance indicated no evidence of Lack of Fit at the 95% confidence (p=0.0938) but revealed a non-significant model (p=0.4932) indicating that the hypothesis that all regression coefficients βi in the model of equation (1) are zero could not be rejected. Thus, the linear model is reduced to the mean.

The GP algorithm has been applied to the DOE data set. The selectivity P_k was selected as the output variable and x_1, x_2, x_3, x_4 were selected as input variables. Several runs were performed with 20 runs, population size between 500 to 900, number of generations between 100 to 500, 40 reproductions per generation, 0.6 probability for function as next node, parsimony pressure between 0.01 and 0.1, and correlation coefficient and sum of squares as optimization criteria. Hundreds of non-linear GP equations were generated. However, no equation was found that accounted for more than 50% of the variation in the data (the maximum correlation coefficient found was 0.5).

This result is potentially significant. In the P_k case, it can be demonstrated through statistical analysis that a statistically significant correlation does not exist between variables and response. The non-linear GP algorithm produced in an independent way the same result from the linear regression analysis by not creating a statistically significant model. This demonstrates that in this case the risk of model over-fitting is low.

4 Conclusions

A novel methodology for empirical modeling based on Design of Experiments and GP has been defined and applied successfully in the Dow Chemical Company. The proposed methodology uses linear regression models and non-linear GP models (statistically designed experiments, linear regression models for undesigned data, and GP-generated symbolic regression). A significant part of this methodology is based on the unique potential of GP algorithms for linearizing regression models in the presence of Lack of Fit and providing additional empirical non-linear functions that can be considered in combination with the linear ones. The proposed methodology has the following advantages:
- increases the options for development of empirical models based on DOE;
- reduces (or even eliminates) the number of additional experiments in the presence of Lack of Fit;
- maximizes the use of available data when model extrapolation is required;
- improves model validation by introducing alternative models.

These advantages are illustrated in an industrial application. The promising results obtained constitute a solid foundation for utilization of linear and non-linear models in industrial applications where extrapolation of an empirical model is often required.

References

1. Box, G.E.P, and Draper, N. R., *Empirical Model Building and Response Surfaces*, John Wiley and Sons, New York, 1987.
2. Koza, J. *Genetic Programming: On the Programming of Computers by Means of Natural Selection*, MIT Press, Cambridge, MA, 1992.
3. Castillo F. A.,Marshall, K. A, Green, J.L., and Kordon, A, "Symbolic Regression in Design of Experiments: A Case Study with Linearizing Transformations, *Proceedings of GECCO'2002*, New York, pp. 1043–1048., 2002
4. Banzhaf W., P. Nordin, R. Keller, and F. Francone, *Genetic Programming: An Introduction*, Morgan Kaufmann, San Francisco, 1998.
5. Myers, R.H., and, Montgomery, D.C., *Response Surface Methodology*, John Wiley and Sons, New York, 1995
6. Montgomery, D.C., and Peck, E.A., *Introduction to Linear Regression Analysis*, John Wiley and Sons, New York, 1992.
7. Cook, R.D., "Detection of Influential Observations in Linear Regression", *Technometrics*, Vol. 19, pp. 15–18, 1977

The General Yard Allocation Problem

Ping Chen[1], Zhaohui Fu[1], Andrew Lim[2], and Brian Rodrigues[3]

[1] Dept of Computer Science, National University of Singapore
3 Science Drive 2, Singapore 117543
{chenp,fuzh}@comp.nus.edu.sg
[2] Dept of IEEM, Hong Kong University of Science and Technology
Clear Water Bay, Kowloon, Hong Kong
iealim@ust.hk
[3] School of Business, Singapore Management University
469 Bukit Timah Road, Singapore 259759
br@smu.edu.sg

Abstract. The General Yard Allocation Problem (GYAP) is a resource allocation problem faced by the Port of Singapore Authority. Here, space allocation for cargo is minimized for all incoming requests for space required in the yard within time intervals. The GYAP is NP-hard for which we propose several heuristic algorithms, including Tabu Search, Simulated Annealing, Genetic Algorithms and the recently emerged "Squeaky Wheel" Optimization (SWO). Extensive experiments give solutions to the problem while comparisons among approaches developed show that the Genetic Algorithm method gives best results.

1 Introduction

The Port of Singapore Authority (PSA) operates this world's largest integrated container port and transshipment hub in Singapore. In year 2000, PSA handled 19.77 million Twenty-foot Equivalent Units (TEUs) of containers worldwide, including 17.04 million TEUs in Singapore. This represents 7.4% of the global container throughput and 25% of the world's total container transshipment throughput.

Typically, storage is a constraining component in port logistics management. Factors that impact on terminal storage capacity include stacking heights, available net storage area, storage density (containers per acre) and dwell times for empty containers and breakbulk cargo. The port in Singapore, faces such storage constraints in heightened way due to scarcity of land available for port activities. As such, the optimization of storage of cargo in its available yards is crucial to its operations. In studying its operations with the view of finding better ways to utilize storage space within the dynamic environment of the port, we have focused on a central allocation process in storage operations which allows for improved usage of space. In this process, requests are made from operations which coordinate ship berthing and ship-to-apron loading as well as apron-to-yard transportation. Each request consists of a set of yard spaces required for a single time interval. If space is allocated to the request, this space cannot be freed until completion of the request, i.e, until the end time. The major reason for such a constraint is that once a container is placed in the yard, it will not be removed until the ship for which it is bound arrives to reduce labor cost. The purpose is to pack all such requirements into a minimum space. The current allocation is made manually, hence it requires a considerable amount of manpower and the yard arrangement generated is not efficient.

Sabria and Daganzo [1] give a bibliography on port operations with the focus on berthing and cargo-handling systems. On the other hand, traffic and vehicle-flow scheduling on landside upto yard points have been studied well (see for example, Bish et al.[2]). Other than studies such as Gambardella et al. [3], which address spatial allocation of containers on a terminal yard using simulation techniques, there has been little direct study on yard space allocation as described in this paper. We propose a basic model to address this port storage optimization problem. The model is applicable elsewhere as it is generic in form. We call it General Yard Allocation Problem (GYAP), which is an extended study of The Yard Allocation Problem (YAP) (see [4] [5]). The yard is treated as a two dimensional space instead of one dimensional in YAP.

This paper is organised as follows: problem definition is given in Sect. 2, followed by a discussion on two-dimensional packing, which is highly related to our problem, in Sect. 3. Different heuristic approaches including Tabu Search, "Squeaky Wheel" Optimization, Simulated Annealing and Genetic Algorithms, as discussed in the next four sections, are applied. Before we conclude in Sect. 9, experimental results are presented and analyzed in Sect. 8.

2 Problem Description and Formulation

The main objective of the GYAP problem is to minimize the container yard space used while satisfying all the space request. The GYAP problem can be described as follows:

A set R of n yard space requests, as described above, and a two-dimensional infinite container yard E are given. We can think of E as the being first quadrant in $\Re^2$. Each request $R_i \in R$ $(i = 1, ..., n)$ has a series of (continuous) space requirements Y_{i_j} with length L_{i_j} and width W_{i_j}, where $j \in [T_{i_{start}}, T_{i_{end}}]$, where the latter time interval is defined by the request R_i.

A mapping, F, such that $F(Y_{i_j}) = (x, y)$, where $(x, y) \in E$ gives the coordinate of the bottom left corner of Y_{i_j} as it is aligned in E with its sides parallel to the X-Y axes. Each map, F, must also satisfy the condition that for all $p, q \in [T_{i_{start}}, T_{i_{end}}]$ such that $p = q - 1$, and for $F(Y_{i_p}) = (x_{i_p}, y_{i_p})$ and $F(Y_{i_q}) = (x_{i_q}, y_{i_q})$, we must have $x_{i_p} \geq x_{i_q}, y_{i_p} \geq y_{i_q}, x_{i_p} + L_{i_p} \leq x_{i_q} + L_{i_q}$ and $y_{i_p} + W_{i_p} \leq y_{i_q} + W_{i_q}$. This constraint provides the fact that the total space requests increase in time, as would be expected in a realistic situation. Our objective is then to minimize, over all possible mappings F, :

$$\max_{i,j,Y_{i_j} \in R_i} [(Proj_X F(Y_{i_j}) + L_{i_j})] \times \max_{i,j,Y_{i_j} \in R_i} [(Proj_Y F(Y_{i_j}) + W_{i_j})]$$

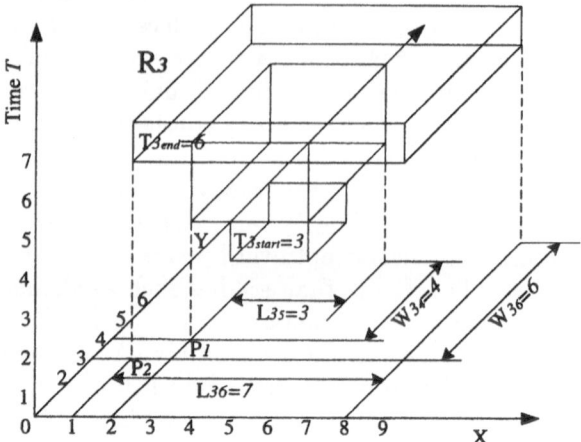

Fig. 1. A *valid* request R_3 in GYAP. The coordinates for P_1 and P_2 are $(2,4)$ and $(1,3)$ respectively. Same amount of space is required at time 4 and 5.

where Proj_X(Proj_Y) denotes the orthogonal projections from $\Re^2$ onto the X (Y) axis. In other words, we would like to minimize the total area used in the yard while satisfying all the space requests.

Figure 1 shows a layout with only one *valid* request, R_3. Time T is taken as a discrete variable with a minimum unit of 1. R_3 has four space requirements, at times 3, 4, 5, and 6, two of which are the same at time 4 and 5, within the time interval $[T_{3_{start}} = 3, T_{3_{end}} = 6]$. The final positions for Y_{3_5} and Y_{3_6} are $F(Y_{3_5}) = (2, 4)$ and $F(Y_{3_6}) = (1, 3)$, respectively, as shown. The corresponding mappings for R_3 will then be $\{(3,4),(2,4),(2,4),(1,3)\}$, where all the constraints imposed can be seen to hold. The maximum value in this case, which is the product of the maximum X−coordinate and the maximum Y−coordinate, is 72. Note that the space requests is look like a inverted pyramid when together. We will call the each space requirement at each time slot a *layer*.

It is clear that GYAP is NP-hard, since (see [4]) we have found YAP is NP-hard, which is a special case of GYAP in the case when each request has the same length (width) which equals to the length(width) of the yard.

3 Two-Dimensional Rectangular Packing Problem

In finding solutions for the GYAP, we confront, at each time slot, a Two-Dimensional Rectangular Packing Problem (2-DRPP) with certain boundary constraints.

The 2-DRPP has been proven to be NP-hard. Various heuristic methods have been proposed. However, most of these are *meta-heuristics* because the complex representation of the 2DRPP makes it difficult to apply other heuristics. Such meta-heuristics usually use a lower-level packing algorithm, for example, greedy packing, and higher-level algorithms, such as Tabu Search.

A lower-level greedy packing routine allocates the objects according to certain given ordering, which is determined by the higher-level heuristics. A greedy packing routine is the Bottom Left (BL) heuristic [6] [7] [8]. Starting from the top right corner each item is slid as far as possible to the bottom and then as far as possible to the left of the object. These successive vertical and horizontal operations are repeated until the item locks in a stable position.

The major advantage of this routine is its simplicity. Its time complexity is only $O(N^2)$, where N is the total number of items to be packed. Due to its low complexity this simple heuristic is favorable in a hybrid combination with a meta-heuristic, since the decoding routine has to be executed every time the quality of a solution is evaluated and hence contributes significantly to the running time of any hybrid algorithm [8].

The BL packing routine uses a deterministic algorithm, hence the input ordering of the objects fully determines the final layout. Heuristics can be applied to optimize the ordering. In fact, such orderings can also be considered as permutations, and hence, in this work, we use the word *permutation* for a solution representation.

As the objective of GYAP is to minimize the total yard space while satisfying all the space requests, A procedure, called *EVALUATE_SOLUTION*, is required. It takes a solution (i.e permutation) and computes the minimum space required by applying BL heuristic on packing those pyramids into $\Re^3_+$ (with two dimensions representing the yard and one dimension for the time axis).

EVALUATE_SOLUTION (P)
1 **for each** $R \in P$
2   **while** the position of R changes
3      $SHIFT(R, R_{end}, 0, 0, bottom)$
4      $SHIFT(R, R_{end}, 0, 0, left)$
5 **return** Minimum area used

SHIFT (R, t, l, b, o)
01 $B :=$ bottom most position to shift all layers (time t')
02 **if** $B < b$
03    $B = b$
04 $L :=$ left most position to shift all layers (time t'')
05 **if** $L < l$
06    $L = l$
07 **if** $o = bottom$
08   **for all** layers r after $t' - 1$
09   shift r bottomwards to B
10      $SHIFT(R, t' - 1, L, B, o)$
11 **else**
12   **for all** layer r after $t'' - 1$
13   shift r leftwards to L
14      $SHIFT(R, t'' - 1, L, B, o)$

4 Tabu Search

Tabu Search (TS) is a local search meta-heuristic that uses the best neighborhood move that is not "tabu" active to move out from local optimum by incorporating *adaptive memory* and *responsive exploration* [9]. According to the different usage of memory, conventionally, Tabu Search has been classified into two categories: Tabu Search with Short Term Memory (TSSTM) and Tabu Search with Long Term Memory (TSLTM) [10] [11].

4.1 Tabu Search with Short Term Memory

The usage of memory of TSSTM is via the Tabu List. Such an adaptation is also known as *recency*-based Tabu Search. The neighborhood solution can be obtained by swapping any two numbers in the permutation. For example: [2, 3, 0, 1, 4] is a neighborhood solution of [1, 3, 0, 2, 4] by interchanging the positions of 1 and 2. Neighborhood solutions that are identical to the original solution after normalization are excluded for efficiency reasons.

4.2 Tabu Search with Long Term Memory

TSLTM uses more advanced Tabu Search techniques including intensification and diversification strategies. It archives total or partial information from all the solutions it has visited. This is also known as *frequency*- based Tabu Search. It attempts to identify certain potentially "good" patterns, which will be used to guide the search process towards possibly better solutions [12]. Two kinds of diversification techniques are used. One is random re-start. The other is randomly picking a sub-sequence and inserting it to a random position. For example, [0, 1, 2, 3, 4] may be changed to [0, 3, 2, 1, 4] if (2, 3) is chosen as the sub-sequence and its inverted order (or original, if random) is inserted back in the position in ahead of 1. Intensification is similar to TSSTM. TSLTM uses a *frequency*-based memory by recording both *residence*-frequency and *transition*-frequency of the visited solutions.

5 "Squeaky Wheel" Optimization

"Squeaky Wheel" Optimization (SWO) is a new heuristic approach proposed in [13]. Until now, this concept can only be found in a few papers: [14], [15] and [16]. In 1996, a "doubleback" approach was proposed to solve the Resource Constrained Project Scheduling (RCPS) problem [17] , which motivated the development of SWO in 1998. The YAP is similar to the RCPS except that it has no precedence constraints and the tasks (requirements) are Stair Like Shapes (SLS). Instead of Left-Shift and Right-Shift in "doubleback", we only use a "drop" routine similar to Left-Shift. We continue using this technique in GYAP.

The ideas in SWO mimic how human beings solve problems by identifying the "trouble-spot" or the "trouble-maker" and trying to resolve problems caused by the latter. In SWO, a greedy algorithm is used to construct a solution according to certain priorities (initially randomly generated) which is then analyzed to find the "trouble-makers", i.e. the elements whose improvements are likely to improve the objective function score. The results of the analysis are used to generate new priorities that determine the order in which the greedy algorithm constructs the next solution. This Construct/Analyze/Prioritize (C/A/P) cycle continues until a certain limit is reached or an acceptable solution is found. This is similar to the Iterative Greedy heuristic proposed in [18]. Iterative Greedy is especially designed for Graph Coloring Problem and may not be directly applicable to other problems, whereas SWO is a more general optimization heuristic.

In our problem, we start with a random solution, which is a random permutation. The Analyzer evaluates the solution by applying the packing routine. If the best known result (yard space) is B, then a threshold T is set to be $B-1$. The blame factor for each request is the sum of the space requirements that exceed this threshold T, i.e. the total area of the pyramid above the cutting line T. All blame information is passed to the Prioritizer, which is a priority queue in our case. When the control has been handed over to the Constructor again, it continuously deletes the elements from the priority queue and immediately drops them into the yard. A tie, i.e. more than one element with the same priority, is broken by considering their relative positions in previous solution. This tie-breaker also helps avoid cycles in our search process.

We also found that the performance of the SWO can be further improved if a "quick" Tabu Search technique TSSTM is embedded in the SWO. We call this modified algorithm SWO+TS, where TS denotes Tabu Search. The Constructor passes its solution to a TSSTM engine, which performs a quick local search and passes the local optimum to the Analyzer. Experiments shows a considerable improvement against the original SWO system. Similar ideas of SWO with "intensification" have been proposed in [14], where solutions are partitioned and SWO is applied to each partition.

6 Simulated Annealing

Simulated annealing [19] stochastically simulates the slow cooling of a physical system. We used the following Simulated Annealing algorithm on our problem:

Step 1. Choose some initial temperature T_0 and a random initial starting configuration θ_0. Set $T = T_0$. Define the Objective function (Energy function) to be $En()$ and the cooling schedule σ.

Step 2. Propose a new configuration, θ', of the parameter space, within a neighborhood of the current state θ, by setting $\theta' = \theta + \phi$ for some random vector ϕ.

Step 3. Let $\delta = En(\theta') - En(\theta)$. Accept the move to θ' with probability

$$\alpha(\theta, \theta') = \begin{cases} 1 & if \delta < 0 \\ exp(-\frac{\delta}{T}) & otherwise \end{cases}$$

Step 4. Repeat Step 2 and 3 for K iterations, until it is deemed to have reached the equilibrium.

Step 5. Lower the temperature by $T = T \times \sigma$ and repeat Steps 2-4 until certain stopping criterion, for our case $T < \epsilon$ (for some small ϵ) is met.

Due to the logarithmic decrement of T, we set $T_0 = 1000$. The Energy function is simply defined as the length of the yard required. The probability $exp(-\frac{\delta}{T})$ is a Boltzmann factor. The number of iterations K is proportional to the input size n. A neighborhood is defined similarly as the one in TS by swapping of any two permutations and re-positioning a random permutation.

7 Genetic Algorithms

Genetic Algorithms [20] (GA) are search procedures based notions from natural selection. It is clear that the classical binary representation is not a suitable in GYAP, in which a permutation $(0, 1, \ldots, n-1)$ is used as the solution representation. The solution space is a permutation of $(0, 1, \ldots, n-1)$. The binary codes of these permutations do not provide any advantage. At times, the situation is even worse: the change of a single bit may not lead to a valid solution. Here, we adopt a vector representation, i.e. by using a permutation directly as the chromosome in the genetic process. We will illustrate the two major genetic operators used in our approach, crossover and mutation.

7.1 Crossover Operator

Using permutations as chromosome, we have implemented three crossover operators: *Classical crossover with repair*, *Partially-mapped crossover* and *Cycle crossover*. All these operators are be tailored to suit our problem domain. A small change in the crossover operator may cause totally different results.

Classical Crossover with Repair. The Classical Crossover operator builds the offspring by appending the head from one parent with the tail from the other parent, where the head and tail come from a random cut of the parents' chromosomes. A repair procedure may be necessary after the crossover [21]. For example, the two parents (with random cut point marked by '|'):

$$p_1 = (0\ 1\ 2\ 3\ 4\ 5\ |\ 6\ 7\ 8\ 9) \ \ and \ \ p_2 = (3\ 1\ 2\ 5\ 7\ 4\ |\ 0\ 9\ 6\ 8)$$

will produce the following two offsprings:

$$o_1 = (0\ 1\ 2\ 3\ 4\ 5\ |\ 0\ 9\ 6\ 8) \ \ and \ \ o_2 = (3\ 1\ 2\ 5\ 7\ 4\ |\ 6\ 7\ 8\ 9)$$

However, the two offsprings are not valid permutations after the crossover. A repair routine replaces the repeated numbers with the missing ones randomly. The repaired offsprings will be:

$$o_1 = (7\ 1\ 2\ 3\ 4\ 5\ |\ 0\ 9\ 6\ 8) \ \ and \ \ o_2 = (3\ 1\ 2\ 5\ 7\ 4\ |\ 6\ 0\ 8\ 9)$$

The classical crossover operator tries to maintain the absolute positions in the parents.

Partially Mapped Crossover. Partially Mapped Crossover (PMX) was first used in [22] to solve the Traveling Salesman Problem. We have made several adjustments to accommodate our permutation representation. The modified PMX builds an offspring by choosing a subsequence of a permutation from one parent and preserving the order and position of as many numbers as possible from the other parent. The subsequence is determined by choosing two random cut points. For example, the two parents:

$$p_1 = (0\ 1\ 2\ |\ 3\ 4\ 5\ 6\ |\ 7\ 8\ 9) \text{ and } p_2 = (3\ 1\ 2\ |\ 5\ 7\ 4\ 0\ |\ 9\ 6\ 8)$$

would produce offspring as follows. First, two segments between cutting points are swapped (symbol 'u' represents 'unknown' for this moment):

$$o_1 = (u\ u\ u\ |\ 5\ 7\ 4\ 0\ |\ u\ u\ u) \text{ and } o_2 = (u\ u\ u\ |\ 3\ 4\ 5\ 6\ |\ u\ u\ u)$$

The swap defines a series of mappings implicitly at the same time:

$$3 \leftrightarrow 5, 4 \leftrightarrow 7, 5 \leftrightarrow 4 \text{ and } 6 \leftrightarrow 0.$$

The 'unknown's are then filled in with numbers from original parents, for which there is no conflict:

$$o_1 = (u\ 1\ 2\ |\ 5\ 7\ 4\ 0\ |\ u\ 8\ 9) \text{ and } o_2 = (u\ 1\ 2\ |\ 3\ 4\ 5\ 6\ |\ 9\ u\ 8)$$

Finally, the first u in o_1 (which should be 0 will cause a conflict) is replaced by 6 because of the mapping $0 \leftrightarrow 6$. Note such replacement is transitive, for example, the second u in o_1 should follow the mapping $7 \leftrightarrow 4, 4 \leftrightarrow 5, 5 \leftrightarrow 3$ and is hence replaced by 3. The final offspring are:

$$o_1 = (6\ 1\ 2\ |\ 5\ 7\ 4\ 0\ |\ 3\ 8\ 9) \text{ and } o_2 = (7\ 1\ 2\ |\ 3\ 4\ 5\ 6\ |\ 9\ 0\ 8)$$

The PMX crossover exploits important similarities in the value and ordering simultaneously when used with an appropriate reproductive plan [22].

Cycle Crossover. Original Cycle Crossover (CX) was proposed in [23], again for the TSP problem. Our CX builds offspring in such a way that each number (and its position) comes from one of the parents. We explain the mechanism of the CX with following example. Two parents:

$$p_1 = (0\ 1\ 2\ 3\ 4\ 5\ 6\ 7\ 8\ 9) \text{ and } p_2 = (3\ 1\ 2\ 5\ 0\ 4\ 7\ 9\ 6\ 8)$$

will produce the first offspring by taking the first number from the first parent:

$$o_1 = (0\ u\ u\ u\ u\ u\ u\ u\ u)$$

Since every number in the offspring should come from one of its parents (for the same position), the only choice we have at this moment is to pick number 3, as

the number from parent p_2 just "below" the selected 0. In p_1, it is in position 3, hence:
$$o_1 = (0\ u\ u\ 3\ u\ u\ u\ u\ u)$$
which, in turn, implies number 5, as the number from p_2 "below" the selected 3:
$$o_1 = (0\ u\ u\ 3\ u\ 5\ u\ u\ u\ u)$$
Following the rule, the next number to be inserted is 4. However, selection of 4 requires the selection of 0, which is already in the list. Hence the cycle is formed as expected.
$$o_1 = (0\ u\ u\ 3\ 4\ 5\ u\ u\ u\ u)$$
The remaining 'u's are filled from p_2:
$$o_1 = (0\ 1\ 2\ 3\ 4\ 5\ 7\ 9\ 6\ 8).$$
Similarly,
$$o_2 = (3\ 1\ 2\ 5\ 0\ 4\ 6\ 7\ 8\ 9).$$

CX preserves the absolute position of the elements in the parent sequence [21].

Our experiments shows Classical crossover and CX have a stable but slow improvement rates, while PMX demonstrates oscillating but fast convergence. In our later experiments, the majority of the crossover is done by PMX. Classical crossover and CX are applied at a much lower probability.

7.2 Mutation Operator

Mutation is another classical genetic operator, which alters one or more genes (part of a chromosome) with a probability equal to the mutation rate. There are several known mutation algorithms which work well on different problems:

- Inversion: invert a subsequence.
- Insertion: select an number and insert it back in a random position.
- Displacement: select a subsequence and insert it back in a random position.
- Reciprocal Exchange: swap two numbers.

In fact the Inversion, Displacement and Reciprocal Exchange are quite similar to our neighborhood solution and diversification techniques used in Tabu Search and Simulated Annealing in previous sessions. We adopt a relatively low mutation rate of 1%.

We use population size $P = 1000$ for most cases. The evolution process starts with a random population. The population is sorted according to the objective function, the better the quality, the higher the probability it will be selected for reproduction. At each iteration, a new generation with population size $2P$ is produced and the better half, which is of size P, survive for the next iteration. The evolution process continues until certain stop criterion are met.

8 Experimental Results

We conducted extensive experiments on randomly generated data [1]. The graph for each test case contains one components, so that the cases cannot be partitioned into more than one independent sub-case.

All the programs implementing various heuristic methods are coded in GNU C++, with an extensive use of Standard Template Library (STL) for efficiency data structures like priority queue, set and map.

Due to the difficulties of finding any optimal solution in the experiments, a trivial *lower bound* is taken to be the sum of the space requirements at each time slot and used for benchmarking purpose.

Table 1. Experimental results for GYAP (Entries in the table show the minimum length of the yard required. Names of Data Sets show the number of pyramids in the file; LB:Lower Bound)

Data Set	Lower Bound	TSSTM	TSLTM	SWO	SWO+TS	SA	GA
P35	47	84	80	112	84	76	76
P86	218	616	583	748	550	572	550
P108	78	205	200	270	200	210	175
P127	221	660	649	891	671	671	550
P137	239	680	670	950	690	600	560
P142	40	135	125	190	140	120	115
P160	249	828	828	996	852	840	780
P167	243	814	792	1012	825	748	649
P187	302	948	912	1032	912	816	768

Table 1 illustrates the results and Table 2 shows the running time for each of the test performed in Table 1. GA is the most cost-effective approach while TSSTM has the simplest implementation for which it is not surprising to see that it achieves the poorest results. Long term memory certainly improves the performance of TS, though the improvement is not very obvious and stable sometimes. We believe one of the major difficulties with long term memory is the fine tuning of parameters, including the assignment of relative weights to yard length, residence frequency and transition frequency in the objective function.

The performance of SWO is poor in our experiments because of two reasons. Firstly, the BL packing routine makes it difficult to identify the bottleneck, or the "trouble makers". The objects are manipulated in a two-dimensional space, and assigning blame factor according to only one dimension may not well reflect the structure of the solution. Secondly, there is the loss of correspondence between physical layout and actual solutions. SWO is more sensitive to the problem domain as it needs to know the exact structure of the solution in order to assign

---

[1] All test data are available on the web with URL:
http://www.comp.nus.edu.sg/~fuzh/GYAP

Table 2. Experiment running time (in seconds) for Table 1.

Data Set	TSSTM	TSLTM	SWO	SWO+TS	SA	GA
P35	783	4521	1033	5462	4512	923
P86	948	6529	2312	5978	2351	1423
P108	2321	8392	4528	8432	8934	3452
P127	2783	9837	4678	10262	11023	6621
P137	2796	9640	4780	9892	9857	7048
P142	893	3428	1532	4582	3275	1094
P160	4781	15327	3085	18539	6793	3583
P167	6063	14294	3769	19852	6832	3781
P187	7806	17327	4085	20542	6673	6849

proper blame values. SA is the second best approach. We believe it is because the cooling schedule is not much affected by the loss of correspondence.

9 Conclusion

In this paper, the General Yard Allocation Problem is studied which involves two-dimensional rectangle packing as a related problem. We adopted a simple Bottom Left packing strategy as a first heuristic. Heuristics like Tabu Search, Simulated Annealing, Genetic Algorithms and "Squeaky Wheel" Optimization were then applied to problem in the extensive experiments and solutions obtained. Our approach sheds light on the use of these meta-heuristics on a set of problems, including those of packing where packing lists can change in time. In comparisons between results obtained, we found that the GA implementation achieves the best results to this problem.

References

1. F. Sabria and C. Daganzo: Queuing systems with scheduled arrivals and established service order. Transportation Research B, vol. 23. (1989) 159–175
2. E.K. Bish, T.-Y. Leong, C.-L. Li, J. W.C. Ng, D. Simchi-Levi: Analysis of a new vehicle scheduling and location problem. Naval Research Logistics, vol. 48. (2001) 363–385
3. L. Gambardella, A. Rizzoli, M. Zaffalon: Simulation and planning of an intermodal container terminal. Simulation, vol. 71. (1998) 107–116
4. P. Chen, Z. Fu, A. Lim: The yard allocation problem. In: Proceedings of the Eighteenth National Conference on Artificial Intelligence (AAAI-02), Edmonton, Alberta, Canada (2002)
5. P. Chen, Z. Fu, A. Lim: Using genetic algorithms to solve the yard allocation problem. In: Proceedings of the Genetic and Evolutionary Computing Conference (GECCO-2002) New York City, USA (2002)
6. B. Baker, E.C. Jr., R. Rivest: Orthogonal packing in two dimensions. SIAM Journal of Computing, vol. 9. (1980) 846–855

7. S. Jacobs: On genetic algorithms for the packing of polygons. European Journal of Operational Research, vol. 88. (1996) 165–181
8. E. Hopper, B. Turton: An empirical investigation of meta-heuristic and heuristic algorithms for a 2d packing problem. European Journal of Operational Research, vol. 128. (2001) 34–57
9. P. L. Hammer: Tabu Search. J.C. Baltzer, Basel, Switzerland (1993)
10. F. Glover, M. Laguna: Tabu Search. Kluwer Academic Publishers (1997)
11. S. Sait, H. Youssef: Iterative Computer Algorithms with Applications in Engineering: Solving Combinatorial Optimization Problems. IEEE (1999)
12. D. Pham, D. Karaboga: Intelligent optimisation techniques: genetic algorithms, tabu search, simulated annealing and neural networks. London and New York, Springer (2000)
13. D. Clements, J. Crawford, D. Joslin, G. Nemhauser, M. Puttlitz, M. Savelsbergh: Heuristic optimization: A hybrid ai/or approach. In: Workshop on Industrial Constraint-Directed Scheduling (1997)
14. D.E. Joslin, D. P. Clements: "squeaky wheel" optimization. Journal of Artificial Intelligence Research, vol. 10. (1999) 353–373
15. D.E. Joslin, D. P. Clements: Squeaky wheel optimization. In: Proceedings of the Fifteenth National Conference on Artificial Intelligence (AAAI-98), Madison, USA (1998) 340–346
16. D. Draper, A. Jonsson, D. Clements, D. Joslin: Cyclic scheduling. In: Proceedings of the Sixteenth International Joint Conference on Artificial Intelligence (1999)
17. J.M. Crawford: An approach to resource constrained project scheduling. In: Proceedings of the 1996 Artificial Intelligence and Manufacturing Research Planning Workshop (1996)
18. J.C. Culberson, F. Luo: Exploring the k-colorable landscape with iterated greedy. Cliques, Coloring, and Satisfiability: Second DIMACS Implementation Challenge, David S. Johnson and Michael A. Trick (eds.). DIMACS Series in Discrete Mathematics and Theoretical Computer Science (1996) 245–284
19. S. Kirpatrick, C. Gelatt, Jr., M. Vecchi: Optimization by simulated annealing. Science, vol. 220. (1983) 671–680
20. J.H. Holland: Adaptation in natural artificial systems. Ann Arbor: University of Michigan Press (1975)
21. Z. Michalewicz: Genetic Algorithms + Data Structure = Evolution Programs. Springer-Verlag, Berlin Heidelberg New York (1996)
22. D. Goldberg, R. Lingle: Alleles, loci, and the traveling salesman problem (1985)
23. I. Oliver, D. Smith, J. Holland: A study of permutation crossover operators on the tsp (1987)